physica status solidi (a)

ISSN 0031-8965 * VOL. 65 · NO. 1 · MAY 1981

AKADEMIE-VERLAG · BERLIN

THEORIA CUM PRAXI

Classification Scheme

1. Structure of Crystalline Solids
   1.1 Perfectly Periodic Structure
   1.2 Solid-State Phase Transformations
   1.3 Alloys. Metallurgy
   1.4 Microstructure (Magnetic Domains See 18; Ferroelectric Domains See 14.4.1)
   1.5 Films
   1.6 Surfaces
2. Non-Crystalline State
3. Crystal Growth
4. Bonding Properties
5. Mössbauer Spectroscopy
6. Lattice Dynamics. Phonons
7. Acoustic Properties
8. Thermal Properties
9. Diffusion
10. Defect Properties (Irradiation Defects See 11)
    10.1 Metals
    10.2 Non-Metals
11. Irradiation Effects (X-Ray Diffraction Investigations See 1 and 10)
12. Mechanical Properties (Plastic Deformations See 10)
    12.1 Metals
    12.2 Non-Metals
13. Electron States
    13.1 Band Structure
    13.2 Fermi Surfaces
    13.3 Surface and Interface States
    13.4 Impurity and Defect States
    13.5 Elementary Excitations (Phonons See 6)
        13.5.1 Excitons
        13.5.2 Plasmons
        13.5.3 Polarons
        13.5.4 Magnons
14. Electrical Properties. Transport Phenomena
    14.1 Metals. Semi-Metals
    14.2 Superconductivity. Superconducting Materials and Devices
    14.3 Semiconductors
        14.3.1 Films
        14.3.2 Surfaces and Interfaces
        14.3.3 Devices. Junctions (Contact Problems See 14.3.4)
        14.3.4 High-Field Phenomena, Space-Charge Effects, Inhomogeneities, Injected Carriers (Electroluminescence See 20.3; Junctions See 14.3.3)
    14.4 Dielectrics
        14.4.1 Ferroelectrics
15. Thermoelectric and Thermomagnetic Properties
16. Photoconductivity. Photovoltaic Effects
17. Emission of Electrons and Ions
    17.1 Field Emission Microscope Investigations
18. Magnetic Properties
    18.1 Paramagnetic Properties
    18.2 Ferromagnetic Properties
        18.2.1 Ferromagnetic Films
    18.3 Ferrimagnetic Properties
    18.4 Antiferromagnetic Properties

(Continued on cover three)

Author Index

physica status solidi (a)

applied research

Volume 65 · Number 1 · Pages 1 to 404, K1 to K102, and A1 to A8

May 16, 1981

PSSA 65(1) 1—404, K1—K102, A1—A8 (1981)

ISSN 0031-8965

THEORIA CUM PRAXI

AKADEMIE-VERLAG · BERLIN

Subscriptions and orders for single copies should be directed
in the GDR:
to the Postzeitungsvertrieb,
to a book-shop, or
to the Akademie-Verlag, DDR-1080 Berlin, Leipziger Straße 3 – 4;
in the other socialist countries:
to a book-shop for foreign language literature or
to the competent news-distributing agency;
in the FRG and BERLIN (WEST):
to a book-shop or
to the wholesale distributing agency Kunst und Wissen, Erich Bieber, D-7000 Stuttgart 1, Wilhelmstr. 4 – 6;
in AUSTRIA:
to the Globus-Buchvertrieb, A-1201 Wien, Höchstädtplatz 3:
in the other Western European countries:
to Kunst und Wissen, Erich Bieber GmbH, CH-8008 Zürich, Dufourstr. 51;
in USA and CANADA:
to Verlag Chemie International, Inc., Plaza Centre. Suite E, 1020 N. W. 6th Street, Deerfield Beach, FL 33441, USA;
in other countries:
to the international book- and journal-selling trade,
to Buchexport, Volkseigener Außenhandelsbetrieb der Deutschen Demokratischen Republik, DDR-7010 Leipzig, Postfach 160, or
to the Akademie-Verlag, DDR-1080 Berlin, Leipziger Straße 3 – 4.

Editorial Note:

"physica status solidi (a)" undertakes that an original paper *accepted for publication* before the 23rd of any month will be published within 50 days of this date unless the author requests a postponement. In special cases there may be some delay between *receipt* and *acceptance* of a paper due to the review and, if necessary, revision of the paper.

Schriftleiter und verantwortlich für den Inhalt:
Professor Dr. Dr. h. c. P. Görlich, DDR-1020 Berlin, Neue Schönhauser Str. 20 bzw. DDR-6900 Jena, Schillbachstr. 24.
Verlag:
Akademie-Verlag, DDR-1080 Berlin, Leipziger Str. 3—4; Fernruf: 2236221 und 2236229; Telex-Nr.: 114420; Bank: Staatsbank der DDR, Berlin, Kto.-Nr.: 6836-26-20712.
Chefredakteur:
Dr. H.-J. Hänsch.
Redaktionskollegium:
Prof. Dr. E. Gutsche, Dr. H.-J. Hänsch, Dr. H. Lange, Dr. S. Oberländer.
Anschrift der Redaktion:
DDR-1020 Berlin, Neue Schönhauser Str. 20; Fernruf: 2823380.
Veröffentlicht unter der Lizenznummer 1620 des Presseamtes beim Vorsitzenden des Ministerrates der Deutschen Demokratischen Republik.
Gesamtherstellung:
VEB Druckerei „Thomas Müntzer", DDR-5820 Bad Langensalza.
Erscheinungsweise:
Die Zeitschrift „physica status solidi (a)" erscheint jeweils am 16. eines jeden Monats. Jährlich erscheinen 6 Bände zu je 2 Heften.
Bezugspreis:
je Band 165,— M zuzüglich Versandspesen (Preis für die DDR: 130,— M).
Bestellnummer dieses Bandes: 1085/65.

Printed in the German Democratic Republic.
AN (EDV) 20735

Contents

Review Article

Original Papers

Short Notes

Pre-Printed Titles

Systematic List

Contents of Volume 65 Continued on Page 407

Review Article

phys. stat. sol. (a) **65**, 11 (1981)

Subject classification: 14.3.1; 1.5; 22.9

Physics Department, Brunel University, Uxbridge[1])

The Electrical and Structural Properties of Thin Films of Polymers Deposited from the Vapour Phase, with Special Reference to Polypropylene

By

C. A. HOGARTH and T. IQBAL

Contents

1. Introduction

1.1 General background

The electrical properties of polymeric materials have been the subject of extensive study [1 to 5]. Due to low production costs and to their excellent electrical and

---

[1]) Uxbridge, Middlesex UB8 3PH, Great Britain.

mechanical properties, polymers have many industrial applications. Thin polymeric films can be used for the investigation of electrical and optical properties and may be obtained by several means including thermal evaporation [4] and deposition by polymerization of monomers on surfaces bombarded with electrons or exposed to other radiations, e.g. a glow discharge [6] or ultra-violet light. Radio-frequency sputtering [7] and dipping of a substrate into an appropriate solution can also produce thin layers of polymers [8]. When polymers are heated in vacuum, one result is the breaking of certain chemical bonds within the molecules, thus leading to the formation of active fragments which sublime from the surface of the polymer material [9]. These fragments when incident on to a substrate react with one another to form a polymer coating on its surface [10 to 12]. Not all polymers release active fragments, when heated in vacuum. Some of them such as the polyurethanes, epoxides, and polyvinylchlorides decompose into gaseous fragments that are not capable of recombining on a substrate. Unlike vinyl polymers, halogen polymers such as polytetrafluorethylene (PTFE) and polychlortrifluorethylene release significant numbers of active fragments, when they are heated in vacuum [13], and these fragments form polymer layers when they condense in vacuum.

The structure of the polymeric materials plays an important role in controlling their electrical properties. Most polymers of practical significance are composed of a fine mixture of small micrometre size crystalline regions, embedded in an amorphous matrix of similar molecular composition [14]. The polymeric chains are thought to be folded in an orderly arrangement within the crystalline regions and these regions have the form of small platelets having thicknesses of the order of a few nanometres. Since the amorphous regions have no structural order it is therefore most likely that they contain most of the impurities present in the polymer. The available theories for the crystalline, polycrystalline, and amorphous inorganic solids are usually used to describe the conduction process of organic materials. The charge carrier trap depth in certain polymeric materials has been investigated by using the technique of thermally stimulated currents [15 to 17]. It has also been found that on exceeding a threshold applied voltage, thin films of polymeric materials sandwiched between metal electrodes may undergo an electroforming process [4, 18].

We have examined the properties of evaporated polypropylene for possible use as a thin dielectric film and were able to produce standard and reproducible results of electrical measurements. The trap depth in the evaporated polypropylene was calculated from the measurements of thermally stimulated currents and was found to be 0.33 eV [15]. A vacuum heat treatment of the films at 110 °C leads to an increase in its resistivity. Examination by electron microscopy and by electron diffraction showed an aggregation of particulate material of more crystalline nature as a result of the heat treatment. The ac conductivity measurements of polypropylene demonstrated the power law dependence of conductivity on frequency, which characterizes a process of hopping conduction [19].

1.2 Thermal decomposition

Two types of change take place when polymers are heated in vacuum or in an inert gas atmosphere, namely, (i) depolymerization and (ii) degradation.

When thermal decomposition gives rise to the evaporation of volatile low molecular weight products, mainly consisting of original monomers, the process is known as depolymerization, while in the case of degradation a considerable decrease in both the chain length and average molecular weight of the polymer are observed. Also some volatile products may appear but these are normally present in negligible amount.

1.2.1 *Degradation*

During the degradation a considerable reduction in the molecular weight and in the chain length of the polymer are observed. Thermal degradation of a number of polymers such as polystyrene, polyisoprene, polybutadiene, and polyethylene was brought about in vacuum at 300 to 400 °C by Madorsky et al. [20]. The resultant volatile products were composed of fragments varying in size from a monomer to material having a molecular weight of about 1000. Madorsky [21] evaporated the polyethylene (PE) under high vacuum conditions at 10^{-6} Torr and the temperature range was from 380 to 475 °C. The average molecular weight of the resultant fragments was found equal to 700, which indicates that the chains were broken at random positions. Grassie [22] pointed out that when the temperature is raised the first sign of thermal degradation in polyethylene which occurs around 290 °C is a decrease in the molecular weight.

Two-stage degradation in many polymers has been observed. Luff and White [23] found that the degradation of polyethylene (PE) starts at 300 °C and the polyethylene undergoes a two-stage degradation process. Fig. 1 shows the evaporation rate plotted against the reciprocal of temperature. According to Amelin et al. [24], the low temperature range having an activation energy of ≈ 1 eV characterises the breaking of the weak bonds and removal of side chains from the main molecule. The second stage which reflects the rupturing of primary bonds has an activation energy of 3.04 eV. Luff and White [23] reported the activation process above 350 °C as having an activation energy of 3.68 eV. Two-stage degradation was also observed in polymethylmethacrylate and in polystyrene [24]. Polystyrene when heated in high vacuum at 325 to 375 °C gives rise to the evaporation of the monomer and of some dimer, trimer, etc. This shows that the chains are broken largely at the ends. Madorsky [21] calculated the activation energy at various temperatures, on the basis of degradation, as 2.95 eV for polyethylene and 2.51 eV for polystyrene. Evaporation of PTFE was studied by Lane [25] under high vacuum at 400 °C. It was found that PTFE degrades to leave a residue of carbon in the evaporation source, while on completion of a similar evaporation of PE no residue of PE was found [26]. Thermal degradation of polyvinyl compounds was described by Grassie and Melville [27]. They put forward the three possible extreme degradation mechanisms as random breaking, stepwise degradation, and depolymerization. Recently Kinstle [28] observed the thermal decomposition of {poly(oxycarbonylimino-1,4 phenylene)} a polymer which undergoes a purely decomposition process.

When polypropylene is heated in vacuum, the decomposition rate at 412 °C is 1000 times that at 320 °C. The gas consists of 34 to 52% propylene, 2.6% butenes, and 39 to 41% amylenes and pentenes. Also the non-volatile residue left in the evaporation source becomes amorphous [29]. The molecular weight of the polypropylene decreases to a value characteristic of normal polyethylene in the temperature range

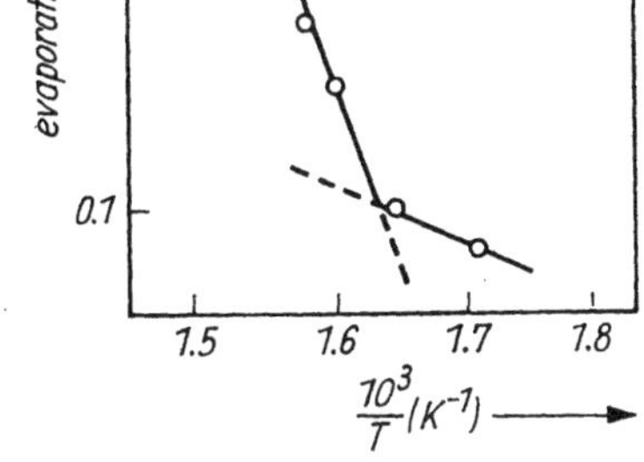

Fig. 1. Evaporation rate as a function of inverse absolute temperature (after Luff and White [23])

of 230 to 300 °C [30]. Decomposition of polypropylene starts by the random breaking of C–C bonds which then leads to the breaking of terminals from macroradicals. Then the hydrogen atoms are removed from the neighbouring molecules. Finally, the chains are broken as shown below.

$$R\text{–}\underset{\displaystyle CH_3}{\underset{|}{CH}}\text{–}CH_2\text{–}\underset{\displaystyle CH_3}{\underset{|}{CH}}\text{–}CH_2\text{–}R' \rightarrow R\text{–}\underset{\displaystyle CH_3}{\underset{|}{CH}}\text{–}\dot{C}H_2 + \underset{\displaystyle CH_3}{\underset{|}{\dot{C}H}}\text{–}CH_2\text{–}R'$$

$$R\text{–}\underset{\displaystyle CH_3}{\underset{|}{CH}}\text{–}\dot{C}H_2 \rightarrow \dot{R} + \underset{\displaystyle CH_3}{\underset{|}{CH}} = CH_2$$

$$R'\text{–}CH_2\text{–}\underset{\displaystyle CH_3}{\underset{|}{\dot{C}H}} \rightarrow \dot{R}' + CH_2 = \underset{\displaystyle CH_3}{\underset{|}{CH}}$$

($\dot{R}$ and $\dot{R}'$ are macroradicals)

2. Electrical Properties

The mechanism of electrical conduction in polymers is quite complicated and there is no satisfactory theory to explain the conduction mechanism in these organic disordered materials, so the available theories for the crystalline, polycrystalline, and amorphous inorganic solids are usually used to describe the conductivity of these materials. The conduction mechanism in polymeric thin films may depend on the thickness and also on the method of producing them. As a consequence different workers have reported different conduction mechanisms for the same nominal material.

2.1 Ohmic conduction

Ohmic conduction in polymer thin films is not very common and most of the polymer films exhibit non-ohmic behaviour at high electric fields. Gregor [31] examined thin films of polydivinylbenzene having thicknesses between 100 and 1000 Å and produced by the glow discharge method. He reported ohmic conduction in 1000 Å films up to the breakdown point which occurred above 10^6 V cm^{-1}. Hogarth and Iqbal [5] obtained their films of polypropylene (PP) by vacuum deposition, having thicknesses between 1200 and 2400 Å. They found systematic differences in the electrical behaviour of these films as a consequence of using different electrode materials, even though the current–voltage characteristics remained ohmic. The current–voltage characteristics of Cu–PP–Cu and Al–PP–Cu samples with three different thicknesses

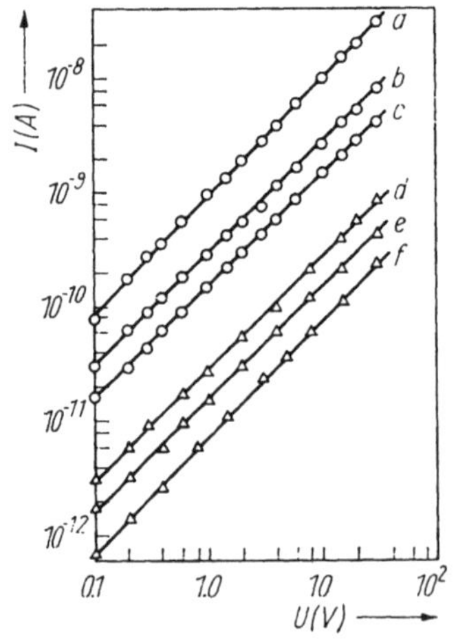

Fig. 2. Current–voltage characteristics of some thin films of propylene. (a) Cu–PP(1200 Å)–Cu, (b) Cu–PP(1800 Å)–Cu, (c) Cu–PP(2400 Å)–Cu, (d) Al–PP(1200 Å)–Cu, (e) Al–PP (1800 Å)–Cu, (f) Al–PP(2400 Å)–Cu (after Hogarth and Iqbal [5])

of polypropylene are shown in Fig. 2. It was found that the polypropylene films were highly resistive with resistivities of the order of 10^{14} Ω cm. The breakdown of the films was not thoroughly investigated; nevertheless the films withstood an applied field of at least 2.48×10^6 V cm^{-1}.

2.2 Super-ohmic and tunnelling characteristics

Super-ohmic and tunnelling effects have been reported as the dominant conduction mechanisms in many thin polymer films. Super-ohmic effects include the conduction by Schottky emission or by the Poole-Frenkel effects. In Schottky emission the electrons are emitted by thermal activation from metal electrodes into the conduction band of the insulator over a potential barrier at the metal/insulator interface. In the Poole-Frenkel effect, the electrons are thermally emitted from randomly distributed trapping centres to the conduction band of the insulator by means of the lowering of the coulombic potential barrier surrounding the trapped electrons, by an external electric field. Thin polymer films prepared from silicone oil and having thicknesses between 50 and 2500 Å were investigated by Christy [2], Emtage and Tantraporn [32], and Mann [33]. Christy obtained these films by electron bombardment and they had thicknesses between 50 and 150 Å. He concluded that the conduction mechanism at 4 K was due to tunnelling, while at room temperature no single mechanism was adequate to describe the conduction process, although the current showed a $\log I \propto U^{1/2}$ dependence on voltage characteristic of either a Schottky or Poole-Frenkel process. Emtage and Tantraporn reported that Schottky emission was the main conduction mechanism in 100 Å films at 300 K formed in the same way, as shown in Fig. 3. For films in the thickness range from 500 to 2500 Å sandwiched between aluminium electrodes, Mann found that in the temperature range from 4 to 300 K the conductivity under high electric fields was related to Schottky emission or tunnelling or to a mixture of both. Gregor and Kaplan [34], who studied the thin films of epichlorohydrin (100 Å thick), produced by electron bombardment, came to the conclusion that the electrical characteristics could be related to a combination of tunnelling and either Poole-Frenkel or Schottky emission, depending upon the temperature range and electric field strength. The electrical properties of thin films of polysilazane having thicknesses between 0.6 and 3.5 μm formed by the glow discharge method were investigated by Tyczkowski et al. [35]. They suggested that electrode-limited (Schottky) conduction was the dominant conduction mechanism. Gregor [31] examined thin films of polydivinylbenzene having thicknesses between 100 and 1000 Å, produced by the glow discharge method, and concluded that Schottky emission was the dominant conduction mechanism in 100 to 500 Å thick films as shown in Fig. 4. Lengyel [36] investigated films of polyethylene terephthalate (Mylar) and polyvinylformal and reported the conduction mechanism as Schottky emission for fields between 20 and 200 kV cm^{-1} and temperatures from 298 to 373 K. Lilly

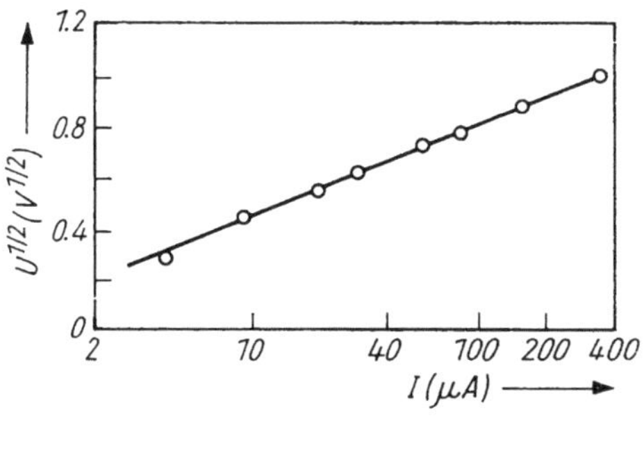

Fig. 3. Current–voltage characteristics of an Au–silicone oil (PSO)–Au sample at 300 K (after Emtage and Tantraporn [32])

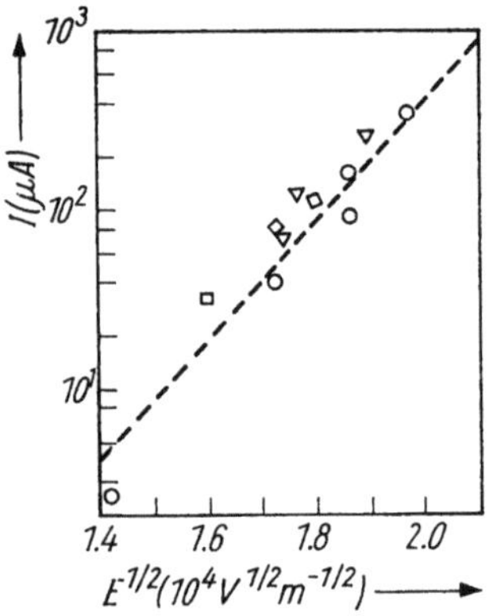

Fig. 4

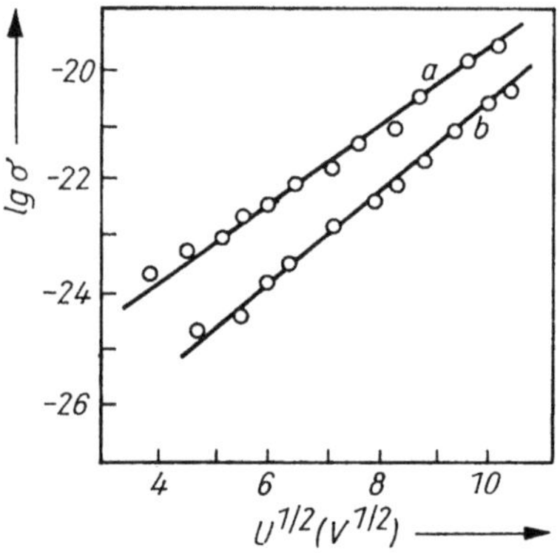

Fig. 5

Fig. 4. Current as a function of electric field for a Pb–polydivinylbenzene–Pb structure at 300 K (after Gregor [31])

Fig. 5. Current–voltage characteristics of P-terphenyl layers at 298 K (a) 2.6, (b) 3.5 µm thick (after Staryga and Swiatek [41])

and Mcdowell [37] worked on Mylar and Teflon (polytetrafluorethylene) films. They also interpreted their results using the theory of Schottky emission. According to Lilly and Lowitz [38] the conduction mechanism in 25 to 125 µm Mylar films was governed by the O'Dwyer [39] theory of space-charge conduction. The space charge extended throughout the thickness of the sample rather than being localised near the cathode. Kamisako et al. [40] measured the steady-state conductivity of polystyrene films and the conduction mechanism was explained in terms of a space charge formed by the injection of electrons into the film over the Schottky barrier and was found

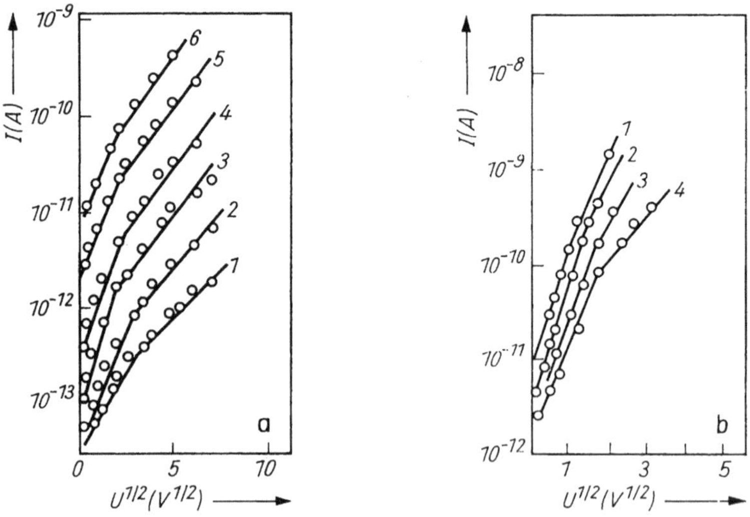

Fig. 6. Schottky plot for a) a polystyrene film of thickness 314 Å at $T = 293$ (1), 312 (2), 326 (3), 336 (4), 356 (5), 373 K (6) and b) a PVK film of thickness 740 Å at $T = 334$ (1), 324 (2), 308 (3), 296 K (4) (after Kryszewski and Swiatek [42])

diffusing towards the anode. They also made a rough estimate of the density of traps and its temperature dependence. The electrical conductivity of thin polycrystalline P-terphenyl films was investigated as a function of temperature, electric field, and film thickness by Staryga and Swiatek [41]. They interpreted their data in terms of a hopping process between localized sites over potential barriers lowered by the electric field according to the Poole-Frenkel formula as shown in Fig. 5. Kryszewski and Swiatek [42] investigated the electrical properties of thin polystyrene (PS) and poly-(N-vinyl carbazole) (PVK) films, obtained by the glow discharge method. The conductivity was described by both electrode-limited and bulk-limited conduction processes. At lower field strengths, $E \leqq 10^5\ \text{V cm}^{-1}$, the electrode-limited conduction mechanism was dominant, while at higher fields ($E \geqq 10^5\ \text{V cm}^{-1}$) the conductivity was related to the volume properties of the polymer, i.e. it was dominated by the Poole-Frenkel effect. Fig. 6 a and b show the current–voltage characteristics of PS and PVK films at different temperatures, respectively.

Electrical conduction through insulating layers of polyethylene has been investigated by many research workers during the last few years. The current–voltage behaviour of thermally evaporated polyethylene films sandwiched between aluminium electrodes was explained in terms of the anomalous Poole-Frenkel effect by Zor [43]. Miyoshi and Chino [44] studied polyethylene films formed by vacuum deposition and also by dipping the substrate into solution. They discussed their results by using the simple theories of tunnelling (in 50 Å films) and Schottky emission (in 500 to 1200 Å films). Rogger [45] observed the current–voltage characteristics of 100 Å thick polyethylene films and found tunnelling as the conduction mechanism. Gupta and Barbarez [46] investigated the conduction process in polyethylene films and concluded that the carriers were introduced into the conduction band of the polymer via defects or impurity levels and the subsequent motion through the insulator was by means of hopping. Suzuki et al. [47] obtained their films of polyacrylonitrile by vacuum deposition. After subjecting the films to heat treatment in nitrogen they suggested that the conduction should be understood in terms of the hopping transport of charge carriers in the localized states present in the forbidden band due to the random networks of conjugate double bond systems of carbon-to-carbon and carbon-to-nitrogen.

2.3 Space-charge-limited conduction

If the electrode makes ohmic contacts to the insulator, the electrons travel from the metal into the conduction band of the insulator with a velocity determined by their mobility in the electric field, and from a space charge similar to that of a vacuum diode, then the conduction process is known as space-charge-limited.

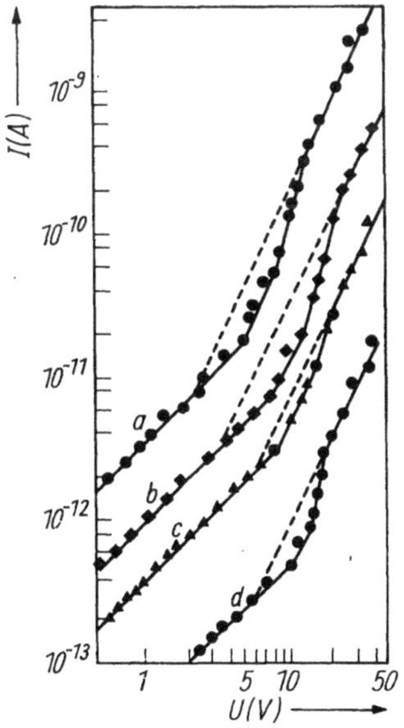

Fig. 7. Current–voltage characteristics of some films of polyferrocene Al–PF–Al. Thickness of PF film (a) 210, (b) 600, (c) 900, (d) 3000 Å (after Phadke et al. [48])

Table 1

Selected results on evaporated polymer films

polymer	method of obtaining films	thicknesses (Å)	temperature (K)	observed conduction mechanism	ref.
silicone oil	electron bombardment	50 to 150	4 to 300	tunnelling at 4 K; lg I vs. $U^{1/2}$ dependence at 300 K. No single mechanism was dominant	[2]
silicone oil	electron bombardment	100	300	Schottky emission	[32]
silicone oil	electron bombardment	500 to 2500	4 to 300	Schottky emission or tunnelling or a mixture of both	[33]
polyepichlorohydrin	electron bombardment	100	160 to 300	tunnelling at low temperature (160 K) and low voltages. Schottky emission or Poole-Frenkel effect at 300 K and high voltages.	[34]
polydivinylbenzene	glow discharge	100 to 1000	300	Schottky emission in 100 to 500 Å films, while ohmic conduction in 1000 Å films up to the breakdown point, which occurred at 10^6 V cm^{-1}.	[31]
polyethylene-terephthalate (Mylar)	commercially supplied samples were used	27.50×10^4	298 to 383	Schottky emission for fields between 20 and 200 kV cm^{-1}	[36]
polyvinylformal	commercially supplied samples were used	$(50 \text{ to } 55) \times 10^4$	298 to 383	Schottky emission for fields between 20 and 200 kV cm^{-1}	[36]
Mylar and Teflon (polytetrafluorethylene)	films were cut from sheets of Mylar and Teflon	$(25 \text{ to } 250) \times 10^4$	343 to 436	Schottky emission	[37]

polyethylene	—	100	298	tunnelling	[45]
Mylar	—	(25 to 125) × 10^4	300 to 425	Dwyer theory of space-charge conduction	[38]
polyethylene	films were supplied by Monsanto	(25 to 75) × 10^4	313	Poole-Frenkel and hopping	[46]
polyacrylonitrile	vacuum deposition	100	160 to 500	hopping through localized states in the forbidden gap	[47]
polycrystalline P-terphenyl	vacuum deposition	(1.7 to 3.5) × 10^4	298	Poole-Frenkel and hopping. Also, space-charge-limited conduction above 300 K	[41]
polyerocene	plasma polymerization	210 to 3000	298	space-charge-limited currents	[48]
polymethyl-methacrylate	isothermal immersion	9646	298 to 333	space-charge-limited currents	[49]
polypropylene	vacuum deposition	1200 to 2400	300	ohmic	[5]
polystyrene	styroflex films were used	(33 to 40) × 10^4	300 to 375	Schottky emission	[40]
polystyrene and poly-N-vinylcarbazole	glow discharge	200 to 800	293 to 373	for $E \leqq 10^5$ V cm^{-1}, electrode-limited	[42]
polysilazane	glow discharge	(0.6 to 3.5) × 10^4	298	Schottky emission	[35]
PTFE	vacuum deposition	400 to 1650	298	no single conduction mechanism	[25]
polyethylene	vacuum deposition	1000 to 2000	298	Schottky emission in PE single crystal of 500 to 1200 Å thickness	[44]
	solution deposition	50 to 1000	200 to 400	tunnelling in 50 Å	
polyethylene	vacuum deposition	1100 to 4200	298	space-charge-limited currents with copper electrodes. Poole-Frenkel effect with aluminium electrodes	[43]

Space-charge-limited currents provide the dominant conduction mechanism in some polymer thin films as has been reported by several workers. Phadke et al. [48] obtained their films of polyerocene sandwiched between Al electrodes and having thicknesses between 210 and 300 Å by a plasma polymerization method. They concluded that the conduction mechanism was governed by space-charge-limited currents as shown in Fig. 7. An investigation was undertaken by Chutia and Barua [49] and the conduction mechanism in polymethyl-methacrylate (PMMA) films was found to be due to space-charge-limited currents. Zor [43] also found the space-charge-limited currents as the dominant conduction mechanism in evaporated polyethylene films, having thicknesses in the range from 1200 to 4200 Å and sandwiched between Cu and Al electrodes. Some influence of space-charge-limited currents above 300 K was found in the current–voltage characteristics of vacuum-deposited polycrystalline P-terphenyl films by Staryga and Swiatek [41]. Some electrical properties of the polymers discussed above are presented in Table 1.

2.4 Electroforming

Electroforming, in which a large reduction in the apparent resistance of the material is observed when a threshold voltage is applied, may occur in the thin layers of many polymeric materials. Associated with the electroforming process are a voltage-controlled negative resistance (VCNR), electroluminescence, electron emission, and a memory effect. Several theories [50 to 57] have been proposed to explain electroforming and associated effects in insulators. Mann [33] worked on thin films of silicone oil prepared by the electron bombardment method and having thicknesses between 500 and 2500 Å sandwiched between Al electrodes, and observed differential negative resistance similar to that reported by Hickmott [50] in some of his samples. The current rose exponentially to its maximum value at about 5 V and was found independent of the film thickness. While applying a decreasing voltage, a similar negative resistance effect was noticed but always with a smaller maximum current. The memory effect was also reported up to 3 V and this was stable and independent of temperature. For voltages greater than 6 V and up to 50 V, tunnelling was observed in these devices as shown in Fig. 8. Gundlach and Kadlec [58] investigated the monomolecular multi-layers of cadmium arachidate sandwiched between Al and Au electrodes. The monolayers were deposited onto previously evaporated aluminium electrodes, by dipping the substrate several times into an aqueous solution of Cd ions and with nonomolecular layers of arachidic acid spread on the surface of the water. Each monolayer had a thickness of 26 Å. To complete the sandwich, top electrodes of gold or aluminium were evaporated. For the electrical measurements the sample

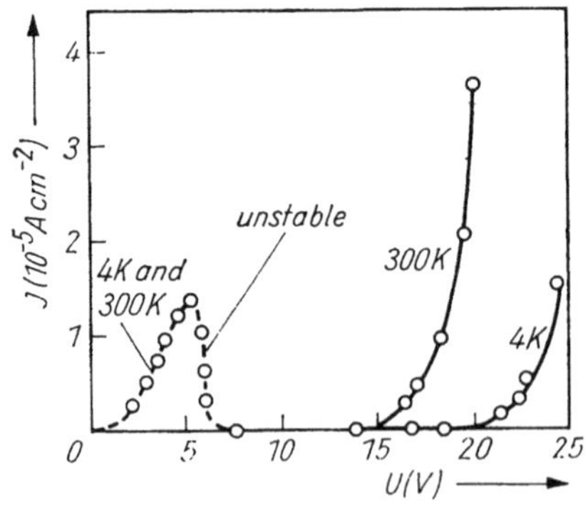

Fig. 8. Differential negative resistance and tunnelling in metal–Dow Corning DC 704–metal structures (after Mann [33])

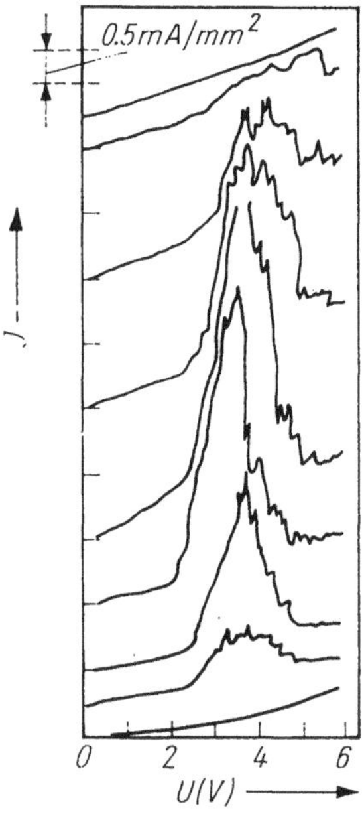

Fig. 9. Current–voltage characteristics of an Al–Al oxide–eleven monolayer–Au sandwich at varisou temperatures. (The vertical scale is displaced for each curve for clarity.) (After Gundlach and Kadlec [58])

was cooled down to 77 K and the top electrode was always positively biased. A sawtooth voltage with an increasing rate of 0.3 V s^{-1} was used. The device was electroformed by applying a bias voltage of 5 to 6 V or more. Now the sandwich was warmed at a rate of 5 K min^{-1} and the first sign of the differential negative resistance (DNR) was observed at about 150 K. The DNR was most pronounced at about 190 K and it disappeared above 210 K. The current–voltage characteristics of an Al–(Al oxide)–eleven monolayer–Al sandwich with increasing temperature are shown in Fig. 9. It was not possible to form the devices having thicknesses less than five monolayers. On decreasing the voltage within 0.02 s the device was left in a memory state which was erased at a threshold voltage of between 2 and 3 V. The electroforming of evaporated polyethylene has been reported by many research workers. Gazso [59] observed the DNR in the evaporated polyethylene films, having thicknesses in the range of 1000 to 2000 Å and sandwiched between aluminium and gold electrodes. The conductivity increased due to the in-diffusion of gold from the top electrode under the influence of the applied electric field. This was supported by the fact that the sandwiches prepared only with aluminium electrodes did not show any electroforming. Miyoshi and Chino [44] also observed the negative resistance similar to Hickmott [50] in their samples of evaporated polyethylene having the same thicknesses as reported by Gazso. Hogarth and Zor [4] evaporated polyethylene at 10^{-4} Torr onto previously deposited aluminium electrodes. Their films were 1140 Å thick. In order to complete the MIM structure, aluminium was evaporated as top electrodes. Unlike Gazso [59] in these films of polyethylene sandwiched between Al electrodes, the electrical process was influenced by the injection of metal from the electrodes and irreversible forming and related effects were also observed. Fig. 10 shows the effect of time when an Al–PE(1140 Å)–Al sample was left under 16 V (minimum forming voltage) for an extended period and significant forming was observed after 313 h. At the end

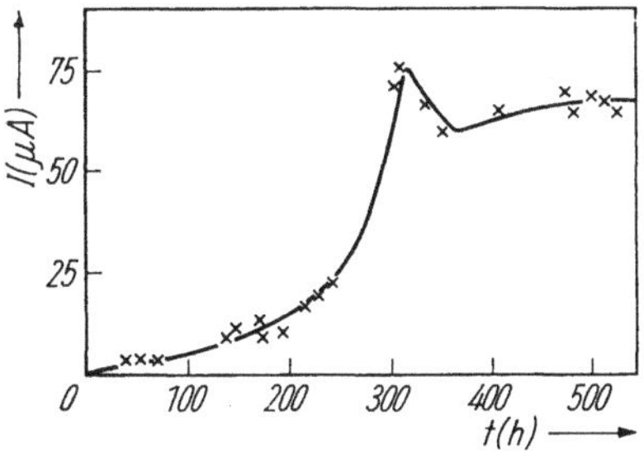

Fig. 10. Current at constant supply voltage as a function of time for an Al–PE (1140 Å)–Al structure (after Hogarth and Zor [4])

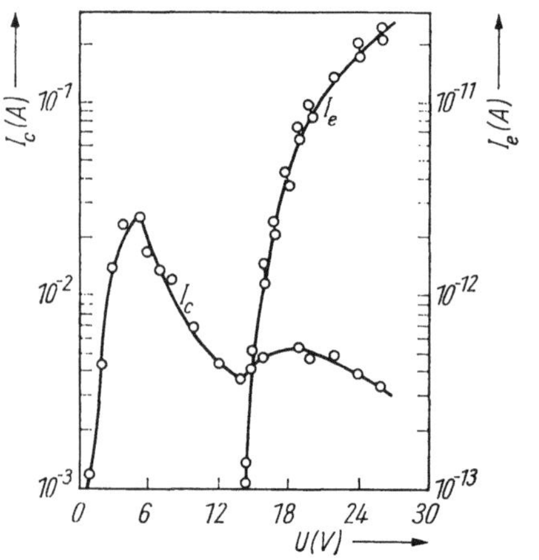

Fig. 11. Circulation (I_c) and emission (I_e) currents as functions of applied voltage for a Cu–PP (600 Å)–Cu sample at room temperature and $\approx 10^{-5}$ Torr (after Hogarth and Iqbal [60])

of this time on increasing the applied voltage up to 25 V, no further forming was observed. Thin layers of evaporated polypropylene sandwiched between metal electrodes (aluminium or copper) having thickness ranging from 600 to 1000 Å showed electroforming effects followed by negative resistance and electron emission into a vacuum under suitable conditions as described by Hogarth and Iqbal [60]. In their measurements the forming voltage varied from 12 to 15 V and the peak current in the formed specimens occurred at between 5 and 7 V. The results were found consistent with a theory based on the generation of conducting filaments during the forming process. Fig. 11 shows the current–voltage characteristics of a Cu–PP(600 Å)–Cu sample after electroforming at room temperature and under a pressure of $\approx 10^{-5}$ Torr. The emission current as shown in Fig. 11 was collected by a subsidiary anode biased to 100 V and positioned about 1.5 cm from the surface of the sample. Unlike the behaviour of many reported MIM samples, the rapid rise of emitted current occurred not at a voltage corresponding to a peak current but at a voltage corresponding closely to the initial forming voltage 14.6 V. The behaviour in polyethylene reported by Hogarth and Zor [4] was quite different.

2.5 Hopping and ac conduction

Hung and Gliessman [61] first observed the impurity conduction in their low-temperature investigation of germanium. Since then, a number of papers [62 to 65] have been published on the same kind of conduction resulting from the application of steady fields. In order to explain the conduction, they all used the model based on the impurity theory of Conwell [62] and Mott [66]. According to this model, which is valid for low-temperature impurity conduction, transport occurs by electrons hopping between states which are essentially localized around acceptor or donor impurities, by exchanging energy with phonons. Conduction may also be explained in terms of tunnelling through the energy barrier from the occupied donor to the unoccupied one. For such transport to take place, it is necessary for some of the localized states to be vacant and hence compensation of the majority impurity becomes an essential feature of impurity conduction, as demonstrated by Fritzsche [67]. Pollack and Geballe [19] have presented a theory relating the hopping conductivity with ac frequency. Their theory is also based upon the original model of impurity conduction.

We have made some ac measurements with evaporated polypropylene sandwiched between copper electrodes. The samples for the ac measurements were prepared in

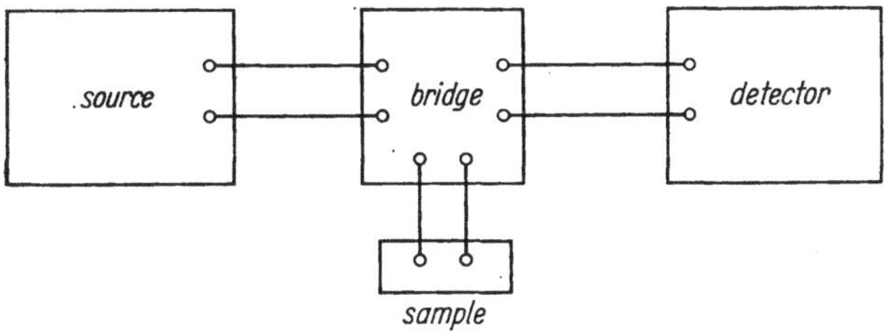

Fig. 12. Schematic circuit for ac measurements

the same way as described by Hogarth and Iqbal [5]. The measurements were carried out under a pressure of 10^{-5} Torr using the frequency range between 10^3 and $2 \times \times 10^4$ Hz at room temperature, unless stated otherwise. An audio-frequency voltage generator was used to apply a signal of 5 V across the sample and a waveform analyser and a universal bridge were used for the detector. A block diagram of the measurement system is shown in Fig. 12. Measurements of capacitance and conductance were carried out before and after the electroforming of the samples. An expected consequence of the Dearnaley filamentary model [68] used to explain the electroforming phenomena in our samples, would be some significant differences in the frequency dependence of capacitance and conductance after forming, compared with the behaviour prior to forming. The hopping conduction at low and room temperatures can be identified by the ac conductivity measurements. Pollack and Geballe [19] have given an equation for this process,

$$\sigma(\omega) = A\omega^s ,$$

where A is a complex constant, ω the angular frequency, $\sigma(\omega)$ the ac conductivity, and $0.5 < s < 1$. For the hopping conduction s is generally found to be equal to 0.8. In the radio-frequency range of the applied frequency, the square-law dependence of ac conductivity upon frequency has been explained in terms of a two-centre hopping process [69]. Fig. 13 a and b show the plot of ac conductance versus frequency for evaporated polypropylene before electroforming at room temperature and at 241 K, respectively. The slopes of the straight lines fall within the limit $0.5 < s < 1$ and are equal to 0.85 in both cases; these results characterize the process of hopping conduction as a dominant process at both low and room temperatures. Fig. 14a shows the variation of capacitance with frequency before electroforming for a Cu–PP(1800 Å)–Cu sample at an ambient pressure less than 10^{-5} Torr and at room temperature. Fig. 14b demonstrates the variation of capacitance with frequency at low temperature (241 K) for the same sample. Before the electroforming a linear decrease of capacitance with increasing frequency is clear from Fig. 14. This could be due to the increasing leakage

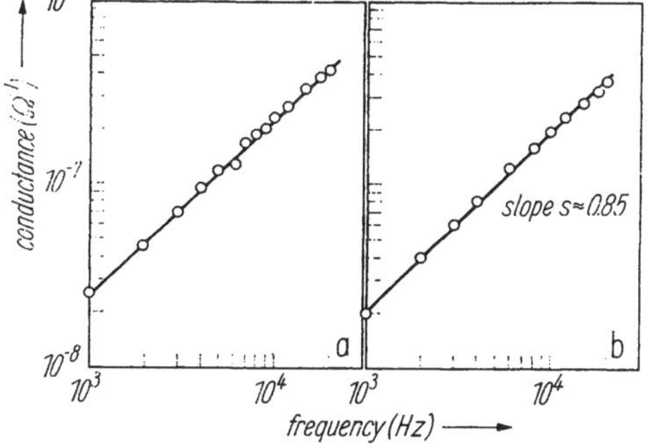

Fig. 13. Frequency variation of ac conductance for Cu–PP(1800 Å)–Cu sample at a) room temperature and b) 241 K and 10^{-5} Torr

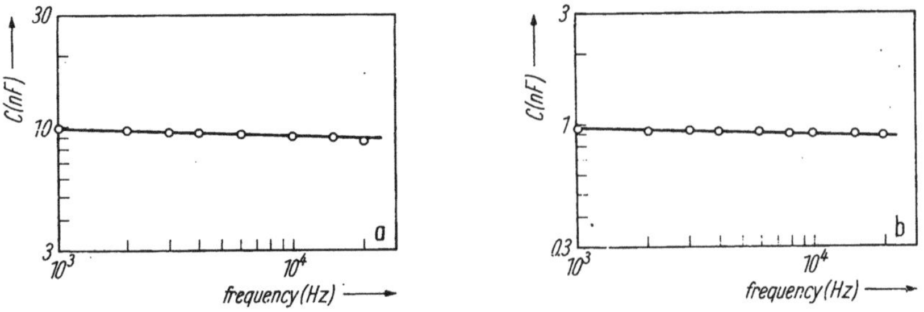

Fig. 14. Variation of capacitance with frequency for a Cu–PP (1800 Å)–Cu sample at a) room temperature and b) 241 K and $\approx 10^{-5}$ Torr

current with frequency and is associated with capacitance reduction. After electroforming a Cu–PP(700 Å)–Cu device an enormous increase in the conductance was noticed. Further the conductance was found to be proportional to the degree of forming and this is consistent with the Dearnaley theory based on the generation of essentially metallic conducting filaments during the forming process. The capacitance of the sample was decreased after electroforming. It was not possible to balance the bridge and to measure the values of conductance and capacitance simultaneously after the electroforming, a result which is explained by the incidence of a high proportion of metallic conduction following the electroforming.

3. Thermally Stimulated Currents

The measurements of thermally stimulated currents (TSC) have been widely used to determine the properties of traps, i.e. the density and energy distribution of charge carrier trapping centres in solids as reported by Bube [70]. Recently Zielinski and Samoć [75] have used this technique to investigate the Poole-Frenkel effect in their samples. In polymers, trapping may take place due to the structural disorder resulting from the unsaturated ions in chains, chain terminations, chain folding, and from the amorphous/crystalline interface [14]. The study of thermally stimulated currents in irradiated polyethylene and in Teflon revealed that in Teflon the $-(CF{=}CF_2)$, $-(CF{=}CF)-$, and $-(CF{-}CF)-$ groups act as traps [71, 72]. The foreign elements present in the polymers are responsible for the deeper traps, while the shallow traps are due to the defects in the chain structure, e.g. the presence of –OH and $\rangle$CO groups or $-C{=}C-$ bonds [73].

3.1 Experimental techniques

In this technique the samples are kept overnight under vacuum in a dark chamber at room temperature with their electrodes short-circuited as described by Slowik [16]. The samples are then cooled rapidly to liquid nitrogen temperature while still in the dark and are illuminated through the top electrode with light of the appropriate wavelength to fill the trapping centres. Then a constant external field is applied and the samples are heated in the dark at a constant heating rate. The current flowing in the external circuit is recorded as a function of the time, which is proportional to the temperature. The emptying of traps when a particular temperature is reached leads to an increase in the current in the form of a peak over and above the normal background current. For the fixed heating rate, Pai [74] reported that the peak temperature shifted to lower values, as the applied field increased. But, when the field was fixed and the heating rate was varied, the peak shifted to higher temperature as the heating rate increased.

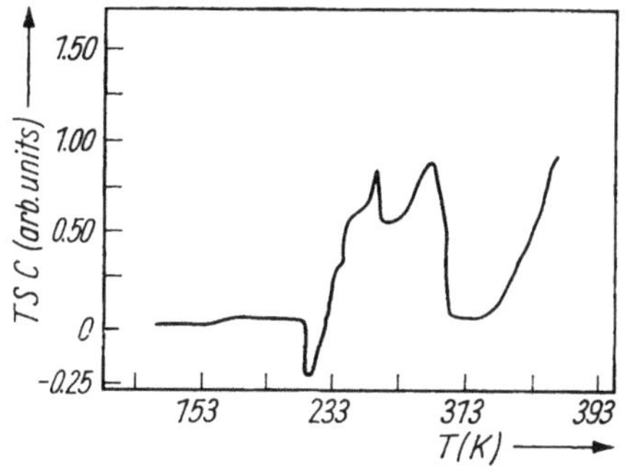

Fig. 15. TSC spectrum of a P-terphenyl layer 2.6 μm thick (after Staryga and Swiatek [41])

3.2 Analysis of TSC curves

The energies of the trap levels may be determined from the experimental data in a number of ways. Bube [70] calculated the trap depth E_t from the curves of the thermally stimulated current by using the conductivity and the peak temperature (T_m). His calculation was based on the assumption that the Fermi level is located at the level corresponding to the traps, when the conductivity attains its maximum value and his expression for E_t includes re-trapping as well. So, according to Bube the mean trap depth is given as

$$E_t = kT_m \ln \frac{N_c e\mu}{\sigma_m}, \tag{1}$$

where E_t is the mean trap depth, N_c the equivalent density of states at the bottom of the conduction band, σ_m the peak conductivity, μ the mobility, and e refers to the electronic charge.

Many papers have been published giving different treatments for the curves of thermally stimulated currents [74 to 82]. Recently Staryga and Swiatek [41] have calculated the trap depth in P-terphenyl layers from the very complicated TSC spectrum as shown in Fig. 15 by using the very simple relation for E_t,

$$E_t = nkT_m, \tag{2}$$

where E_t is trap depth, k the Boltzmann constant, T_m the peak temperature of the TSC, and n a constant which was taken equal to 22 for P-terphenyl films. They calculated the value of n by comparing equation (2) with the relation proposed by Grossweiner [77] for a well-isolated TSC peak. His equation is

$$E = 1.51k \frac{T^* T'}{T^* - T'}. \tag{3}$$

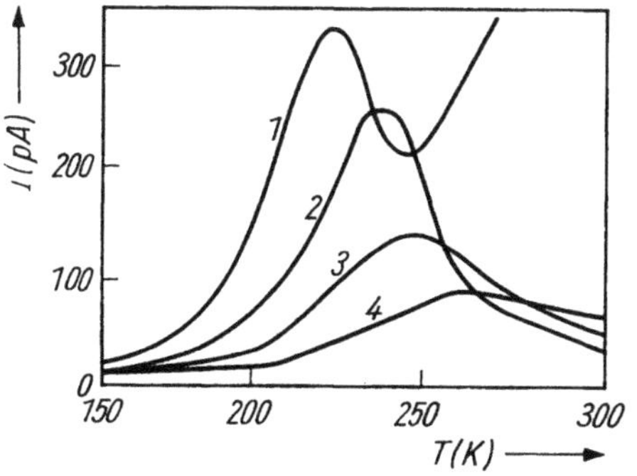

Fig. 16. TSC curves for PVK for heating rate fixed at 0.182 K s^{-1} but for different values of applied field: 18.4 × 10^4 (1), 9.18 × 10^4 (2), 4.23 × 10^4 (3), 1.69 × 10^4 V cm^{-1} (4) (after Slowik [16])

In (3) E is the trap depth, k the Boltzmann constant, T^* the peak temperature, and T' the temperature at which the low-temperature side of the TSC peak attains one-half of its maximum value. Staryga and Swiatek neglected the heating rate dependence of T_m due to the uncertain value of the capture cross-section and suggested that equation (2) can be used to determine the relative trap distribution when the TSC spectra are complicated and isolated TSC peaks cannot be obtained as given in Fig. 15. Slowik [16] studied the thermally stimulated currents in poly-N-vinylcarbazole (PVK). Fig. 16 shows a series of current–temperature curves for poly-N-vinylcarbazole at a constant heating rate but for different applied electric fields. Iqbal and Hogarth [15] have analysed the thermally stimulated conductivity curves of evaporated polypropylene by using the expression given by Haine and Carley-Read [83],

$$\exp\left(\frac{-E_t}{kT}\right) = \frac{n_{c_2}X - n_{c_1}}{N_c(1 - X)}. \tag{4}$$

X is defined in the following equation:

$$X^2 = \frac{\dot{T}_1(1 - n'_{c_1}kT^2/n_{c_1}E_t)}{\dot{T}_2(1 - n'_{c_2}kT^2/n_{c_2}E_t)}, \tag{5}$$

$$N_t(\tau_t + \tau) = \frac{\{n_c + N_c \exp(-E_t/kT)\}^2}{\dot{T}N_c\{E_t/kT^2 - n'_c/n_c\} \exp(-E_t/kT)}. \tag{6}$$

The symbols are defined as follows. n_c is the density of electrons in the conduction band, E_t the trap depth, τ_t is the electron trapping time, τ the recombination lifetime, N_t the density of trapping centres for the particular trapping level, $\dot{T}_1$ and $\dot{T}_2$ are two heating rates, T is the absolute temperature, k the Boltzmann constant, n_{c_1} refers to the carrier density at heating rate $\dot{T}_1$ and similarly for n_{c_2}; n'_{c_1} is defined as dn_{c_1}/dT and similarly for n'_{c_2}. The trap depth calculated from equation (4) was found consistent with the trap depth obtained from the Bube equation and its value was found equal to 0.33 eV. Fig. 17a shows a series of current–temperature curves for evaporated polypropylene at a fixed applied field but at different heating rates, while Fig. 17b shows the dark current measured on the same sample and demonstrates the form of the TSC peak.

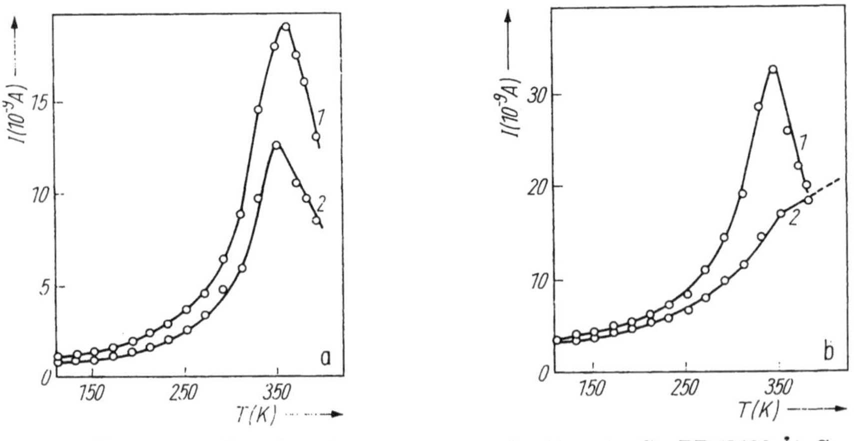

Fig. 17. Current as a function of temperature on heating a) a Cu–PP (2400 Å)–Cu sandwich structure at an applied field of 1.6×10^5 V cm^{-1}, (1) heating rate 0.217, (2) 0.095 K s^{-1}; b) a Cu–PP 1800 Å)–Cu structure after illumination (1) and in the dark (2) (after Iqbal and Hogarth [15])

4. Some Recent Work on the Structure of Vacuum-Evaporated Polypropylene Films

In our previous work we have exmined the properties of evaporated polypropylene for possible use as thin dielectric films and were able to produce standard and reproducible results of electrical measurements [5]. In this section we look at the effect of vacuum heat treatment on the structure and on the relevant electrical properties of these films.

4.1 Experimental techniques

Polypropylene was evaporated in a vacuum of about 10^{-5} Torr from a stainless steel boat on to electron microscope specimen support grids. Other specimens prepared at the same time were deposited on slides with copper electrodes as described previously [5]. Annealing treatments were carried out in a stainless steel container in a subsidiary vacuum system. The electrical measurements were of the kind described previously by the authors. Structural measurements were made using a JEM 7 electron microscope both for microscopy and for electron diffraction.

4.2 Annealing

There is always the possibility of producing structural defects during the deposition of the films, unless a great deal of special care is taken. Some of these defects may be annealed out of the film by heat treatment. As a general rule the resistivity of evaporated films, particularly metallic films, decreases with annealing but for some materials heat treatment may lead to an increase in film resistivity. This increase may be due to the effects of oxidation and/or agglomeration [84]. In some cases the resistance changes are caused by a recombination of the interstitial atoms and lattice vacancies as a result of the heat treatment [85]. Agglomeration is associated with the reverting of a continuous structure of a very thin film to an island structure when subjected to the vacuum heat treatment. This process is detected by the large increase in the resistance of the film. Kane et al. [86] have reported three-stage agglomeration due to vacuum annealing for the gold film deposited on a zinc sulphide substrate. Initially, small holes are formed in the film, then they are enlarged, and finally the continuous film separates into islands. Any change in the resistance of the film occurs during the second and the third stages. They also mentioned the migration of gold atoms which is necessary for agglomeration, as it is a thermally activated process. Skofronick and Phillips [87] studied the additional agglomeration in the films which were

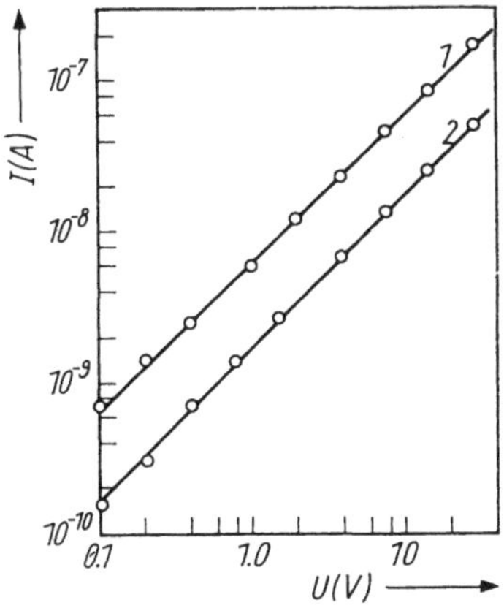

Fig. 18. Current–voltage characteristics before (1) and after (2) annealing of a Cu–PP(900 Å)–Cu thin film sample

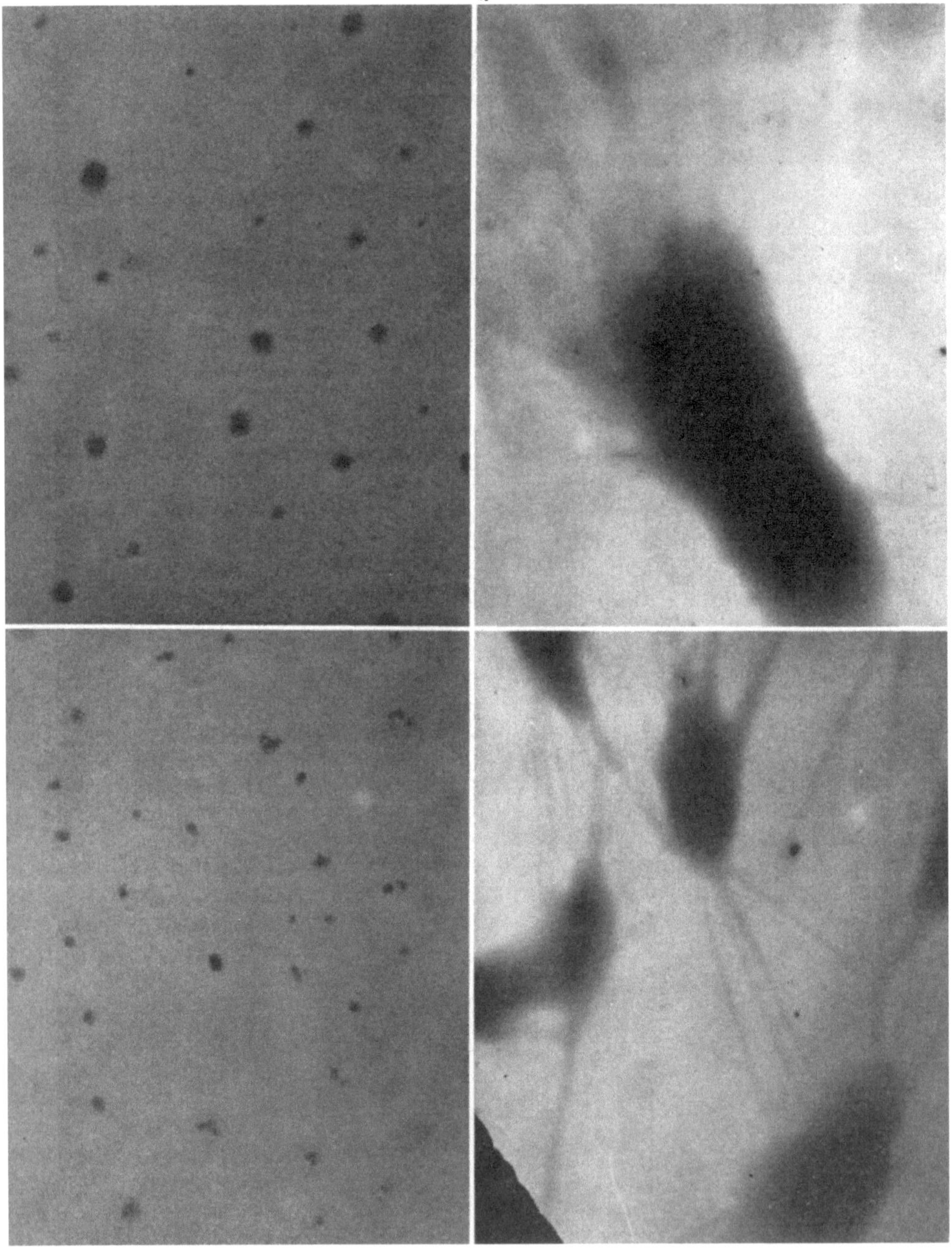

Fig. 19. PP films before (left) and after (right) heat treatment. a) 150 Å (6000 ×), b) 600 Å (1320 ×), c) 1000 Å (1320 ×), d) 1600 Å (1320 ×) thick

c d

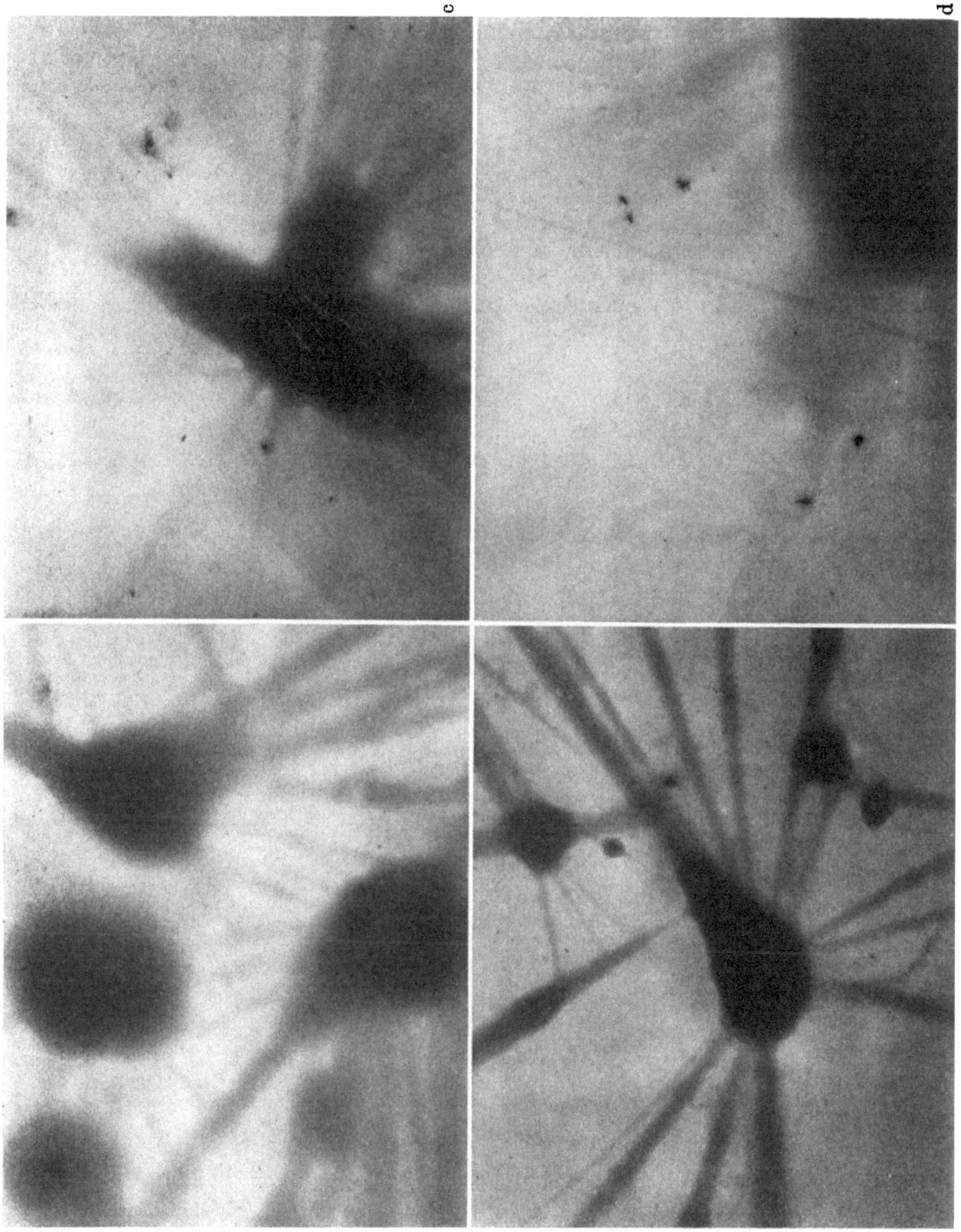

discontinuous when first deposited. Initially, on annealing, a slight decrease in the resistance of the film was observed, but then it increased irreversibly by several orders of magnitude. They explained their results by suggesting that the heat treatment caused the motion of the small islands which then coalesced with the large ones and the mean spacing between them increased. This explains the high value of the resistance. Thin films when heated in air can also undergo an increase in resistance because of oxidation [88].

The current–voltage characteristic of a freshly deposited Cu–PP(900 Å)–Cu sample at a pressure of less than 10^{-5} Torr and at room temperature is shown in curve 1 of Fig. 18. After a vacuum heat treatment at 110 °C for about 30 min and allowing the specimen to cool to room temperature the characteristic was again measured as shown in curve 2. The sample resistivity was found to increase from 3.4×10^{12} to $1.2 \times \times 10^{13}$ Ω cm. Such changes are typical of all the specimens examined. As mentioned above such an increase in the film resistivity is believed to be due to agglomeration.

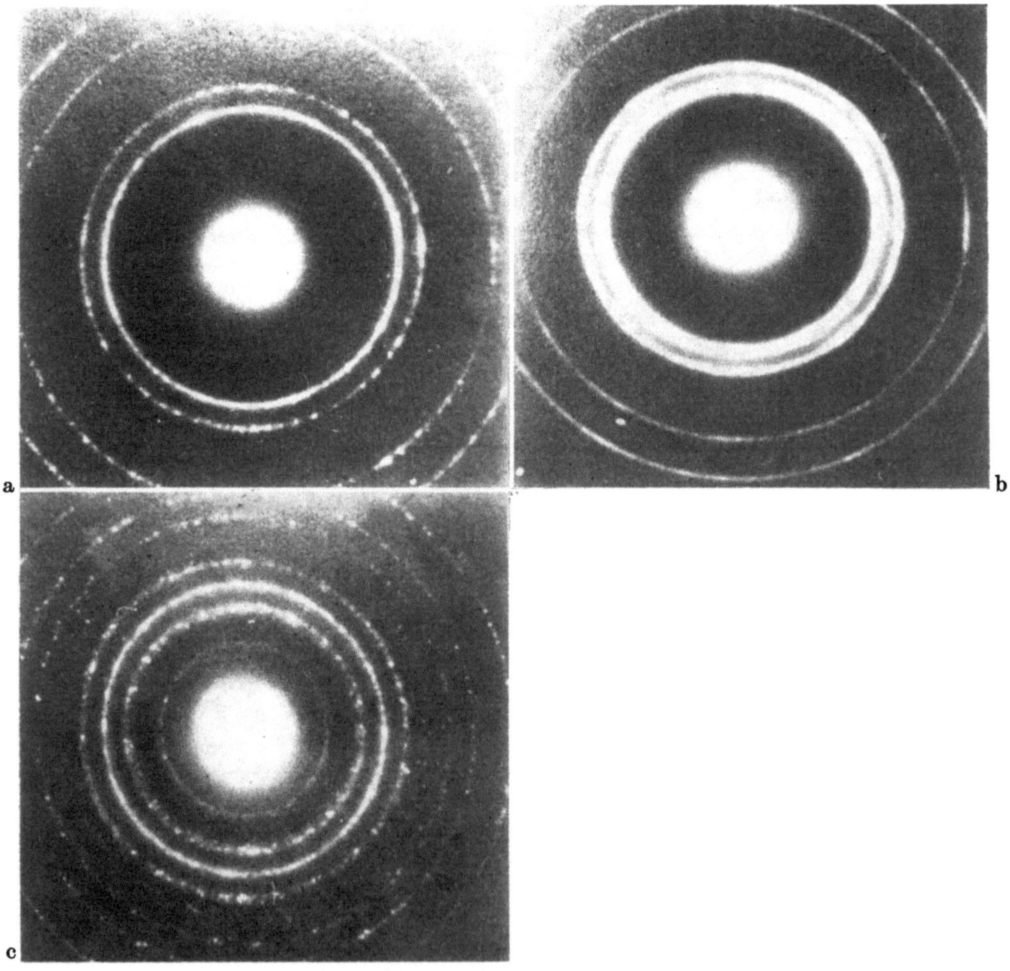

Fig. 20. Selected area electron diffraction patterns of a rhodium film a) without heating and heated in air at b) 300 °C and c) 350 °C (after Koshy [89])

4.3 Microscopy

The resistance measurements can be correlated with electron micrographs to establish the correspondence between agglomeration and resistance increase. A series of layers of polypropylene from 150 to 1600 Å thick were examined in the electron microscope before and after a vacuum heat treatment. By considering the micrographs in Fig. 19a there are two main effects of the heat treatment, an increase of particle size and an increase in the distance between particles. These effects occur as a result of particle migration and agglomeration. If electrical conduction occurs by electron hopping between particles, then an increased spacing would lead to a reduced electrical conductivity [87]. The original chain structure of the polymer is broken during the heat treatment and the atomic clusters are re-arranged. Similar changes were found in layers of thickness 600, 1000, and 1600 Å. Many links between small aggregated particles form during the deposition of the film as may be seen in Fig. 19 b to d (left).

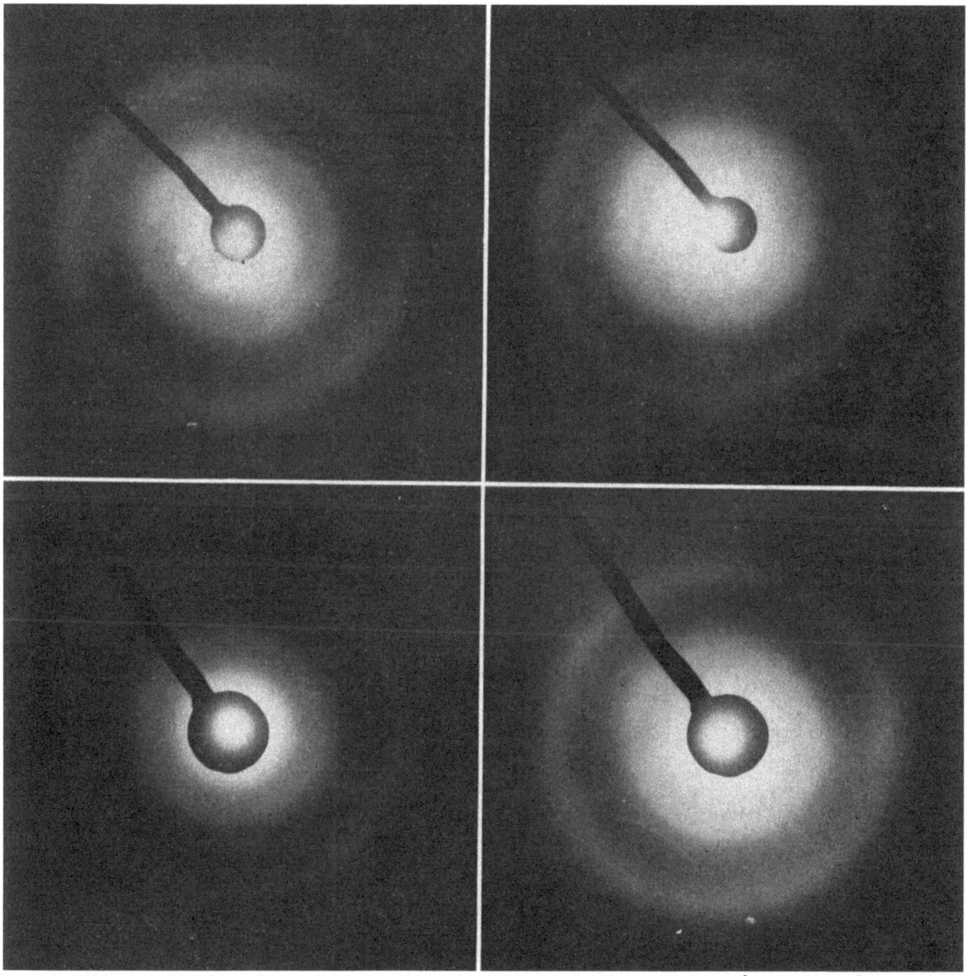
a

b

Fig. 21. Electron diffraction patterns of polypropylene films a) 150 Å, b) 600 Å, c) 1000 Å, d) 1600 Å thick before (left) and after (right) heat treatment to 110 °C approaching the depolymerization temperature

These may provide diffusion paths for the copper introduced from the electrodes under the influence of the applied field as suggested previously [5]. After heat treatments the micrographs Fig. 19 b to d (right) show that the links are reduced in number so that a further reason for the increase in resistivity is given.

4.4 Structure

Apart from the change in resistivity, heat treatment may also bring some microstructural changes in thin films. Koshy [89] has studied the effect of heat treatment on the structure of thin rhodium films in vacuum and in air at different temperatures. Some of his results are shown in Fig. 20. The transmission electron diffraction patterns of evaporated polypropylene shown in Fig. 21 a to d indicate that the samples are essentially amorphous although the diffuse patterns could be in part due to the presence of micro-crystallites. Although the full structural investigation would require very elaborate instrumentation in order to determine the radial distribution function, nevertheless some estimate of changes in the crystalline character of the layers can

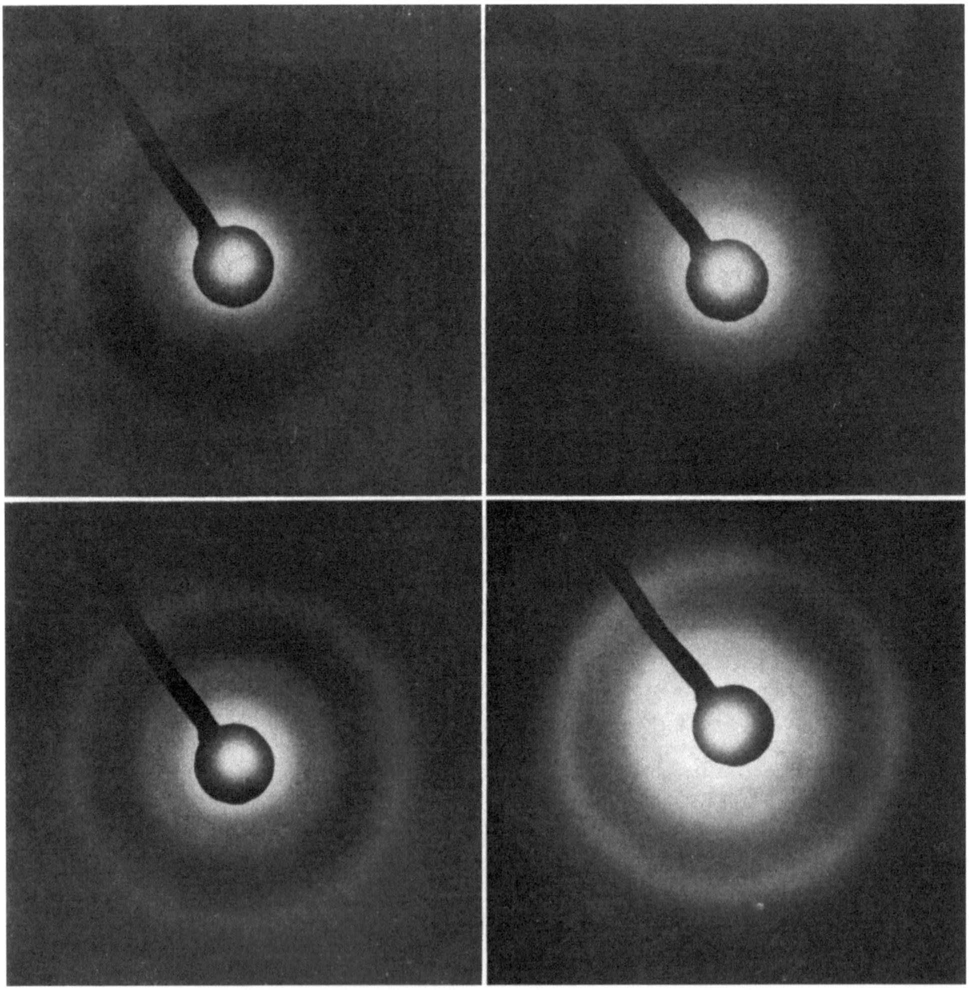

c

d

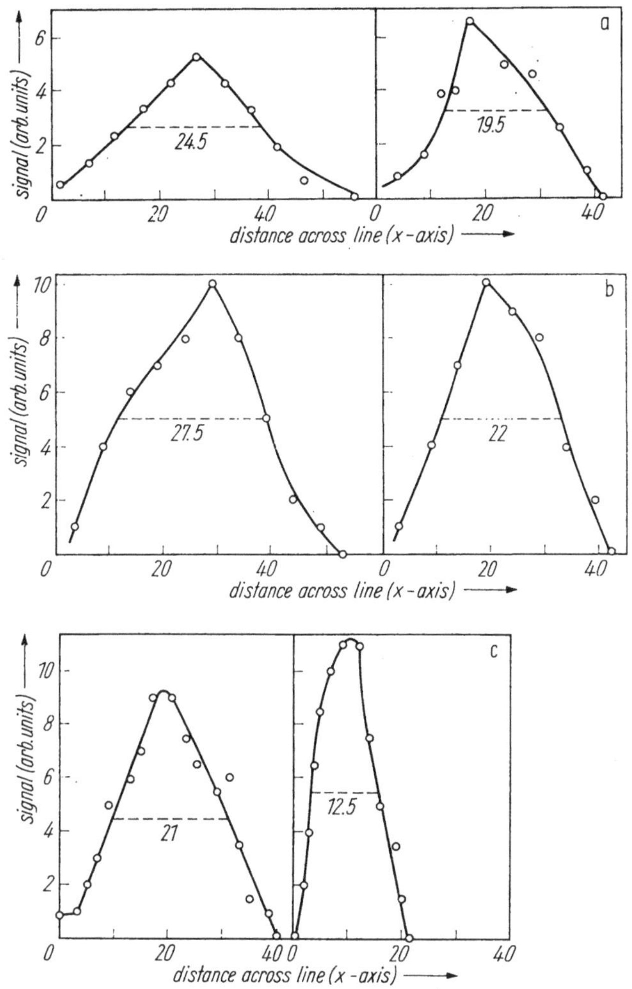

Fig. 22. Microphotometer curve of diffraction linewidth of a PP layer a) 150 Å, b) 1000 Å, c) 1600 Å thick before (left) and after (right) heat treatment

be made by a microphotometer examination of the diffraction patterns before and after the heat treatment. A narrowing of the linewidth indicates an increase in the crystallite size and all the diffraction patterns show this feature as a result of heat treatment. Fig. 22 a to c (left) show the microphotometer traces of non-annealed films and the right parts of Fig. 22 indicate the microphotometer traces for the same rings after the samples had been annealed. These results are entirely consistent with the results of the micrographic examination and of the electrical measurements.

5. Discussion and Conclusions

The ability of an insulator to resist the passage of direct current is not necessarily the same as its ability to resist alternating currents. In the former case the resistivity of the material controls the current while, in the latter, the power factor and permittivity are relevant for determining the power loss occurring in alternating fields. Most common polymer materials can be divided into two groups: (i) polymers having high resistivity, low permittivity, and negligible power factor and usually known as non-polar polymers and (ii) polymers with low resistivity and higher permittivity and power factor. They are generally referred to as polar polymers. Polypropylene belongs to the former group. It has a very high value of electrical bulk resistivity of order 10^{17} Ω cm and a low value of relative permittivity of 2.13. Polypropylene was recognised by Buckingham and Reddish [90] as being a superior insulator to low density polyethylene probably due to the method of polymerization which enabled it to be purified more effectively. The power factor of polypropylene is very low, so that power losses in large capacitors with polypropylene as dielectric are much smaller than for many other dielectrics. The breakdown fields of polypropylene are very high (up to 700 V/μm).

From a study of previously published work it is clear that some vacuum-evaporated polymers have characteristics similar to those of the original bulk material. As an example of this proposition we may consider again the work of White [26] who obtained thin films of polyethylene by vacuum evaporation on to glass microscope slides. For his electrical measurements, Al–PE–Al sandwiches were prepared and Schottky emission was found to be the dominant conduction mechanism. Capacitance measurement led to a value for the permittivity equal to 2.02 to be compared with a value of 2.25 for bulk polyethylene. So, we can say that the deposited material on the surface of the glass slide is similar to the material in the evaporation boat. This result is consistent with the idea that the vacuum evaporation of polymers results in the breaking of certain chemical bonds within their molecules, thus leading to the formation of acitve fragments, and these fragments when they reach the substrate, react with one another to form a polymer coating on its surface. The electrical conductivity is a volume property of a polymer and is directly related to the structure and to the nature of the conduction process [91, 92]. Scott [93] pointed out that changes in the composition or in the conduction process are usually reflected by changes in the resistivity before other properties exhibit significant changes. The exact relationship between the observed resistivity parameters and the structural parameters of the polymer, which determine the magnitude of the resistivity parameters are not fully known. However, once these relations are understood, such measurements will become more significant.

The permittivity of the film can be calculated from the following relation:

$$\varepsilon = \frac{Cd}{A\varepsilon_0}, \tag{7}$$

where ε is the permittivity or dielectric constant of the material, C the capacitance, and A the area of the sample. ε_0 is the permittivity of free space (8.85×10^{-12} F m^{-1}). A polypropylene film sandwiched between copper electrodes and having a thickness of 1800 Å gave the capacitance value equal to 0.9 nF (at 20 kHz). Using $A = 0.1 \times \times 10^{-4}$ m^2 we obtain $\varepsilon = 1.83$, whereas for the bulk polypropylene $\varepsilon = 2.13$. This shows that the permittivity of the evaporated film is close to the value for bulk polypropylene. Therefore we can say that the films formed by the vacuum evaporation are reasonably similar to the starting material. Thin films of polyethylene formed by vacuum evaporation have been studied extensively and it was reported [23] that the vacuum-evaporated films are amorphous before annealing and become polycrystalline after annealing. Also, a large amount of trapped gas and other gaseous hydrocarbons is expelled from the film during annealing, thus leading to a reduction in its volume. Annealing in vacuum increased the resistivity of the polypropylene film, so it seems to be important to measure the permittivity of the sample after annealing. The capacitance of an evaporated layer of polypropylene (2589 Å thick) was found to increase slightly from 0.59 to 0.62 nF after annealing. An increase in the capacitance of evaporated polyethylene after annealing has been reported previously. As mentioned above, during annealing the polyethylene films release some gaseous hydrocarbons, trapped within the film. This causes a decrease in the thickness of the film and therefore the capacitance increases after annealing. The same thing is believed to be happening in the case of polypropylene. Although, there is evidence of rearrangement of the atomic clusters after annealing in the evaporated polypropylene films, nevertheless, the permittivity remains almost the same. For example the permittivity values of a sample (2589 Å) were 1.75 and 1.81 before and after annealing, respectively. So, the annealing does not change the nature of the material; the only effect it has is to break the links between the particles, and the distance between them is increased. As is clear from the ac measurement, the electrical conduction takes place due to the electron hopping between the particles and an increased spacing would lead to a reduced electrical conductivity.

In polymers the similar structures of the monomers which are closely packed in the chains may provide some partly ordered crystalline regions (crystallites) but due to the long chains, the polymer molecules become entangled with one another. The existence of these entanglements prevents the complete crystallization; therefore there must exist disordered or amorphous regions between the crystallites. The structural results do provide for the first time evidence of the microscopic nature of the evaporated polypropylene films. The transmission electron diffraction patterns of the several evaporated layers of polypropylene, having different thicknesses certainly indicated the amorphous structure of these films, but after annealing there was evidence of some micro-crystallites. However, as a whole the structure remained essentially non-crystalline.

References

[1] R. W. Christy, J. appl. Phys. **31**, 1680 (1960).
[2] R. W. Christy, J. appl. Phys. **35**, 2179 (1964).
[3] J. R. Hanscomb and Y. Kaahwa, J. Phys. D **12**, 579 (1979).
[4] C. A. Hogarth and M. Zor, Thin Solid Films **27**, L5 (1975).
[5] C. A. Hogarth and T. Iqbal, Thin Solid Films **51**, L45 (1978).
[6] A. M. Mearns, Thin Solid Films **3**, 201 (1969).
[7] M. White, Thin Solid Films **18**, 157 (1973).
[8] A. C. Rastogi and K. L. Chopra, Thin Solid Films **18**, 187 (1973).
[9] S. L. Madorsky, Thermal Degradation of Organic Polymers, Interscience, New York 1975.
[10] A. G. Cooper, A. B. Keller, and J. R. Waring, J. Polymer Sci. **11**, 215 (1953).
[11] J. H. Correl, J. Plast. Techn. **10**, 45 (1969).
[12] G. L. Slominskii, V. A. Belyi, and A. M. Krasnovskii, J. High-Molecular Compounds (USSR) B **4**, 427 (1968).
[13] I. L. Rock, D. M. Rafalovick, and V. G. Zadorozhnyi, Instrum. Exp. Tech. (USSR) **19**, 1215 (1976).
[14] A. K. Jonscher, Internat. Meeting Electrostatic Charging, Frankfurt 1972, Verlag Chemie, 1974 (p. 29).
[15] T. Iqbal and C. A. Hogarth, Thin Solid Films **61**, 23 (1979).
[16] J. H. Slowik, J. appl. Phys. **47**, 2982 (1976).
[17] I. Chen, J. appl. Phys. **47**, 2988 (1976).
[18] J. Kevorkian, M. M. Labes, D. C. Larson, and D. C. Wu, Disc. Faraday Soc. **51**, 139 (1971).
[19] M. Pollack and T. H. Geballe, Phys. Rev. **122**, 1742 (1961).
[20] S. L. Madorsky, S. Strauss, D. Thompson, and L. Williamson, J. Polymer Sci. **4**, 639 (1949).
[21] S. L. Madorsky, J. Polymer Sci. **9**, 133 (1952).

[22] N. Grassie, in: A Handbook for Polymer Science, Vol. 2, Ed. A. K. Jenkins, North-Holland Publ. Co., 1972 (p. 1443).
[23] P. P. Luff and M. White, Thin Solid Films 6, 175 (1970).
[24] A. V. Amelin, O. F. Pozdnyakov, V. R. Regel, and T. R. Sanfirana, Soviet Phys. — Solid State 12, 2034 (1970).
[25] J. A. Lane, Brighton College of Tech., Project Report No. 156 (1967).
[26] M. White, Vacuum 15, 449 (1965).
[27] N. Grassie and H. W. Melville, Proc. Roy. Soc. A199, 14 (1949).
[28] J. F. Kinstle, J. Polymer Sci. 15, 467 (1977).
[29] V. D. Moiseev, M. B. Neman, and A. N. Kryukova, Vysokomolekulyarnye Soedineniya 1, 1352 (1956).
[30] D. Rysovy and L. Balaban, Polymer Sci. (USSR) 3, 147 (1962).
[31] L. V. Gregor, Thin Solid Films 2, 235 (1968).
[32] P. R. Emtage and W. Tantraporn, Phys. Rev. Letters 8, 267 (1962).
[33] H. T. Mann, J. appl. Phys. 35, 2173 (1964).
[34] L. V. Gregor and L. H. Kaplan, Thin Solid Films 2, 95 (1968).
[35] J. Tyczkowski, M. Zielinski, and M. Kryszewski, Thin Solid Films 55, 253 (1978).
[36] G. Lengyel, J. appl. Phys. 37, 807 (1966).
[37] A. C. Lilly and J. R. Mcdowell, J. appl. Phys. 39, 141 (1968).
[38] A. C. Lilly and D. Lowitz, J. appl. Phys. 39, 4360 (1968).
[39] J. J. O'Dwyer, J. appl. Phys. 37, 599 (1966).
[40] K. Kamisako, S. Akiyama, and K. Shinohora, Japan. J. appl. Phys. 13, 1780 (1974).
[41] E. Staryga and J. Swiatek, Thin Solid Films 56, 311 (1979).
[42] M. Kryszewski and J. Swiatek, Acta phys. Polon. A45, 119 (1974).
[43] M. Zor, Ph. D. Thesis, Brunel University, 1977.
[44] Y. Miyoshi and K. Chino, Japan. J. appl. Phys. 6, 181 (1967).
[45] A. V. Rogger, Phys. Rev. Letters 9, 368 (1962).
[46] D. K. Gupta and M. K. Barbarez, J. Phys. D 6, 867 (1973).
[47] M. Suzuki, K. Takahashi, and S. Mintani, Japan. J. appl. Phys. 14, 741 (1976).
[48] S. D. Phadke, K. Sathianandan, and N. Karekar, Thin Solid Films 51, L9 (1978).
[49] J. Chutia and K. Barua, Thin Solid Films 55, 387 (1978).
[50] T. W. Hickmott, J. appl. Phys. 33, 2669 (1962).
[51] T. W. Hickmott, J. appl. Phys. 34, 1569 (1963).
[52] J. G. Simmons and R. R. Verderber, Proc. Roy. Soc. A301, 77 (1967).
[53] R. R. Verderber, J. G. Simmons, and B. Eales, Phil. Mag. 16, 1049 (1967).
[54] P. D. Greene, E. L. Bush, and I. R. Rawlings, Proc. Symp. Deposited Thin Film Dielectric Materials, Ed. F. Vratny, The Electrochemical Society, New York 1968 (p. 167).
[55] C. Barriac, P. Pinard, and F. Davoine, phys. stat. sol. 34, 621 (1969).
[56] G. Dearnaley, D. V. Morgan, and A. M. Stoneham, J. non-crystall. Solids 4, 593 (1970).
[57] J. E. Ralph and J. M. Woodcock, J. non-crystall. Solids 7, 236 (1972).
[58] K. H. Gundlach and J. Kadlec, phys. stat. sol. (a) 10, 371 (1972).
[59] J. Gazso, Thin Solid Films 21, 43 (1974).
[60] C. A. Hogarth and T. Iqbal, Internat. J. Electronics 47, 349 (1979).
[61] C. S. Hung and J. R. Gliessman, Phys. Rev. 79, 726 (1950).
[62] E. M. Conwell, Phys. Rev. 103, 51 (1956).
[63] S. H. Koeing and G. G. Mohr, J. Phys. Chem. Solids 2, 268 (1957).
[64] A. Miller and E. Abrahans, Phys. Rev. 120, 745 (1960).
[65] K. R. Atkins, R. Donovand, and R. H. Walmsley, Phys. Rev. 118, 411 (1960).
[66] N. F. Mott, Canad. J. Phys. 34, 1356 (1956).
[67] H. Fritzsche, J. Chem. Phys. Solids 6, 69 (1958).
[68] G. Dearnaley, Thin Solid Films 3, 161 (1969).
[69] F. Argall and A. K. Jonscher, Thin Solid Films 2, 185 (1968).
[70] R. H. Bube, Photo-Conductivity of Solids, Wiley, 1960 (p. 292).
[71] M. M. Perlman and S. Unger, J. Phys. D 5, 2115 (1972).
[72] M. M. Perlman and S. Unger, in: Electrets, Charge Storage, and Transport in Dielectrics, Ed. M. M. Perlman. The Electrochemical Society, New York 1973 (p. 105).
[73] M. Kryszewski, J. Polymer Sci. (Symp.) 50, 359 (1975).

[74] D. M. Pai, J. chem. Phys. **52**, 2285 (1970).
[75] M. Zielinski and M. Samoć, J. Phys. D. **10**, L105 (1977).
[76] G. F. J. Garlick and A. F. Gibson, Proc. Phys. Soc. **60**, 574 (1948).
[77] L. I. Grossweiner, J. appl. Phys. **24**, 1306 (1953).
[78] R. H. Bube, J. chem. Phys. **23**, 18 (1955).
[79] W. Hoogestraaten, Philips Res. Rep. **13**, 515 (1958).
[80] P. N. Keatings, Proc. Phys. Soc. 78, 1408 (1961).
[81] G. A. Dussel and R. H. Bube, Phys. Rev. **155**, 764 (1967).
[82] A. R. McGhie, G. Megibbon, A. Sharples, and E. J. Stanley, Polymer (London) **13**, 371 (1972).
[83] M. E. Haine and R. E. Carley-Read, Brit. J. appl. Phys. **1**, 1257 (1968).
[84] L. I. Maissel, Handbook of Thin Film Technology, McGraw-Hill, 1970 (pp. 13 to 26).
[85] V. Vand, Proc. Phys. Soc. **55**, 222 (1943).
[86] W. M. Kane, J. P. Spratt, and L. W. Hershinger, J. appl. Phys. **37**, 2085 (1966).
[87] J. E. Skofronick and W. B. Phillips, Proc. Internat. Symp., 1966 (p. 591).
[88] H. Basseches, J. Electrochem. Soc. **109**, 475 (1962).
[89] J. Koshy, Thin Solid Films **51**, L17 (1978).
[90] K. A. Buckingham and W. Reddish, Proc. IEE **144**, 1810 (1967).
[91] R. W. Warfield and M. C. Petree, Soc. Plast. Eng. Trans. **1**, 80 (1961).
[92] R. W. Warfield and M. C. Petree, Makromol. Chem. **58**, 139 (1962).
[93] A. H. Scott, Soc. Plast. Eng. J. **18**, 1375 (1962).

(Received September 4, 1980)

phys. stat. sol. (a) **65**, 39 (1981)

Subject classification: 11 and 14.3; 13.4; 22.2.1

A. N. Sevchenko Scientific-Research Institute for Applied Physics Problems, Minsk[1])

Simulation of Electrical Characteristics and Properties of Defect Clusters in Irradiated Semiconductors

II. Self-Consistent Solution[2])

By

V. M. Lomako and P. Ya. Starostin

A simulation is made of electrical characteristics of defect clusters (DC) using a self-consistent expression of the potential barrier height with the electric charge density. A power dependence of the initial charge-carrier removal rate on the semiconductor doping level is found. The regularities of changes in electrical characteristics with core radius, concentration, and defect level energy positions in the core are obtained. The dependence of the initial charge-carrier removal rate is shown to be defined by the energy positions of defect levels in the core and their concentrations. A comparison is made of the calculation results with the experimental data.

В работе проведено моделирование электрических характеристик областей скопления дефектов в самосогласованном выражении высоты потенциального барьера с плотностью электрического заряда. Установлена степенная зависимость начальной скорости удаления носителей заряда от уровня легирования полупроводника. Определены закономерности изменения электрических характеристик от радиуса ядра, концентрации и энергетического положения уровней дефектов в ядре. Показано, что зависимость начальной скорости удаления носителей заряда от уровня легирования полупроводника определяется энергетическим положением уровней дефектов в ядре и их концентрацией. Проводится сопоставление результатов расчета с экспериментальными данными.

1. Introduction

The analysis of the electrical properties of semiconductors irradiated with high-energy particles, is closely related to the simulation of DC characteristics [1 to 7]. The outer radius of the DC and the space-charge layer width are defined in [2], where the distribution of the electrical charge density in a first approximation is chosen as constant. The range of values for the DC radius and the potential barrier height, for which the dependence of the initial charge-carrier removal rate on doping level has a form which is qualitatively consistent with the similar experimental dependence, is obtained. However, the comparison of the experimental results with those of the simulation [2] gives no possibility to determine the energy positions of defect levels in the core and their concentrations.

The analysis of the self-consistent effect of the barrier field on the electrical charge density of DC is the basic task of the present paper. Such an effect should increase the sensitivity of the simulation results, depending on defect parameters in the core, since this approach is based on the relation between the electrical charge density and these parameters. The self-consistent solution is also important for the determination of application limits for the results of the simulation in a first approximation.

[1]) Minsk 106, USSR.

[2]) Part I see phys. stat. sol. (a) **63**, 585 (1981).

2. DC Simulation

A DC of spherical form is considered. The details of the energetic and spatial schemes are given in [2]. The coordinate functions of the potential barrier height $\varphi(r)$ and the electrical charge density $\varrho(r)$ are related self-consistent by through the Poisson equation where

$$\varrho(\varphi) = e[n(\varphi) - p(\varphi) + N_a^-(\varphi) + M_a^-(\varphi) - N_d^+(\varphi) - M_d^+(\varphi)] ,$$

$$n(\varphi) = n_0 \exp\left(-\frac{e\varphi}{kT}\right), \qquad p(\varphi) = p_0 \exp\left(\frac{e\varphi}{kT}\right), \tag{1}$$

$$N_a^-(\varphi) = N_a \left\{1 + \frac{N_c}{n_0} \exp\left[(e\varphi + \Delta E_a - E_g)/kT\right]\right\}^{-1},$$

$$M_a^-(\varphi) = M_a \left\{1 + \frac{N_c}{n_0} \exp\left[(e\varphi + \Delta E_{aM} - E_g)/kT\right]\right\}^{-1},$$

$$N_d^+(\varphi) = N_d \left\{1 + \frac{n_0}{N_c} \exp\left[(\Delta E_d - e\varphi)/kT\right]\right\}^{-1},$$

$$M_d^+(\varphi) = M_d \left\{1 + \frac{n_0}{N_c} \exp\left[(\Delta E_{dM} - e\varphi)/kT\right]\right\}^{-1}.$$

Integration of the Poisson equation is performed at the following boundary conditions:

$$\varphi(r)|_{r \geqq r_2} = 0 , \qquad \varphi(r)|_{r \leqq r_0} = \varphi_0 , \qquad \left.\frac{d\varphi}{dr}\right|_{r \geqq r_2} = \left.\frac{d\varphi}{dr}\right|_{r \leqq r_0} = 0 . \tag{2}$$

Moreover, it is necessary that the charge of the total DC volume be equal to zero, i.e. the condition of electroneutrality is valid,

$$4\pi \int_0^{r_2} r^2 \varrho(\varphi)\, dr = 0 , \tag{3}$$

where n_0, p_0 are the electron and hole concentrations in the original material, N_a, N_d acceptor- and donor-type defect concentrations in the core, E_a, E_d the energy positions of defect levels in the core, M_a, M_d the atom concentrations of acceptor and donor impurities in the original material, E_{aM}, E_{dM} the energy positions of the levels of these impurities, ΔE_a, ΔE_d, ΔE_{aM}, ΔE_{dM} the ionization energies for defect levels and impurities, respectively; (the donor levels are counted from the C-band, and the acceptor ones from the V-band); r_1 is the DC core radius, r_0 the radius of the neutral part of the core, r_2 the outer radius of the DC, φ_0 the potential barrier height, E_g the band gap, eN^- the electrical charge density in the core.

Since the real distribution of defects in the DC core is unknown, we consider the case of a uniform defect distribution. The choice of a uniform distribution reduces the number of initial parameters, that leads to the simplification of the problem. Within the frame of such a distribution two cases may be singled out:

A. Concentration of acceptor-type defects is not equal to the donor-type defect concentration.

B. Defect concentrations of different types are equal. (This is also the case, when the DC core is formed by an amphoteric defect, which has one acceptor state and one donor state.)

As in [2], the initial charge-carrier removal rate is determined by the expression

$$\left.\frac{\Delta n}{\Delta \Phi}\right|_{\Phi\to 0} = \sigma_r \int_V [n_0 - n(\varphi)]\, \mathrm{d}V \equiv \sigma_r R_T\,, \tag{4}$$

where σ_r is the effective cross-section for DC formation.

Integration of the Poisson equation is performed by numerical methods. The DC core radius r_1, the potential barrier height φ_0, and N^- were the initial parameters. The outer DC radius r_2, the space-charge layer width $L = r_2 - r_0$, and the initial charge-carrier removal rate R_T were determined. The dependence of the potential barrier height on the semiconductor doping level, concentration, and the energy positions of defect levels in the core was determined in [2].

3. Results

The values of the initial charge-carrier removal rate R_T are calculated for n-GaAs at $T = 300$ K. The functional dependence of R_T on charge-carrier concentration in a semiconductor has the form

$$R_T(n_0) \sim n_0^\beta \tag{5}$$

and is given in Fig. 1.

When the ionization energy of defects in the core is small, a weak dependence of the charge-carrier removal rate on doping level is observed ($\beta \approx 0.07$), the behaviour

Table 1

No.	ΔE_a (eV)	ΔE_d (eV)
1	0.3	0.3
2	0.7	0.7
2	0.73	0.73
3	1.0	0.2
4	1.2	0.2

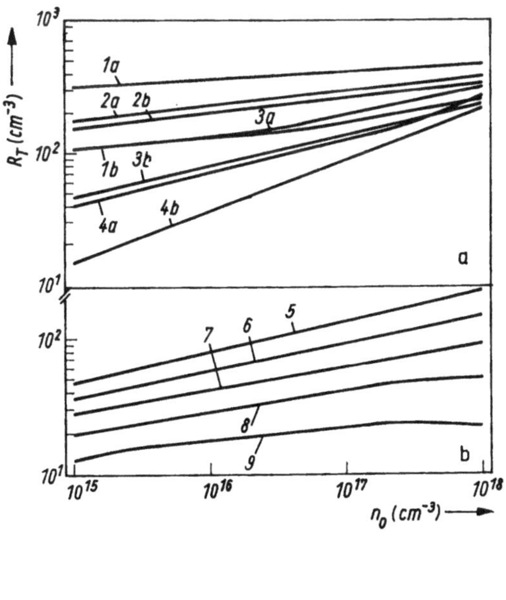

Fig. 1. The values of the initial charge-carrier removal rates against semiconductor doping level. a) Change in the $R_T(n_0)$ dependence with the energy position of defect levels in the core; indices a and b correspond to defect concentrations a $N_a = 10^{21}$ cm^{-3}, $N_d = 10^{20}$ cm^{-3}; b $N_a = N_d = 10^{20}$ cm^{-3} (the numbers of the dependences correspond to the numbers given in Table 1. $N^- = 10^{19}$ cm^{-3}. b) Change in the $R_T(n_0)$ dependence with the core radius r_1. (5) $r_1 = 300$, (6) 250, (7) 200, (8) 150, (9) 100 Å. ($\Delta E_a = 1.0$ eV, $\Delta E_d = 0.2$ eV; $N_a = N_d = 10^{20}$ cm^{-3}; $N^- = 10^{19}$ cm^{-3})

of $R_T(n_0)$ for cases A and B being qualitatively different. If $N_a \neq N_d$, (5) is valid in a wide range of concentrations n_0 (line 1a). If $N_a = N_d$, (5) is valid only for some values of n_0 (line 1b).

Shifting the acceptor defect level to the bottom of the conduction band (the donor defect level being in the upper half of the gap), results in an increase of index β (curves 3 and 4). The character of the difference between the $R_T(n_0)$ dependences for cases A and B is contrary to that stated above.

When the defect levels are near the middle of the band gap, $\beta \approx 0.12$. This case is given in Fig. 1a by lines 2a and 2b. It should be noted that changes in the electrical charge density in the core eN^- (by two orders of magnitude for heavily doped samples and by four orders of magnitude for lightly doped samples) do not affect significantly the R_T values ($< 4\%$).

Fig. 1b gives the change in $R_T(n_0)$ depending on the DC core radius r_1. It is seen that the applicability of (5) is reduced with decrease in r_1. At higher doping levels this deviation increases. This may be explained by the increase in the ratio of the core volume to the total DC volume.

From Fig. 1a and b it is seen that the energy position of defect levels in the core defines the value of β in [5]. The change in the core radius r_1 results only in a parallel shift of $R_T(n_0)$ and a decrease in the concentration interval for which (5) is valid.

Fig. 2 gives the dependences of r_0, L, and r_2 on n_0. It is seen, that the dependences $r_2(n_0)$ and $L(n_0)$ are analogous to those obtained in [2]. The size of the neutral part of the core r_0 is related to the value of the electrical charge density in the core. For $N^- \geqq 0.1 N_a$ the electrical charge is distributed in a thin near-surface layer of the core. In this case the value of r_0 is practically independent of doping level. The width of the charged layer of the core $r_1 - r_0$ increases with decrease in the N^- value.

The results obtained allow to make an independent estimate of both the energy position of defect levels in the core and the core radius by a comparison of the theoretical data with experimental ones.

Fig. 3 gives the experimental results obtained on n-GaAs irradiated with fast neutrons. The theoretical dependences, which allow to obtain the best fit to the expermental data, are given in the same figure. It is found that a good fit of the theoretical

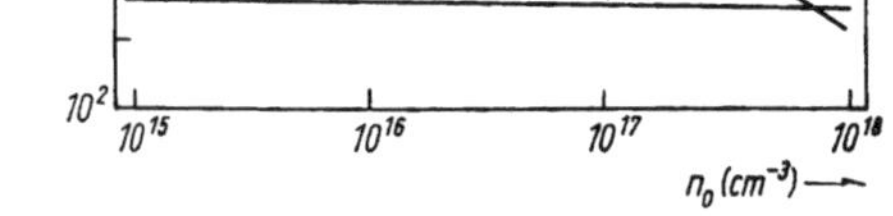

Fig. 2

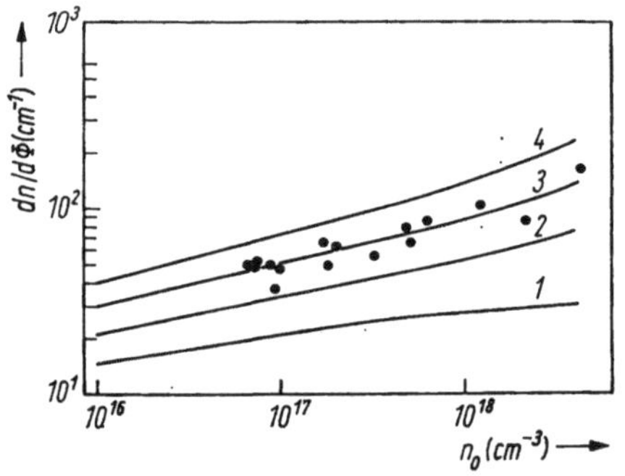

Fig. 3

Fig. 2. The DC size r_2 (curve 3), the space-charge layer width L (curve 2), and the neutral part of the core r_0 (curve 1) against the semiconductor doping level. ($\Delta E_a = 1.0$ eV, $\Delta E_d = 0.20$ eV, $N_a = N_d = 10^{20}$ cm^{-3}, $N^- = 10^{19}$ cm^{-3})

Fig. 3. The calculated values of the initial charge-carrier removal rates compared with the experimental ones for n-GaAs [5]. $T = 300$ K. (1) $r_1 = 150$, (2) 200, (3) 250, (4) 300 Å ($\Delta E_a = 1.0$ eV, $\Delta E_d = 0.2$ eV; $N_a = N_d = 10^{20}$ cm^{-3}; $N^- = 10^{19}$ cm^{-3})

data to the experiment may be obtained only when the defect levels in the core are in the upper half of the gap, and the DC core radius $r_1 \approx 250$ Å; the potential barrier height $e\varphi_0 \approx 10kT$.

4. Conclusions

Simulation of electrical characteristics of DC in irradiated semiconductors allows to obtain the general regularities of changes in the initial charge-carrier removal rate, the outer DC radius r_2, the space-charge layer width L, and the neutral part of the core r_0, depending on defect concentration in the core, the energy position of defect levels, the DC core size r_1, and the semiconductor doping level.

It is shown that the dependence of the initial charge-carrier removal rate on the semiconductor doping level is determined by both defect concentration in the core and energy positions of the levels in the gap. The results obtained allow to make a quantitative estimate of the energy positions of the defect levels in the core and the DC size (r_2, r_1, r_0) by a comparison of experimental and theoretical data.

It is shown that in n-GaAs irradiated with fast neutrons, the energy levels of the acceptor- and donor-type defects in the core are in the upper half of the gap. These defects have equal concentrations $\approx 10^{20}$ to 10^{21} cm^{-3}, the core radius $r_1 \approx 250$ Å and the potential barrier height $\approx 10kT$.

References

[1] T. I. Kolchenko, V. M. Lomako, A. M. Novoselov, and P. Ya. Starostin, Internat. Conf. Radiation of Semiconductors and Related Materials, September, 13 to 19, Tblissi 1979 (p. 106).
[2] V. M. Lomako and P. Ya. Starostin, phys. stat. sol. (a) **63**, 585 (1981).
[3] G. H. Marcus and H. P. Bruemmer, IEEE Trans. Nuclear Sci. 17, 230 (1970).
[4] A. N. Zchevno, N. A. Poklonskii, and V. F. Stelmakh, Radiation Defects in Semiconductors, Izd. V. I. Lenin, Belorussian State Univ., Minsk 1972 (p. 58).
[5] T. I. Kolchenko and V. M. Lomako, phys. stat. sol. (a) **48**, 263 (1978).
[6] R. R. Holmes, IEEE Trans. Nuclear Sci. 17, 137 (1970).
[7] V. F. Konopleva, S. R. Novikov, and E. E. Rubinova, Fiz. Tekh. Poluprov. **9**, 330 (1975).

(Received October 16, 1980)

phys. stat. sol. (a) 65, 45 (1981)

Subject classification: 18.2.1; 21.1.1

Institute of Control Sciences, Moscow[1])

Structure and Magnetic Properties of Etched NiFeCo Films

By

V. I. MALYUTIN, V. E. OSUKHOVSKII, YU. D. VOROBIEV, A. G. SHISHKOV, and V. V. YUDIN

The effect of chemical etching on the surface relief of NiFe and NiFeCo films and their magnetic parameters is studied. Electron micrograph analysis shows that during the etching inhomogeneities in the film thickness appear with a period of about 1000 Å and an amplitude of the thickness fluctuations of about 100 Å. This process goes along with the increase of the angular amplitude of the magnetization ripple and coercivity and is well described by Hoffmann's theory. The coercive force component stipulated by pinning of the domain walls on localized defects does not change significantly during the etching.

Der Einfluß des chemischen Ätzvorgangs auf das Oberflächenrelief und die magnetischen Parameter von NiFe- und NiFeCo-Schichten wird untersucht. Elektronenmikroskopische Untersuchungen zeigten, daß beim Ätzen Schichtdickeninhomogenitäten mit einer Periode von ca. 1000 Å und Dickenänderungen von etwa 100 Å entstehen. Dieser Vorgang ist begleitet von einem Anwachsen der Ripple-Winkelamplitude und der Koerzitivität. Er läßt sich mit der Hoffmannschen Theorie gut beschreiben. Die durch das Haften der Bereichsgrenzen an ortsfesten Fehlstellen bedingte Koerzitivkraftkomponente zeigt während des Ätzens keine merkliche Veränderung.

1. Introduction

Chemical etching is used at present to create a hard magnetic background in flat magnetic domain devices [1]. Etching thin films [1] and micron particles [2] in weak acid solutions changes significantly the coercive force H_c and other magnetic parameters. The coercive force of domain wall (DW) motion is known to include two components, $H_c = H_0 + H_D$ [3]. The component H_0 is determined by a quasicontinuous distribution of inhomogeneities (dimension less than the width of the domain wall) which provide continuous pinning of DW. Such quasicontinuous inhomogeneities in films are grains, intergrain layers, micropores, etc. The presence of such inhomogeneities results in magnetization ripples. The component H_D is caused by large defects (with dimensions of the order of DW width and larger) which lead to a local pinning of the domain wall, bulging and uneven motion of its sections. Such defects may result from magnetic and nonmagnetic inclusions, macropores, holes, regions with increased local strains, inhomogeneities of chemical composition, etc.

This paper studies the surface relief of etched $Ni_{80}Fe_{20}$ and $Ni_{63}Fe_{12}Co_{25}$ films and the effect of this relief upon fine magnetic structure, integral coercive force of the DW motion H_c, and its components, H_0 and H_D.

2. Experimental

The film were prepared by means of evaporation of $Ni_{80}Fe_{20}$ and $Ni_{63}Fe_{12}Co_{25}$ alloys with magnetostriction constant near zero in 10^{-6} Torr vacuum in the presence of a

[1]) Profsoyuznaya 65, 117342 Moscow, USSR.

magnetic field onto carefully cleaned glass substrates. The initial thickness of the films d_0 was between 200 and 1000 Å. Chemical etching was carried out in a 1% aqueous solution of sulphuric acid at a temperature of (20 ± 1) °C. The coercive force H_c was measured from the hysteresis loop; the anisotropy field H_k was found by improved Kobelev technique [4], and the angular dispersion α_{90} by the Crowther technique [5]. The effect of etching upon the surface relief of the film was studied with the help of transmission and scanning electron microscopy and the method of replicas. To separate the film from the substrate (in the case of observation of a grain or magnetic fine structure) the glass was covered with a photoresistive sublayer easily solvable in acetone.

3. The Effect of Etching upon the Magnetic Parameters of the Films

Fig. 1 presents typical dependences of the coercive force, anisotropy field, anisotropy angular dispersion, and the width of the ferromagnetic resonance (FMR) line measured at 9300 MHz upon the etching time t for a film of an original thickness of $d_0 = 600$ Å. It may be seen that after the etching within minutes H_c increases by ten times, H_k decreases insignificantly, and α_{90} increases rather noticeably.

Coercive force versus time of etching for films of different initial thickness d_0 is shown in Fig. 2a. The thinner the film in its initial state, the greater the rate of $H_c(t)$ increase. The shape of the $H_c(t)$ curve cannot be attributed to the decrease of their thickness in etching. Fig. 2b shows the dependence of H_c on the mean thickness of the film d during the etching. The thickness d was determined from the magnetic moment variation. For all the films the rate of etching was approximately the same and was about 1 Å/s. The dashed line in Fig. 2b shows the relationship H_c versus d_0 for non-etched films. Within the range of thickness between 200 and 1000 Å the type of DW changes significantly [6]. This, however, has a minor effect on the shape of the curve $H_c(d_0)$.

As one may see from Fig. 1, the curve $H_c(t)$ is correlated with the FMR linewidth curve $\Delta H(t)$ almost throughout the range of variation. Within this range, the relationship $H_c = 2.5H_{co}(\Delta H/\Delta H_0)$ holds where H_{co} and ΔH_0 are the coercive force and the FMR linewidth for a non-etched film, respectively. The width of the line ΔH is well known to be proportional to the relative volume of pores; therefore, the rise of $H_c(t)$ may be explained by the increase of film porosity after etching.

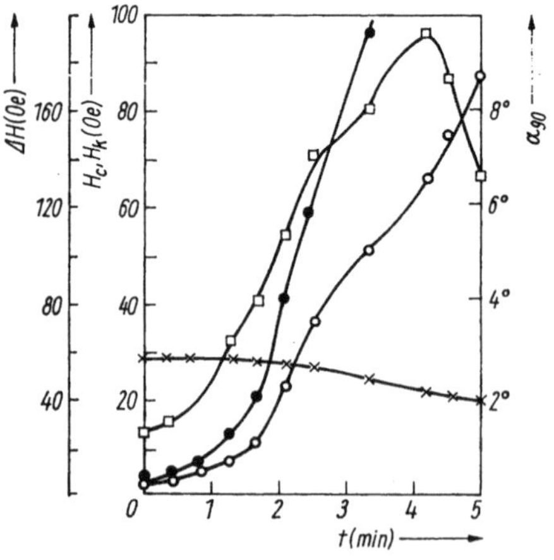

Fig. 1. Typical dependences of H_c (○), H_k (×), ΔH (□), and α_{90} (●) on the time of etching. The composition of the film is $Ni_{63}Fe_{12}Co_{25}$; $d_0 = 600$ Å

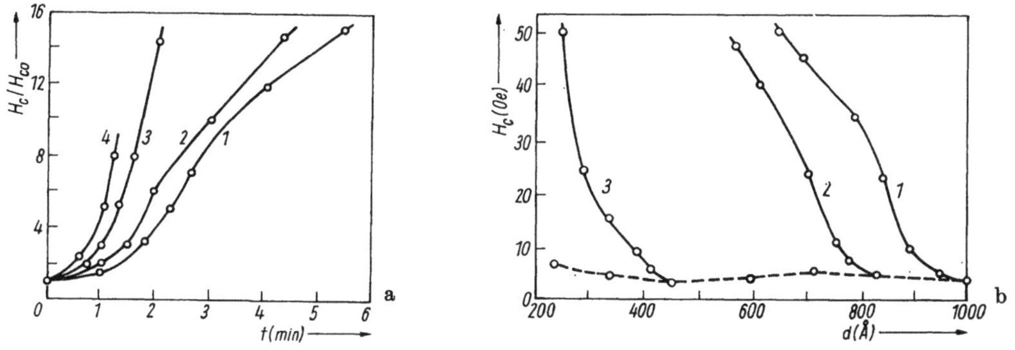

Fig. 2. The coercive force for $Ni_{63}Fe_{12}Co_{25}$ films of different thickness vs. a) the etching time and b) the averaged thickness. (1) $d_0 = 1000$, (2) 840, (3) 450, (4) 220 Å. The dashed line presents the relationship $H_c(d_0)$

4. The Effect of Etching upon the Surface Relief and Magnetic Fine Structure

Fig. 3a, b, and c show TEM microgarphs of the film structure of the composition $Ni_{80}Fe_{20}$ of an initial thickness $d_0 = 500$ Å at different stages of etching. During the first minute of etching no noticeable change of the grain structure occurs. In further etching hollows occur at the film surface the density of which rapidly increases and at the end of the second minute of etching it amounts to about 10^{10} cm^{-2}. In further etching these hollows coagulate (see Fig. 3c). This time is associated with an unusual drop of the FMR linewidth ΔH (Fig. 1).

To reveal the structural details of the morphology of films in etching, system analysis was performed of the TEM micrographs (Fig. 3a, b, and c), with the use of an optical digital complex [7]. Analysis was based on spectrum estimates of the structure images (Fraunhofer's diffraction pictures) in various stages of etching with subsequent minicomputer processing (filtering and smoothing). Fig. 4 presents the envelopes of the radial intensity profiles $I(\lambda)$ averaged in two orthogonal directions over the 200 to 3000 Å range. The non-etched film (curve 1) features two regions of spatially periodic areas, 250 to 300 Å and 600 to 1200 Å. The first region is caused by the grain structure of the film and the elements ("light" grain–"dark" grain) are easily distinguishable on the micrograph. The second region reflects certain clusters of grains and its elements are not visible. During the etching the high-frequency component of the grain structure is gradually smoothing (curve 2), the maximum is achieved in the vicinity of 800 to 1000 Å, and a longwave mode is slightly outlined around 2500 Å. With the etching time exceeding three minutes (curve 3) the mode on 800 Å is clearly seen. The next in the representative statistics of structural elements in curve 3 is the longwave mode 2500 Å. A high-frequency intensity region is weakly presented.

To determine the depth of the hollows which result from the etching of films, the method of replicas was employed. After the fabrication of a carbon replica, its surface, which was in a direct contact with the film, was covered by the test particles (small cubic monocrystals of MgO). The replica then was shadowed by platinum at an angle of $\theta = 22°$ to the surface; the local angle θ was, however, changing at the surface of the replica from 15° to 30°. Comparison of the shadows from roughnesses of the replicas surface and the test-particles allowed obtaining the depth of the hollows. Fig. 3d, e, and f show the micrographs of replicas in different stages of etching. The mean

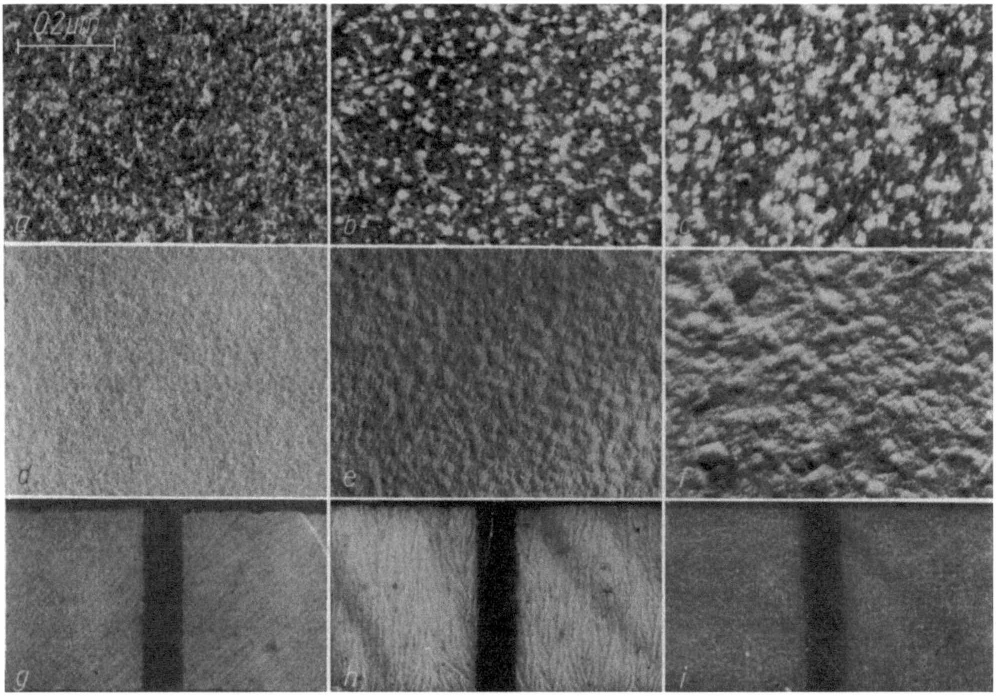

Fig. 3. Electron micrographs of the grain structure (a, b, c), replicas (d, e, f), and magnetic fine structure (g, h, i) of a $Ni_{80}Fe_{20}$ film with $d_0 = 500$ Å in different stages of etching: $t = 0$ (a, d, g); $t = 2$ min (b, c, h); $t = 3$ min (c, f, i). The size of a cell in micrographs of the magnetic structures is 50 μm; (d) $\theta = 23°$; e) $\theta = 17°$; f) $\theta = 25°$)

depth h of the hollows of the non-etched film was 40 Å. After the first minute of etching h keeps practically constant and equals (120 ± 30) Å.

Opto-digital analysis of replica micrographs shows rather distinctly the long-wave mode 2500 Å generated by the fluctuations of the etched hollow density. These inhomogeneities are well exposed on the SEM micrographs (Fig. 5), when scanning the film by an electronic beam at small angles to the surface (8° to 10°).

The appearance of the essential relief of the surface in etching with the amplitude of oscillations commensurable with the film thickness leads to an increase of the

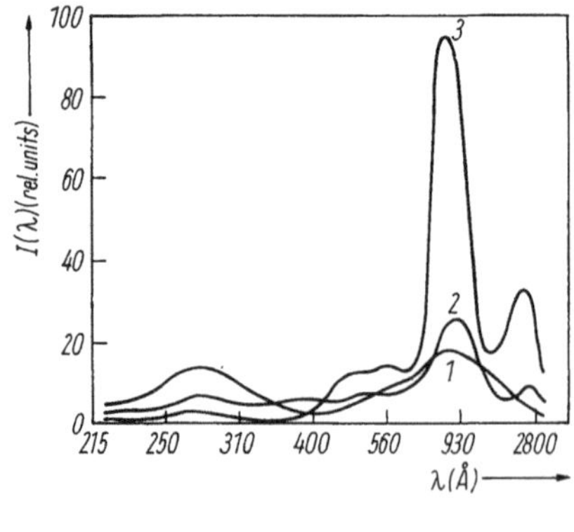

Fig. 4. Envelopes of the radial intensity profiles $I(\lambda)$

Fig. 5. Scanning electron micrograph of the surface of the film. The upper part of the picture shows a non-etched surface, the lower that with an etching time of 3 min. The angle of scanning is 9° to the surface

magnetostatic energy of the film [8]. This should be accompanied by an increase in the local anisotropy constant K_s and angular dispersion of the local easy axes due to the anisotropy magnetostatic component K_m. According to Hoffmann's theory [9], the angular amplitude of the magnetization ripple $\bar{\varphi}$ is determined as

$$\bar{\varphi} = \frac{K_s a}{4 \sqrt[4]{2} \sqrt{\pi N}\, M_s d^{1/2} A^{3/8} K_u^{3/8}}, \tag{1}$$

where M_s is the saturation magnetization, a the grain size, N the number of grains within the thickness of the film d, A the exchange constant, and K_u the uniaxial anisotropy constant.

In the presence of roughness of the film surface with a period of $2a_m$ ($2a_m > 2a$) these may be considered as the easy axis skew area. The expression for the resulting angular amplitude of the ripple may then be written in the form [10]

$$\bar{\varphi} = 1.7 \left\{ \left(\frac{\bar{\varphi}_1}{2.7} \right)^2 + \bar{\varphi}_m^2 \right\}^{1/2} + \frac{\bar{\varphi}_1}{2.7}, \tag{2}$$

where $\bar{\varphi}_1 = K_1 a/B$ is the angular amplitude of the ripple caused by the grain structure, $\bar{\varphi}_m = K_m a_m/B$ is the angular amplitude of the ripple, caused by the surface relief of the film, K_1 the crystalline anisotropy constant, and B the denominator of expression (1).

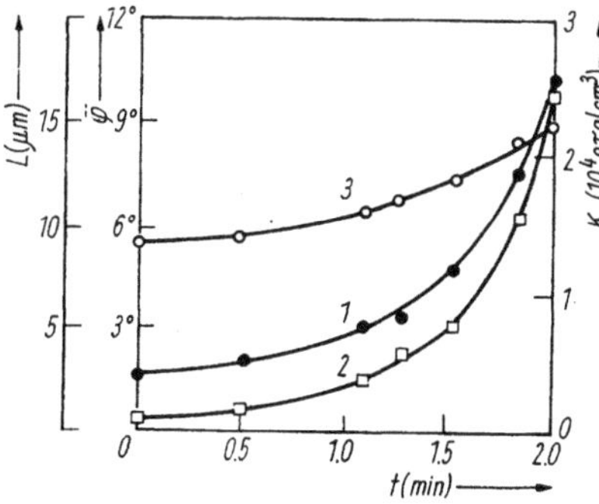

Fig. 6. The ripple angular amplitude $\bar{\varphi}$ (curve 1), the distance between the cross-ties L (curve 3) and the magnetostatic anisotropy constant K_m (curve 2) as functions of the $Ni_{80}Fe_{20}$ film etching time; $d_0 = 500$ Å

Experimentally, the angle $\bar{\varphi}$ was found from optical spectra of magnetic fine structure micrographs, which are shown in Fig. 3g, h, and i; $\bar{\varphi}$ versus the time of etching is presented in Fig. 6 (curve 1).

The value of the local anisotropy magnetostatic component K_m may be obtained from the formula (2) by experimental values of $\bar{\varphi}(t)$, $d(t)$, $K_u = 2 \times 10^3$ erg cm^{-3}, $a = 1 \times 10^{-6}$ cm, $a_m = 4 \times 10^{-6}$ cm (the value of a_m was found from the spectra in Fig. 4), $M_s = 760$ G. The exchange constant was assumed to be $A = 10^{-6}$ erg cm^{-1}, $K_1 = 7 \times 10^3$ erg cm^{-3}, $N = 1$. During the etching, the component K_m of the local anisotropy increases in 2 min from 1×10^3 erg cm^{-3} to 2.5×10^4 erg cm^{-3} (Fig. 6 curve 2).

5. Domain Wall Deformation in the Applied Magnetic Field and the Threshold Field H_0

The effect of etching upon the shape of DW deformation caused by the applied magnetic field was studied by the method of Bitter patterns. Microphotometry of DW photographs allowed the determination of the wall centre in the film with an accuracy of ± 0.05 μm [11].

Fig. 7 shows the dependence of the averaged displacement $\bar{S}$ on the magnitude of the applied magnetic field H for non-etched (curve 1) and etched (curves 2, 3, and 4) films. One may see the field of the total wall pinning H_0 [3] increases during the etching while the deflection magnitude of DW (dotted curve) decreases by breaking loose from localized defects. The dependences of H_0, H_c, and H_k on the etching time are presented in Fig. 8. It follows from this figure that the rise of the coercive force at etching is due to the increase of the threshold field H_0 while the component $H_D = H_c - H_0$ is practically unchanged. At further etching, the film is inversed ($H_c > H_k$) and speaking of the DW motion coercive force and the threshold field H_0 makes no sense.

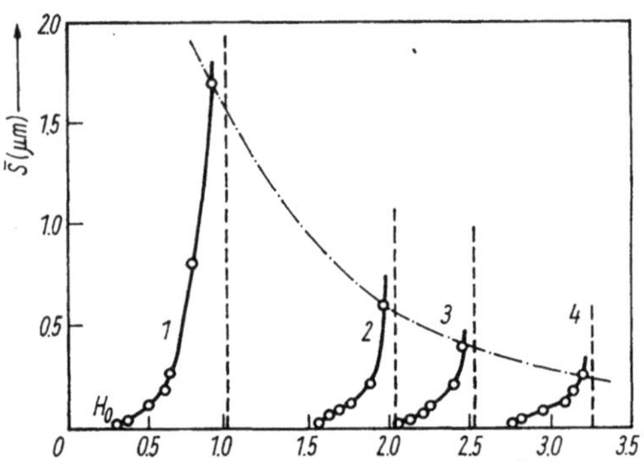

Fig. 7

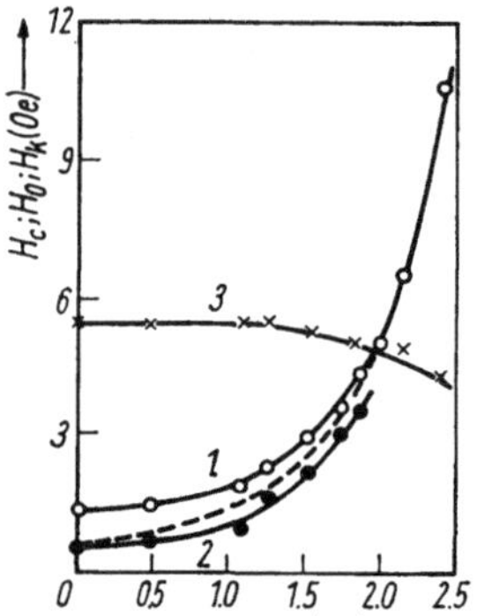

Fig. 8

Fig. 7. The averaged deflection $\bar{S}$ vs. the applied magnetic field magnitude for the $Ni_{80}Fe_{20}$ film; $d_0 = 500$ Å. The time of etching: (1) 0, (2) 1.2, (3) 1.6, (4) 1.8 min. The vertical dashed lines represent the local field H_c

Fig. 8. The dependences of H_c (curve 1), H_0 (curve 2), and H_k (curve 3) on the time of etching $Ni_{80}Fe_{20}$ film; $d_0 = 500$ Å. The dashed curve ($H_{0m}(t)$) was computed by formula (4)

According to Hoffmann's theory [12, 13] the coercive force H_0 caused by magnetization ripple is determined as

$$H_0 = D\sqrt{\frac{2}{\pi}}\,\frac{K_u^{1/4}a^{3/2}K_s}{M_s A^{1/4}\bar{L}}\,, \tag{3}$$

where D is a slowly changing factor and $\bar{L}$ the coupling length parallel to the wall. For a thin film with the roughness period $2a_m$ the expression for the coercive force H_{0m} stipulated by the surface relief may, in analogy with (3), be written in the form

$$H_{0m} = D\sqrt{\frac{2}{\pi}}\,\frac{K_u^{1/4}a_m^{3/2}K_m}{M_s A^{1/4}\bar{L}}\,. \tag{4}$$

Assuming $\bar{L}$ to be the distance between the cross-ties (curve 3 in Fig. 6) for the cross-tie DW and using the values K_m determined through the relationship $\bar{\varphi}(t)$ (curve 2 in Fig. 6) we may obtain the relationship $H_{0m}(t)$ from formula (4) (dashed line in Fig. 8). The factor D was assumed equal to 192.5. As one may see in Fig. 8 the computed curve describes accurately enough the experimental relationship $H_0(t)$.

6. Magnetic Potential Relief of the Localized Defects and the Field H_D

To study the distribution of the localized defects responsible for the coercive force component H_D, a DW was used which moved in the creep mode (the direct field along the easy axis $H_L = 0.6H_c$ and the alternate field along the hard axis $H_T = 0.2H_k$). After shooting the position and the density of "point" defects were found by studying every shot.

Fig. 9 shows the position of DW before breaking loose from the large defects. The mean break-loose angle $\bar{\alpha} = (\alpha_1 + \alpha_2)/2$ represents the height of the local potential barrier. Each film was experimentally found to feature its maximum (critical) break-loose angle α_k, which represents the critical defects. The value of the angle α_k depends upon H_k, α_{90}, the thickness, and the composition of the film. The map of critical defect locations (black points) corresponding to the angle $\alpha_k = (21 \pm 3)°$ for a non-etched film $Ni_{80}Fe_{20}$ with a thickness of $d_0 = 600$ Å is shown in Fig. 9. The mean distance between the critical defects was determined as $\bar{l}_k = \sqrt{\Delta S/\Delta n}$, where ΔS is the area of the region of the film and Δn is the number of critical defects in this area. For the film considered $\bar{l}_k = 130$ µm and the density of critical defects $n_k = 6 \times 10^3$ cm^{-2}, respectively. The mean distance $\bar{l}$ between the points of DW pinning on critical defects is $\bar{l} \approx 2\bar{l}_k$.

The dependence of the deflection magnitude S_m before breaking loose from critical defects on the distance between the point of the wall pinning l is shown in Fig. 10.

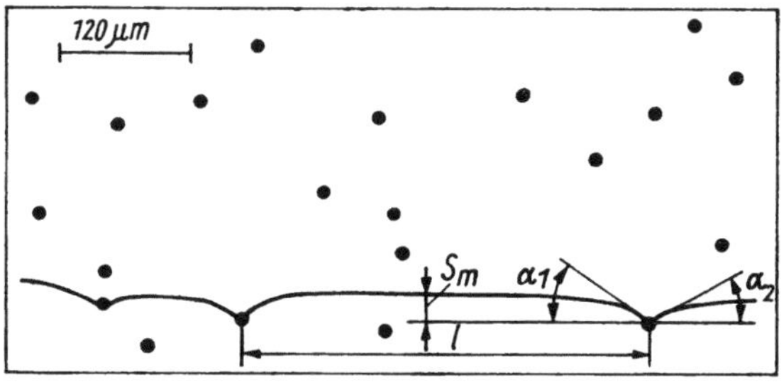

Fig. 9. The map of critical defects (black points) and DW locations before breaking loose from large defects. The composition of the film $Ni_{80}Fe_{20}$; $d_0 = 600$ Å

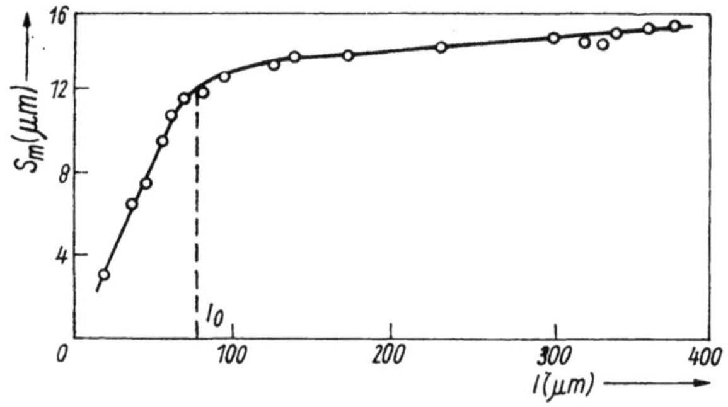

Fig. 10. The relationship between S_m and the distance between the wall pinning points

One may see that there is a certain critical distance $l_0 = 80\ \mu m$. With $\bar{l} < l_0$ a nearby linear relationship $S_m(l)$ is observed. In this case the entire section of the deflected DW is bent. With $\bar{l} > l_0$ the deflection S_m depends on l much weaker and the central part of the deflected DW is almost unbent (Fig. 9).

Thus at a small density of defects n_k ($\bar{l} > l_0$) the DW interact with each defect individually as though there are no other critical defects. The coercive force of the localized defects H_D in this case does not strongly depends on n_k. If, however, the defect density is high and, consequently, $\bar{l} < l_0$, the DW interacts simultaneously with many defects. In this case H_D should be strongly dependent upon n_k.

For the mean distance between the DW pinning points $\bar{l} = l_0 = 80\ \mu m$ the density of critical defects (with regard for $\bar{l} \approx 2\bar{l}k$) $n_{k0} \approx 10^5\ cm^{-2}$.

The critical defect density in our films, $n_k = 6 \times 10^3\ cm^{-2}$ is much less than n_{k0} and is practically constant in etching. This accounts for the weak correlation between the localized defect coercive force H_D and the time of etching.

Acknowledgement

The authors wish to thank Mrs. B. Bey (University Regensburg, FRG (DFG, Al 56/6)) for her great help in carrying out the SEM investigations.

References

[1] N. P. Vasilieva, V. I. Malyutin, Yu. S. Kulicov, and F. A. Gal, IEEE Trans. Magnetics **15**, 1339 (1979).
[2] R. A. Currie and L. E. Willmore, J. appl. Phys. **50**, 3560 (1979).
[3] R. V. Telesnin, A. G. Shishkov, V. E. Osukhovskii, A. S. Sigov, and L. P. Osukhovskaya, phys. stat. sol. (a) **22**, 749 (1974).
[4] V. V. Kobelev, Fiz. Metallov i Metallovedenie **13**, 467 (1962).
[5] T. S. Crowther, J. appl. Phys. **34**, 580 (1963).
[6] S. Methfessel, S. Middelhoek, and H. Thomas, J. appl. Phys. **31**, 302 (1960).
[7] L. I. Antonov, V. V. Veter, L. A. Yudina, G. G. Kozodoi, E. L. Fedorova, and V. V. Yudin, Fiz. Metallov i Metallovedenie **43**, 518 (1977).
[8] A. G. Lesnik, G. I. Levin, and S. I. Kaverina, Izv. Akad. Nauk SSSR, Ser. fiz. **29**, 591 (1965).
[9] H. Hoffmann, phys. stat. sol. **5**, 187 (1964).
[10] G. Bayreuther and H. Hoffmann, Czech. J. Phys. **B21**, 541 (1971).
[11] R. V. Telesnin, E. H. Ilyicheva, N. G. Kanavina, V. E. Osukhovskii, and A. G. Shishkov, phys. stat. sol. **34**, 443 (1969).
[12] H. Hoffmann, IEEE Trans. Magnetics **9**, 17 (1973).
[13] H. Hoffmann and T. Fujii, Proc. Internat. Conf. Magnet., Vol. **3**, Moscow 1974 (p. 455).

(Received October 17, 1980)

phys. stat. sol. (a) 65, 53 (1981)

Subject classification: 10.1; 21

Physico-Technical Institute of Low Temperatures, Academy of Sciences of the Ukrainian SSR, Kharkov[1])

Plasticity of Molybdenum Single Crystals at Helium Temperature

By

H.-J. KAUFMANN[2])

The plastic behaviour of molybdenum single crystals of different high purity in several crystal orientations at and below helium temperature is investigated. Up to the lowest applied temperature (0.48 K) this material deforms plastically with more or less large fracture elongations, depending on orientation, purity, and deformation rate. The fracture mode is partly ductile and partly brittle. No indication of an anomalous influence of very low temperatures on the critical stress (proportional limit) can be established.

Die plastische Verformung unterschiedlich hochreiner Mo-Einkristalle verschiedener Kristallorientierungen wird bei Heliumtemperatur und darunter untersucht. Bis zur niedrigsten angewendeten Temperatur (0,48 K) verformt sich dieses Material plastisch bis zu mehr oder weniger großen Bruchdehnungen, je nach Orientierung, Reinheit oder Verformungsrate. Der Bruchmodus ist teils duktil, teils spröde. Anzeichen eines anomalen Einflusses tiefster Temperaturen auf die kritische Spannung (Proportionalitätsgrenze) können nicht festgestellt werden.

1. Introduction

The plastic deformation of body-centred-cubic (b.c.c.) metals at low temperatures is still far from being clearly understood [1]. Most of the systematic studies on the deformation behaviour of single crystals have been made on iron and iron-base alloys [2]. A complete understanding of the deformation modes, work hardening behaviour, and fracture mechanisms of b.c.c. metals is not possible without detailed studies at very low temperatures.

Molybdenum single crystals below room temperature in general show parabolic stress–strain curves characterized by high initial work hardening rates [3 to 7]. Up to now, no systematic studies of deformation behaviour of molybdenum at very low temperatures have been made. The present paper describes the flow stress behaviour of molybdenum single crystals of high purity and different orientations at and below helium temperature.

2. Experimental Procedure

The objects of investigation were molybdenum single crystals of two different purities, grown by electron-beam zone melting with the crystal axis orientations [100], [110], [111], [210], and [941]. (The as-grown states of the crystals are tabulated in Table 1.) The residual resistance ratio (RRR) ($R_{300\,\mathrm{K}}/R_{0\,\mathrm{K}}$) ranged from 1×10^5 to 2×10^5; for several [110] crystals additionally investigated RRR was of about 3×10^3.

From the cylindrical as-grown crystals tensile specimens were prepared by a sequence of centreless grinding, electrolytic shaping, and final electropolishing. The specimen dimensions were gauge length 4 mm, gauge diameter 1 mm, head diameter

[1]) Leninskii prospekt 47, 310164 Kharkov, USSR.

[2]) Present address: Akademie der Wissenschaften der DDR, Otto-Nuschke-Str. 22/23, DDR-1080 Berlin, DDR.

Table 1

crystal quality	axis orientation	RRR $\frac{R_{300\,K}}{R_{0K}}$	main impurity (wt ppm)	as-grown disloc. density (cm^{-2})	producer	ref.
DC-T-Mo	[100] [110] [111] [941] [210]	2×10^5	W < 1, C 3, O 3, N 0.4	10^5	ZFW, Acad. Sci. GDR, Dresden	[13, 14]
TC-Ch-Mo	[110]	3×10^3	W 50, Nb 10, C 10, O 30, N 20	10^6	IFFT, Acad. Sci. USSR, Chernogolovka	[15]

4.5 mm. The specimens were predominantly deformed by tension in liquid helium at temperatures of 1.5 and 4.2 K and cross-head velocities of 6.66×10^{-4} to $9.75 \times \times 10^{-5}$ mm s^{-1}. The temperature of 1.5 K was obtained by pumping off helium vapour. (The used special material testing machine was developed by FTINT, Kharkov.)

Some individual tests were carried out at 0.5 K with another special material testing machine [8] working in the environment of liquid ^{3}He. (This temperature is somewhat lower than the transition temperature to superconductivity of molybdenum, $T_t = 0.92$ K.) Additional straining experiments at temperatures of 77 and 300 K served to get a survey over the whole range below room temperature.

It is well known that slight predeformation at higher temperatures significantly improves the ductility of molybdenum at 4.2 K or lower temperatures (cf. [10]). Therefore, all the following results refer to crystals deformed at low temperatures after about 1% pre-straining at room temperature or 77 K. In order to remove any deformation traces the pre-strained samples were electropolished once more.

3. Experimental Results

3.1 Influence of temperature on stress–strain behaviour and flow stress

Fig. 1 shows typical stress–strain curves for middle oriented [941] crystals (Schmid factor ≈ 0.5 on the (101) [11$\bar{1}$] slip system) at different temperatures (crystal quality DC). Because no yield drop is observed, the onset of plastic deformation may be characterized by the proportional limit $\sigma_0(T)$. The respective values for the pre-strained crystals are $\sigma_0(1.5\text{ K}) = 1.496$ GPa, $\sigma_0(4.2\text{ K}) = 1.427$ GPa. Until fracture, this orientation exhibits a plateau, quite similar to that observed at higher temperatures. Evidently, the macroscopic deformation takes place with a very low work hardening

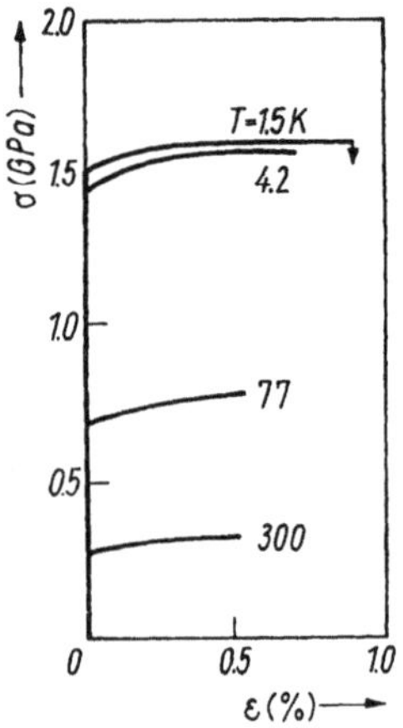

Fig. 1. Stress–strain curves of Mo single crystals with deformation axis in the [941] direction at a strain rate of 3.2×10^{-5} s^{-1}

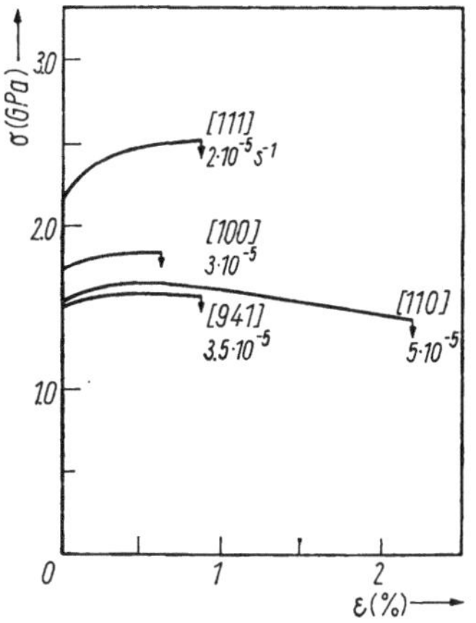

Fig. 2. Stress–strain curves of Mo single crystals of different orientations at 1.5 K at a strain rate of 2 to 5 × 10^{-5} s^{-1}

rate. A preliminary test showed: The [941] orientation fractures brittly without any detectable prior slip deformation or necking at 4.2 K. Also no mechanical twins could be observed at the surface of the samples as it is the case for molybdenum wires, niobium or niobium–molybdenum single crystals [4, 9, 10]. At about 0.7% strain the 4.2 K sample related to Fig. 1 was unloaded and then restrained at 1.5 K. Under these conditions the crystal fractured also in a brittle manner at about 1% strain.

3.2 Influence of crystal orientation on the stress–strain curves

Fig. 2 shows typical stress–strain curves of all strain orientations (quality DC) at 1.5 K. The strain rate amounted to the order of magnitude of 10^{-5} s^{-1}, as indicated within the diagram. The [100] and [111] crystals exhibit parabolic stress–strain curves, whereas the [110] curve goes through a broad maximum, and then it softens until fracture at a relatively high strain (2.1%). (This effect may be considered as a "truncated" yield drop, interrupted by fracture before going through the typical minimum.) No indications for stages of low work hardening rates are present. Pronounced yield drops are also absent in all cases. As a rule, the proportional limits of all corner orientations of the stereographic triangle are somewhat higher than those of the single slip orientation [941]. The fracture mode of the multislip orientations is always brittle without any observable slip traces or twins. Supplementary experiments with [210] crystals showed only very weak plastic deformation in the temperature range below 4.2 K. This orientation shows also fracture in creep tests.

3.3 Effect of purity on the flow stress

The influence of purity on the plasticity of [110] crystals over the whole temperature range is plotted in Fig. 3. (Note, here the minimum temperature $T = 0.48$ K is lower than the transition temperature of Mo.) The diagram compares the two crystal qualities DC and TC, with RRR's of the order of magnitude of 10^3 or 10^5. Within the whole temperature range (0.5 to 300 K) the proportional limits σ_0 of the crystals of TC quality are significantly higher than those of DC quality.

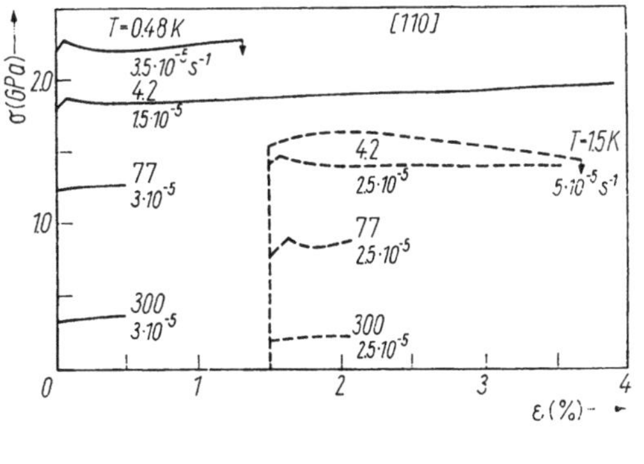

Fig. 3. Stress–strain curves of Mo single crystals of [110] orientation depending on purity and strain rate; ——— crystal quality TC, $R/R_0 \approx 3 \times 10^3$; – – – crystal quality DC, $R/R_0 \approx 2 \times 10^5$

Under the given conditions of temperature and strain rate the [110] crystals reveal typical yield drop-effects.[3]) They are more pronounced for DC quality at 4.2 and 77 K than for TC quality at 0.48 and 4.2 K. Also these phenomena reflect the significant influence of the different purities of the two crystal qualities on the stress–strain behaviour, especially at low temperatures.

The plasticity of crystals of higher purity is smaller and fracture occurs at lower applied stresses than for less pure samples. The influence of the superconductive state on plastic deformation [10, 12] is in general the smaller the purer the crystals are. Sometimes, at high stresses and at the lowest temperatures, crystals of the lower quality show the effect of "continuous yielding", accompanied by homogeneous slip and macroscopic tapering of the cross-section. No indication of any localized plastic deformation was observed along the stress–strain curves. The maximum elongation is slightly influenced by the deformation rate. Deformation of such crystals ends by chisel fracture; the maximum strain observed amounted to 8%.

3.4 Temperature dependence of the proportional limit

The temperature dependence of the proportional limit σ_0 (as it was defined above) is drawn in Fig. 4a for all investigated crystal orientations (see inset). Within the whole temperature range investigated no temperature-influenced anomalies could be observed under the given experimental conditions: σ_0 strongly increases with decreasing temperature down to the superconducting state.

The stress–strain curves as well as the temperature dependence of σ_0 of this investigation correspond to those obtained in the constant strain rate tests within the temperature range partly down to 4.2 K, reported in [3 to 5, 11]. Some differences in characteristic quantities may possibly be due to differences in specimen geometry and the as-grown states of the crystals.

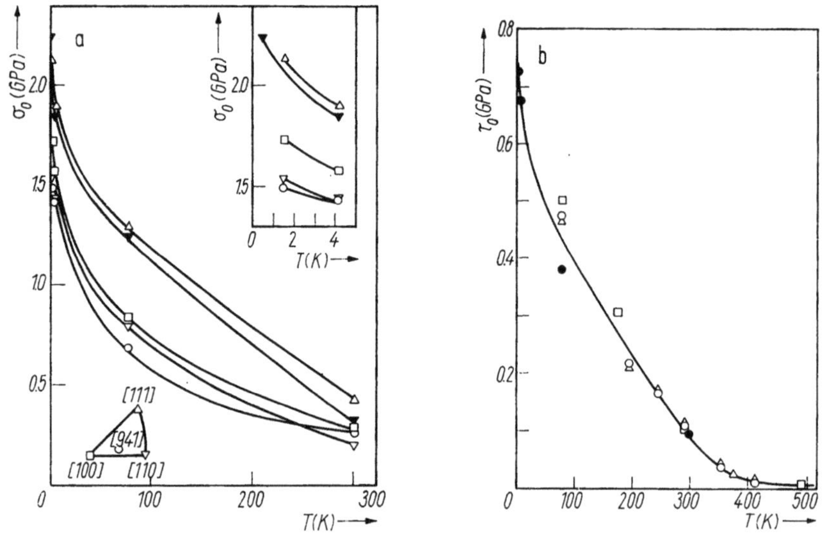

Fig. 4. a) The effect of temperature on the critical stress of Mo single crystals of different purities and orientations (the black symbol represents TC quality, the others DC quality). b) Temperature dependence of the critical resolved shear stress of [110] oriented Mo single crystals. △ [4], ○ [5], □ [11], ● this work

[3]) Also in this special case we used the proportional limit σ_0 as a measure for the onset of plastic deformation.

Fig. 4b shows the temperature dependence of the critical resolved shear stress (crss) of [110] crystals combined with corresponding results on high-purity Mo single crystals of Guiu and Pratt [4], Sherwood et al. [5], and Schaper [11] (Schmid factor 0.472, according to slip line observations in the system (112) [11$\bar{1}$] [16]). The plot clearly exhibits the existence of three distinguishable temperature ranges within the low-temperature rise of the crss, as it was observed previously by Diehl et al. [17] on Fe and Nb single crystals and, more detailed, by Zwiesele and Diehl [18] on mono- and polycrystalline Fe. For [110] Mo single crystals the weakly curved intermediate region ranges between about 80 and 300 K and hence is enlarged in comparison with the analogous region in the crss versus T plot of iron (120 to 250 K [18]). Accordingly, the stronger curved low-temperature region of Mo extends at lower temperatures down to about 0.5 K. To derive a "knee temperature" for comparison with current theories, as it was done by Zwiesele and Diehl [18], is impossible at the moment because of the small number of points in the sequence of deformation temperatures.

3.5 Activation volume

Starting from the yield point, with growing strain the relaxation curves $\sigma(t)$ were registered. At all temperatures the stress relaxation can be described by a logarithmic time law. The activation volume has been calculated for [110] crystals of DC and TC quality by the relation $V = kT\ (\partial \ln \gamma/\partial\tau)_T$ (Table 2). At and below room temperature the results fairly correspond with data from Guiu and Pratt [4] and down to 77 K with those from Schaper [11]. The values of the activation volume are slightly influenced by the strain rate (cf. Table 2, DC crystals at 300 K).

Table 2

Temperature dependence of the activation volume

crystal quality	axis orientation	temperature (K)	strain rate (10^{-5} s^{-1})	activation volume (10^{-23} cm^3)	activation volume (b^3)
DC-T-Mo	[110]	4.2	2.5	2.17	1
		77	2.5	25.63	13
		300	2.5	58.76	29
		300	25	78.06	38
TC-Ch-Mo	[110]	4.2	1.5	1.83	0.9
		77	3	11.26	6
		300	3	50.6	25

As the temperature is reduced, the activation volume thus obtained decreases to very low values (lower than $1b^3$) at helium temperature. Measurements of internal friction of the same molybdenum qualities resulted in different values of the activation volume [19]. It decreases from $1.48 \times 10^4 b^3$ at 300 K to $17b^3$ at 6.5 K for DC and TC quality, respectively.

4. Discussion

The conception that the thermally activated formation of kink pairs in a screw dislocation is the prevailing process, has found wide acceptance in the theoretical explanations of the low-temperature increase of the crss. The temperature dependences of crss and activation volume observed here within certain limits are consistent with this idea. But the stronger increase of the crss in the low-temperature range has not been explained so far [18]. Seegers "line tension approximation" fits the experimental crss versus T data on iron only down to about 100 K (cf. [18]). The region below 100 K especially interesting here obviously remains unexplained within the framework of this approach. Moreover, in the present investigation the activation volume tends to values lower than $1b^3$ at helium temperature. This observation can hardly be understood in the picture of two separated kinks. On the other hand, under the conditions of internal friction or stress relaxation experiments, where the applied stresses are relatively small, the critical kink separation corresponds more or less to the interatomic distance [20].

Deforming high-purity iron single crystals, Kuramoto et al. [2] with conventional procedures obtained also very small values of the activation parameters below 30 K. This deviation from the Arrhenius rate equation of deformation has been interpreted by them as a quantum effect of the vibrational mode of dislocations in overcoming the Peierls potential.

Finally, one possible generalization of the present investigation should be mentioned here: The less pure the crystals and the lower the deformation temperature, the higher is the ductility, weakly influenced by the strain rate. It seems that an expanded correlation with the influence of superconductivity cannot be excluded.

Concluding, the low-temperature dependence of the crss, i.e. the yield phenomenon in different temperature ranges cannot be explained simply in terms of a single rate controlling process. A more comprehensive study would be required.

Acknowledgements

The author is indebted to Prof. D. Schulze (ZFW), Prof. V. I. Startsev (FTINT), and Prof. V. V. Pustovalov(FTINT) for helpful discussions and a critical revision of the manuscript. He expresses his gratitude to Dr. M. Jurisch and Ing. R. Petri (both ZFW) for the molybdenum crystals and I. N. Kuzmenko (FTINT) for experimental assistance during special tests with liquid ^{3}He.

References

[1] B. Šesták and A. Seeger, Z. Metallk. **69**, 195, 355, 425 (1978).
[2] E. Kuramoto, Y. Aono, and K. Kitajima, Phil. Mag. **39**, 717 (1979).
[3] A. Lawley and H. Gaigher, Phil. Mag. **10**, 15 (1964).
[4] F. Guiu and P. L. Pratt, phys. stat. sol. **15**, 539 (1966).
[5] P. J. Sherwood, F. Guiu, H. C. Kim, and P. L. Pratt, Canad. J. Phys. **45**, 1075 (1967).
[6] J. Richter, phys. stat. sol. **40**, 565 (1970); phys. stat. sol. (b) **46**, 203 (1971).
[7] Ch. Ritschel, A. Luft, and D. Schulze, Kristall und Technik **13**, 791 (1978).
[8] I. N. Kuzmenko, T. A. Parkhomenko, and V. V. Pustovalov, Fiz. nizk. Temp. **4**, 1340 (1978).
[9] L. P. Kubin and B. Jouffrey, Phil. Mag. **24**, 437 (1971).
[10] G. Kostorz, Acta metall **21**, 813 (1973).
[11] M. Schaper, Dissertation, AdW der DDR, Dresden 1972.
[12] V. V. Pustovalov, V. I. Startsev, and V. S. Fomenko, phys. stat. sol. **37**, 413 (1970).
[13] K. Friedrich, J. Barthel, and J. Kunze, J. less-common Metals **14**, 55 (1968).
[14] A. Luft and L. Kaun, phys. stat. sol. **34**, 135 (1969).
[15] E. B. Leiko, private communication (1979).
[16] A. Luft and L. Kaun, phys. stat. sol. (a) **18**, 109 (1973).
[17] J. Diehl, M. Schreiner, S. Staiger, and S. Zwiesele, Scripta metall. **10**, 949 (1976).
[18] S. Zwiesele and J. Diehl, Proc. V. Internat. Conf. ICSMA, Aachen 1979 (p. 59).
[19] H.-J. Kaufmann and P. P. Pal-Val, phys. stat. sol. (a) **62**, 569 (1980).
[20] A. Seeger, phys. stat. sol. (a) **55**, 457 (1979).

(Received January 16, 1981)

phys. stat. sol. (a) 65, 59 (1981)

Subject classification: 5

Institute of Physics, Jagiellonian University, Cracow[1])

Observation of the Double Resonance Scattering of γ-Rays

By

J. J. BARA

The double resonance scattering process is observed in an enriched hematite by detecting the 6.47 keV conversion X-rays following the nuclear resonance absorption of 14.4 keV γ-rays which were previously recoilless–recoillessly scattered by an enriched metallic iron foil.

Измерен спектр двойного мессбауэровского рассеяния. γ-кванты с энергей 14,4 keV впервые были бесотдачно рассеянны в обогащенном металлическом железе, а затем они поглащались в обогащенном гематите. Использован метод регистрации рентгеновского конверсионного излучения с энергей 6,47 keV, исспускаемого в втором мессбауэровском резонансе.

1. Introduction

The wide applications of Mössbauer spectroscopy are due to its unique properties. Three of them, that is, high-energy resolution, large value of the nuclear resonance cross-section, and short observation time, are very well known. They are attributed to both the transmission and scattering methods of Mössbauer spectroscopy. The fourth property, as yet not commonly known, is the filtration property of the nuclear resonance scattering process [1]. It can be observed most directly in double Mössbauer resonance experiments [2]. The filtration property of the nuclear resonance scattering process enables one to reduce the width of a Mössbauer line [3]. Such a line narrowing may be even more effective when multiple resonance scattering is considered.

We have made an attempt to observe the double resonance scattering process. The purpose of this paper is to present the results of these investigations.

2. The Shape of the Double Resonance Scattering Line

The method of calculation, developed previously [4] for single resonance scattering of γ-rays, was extended to the double resonance scattering process. The double resonance scattering geometry shown in Fig. 1 was considered.

In general, there are many contributions to the total double resonance scattered intensity. However, not all of them have to be considered, especially when thin scatterers are used. It is sufficient to include the recoilless–recoilless and recoilless–nonrecoilless nuclear resonance scattering processes in the lineshape calculations. In this case the shape of the background-corrected double resonance scattering line is described by the formula

$$\varepsilon(S_0, S_1) = \frac{a_\mathrm{r}^2}{f_0} \frac{1}{\pi[1 - \exp(-T)]^2} \int_{-\infty}^{\infty} \frac{\mathrm{d}y \; T_{\varkappa\mathrm{r}}^2(y, S_1)}{(y - S_0)^2 + 1} +$$

[1]) Reymonta 4, 30-059 Cracow, Poland.

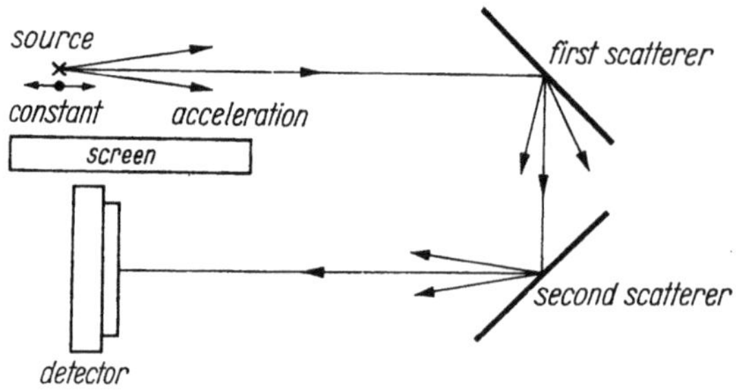

Fig. 1. Double resonance scattering geometry

$$+\frac{a_r a_n}{f_0}\frac{1}{\pi[1-\exp(-T)]^2}\int\limits_{-\infty}^{\infty}\frac{dy\, T_{\varkappa_r}(y, S_1)\, T_{\varkappa_n}(y, S_1)}{(y-S_0)^2+1}, \tag{1}$$

where

$$T_\varkappa(y, S_1) = \frac{1-\exp\left[-T\frac{(y-S_1)^2+\varkappa^2}{(y-S_1)^2+1}\right]}{(y-S_1)^2+\varkappa^2}.$$

The isomer shift S_1 of the scatterer with respect to the source and the Doppler energy shift of the moving source S_0 are expressed in $\Gamma/2$ units. All other symbols are the same as those used in [4].

In the very thin sample limit approximation (1) has the analytical form

$$\varepsilon(S_0, S_1) = \frac{f_1}{2f_0}\left[\frac{a_r+a_n}{2}\right]^2\left\{\frac{4}{4+(S_0-S_1)^2}+\left[\frac{4}{4+(S_0-S_1)^2}\right]^2\right\}. \tag{2}$$

By comparing (2) with (21) of [4] obtained for single resonance scattering it can be noticed that the double resonance scattering process reduces the width and amplifies the magnitude of a Mössbauer line. These effects are even more pronounced in higher-order scattering processes [5].

3. Observation of Double Resonance Scattering Process

The double resonance scattering process may be observed either by detecting doubly scattered γ-rays or by detecting conversion X-rays or conversion electrons which follow the resonance absorption in the second scatterer of γ-rays recoilless–recoillessly scattered by the first scatterer. The first method of detection enables one to get a very high amplitude of the background-corrected double resonance scattering line. However, a very good screening is necessary for this case. Moreover, the two small solid angles used in double scattering geometry greatly reduce the efficiency of γ-ray detection. The application of the second or third method of detection makes it possible to utilize half of the solid angle in the second scattering process and hence greatly enhances the detection efficiency. The detection of conversion X-rays was used in our double resonance scattering experiment since the penetration depth of X-rays is much higher than that of conversion electrons.

Before the double resonance scattering experiment was performed the single resonance X-ray scattering spectrum was measured by detecting X-rays emitted mainly in the forward direction by an enriched hematite scatterer irradiated through a 4.5 mm thick plexiglass plate by 14.4 keV γ-rays from the $^{57}Co(Cr)$ source moving with a constant acceleration. The hematite scatterer was placed directly on a beryl-

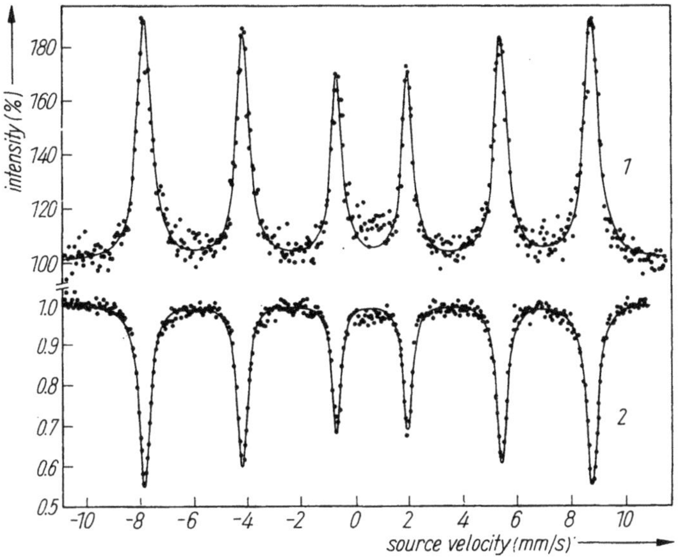

Fig. 2. Single resonance X-rays scattering (1) and γ-ray absorption (2) spectra for enriched hematite (The intensity scale for (2) has to be multiplied by a factor 100)

lium window of a proportional counter which was filled to 90% with argon and 10% with methane under a preasure of 450 Torr. This conversion X-ray single resonance scattering Zeeman spectrum of the hematite sample is shown in Fig. 2 together with the ordinary γ-ray absorption spectrum measured for the hematite sample. The rather high line amplitudes of the scattering spectrum explain the usefulness of the application of X-ray detection in double resonance scattering experiments. The resonance absorption in the hematite sample used was quite large as can bee seen from the amplitudes of the absorption spectrum.

We have succeeded in observing a double resonance scattering process. The double resonance scattering spectrum was measured by detecting conversion X-rays following resonance absorption in an 80% enriched hematite scatterer of γ-rays after being recoilless-recoillessly scattered by a 90.6% enriched 2.28×10^{-3} mm thick metallic iron foil. The double scattering geometry used in the experiment is shown in Fig. 3. The scattering angle for the first process was 90°, whereas the detected conversion X-rays were emitted mainly in the forward direction. The hematite scatterer was placed directly on the beryllium window of a proportional counter. The X-rays

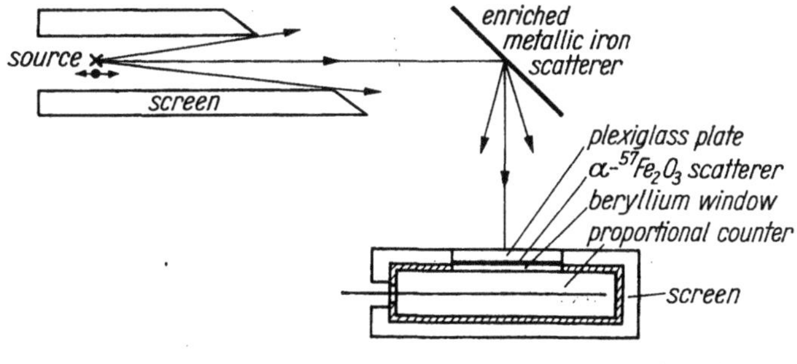

Fig. 3. The double resonance scattering geometry used in the experiment

scattered by the metallic iron foil towards the detector were absorbed by a 4.5 mm thick plexiglass plate. Attention was paid to enhancing the X-ray intensity rather than to minimize noise created in the surrounding screens. The 25 mCi $^{57}Co(Cr)$ source was driven with a constant acceleration. A Polon-type Mössbauer spectrometer was used. X-rays and γ-rays were detected and the two corresponding scattering spectra were stored simultaneously in a Mera 303 minicomputer. The observed X-ray (1) and γ-ray (2) scattering spectra are shown in Fig. 4, where the positions of the Zeeman lines in the metallic iron and hematite scatterers are also indicated.

The single resonance γ-ray scattering spectrum (2) of the metallic iron foil is distorted mainly by the resonance absorption of the recoilless–recoillessly scattered γ-rays which takes place in the hematite sample. The influence of doubly scattered γ-rays on the spectrum is negligible. It can be noticed that the amplitudes of the sixth, fifth, and third lines of the metallic iron Zeeman spectrum (2) are smaller than those of the first, second, and fourth lines of the double resonance spectrum respectively. Nearly the same positions of 3F and 3H and also 6F and 5H lines on the velocity scale (Fig. 4) explain the reduction of the amplitudes of the sixth and third Zeeman lines of the metallic iron scattering spectrum. The γ-rays absorbed by the fifth Zeeman line of the metallic iron foil are re-emitted in two fractions with energies corresponding to the fifth and third Zeeman lines, respectively. This explains the reduction in amplitude of the fifth line.

The influence of the double resonance scattering process on the X-ray scattering spectrum (1) can be clearly noticed by comparing the amplitudes of the first and sixth, second and fifth, and also third and fourth Zeeman lines of this spectrum. The enhancement of the sixth, fifth, and third line amplitudes as compared with the corresponding ones is due to the emission of conversion X-rays which followed the resonance absorption in the hematite sample of γ-rays recoilless–recoillessly scattered by the metallic iron foil. The attenuation of the sixth, fifth, and third Zeeman lines

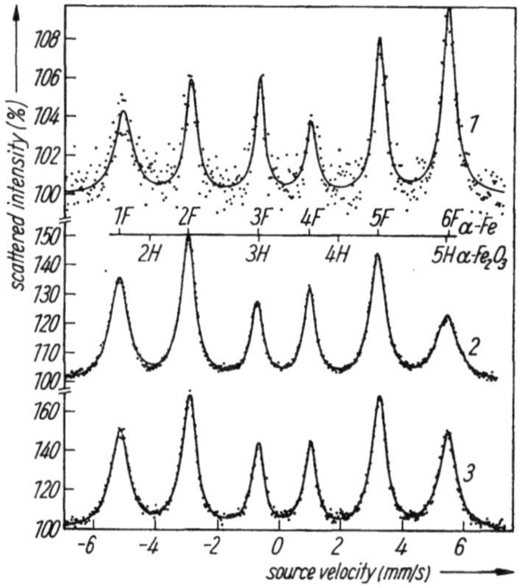

Fig. 4

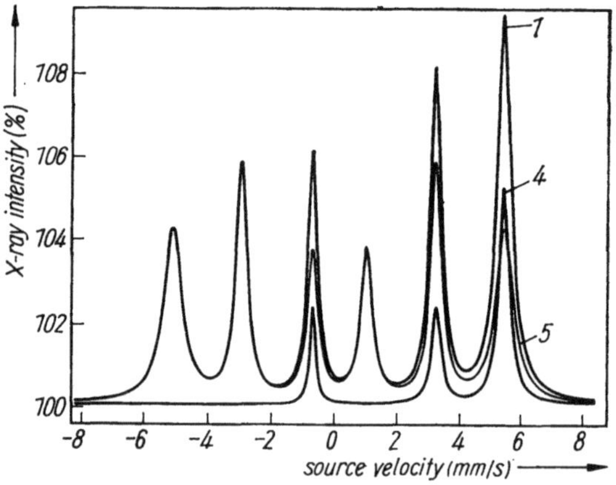

Fig. 5

Fig. 4. The double resonance scattering spectrum (1) and the corresponding single resonance distorted (2) and undistorted (3) γ-ray scattering spectra

Fig. 5. The decomposition of the double resonance scattering spectrum (1) into the pure double resonance scattering lines (4) and the single resonance X-ray scattering spectrum (5)

of the single resonance γ-ray scattering spectrum (2) is, of course, reflected in the form of the enhancement of the sixth, fifth, and third Zeeman lines of the X-ray scattering spectrum (1).

The computer decomposition of the spectrum (1) is shown in Fig. 5. The pure double resonance scattering lines are superimposed at the positions of the sixth, fifth, and third lines on the single scattering X-ray spectrum. The single resonance scattering X-ray spectrum (5) is mainly created by the X-rays which follow photoabsorption in the hematite scatterer of γ-rays resonantly scattered by the metallic iron foil. Its linewidths and relative line amplitudes are nearly the same as those of the γ-ray single resonance scattering spectrum (3) of the metallic iron foil measured with the same geometry without the hematite scatterer (Fig. 4). The subtraction of spectrum (5) from the spectrum (1) gives the pure double resonance scattering spectrum (4) composed of three lines. Their widths are smaller than those of corresponding lines of spectrum (5). In order to get more efficient linewidth reduction, thin samples and small solid angles have to be used.

References

[1] J. J. Bara, in: Proc. Winter Physics School Hyperfine Interactions, Bielsko-Biała (Poland) 1979, Ed. R. Kulessa and K. Królas, Rep. 1066/PL, Institute of Nuclear Physics, Cracow (p. 226).
[2] J. J. Bara, Rep. No. 1011/Pl, Institute of Nuclear Physics, Cracow 1978.
[3] J. J. Bara, phys. stat. sol. (a) **59**, 155 (1980).
[4] J. J. Bara, phys. stat. sol. (a) **58**, 349 (1980).
[5] J. J. Bara, phys. stat. sol. (a) **64**, 509 (1981).

(Received September 24, 1980)

phys. stat. sol. (a) 65, 65 (1981)

Subject classification: 5 and 18.2; 22.8

Solid State Laboratory, Department of Chemistry (a) and Department of Physics (b), Indian Institute of Technology, Bombay[1])

Mössbauer Spectroscopic Studies of the $Ba_xLa_{1-x}CoO_3$ System

By

S. B. PATIL (a), D. K. CHAKRABARTY[2]) (a),
M. V. BABU (b), and S. N. SHRINGI (b)

Mössbauer spectra of the system $Ba_xLa_{1-x}CoO_3$ ($0 \leqq x \leqq 0.4$) are studied at various temperatures. The results show an equilibrium between low spin and high spin states of Co^{3+} at room temperature. The compounds with $x > 0.1$ are largely oxygen deficient. Above room temperature tetravalent cobalt is formed by electron transfer from one trivalent cobalt to another. The Mössbauer spectral data together with initial susceptibility and hysteresis properties suggest the formation of a number of single domain magnetic clusters in a paramagnetic matrix.

Die Mößbauerspektren des Systems $Ba_xLa_{1-x}CoO_3$ ($0 \leqq x \leqq 0{,}4$) werden bei verschiedenen Temperaturen untersucht. Die Ergebnisse zeigen ein Gleichgewicht zwischen Low-Spin- und Hoch-Spin-Zuständen des Co^{3+} bei Zimmertemperatur. Die Verbindungen mit $x > 0{,}1$ haben starken Sauerstoffunterschuß. Oberhalb Zimmertemperatur wird tetravalentes Kobalt durch Elektronentransfer von einem trivalenten Kobaltatom zum anderen gebildet. Die Mößbauerspektralwerte zusammen mit der Anfangssuszeptibilität und den Hysterese-Eigenschaften weisen auf die Bildung magnetischer Eindomänencluster in einer paramagnetischen Matrix hin.

1. Introduction

The compound lanthanum cobalt oxide ($LaCoO_3$) has been studied extensively for its crystallographic [1, 2], electrical [3, 4], and magnetic properties [3 to 5]. The rhombohedrally distorted perovskite undergoes a first-order transition at 1210 K to pseudo-cubic structure [2]. The ground state of cobalt is the low spin t_{2g}^6 and this differs only by ≈ 0.05 eV from the high spin trivalent state $t_{2g}^4e_g^2$ of cobalt [5] and at room temperature, both the states coexist. Following the notation of Raccah and Goodenough [2], they will be referred to as Co^{III} and Co^{3+}, respectively. Substitution of lanthanum by divalent strontium gives rise to some interesting properties. It was observed, for example, that compounds $Sr_xLa_{1-x}CoO_3$ are ferromagnetic at $x \geqq 0.15$ [3] and the composition $x = 0.5$ is metallic [6].

Bhide et al. [7] have shown the coexistence of Co^{III} and Co^{3+} in $LaCoO_3$ by Mössbauer spectroscopy at low temperature. In order to account for the anomalous changes in the areas under resonance near room temperature and above, they have proposed that the change Co^{III} to Co^{3+} begins at a low temperature and continues up to 200 K. Above this temperature, an electron transfer from Co^{3+} to Co^{III} takes place leading to the formation of low spin divalent Co^{II} ($t_{2g}^6e_g^1$) and intermediate spin tetravalent Co^{IV} ($t_{2g}^4e_g^1$). These authors have also studied the Mössbauer spectra of the system $Sr_xLa_{1-x}CoO_3$ [8]. The spectra of the paramagnetic samples ($x < 0.125$) were qualitatively similar to that of $LaCoO_3$. For magnetic samples ($0.125 \leqq x \leqq 0.5$), they observed at 78 K an intense central peak superimposed on the six-line ferromagnetic pattern. The authors interpreted the results on the assumption of a model originally proposed by Raccah and Goodenough [9].

[1]) Bombay 400076, India.

[2]) To whom all correspondence should be addressed.

According to the model proposed by Raccah and Goodenough [9], the 3d holes introduced by substitution of lanthanum by divalent strontium is not localized at a single cobalt ion, but is shared by eight cobalt ions nearest to the strontium forming an acceptor level. As x increases, the acceptor complex grows in size to form an impurity band giving rise to itinerant ferromagnetism.

Earlier [10], we reported the structural, electrical, and magnetic properties of the system $Ba_xLa_{1-x}CoO_3$ for $0 \leqq x \leqq 0.5$. It was shown that the results could be explained by the itinerant electron model as well as by Zener double exchange model. Zener double exchange requires the existence of the same metal atom in two different oxidation states differing by one electron to be linked by an oxygen atom. A simultaneous transfer of an electron from the metal orbital to the p orbital of oxygen and an oxygen p electron to the orbital of the other metal atom gives rise to ferromagnetic interaction [11]. Our interpretation was based on the assumption that for higher values of x, the compounds were highly oxygen deficient. We have now studied the Mössbauer spectra of these compounds. The results confirm that our assumption on oxygen deficiency was correct. Besides giving us an opportunity to compare our results with those reported for $Sr_xLa_{1-x}CoO_3$, they suggest the formation of magnetic clusters in a crystallographically homogeneous phase. The formation of magnetic clusters is supported by the measurement of initial susceptibility and hysteresis as a function of temperature. These results are also reported here.

2. Experimental

Preparation of barium lanthanum cobaltates has been described elsewhere [10]. For preparing the Mössbauer sources, a few drops of $^{57}CoCl_2$ in hydrochloric acid solution were added to the sample in the form of 1 to 2 mm thick pellet and dried with an IR lamp. Activity was diffused into the sample by heating in UHP argon at 1200 K for 4 h and then heating in air at 1250 K for 4 h. The barium containing samples were further heated for 2 h at 1400 K and slowly cooled. The surface was then polished to remove any undiffused surface radioactivity.

The Mössbauer spectra were obtained using a electro-mechanical transducer drive operating in a constant acceleration mode in conjunction with a 1024 channel analyser (ECIL MCA-38A). For the spectrum of standard iron foil, the linewidth obtained was 0.3 mm/s. The spectra were obtained at several temperatures in the range 300 to 725 K ($\pm$ 5 K). The standard absorber used was 310 stainless steel. The 300 K data were also repeated using $K_4Fe(CN)_6 \cdot 3H_2O$ absorber to get a better resolution. The spectra were analysed by least square fitting technique.

Initial susceptibility was measured in a field 0.5 Oe by a double coil method of Radhakrishnamurty and Likhite [12]. Hysteresis loops were obtained with an improved version of the apparatus described by Likhite et al. [13].

3. Results

Room temperature Mössbauer spectra of the various samples with $K_4Fe(CN)_63H_2O$ absorber are shown in Fig. 1. The asymmetric spectra are resolved into two Lorentzians. The relative areas under resonance, isomer shift (IS), and linewidth of these spectra are given in Table 1. Fig. 2a and b shows the spectra for the compositions $x = 0.2$ and $x = 0.4$ at various temperatures. The variation of the various Mössbauer parameters with temperature are shown in Table 2. Variation of initial susceptibility with temperature are given in Fig. 3. Initial susceptibility increases rapidly with temperature to attain a maximum value and then sharply falls to zero. Curie temperatures (T_C) were obtained from these curves by extrapolating the part with highest slope to the temperature axis.

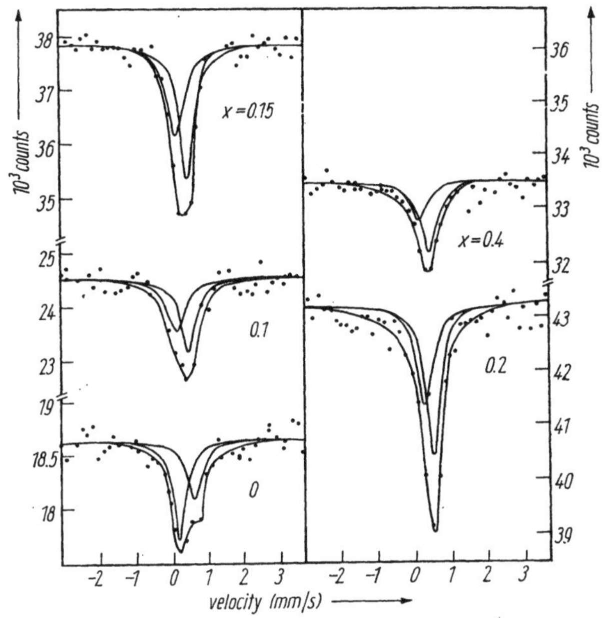

Fig. 1. Mössbauer spectra of $Ba_xLa_{1-x}CoO_3$: ^{57}Co matched against $K_4Fe(CN)_6 \cdot 3\,H_2O$ absorber at room temperature

Table 1

Mössbauer parameters for $Ba_xLa_{1-x}CoO_3$ at 300 K

composition x	high energy resonance line			low energy resonance line		
	(% area)	linewidth+) (mm/s)	IS*)+) (mm/s)	(% area)	linewidth+) (mm/s)	IS*)+) (mm/s)
0.0	40	0.47	0.68	60	0.42	0.23
0.1	55	0.47	0.61	45	0.56	0.31
0.15	56	0.47	0.61	44	0.56	0.31
0.2	55	0.47	0.53	45	0.56	0.31
0.4	66	0.56	0.53	34	0.56	0.23

The absorber was $K_4Fe(CN)_6 \cdot 3\ H_2O$ single crystal.

*) IS are measured with respect to Fe.

+) Estimated error is ± 0.05 mm/s.

4. Discussion

Our interpretation of the Mössbauer spectra are based on the assumption that the presence of high spin Co^{3+} will show the resonance of high spin Fe^{3+} and that of low spin Co^{III} will show the resonance of low spin Fe^{III} in the Mössbauer spectra. This assumption has been justified by several authors [7, 14, 15]. Based on isomer shift systematics, the high resonance line can be assigned to high spin Fe^{3+} arising from Co^{3+}. The low resonance line may be due to low spin Fe^{III} formed by electron capture from Co^{III}. The barium containing samples must contain some tetravalent cobalt most likely to be present in the low spin state (Co^{IV}), the resonance of which will lie in the same region as that of Co^{III}. Hence the low resonance line in these compounds is assigned to a mixture of Co^{III} and tetravalent cobalt.

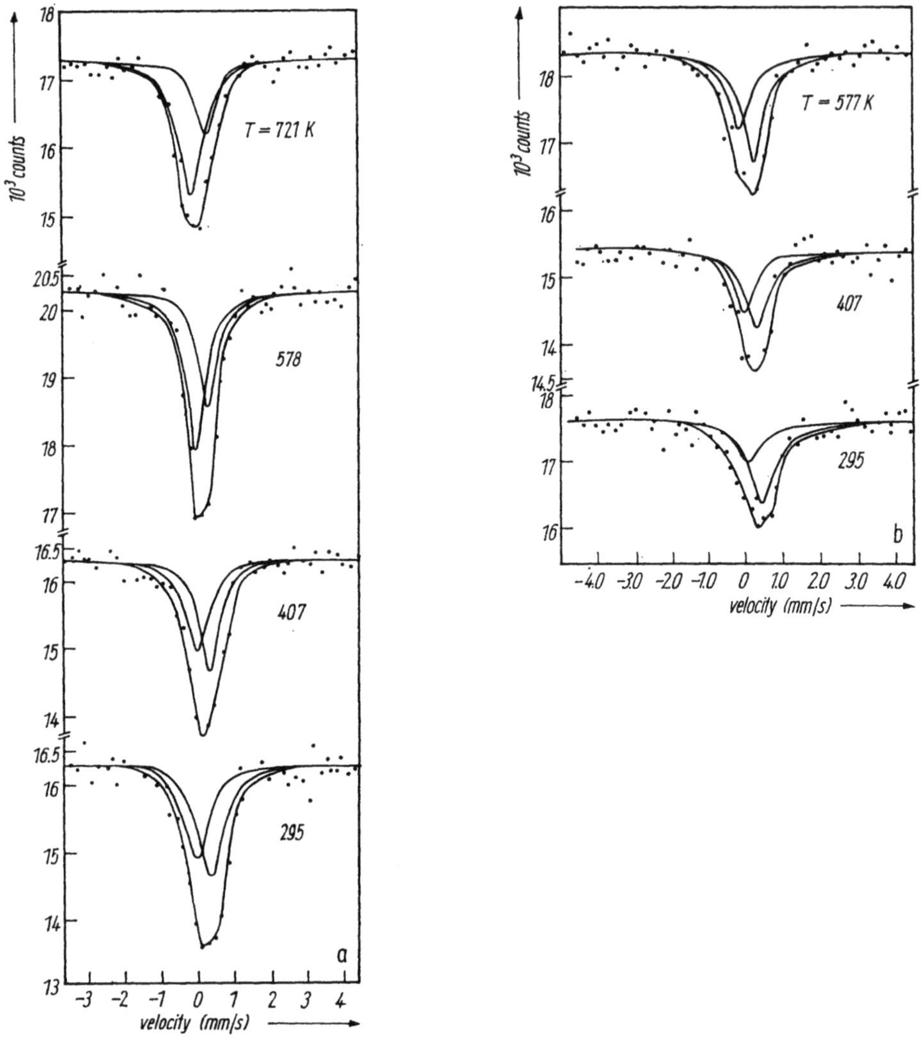

Fig. 2. Mössbauer spectra of a) $Ba_{0.2}La_{0.8}CoO_3$ and b) $Ba_{0.4}La_{0.6}CoO_3$ matched against 310 stainless-steel absorber at various temperature

The room temperature spectra of these compounds show some interesting features. Although IS due to Co^{3+} remains unchanged with composition, the IS values for the low resonance shift slightly towards higher values as x increases. This may suggest that Co^{IV} has a slightly higher isomer shift than Co^{III} and as the concentration of Co^{IV} should increase with x, the resonance shifts towards the positive side although the spectrum does not get resolved into separate resonances due to Co^{III} and Co^{IV}.

Another important aspect of the room temperature spectra is the change in the relative area under resonance with composition. It is seen from Table 1 that as x changes from zero to 0.1, the intensity of the resonance due to Co^{3+} decreases, obviously due to its conversion to Co^{IV}. With further increase in x, however, the change is much less pronounced. Assuming that the relative area under resonance is indicative of the

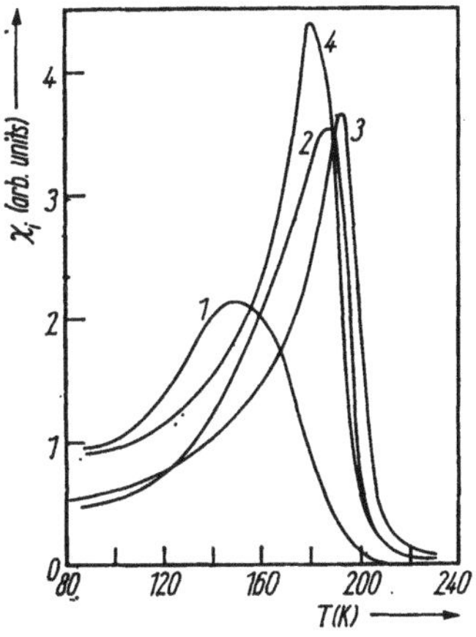

Fig. 3. Variation of initial susceptibility with temperature for samples $Ba_xLa_{1-x}CoO_3$: (1) $x = 0.2$, $T_C = 194$; (2) 0.3, 203; (3) 0.4, 208; (4) 0.5, 200 K

Co^{3+}/Co^{III} ratio, such a behaviour can only mean that by increasing x beyond 0.1, one does not change Co^{3+} to Co^{IV} appreciably. This can only happen if a large concentration of oxygen vacancies is present in these compounds, which was suggested by us earlier to explain the variation of electrical resistivity of these samples with composition [10].

It is well known that lanthanum cobaltate is oxygen deficient and should be written as $LaCoO_{3-\lambda}$. Some authors have suggested a volumetric method to obtain the oxidation states of cobalt in these cobaltates from which λ can be calculated [16]. This method did not give us any consistent results. Although we could not determine λ

Table 2

Mössbauer parameters for $Ba_xLa_{1-x}CoO_3$ obtained at various temperatures

composition x	T (K)	high energy resonance line			low energy resonance line		
		(% area)	linewidth⁺) (mm/s)	IS⁺)*) (mm/s)	(% area)	linewidth⁺) (mm/s)	IS⁺)*) (mm/s)
	295	55	0.56	0.43	45	0.56	0.21
	407	48	0.56	0.38	52	0.56	0.12
0.1	574	37	0.68	0.43	63	0.56	0.12
	720	34	0.85	0.29	66	0.85	0.16
	295	58	0.68	0.56	42	0.68	−0.16
	405	49	0.56	0.48	51	0.56	0.25
0.15	575	40	0.68	0.39	60	0.68	0.12
	719	33	0.68	0.39	67	0.85	−0.07
	295	55	0.85	0.47	45	0.85	0.12
	407	49	0.68	0.39	51	0.85	0.03
0.2	578	42	0.68	0.39	58	0.68	0.03
	721	29	0.68	0.39	71	0.85	−0.06
	295	66	0.85	0.48	34	0.85	0.12
0.4	407	61	0.85	0.34	39	0.68	0.03
	577	59	0.68	0.30	41	0.68	−0.11

The absorber used was 310 stainless steel.

⁺) Estimated error is ± 0.05 mm/s.

*) IS are measured with respect to Fe.

it is supported by other experiments that the compositions beyond $x = 0.1$ are more oxygen deficient. Differential thermial analysis and thermogravimetric experiments have shown that the oxygen uptake by heating the samples at constant temperature (900 to 1000 K) is negligible for $x = 0$ and 0.1, but increases sharply for $x \geqq 0.2$ [17].

It was shown earlier that the compounds $M_xLa_{1-x}CoO_3$ where ($M = Sr^{2+}$ or Ba^{2+}) are paramagnetic for $x \leqq 0.125$ and ferromagnetic beyond this [8, 10]. Bhide et al. [8] have reported two resonances in the Mössbauer spectra of the paramagnetic compositions in the system $Sr_xLa_{1-x}CoO_3$, but noticed a single peak in the ferromagnetic compositions above Curie temperature (T_C). Our results for the $Ba_xLa_{1-x}CoO_3$ system are distinctly different in that we observed two resonance peaks in the spectra of both paramagnetic and ferromagnetic samples. The assignment of the two resonances in the paramagnetic samples by Bhide et al. [7, 8] are similar to that by us. However, the single resonance in the Mössbauer spectra of the ferromagnetic samples was interpreted by them as due to the itinerant electron which gives the cobalt ion a time-averaged configuration between Co^{3+} and Co^{IV}. It is difficult to comment on these results as the authors do not provide complete tables of isomer shift and linewidth for the various compositions. For the $Ba_xLa_{1-x}CoO_3$ system studied by us, the spectra are distinctly asymmetric and could be resolved into two Lorentzians in case of the paramagnetic as well as the ferromagnetic samples.

We shall now comment on the temperature dependence of the Mössbauer spectra. It is clear from Table 2 that the intensity of the resonance due to Co^{3+} decreases at the cost of the low resonance peak. Normally, one would expect an increase in Co^{3+} with temperature, as it has slightly higher energy than Co^{III}, which along with Co^{IV} constitutes the low resonance peak. In order to explain the changes in electrical conductivity of $LaCoO_3$ with temperature, Heikes et al. [4] suggested that there is an electron transition from one trivalent cobalt to another resulting in a pair of divalent and tetravalent cobalt. This exchange, needless to say, takes place through the p orbital of the intermediate oxygen atom. The presence of high spin Co^{2+} can be ruled out as it should show a high isomer shift (≈ 1 mm/s with respect to iron). This would mean that divalent cobalt is in a low spin state (Co^{II}), that will have Mössbauer resonance near zero velocity and would merge with the low resonance peak in the spectra. The electron jump can be represented as

$$Co^{3+}(t_{2g}^4e_g^2)\text{–O–}Co^{III}(t_{2g}^6e_g) \rightarrow Co^{IV}(t_{2g}^4e_g^1)\text{–O–}Co^{II}(t_{2g}^6e_g^1)\ ,$$

where $Co^{IV}(t_{2g}^4e_g^1)$ is an intermediate spin state of tetravalent cobalt. Such a transition that does not need any spin reversal can occur easily and has been suggested by Bhide et al. to explain the Mössbauer spectra of $LaCoO_3$ [7]. This explains very well why the intensity of the resonance due to Co^{3+} decreases with temperature.

Bhide et al. [8] have considered several models to explain ferromagnetism in the system $Sr_xLa_{1-x}CoO_3$ ($x \geqq 0.125$). Their Mössbauer results are compatible with itinerant electron ferromagnetism [9] as well with Zener double exchange [11]. The initial susceptibility (χ_i) versus temperature plots (Fig. 3) of our samples are typical of single domain materials [18, 19]. χ_i versus T curves may show three distinct types: (i) Hopkinson type, (ii) single domain (SD) type, and (iii) multidomain (MD) type. Single domain character of the material is also reflected in the low temperature Mössbauer results [8]. As the temperature is raised from 80 K the six-fingered spectrum of $Sr_{0.5}La_{0.5}CoO_3$ gradually changes showing an intense central peak superimposed on the six-line spectrum. This is typical of SD material changing over to the superparamagnetic state before reaching T_C [20]. The change in the hysteresis loop characteristics with temperature shows the following featurse. The coercive field (H_c)

decreases with rise in temperature and the relative remanence also decreases. This is suggestive of randomly oriented SD particles.

All these results suggest that the ferromagnetic samples consists of grains having a number of single domains. In a chemically and crystallographically homogeneous grain, there may be a statistical fluctuation of compositions. This may give rise to magnetic clusters in the barium rich regions in a otherwise paramagnetic matrix. Formation of such clusters has been reported for $BaFe_4Ti_2O_{11}$ [21]. Strong ferromagnetic interaction may arise in the barium rich region due to exchange interaction between intermediate spin tetravalent cobalt ($t_{2g}^4e_g^1$) and high spin $Co^{3+}(t_{2g}^4e_g^2)$ lying across an oxygen atom. The itinerant model, on the other hand, would require the impurity band (due to tetravalent cobalt) to extend throughout the lattice. This is not likely to give rise to single domain grains. It thus appears that a large number of single domain clusters formed in a paramagnetic host determines the magnetic properties of these compounds.

5. Conclusions

Mössbauer spectral studies of the system $Ba_xLa_{1-x}CoO_3$ ($0 \leqq x \leqq 0.4$) confirm the coexistence of high spin and low spin trivalent cobalt at room temperature. At higher temperature, electron transfer occurs from high spin to low spin cobalt. The Mössbauer results along with initial susceptibility and hysteresis loop properties suggests the formation of single domain magnetic clusters in compositions with $x \geqq 0.1$.

Acknowledgement

We thank Dr. C. Radhakrishnamurty of the Tata Institute of Fundamental Research, Bombay for kindly extending experimental facilities and helpful discussions.

References

[1] H. L. Yakel, Acta cryst. **8**, 394 (1955).
[2] P. M. Raccah and J. B. Goodenough, Phys. Rev. **155**, 932 (1967).
[3] G. H. Jonker and J. H. van Santen, Physica (Utrecht) **19**, 120 (1953).
[4] R. R. Heikes, R. C. Miller, and R. Mazelsky, Physica (Utrecht) **30**, 1600 (1964).
[5] G. H. Jonker, J. appl. Phys. **37**, 1424 (1966).
[6] N. Menyuk, P. M. Raccah, and K. Dwight, Phys. Rev. **166**, 510 (1968).
[7] V. G. Bhide, D. S. Rajoria, G. Rama Rao, and C. N. R. Rao, Phys. Rev. B **6**, 1021 (1972).
[8] V. G. Bhide, D. S. Rajoria, C. N. R. Rao, G. Rama Rao, and V. G. Jadhao, Phys. Rev. B **12**, 2832 (1975).
[9] P. M. Raccah and J. B. Goodenough, J. appl. Phys. **39**, 1209 (1968).
[10] S. B. Patil, H. V. Keer, and D. K. Chakrabarty, phys. stat. sol. (a) **52**, 681 (1979).
[11] C. L. Zener, Phys. Rev. **82**, 403 (1951).
[12] C. Radhakrishnamurty and S. D. Likhite, Earth Planet Sci. Letters **7**, 389 (1970).
[13] S. D. Likhite, C. Radhakrishnamurty, and P. W. Sahasrabudhe, Rev. sci. Instrum. **36**, 1558 (1965).
[14] J. G. Mullen and H. N. Ok, Phys. Rev. Letters **17**, 287 (1966).
[15] H. N. Ok and J. G. Mullen, Phys. Rev. **168**, 563 (1966).
[16] H. Taguchi, J. Solid State Chem. **29**, 221 (1979).
[17] D. Y. Rao, S. B. Patil, and D. K. Chakrabarty, to be published.
[18] C. Radhakrishnamurty, S. D. Likhite, and R. Nagarjan, J. Magnetism magnetic Mater. **15/18**, 195 (1980).
[19] C. Radhakrishnamurty and N. G. Nanadikar, Pramana **13**, 413 (1979).
[20] U. Gonser, H. Wiedersich, and R. W. Grant, J. appl. Phys. **39**, 1004 (1968).
[21] E. Kneller, M. Velicescu, and F. Habery, J. Magnetism magnetic Mater. **7**, 49 (1978).

(Received November 5, 1980)

phys. stat. sol. (a) **65**, 73 (1981)

Subject classification: 1.1 and 1.4; 21.6

Rijksuniversitair Centrum Antwerpen[1])

Electron Diffraction and High Resolution Electron Microscopy Study of Ordering in the Gold–Manganese System[2])

I. Lozenge[3]) Shaped Island Structures Derived from the DO_{22} Structure $Au_{3+x}Mn$

By

G. VAN TENDELOO and S. AMELINCKX[4])

Ordered alloys with a composition around $Au_{3+x}Mn$ ($0.3 < x < 0.5$) are analysed by means of electron diffraction and dark field as well as bright field high resolution electron microscopy. Under suitable conditions the configuration of Mn columns can be imaged as bright dots and different superstructures of the DO_{22} basic structure can be identified in this way. The superstructures consist of DO_{22} islands of different sizes separated by two families of antiphase boundaries enclosing an angle of 71° to 77° depending upon the island size. The formation of such two-dimensional superstructures is clearly a way to accomodate changes in composition when deviating from the ideal 3–1 composition towards the Au_4Mn composition. Along the non-conservative antiphase boundaries, atoms adopt the Au_4Mn configuration. Very small changes in composition probably alter the island size or the frequency of occurrence of a given island type since very often different island sizes are present in the same area.

Geordnete Legierungen mit einer Zusammensetzung um $Au_{3+x}Mn$ ($0,3 < x < 0,5$) werden mittels Elektronenbeugung sowie Dunkel- als auch Hellfeld-Hochauflösungselektronenmikroskopie analysiert. Unter geeigneten Bedingungen kann die Konfiguration der Mn-Spalten als helle Punkte abgebildet werden, und verschiedene Superstrukturen der DO_{22}-Basisstruktur lassen sich auf diese Weise identifizieren. Die Superstruktur besteht aus DO_{22}-Inseln unterschiedlicher Größe, getrennt durch zwei Familien von Antiphasengrenzen, die einen Winkel von 71° bis 77° in Abhängigkeit von der Inselgröße bilden. Die Erzeugung solcher zweidimensionaler Superstrukturen ist ein sicherer Weg, Änderungen in der Zusammensetzung anzupassen, wenn sie von der idealen 3–1-Zusammensetzung hin zur Au_4Mn-Zusammensetzung abweicht. In Richtung der nichtkonservativen Antiphasengrenzen nehmen die Atome die Au_4Mn-Konfiguration an. Sehr geringe Änderungen in der Zusammensetzung ändern wahrscheinlich die Inselgröße oder die Frequenz des Auftretens eines gegebenen Inseltyps, da sehr oft verschiedene Inselgrößen auf derselben Fläche anwesend sind.

1. Introduction

In the course of a systematic study of the structure of gold–manganese alloys with compositions in the vicinity of Au_4Mn we discovered a new structure type derived from the DO_{22} structure [1]. The structure images, obtained in dark field, using superstructure spots only, suggested a non-ideal structure, consisting of islands of DO_{22} structure of varying sizes, i.e. containing arrays of 3×3, 3×4, and 4×4 manganese columns, separated by systematically ledged antiphase boundaries composed of segments of $(101)_{DO_{22}}$ planes, along which the atom configuration is that of Au_4Mn.

[1]) Groenenborgerlaan 171, B-2020 Antwerpen, Belgie.

[2]) Work performed under the auspices of the association RUCA-SCK with the financial support of the IIKW-NFWO (Brussels).

[3]) We shall use the term "lozenge" shaped islands rather loosely denoting also parallelogram shaped islands.

[4]) Also at SCK/CEN, B-2400 Mol, Belgium.

Fig. 1. Specimen area containing the following different phases: (1) and (1') two different variants of the Au_4Mn phase with the Ni_4Mo structure. (2), (2'), and (2'') three different variants of a one-dimensional long period antiphase boundary structure derived from the Au_4Mn structure. (3) one variant of the square island superstructure derived from the Au_4Mn structure, and containing 3 × 3 manganese columns. (4) one variant of a lozenge shaped island superstructure of the DO_{22} structure consisting of islands containing 3 × 4 and 3 × 3 manganese columns. The different structures are presumably formed because of small local differences in composition

In the idealized structure, consisting only of 3×3 islands, the ideal composition is $Au_{31}Mn_9$ and the average directions of the antiphase boundaries enclose angles of $\approx 77°$. The model, suggested by the images, could so far only be verified through a comparison of observed and calculated multidomain diffraction patterns, taking all six orientation variants into account. This was due to the fact that in the alloy with average composition Au_4Mn (20 at% Mn) it proved to be difficult to produce a domain size large enough to obtain monodomain patterns, especially since at this composition the new phase is metastable and after long annealing the final stable Au_4Mn structure is formed.

We have now studied alloys with a macroscopic composition in the range 20.5 to 23 at% Mn. After heat treatment (one week at 300 °C) domains are formed in some alloys which are large enough to allow monodomain patterns to be produced.

However, the complex microdomain structure in most of the alloys is well illustrated on the high resolution micrograph of Fig. 1. Apart from two orientation variants of the Au_4Mn structure (marked by 1 and 1′) three more superstructures are present in

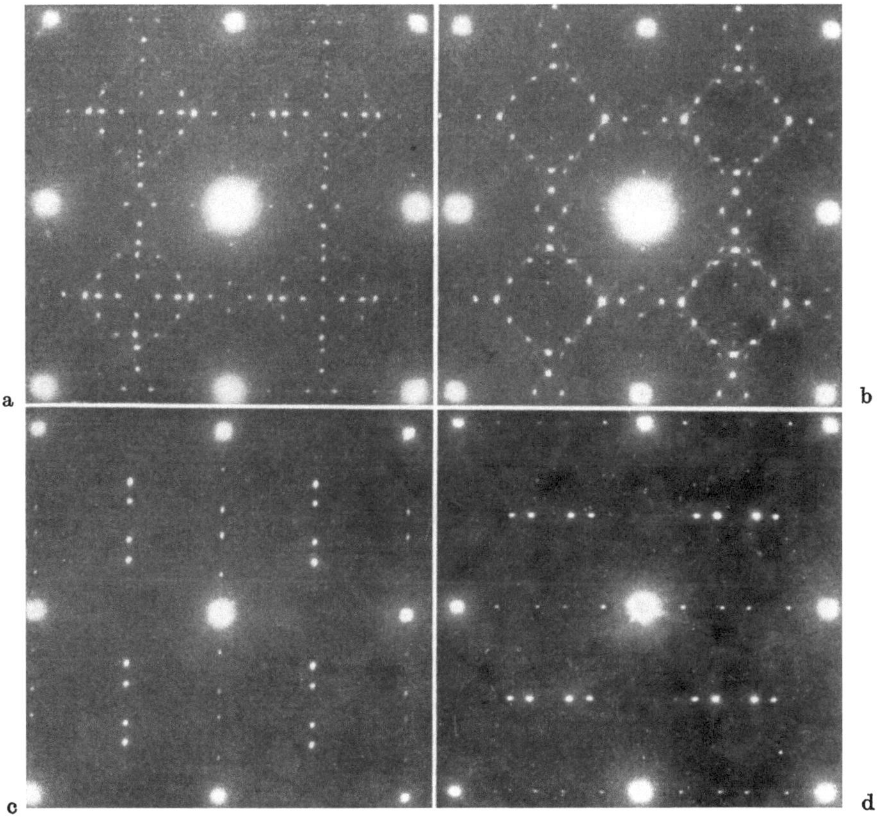

Fig. 2. Poly-domain diffraction patterns due to parallelogram and lozenge shaped island structures. a) All six orientation variants contribute to the diffraction pattern. b) Only the two coaxial variants with their shortest lattice vector along the zone axis contribute to the image. c) Two coaxial variants with the island plane vertical. d) Two coaxial variants with the island plane horizontal. Note that the lattice parameter perpendicular to the island plane is equal to the cube edge of the basic f.c.c. lattice

this restricted area of a few hundred Å wide: a one-dimensional superstructure of Au_4Mn (areas 2, 2′ and 2″), a two-dimensional superstructure based on Au_4Mn (region 3) and an island superstructure based on a DO_{22} configuration (region 4). This last type of superstructure will be the main subject of this part I, the one- and two-dimensional superstructures of Au_4Mn will be treated in part II.

The new observations presented here support our previous model and have moreover allowed to establish other regular structures built on the same principle. We already reported very briefly on these new observations in [2]. Our initial results [1] have since been extended by Terasaki et al. [3] who have noted also the existence of islands of different sizes.

2. Alloy and Specimen Preparation

Ten different gold–manganese alloys with macroscopic compositions in the range 17 to 25 at% Mn were prepared by arc melting. The weight loss upon melting and homogenizing the material was such that if all losses would be attributed either to gold or to manganese the final composition would not differ from the original one by more than 0.50 at%. All alloys containing more than 77 at% of gold could be rolled. Circular discs with a diameter that would allow to fit them in the specimen holder were obtained by punching. All specimens were first heat treated during three days at 500 °C and then water quenched; they were subsequently ordered at 300 °C during at least one week. Final electropolishing of the discs was done in an electrolyte consisting of water, sulphuric acid, and hydrochloric acid [4].

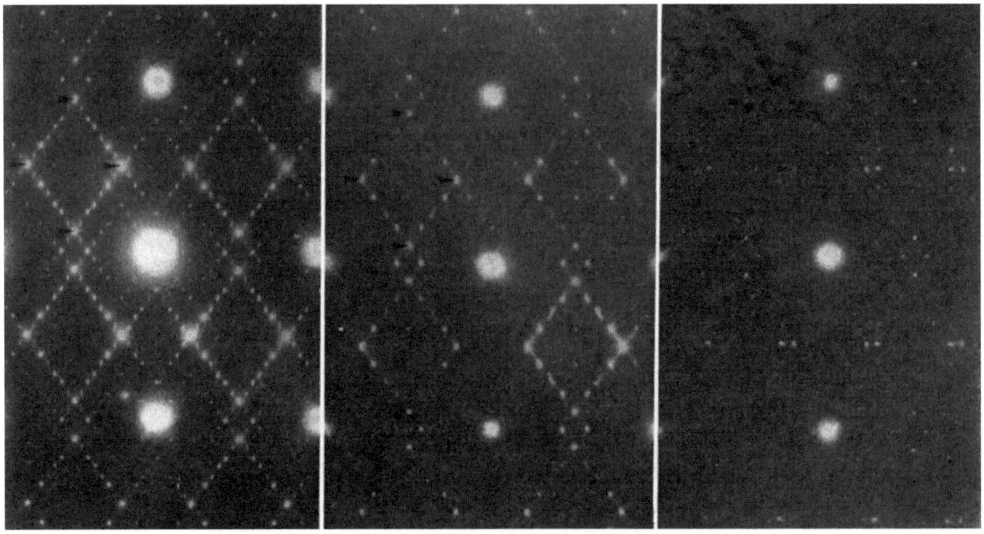

a b c

Fig. 3. Mono-domain diffraction patterns of the "lozenge" shaped island superstructure of the DO_{22} structure. Note that the most intense spots or groups of spots are situated in the immediate vicinity of DO_{22} spots. a) Well ordered structure containing a mixture of 3 × 3 and 3 × 4 islands (see also Fig. 1, area 4). b) Less well ordered structure containing predominantly 4 × 4 islands. c) Structure with variable island sizes

3. Diffraction Patterns

3.1 Observations

The diffraction pattern of a polydomain single crystal with nominal composition Au_4Mn was reproduced in [1] as Fig. 2a and compared with a polydomain calculated pattern. A representative polydomain pattern of the present alloy, with 22 at% Mn, is shown in Fig. 2a. Since the domain size could be made larger in this new alloy we have now been able to unravel such diffraction patterns, due to a superposition of different variants (Fig. 2) and to obtain monodomain patterns by selecting a single orientation variant on the image. Apparently different types of patterns were obtained from different areas of the specimen. Three typical patterns are reproduced in Fig. 3a, b, c and schematically represented in Fig. 4. It should be noted that the most intense superlattice spots are located close to the positions of DO_{22} spots marked by open circles in Fig. 4b, suggesting that the basic structure is the DO_{22} structure.

In particular the pattern of Fig. 3c can be interpreted as being derived from the DO_{22} diffraction pattern by splitting of each superlattice spot in two components, connected by a weak diffuse streak. This pattern corresponds to a structure with a variable island size, the distribution being less sharply peaked than in the case of Fig. 3a. Diffraction patterns of this type have been analysed in detail in relation with the digenite structure [5].

In Fig. 3b the distance between the most intense superlattice spots, marked by arrows, is divided roughly in four parts. In Fig. 3a, on the other hand, the distance between the most intense spots is divided in seven visibly unequal intervals. The array consists in fact of two interpenetrating equidistant subarrays each associated with one of the intense spots, close to the DO_{22} spots (Fig. 4a). The spacing in these equidistant arrays is now $\approx 1:3.5$ of the distance between the intense superlattice spots (Fig. 4a). Moreover the arrays of superlattice diffraction spots are not aligned precisely along the line connecting these two intense superlattice spots. We conclude that there are small spacing anomalies as well as small orientation anomalies, leading to a pseudo-incommensurate diffraction pattern.

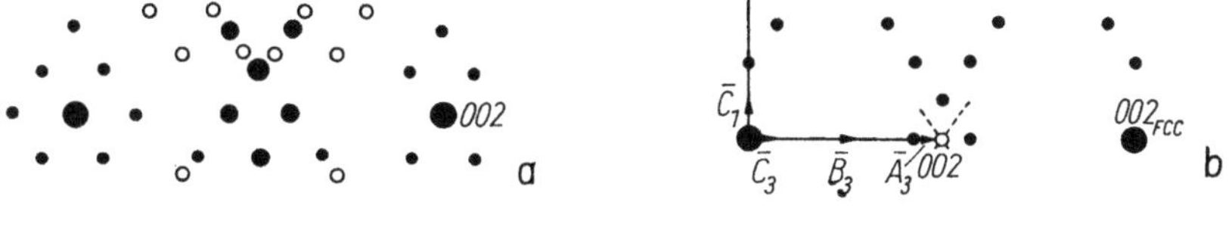

Fig. 4. a) Schematic representation of a diffraction pattern of type (a) of Fig. 3. Superstructure spots derived from different basic spots are represented by a different symbol (full or open dot). b) Schematic representation of a diffraction pattern of the ideal island structure containing 4×4 islands only. The spots due to the basic DO_{22} spots are represented as open circles. Note the fractional displacements of the sequences of superlattice spots with respect to the DO_{22} spots

3.2 *Analysis of the diffraction patterns*

As noted above the configurations of diffraction spots can be analysed in terms of linear arrays of equidistant spots of which the positions are determined with respect to the basic spots, i.e. the DO_{22} spots in the manner shown schematically in Fig. 4b.

The observations indicate that the structure contains pseudo-periodic sequences of interfaces with an average distance of 14 Å and with average directions which are not simple crystallographic directions; these directions enclose a small angle with the [101] and [10$\bar{1}$] directions as referred to the DO_{22} structure. From the diffraction pattern we can furthermore deduce that the angle enclosed by the average directions of the APB's is $\approx 74°$.

The arrays of superlattice spots are shifted with respect to the positions of the DO_{22} spots, indicated by open circles in Fig. 4b. The fractional shifts are indicated as well. It has been shown [6] that the displacement vectors of the families of parallel interfaces can be deduced from such fractional shifts. In Table 1 we have indicated the observed values of the fractional shifts associated with the different DO_{22} reflections. These fractional shifts are the same in all diffraction patterns. The calculated fractional shifts, assuming for both families of APB's suitable displacement vectors of the type $1/2\ [110]_{\text{f.c.c.}}$ are indicated in the table as well. The displacement vectors $\boldsymbol{R}_1$ and $\boldsymbol{R}_2$ are referred to the DO_{22} unit cell.

The projections of the displacement vectors along the zone of observation can also be derived in a more systematic manner. Let $\boldsymbol{R}_1$ be represented by $[uvw]$ then we have from Table 1 (4): $2w = 1/2$ i.e. $w = 1/4$ or $3/4$ (mod 1). Furthermore from Table 1 (2) we deduce that $u + w = 1/4$ or $3/4$ i.e. $u = 0$ or $1/2$ (mod 1). We have no information on the v component; however since it is reasonable to assume that the vector has to be of the type $1/2\ [110]_{\text{f.c.c.}}$ the only acceptable proposal is $v = 0$ i.e. $\boldsymbol{R}_1 = 1/4\ [201]_{DO_{22}}$. The same type of reasoning applies to $\boldsymbol{R}_2$, which is found to be $\boldsymbol{R}_2 = 1/4\ [20\bar{1}]_{DO_{22}}$.

The distance between interfaces, and hence the island size, is related to the spacing in the arrays of superlattice spots. The pseudo-incommensurate character of the diffraction pattern suggests the presence of a regular mixture of different integral spacings between interfaces, leading to a mixture of island sizes of 3×3, 3×4, and 4×4 manganese columns.

3.3 *Models*

The models suggested by these observations are closely related to the idealized model proposed previously for the transition structure of the alloy with nominal composition Au_4Mn [1]. Already in this paper we pointed out that the island size was not uniform

Table 1

$\boldsymbol{g}$ (DO_{22})		observed displacement (mod 1)		calculated $\boldsymbol{g} \cdot \boldsymbol{R}$	
		array 1	array 2	array 1 $\boldsymbol{R}_1 = \frac{1}{4}$ [201]	array 2 $\boldsymbol{R}_2 = \frac{1}{4}$ [20$\bar{1}$]
202	(1)	1/2	1/2	1/2	1/2
101	(2)	1/4(3/4)	1/4(3/4)	3/4	1/4
103	(3)	3/4(1/4)	3/4(1/4)	1/4	3/4
002	(4)	1/2	1/2	1/2	1/2

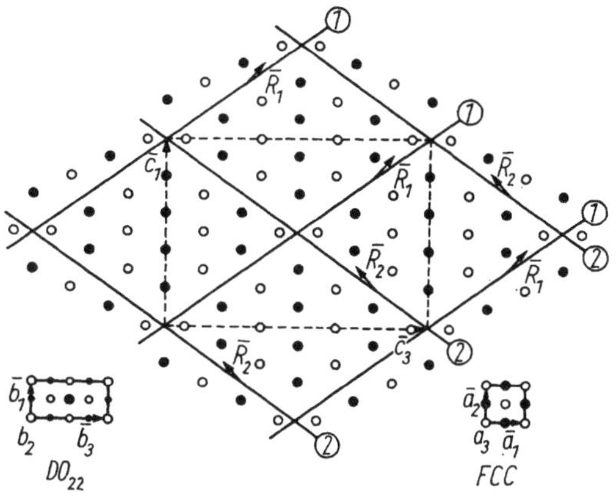

Fig. 5. "Lozenge" shaped island superstructure as derived from the diffraction pattern of Fig. 3b. The directions of the APB's are perpendicular to the rows of superstructure spots, whereas their spacing follows from the interspot distance. The displacement vectors are derived from the fractional shifts in the diffraction pattern of Fig. 4

and that islands containing 3 × 3, 3 × 4, and 4 × 4 columns occurred. We have represented in Fig. 5 the idealized 4 × 4 island structure as deduced from the diffraction pattern.

In the present case diffraction patterns such as 3a suggest a regular mixture of these islands leading to an average size of around 3.5 intercolumn distances in one or in both directions. This would lead to the possible, idealized models represented in Fig. 6, where also the angle φ enclosed by the average directions of the ledged antiphase boundaries have been indicated. We shall see that these models are consistent with the high resolution images to be discussed below. Also it will be clear that in each area of a given variant, mixing of island sizes seems to occur preferentially along one direction only, i.e. in one family of interfaces. This cannot be concluded from the diffraction pattern alone, but is suggested by the high resolution images as we shall see.

Occasionally mixing of three island sizes occurs as is represented in Fig. 6g.

4. High Resolution Images

4.1 Imaging technique

As already described in [1] we obtained dark field images by using all possible superlattice spots within a mesh of the reciprocal lattice of the basic f.c.c. lattice. Care has to be taken to make sure that the intensities of the superlattice diffraction spots are as large as possible as compared to those of the basic spots and that their configuration reflects the symmetry of the crystal. Alternatively formulated, the projection of the centre of Ewald's sphere should coincide with the centre of the unit mesh of the basic f.c.c. lattice. In this way all beams, producing spots located on a circle centered on the centre of the unit mesh, undergo the same phase shift as a result of instrumental factors since they enclose the same angle with the optical axis of the microscope. It was found previously that under these conditions the lighter minority atoms can be revealed as bright dots in Au–Mn alloys with a column structure [7]. Using this technique the information concerning the basic f.c.c. lattice is rejected intentionally as being irrelevant and redundant for the problem under consideration.

We also produced bright field images using a more recent instrument with a better instrumental resolution; such images are reproduced in Fig. 8. Also in this case the manganese columns are represented as white dots under suitable defocus conditions. The white dot contrast is most pronounced in the first dark thickness fringe associated with the fundamental reflections, i.e. again under conditions which minimize the intensity of the basic reflections, which are rejected by the aperture, except for the

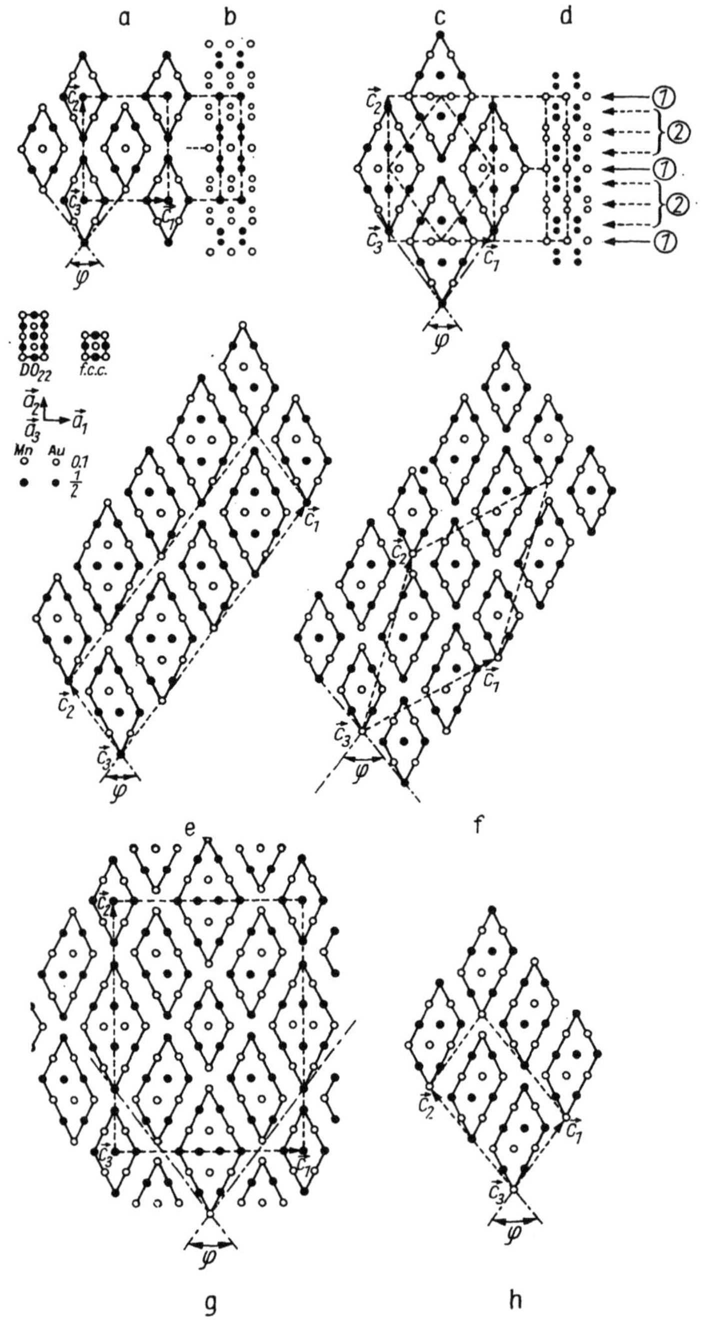

Fig. 6. Different "lozenge" shaped island superstructures of the DO_{22} structure a) and b) structure consisting of 3×3 islands only; a) is a view along the normal to the island plane whereas b) is a side view along another cube direction, c) and d) structure consisting of 4×4 islands only; c) is again a "top" view and d) a side view. e) Mixture of 3×4 and 4×4 islands. f) Mixture of 3×4 and 3×3 islands. g) Mixture of three types of islands: 3×3, 4×4, and 3×4. h) Structure consisting of 4×3 islands only

Table 2

Lozenge shaped island structures derived from the DO_{22} structure

model (Fig. 6)	unit cell size*)		Bravais lattice	point group	number of orientation variants	angle between average direction of APB	ideal composition		
							chemical formula	at% Mn	atom ratio Au/Mn
(a)	$\begin{bmatrix} 4 & 0 & 0 \\ 0 & 5 & 0 \\ 0 & 0 & 1 \end{bmatrix}$	$c_1 = 4a$, $c_2 = 5a$, $c_3 = a$	orthorhombic body centered	2/m 2/m 2/m	6	$\approx 77°$	$Au_{31}Mn_9$	0.2250	3.44
(c)	$\begin{bmatrix} 5 & 0 & 0 \\ 0 & 7 & 0 \\ 0 & 0 & 1 \end{bmatrix}$	$c_1 = 5a$, $c_2 = 7a$, $c_3 = a$	orthorhombic one face centered	2/m 2/m 2/m	6	$\approx 71°$	$Au_{27}Mn_8$	0.2285	3.37
(e)	$\begin{bmatrix} 9 & 12 & 0 \\ -5/2 & 7/2 & 0 \\ 0 & 0 & 1 \end{bmatrix}$	$c_1 = 15a$, $c_2 = \frac{a}{2}\sqrt{74}$, $c_3 = a$	monoclinic one face centered	2/m	12	$\approx 74°$	$Au_{95}Mn_{28}$	0.2276	3.39
(f)	$\begin{bmatrix} 13/2 & 7/2 & 0 \\ 5/2 & 17/2 & 0 \\ 0 & 0 & 1 \end{bmatrix}$	$c_1 = \frac{a}{2}\sqrt{218}$, $c_2 = \frac{a}{2}\sqrt{314}$, $c_3 = a$	monoclinic primitive	2/m	12	$\approx 77°$	$Au_{72}Mn_{21}$	0.2258	3.42
(g)	$\begin{bmatrix} 9 & 0 & 0 \\ 0 & 12 & 0 \\ 0 & 0 & 0 \end{bmatrix}$	$c_1 = 9a$, $c_2 = 12a$, $c_3 = a$	orthorhombic body centered	2/m 2/m	6	$\approx 74°$	$Au_{167}Mn_{49}$	0.2268	3.41
(h)	$\begin{bmatrix} 5/2 & 7/2 & 0 \\ -4 & 5 & 0 \\ 0 & 0 & 1 \end{bmatrix}$	$c_1 = \frac{a}{2}\sqrt{74}$, $c_2 = a\sqrt{31}$, $c_3 = a$	monoclinic one face centered	2/m	12	$\approx 74°$	$Au_{41}Mn_{12}$	0.2264	3.416

*) The transformation matrix relating the base vectors $\boldsymbol{c}_1$, $\boldsymbol{c}_2$, and $\boldsymbol{c}_3$ of the superstructure with those of the f.c.c. basic lattice $\boldsymbol{a}_1$, $\boldsymbol{a}_2$, and $\boldsymbol{a}_3$. The lengths of the vectors are expressed in units $a = |\boldsymbol{a}_1| = |\boldsymbol{a}_2| = |\boldsymbol{a}_3|$.

direct beam. Similar thickness effects have been observed and computer simulated by Hiraga et al. [8] for a two-dimensional superstructure based on Au_4Mn.

The images of antiphase boundaries are less prominent than in the dark field images, but the atomic positions along the antiphase boundaries are better defined in the bright field images.

4.2 Observation of structures

A structure image of the present material, taken under dark field conditions is reproduced in Fig. 7. It is immediately evident that the structure is not strictly periodic except for small areas:

Some parts, marked A, in Fig. 7 consist of 4 × 4 islands. This structure is the same as the idealized structure in [1] except for the island size. Like this structure it is orthorhombic, one face centered, with a monoclinic primitive cell and the angle enclosed by the average directions of the APB's is 71° in agreement with the model of Fig. 6c.

Other parts, marked B in Fig. 7 consist of alternating rows of 3 × 4 and 4 × 4 islands. This structure is shown schematically in Fig. 6e; it is monoclinic. The measured angle enclosed by the average directions of the APB's is now 74°. A bright field image of such an area is reproduced in Fig. 8; the unit cell has been indicated. There is

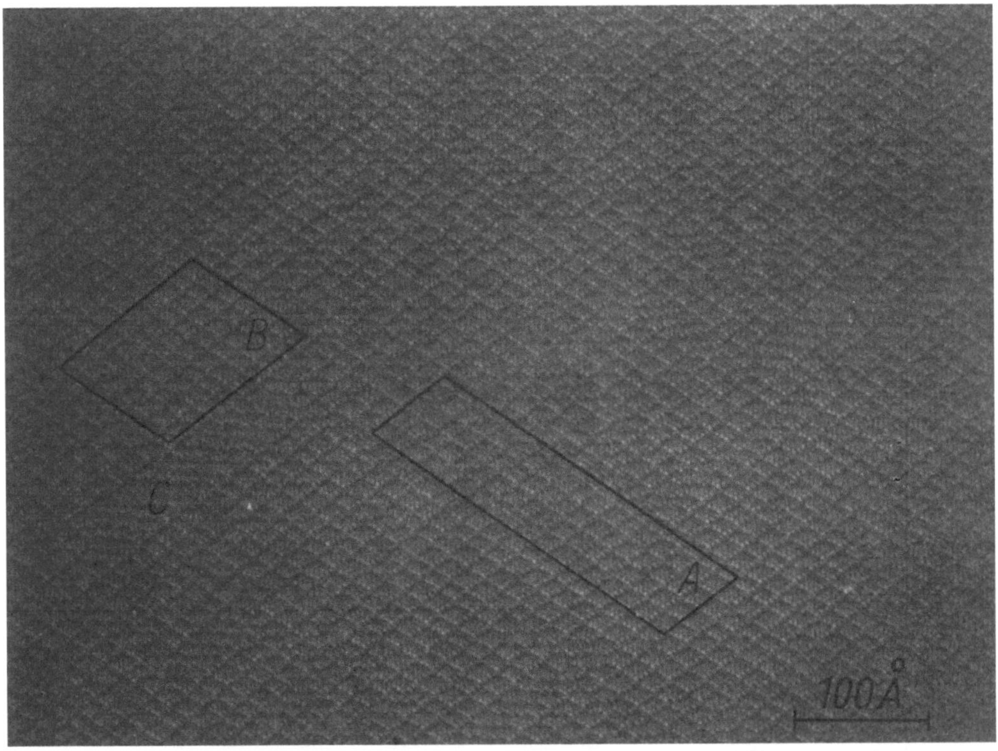

Fig. 7. Dark field image in superlattice spots of a specimen area containing different types of island structure derived from the DO_{22} structure. In (A) the structure represented schematically in Fig. 6c; in (B) the structure represented schematically in Fig. 6e; in (C) a small area of one-dimensional structure is present as well. Note that most areas contain a mixture of different island sizes

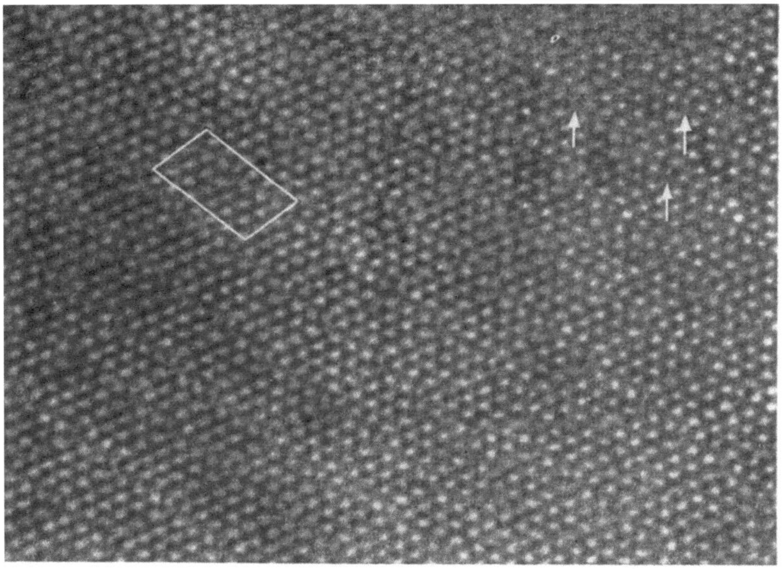

Fig. 8. Bright field image of an area containing a structure of the type of Fig. 6e, i.e. containing only 3 × 4 islands

a close relationship between the bright dots in the image and the Mn-configuration of Fig. 6e. Along the intersection line of two antiphase boundaries in the ideal island structure the same configuration of four manganese atoms as in the DO_{22} structure is present. This configuration is usually not clearly resolved but one sharp corner of

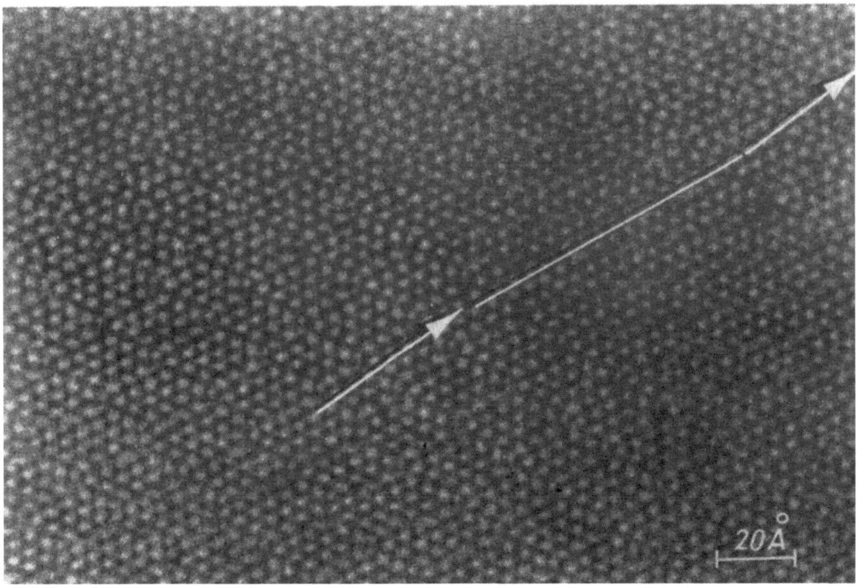

Fig. 9. Bright field image in alloy with composition Au–22 at% Mn. Areas having a one-dimensional long period superstructure are generated within an area consisting of two-dimensional DO_{22} islands

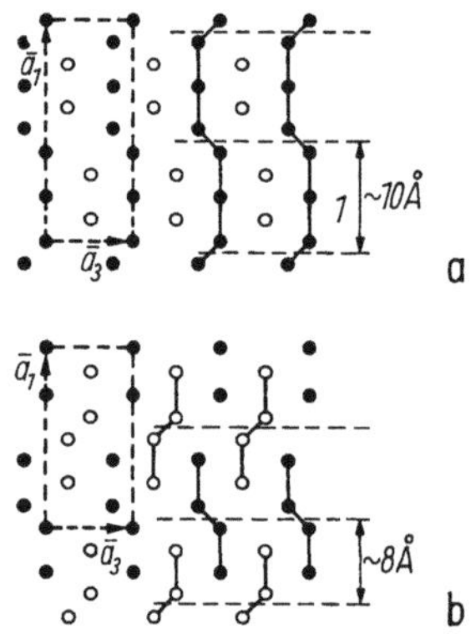

Fig. 10

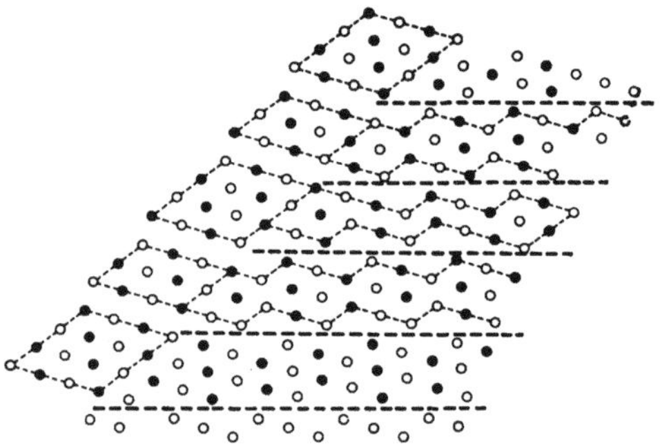

Fig. 11

Fig. 10. Comparison between a) the orthorhombic structure [7] and b) the Watanabe-type structure [9]. Both are one-dimensional superstructures of the DO_{22} structure

Fig. 11. Generation of a one-dimensional long period structure by a change in orientation of the APB's. This scheme should be compared with Fig. 7, area C and Fig. 9

the island is usually truncated, i.e. not marked by a white dot (indicated by arrows in Fig. 8). It is possible that this has some structural significance and is due to the fact that these corresponding columns consist partly of maganese and partly of gold.

The models suggested by the images are clearly consistent with the diffraction data. It should be noted, however, that the structures derived from the diffraction patterns can only be idealized since they ignore the imperfections in the arrangements.

In other small areas (Fig. 7 part marked C) the antiphase boundaries become parallel and equidistant; their direction encloses a small angle with the $[100]_{cube}$ direction. The idealized structure in which the APB's would be in (100) or $(010)_{DO_{22}}$ planes would be closely related to either the so-called Watanabe structure [9] or to the orthorhombic structure described in [10] the difference between both only consisting in the spacing of the APB's (Fig. 10). The small deviation from the simple cube plane changes the composition and moreover creates a structure of the monoclinic II type described in [11].

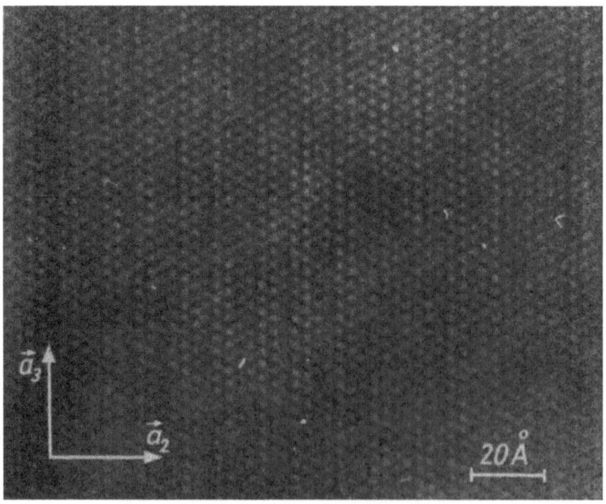

Fig. 12. Side view of an island structure consisting mainly of 4 × 4 islands. This image can be compared with the schematic side view of Fig. 6d

In the bright field mode such areas are hard to recognize (Fig. 9). When looking along the direction of the antiphase boundaries indicated by arrows the orientation change is clear however. Broad, almost one-dimensional areas have the DO_{22} structure but only very locally a known structure can be identified. Fig. 11 shows schematically how such a structure can be generated from the two-dimensional island structure.

As already noted by Terasaki et al. [3] also regular arrangements of three types of islands 3×3, 3×4, and 4×4 occur. We found this arrangement in small parts of our specimen only.

All the island arrangements result from the intersection of two sets of parallel boundaries, forming an angle in the range 71° to 77°.

In regions containing predominantly 4×4 islands, the structure could also be imaged along the cube zone which is parallel with the small diagonal of the 4×4 lozenge shaped islands (Fig. 12). The period in the direction perpendicular to the plane of the islands is equal to the length of one cube edge of the f.c.c. structure, as already proposed in [1] on the basis of the multidomain diffraction pattern. We have now also been able to obtain a mono-domain pattern along this zone confirming independently the prediction of Fig. 6 and the assumption of [1]. The periodicity of the dot pattern along the island plane can be understood on the basis of the schematic drawing of Fig. 6d. Certain atom rows such as those marked (1) in Fig. 6d are con tinuous across the islands; they form columns in the sense discussed previously in [7]. Such columns produce well defined intense dots, separated by three rows of less intense and more diffuse dots marked (2) in Fig. 6d. The latter are produced by unresolved pairs of atom rows, which are periodically interrupted and shifted from one position to the other. This is shown schematically in Fig. 6d. The ratio of the distance between rows of the type 1 to the distance between spots along such rows is 3.5 according to the model; the measured ratio is 3.45. This difference may be due to the fact that the c/a ratio in the DO_{22} structure is not ideal, i.e. slightly smaller than 2.

5. Discussion

Ordered structures in the Au–Mn system with compositions Au_xMn, where $3 \leqq x \leqq 3.5$, can be described in terms of two-dimensional periodic antiphase boundary structures derived from the DO_{22} structure, which itself corresponds with $x = 3$. For $x = 4$ the structure is the tetragonal Ni_4Mo type structure, which can also be generated as a one-dimensional periodic antiphase boundary structure of the DO_{22} structure [12]. The Watanabe structure [9] and the previously described orthorhombic and monoclinic structures in alloys with $x \leqq 3$ [10, 11] are also one-dimensional periodic antiphase boundary structures derived from the DO_{22} structure.

The DO_{22} structure is built on tetrahedral clusters of atoms containing three gold and one manganese atom; they thus have the macroscopic composition, such clusters are known to subsist even in the as quenched state of alloys within the above-mentioned composition range [1]. Also the Cu_3Au structure, of which the DO_{22} is a periodic one dimensional APB structure, is built on such clusters. However, the average distance between the repelling Mn atoms is larger in the DO_{22} structure than in the Cu_3Au structure.

We conclude that the DO_{22} structure is a particularly stable structure in this composition range; it is therefore justified to refer our description of structures to this basic structure.

For compositions close to Au_3Mn with $x \leqq 3$ the one-dimensional periodic APB structures prevail; adjustment of the structure to the composition can be achieved by changing the spacing between non-conservative APB's and by changing the average orientation of the APB's as a result of systematic ledging [11, 12].

For x-values exceeding 3 the structures consist of two-dimensional islands with the DO_{22} structure (shown in Fig. 6) and separated by APB's along which locally the Au_4Mn configuration is realized. The size of the islands becomes larger as the composition approaches Au_3Mn for which the island size is infinite. As the manganese contents decreases the island size also decreases and the concentration of antiphase boundaries, i.e. of elements of Au_4Mn structure increases.

At some concentration, rows of Au_4Mn elements i.e. of antiphase boundary strips would have to become neighbours and now the structure could more conveniently be described as consisting of islands of Au_4Mn structure separated by APB's along which the DO_{22} structure is locally realised; the size of such islands again increases as the composition approaches Au_4Mn; at this composition the island size becomes infinite. This type of structures will be described in Part II of this paper.

In the alloy with macroscopic composition Au_4Mn (20 at% Mn) the lozenge shaped island structure was found to occur as a transition structure after slow quenching or very short annealing treatment, i.e. as an intermediate phase before the final Au_4Mn structure is formed on long annealing at 300 °C. The occurrence of the island structure in this alloy may be related to inhomogeneities in composition, which are levelled out on sufficiently long annealing and thus cause the disappearance of the structure.

In the alloy with composition 21.5 to 23 at% Mn, on the other hand, the island structure is better defined and it still occurs after long annealing treatments at 300 °C. The average size of the islands is larger which is consistent with the larger manganese contents. In alloys with this composition it is apparently a stable structure, which an adapt itself to local variations in compositions by changes in island size.

Perhaps the most striking result of this study is the extreme sensitivity of island configurations to very small composition changes as is evident from Table 1 and Fig. 3.

Acknowledgement

Thanks are due to L. Rossou for careful preparation of the different alloys.

References

[1] G. van Tendeloo, R. de Ridder, and S. Amelinckx, phys. stat. sol. (a) **49**, 655 (1978).
[2] G. van Tendeloo, J. van Landuyt, and S. Amelinckx, in: Electron Microscopy 1980, Vol. 1 Ed. P. Brederoo and G. Boom, VII. Europ. Congr. Electr. Microscopy Foundation, Leiden 1980 (p. 226).
[3] O. Terasaki, D. Watanabe, K. Hiraga, and M. Hirabayashi, Acta cryst., to be published.
[4] R. Wolf, Micron **8**, 171 (1977).
[5] C. Conde, C. Manolikas, D. van Dyck, P. Delavignette, J. van Landuyt, and S. Amelinckx, Mater. Res. Bull. **13**, 1055 (1978).
D. van Dyck, C. Conde, and S. Amelinckx, phys. stat. sol. (a) **56**, 327 (1979); **58**, 451 (1980).
[6] J. van Landuyt, R. Gevers, and S. Amelinckx, Mater. Res. Bull. **5**, 353 (1970).
[7] G. van Tendeloo, R. Wolf, D. van Dyck, and S. Amelinckx, phys. stat. sol. (a) **47**, 105 (1978).
[8] K. Hiraga, D. Shindo, and M. Hirabayashi, in: Electron Microscopy 1980, Vol. 4, Ed. P. Brederoo and J. van Landuyt, VII. Europ. Congr. Electr. Microscopy Foundation, Leiden 1980 (p. 170).
[9] D. Watanabe, J. Phys. Soc. Japan **15**, 1030 (1960).
[10] G. van Tendeloo, R. Wolf, J. van Landuyt, and S. Amelinckx, phys. stat. sol. (a) **47**, 539 (1978).
[11] R. Wolf, G. van Tendeloo, J. van Landuyt, and S. Amelinckx, phys. stat. sol. (a) **48**, 39 (1978).
[12] G. van Tendeloo and S. Amelinckx, phys. stat. sol. (a) **43**, 553 (1977).

(Received December 15, 1980)

phys. stat. sol. (a) **65**, 87 (1981)

Subject classification: 1.2; 21.1

Institute of Metal Physics, Academy of Sciences of the Ukrainian SSR, Kiev[1])

Shape Memory Effect and Phase Transitions in Ti–Ni Alloys

Electron Microscopy Studies of the Phase Transitions after Thermomechanical Treatment

By

V. I. Kolomytsev, V. A. Lobodyuk, and L. G. Khandros

Diffuse scattering is observed in Ti–53 to 56 wt% Ni alloys after high-temperature treatment (within temperature range 650 to 800 °C) which can be explained as the result of the short-range order arrangement of the structural vacancies. During aging within the temperature range 400 to 650 °C this short-range order state transforms to a long-range order state which can be interpreted as a long-period superstructure with lattice constant $a = 1.58$ nm (x-phase). The orientation relationships between β-phase and x-phase are: $\{111\} \langle 321 \rangle_\beta \parallel \{111\} \langle 110 \rangle_x$. The quantity of the x-phase formed during aging at intermediate temperatures is limited ($\approx 15\%$). The x-phase particles arrange in two to three directions and look like thin elongated lentiform plates coherent with the parent β-phase matrix. These particles are inherited partly by the martensite formed near room temperature.

В области температур 650 до 800 °С обнаружено возникновение ближнего порядка по вакансиям, который при выдержках при более низких температурах (400 до 600 °С) переходит в дальний порядок. Состояние с дальним порядком можно представить как длиннопериодную структуру с ГЦК решеткой и параметром $a = 1{,}58$ nm (x-фаза). Между исходной и x-фазой существует следующая взаимосвязь: $\{111\} \langle 321 \rangle_\beta \parallel \{111\} \langle 110 \rangle_x$. Общее количество возникающей при высокотемпературных выдержках x-фазы ограничено. Частицы x-фазы имеют вид вытянутых тонких пластин, когерентных с матрицей и расположенных в 2 до 3 направлениях. При последующем мартенситном превращении x-фаза частично наследуется мартенситом.

1. Introduction

The influence of various thermomechanical treatments (TMT) on the mechanical properties of Ti–Ni alloys, the phase transitions which take place in them, the crystal structures of both the parent and the martensite phases, and the martensite substructure has been the subject of a number of studies [1 to 7]. There is a disagreement between the experimental results obtained by different authors. Taking into account the fact that TMT leads to the corresponding structural changes in the parent phase (e.g., the redistribution of the various types of defects, the decomposition of the initial solid solution) and to the appearance of the new phases one might attribute this disagreement to the difference in TMT used.

The aim of the present work is to investigate the influence of TMT on both the structure and the phase changes of Ti–Ni alloys with nearly equiatomic compositions. The experimental data concerning the influence of TMT on the transformation temperatures, the hysteresis, and the shape memory effect will be presented elsewhere [8].

---

[1]) Vernadskogo ul. 36, Kiev, USSR.

2. Experimental Technique and Results

The alloys under consideration were prepared by arc-melting the constituent elements (Ti and Ni of 99.9 and 99.99% purity, respectively) in vacuum. The composition of the alloys (according to the chemical analysis) and the employed TMT are given in Table 1. The melted ingots of Ti–Ni alloys were hot-rolled at 600 to 800 °C to 0.6 to 1.0 mm strips. TMT was conducted according to two schemes which were per-

Table 1

composition	M_s (°C)	thermomechanical treatment
Ti–54.2 wt% Ni	0	1st scheme: Preliminarily hot-rolled and quenched from 1100 °C, samples were tempered in the 200 to 800 °C range.
Ti–55.5 wt% Ni	10	
Ti–56.4 wt% Ni	−30	2nd scheme: Preliminarily hot-rolled and heated to 1100 °C, samples were quickly cooled to the temperature of aging in the 200 to 800 °C range.

formed in argon atmosphere or by placing the sample between two Ti–Ni plates. Under the first scheme the samples were preliminarily hot-rolled and quenched from 1100 °C and then annealed within the temperature range 200 to 800 °C during times from 1 min to 24 h.

Under the second scheme the hot-rolled samples were heated to the temperature 1100 °C and then quickly cooled to the aging temperature within the temperature range 200 to 800 °C.

The phase composition was checked by the X-ray photographic method. The electron diffraction and the electron microscopy studies were carried out in an EM-200 electron microscope operated at 100 and 175 kV. The thin foils were prepared by using the two-jets electropolishing method.

After hot-rolling a high density of dislocations and elongated $((0.3 \text{ to } 0.5) \times \times (3 \text{ to } 5)\ \mu m^2)$ subgrains is observed (Fig. 1a). This substructure leads to the arc-broadening to the spots on the SAD patterns. Aging at 400 °C during 5 min, on the other hand, results in the initial stage of recrystallization (Fig. 1b). The high-temperature β-phase is ordered at room temperature: $\{100\}_\beta$, $\{210\}_\beta$, and other superstructure reflections are seen on the SAD patterns (Fig. 1c). Diffuse spikes and diffuse spots appear after aging within the temperature ranges 200 to 400 or 800 to 700 °C (Fig. 2a). They appear independently of the way by which the temperature of aging is reached: whether by heating from room temperature or by cooling from a high temperature ($\approx$ 1000 °C).

The SAD pattern in Fig. 2b (orientation is near $\langle 111\rangle_\beta$) shows a number of diffuse spots in addition to the strong reflections of the parent phase. The diffuse spots are similar to those of the ω-phase observed in Ti, Zr, and Nb [9]. Similar spots are also observed at other foil orientations ($\langle 100\rangle_\beta$, $\langle 310\rangle_\beta$, $\langle 321\rangle_\beta$). All basic reflections are connected by the spikes in the $\langle 110\rangle_\beta$, $\langle 211\rangle_\beta$, and $\langle 321\rangle_\beta$ directions and diffuse spots spread in $\langle 321\rangle_\beta$ direction. It is necessary to mention that at the exact Bragg orientation both diffuse spikes and diffuse spots are practically invisible. To observe these spots and spikes one needs an inclination of about 3° to 4°.

The SAD patterns of greatest interest are observed near $[001]_\beta$ foil orientation: four pairs of reflections weaker than the basic one appear between $\{110\}_\beta$ and $\{200\}_\beta$

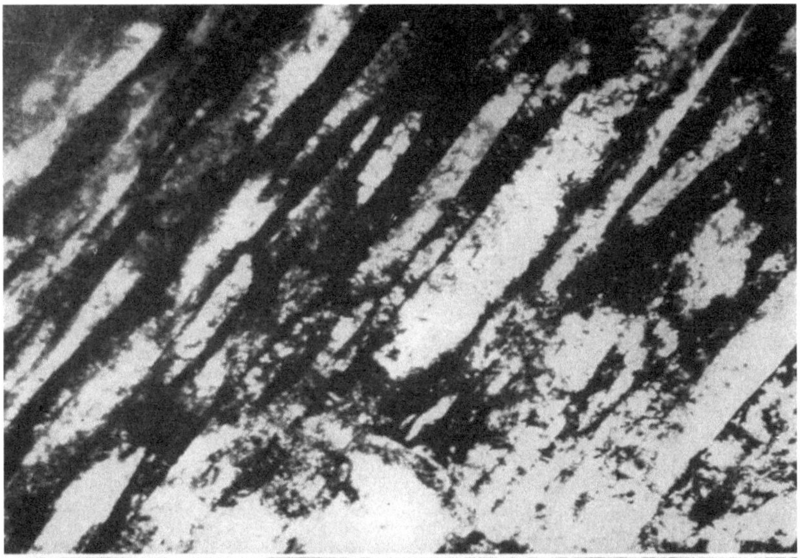

a

b

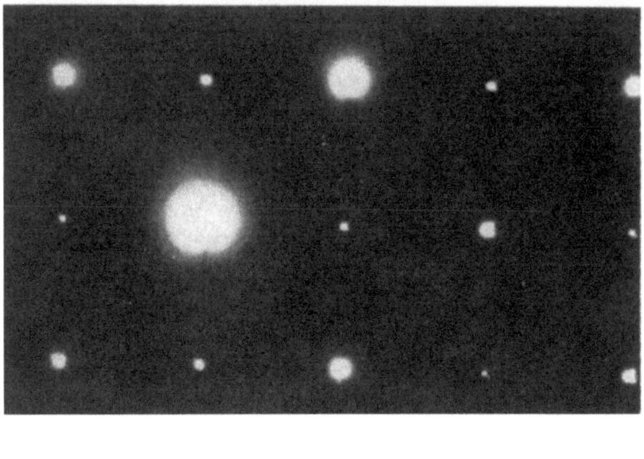

c

Fig. 1. a) Elongated subgrains of the parent β-phase after rolling at high temperature; b) initial stage of the recrystallization in the β-phase (after tempering at 400 °C for 5 min); c) SAD pattern of the high-temperature β-phase, orientation near $[100]_\beta$

a b c d e f

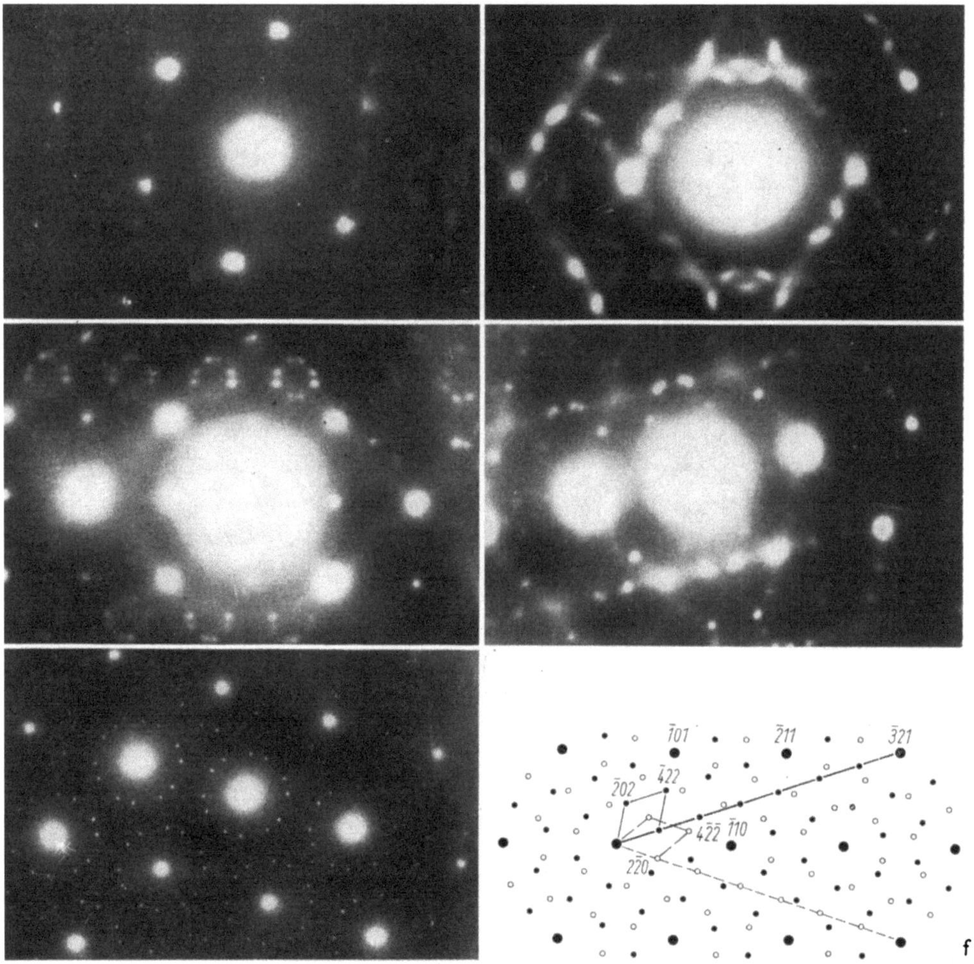

Fig. 2. SAD patterns of heat-treated alloys. a) Tempering at 200 °C for 5 min, orientation near $[111]_\beta$; weak diffuse spikes and diffuse β-spots are seen. b) Tempering at 250 °C for 5 min, orientation near $[111]_\beta$; diffuse spots are visible between $\{110\}_\beta$ reflections. c) Tempering at 350 °C for 5 min, orientation near $[001]_\beta$; additional spots situated on circles with the radius $r = 2.3a_0^*$ nm^{-1} centered between the basic β-phase reflections in $\langle 100\rangle_\beta$ direction are observed. d) Tempering at 350 °C for 60 min, orientation near $[111]_\beta$; splitting of the additional spots is visible. e) Tempering at 550 °C for 5 min, orientation near $[111]_\beta$ ($[111]_x$); x-phase reflections (two variants) are situated on the circles with the radius $r = 5.3a_0^*$ nm^{-1} around each β-reflection. f) Schematic representation of the SAD patterns

or $\{110\}_\beta$ and $\{220\}_\beta$ reflections (Fig. 2c). They are arranged on circles with a radius of about 2.3^*_0 nm^{-1}. The centres of these circles are situated between the corresponding reflections of the β-phase. The intensity of these additional reflections depends on the deviation from the Bragg orientation.

When the temperature is increased the diffuse spikes become weaker and then disappear. On the other hand, the diffuse spots become more distinct (Fig. 2d). Each of these diffuse spots breaks up into two separate reflections, one of them being shifted to the new position. As a result of this process a system of new reflections appears which can be attributed to the new phase. These reflections appear after tempering within the temperature range 400 to 600 °C. They are situated on circles with the radius 5.3_0^* nm^{-1} centred on the β-phase reflections (Fig. 2e). These additional reflections belong to an f.c.c. structure (x-phase) with lattice constant $a = 1.58$ nm [10]. Schematic representation of the SAD pattern is shown in Fig. 2f. The $\{321\}_\beta$ and $\{14, 14, 0\}_x$ reflections coincide on this SAD pattern. $\{100\}_\beta$ and $\{511\}_x$ reflections coincide at the other foil orientations. The two systems of the x-phase reflections of Fig. 2e coincide when one of them is rotated relative to another around the common $\langle 111 \rangle$ axis by the angle of 19°. The orientation relationship between the parent β-phase and the new x-phase is

$$\{111\} \langle 321 \rangle_\beta \parallel \{111\} \langle 110 \rangle_x .$$

It is necessary to mention the rapid decrease in the "diffuseness" of both the β-phase and the x-phase reflections during tempering within the 400 to 600 °C temperature range. Outside this range the reflections remain broadened.

The observed diffraction effects (diffuse spikes, diffuse spots, new systems of reflections) are associated with definite structural changes. After quenching from high temperature in the material only dislocations, mottling, and Ti_2Ni or $TiNi_3$ particles are observed. Tempering at 200 to 300 or 700 to 800 °C results in the more prominent mottling caused by fine particles (3 to 10 nm) of the other phase. But the dark-field images taken in the β-phase reflections or in the diffuse spikes show no clear evidence for the presence of these particles.

A peculiar microstructure appears with rising temperature from 200 to 400 °C (or lowering from 700 to 650 °C): there are often seen dislocation loops (Fig. 3a) and fine (0.02 to 0.03 μm) particles elongated in the $\langle 110 \rangle_\beta$ direction on the dark-field images taken in the x-phase reflections (Fig. 3b).

Annealing at higher temperatures (500 °C) results in the growing of x-phase particles. They have the form of elongated thin lentil plates arranged in two or three directions (Fig. 3c). Usually the particles consist of two to three "subgrains": along them the contrast is changed from dark to light and again to dark. A distinct boundary is often seen between the dark and the light parts but a "spread" boundary also occurs. The latter can be caused by a local foil bending. The fringe contrast may often be seen on the phase boundary of thin x-phase particles.

The particles of the x-phase continue to grow up with tempering, lose their coherency with the parent phase, and on the boundaries between x-phase and β-phase there appear misfit dislocations with distances of 750 to 800 nm between them (Fig. 3d). Usually the length of the particles increases to 3 to 4 μm after annealing for 6 h at 550 °C. Then the particles do no longer grow with annealing during the period of time up to 24 h. In this case the particles are homogeneous. In the same part of the foil one can observe 2 to 3 sets of x-phase particles. By meeting another particle the growing particles stop to grow, but can become a little thicker.

It is necessary to attract attention to another type of SAD patterns which appear after aging within the 200 to 400 °C temperature range. Between the strong reflections of the β-phase two additional spots located at 1/3 and 2/3 positions are seen. These reflections belong to the "b.c.c. structure" with lattice constant three times greater than the parameter of the parent β-phase: $a = 3a_0$ [1, 2, 5, 11]. The bright- and dark-field images show no peculiar microstructure responsible for this type of reflections. Probably we cannot observe any microstructure in view of the small size of the particles.

The high-temperature β-phase transforms partly to the monoclinic martensite (with lattice parameters $a = 0.29$, $b = 0.411$, $c = 0.463$ nm; $\gamma = 96.7°$ [6]) when the specimens containing x-phase particles are cooled to or below room temperature.

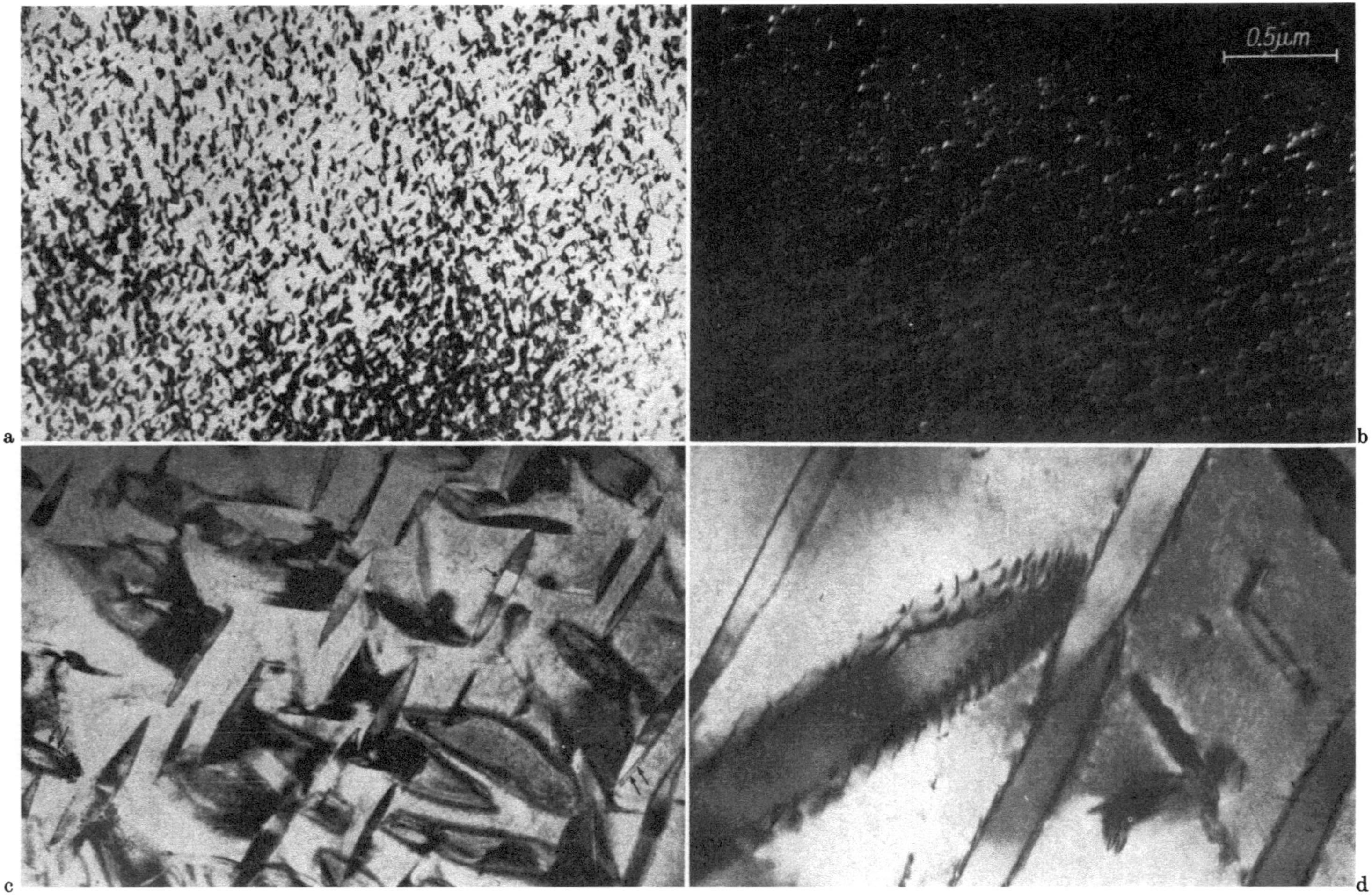

Fig. 3. Microstructure of heat-treated Ti–Ni alloys. a) After tempering at 450 °C for 120 min. b) Dark-field image (reflection $(\bar{4}04)_x$) of the x-phase particles which are elongated in $\langle 110 \rangle_\beta$ direction. c) After tempering at 550 °C for 30 min; distinct and "spread" boundaries of the x-phase are visible. d) After tempering at 550 °C for 360 min; misfit dislocations are observed on the interphase boundaries

Fig. 4. Microstructure of the martensite which are formed after tempering the β-phase at 450 °C for 90 min

On a number of the SAD patterns both the x-phase and martensite reflections are observed, i.e. the x-phase remains unchanged during the martensitic transformation. Reflections at 1/3 and 2/3 positions between the reflections of the β-phase coexist sometimes with the martensite reflections in the M_s to M_f temperature range.

The martensite which is formed on cooling looks like elongated alternative plates (Fig. 4). The substructure and the morphology of this martensite will be reported (elsewhere [8].

3. Discussion

Thus on the SAD patterns one can observe the following sequence of changes which are caused by TMT. We have observed diffuse scattering (diffuse spikes in the $\langle 110 \rangle_\beta$, $\langle 211 \rangle_\beta$, $\langle 321 \rangle_\beta$ directions) on SAD patterns after annealing at temperatures above 700 or below 400 °C. The mottling on the bright-field images corresponds to such SAD patterns.

After annealing within the temperature range 400 to 600 °C the intensity of the diffuse scattering decreases sharply and there appear new systems of distinct spots which belong to an f.c.c. structure with lattice parameter $a = 1.58$ nm (x-phase). On the micrograph usually sets of x-phase particles are observed located in two to three directions of the type $\langle 110 \rangle_\beta$ (or $\langle 321 \rangle_x$). The amount of the particles of the x-phase and the sizes of these particles increase with the duration of the tempering. After annealing near the optimum temperature (550 to 600 °C) for a period up to 24 h the total amount of the particles is limited and remains the same (about 10 to 15%). It is necessary to emphasize that weak diffuse rel-rods (spikes) are observed irrespectively which TMT was used.

On cooling to room temperature the x-phase remains unchanged. Moreover at least some part of it is conserved in the martensite. After heating above 700 °C the particles of the x-phase disappear. At room temperature the homogeneous β-phase is again observed.

The amount and the sizes of Ti_2Ni particles are not changed during TMT. This means that there are probably no considerable changes in the solid solution composition.

Plate-like segregations of f.c.c. structure with the lattice parameter $a = 1.52$ to 1.58 nm were also observed after furnace cooling from 1050 to 300 °C [12] and after slow cooling from 900 °C or prolonged aging at intermediate temperatures [13]. The authors of [12] have suggested that the dispersion of the lattice constant of the f.c.c. phase may be explained by an inaccuracy of the electron diffraction method and/or by the strains of the particles which are arranged parallel or perpendicular to the plane of the foil. However, there is no dispersion if the second- or the third-order reflections were used to determine the lattice parameter. The parameter of the f.c.c. structure lattice was always found in our case to be equal to $a = (1.58 \pm 0.01)$ nm. It was supposed also [12] that the f.c.c. phase is intermediate between the high-temperature β-phase and the stable equilibrium phases: below the eutectic temperature (≈ 650 °C) the β-phase transforms to the intermediate f.c.c. phase and the latter decays into Ti_2Ni and $TiNi_3$ (e.g. after aging at 600 °C for 1 month). In our case the particles of x-phase are always present after tempering during 150 h at 600 °C and the amount and the sizes of the particles like those of Ti_2Ni or $TiNi_3$ remain unchanged.

The homogeneous β-phase of Ti–Ni alloys exists in a very narrow (about 5%) range near the equiatomic composition and this range is reduced at lower temperatures [6]. In the neighbourhood of this range a mixture of phases exists, TiNi + Ti_2Ni or TiNi + $TiNi_3$. Alloys used in our work were of near equiatomic composition (see Table 1). But there is always a small deviation from exact equiatomic composition. Thus structural vacancies can arise as a result of deficiency of atoms of one or the other sort. In [14] a high strengthening was observed in Ti–Ni alloys for compositions on the phase boundary between TiNi and TiNi + $TiNi_3$. The maximum of the strengthening corresponds to the Ti–54.5 at% Ni alloy and this was explained by the excess of the point defects (structural vacancies).

If we suppose that the appearance of the new phase is caused by the structural vacancies [15] then we can imagine the mechanism of the formation of the phase in the following way. Annealing at temperatures higher than 700 or lower than 400 °C induces a short-range order arrangement of structural vacancies that causes the diffuse scattering observed on the SAD patterns. At aging within the 400 to 650 °C temperature range this short-range order arrangement transforms to the long-range order arrangement of the structural vacancies. It can be represented as a new phase state. One can easily show that 1.35 at% difference in atomic composition of any alloying element is sufficient to produce the f.c.c. structure with the lattice parameter $a = 1.58$ nm and with vacancies at the lattice nodes and faces.

An increase in the degree of order under aging was also observed in the Fe–24 at% Pt alloy [16]. The disordered state remains in this alloy after quenching from 1000 °C. However, aging at temperatures below 650 °C results in ordering. The degree of order increases from zero after quenching and aging during 1 h to 0.8 after aging during 10 h.

Similar diffraction effects and a similar succession in which they appeared were also observed after aging of AlCuNi [17], InTl [18], NiAl [19], and other alloys. The short-range order arrangement of the vacancies and the transition to the long-range order arrangement under aging is an adequate interpretation of these effects. The dependence of free energy versus degree of order has been calculated [20]. It was shown that there exists a temperature range where both the ordered and the disordered states coexist. The disordered state is stable above this range and only the ordered state is stable below this range.

It is convenient to use the symbolic notation and scheme proposed in [21] for the interpretation of the SAD patterns in tempered Ti–Ni alloys. In the [111] foil orientation the distance between origin and the $\{321\}_\beta$ reflections on the SAD patterns is divided into seven equal parts. According to [21] we have to proceed two steps ($n = 2$) in one close-packed direction (e.g. $[\bar{1}01]_\beta$) and then one step ($m = 1$) in another ($[1\bar{1}0]_\beta$) close-packed direction. Then the number of equal parts between origin and one of the corresponding spots is equal to $p = n^2 + nm + m^2 = 7$.

There were observed reversible long-period superstructures with lattice parameters equal to multiples of $\sqrt{3}$, $\sqrt{7}$, $\sqrt{13}$, $\sqrt{19}$, ... in transition metal dichalcogenides [21, 22]. As a result of symmetrical shifts of atoms surrounding metal ions there appear clusters which lead to the formation of long-period superstructures. These superstructures can be also represented as a structural modulation by the charge density waves [22]. Clusters of this type appear as a result of a redistribution of outer electrons among atoms. The distance between centres of the clusters, which determines the new lattice parameter becomes equal to $a_0 \sqrt{p}$.

The formation of the new phase and structural states in Ti–Ni alloys due to the charge density waves is not also forbidden. The $K_{\beta_{2,5}}$ band of Ti and Ni is shown to be changed when 50Ti50Ni is formed [23]. This change can be easily explained by the assumption that electrons are transported from Ni to Ti. Thus the formation of the long-period superstructures in Ti–Ni alloys could be also interpreted in terms of the periodic structural distortion caused by the charge density waves. The latter can appear as a result of redistribution of the outer electrons between Ti and Ni atoms.

References

[1] F. E. Wang, W. J. Buehler, and S. J. Pickart, J. appl. Phys. **36**, 3232 (1965).
[2] G. D. Sandrock, A. J. Perkins, and R. F. Hehemann, Metallurg. Trans. **2**, 2769 (1971).
[3] R. J. Wasilewski, Metallurg. Trans. **2**, 2973 (1971).
[4] C. M. Wayman, T. Cornelis, and K. Shimizu, Scripta metall. **6**, 115 (1972).
[5] S. P. Gupta, Mater. Sci. Engng. **14**, 157 (1974).
[6] I. I. Kornilov, O. K. Belousov, and E. V. Kachur, Nickel–Titanium and Other Alloys with the Shape Memory Effect, Izd. Nauka, Moscow 1977.
[7] Yu. N. Koval, V. I. Kolomytsev, V. A. Lobodyuk, V. V. Martinov, and L. G. Khandros, Mater. Internat. Conf. ICOMAT-77 Martensitic Transformations in Metals and Alloys, Izd. Naukova Dumka, Kiev 1979 (p. 145).
[8] V. M. Ermakov, V. I. Kolomytsev, V. A. Lobodyuk, and L. G. Khandros, to be published.
[9] S. L. Sass, Acta metall. **17**, 813 (1969).
[10] V. I. Kolomytsev, V. A. Lobodyuk, and L. G. Khandros, phys. stat. sol. (a) **50**, K127 (1978).
[11] A. Nagasawa, J. Phys. Soc. Japan **29**, 1386 (1970).
[12] D. Koskimaki, M. J. Marcincowski, and A. Sastri, Trans. MS AIME **245**, 1883 (1969).
[13] S. P. Gupta, K. Mukherjee, and A. A. Johnson, Mater. Sci. Engng. **11**, 283 (1973).
[14] T. Suzuki, Trans. Japan. Inst. Metals **14**, 31 (1973).
[15] D. Watanabe, J. R. Castles, A. A. Johnson, and A. S. Malin, Acta cryst. **23**, 307 (1967).
[16] T. Tadaki, K. Katsuki, and K. Shimizu, Trans. Japan. Inst. Metals, Suppl. **17**, 187 (1976).
[17] M. van Sande, R. de Ridder, J. van Landuyt, and S. Amelinckx, phys. stat. sol. (a) **50**, 587 (1978).

[18] G. L. Bleris, Th. Karakostas, J. Stoemenos, and N. A. Economou, phys. stat. sol. (a) **34**, 243 (1976).
[19] R. de Ridder, G. van Tendeloo, and S. Amelinckx, phys. stat. sol. (a) **43**, 133 (1977).
[20] G. L. Bleris, Th. Karakostas, N. A. Economou, and R. de Ridder, phys. stat. sol. (a) **50**, 579 (1978).
[21] J. van Landuyt, G. A. Wiegers, and S. Amelinckx, phys. stat. sol. (a) **46**, 479 (1978).
[22] J. A. Wilson, F. J. DiSalvo, and S. Mahajan, Adv. Phys. **24**, 17 (1975).
[23] V. V. Nemoshkalenko, V. Ya. Nagornyi, and G. P. Gigolashvili, Dokl. Akad. Nauk SSSR **203**, 79 (1972).

(Received April 8, 1980)

phys. stat. sol. (a) **65**, 97 (1981)

Subject classification: 14.3.1 and 14.3.4; 22.1.3

Laboratoire de Physique des Matériaux et Composants de l'Electronique, Institut de Physique Université de Nantes[1])

Commutations dans les couches minces de sélénium

Influence des électrodes

Par

J.-C. BERNÈDE

Tandis que l'étude des caractéristiques courant–tension (I–U) des structures métal–sélénium–pointe (M–Se–P) montre que les commutations obtenues sont peu reproductibles, l'expérience prouve que celles des structures Ag_2Se–Se–P sont stables et relativement fiables. La pointe étant polarisée négativement, après une première commutation pour une tension de formation U_F qui augmente avec l'épaisseur de Se, la tension de seuil U_S se stabilise après quelques cycles. U_S est indépendante de l'épaisseur de Se mais est fonction de la tension inverse maximale et de la fréquence. Le cycle ($I-U$) est expliqé en terme de diffusion d'ions Ag^+. A U_F correspond la formation de dendrites riches en Ag à travers Se, puis à U_S correspond une zone vidée de porteurs à l'interface P–Se lors du passage en polarité inverse.

While (I–U) characteristics of metal–selenium–test needle (M–Se–P) show instable switchings, those of Ag_2Se–Se–P structures are reproducible. With the test needle negatively biased, after a first commutation at the formation voltage U_F (U_F increases with the selenium thickness), the threshold voltage U_S is stabilized after some switchings. U_S is independent of the selenium thickness but U_S increases with the maximum reverse applied voltage and with the frequency. This is explained by Ag^+ ion diffusion.

1. Introduction

Le sélénium étant un chalcogène simple à obtenir à l'état amorphe en couche mince, plusieurs auteurs [1 à 5] ont étudié les possibilités d'obtenir des commutations dans des structures à base de Se.

Les études ont montré que les commutations étaient peu fiables [6]. Aussi avons-nous entrepris au laboratoire, parallèlement à celle des structures métal–Se–pointe (M–Se–P), l'étude des structures semi-conducteur–Se–pointe (SC–Se–P). Le semi-conducteur choisi est Ag_2Se. Ce choix est guidé par la faible résistivité [7] de Ag_2Se et par le fait que l'argent semble jouer un rôle favorable à l'obtention des commutations dans le sélénium [3]. Le choix d'une pointe comme contre-électrode est justifié par les résultats obtenus au laboratoire sur les structures M–Ag_2Se–P [8]: Nous avons montré que ce type de géométrie rend possible l'obtention de cycles commutant reproductibles impossibles à obtenir dans le cas d'électrodes symétriques.

Les expériences réalisées au laboratoire d'une part confirment la faible reproductibilité des caractéristiques (I–U) dans le cas M–Se–P, d'autre part, dans le cas Ag_2Se–Se–P, elles mettent en évidence l'existence de commutations sensibles à la polarité mais stables et relativement fiables.

Ces résultats sont reliés aux propriétés ioniques de Ag_2Se et à la dissymétrie des électrodes qui rendent possible la diffusion d'ions Ag^+ dans Se, formant ainsi des canaux conducteurs entre Ag_2Se et la pointe.

---

[1]) 2, rue de la Houssinière, 44072 Nantes Cédex, France.

2. Conditions expérimentales

Les couches superposées Ag_2Se–Se ou M–Se (M = Au, Cr) sont obtenues par évaporations thermiques classiques successives, sans casser le vide (10^{-6} Torr), sur un substrat à température ambiante.

Les épaisseurs des couches, mesurées par alphastep, sont, pour les électrodes métalliques, de l'ordre de 5000 Å, pour Ag_2Se 2000 Å, tandis que celle de Se varie de 500 à 10000 Å. Le composé Ag_2Se évaporé a été obtenu de la façon suivante: après pondération dans le rapport stoechiométrique des poudres d'argent et de sélénium (99,999%), le mélange est introduit dans un tube de silice qui est ensuite scellé sous vide, chauffé à 900 °C (± 3 °C) pendant 50 h, puis refroidi à 600 °C et enfin trempé à la température de la glace fondante. Le composé récupéré sous forme d'aggloméràt est broyé en poudre. Une analyse au goniomètre à rayons X nous permet de contrôler la nature du produit obtenu.

Lors de l'évaporation de Ag_2Se, dans un creuset de tantale, une décomposition sensible de la poudre se produit. Toutefois, il y a recomposition immédiate au niveau du substrat, comme le montrent les spectres obtenus avec le goniomètre à rayons X (Fig. 1).

La contre-électrode est une aiguille de test de différents métaux (bronze de tungstène, Au, Al, Cu, Ag). Le positionnement de cette électrode est réalisé à l'aide d'un Microprober Wentworth MP 0705. Le contact sur l'électrode inférieure est pris à l'aide d'un fil de cuivre fixé à la laque d'argent.

L'étude des caractéristiques (I–U) est faite à l'aide d'un générateur de signaux IEC F 72 (dents de scies isocèles −15 à +15 V) ou d'un générateur de pulses Hewlett Packard 219 A (pulses de 0 à 120 V, larges de 0,05 μs à 10 ms). Les courbes sont recueillies sur un oscilloscope Téléquipement DM 64 ou/et sur une table traçante X–Y Hewlett-Packard 7044 A.

3. Résultats expérimentaux

3.1 *Structure Ag_2Se-a-Se–P*

Si la structure est soumise à des dents de scies isocèles de polarité négative, la pointe étant prise comme référence, l'état off initial déterminé par l'épaisseur de la couche de sélénium amorphe se conserve ou même s'amplifie. Si la polarité est inversée pour une tension U_F, il y a commutation dans un état on conducteur (R_{on} est de l'ordre de quelques centaines d'ohm). Cet état on et ceux des cycles ultérieurs peuvent s'effacer de deux façons:

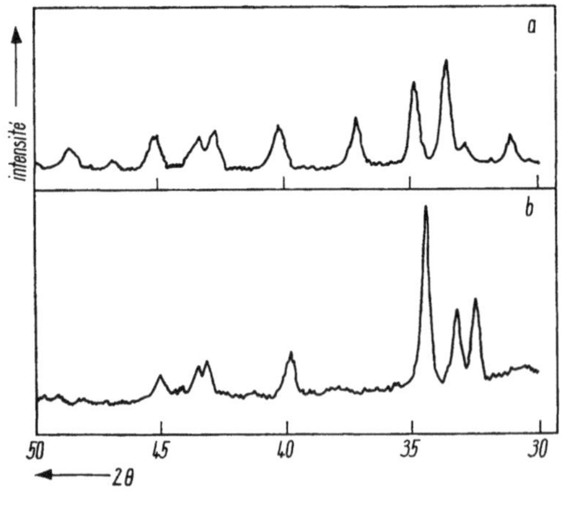

Fig. 1. Spectre au goniomètre à rayons X pour Ag_2Se. a) En poudre cristalline, b) en couche mince

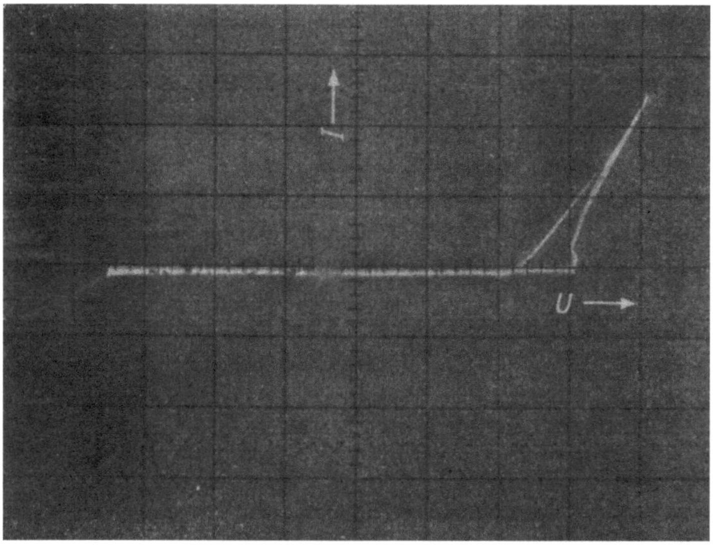

Fig. 2. Caractéristiques (I–U) d'une structure Ag_2Se-a-Se–P (W). $F = 0{,}05$ Hz (0,5 mA/div., 0.2 V/div.)

1. Dans le temps, ils s'effacent progressivement au bout de quelques heures, la tension aux bornes de l'échantillon étant nulle.

2. En inversant la polarité; le retour à l'état off est, dans ce cas, très rapide. Il est alors possible de décrire le cycle de la Fig. 2, avec une tension de seuil $U_S < U_F$. Il y a donc eu formation de la structure (Fig. 3). Cette formation peut être progressive et U_S ne se stabilise qu'au bout de quelques cycles. Contrairement au cas M–Ag_2Se–P [8], la formation s'avère possible de 0,001 Hz à 10 kHz. L'étude de la formation en fonction de l'épaisseur montre que U_F croit rapidement avec l'épaisseur e_{Se} de sélénium (Tableau 1).

Tableau 1

Variation de U_F et U_S en fonction de l'épaisseur de sélénium

e_{Se} (Å)	2500	3200	4200
U_F (V)	1	5	15
U_S (V)	0,5	0,5	0,5

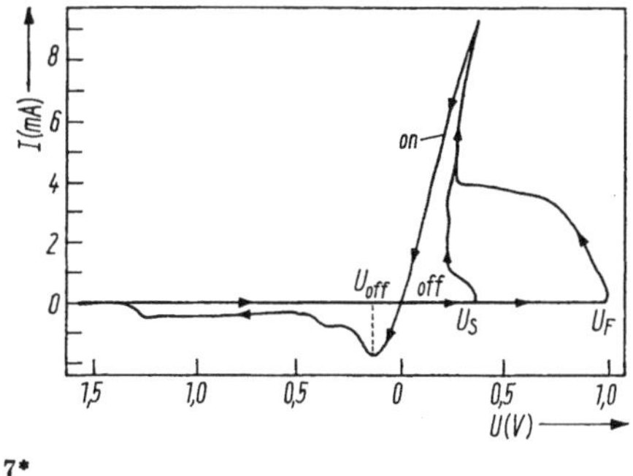

Fig. 3. Formation d'une structure Ag_2Se-a-Se–P (W). $F = 0{,}05$ Hz

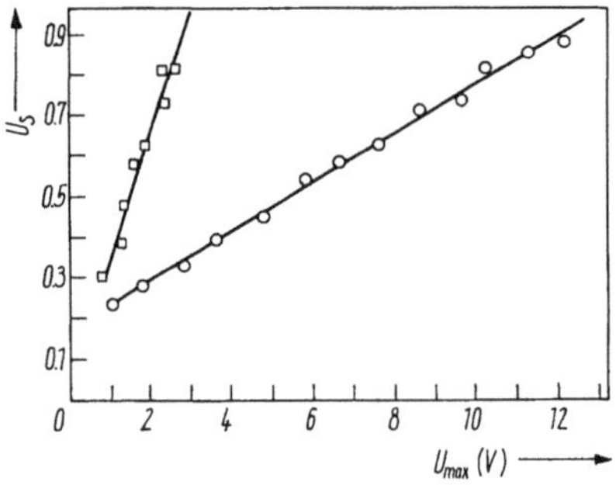

Fig. 4. Variation de la tension de seuil U_S en fonction de la tension inverse maximum pour différentes fréquences. ○ $F = 0{,}05$ Hz, □ 5 kHz

Tous ces résultats sont valables, que la pointe soit en bronze de tungstène, en or ou en aluminium. Par contre, si la pointe est en argent ou en cuivre, la formation est possible, quelle que soit la polarité et pour des tensions inférieures à celles exposées précédement. De plus, l'état off est peu isolant et le cycle peu reproductible avec rapidement mémorisation définitive à l'état on.

En dehors du cas du cuivre ou de l'argent, le cycle de caractéristiques (I–U) est très reproductible (10^8 cycles). La tension de seuil U_S croit avec la tension maximum en inverse, ainsi qu'avec la fréquence (Fig. 4). Mais, contrairement à U_F, U_S est peu sensible à l'épaisseur de a-Se (Tableau 1). Plus la fréquence est élevée, de 0,001 Hz à 10 kHz, le cycle devenant instable aux fréquences plus élevées, moins les états on et off sont différenciés (Fig. 5).

Dernier cas, celui d'une pointe en aluminium. Ici, U_S est plus élevé et plus aléatoire que dans les cas précédents. De plus, la commutation n'intervient plus dès que la fréquence dépasse 50 Hz. Pour une structure Ag_2Se–a-Se recuite 15 h sous atmosphère d'azote à 70 °C, le spectre de rayons X montre que le sélénium s'est cristallisé suivant une orientation préférentielle. Le cycle de caractéristiques (I–U)reste le même avec

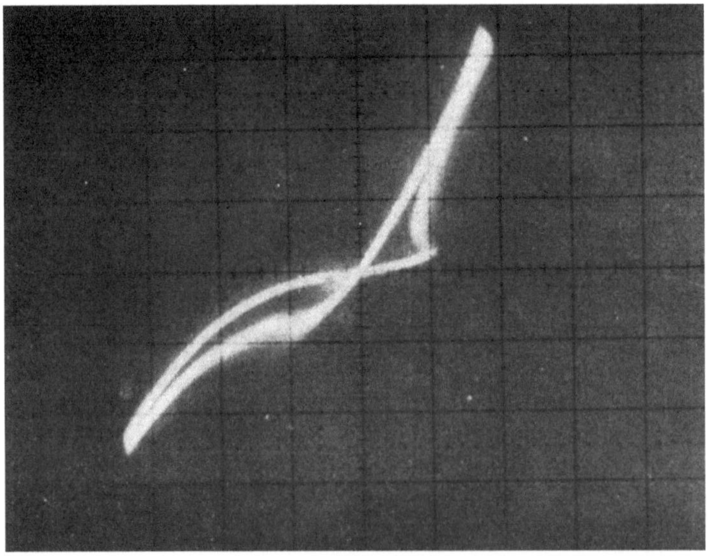

Fig. 5. Caractéristiques (I–U) d'une structure Ag_2Se–a-Se–P (W). $F = 5$ kHz (0,5 mA/div., 0,5 V/div.)

toutefois, dès que la fréquence dépasse quelques Hertz, des états off peu isolants et des commutations irréversibles dês que $U > 10$ V.

3.2 Structure M-a-Se-P

Les résultats sont identiques, que l'électrode inférieure soit en or ou en chrome. Ils confirment ceux précédemment exposés par Jones et Collins [6], à savoir: que la pointe soit en bronze de tungstène, en or, en aluminium, la commutation intervient pour la même tension de seuil U_S et ceci quelle que soit la polarité. U_S augmente avec l'épaisseur e_{Se} de sélénium ($e = 3000$ Å, $U_S = 25$ V; $e = 700$ Å, $U_S = 5$ V). Jones et Collins montrent que la commutation est irréversible, sauf pour $e_{Se} > 3$ µm.

De fait, nos résultats confirment que la reproductibilité du phénomène est fonction de l'épaisseur de Se, mais de plus qu'elle dépend aussi de la résistance de protection R_S incluse en série dans le circuit, Ainsi, si pour $e_{Se} = 1000$ Å on a quelque soit $0 < R_S < 10^4\ \Omega$ mémorisation définitive dès la première commutation, pour $e_{Se} =$ $= 3000$ Å ce ne sera vrai que si $R < 10^3\ \Omega$, par contre il y aura une certaine reproductibilité pour $R = 10^4\ \Omega$.

Après recuit à 70 °C et cristallisation avec orientation préférentielle du sélénium, les caractéristiques sont toujours les mêmes; toutefois U_S est passé de 25 avant à 10 V après le recuit.

Dans le cas d'une pointe Ag ou Cu, U_S est plus faible et plus aléatoire.

4. Discussion

Dans le cas des structures Ag_2Se–Se–P, la tension de formation U_F est fonction de l'épaisseur de la couche de sélénium; il s'agit donc d'un phénomène volumique. La tension de seuil est, elle, indépendante de l'épaisseur de Se; il s'agit donc plutôt d'un phénomène d'interface.

L'étude de la formation a montré que dans le cas où l'électrode supérieure est en bronze de tungstène, en or ou en aluminium, la formation n'est possible que lorsque la pointe est prise comme cathode. Cette influence de la polarité permet d'écarter l'hypothèse d'une formation par effet thermique. Il est plus probable que la formation corresponde à une diffusion d'argent sous forme d'ions Ag^+, Ag_2Se étant un semi-conducteur ionique [9]. Les ions Ag^+ issus de la couche d'Ag_2Se migrent vers la pointe polarisée négativement. Leur mouvement est facilité par la dissymétrie géométrique des électrodes [8]. Lorsque les dendrites (ou le filament) ainsi formés atteignent la pointe, il y a commutation à l'état on. Ainsi, plus la couche de sélénium est épaisse, plus U_F est élevée.

Cette hypothèse est confortée par les résultats obtenus avec les pointes Ag et Cu, métaux dont la grande réactivité avec le sélénium est connue [10]. Ces pointes servent de pourvoyeuses d'ions Ag(Cu) d'où la possibilité de formation, quelle que soit la polarité, la faiblesse de l'état off et la faible reproductibilité des cycles. Quand les aiguilles de test ne sont ni en cuivre, ni en argent, la mémorisation de durée limitée dans le temps s'explique de la façon suivante: Le gradient de concentration Ag impose un retour progressif des ions vers la couche de Ag_2Se, d'où une diminution progressive de la conductivité, puis une rupture des dendrites faiblement résistives et le retour à l'état off. Il est une autre méthode pour rompre le pont conducteur entre les électrodes: il suffit d'inverser la polarité. La pointe devenue positive repousse les ions Ag^+ et reforme au niveau de l'interface P–Se une zone de Se hautement résistive, d'où le retour à l'état off [11]. Il découle de ce qui précéde que U_S sera indépendant de l'épaisseur de a-Se. Par contre, elle dépendra de l'épaisseur de la zone vidée en Ag au niveau du contact, qui elle est fonction de la tension maximum en inverse U^-_{max}. Plus U^-_{max} augmente, plus U_S est élevée. A fréquence élevée, les ions se déplaçant lentement, ils sont repoussés moins massivement, d'où un état off peu résistif.

Leur retour vers la pointe, la vitesse de description du cycle étant élevée, nécessitera une tension de seuil plus élevée, U_S augmente avec la fréquence.

Pour ce qui est de la pointe aluminium, la valeur plus élevée du U_S, le manque de reproductibilité du cycle et l'inexistence de la commutation dès que $F > 50$ Hz doivent probablement être reliés à la présence d'une couche d'Al_2O_3 sur la pointe [12].

Pour les structures M–Se–P, la faible reproductibilité du phénomène limite les possibilités d'investigation. Toutefois, l'influence de la résistance série R_S du circuit sur la reproductibilité du cycle montre que la mémorisation définitive est liée à un phénomène thermique. Cependant, le maintien de la commutation après cristallisation du sélénium montre qu'un changement de structure du Se ne suffit pas à expliquer la mémorisation. On peut alors supposer que le phénomène est lié à une diffusion métallique interélectrode stimulée thermiquement et non sous champ, d'où la symétrie du phénomène malgré la forte dissymétrie des électrodes.

Ici, l'initiation de la commutation à l'état on est probablement d'origine électronique et c'est la valeur du courant I_{on} (fixée par la valeur de R_S) qui détermine l'échauffement du canal conducteur et donc l'existence ou non de la diffusion des électrodes.

5. Conclusion

Si les structures M–Se–P présentent des commutations peu exploitables liées à des phénomènes de diffusion thermique, les structures Ag_2Se–Se–P commutent, elles, de façon reproductible. Leur formation fait intervenir une diffusion d'argent à travers la totalité de la couche de a-Se, tandis que lors des cycles suivants, seule intervient la zone vidée en argent autour de la pointe, lors de l'inversion de la polarité.

Le maintien du phénomène après recuit écarte le problème de la stabilité du sélénium amorphe à la température ambiante. Toutefois, le peu de maniabilité des structures sous pointes fait que l'utilisation de telles structures est difficilement envisageable sous leur forme actuelle. C'est pourquoi il est envisagé au laboratoire de réaliser des structures Ag_2Se–Se–M entièrement évaporées, mais conservant la dissymétrie des électrodes, ce qui les rendra plus souples de manipulation. Nous pourrons alors réaliser l'étude $R_{on} = f(T)$ afin de déterminer la nature des dendrites, Ag ou Ag_2Se.

Bibliographie

[1] T. Matsushita, T. Yamagami et M. Okuda, Japan. J. appl. Phys. **11**, 1657 (1972).
[2] A. Prikhodko, A. P. Desksnis et A. Chesnis, Soviet. Phys. — Semicond. **13**, 111 (1979).
[3] M. I. Korsunskii, S. Ya. Maksimova, Sh. Sh. Sarsembinov, V. V. Smetannitov et V. P. Solodukhin, Soviet Phys. — Solid State **19**, 753 (1977).
[4] A. I. Popov, I. Kh. Geller et V. K. Shemetova, phys. stat. sol. (a) **44**, K71 (1977).
[5] T. Hayashi et Y. Ono, Rep. Fac. Engng., Shizuoka Univ. **26**, 25 (1975).
[6] G. Jones et R. A. Collins, phys. stat. sol. (a) **53**, 339 (1979).
[7] P. Junod, H. Hediger, B. Kilchor et J. Wullschleger, Phil. Mag. **36**, 941 (1977).
[8] J.-C. Bernède, phys. stat. sol. (a) **57**, K101 (1980).
[9] J. Miyatani, J. Phys. Soc. Japan **13**, 317 (1958).
[10] S. K. Sharma, J. Mater. Sci. **4**, 189 (1969).
[11] J.-C. Bernède et Y. Tregouët, Rev. Phys. appl. **14**, 551 (1979).
[12] Y. Tregouët, Thèse d'état, Nantes 1980.

(Received December 22, 1980)

phys. stat. sol. (a) 65, 103 (1981)

Subject classification: 11 and 20.3; 14.3; 22.2.1

Laboratorio de Semiconductores, Instituto de Fisica del Estado Sólido, C.S.I.C.-U.A.M., Universidad Autónoma de Madrid[1])

Photoluminescence in Gallium Arsenide Irradiated with Thermal Neutrons[2])

By

J. GARRIDO, J. L. CASTAÑO, and J. PIQUERAS

Photoluminescence and Hall effect measurements are made on GaAs irradiated with thermal neutrons. In addition to the standard peaks at 1.20 and 1.55 eV, three new peaks appeared in the most heavily irradiated samples, at 1.30, 1.40, and 1.46 eV associated with Cu-contamination, As-vacancies, and T_{AS}–BV transitions, respectively. The radiation damage introduced may be fully recovered by a slight annealing, at 480 °C for 2 min.

On a fait des mesures d'effect Hall et de photoluminescence sur des echantillons d'AsGa préalablement irradiés avec des neutrons termiques. De plus aux pics standard placés à 1,20 et 1,55 eV, on a trouvé trois nouveaux pics dans l'echantillon le plus fortement irradié, et placés à 1,30, 1,40 et 1,46 eV, associés avec Cu-contamination, As lacunes et T_{AS}–BV transitions respectivement. L'endommagement produit par l'irradiation peut disparaitre complètement avec une faible traitement termique à 480 °C pendant 2 min.

1. Introduction

Transmutation doping is a well established technique for obtaining low doping concentrations in silicon for rectifier and thyristor technology [1]. Also, in GaAs, this technique could be very useful in order to introduce low doping without compensation. As it is well known the usual low concentration dopants, like tin, have an amphoteric behaviour causing strong compensation in low resistivity material and consequently low lifetime and mobilities. On the contrary, transmutation doping in GaAs gives rise to stable isotopes, Ge on Ga-sites and Se on As-sites, both acting as donors [2 to 4].

However, there is a lack of data concerning the damage introduced by thermal neutron irradiation and recovery of the different defects appeared during the irradiation.

In this work we have studied the defects introduced with different irradiation doses by means of photoluminescence at 77 K and room temperature [5 to 9]. Hall mobilities were also measured after a heat treatment at which the different photoluminescence peaks caused by the irradiation have completely disappeared.

2. Experimental

The starting samples were (111) oriented GaAs slices doped with Te in the 10^{17} atoms/cm^3 range. Before irradiation the samples were mechanically lapped and polished with alumina powder up to a grain size of 0.3 μm, and later chemically etched with a 0.5% bromine–methanol solution during a few minutes.

The irradiation was carried out in the experimental reactor JEN-1 of the Spanish Institute of Nuclear Studies (JEN). The different samples were subjected to a neutron flow of around 10^{11} cm^{-2} s^{-1} during times of 3, 6, 12, and 24 h. An additional sample was left unirradiated to be used as a test.

---

[1]) Cantoblanco, Madrid, Spain.

[2]) Work partially supported by the Spanish-USA Commitee COOP. 793031.

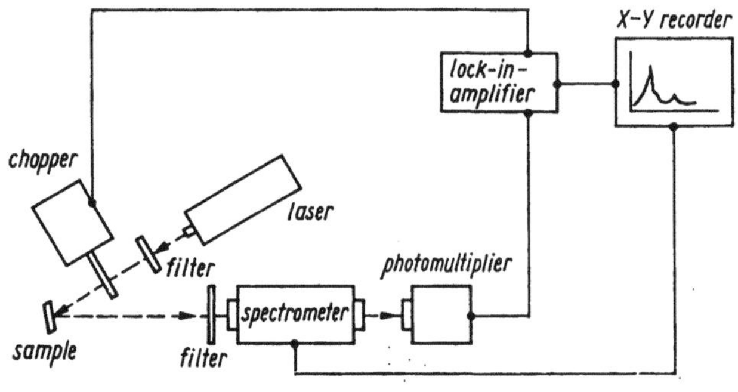

Fig. 1. Scheme of the photoluminescence system

The so irradiated samples were stored for more than 21 days, after which they had negligible radioactivity, less than 1 μCi/g.

The experimental set-up for the photoluminescence measurements is schematically shown in Fig. 1, The excitation light source was a 5 W argon laser modulated at a frequency of about 1.2 kHz. The emitted radiation was focused onto the entrance slit of the monochromator, taking care that the reflected beam comes out of the slit. Furthermore, a convenient filter was situated between the sample and the monochromator in order to absorb the light scattered by the sample surface. Finally, the light emitted by the sample was synchronously detected with the aid of a lock-in amplifier and an S-1 photomultiplier located at the output slit of the monochromator.

After the photoluminescence measurements Au–Ge contacts were alloyed at 480 °C in a pure H_2 atmosphere for 2 min and Hall mobilities were measured in all the samples by the van der Pauw method.

3. Results and Discussion

The photoluminescence measurements taken at room temperature were made with an excitation power of 200 mW. The spectra obtained were the standard ones in n-type GaAs, that is, they consist of a peak around 1.42 eV corresponding to the valence band to conduction band transition and a broad band at about 1.14 eV generally assigned to transitions (Te_{As}–V_{Ga}). The intensity of both peaks decreases with irradiation time disappearing below the sensitivity of our experimental arrangement for times of 12 and 24 h as shown in Fig. 2

The photoluminescence spectra were also taken at 77 K but with an excitation power lower than in the preceding case, 20 mW, to prevent sample heating. The intensity of the Te_{As}–V_{Ga} band (1.2 eV) also decreases with irradiation time as can be seen in Fig. 2.

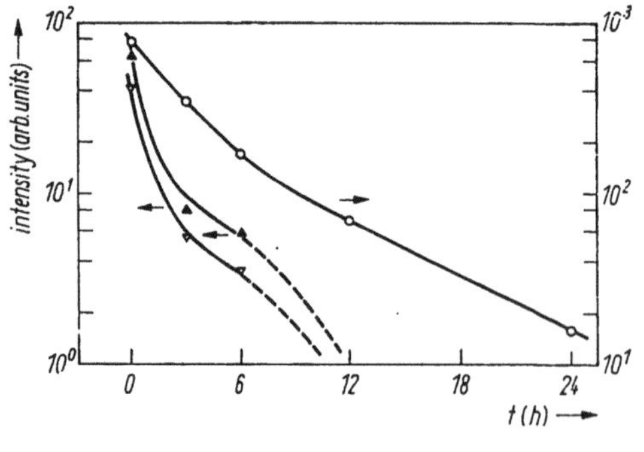

Fig. 2. Photoluminescence intensities versus irradiation time for (BV–BC) and (Te_{As}–V_{Ga}) transitions at 77 K and room temperature before annealing. ○ $E \approx 1.24$ eV, $T \approx 77$ K, ▲ $E \approx 1.42$ eV, $T \approx$ ambient temperature, ▽ $E \approx 1.14$ eV, $T \approx$ ambient temperature

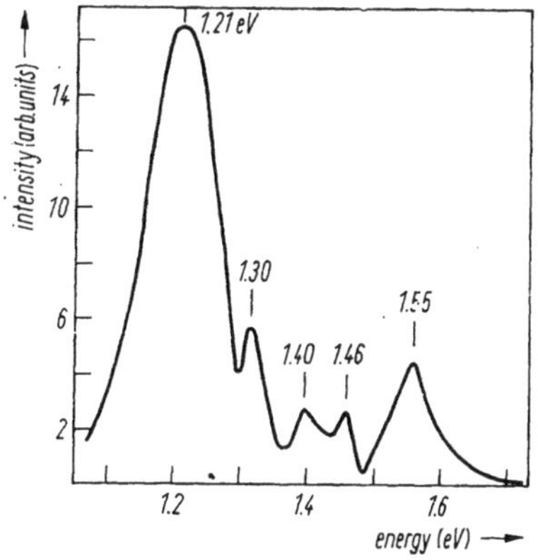

Fig. 3. Photoluminescence spectra for the sample irradiated for 24 h at 77 K before annealing

In the sample irradiated for 24 h the spectrum becomes rather complex and three new peaks appear (see Fig. 3). One of them, at 1.3 eV, is generally assumed to be related to Cu contamination [9]. A second peak at around 1.40 eV could be associated with the As vacancy [5]. A peak at about 1.46 eV has been also found by other authors [9] in Te-doped GaAs irradiated with electrons and it has been assigned to a Te_{As}–BV transition.

After alloying of the Au–Ge ohmic contacts for the Hall measurements, the photoluminescence measurements were repeated at both temperatures (77 and 300 K) in the same conditions.

The salient aspect of the different spectra was that the additional peaks (1.30, 1.40, and 1.46 eV) found in the most heavily irradiated samples (24 h) had fully disappeared. The peaks corresponding to the BC to BV transition and the band associated with the Te_{As}–V_{Ga} complexes show small variations in their intensities, slightly decreasing with the irradiation time as can be seen in Fig. 4, but in general, their intensities had recovered approximately to the value obtained in the unirradiated sample.

Once the ohmic contacts have been alloyed Hall measurements by the van der Pauw method were carried out at 77 and 300 K. The mobilities, resistivities, and carrier concentrations are shown in Table 1 for the different irradiation times. As it can be seen no significant change, except for a slight decrease in mobility and concentration with irradiation time, has been observed.

Finally, by means of the transient capacitance technique [10] we have looked for the possible existence of deep centres introduced by the irradiation process and no appreciable concentration of deep levels has been found at least over the sensitivity limit of our system, in the whole temperature range (77 to 300 K) explored.

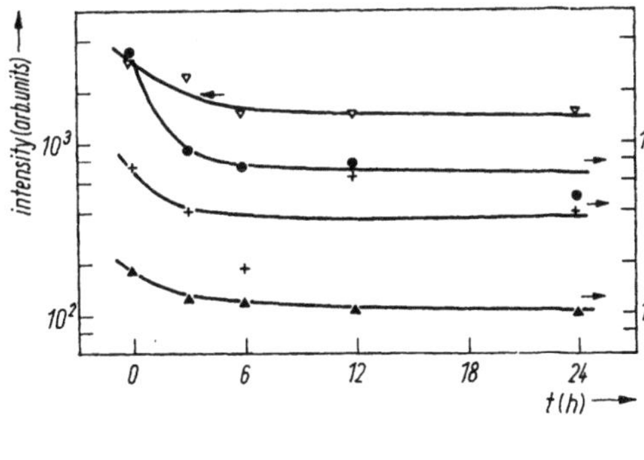

Fig. 4. Photoluminescence intensities versus irradiation time for (BV–BC) and (Te_{As}–V_{Ga}) transitions at 77 K and room temperature after annealing. ● $E \approx 1.42$ eV, + $E \approx 1.14$ eV, $T \approx$ ambient temperature; ▲ $E \approx 1.50$ eV, ▽ $E \approx 1.29$ eV, $T \approx 77$ K

Table 1
Mobilities, resistivities, and carrier concentrations for annealed samples

		unirradiated	irradiated			
			3 h	6 h	12 h	24 h
	μ (10^{-3} Ω cm)	4.83	—	8.63	7.98	8.4
T = amb.	ϱ (10^3 cm^2 V^{-1} s^{-1})	4.87	—	2.94	4.65	4.6
	N (10^{17} cm^{-3})	2.66	—	2.53	1.68	1.59
	μ (10^{-3} Ω cm)	5.54	7.51	5.38	9.62	1.04
T = 77 K	ϱ (10^3 cm^2 V^{-1} s^{-1})	4.23	4.74	4.40	4.04	3.77
	N (10^{17} cm^{-3})	2.66	1.75	2.64	1.61	1.59

4. Conclusions

In view of the results obtained, it seems that, at least for the irradiation doses used, the only observed effect is a decrease of the luminescence peaks with irradiation time, except for the most heavily irradiated samples in which the spectra show new peaks besides the standard ones. These new peaks have been identified with those reported in the literature. that is, one at 1.40 eV related with the As-vacancy, another peak at 1.46 associated with the Te_{As}–BV transition, and that at 1.30 eV generally assumed to be due to Cu-contamination.

In all cases the radiation damage was not too severe and it could be fully recovered by a slight annealing, say, of 2 min at 480 °C. For such low-temperature heat treatment it can be assumed that no migration of the created impurities (Te and Ge) from their rest sites occurs, making this method a valuable tool for introducing n-type impurities in low concentrations and without compensation in GaAs.

Acknowledgements

The authors wish to thank Dr. V. Alcober from the Instituto de Estudios Nucleares for the irradiation facilities.

References

[1] J. M. Meese (Ed.), Proc. 2nd Internat. Conf. Transmutation Doping in Semiconductors, Missouri 1978, Plenum Press, New York 1979.
[2] Sh. M. Mirianshvili and D. I. Nanobashvili, Soviet Phys. – Semicond. 4, 1612 (1971).
[3] J. H. M. Stoelinga, D. M. Larsen, W. Walukiewicz, R. L. Aggarwall, and C. O. Bozler, J. Phys. Chem. Solids 39, 873 (1978).
[4] J. E. Mueller, W. Kellner, and H. Kniepkamp, J. appl. Phys. 51, 3178 (1980).
[5] L. L. Chang, L. Esaki, and R. Tsu, Appl. Phys. Letters 19, 143 (1971).
[6] R. Romano-Moran and K. L. Ashley, J. Phys. Chem. Solids 34, 427 (1973).
[7] T. Itoh and J. Kajahara, J. appl. Phys. 45, 4915 (1974).
[8] H. Birey and J. Sites, J. appl. Phys. 51, 619 (1980).
[9] M. U. Jeong, J. Shirafuji, and Y. Invishi, Proc. Conf. Radiation Eff. in Semicond., Ed. J. W. Corbett and G. D. Warkins, State University of New York, Albany 1970 (p. 287 to 292).
[10] M. Bleicher and E. Lange, Solid State Electronics 16, 375 (1973).

(Received December 2, 1980)

phys. stat. sol. (a) 65, 107 (1981)

Subject classification: 11; 22.1.2

Institute of Nuclear Physics, Moscow State University[1])

Spatial Distribution of Channeled Ions

By

M. A. Kumakhov and V. A. Muralev

An investigation is made of the dependence of distributions of implanted heavy ions, with the inclusion of the channeling effect, upon the initial ion energy, the thickness of the amorphous layer on the crystal surface, the direction of target irradiation, the target temperature, the angle of orientation of the ion beam with respect to the channel axis, the type of irradiating ions, and the radiation dose. Problems of capture of ions into channels are also considered.

Теоретически изучаются пространственные распределения каналированных ионов в зависимости от начальной энергии ионов, от толщины аморфного слоя на поверхности кристалла, от направления облучения мишени, от температуры мишени, от угла между осью канала и направлением пучка, от сорта ионов, от дозы облучения. Обсуждается также проблема захвата ионов в канал.

1. Introduction

In the investigation of heavy ion channeling the problem of the spatial distribution of implanted ions in crystal lattices, which is dependent on the target–ion beam orientation is of particular concern [1]. Thus, if a crystal is irradiated by ions along the channeling directions the final distribution depends upon a great number of parameters: the amorphous layer thickness, the initial energy of ions, the type of ions, the target temperature, the type of target, the crystallographic direction, the angle of beam disorientation with respect to the channel axis, the radiation dose.

For the disoriented irradiation, in order to calculate the implanted ion profile in the Gaussian approximation it is sufficient to know the projective range, the straggling and the parameters characterizing the implanted ions and the target (when secondary processes are neglected) [2].

The theoretical and experimental results of the channeling effect upon the implanted ion profile are discussed in reviews by Gibbons [3], Mayer et al. [4], and in [5 to 31]. The present paper gives the results of a theoretical investigation of heavy ion channeling in crystals which were obtained by the authors in the course of several years at the Institute of Nuclear Physics, Moscow State University.

2. Theoretical Model

The general approach to the formulated problem is based on the derivation of the distribution function depending upon the transverse and total energy of channeled ions from the Focker-Planck two-dimensional kinetic equation [30]. A study of this equation and also its numerical solutions even with an electronic computer presents considerable technical difficulties. This paper gives a very simple method for solving the above-noted problem.

---

[1]) 117234 Moscow, USSR.

Let us proceed from the main assumptions implied in our model.

1. When considering heavy ion channeling we use the classical approximation. The discrete atomic string potential is substituted by the Lindhard continuous potential [31]. In this approach the potential in the axial channel is equal to the sum of string potentials forming it.
2. Consideration is given to axial channeling only.
3. Ion transitions from axial to planar channeling are not considered.
4. Secondary processes (diffusion etc.) are neglected.
5. Rechanneling is also neglected.
6. The condition of ion motion in the channel is that the angle between the channel axis and the ion motion should not exceed a critical angle of channeling, i.e. the transverse energy of ions in a channel is lower than the Lindhard critical energy.
7. If the transverse ion energy is higher than the critical one, it is assumed that the ions leave the channeling regime. As the ions leave the channel their paths are calculated as in amorphous targets [32], allowing for the fact that the ions have lost a certain amount of energy in the channel.
8. For the oriented ion irradiation incident upon the crystal target the ion beam is assumed to be divided into two parts, random and channeled. In addition, the ions forming these beams are assumed to be slowed down independently of each other. The random beam for which the crystal behaves as an amorphous target, suffers stopping due to nuclear collisions and to electronic interactions. The main energy loss mechanism for the ions forming the channeled beam is electronic stopping. Therefore, the channeled ions penetrate deeper into the target than the non-channeled ones. In this model the superposition of distributions of ions which initially did not enter the channel or became channeled and perhaps dechanneled, leads to distributions with two maxima.
9. The transverse energy of an ion is assumed to be monotonously varying, i.e. diffusion in the transverse momentum space is neglected.

In order to calculate the ion distributions in a crystal target it is necessary to know the energy losses of ions in the channel, the initial distribution in transverse energy, and also the transverse energy distribution for the particles that passed through the amorphous layer Δx. We also shall assume that the crystal thickness is sufficient for averaging over all impact parameters of ion collisions with an atomic string.

When travelling along the axial channel, the ion is influenced by four types of interactions leading to a change in its transverse energy $E_\perp$ (in terms of dimensionless units $\varepsilon_\perp = E_\perp d/Z_1Z_2e^2$).

(i) Electronic stopping

In the first case referred to as the square-well approximation the mean kinetic transverse energy will be assumed to be equal to the total transverse energy. Therefore, after averaging the relation over the accessible region we obtain the expression

$$\left\langle \frac{\Delta\varepsilon_\perp}{\Delta x} \right\rangle_e = \frac{\varepsilon_\perp}{E} \left\langle \frac{\Delta E}{\Delta x} \right\rangle_e, \tag{1}$$

where $\langle \Delta E/\Delta x \rangle_e$ is the change in the total energy due to electronic stopping; here and henceforth the brackets $\langle ... \rangle$ mean the averaging over the region accessible for an ion. When the potential in the channel is considered as harmonic in accordance with the virial theorem, we have

$$\left\langle \frac{\Delta\varepsilon_\perp}{\Delta x} \right\rangle_e = \frac{1}{2} \frac{\varepsilon_\perp}{E} \left\langle \frac{\Delta E}{\Delta x} \right\rangle_e. \tag{2}$$

(ii) Electronic scattering

Lindhard's formula including close and distant collisions is used for this scattering

$$\left\langle\frac{\Delta\varepsilon_\perp}{\Delta x}\right\rangle_s = \frac{d(5mv^2\hbar\omega_p)^{1/2}}{E4Z_1Z_2e^2}\left\langle\frac{\Delta E}{\Delta x}\right\rangle_e . \tag{3}$$

(iii) Scattering at thermally vibrating lattice nuclei

$$\left\langle\frac{\Delta\varepsilon_\perp}{\Delta x}\right\rangle_{th} = \frac{\pi N d Z_1 Z_2 e^2 u_\perp^2}{EC^2a^2}\left(\frac{2}{3} + e^{\varepsilon_\perp}\right)(1 - e^{-\varepsilon_\perp})^3 . \tag{4}$$

(iv) Scattering by implanted ions

Here we consider a change in transverse energy of channeled ions due to scattering on defects present in channels after irradiation. Then the total change in transverse energy is equal to electronic stopping, scattering by electrons, by thermally vibrating lattice nuclei, and by lattice defects. The first three interactions were considered above.

The change in transverse energy due to multiple scattering on defects is determined by the formula

$$\delta E_\perp = \langle\theta^2\rangle E = \frac{m_2}{m_1}\frac{\delta E}{E}E = \frac{m_2}{m_1}\left(\frac{\delta E}{\delta x}\right)\delta x , \tag{5}$$

m_1 is the mass of an incident ion, m_2 the target mass, $\langle\theta^2\rangle$ the mean square angle of scattering. If the defect density makes up N_1, then

$$\frac{\delta E}{\delta x} = S_n N_1 , \tag{6}$$

S_n is the nuclear stopping power according to Firsov [34]. Then for $(\delta E_\perp/\delta x)$ we obtain the formula

$$\frac{\delta E_\perp}{\delta x} = \frac{m_2}{m_1} S_n N_1 . \tag{7}$$

Passing over to the dimensionless transverse energy we obtain

$$\left.\begin{aligned}
\left(\frac{\delta\varepsilon_\perp}{\delta x}\right)_d &= \frac{d}{Z_1Z_2e^2}\left(\frac{\delta E_\perp}{\delta x}\right) = \frac{d}{Z_1Z_2e^2}\frac{m_2}{m_1}S_nN_1 = \\
&= \frac{2\pi d}{E}Z_1Z_2e^2N_1 \ln\left|1 + \frac{E\sqrt{\Lambda}}{E_1}\right| , \\
E_1 &= \frac{8}{\pi}e^2\frac{Z_1Z_2}{r_1}\psi\sqrt{\frac{m_1}{m_2}} ; \qquad r_1 = 0.47\ \text{Å} , \\
\psi &= (Z_1^{1/2} + Z_2^{1/2})^{2/3} , \qquad \Lambda = \frac{4m_1m_2}{(m_1 + m_2)^2} .
\end{aligned}\right\} \tag{8}$$

In formulas (1) to (8) N is the number of atoms in the target per unit volume, d the lattice parameter, e the charge of an electron, a the screening constant, $C^2 = 3$, $u_\perp = \frac{2}{3}\langle u^2\rangle$, $\langle u^2\rangle$ the mean square amplitude of thermal vibrations, Z_1e the charge of an incident ion, Z_2e the target atom charge, m the electron mass, v the ion velocity, ω_p the plasma oscillation frequency, $\hbar$ Planck's constant divided by 2π.

It should be noted that the scattering by electrons, thermal vibrations, and implanted ions leads to an increasing transverse energy of channeled particles. Although electron scattering is small for heavy ions, it affects the profile which is formed by the ions remaining in the channel nearly to the end. Electronic stopping is, on the contrary, focusing and taking it into account leads to a decreasing transverse energy. Thus, the total change in ion transverse energy is determined by the formula

$$\frac{d\varepsilon_\perp}{dx} = \left\langle\frac{\Delta\varepsilon_\perp}{\Delta x}\right\rangle_e + \left\langle\frac{\Delta\varepsilon_\perp}{\Delta x}\right\rangle_s + \left\langle\frac{\Delta\varepsilon_\perp}{\Delta x}\right\rangle_{th} + \left\langle\frac{\Delta\varepsilon_\perp}{\Delta x}\right\rangle_d . \tag{9}$$

Equation (9) in the square-well approximation may be rewritten as

$$\frac{d\varepsilon_\perp}{dx} = \frac{\varepsilon_\perp}{E}\left\langle\frac{\Delta E}{\Delta x}\right\rangle_e - \left\langle\frac{\Delta\varepsilon_\perp}{\Delta x}\right\rangle_{th} - \frac{d(5mv^2\hbar\omega_p)^{1/2}}{E4Z_1Z_2e^2}\left\langle\frac{\Delta E}{\Delta x}\right\rangle_e - \left\langle\frac{\Delta E_\perp}{\Delta x}\right\rangle_d . \tag{10}$$

In the harmonic potential approximation

$$\frac{d\varepsilon_\perp}{dx} = \frac{\varepsilon_\perp}{2E}\left\langle\frac{\Delta E}{\Delta x}\right\rangle_e - \left\langle\frac{\Delta\varepsilon_\perp}{\Delta x}\right\rangle_{th} - \frac{d(5mv^2\hbar\omega_p)^{1/2}}{E4Z_1Z_2e^2}\left\langle\frac{\Delta E}{\Delta x}\right\rangle_e - \left\langle\frac{\Delta\varepsilon_\perp}{\Delta x}\right\rangle_d . \tag{11}$$

The notation in (10) and (11) follows that of (1) to (5).

Dividing (10) and (11) by $\langle\Delta E/\Delta x\rangle_e$, we obtain

$$\left\langle\frac{\Delta E}{\Delta x}\right\rangle_e^{-1}\frac{d\varepsilon_\perp}{dE} = \frac{\varepsilon_\perp}{E} - \left\langle\frac{\Delta\varepsilon_\perp}{\Delta x}\right\rangle_{th}\left\langle\frac{\Delta E}{\Delta x}\right\rangle_e^{-1} - \frac{d(5mv^2\hbar\omega_p)^{1/2}}{E4Z_1Z_2e^2} - \left\langle\frac{\Delta\varepsilon_\perp}{\Delta x}\right\rangle_d\left\langle\frac{\Delta E}{\Delta x}\right\rangle_e^{-1}, \tag{12}$$

$$\left\langle\frac{\Delta E}{\Delta x}\right\rangle_e^{-1}\frac{d\varepsilon_\perp}{dE} = \frac{\varepsilon_\perp}{2E} - \left\langle\frac{\Delta\varepsilon_\perp}{\Delta x}\right\rangle_{th}\left\langle\frac{\Delta E}{\Delta x}\right\rangle_e^{-1} - \frac{d(5mv^2\hbar\omega_p)^{1/2}}{E4Z_1Z_2e^2} - \left\langle\frac{\Delta\varepsilon_\perp}{\Delta x}\right\rangle_d\left\langle\frac{\Delta E}{\Delta x}\right\rangle_e^{-1}. \tag{13}$$

The ion cannot be in the channeling regime at any value of transverse energy. Lindhard has estimated the admissible values of transverse energy for a channeled ion [31],

$$3a^2\pi Nd \lessapprox \varepsilon_\perp \lessapprox \varepsilon_\perp^* = \frac{2a\sqrt{E}}{\sqrt{dZ_1Z_2re^2}} . \tag{14}$$

The total range of an ion R is composed of its range in the channel R_c and beyond it, after its leaving the channel, and is determined by the formula

$$\left.\begin{aligned} R(\varepsilon_\perp^0) &= \int_0^{E^*}\left(\frac{dE}{dx}\right)_a^{-1} dE + \int_{E^*}^{E}\left\langle\frac{dE}{dx}(E, E_\perp(E, \varepsilon_\perp^0))\right\rangle_e^{-1} dE = R_a + R_c , \\ R_c(\varepsilon_\perp^0) &= \int_{E^*}^{E}\left\langle\frac{dE}{dx}(E, E_\perp(E, \varepsilon_\perp^0))\right\rangle_e^{-1} dE , \end{aligned}\right\} \tag{15}$$

where $(dE/dx)_a$ are the energy losses in an amorphous medium, E^* is the energy of an ion leaving the channel, and the second integral gives the value of the ion range in the channel, R_c.

It should be noted that due to resonance motion [33] the ions leave the channel near the end of the range at an energy roughly equal to several keV (depending upon the type of ion and crystallographic direction of the target). This means that a channeled ion has minimum threshold energy. Having an energy below such a critical value, the ion leaves the channel. The total range in (15) is obtained in the following way. For some initial value $\varepsilon_\perp^0$ in the interval given by (14), equations (12) or (13) are solved by the Runge-Kutta method. Integration is performed until $\varepsilon_\perp$ exceeds $\varepsilon_\perp^*\sqrt{E/E_0}$ where E is the variable of (12) and (13) and E_0 is the initial energy of an ion. When this condition is fulfilled the value of the argument $E = E^*$ at which it is fulfilled, is fixed. Thereby one obtains the function $E_\perp = E_\perp(E, \varepsilon_\perp^0)$ and the energy of an ion leaving the channel E^*. Substituting the obtained values into (15) we obtain, after numerical integration, $R(\varepsilon_\perp^0)$. Now it is not difficult to obtain the range distribution,

$$\frac{dN}{dR} = \frac{dN}{d\varepsilon_\perp^0}\frac{d\varepsilon_\perp^0}{dR} = \frac{dN/d\varepsilon_\perp^0}{dR/d\varepsilon_\perp^0}, \tag{16}$$

where $dN/d\varepsilon_\perp^0$ is the initial distribution in transverse energies which is found by the use of the Lindhard potential. The function $R(\varepsilon_\perp^0)$ is known from (15).

To obtain the dechanneling function or the fraction of ions leaving the channel in the interval x_i; $x_i + \Delta x_i$, it is sufficient to calculate only the second integral in (15). Ions leaving the channel undergo strong nuclear scattering, therefore it is necessary to take into account range straggling connected with this scattering. In this case we obtain for the range distribution of a dechanneled ion

$$\overline{n(x)} = \int_0^\infty \varphi(x_i)\, f(x - x_i - R_p(x_i))\, dx_i\,, \tag{17}$$

where $\varphi(x_i)\, dx_i$ is the portion of ions leaving the channel in the interval x_i; $x_i + dx_i$. The distribution function is Gaussian,

$$f(x - x_i - \overline{R_p(x_i)}) = \frac{1}{\sqrt{2\pi\, \overline{\Delta R_p^2}(x_i)}} \exp\left\{-\frac{[x - x_i - \overline{R_p(x_i)}]^2}{\overline{\Delta R_p^2}(x_i)}\right\}. \tag{18}$$

Here $\overline{R_p(x_i)}$ is the projective range of an ion in the amorphous medium if it leaves the channel at the depth x_i; $\overline{\Delta R_p}(x_i)$ is the corresponding range straggling. The final profile is composed of three regions. The first region is formed by the ions which do not enter the channel at all. Their distribution is practically the same as the distribution of ions implanted into the amorphous target which is multipled by the fraction of ions which did not enter the channel. The second profile is formed by the ions dechanneled during their motion along the channel. The third one is formed by the ions remaining in the channel to the end.

Thus, in this section we have considered the main principles underlying our model. It follows from (16) that in order to obtain the distributions in the ranges dN/dR one has to know the initial distribution in transverse energies $dN/d\varepsilon_\perp^0$. It is seen from (12) and (13) that one should know how the energy losses of channeled ions are calculated.

3. Energy Losses of Channeled Ions

In order to obtain the ion energy losses in a channel one should know the energy transferred by the ions to a target atom depending upon the impact parameter ϱ. The statistical theory underlying Firsov's treatment [34] provides not very accurate values of energy losses in the channel. For well-channeled ions it overestimates ranges. Therefore, in order to use the formula

$$\left\langle\frac{dE}{dx}(E_\perp)\right\rangle = \frac{1}{d}\left[\frac{1}{S(E_\perp)} \sum_{i=1}^{N} \int_{S(E_\perp)} T(x_i y_i)\, dS\right] \tag{19}$$

and the result not to exceed the experimental data, the ion ranges for a given direction were normalized to the experimental data of Eisen [7] only for one energy at the maximum range in the channel. Ranges at other energies were not corrected with experimental data. In (19) the summation is performed over N nearest-neighbour strings, $S(E_\perp)$ is the accessible region for a given ion, d the lattice parameter, and T

the energy transferred to an atom. We succeeded in approximating the ion ranges $R(\varepsilon_t)$ by simple analytical expressions [33]. The energy losses of ions in the channel were also theoretically estimated with the use of the Hartry-Fock-Slater model for calculating the electron density in the channel. The two approaches have comparable results [35 to 37].

4. Initial Particle Distribution in Transverse Energy

Now, let us define $dN/d\varepsilon_\perp^0$, the initial distribution in transverse energy. We are mainly interested in ions with energy from tens to hundreds of keV. In this energy region one may use the potential $\sim r^{-2}$. Then, the angular distribution of particles which passed through the layer Δx, is equal to

$$\frac{dN}{\theta\, d\theta} = \frac{\overline{P}}{(\theta^2 + \overline{P}^2)^{3/2}}, \tag{20}$$

where $\overline{P} = N\, \Delta x a Z_1 Z_2 e^2 \left((M_1 + M_2)/M_2 E\right)$; a is the screening parameter. Equation (20) is derived from the Gott formula [38] when $\overline{P} < 1$. From (20) we obtain [39]

$$\frac{dN}{d\varepsilon_\perp^0} = \frac{0.25\overline{P}/\psi_1}{[(\overline{P}/\psi_1)^2 + 0.5\varepsilon_\perp^0]^{3/2}}. \tag{21}$$

If the beam falls upon the surface at an angle $\psi_{in} \neq 0$, it is evident that [39]

$$\frac{dN}{d\varepsilon_\perp'} = \frac{0.25\overline{P}/\psi_1}{[(\overline{P}/\psi_1)^2 + 0.5\varepsilon_\perp']^{3/2}}, \tag{22}$$

where $\varepsilon_\perp' = \varepsilon_\perp^0 + 2\psi_{in}^2/\psi_1^2$, $\psi_1 = \sqrt{2Z_1Z_2e^2/Ed}$ is Lindhard's critical angle. The channeled ion acquires additional energy due to its interaction with the crystal lattice field. This fact is taken into account by the transmission factor

$$T(\varepsilon_\perp, \varepsilon_\perp') = \int \frac{dS}{S_0}\, \delta\left(\varepsilon_\perp - \varepsilon_\perp' - \frac{2U_T}{E\psi_1^2}\right), \tag{23}$$

where S_0 is the square per channel, U_T the potential in the channel which is produced by the nearest-neighbour atomic strings. The final distribution $g(\varepsilon_\perp, 0)$ in $\varepsilon_\perp$ for the ions that passed through the layer Δx and entered the channel has the form

$$g(\varepsilon_\perp, 0) = \int \left(\frac{dN}{d\varepsilon_\perp'}\right) T(\varepsilon_\perp, \varepsilon_\perp')\, d\varepsilon_\perp'. \tag{24}$$

5. Channeled Ion Ranges. Discussion of the Results

In this section we shall study the influence of various factors upon the total ranges of ions R and the ranges of ions in channels R_c which are calculated by (12), (13), (15). It should be mentioned that by the range of an ion we mean the value equal to the sum of ranges which an ion has traveled in the channel and after dechanneling.

Results of a calculation of the dependence of ranges of phosphorus ions in the ⟨110⟩ silicon channel upon transverse energy at the target temperature 25 °C for various initial values ion energy are given in Fig. 1a. In calculating the range of a channeled ion by the second formula (15) we use the solution of (12) written for the square-well potential. It follows from Fig. 1a that curves 1 to 4 have common features. Thus, at $\varepsilon_\perp \to 0$ the ranges in the channel tend to their maximum values which are equal at initial energies of 200, 150, 100, 50 keV to 2.1, 1.8, 1.5, 1 μm, respectively. With increasing $\varepsilon_\perp$ the ranges of the channeled ions decrease almost

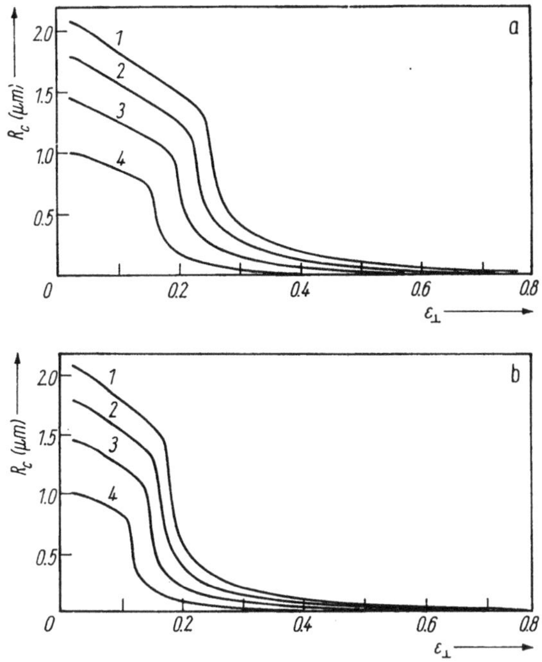

Fig. 1. Dependence of the ranges of phosphorus ions in the Si ⟨110⟩ channel upon $\varepsilon_\perp$. (1) E_0 = 200, (2) 150, (3) 100, (4) 50 keV, square-well potential, a) at a target temperature of 25 °C and b) at a target temperature of 300 °C

linearly until the influence of thermal vibrations plays a small part as compared with electronic stopping. As it follows from the figure, at any values of $\varepsilon_\perp$ the ranges of the channeled ions increase with energy.

Fig. 1b presents the results of calculations of ranges R_c of phosphorus ions in a ⟨110⟩ silicon channel at the same energies as in Fig. 1a but at a higher target temperature 300 °C. It is seen that the increase in target temperature does not affect the maximum ranges of well-channeled ions. However, with increasing temperature one observes a reduction and considerable steepening of the section in which the ranges decrease linearly with increasing $\varepsilon_\perp$ which is indicative of an increased contribution by thermal vibrations to the increase of $\varepsilon_\perp$. As it follows from a comparison of Fig. 1a and b at E_0 = 200 keV a sharp decrease in the range is observed starting from $\varepsilon_\perp = 0.17$ (at the target temperature 25 °C starting from $\varepsilon_\perp = 0.248$), at E_0 = 150 keV from $\varepsilon_\perp = 0.16$ (at 25 °C starting from $\varepsilon_\perp = 0.19$), at E_0 = 50 keV from $\varepsilon_\perp = 0.11$ (at 25 °C from $\varepsilon_\perp = 0.15$). Thus, the boundary values of $\varepsilon_\perp$ decrease with increasing temperature.

The results of range calculations of phosphorus ions as a function of $\varepsilon_\perp$ at target temperatures 25 and 300 °C are given in Fig. 2a. The present calculation was made with the use of the square-well potential for obtaining $R(\varepsilon_\perp)$. As was noted, the total range R is the sum of the ion ranges in the channel and the amorphous medium. Therefore, the difference between Fig. 2a and Fig. 1a, b is that with increasing $\varepsilon_\perp$, when the ranges in channels are small and the ions are rapidly dechanneled, R tends to the values of ranges R_a in the amorphous medium, whereas the ranges in channels R_c tend to zero. At small $\varepsilon_\perp$ the range of ions in the amorphous medium is small because these ions leave the channel with low energies. Therefore, the values do practically not differ from the ranges R_c (Fig. 1 and 2a). Let us recall the ranges of phosphorus ions in amorphous silicon at energies of 50, 100, 200 keV. They are equal to 0.073, 0.14, 0.213, 0.25 µm, respectively. Exactly to these values tend the total ranges with increasing $\varepsilon_\perp$ for the above-noted energies.

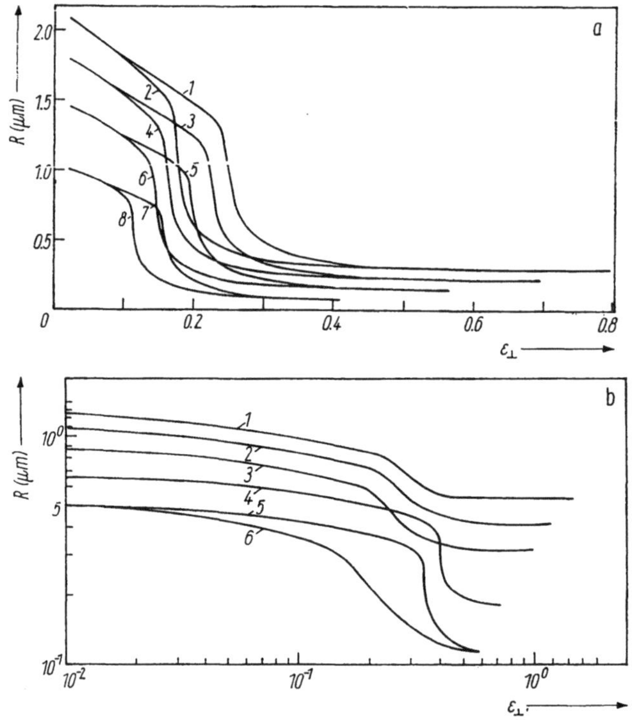

Fig. 2. Dependence of the total ranges R a) of phosphorus ions upon $\varepsilon_\perp$ in the Si ⟨110⟩ channel, square-well potential; $E_0 = 200$ keV, (1) $T = 25$, (2) 300 °C; $E_0 = 150$ keV, (3) $T = 25$, (4) 300 °C; $E_0 = 100$ keV, (5) $T = 25$, (6) 300 °C; $E_0 = 50$ keV, (7) $T = 25$, (8) 300 °C; b) of boron ions upon $\varepsilon_\perp$ in the ⟨111⟩ channel. Target temperature 25 °C; (1) $E_0 = 200$, (2) 150, (3) 100, (4) 50, (5) 30, (6) 30 keV, respectively; (1), (2), (3), (6) harmonic potential, (4), (5) square-well potential

From experiments [17], and also from semiempirical estimates by Eisen [7], it is known that the ranges of channeled boron ions are smaller than those of phosphorus ions. The calculation of the dependence of total ranges R of boron ions upon $\varepsilon_\perp$ is illustrated in Fig. 2b. The behaviour of the total ranges of boron ions is similar to that of total ranges of phosphorus ions. Thus, for example, with increasing $\varepsilon_\perp$ the total ranges R tend to the ranges in the amorphous medium, R_a, which are equal, for the noted energies, to 0.11, 0.18, and 0.32 μm, respectively. With decreasing $\varepsilon_\perp$ the total ranges tend, as in the case of phosphorus ions, to the maximum ranges in the channel.

In addition to the energy dependence of total ranges, Fig. 3 presents their dependence upon the potentials. Comparison of the behaviour of curves 5 and 6 shows that the curves obtained on the basis of the harmonic potential (curve 6) show practically no sharp decrease in the value of the total range. The range is smoothly decreased with increasing $\varepsilon_\perp$. It should be noted that the total range is here dropping more quickly with increasing $\varepsilon_\perp$ than in the model with the square-well potential. The absence of a region of abrupt drop is due to a greater tendency to dechanneling at the same $\varepsilon_\perp$ in the model with the harmonic potential. Fig. 2a and b show that the ranges of boron ions in the amorphous medium and in the channel differ not so drastically as the corresponding ranges of phosphorus ions at the same energies. For example, the range of a boron ion with an energy of 150 keV in the ⟨110⟩ silicon

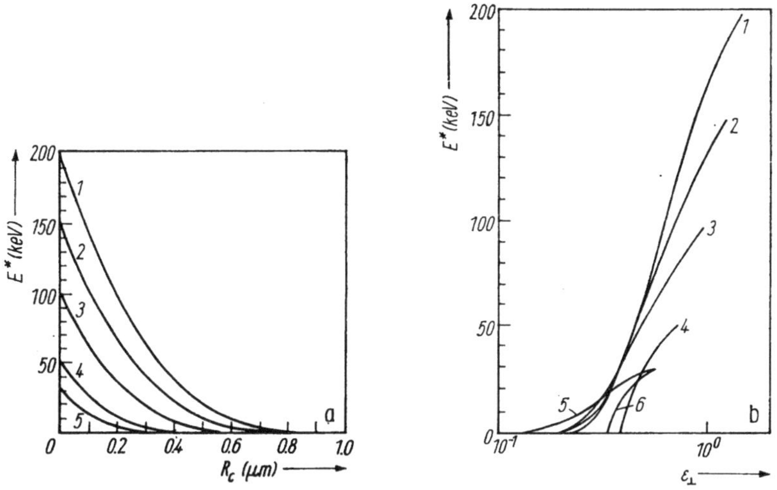

Fig. 3. Dependence of E^* a) upon R_c for boron ions in the ⟨111⟩ channel, harmonic potential; (1) $E = 200$, (2) 150, (3) 100, (4) 50, (5) 30 keV, respectively; b) upon $\varepsilon_\perp$ for boron ions in the ⟨111⟩ Si channel, (1) $E_0 = 200$, (2) 150, (3) 100, (4) 50, (5) 30, (6) 30 keV, respectively; (1), (2), (3), (5) harmonic potential; (4) to (6) square-well potential

channel is 1.6 µm, and in the amorphous medium it is 0.50 µm, and the range of a phosphorus ion with the same energy and in the same channel is 1.75 µm and in the amorphous medium 0.25 µm.

The dependence of the energy of ions leaving the channel upon the range in the channel is of interest for studying the dynamics of ions in the channel. The results of a calculation of these dependences for various initial energies of boron ions in the ⟨111⟩ silicon channel at the target temperature 25 °C are given in Fig. 3a. It is seen from the figure that if the range in the channel R_c is small, then the energy with which the ions leave the channel E^* is large, which evidences strong ion dechanneling; their further path is in the non-oriented target. With increasing R_c one observes a reduction in E^* which is natural since the ions cover large distances in the channel thereby loosing a greater amount of energy.

The results of calculations of the dependence of energy of an ion leaving the ⟨111⟩ channel, E^*, upon $\varepsilon_\perp$ for boron ions with initial energies 30, 50, 100, 150, 200 keV are shown in Fig. 3b. This figure accounts for the presence of peculiarities in the behaviour of the ranges of channeled ions which are calculated by the models using the square-well potential (curve 6) and the harmonic potential (curve 5). It is seen from the figure that E^* is strongly dependent upon the potential in the channel. Thus, in the model with the square-well potential for the initial energy $E_0 = 30$ keV, $E^* = 0$ at $\varepsilon_\perp < 0.34$, which means absence of dechanneling within $0 < \varepsilon_\perp < 0.34$. However, at $\varepsilon_\perp > 0.34$, E^* rapidly increases. It is exactly that this leads to a sharp change in the range in the region of $\varepsilon_\perp = 0.34$. In the model with the harmonic potential there is no such boundary value of $\varepsilon_\perp$, E^* changes smoothly with $\varepsilon_\perp$. The physical difference of the models is that if the square-well potential is used in (10) the focusing term in the right-hand side of the equation, which is due to electronic stopping and which leads to the decrease in transverse energy of an ion, exceeds the defocusing one, the thermal vibrations, up to the boundary value of $\varepsilon_\perp^*$. At $\varepsilon_\perp > \varepsilon_\perp^*$ one observes strong dechanneling due to thermal vibrations because here $\varepsilon_\perp$ acquires values at which the contribution from the defocusing factor exceeds that from the focusing one. If the harmonic

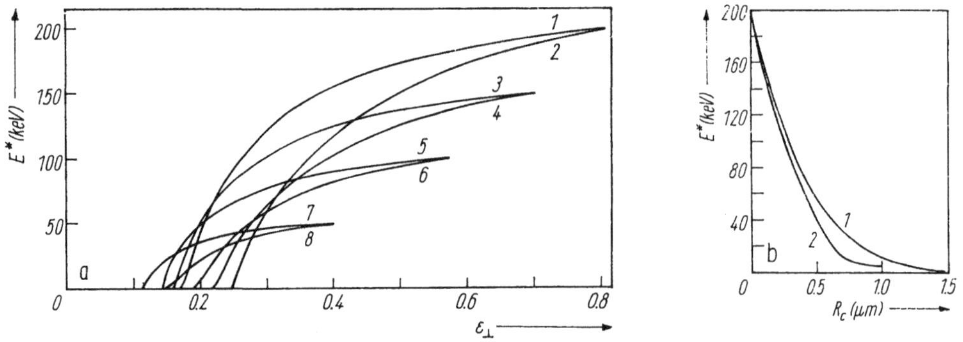

Fig. 4. Dependence E^* a) upon $\varepsilon_\perp$ for phosphorous in the ⟨110⟩. Si channel, square-well potential; $E_0 = 200$ keV, (1) $T = 300$, (2) 25 °C; $E_0 = 150$ keV, (3) $T = 300$, (4) 25 °C; $E_0 = 100$ keV, (5) $T = 300$, (6) 25 °C; $E_0 = 50$ keV, (7) $T = 300$, (8) 25 °C; b) upon R_c for phosphorus ions in the Si ⟨110⟩ channel, square-well potential; (1) $T = 300$, (2) 25 °C

potential is used in (11) the focusing factor is damped and, consequently, the influence of the thermal vibrations becomes more significant. Therefore, at $\varepsilon_\perp < \varepsilon_\perp^*$ dechanneling is also observed, i.e. $E^*(\varepsilon_\perp < 0.34) \neq 0$ (curve 5). The initial energy of ions $E_0 = 50$ keV corresponds to curve 4. The calculation is made with the use of the square-well potential, $E^* = 0$ at $\varepsilon_\perp \leqq 0.4$. A sharp boundary at $\varepsilon_\perp = 0.4$ is also observed here as in curve 6. The remaining curves 1, 2, 3 are calculated in the harmonic potential approximation and so they exhibit no sharp boundary $\varepsilon_\perp$. E^* takes maximum values equal to 30, 50, 100, 150, 200 keV at $\varepsilon_\perp^{**} = 5.5$, 7.5, 1, 1.2, 1.4, respectively. At $\varepsilon_\perp \geqq \varepsilon_\perp^{**}$ the ions instantly leave the channel.

Fig. 4a gives the results of a calculation of the dependence E^* upon $\varepsilon_\perp$ for phosphorus ions in the ⟨110⟩ silicon channel at initial values of energy 50, 100, 150, 200 keV. In addition to the energy dependence, the figure shows the temperature dependence at 25 and 300 °C. All calculations in Fig. 4a are made for the square-well potential. The dependence upon transverse energy of phosphorus ions is similar to that for boron ions. However, the boundary value of $\varepsilon_\perp^*$ at which $E^* = 0$, is smaller than in the case of boron ions for the same energies. Thus, if the target temperature is 25 °C (curves 2, 4, 6, 8) then, for example, at $E_0 = 50$ keV ($\varepsilon_\perp^* = 0.15$) at $E_0 = 200$ keV ($\varepsilon_\perp^* = 0.25$). It is seen that at $\varepsilon_\perp \gtrapprox \varepsilon_\perp^*$, E^* sharply increases and reaches the maximum values equal to 50, 100, 150, 200 keV and at $\varepsilon_\perp^{**} = 0.40$, 0.57, 0.70, 0.80, respectively. A comparison between Fig. 3b and 4a shows that $\varepsilon_\perp^*$ for boron ions is larger than that for phosphorus ions at equal initial energies. Consequently, the contributions from thermal vibrations to the change in $\varepsilon_\perp$ in channeling of boron and phosphorus ions are different. It follows from the figure that with increasing temperature the boundary values of transverse energy at which $E^* = 0$, are shifted to the left.

For example, at 300 °C for the initial energy of ions $E_0 = 50$ keV, $\Delta\varepsilon_\perp = 0.04$ and at $E_0 = 100$ keV $\Delta\varepsilon_\perp = 0.08$. From the value of the shift for a given initial energy one may judge the target temperature. A calculation of the dependence of E^* upon the value of ranges R_c of channeled phosphorus ions with energy of 200 keV is given in Fig. 4b. It is seen from a comparison of curves 1 and 2 that for the same ranges R_c, E^* increases with target temperature due to the fact that at higher temperatures dechanneling is stronger and, consequently, the energy of ions leaving the channel is higher than found under otherwise equal conditions.

The present discussion shows that our approach provides the possibility of considering the dynamics of the motion of ions in crystals, producing not only qualitative but also quantitative results.

References

[1] M. A. Kumakhov and V. A. Muralev, Izv. Akad. Nauk SSSR, Ser. fiz. 37, 2585 (1973).
[2] M. A. Kumakhov and V. A. Muralev, Fiz. Tekh. Poluprov. 6, 1564 (1972).
[3] J. F. Gibbons, Proc. IEEE 56, 2204 (1968).
[4] J. Mayer, L. Erickson, and J. Davis, Ion Alloyage of Semiconductors, Mir, Moscow 1973.
[5] A collection of papers on ion implantation. Atomizdat, Moscow 1971.
[6] P. V. Pavlov, D. I. Tetelbaum, E. I. Zorin, and B. I. Alekseev, Fiz. tverd. Tela 8, 2679 (1966).
[7] F. M. Eisen, Canad. J. Phys. 46, 561 (1968).
[8] W. J. Kleinfelder, W. S. Johnson, and J. F. Gibbons, Canad. J. Phys. 46, 597 (1968).
[9] G. Dearnaley, J. H. Freeman, G. A. Gard, and M. A. Wilkins, Canad. J. Phys. 46, 587 (1968).
[10] L. Erickson, J. A. Davies, and P. Jespersgaard, Phys. Rev. 161, 219 (1967).
[11] P. D. Goode, M. A. Wilkins, and G. Dearnaley, Radiat. Eff. 6, 237 (1970).
[12] R. A. Moline, J. appl. Phys. 42, 3553 (1971).
[13] L. Howe and D. A. Channing, Canad. J. Phys. 45, 2467 (1967).
[14] G. Dearnaley, B. W. Famery, J. V. Mitchell, R. S. Nelson, and M. W. Thompson, Phil. Mag. 18, 985 (1968).
[15] K. B. Alexander, E. Dearnaley, D. V. Morgan, and J. M. Poate, Phys. Letters A 32, 365 (1970).
[16] V. M. Gusev, N. P. Busharov, K. D. Demakov, and Yu. Kozlov, Dokl. Akad. Nauk SSSR 197, 319 (1971).
[17] B. L. Crouder, J. Electrochem. Soc. 48, 943 (1971).
[18] V. G. K. Reddi and J. D. Sansbury, J. appl. Phys. 44, 2951 (1973).
[19] N. N. Busharov, V. M. Gusev, M. U. Guseva, M. D. Demakov, Yu. G. Kozlov, Yu. V. Martynenko, A. N. Mansurova, and K. V. Starinin, Proc. IV. All-Union Conf. Physics of Interaction between the Charged Particles and Single Crystals, Izd. MGU, Moscow 1973 (p. 337).
[20] I. G. Gverdtsitelly, A. I. Guldamashvilly, E. M. Diasamidze, S. A. Zaslavsky, A. N. Kalinin, M. A. Kumakhov, and V. A. Muralev, Radiat. Eff. 19, 171 (1973).
[21] A. Desalvo, R. Rosa, and F. Zignani, J. appl. Phys. 43, 3755 (1972).
[22] A. Desalvo, R. Rosa, and F. Zignani, Radiat. Eff. 27, 89 (1975).
[23] A. Desalvo and R. Rosa, Radiat. Eff. 31, 41 (1976).
[24] R. G. Wilson, Appl. Phys. Letters 29, 770 (1976).
[25] F. Burkhardt and C. Wagner, phys. stat. sol. (a) 39, K63 (1977).
[26] M. Huez, F. Cappellani, and G. Restelli, Radiat. Eff. 32, 147 (1977).
[27] M. Meyer, C. Dolin, R. Talon, and C. Zacharis, J. Phys. C 10, 2039 (1977).
[28] K. Tsukamoto, Y. Akasaka, and K. Horie, Japan. J. appl. Phys. 16, 663 (1977).
[29] D. Lecrosnier, J. Paugam, and J. Gallou, Appl. Phys. Letters 30, 323 (1977).
[30] M. A. Kumakhov, V. A. Muralev, A. S. Rudnev, and E. I. Sirotinin, Radiat. Eff. 24, 273 (1975).
[31] J. Lindhard, Kong. Danske Vid. Selsk., mat.-fys. Medd. 34, No. 14 (1965).
[32] M. A. Kumakhov, V. A. Muralev, E. G. Averyanov, V. A. Simonov, and L. P. Khavkii, Projective Ranges Straggling for 1240 Combinations of Ion–Target within 20 keV $\leqq E \leqq$ $\leqq$ 1000 keV, VINITI, Moscow 1974.
[33] V. V. Beloshitsky, M. A. Kumakhov, and V. A. Muralev, Radiat. Eff. 13, 9 (1972).
[34] O. B. Firsov, Zh. eksper. teor. Fiz. 36, 1517 (1959); 32, 1465 (1957); 33, 696 (1957).
[35] M. A. Kumakhov, Proc. III. All-Union Conf. Physics of Interaction between the Charged Particles and Single Crystals, Izd. MGU, Moscow 1972 (pp. 41, 54).
[36] M. A. Kumakhov and V. A. Muralev, Theses of Rep. All-Union Conf. Physical Fundamentals of Ion-Beam Alloyage, Gorkii 1972 (p. 6).
[37] M. A. Kumakhov and V. A. Muralev, see [35] (p. 81).
[38] Yu. Gott, Zh. eksper. teor. Fiz. 60, 1291 (1971).
[39] M. A. Kumakhov, V. A. Muralev, and V. A. Simonov, see [19] (p. 118).

(Received January 26, 1981)

phys. stat. sol. (a) **65**, 119 (1981)

Subject classification: 10.1; 14.2

Physico-Technical Institute of Low Temperatures, Academy of Sciences of the Ukrainian SSR, Kharkov[1])

The Effect of Dislocation Inertia on the Thermally Activated Low Temperature Plasticity of Materials

II. Methods of Calculations and Rate Dependences[2])

By

A. I. LANDAU

The technique of numerical calculations based on the thermoinertial theory developed earlier is considered in detail. The former theory takes into account the influence of dislocation inertial properties on thermally activated ductility of crystals. The values of flow stress σ_{0N} and σ_{0S} are calculated for normal (N) and superconducting (S) states, respectively, and the change of flow stress $\Delta\sigma_{NS} = \sigma_{0N} - \sigma_{0S}$ at the N → S transition. The dependence of the value on strain rate v_{fix} is investigated and it is shown that in a wide range of v_{fix} values $\Delta\sigma_{NS}$ is independent of v_{fix} in terms of the proposed theory. The theoretical results are in rather good qualitative agreement with experimental data known from the literature.

Die Technik der numerischen Berechnung, die auf der früher entwickelten Thermo-Trägheitstheorie beruht, wird ausführlich betrachtet. Die frühere Theorie berücksichtigt den Einfluß der Trägheitseigenschaften von Versetzungen auf die thermisch aktivierte Duktibilität der Kristalle. Die Werte der Fließspannungen σ_{0N} und σ_{0S} für normale (N) bzw. supraleitende (S) Zustände und die Änderung der Fließspannung $\Delta\sigma_{NS} = \sigma_{0N} - \sigma_{0S}$ am N → S-Übergang werden berechnet. Die Abhängigkeit des Wertes von der Verformungsgeschwindigkeit v_{fix} wird untersucht und es wird gezeigt, daß in einem weiten Bereich von v_{fix}-Werten $\Delta\sigma_{NS}$ in der vorgeschlagenen Theorie unabhängig von v_{fix} ist. Die theoretischen Ergebnisse befinden sich in guter qualitativer Übereinstimmung mit experimentellen Werten aus der Literatur.

1. Introduction

In Part I of the present paper [1] a thermoinertial theory has been developed allowing for the effect of dislocation inertia on the thermally activated plasticity of crystalline materials. It includes equations governing the average effective velocity $\bar{v}_{eff}$ of thermally activated dislocation motion that contain explicitly the inertial parameter $\gamma = \tan(\alpha^*_{dyn,max})/\tan(\alpha^*_{stat})$. The parameter describes the increase in the dynamic maximum value $\alpha^*_{dyn,max}$ of the angle at the base of a dislocation segment performing undamped vibrations of decreasing amplitude in comparison with the corresponding static angle α^*_{stat} [1]. From the knowledge of $\bar{v}_{eff}$ one is able to calculate the rate $\dot{\varepsilon}$ of macroscopic plastic deformation using the familiar relation $\dot{\varepsilon} = b\varrho\bar{v}_{eff}$ where b is the length of the Burgers vector and ϱ the density of mobile dislocations in the crystal.

The equations suggested in [1] permit a numerical solution and can be compared with experiment. An example was given in [1] itself. As shown by the calculations, dislocation inertia can be the cause of specific low temperature effects, such as a non-monotonic (i.e. anomalous) temperature dependence of the yield stress in crystals or the change in metal plasticity at the transition from the normal to the superconducting state.

---

[1]) Leninprospekt 47, 310164 Kharkov, USSR.

[2]) Part I see phys. stat. sol. (a) **61**, 555 (1980).

The present paper is devoted to the detail description of the numerical calculation technique made in terms of the thermoinertial theory. Then some results of the calculations are given, i.e. those related to the investigation of the influence of the dislocation inertial properties on the rate dependences of the strain and on the effective value of the activation volume.

2. Sequence of Calculations

Using the formulas obtained in [1] calculations of the temperature dependences of the yield stress of metals in the normal (N) and superconducting (S) state have been performed, i.e. σ_{0N} and σ_{0S}. The calculations require the knowledge of parameters as a the crystal lattice constant, G the shear modulus; v_s sound velocity in the crystal, v the velocity of dislocation motion in the lattice between local obstacles, σ the effective load acting on the dislocation, σ_c the Foreman-Makin critical load [1, 2], B the coefficient of total dynamical drag force, k the Boltzmann constant, T temperature, T_c critical temperature of the N $\rightarrow$ S transition, ν the frequency of thermally activated surmounting attempts, H the activation enthalpy corresponding to the thermally activated motion of dislocations, c volume density of obstacles in the crystal,[3]) S_0 the average area per one obstacle in the glide plane of the dislocation, α the angle of attack [1, 3], i.e. at which the dislocation collides with the point obstacle encountered, α_c the critical value of attack angle defined from the condition that at $\alpha \geqq \alpha_c$ the dislocation should be able to surmount the obstacle without thermal activation, g the fraction of obstacles in the glide plane that are surmounted in an unactivated way [1], $\bar{\lambda}$ the average path covered by the dislocation in the direction of its motion per one thermal activation event, and finally, $\bar{l}$ the mean dislocation segment length.

The yield stress σ_0 is obtained as a solution of the transcendental equation

$$v_{\text{fix}} = \frac{\dot{\varepsilon}_{\text{fix}}}{b\varrho} = \frac{\bar{\lambda}}{\bar{\lambda} v_{\max}^{-1} + \nu^{-1} \exp\left[H(\sigma)/kT\right]}, \tag{1}$$

where

$$\bar{\lambda} = \frac{S_0}{ac\bar{l}(\sigma)\left[1 - g(\gamma^{3/2}\sigma, \sigma_c)\right]} \tag{2}$$

and $\dot{\varepsilon}_{\text{fix}} = v_r/L_s$. Here v_r is the prescribed speed of the straining machine rod in the mode of active straining and L_s the length of the specimen. The value of ϱ is assumed to remain constant in (1). The values g, $\bar{l}$, γ, and σ_c participating in the right-hand side of (1) are found from the formulas presented in [1], namely (1) to (3) and (9) to (11). $H(\sigma)$ is calculated according to the relation [1, 4]

$$H = H_0 \ln\left(\sigma_m/\sigma\right), \tag{3}$$

where we assume $\sigma_m = \sigma_c$. The value of $v_{\max}$ is obtained either from the formula $v_{\max} = b\sigma/B$ (for B independent of v) or from the transcendental equation of the same appearance, $v_{\max} = b\sigma/B(v_{\max})$ (B is a function of v in the superconducting state). Since all the relations involve the effective rather than the applied stress, the values σ_{0N} and σ_{0S} resulting from (1) are not the total yield stress but the effective component only.

[3]) It should be emphasized that c denotes the concentration of obstacles rather than impurities in the crystal. Obstacles to the dislocation motion can be of other nature than impurities, e.g. in a number of cases they are dislocations or dislocation loops of other glide systems intersecting with the glide plane of mobile dislocations considered. On the other hand, impurity-produced obstacles may be clustered rather than single atoms. However, the existence of some correlation between the value of c and the concentration of impurities, c_{imp}, cannot be neglected.

The magnitude of B is obtained from the formula

$$B = \Gamma B_{eN} + B_{cryst}\,, \tag{4}$$

where B_{eN} and B_{cryst} are the normal state electron and non-electron components, respectively. The factor Γ is equal to unity in the N state while in the S state $\Gamma \leqq 1$ [5 to 7]. B_{cryst} is the net effect of several mechanisms dissipating the energy of a mobile dislocation [8]. Its temperature dependence is described by the relation

$$B_{cryst} = \beta \left(\frac{T}{T_0}\right)^3 \left[1 + \left(\frac{T}{T_0}\right)^2\right]^{-1}. \tag{5}$$

To represent Γ we have constructed a special analytic function $\Gamma(v, T)$ capable of approximating, to a good accuracy, the numerical values of Γ as obtained in [5] for various v and T within the range $0.1T_c \leqq T \leqq T_c$. Specifically, the function $\Gamma(v, T)$ is

$$\Gamma(v, T) \approx \Gamma_0(v, T) + [1 - \Gamma_0(v, T)]\,(T/T_c)^{4.5} \tag{6}$$

with

$$\Gamma_0(v, T) \approx \begin{cases} 0 & \text{at } v < v_c\,, \\ 1 - \exp\left[-0.6\left(\dfrac{v}{v_c} - 1\right)\right] & \text{at } v \geqq v_c\,, \end{cases} \tag{7}$$

and

$$v_c = 2b\Delta(T)/\hbar\,. \tag{8}$$

Here $\hbar$ is Planck's constant and $\Delta(T)$ the energy gap of the superconductor. In its turn, the temperature dependence of Δ is represented within $0 \leqq T \leqq T_c$ by the function

$$\Delta(T) \approx 1.76kT_c\left[1 - \exp\left(-1.74\,\frac{T_c}{T}\sqrt{1 - \frac{T}{T_c}}\right)\right]. \tag{9}$$

The computer program allowed $\Gamma(v, T)$ of (6) to (8) to be replaced by any other analytical function that would approximately represent the dependence of Γ upon v and T as found in [5 to 7].

Equations (9) to (11) of [1] employed in the present paper for calculating γ are rigorously valid only when B is independent of v. But in view of the monotonic variation of B associated with the increase of v in the S state, the use of these formulas for finding γ in the S state can result only in underestimated (never overestimated!) values of $\Delta\sigma_{NS} = \sigma_{0N} - \sigma_{0S}$, with qualitatively correct dependences of the value upon T, H_0, c, α_c, and other parameters.

To simplify the calculations, we have chosen some average crystal parameters rather than values relating to a specific material. E.g., we have taken $a = b = 10^{-8}$ cm, $v_s = 10^5$ cm s^{-1}, $G = 10^{11}$ g cm^{-1} s^{-2}, and $T_c = 10$ K. The values β and T_0 involved in (5) have been assumed $\beta = 5 \times 10^{-5}$ g cm^{-1} s^{-1} and $T_0 = 15$ K. We have assumed ν constant, of numerical value equal to 10^{11} s^{-1} in accordance with [9 to 11]. Hence, we ignore the dependence of ν on the dynamical drag coefficient B [11 to 15] and the possible change in ν at the N $\rightarrow$ S transition. This point has been especially checked on the computer, namely the contribution of this change to the magnitude of the yield stress jump, $\Delta\sigma_{NS}$, has been calculated.[4]) As it turns out, the corresponding change in

[4]) To allow for the change in ν during the N $\rightarrow$ S transition we have employed the formula [11 to 15] $\nu = (B_0^2/4M^2 + \omega_0^2)^{1/2} - B_0/2M$ where $M = Gb^2/2v_s^2$ is the effective dislocation mass, $\omega_0 = 2\pi v_s/\bar{l}$ the fundamental frequency of the dislocation segment characteristic vibrations, and B_0 the coefficient of the total (i.e. electronic plus non-electronic) dynamical drag force at $v = 0$. It has been also assumed that the electron component of the dynamical friction force is reduced at the N $\rightarrow$ S transition [5 to 7] by $2\{1 + \exp[\Delta(T)/kT]\}^{-1}$ times if $v = 0$.

$\Delta\sigma_{NS}$ is below 10^4 Pa. With any set of parameters this is about two orders of magnitude lower than the change in $\Delta\sigma_{NS}$ due to the thermoinertial mechanism.

As is well known, a decrease of temperature in the low temperature range brings about a monotonous diminishing of H_0 (whereas at higher temperatures $H_0 = \text{const}$). One of the possible physical explanations was pointed out in [4]. However, all the mechanisms of the phenomenon so far have not been investigated and the analytical form of $H_0(T)$ is not known. For this reason we have assumed in our calculations $H_0 = \text{const}$ at all T.

3. Calculated Results

The transcendental equation (1) and equation (9) of [1] were solved on a computer using the bisection technique. The preset relative accuracy provided for the correctness of six first significant characters of the value computed (i.e. σ_0). Let us designate the right-hand side of (1) by $\bar{v}_{eff}(\sigma)$. Then the graphical solution of the equation would take the form of Fig. 1.

As follows from the series of numerical calculations performed, with every fixed set of H_0, c, α_c, and other parameters one can always indicate a low enough temperature at which $\bar{v}_{eff}(\sigma)$ assumes the form represented schematically by curve 1 in Fig. 1. At $v_{fix} > v_{c1} = v_{max}(\sigma_{01})$ where $\sigma_{01} = \gamma^{-3/2}\sigma_c$ the value of g is equal to unity, i.e. the process of plastic straining is in its visco-dynamical regime where the yield stress can be determined from the equation $v_{fix} = v_{max}(\sigma)$.

At $v_{c2} < v_{fix} < v_{c1}$ the process of straining turns to the regime that can be called quasidynamical, i.e. $g \approx 1$ and virtually all the pinning obstacles are surmounted without the aid of thermal activation but the yield stress σ_0 does not depend on v_{fix} to the accuracy of computer calculations. Finally, at $v_{fix} < v_{c2}$ the straining process represented by curve 1 acquires an unambiguously thermally activated nature when $g < 1$, hence either part or all the obstacles are surmounted by dislocations through thermal fluctuations and the $\sigma_0(v_{fix})$ dependence becomes close to the logarithmic law.

As the temperature is increased, the value of v_{c2} gradually approaches v_{c1} until they coincide at $v_c = v_{max}(\gamma^{-3/2}\sigma_c)$ for some temperature T^*. Then for all $T > T^*$ the dependence $\bar{v}_{eff} = \bar{v}_{eff}(\sigma)$ becomes of the form shown schematically by curve 2 of Fig. 1. At $v_{fix} > v_c$ the straining process is visco-dynamical and $g = 1$, while at $v_{fix} < v_c$ it is of thermally activated nature with $g < 1$.

This analysis of the $\bar{v}_{eff}(\sigma)$ dependences characteristic of different temperatures might prove useful for understanding the physical meaning of the numerical results presented below.

3.1 Rate dependences

The rate dependences of the yield stress, i.e. upon $\dot{\varepsilon}/b\varrho$ or, which is the same, on v_{fix}, as obtained numerically for several temperatures, are given in Fig. 2. The course of

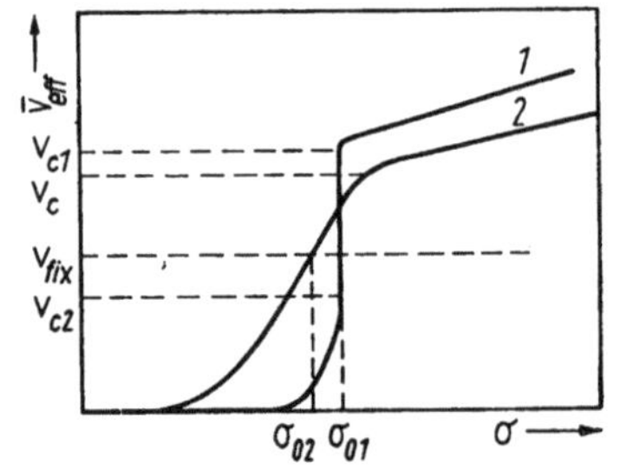

Fig. 1. Graphical solution of the transcendental equation (1) at various temperatures (schematic)

these curves corresponds to the behavior of $\bar{v}_{\text{eff}}(\sigma)$ that has been discussed in detail above (see Fig. 1). The value v_{c2} is smaller than 10^{-10} cm/s for curves 1 to 4 of Fig. 2 while for curves 5 and 6 it is, respectively, about 10^{-2} and 10 cm/s. Curves 7 and 8 of Fig. 2 are of the same general appearance as curve 2 of Fig. 1. The theoretical results obtained for the rate dependences qualitatively agree with similar results of [16].

The rate dependence of $\Delta\sigma_{\text{NS}}$ corresponding to curve 1 of Fig. 2 (i.e. $T = 0.1T_c$) is given in Fig. 3. That $\Delta\sigma_{\text{NS}}$ is practically independent of strain rate in a wide range of rates was earlier reported in [17] and a number of other experimental papers (see the review paper [18]). From the point of view of the theory developed in this paper, the effect is due to the fact that for the range of $\dot{\varepsilon}/b\varrho$ values in question the process of plastic straining is quasi-dynamical both in the N and S states. The low peak observed on the $\Delta\sigma_{\text{NS}}(v_{\text{fix}})$ curve of Fig. 3 appears because the transition from the quasi-dynamical to the visco-dynamical dislocation motion at the increase of $\dot{\varepsilon}/b\varrho$ occurs in the N state earlier than in the S state. At a further increase of $\dot{\varepsilon}/b\varrho$ the mobile dislocations get more and more efficient in breaking Cooper electron pairs in the superconductor. As a result, the growth of v at $v > v_c = 2b\Delta(T)/\hbar$ leads to an increase in Γ and decrease in $\Delta\sigma_{\text{NS}}$. Hence, there is a range of strain rates where $\Delta\sigma_{\text{NS}}$ can decrease at the increase of $\dot{\varepsilon}/b\varrho$. A similar effect was observed experimentally by Iida et al. [19].

Fig. 4 illustrates the variation of the $\sigma_0(T)$ dependence in the normal state caused by changes in v_{fix}.

3.2 Activation volume

Formally the activation volume V^* can be defined as $V^* = kT\,\partial \ln \dot{\varepsilon}/\partial\sigma$. A definition like this conforms to the experimental conditions in that the experimentalist does not know a priori whether the dislocation motion is of thermally activated nature, whereas the derivative $\partial \ln \dot{\varepsilon}/\partial\sigma$ can be measured in almost all cases.

One can see in Fig. 1 that the derivative is not a single-valued function for dislocations moving in the quasi-dynamical regime, its magnitude depending on the measurement procedure. E.g., if the measurement consists in a forced variation of $\dot{\varepsilon}$ by a small value $\delta\dot{\varepsilon}$, then the corresponding $\delta\sigma$ is equal to zero, hence $\partial \ln \dot{\varepsilon}/\partial\sigma \to \infty$. In case the

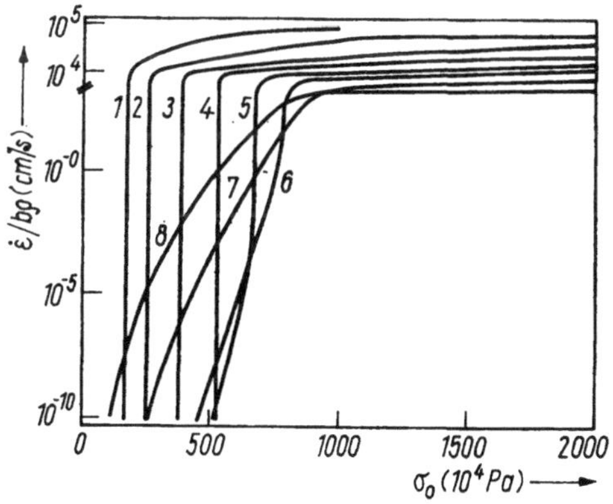

Fig. 2

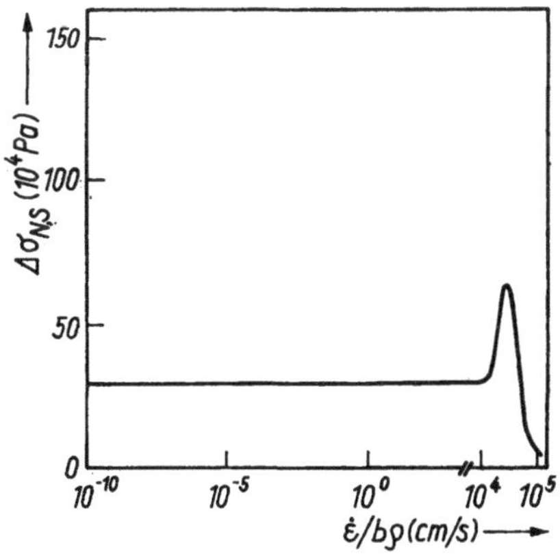

Fig. 3

Fig. 2. Rate dependence of the yield stress corresponding to different temperatures. Curves 1 to 8 are, respectively, for $T = 1$, 10, 15, 20, 25, 30, 50, and 80 K. The other parameters are $B_{\text{eN}} = 10^{-5}$ g cm^{-1} s^{-1}, $H_0 = 0.1$ eV, $\alpha_c = \pi/6$, and $a/\sqrt{S_0} = 10^{-2}$

Fig. 3. The rate dependence of $\Delta\sigma_{\text{NS}}$ at $T = 0.1\,T_c$ corresponding to curve 1 of Fig. 2

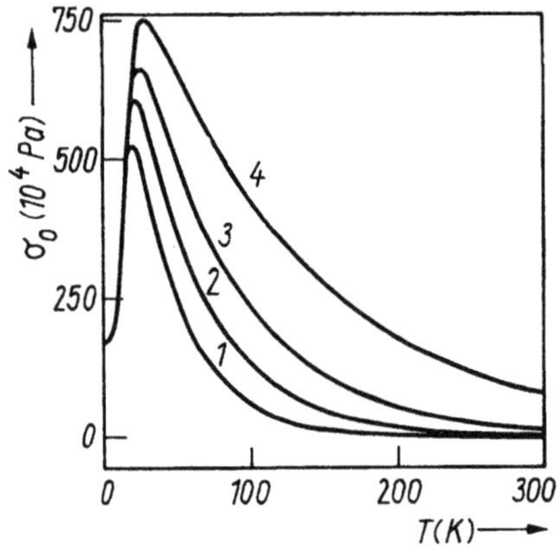

Fig. 4

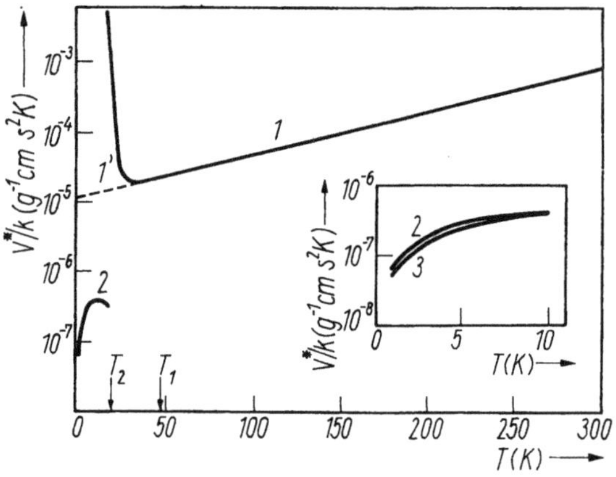

Fig. 5

Fig. 4. Variations in the temperature dependence of the normal state yield stress owing to changes in v_{fix}. Curves 1 to 4 are, respectively, for $v_{\text{fix}} = 10^{-10}$, 10^{-6}, 10^{-3}, and 1 cm s^{-1}. The other parameters are $B_{\text{eN}} = 10^{-5}$ g cm^{-1} s^{-1}, $H_0 = 0.1$ eV, $\alpha_c = \pi/6$, and $a/\sqrt{S_0} = 10^{-2}$

Fig. 5. Temperature dependence of the activation volume. Curve 1 corresponds to the normal state, with account of the dislocation inertia and an arbitrary method of measuring V^*, curve 1' is for the same conditions as curve 1 but with no account of the dislocation inertia. Curve 2 corresponds to the normal state, the dislocation inertia has been taken into account and V^* measured in the incremental loading regime. Curve 3 is the same as curve 2 but for the superconducting state. The parameters are $B_{\text{eN}} = 10^{-5}$ g cm^{-1} s^{-1}, $H_0 = 0.1$ eV, $\alpha_c = \pi/6$, $v_{\text{fix}} = 10^{-3}$ cm s^{-1}, and $a/\sqrt{S_0} = 10^{-2}$

procedure involves incremental loading, i.e. σ_{01} is changed forcedly by $\delta\sigma_{01}$ (i.e., in creep measurements), then $\delta\dot{\varepsilon} \neq 0$ and $\partial \ln \dot{\varepsilon}/\partial\sigma < \infty$. In other cases (i.e. during visco-dynamical or thermally activated dislocation motion) the derivative $\partial \ln \dot{\varepsilon}/\partial\sigma$ can be determined unambiguously.

The temperature dependences of $V^* = kT\, \partial \ln \dot{\varepsilon}/\partial\sigma$ that have been obtained numerically are shown in Fig. 5. At temperatures below T_1 the inertial properties of dislocations are evident ($\gamma > 1$) while below T_2 the dislocation motion is quasi-dynamical ($g \approx 1$). Curves 2 and 3 of Fig. 5 have been calculated for the incremental loading regime. More details of curves 1, 1', 2, and 3 can be found in the caption.

4. Concluding Remarks

The above-described results have been obtained on the assumption that all obstacles in the crystals are identical. In fact some scatter in the obstacle parameters, such as α_c and the potential barrier height U_0, is inevitable. The scatter might give rise to a rate dependence of such quantities as σ_{0N}, σ_{0S}, and $\Delta\sigma_{\text{NS}}$ in the range of quasi-dynamical dislocation glide. It would allow an unambiguous determination of $\partial \ln \dot{\varepsilon}/\partial\sigma$ within the range independent of the measurement technique. In its turn, this would eliminate the gap between curves 1 and 2 in Fig. 5 and cause some other effects.

Acknowledgements

The author is indebted to Dr. S. N. Komnik and V. V. Demirskii for discussion of the obtained results.

References

[1] A. I. Landau, phys. stat. sol. (a) **61**, 555 (1980).
[2] A. J. E. Foreman and M. J. Makin, Phil. Mag. **14**, 911 (1966).
[3] A. I. Landau, phys. stat. sol. (a) **30**, 659 (1975).
[4] A. I. Landau and V. I. Dotsenko, phys. stat. sol. (a) **37**, 709 (1976).
[5] G. P. Huffman and N. Louat, Phys. Rev. Letters **24**, 1055 (1970).
[6] M. I. Kaganov and V. D. Natsik, Zh. eksper. teor. Fiz., Pisma **11**, 550 (1970).
[7] V. G. Baryakhtar, E. I. Druinskii, and I. I. Falko, Fiz. Metallov i Metallovedenie **33**, 5 (1972).
[8] V. I. Alschitz and V. L. Indenbom, in: Dislocation Dynamics, Izd. Naukova Dumka Kiev 1975 (p. 232 to 275) (in Russian).
[9] A. V. Granato, K. Lücke, J. Schlipf, and L. J. Teutonico, J. appl. Phys. **35**, 2732 (1964).
[10] V. M. Chernov, Fiz. tverd. Tela **15**, 323 (1973).
[11] A. I. Landau and Yu. I. Hoffman, Fiz. tverd. Tela **16**, 3427 (1974).
[12] V. L. Indenbom and Yu. Z. Estrin, phys. stat. sol. (a) **4**, K37 (1971).
[13] V. D. Natsik, Zh. eksper. teor. Fiz. **61**, 2540 (1971).
[14] Yu. Z. Estrin, phys. stat. sol. (a) **26**, K161 (1974).
[15] V. D. Natsik, Fiz. nizk. Temp. **1**, 240 (1975).
[16] V. I. Kononenko and V. V. Pustovalov, Problemy Prochnosti **7**, 55 (1974).
[17] G. A. Alers, O. Buck, and B. R. Tittman, Phys. Rev. Letters **23**, 290 (1969).
[18] V. V. Pustovalov, in: Physics of Low Temperature Plastic Deformation, Izd. Naukova Dumka, Kiev 1974 (p. 152 to 210) (in Russian).
[19] F. Iida, T. Susuki, E. Kuramoto, and S. Takeuchi, Acta metall. **27**, 637 (1979).

(Received February 5, 1981)

phys. stat. sol. (a) 65, 127 (1981)

Subject classification: 14.4

Institute of Electronic Technology, Moscow[1])

The Scattering Coefficient and Velocity of Propagation of Electromagnetic Waves in Dielectric Polycrystals

By

O. A. Grigorev and T. D. Shermergor

The scattering coefficient and velocity of propagation of electromagnetic waves are calculated for polycrystals up to triclinic symmetry. The calculation is made in the Bourret approximation by the method of statistical perturbation theory taking account of multiple scattering. The corresponding numerical calculation is presented.

Проведен расчет коэффициента рассеяния и скорости распространения электромагнитных волн в поликристаллах вплоть до триклинной симметрии. Расчет выполнен в приближении Бурре методом статистической теории возмущений, который учитывает многократное рассеяние. Представлен соответствующий численный пример.

1. Introduction

A number of scientific papers published in recent years are devoted to research on the propagation of a plane wave in randomly inhomogeneous media and, particularly, to the calculation of the scattering coefficient and the velocity of propagation throughout the frequency range in the single-scattering approximation [1 to 3]. The present paper offers a calculation of these values for polycrystalline dielectrics over the entire range of frequencies of an incident electromagnetic wave. The calculation has been accomplished in the Bourret approximation by the method of renormalization of the wave equation and re-expansion taking account of multiple scattering. This problem is of interest since the single scattering approximation in the region of short waves proves to be inadequate [3]. The analogous problem for ultrasonic waves has been solved in paper [1].

2. Basic Equations

Let us consider an unlimited polycrystalline dielectric in which a plane monochromatic wave is propagating. The local value of the electric field strength $E_n(\boldsymbol{r})$ will be defined by the wave equation

$$L_{ns}(\boldsymbol{r})\, E_s(\boldsymbol{r}) = 0\,, \qquad L_{ns}(\boldsymbol{r}) = \nabla^2 \delta_{ns} - \nabla_n \nabla_s + (\omega/c_0)^2\, \varepsilon_{ns}(\boldsymbol{r}) \tag{1}$$

and the mean value, $\langle E_n \rangle$, by the wave equation

$$L^*_{ns} \langle E_s \rangle = 0\,, \qquad L^*_{ns} = \nabla^2 \delta_{ns} - \nabla_n \nabla_s + (\omega/c_0)^2\, \varepsilon^*_{ns}\,, \tag{2}$$

$$\langle E_n(\boldsymbol{r}) \rangle = E_n(\boldsymbol{k}) \exp\,(-i\boldsymbol{k} \cdot \boldsymbol{r})\,.$$

Here the components $\varepsilon_{ns}(\boldsymbol{r})$ and ε^*_{ns} are tensors of the local and effective permittivities, respectively. The latter connects the average inductance $\langle D_n \rangle$ with the average electric field strength $\langle E_s \rangle$ through the known equation

$$\langle D_n \rangle = \varepsilon^*_{ns} \langle E_s \rangle\,; \tag{3}$$

c_0 is the velocity of light in vacuum, and ω is the cyclic frequency.

[1]) 103498 Moscow K-498, USSR.

It is assumed that there are sharp boundaries between different grains. The corner-shaped brackets are used to denote the operation of averaging in the region whose size is much less than the degree of inhomogeneity of the field $\langle E_n \rangle$, but materially greater than the size of the crystallites. The nonlocality ε^*_{ns} for such fields does not show up, and averaging in bulk coincides with statistical averaging [4].

In addition to the real field of the dielectric constant tensor $\varepsilon_{ns}(\boldsymbol{r})$ we shall introduce a homogeneous and isotropic comparison field $\varepsilon_0\delta_{ns}$, the selection conditions of which were researched by variation methods in [5], where also were obtained the optimum bounds for ε_0 (lower bound ε_0^- and upper bound ε_0^+). The Hill condition for ε_0 being a good approximation:

$$\varepsilon_0 = \tfrac{1}{2}\,(\varepsilon_0^- + \varepsilon_0^+)\,. \tag{4}$$

For a medium with permittivity $\varepsilon_0\delta_{ns}$ the Green function of the wave equation is known [3], and can be written in the following manner:

$$G_{ns}(\boldsymbol{r}) = \frac{c_0^2}{\omega^2}\left[F_{ns}(\boldsymbol{r}) - \frac{1}{3\varepsilon_0}\,\delta_{ns}\,\delta(\boldsymbol{r})\right], \tag{5}$$

$$F_{ns}(\boldsymbol{r}) = \frac{1}{4\pi\varepsilon_0 r^3}\left[\left(3 + 3\,\frac{i\omega r}{c} - \frac{\omega^2 r^2}{c^2}\right)\frac{r_n r_s}{r^2} - \right.$$
$$\left. -\left(1 + \frac{i\omega r}{c} - \frac{\omega^2 r^2}{c^2}\right)\delta_{ns}\right]\exp\left(-\frac{i\omega r}{c}\right), \tag{6}$$

$$c = c_0\varepsilon_0^{-1/2}\,. \tag{7}$$

After introducing the tensor h_{ns} and integral operator b_{ns} we obtain

$$h_{ns} = (-3\varepsilon_0)^{-1}\,\delta_{ns}\,, \qquad b_{ns}f = F_{ns}(\boldsymbol{r}) * f\,, \tag{8}$$

where f is an arbitrary function, $\delta(\boldsymbol{r})$ is the Dirac delta function. The asterisk indicates the integral convolution operation.

Upon renormalizing (1) with the aid of the Green function (5) we shall obtain an operator series for the effective permittivity. Then we re-expand this series by means of the tensor employing the method suggested in [4]. Then similarly to [4] we derive

$$\left.\begin{aligned}
&\varepsilon = \varepsilon_0 + (1 + d^*h)^{-1}\,d^*\,, \qquad d^* = \langle dR\rangle\,,\\
&R = (1 - bd)^{-1}\,\langle(1 - bd)^{-1}\rangle^{-1} = \lim_{n\to\infty} R_n\,,\\
&R_n = \sum_{k=0}^{n} (Hd)^k\,, \qquad H = (1 - T)\,b\,, \qquad Tf = \langle f\rangle\,,\\
&d = \varepsilon'(1 - h\varepsilon')^{-1}\,, \qquad \varepsilon' = \varepsilon - \varepsilon_0\,,
\end{aligned}\right\} \tag{9}$$

where the tensor suffixes have been omitted for the sake of simplicity.

In the Bourret approximation [1] ($d^* = \langle dbd\rangle - \langle d\rangle\, b\langle d\rangle + \langle d\rangle$) with exponential coordinate dependence of the binary correlation function of the permittivity for fields of the mode $\exp(-i\boldsymbol{k}\cdot\boldsymbol{r})$ we find

$$\langle d''_{nr}(\boldsymbol{r})\, d''_{ms}(\boldsymbol{r}_1)\rangle = A^{ms}_{nr}\exp\left(-\,|\boldsymbol{r} - \boldsymbol{r}_1|/r_0\right), \qquad d'' = d - \langle d\rangle\,,$$

$$d^*_{ns}(\boldsymbol{k}) = \langle d_{ns}\rangle + A^{ms}_{nr}\int F_{ns}(\boldsymbol{r})\exp\left(-\frac{r}{r_0}\right)\cos \boldsymbol{k}\cdot\boldsymbol{r}\,\mathrm{d}\boldsymbol{r}\,, \tag{10}$$

where r_0 is the spatial scale of correlations proportional to the grain size, and $\boldsymbol{k}$ is the wave vector.

3. Attenuation and Lag of the Incident Wave

The renormalized wave equation $L^*_{ns}\langle E_s\rangle = 0$ allows a solution in the form of a plane wave with complex wave number k^* defined by the formula

$$k^* = \frac{k}{\sqrt{1+M}} \approx k\left(1 - \frac{M}{2}\right), \tag{11}$$

where the correlation addition due to scattering M is considered small (i.e. $|M| \ll 1$). Using the equation $k^*(\omega) = \beta(\omega) - i\alpha(\omega)$ we shall find the scattering amplitude coefficient $\alpha(\omega)$ and the electromagnetic wave propagation velocity $c(\omega)$:

$$\alpha(\omega) = \frac{1}{2}\frac{\omega}{c} \operatorname{Im} M, \tag{12}$$

$$c(\omega) = \left[\frac{\mathrm{d}\beta(\omega)}{\mathrm{d}\omega}\right]^{-1} = c\left[1 + \frac{1}{2}\left(1 + \omega\frac{\mathrm{d}}{\mathrm{d}\omega}\right) \operatorname{Re} M\right]. \tag{13}$$

We presume that the local value of E_n in a polycrystalline dielectric is described by the usual equation $\operatorname{div} \varepsilon E_n = 4\pi\varrho$, where ϱ is the density of random charges and $\langle\varrho\rangle = 0$. Then the correlation addend M is equal to the isotropic component of the tensor $\varepsilon_0^{-1}\varepsilon^*_{ns}(\boldsymbol{k})$ [3].

After integrating (10) in a manner similar to [1] and subsequently calculating by use of formulas (9) to (13) we come to the following results:

$$\alpha(x) = \frac{x}{2r_0}\frac{3a_2}{a_1^2 + a_2^2}, \qquad x = \frac{\omega r_0}{c}, \tag{14}$$

$$c(x) = c\left[1 + \frac{3}{2}\left(\frac{a_1 + a_2}{a_1^2 + a_2^2} - 2a_1\frac{a_1a_3 + a_2a_4}{(a_1^2 + a_2^2)^2} + 1\right)\right], \tag{15}$$

where

$$a_1 = \sigma\left[\frac{1}{3} + \frac{1}{2x^2} - \frac{1+2x^2}{4x^3}\arctan 2x + 7\frac{x^2}{1+4x^2}\right] - 1, \tag{16}$$

$$a_2 = \sigma\left[\frac{1}{2x} - \frac{1+2x^2}{8x^3}\ln(1+4x^2) + 7\frac{2x^3}{1+4x^2}\right], \tag{17}$$

$$a_3 = \sigma\left[\frac{-3-10x^2}{2x^2(1+4x^2)} + \frac{3+2x^2}{4x^3}\arctan 2x + 7\frac{2x^2}{(1+4x^2)^2}\right], \tag{18}$$

$$a_4 = \sigma\left[\frac{-3-8x^2}{2x(1+4x^2)} + \frac{3+2x^2}{8x^3}\ln(1+4x^2) + 7\frac{6x^3+8x^5}{(1+4x^2)^2}\right]. \tag{19}$$

Here σ is the coefficient in the covariant tensor of relative permittivities $\tilde{d}_{ns}$:

$$A^{ms}_{nr} = \langle d''_{nr}(0)\, d''_{ms}(0)\rangle = \sigma(\delta_{nm}\delta_{rs} + \delta_{ns}\delta_{rm} - \tfrac{2}{3}\delta_{nr}\delta_{ms}), \tag{20}$$

$$\sigma = \tfrac{1}{15}(\tilde{d}_{11}^2 + \tilde{d}_{22}^2 + \tilde{d}_{33}^2 - \tilde{d}_{11}\tilde{d}_{22} - \tilde{d}_{22}\tilde{d}_{33} - \tilde{d}_{33}\tilde{d}_{11}) + \tfrac{1}{5}(\tilde{d}_{12}^2 + \tilde{d}_{13}^2 + \tilde{d}_{23}^2),$$

$$\tilde{d}_{ns} = (\varepsilon_{nr} - \varepsilon_0\delta_{nr})(\varepsilon_{rs} + 2\varepsilon_0\delta_{rs})^{-1}, \tag{21}$$

where ε_{nr} are the permittivity components of a single crystal.

In the extreme case $a_2 \ll 1$, for $\alpha(x)$ and $c(x)$ the general and limiting (with $x \ll 1$ and $x \gg 1$) formulas for single scattering [3] are obtained from the expressions (14) to (21). The scattering coefficient in another extreme case ($a_2 \gg 1$) is independent of frequency, and velocity dispersion is absent.

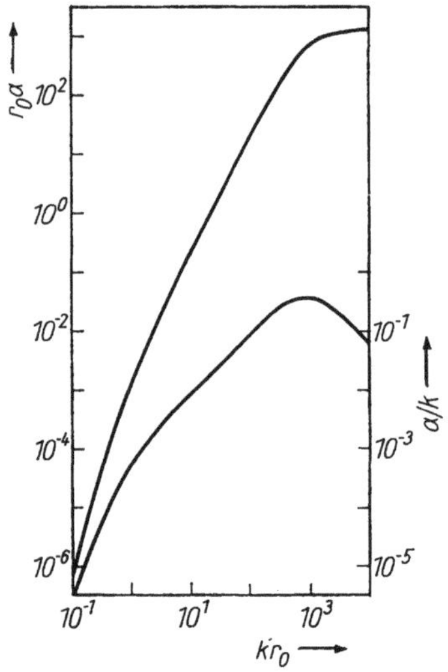

Fig. 1. Dimensionless attenuation coefficients of electromagnetic waves in polycrystalline α-corundum as a function of dimensionless frequency and average grain diameter (with $\overline{D} = r_0$)

4. Conclusion

The results obtained may find application for calculating $\alpha(x)$ and $c(x)$ in dielectric and piezoelectric polycrystals, as well as for structural measurements [3].

Fig. 1 illustrates the dimensionless scattering coefficient in a polycrystal as a function of the dimensionless frequency and the inhomogeneity grain size calculated on the basis of the formulas derived. The correlation radius, by analogy with the acoustic problem [1], is taken to be equal to the average grain diameter, $\overline{D}$. The calculation has been accomplished for α-corundum whose permittivity components $\varepsilon_1 = \varepsilon_2 = 8.6$; $\varepsilon_3 = 10.55$ [6].

It is evident that the scattering coefficient for long waves, intermediate wave range, and short waves associates with the Rayleigh $r_0^3\omega^4$, stochastic $r_0\omega^2$, and diffuse $r_0^{-1}\omega_0$ dependences, respectively.

References

[1] O. A. GRIGOREV and T. D. SHERMERGOR, phys. stat. sol. (a) **64**, 385 (1981).
[2] R. RANGANATHAN and S. DATTAGUPTA, phys. stat. sol. (a) **54**, 537 (1979).
[3] A. A. USOV and T. D. SHERMERGOR, Soviet Phys. — J. tech. Phys. **48**, 1132 (1978).
[4] A. G. FOKIN, Soviet Phys. — J. tech. Phys. **43**, 71 (1973).
[5] G. A. ERMAKOV, A. G. FOKIN, and T. D. SHERMERGOR, Soviet Phys. — J. tech. Phys. **44**, 249 (1974).
[6] International Critical Tables, McGraw-Hill Publ. Co., New York 1929.

(Received January 14, 1981)

phys. stat. sol. (a) 65, 131 (1981)

Subject classification: 14.3; 22.2.3

Institute of Semiconductors, Academy of Sciences of the Ukrainian SSR, Kiev[1])

Magnetoconcentration Effect at Auger Recombination of Current Carriers

By

V. K. MALYUTENKO, K. YU. GUGA, and YU. M. MALOZOVSKII

A theory of the magnetoconcentration effect under conditions when two mechanisms of current carrier recombination — interband Auger and radiative recombination — take place, is described. The spatial distribution of current carriers in the cross-section of the specimen found in electric and magnetic crossed fields and the type of current–voltage characteristics are analysed. The theory can be used for the investigation of the peculiarities of this effect in narrow-band semiconductors. Experiments on InSb samples at room temperature, are in a qualitative agreement with the theory. From the analysis of the current–voltage characteristics of the crystals, the values of surface recombination velocities on the faces as well as the magnetic sensitivity of the crystals are estimated.

Eine Theorie des Magnetokonzentrationseffekts wird unter der Bedingung, daß zwei Mechanismen für die Stromträgerrekombination wirksam sind — Auger-Interbandrekombination und strahlende Rekombination — beschrieben. Die räumliche Verteilung der Stromträger über den Querschnitt der Probe, die aus Untersuchungen in gekreuzten elektrischen und magnetischen Feldern erhalten wird, sowie der Typ der Strom–Spannungscharakteristiken werden analysiert. Die Theorie kann für die Untersuchung der Besonderheiten dieses Effekts in schmalbandigen Halbleitern benutzt werden. Experimente an InSb-Proben bei Zimmertemperatur befinden sich in qualitativer Übereinstimmung mit der Theorie. Aus der Analyse der Strom–Spannungscharakteristiken der Kristalle werden sowohl die Werte der Oberflächenrekombinationsgeschwindigkeiten als auch der magnetischen Empfindlichkeit der Kristalle berechnet.

1. Introduction

The possibility of controlling the total number of current carriers in a semiconductor plate with intrinsic conductivity using the electric ($\boldsymbol{E}$) and magnetic $\boldsymbol{H}$ crossed fields (the fields lie within the plane of the plate, magnetoconcentration or Welker effect [1]) presents interest from scientific as well as practical point of view. The study of plasma reaction to the external influence ($\boldsymbol{E} \times \boldsymbol{H}$) allowed to suggest several methods of determining its most important parameters and also new types of semiconductor devices (see [2, 3]).

The magnetoconcentration effect (MCE) as a method of creating a non-equilibrium plasma of high density is of great interest.[2]) It turns out that in narrow-band semiconductors MCE may in some cases compete with such traditional methods as laser excitation and contact injection [4]. Under such conditions the non-linear mechanisms of recombination must be taken into account for adequate description of the effect and proper estimation of the parameters of the corresponding devices. The quadratic recombination was considered in [3, 5 to 7] and an attempt to take into account the peculiarities of Auger recombination was made in [8]. But in [8] the theory is developed

[1]) Prospekt Nauki 115, 253028 Kiev-28, USSR.

[2]) If surface recombination velocities on the plate faces differ essentially ($s_+ \gg s_-$), under the conditions of transverse carrier drift due to Lorentz force the pre-surface concentration at the face s_- can be increased considerably at the expense of thermal generation of pairs by the face s_+.

only for the case of small deviations from equilibrium when non-linearity of the Auger recombination (the dependence of lifetime on concentration) does not occur.

Considering the expediency of MCE usage for investigating the properties of the high-density electron hole plasma in narrow-band semiconductors (InSb, $Cd_xHg_{1-x}Te$, and others) we report here the theory of the effect under the conditions of Auger recombination. In order to make the model under investigation similar to the existing physical objects, we consider the case when two types of non-linear recombination — interband Auger and quadratic — take place simultaneously. Such an approach allows to use the theoretical results in experimental analyses in a wide range of temperature and carrier concentrations. The results of the calculation are used for analysing experiments performed on indium antimonide plates.

2. Theory

Let us consider the peculiarities of the magnetoconcentration effect in a limited plate ($-d \leq y \leq d$) of a non-degenerate semiconductor with intrinsic conductivity and non-linear mechanism of carrier recombination. The electric field $\boldsymbol{E}$ is applied along the x-axis, the external magnetic field $\boldsymbol{H}$ along the z-axis, the carrier extraction to the surface $y = -d$ corresponds to positive values of $\boldsymbol{E}$ and $\boldsymbol{H}$. The values $\boldsymbol{E}$ and $\boldsymbol{H}$ must be chosen small in order to ignore the carriers heating, internal magnetic field of the current, and magnetic resistance. Suppose that in the semiconductor volume two mechanisms of current carrier recombination take place simultaneously: radiative interband recombination with pair lifetime τ_R and interband Auger recombination of first order with lifetime τ_A for small deviations of the local carrier concentration $p(y)$ from the equilibrium value p_i. The lifetimes τ_R and τ_A can be easily expressed through material constants using the results of [9] and [10], respectively. Let us present the contribution of both mechanisms (supposing they are additive [11]) to the total recombination flow in the form of the quantum efficiency of radiative recombination

$$\eta = \frac{\tau_A}{\tau_A + \tau_R} = \frac{\alpha}{1 + \alpha} \,. \tag{1}$$

Then, if $\alpha = \tau_A/\tau_R \gg 1$, the radiative mechanism is the main recombination mechanism at $\alpha \ll 1$ — Auger recombination. Under these conditions the time-independent continuity equation in quasi-neutral approximation, similarly to [3], can be presented in the form

$$D \frac{d^2p}{dy^2} + \frac{eaD}{2kT} E_x \frac{dp}{dy} - \frac{p^2 - p_i^2}{2p_i\tau_R} - \frac{p}{p_i} \frac{p^2 - p_i^2}{2p_i\tau_A} = 0 \,, \tag{2}$$

where

$$D = \frac{2D_nD_p}{D_n + D_p} \,, \qquad a = \frac{U_n + U_p}{c} H_z \,.$$

D_n, D_p and U_n, U_p are the diffusion coefficients and the electron and hole mobility, k the Boltzmann constant, T the absolute temperature. Let us suppose, then, that the surface recombination velocity S on the plate faces does not depend on the carrier concentration. Then, the solution of (2), which we are interested in, must satisfy the following boundary conditions:

$$D \frac{dp}{dy} + \frac{eaD}{2kT} E_x p = \begin{cases} -s_+(p - p_i) & y = d \,, \\ s_-(p - p_i) & y = -d \,. \end{cases} \tag{3}$$

Let us introduce the symbols

$$\xi = y/d\,,\quad \beta_R = d/\sqrt{D\tau_R} = d/L_R\,,\qquad \beta_A = d/\sqrt{D\tau_A} = d/L_A\,,$$

$$\beta_R = \beta_A\sqrt{\alpha}\,,\qquad f(\xi) = p(y)/p_i\,,\qquad \gamma = \frac{ead}{2kT}E_x\,,\qquad S_\pm = \frac{s_\pm d}{D}\,. \tag{4}$$

Thus expressing the task parameters by dimensionless values: the coordinate ξ, the crystal thickness β_R, β_A, local concentration $f(\xi)$, controlling field γ, and surface recombination velocity $S_\pm$. Then (2), (3) take the form

$$\frac{d^2f}{d\xi^2} + \gamma\frac{df}{d\xi} - \frac{\beta_A^2}{2}(f^2-1)(\alpha+f) = 0\,, \tag{2'}$$

$$\frac{df}{d\xi} + \gamma f = \begin{cases} -S_+(f-1) & \xi = 1,\\ S_-(f-1) & \xi = -1\,. \end{cases} \tag{3'}$$

We obtain the approximate solution of (2′) dividing the crystal in two regions [3, 5] according to the thickness. In the first region bordering the plane $\xi = -1$ and enriched with carriers, we solve (2′) without the term describing the generation–recombination of current carriers, and in the second one, bordering the plane $\xi = 1$, we ignore the diffusion term. We join the solutions for the two parts of the crystal with the help of the balance equation for the volume and surface recombination–generation flows:

$$S_+[f(1)-1] + S_-[f(-1)-1] + \frac{\beta_A^2}{2}\int\limits_{-1}^{1}(f^2-1)(\alpha+f)\,d\xi = 0 \tag{5}$$

which is obtained by integrating (2′) over ξ within the limits $(-1, 1)$ and using (3′). The obtained approximate solution for the region of fields $\gamma/(\beta_R^2+\beta_A^2) \gg 1$, $\gamma \gg 1$ can be expressed in the form

$$f(\xi) = C\,e^{-\gamma(1+\xi)} + f_1 + \frac{(\alpha+f_1)(1-f_1^2)}{2\gamma}\beta_A^2(1-\xi)\,, \tag{6}$$

where $C = x_0 - A$;

$$A = \tfrac{1}{2}(\alpha+3\Phi),\ \Phi = f_1 + \frac{(\alpha+f_1)(1-f_1^2)}{\gamma}\beta_A^2\,,\quad f_1 = \frac{\gamma S_+}{\gamma+S_+} \tag{7}$$

and x_0 is the positive root of the equation

$$x^3 + 3px + 2q = 0\,, \tag{8}$$

$$p = \tfrac{1}{3}B - A\,,$$

$$q = A^3 - \tfrac{1}{2}AB + \tfrac{1}{2}G\,,$$

$$B = 3\left[\Phi(3\Phi+2\alpha) - 1 + \frac{2\gamma S_-}{\beta_A^2}\right],$$

$$G = \frac{6\gamma}{\beta_A^2}\Big\{S_+(f_1-1) + S_-(\Phi-1) + \frac{\beta_A^2}{2}[\tfrac{1}{2}(f_1+\Phi)(f_1^2+\Phi^2) -$$

$$-(f_1+\Phi) + \alpha\{\tfrac{2}{3}(f_1^2+f_1\Phi+\Phi^2) - 2\}]\Big\}.$$

For the boundary conditions the expressions for the spatial distribution of current carriers $f(\xi)$ are simplified. We arrange these formulas in Table 1.

Table 1

S_+	S_-	$f(\xi)$	range of application	
∞	0	$C\,e^{-\gamma(1+\xi)} + 1$	$C \approx 2\left(\frac{\gamma}{\beta_R} - 1\right)$	$C(\gamma) \ll \alpha$
			$C \approx (6\gamma^2/\beta_A^2)^{1/3}$	$C(\gamma) \gg 1, \alpha$
0	0	$C\,e^{-\gamma(1+\xi)} + \frac{\beta_R^2}{2\gamma}(1-\xi)$	$C \approx 2\sqrt{\gamma}$	$C(\gamma) \ll \alpha$
			$C \approx (6\gamma\alpha)^{1/3}$	$C(\gamma) \gg 1, \alpha$
			$C \approx \sqrt{3}$	$\alpha = 0$
0	∞	$\left(1 - \frac{\beta_R^2}{\gamma}\right) e^{-\gamma(1+\xi)} + \frac{\beta_R^2}{2\gamma}\left(1 - \xi\right)$		
∞	∞	1		

Knowing the dependence of $f(\xi)$ on the values of controlling fields (dimensionless field γ) it is easy to obtain expressions for current–voltage characteristics (CVC) (the product of the dimensionless field and the dimensionless total number of carriers). Taking into consideration (6), (7) we obtain

$$I(\gamma) = \gamma \int_{-1}^{1} f(\xi)\,d\xi = C + \gamma(f_1 + \Phi)\,. \tag{9}$$

Strictly speaking the expression for the current in the general form contains, apart from the field, also the diffusion term, proportional to the difference between the surface concentrations on the faces. However, in weak magnetic fields $U_n U_p (H/C)^2 \ll 1$ the diffusion term can be neglected. The dimensional value of the current j is related to the dimensionless value I by the ratio [5]:

$$j = \frac{eD^* p_i}{ad}\,\Sigma\, I(\gamma)\,, \tag{10}$$

where Σ is the crystal section and D^* is determined in [5].

The expressions for CVC and the range of their application are shown in Table 2 for some interesting limiting cases.

3. Material and Experimental Method

Indium antimonide was chosen as the material for the experiment. The small width of the forbidden band and the high carrier mobilities allow to achieve in this material essential concentration deviations from the equilibrium value [4] under the conditions of the magnetoconcentration effect, and thus to study the non-linear mechanisms of the current carrier recombination.

As it is shown in [12], in pure InSb crystals the recombination processes are controlled by two non-linear mechanisms of interband recombination in the temperature range $300 \geq T \geq 200$ K. In the low temperat ureregion the radiative recombination manifests itself more strongly, but around room temperatures the Auger recombination prevails.

Table 2

S_+	S_-	$I(\gamma)$	range of application
0	0	$2\sqrt{\gamma} + \beta_R^2$	$C(\gamma) \ll \alpha$
		$\sqrt[3]{6\gamma\alpha} + \beta_R^2$	$C(\gamma) \gg 1, \alpha$
		$\sqrt{3}$	$\alpha = 0$
∞	0	$2\gamma(1 + \beta_R^{-1})$ $2\gamma + (6\gamma^2/\beta_A^2)^{1/3}$	$C(\gamma) \ll \alpha$ $C(\gamma) \gg 1, \alpha$
0	∞	$1 + \beta_R^2 - \beta_R^2/\gamma$	
∞	∞	2γ	

Naturally, the competition of these mechanisms is determined not only by temperature, but also by the level of excitation. It is interesting that in almost all cases (except the levels of extremely high excitations) it is necessary to take into account both mechanisms simultaneously. Thus, the conditions of investigation of the magnetoconcentration effect in InSb at $T > 200$ K correspond to the situation considered above (see (2)).

The experiments were performed on p-InSb ($N_a - N_d = 7 \times 10^{14}$ cm^{-3}) specimens having the form of thin plates ($4 \times 1 \times 0.015$ mm^3) with asymmetric and symmetric treatment of opposite faces. In the case of asymmetric treatment the specimens were polished mechanically and after that one of the faces was etched in the standard etch CP-4. Such treatment allowed to obtain values of surface recombination velocities of $s_+ \approx 10^6$ cm/s and $s_- \approx 10^3$ cm/s on polished and etched faces, respectively. During symmetric treatment the specimens were polished mechanically and then etched in CP-4 ($s_+ \approx s_- \approx 10^3$ cm/s).

To avoid heating, the electric field was applied to the specimens in the form of impulses lasting no more than 1 μs, the magnetic field was steady. The current–voltage characteristics were investigated in a large range of electric ($E \leqq 80$ V/cm) and magnetic fields ($H \leqq 6$ kOe). Under such conditions [4] one may expect an increase of the surface concentration of current carriers up to 10^{17} cm^{-3}.

4. Discussion of the Results

The magnetoconcentration effect has a peculiar feature — the possibility to control the profile of the carrier distribution in the crystal $f(\xi)$ and their total number $F(\gamma) = \int_{-1}^{+1} f(\xi)\, d\xi$ within wide limits.

From this point of view we shall consider the influence of the Auger recombination, taking different values of the quantum efficiency η (parameter α, see (1)).

The spatial distribution of current carriers in the crystal with different relations between τ_R and τ_A is presented in Fig. 1. As one can see, when α decreases (the contribution of the Auger recombination increases) at fixed parameters γ, β_R, $S_\pm$, the degree of enrichment (the deviation of $f(\xi)$ from its equilibrium value equals unity) decreases (see Fig. 1 a, b). Thus, for example, the carrier concentration on the faces of the crystal $\xi = -1$ with asymmetrically treated surfaces ($S_+ = \infty$, $S_- = 0$) when

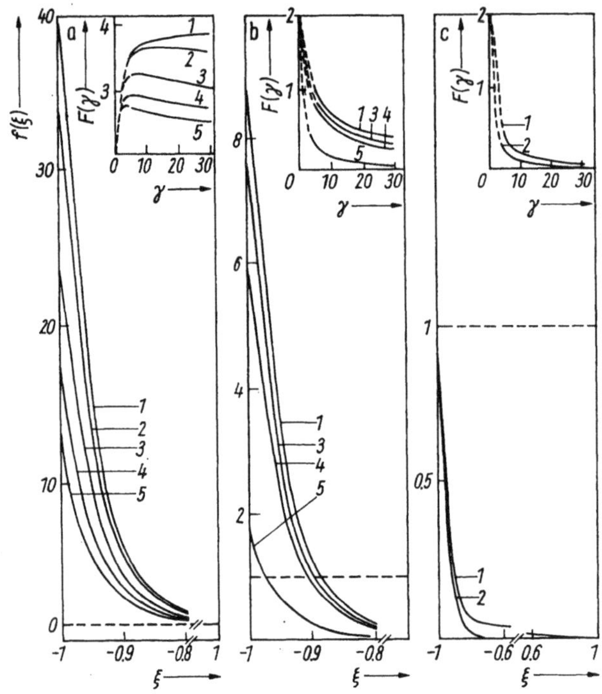

Fig. 1. The spatial distribution of the current carriers $f(\xi)$ at $\gamma = 20$, $\beta_R = 1$ (curves 1 to 4), $\beta_A = 1$ (curve 5) and different values of the parameter α. (1) $\alpha = \infty$, (2) 100, (3) 10, (4) 3, (5) 0. At the top the field dependence of the total carrier number $F(\gamma) = \int_{-1}^{1} f(\xi)\, d\xi$ is plotted a) $S_+ = \infty$, $S_- = 0$; b) $S_+ = S_- = 0$; c) $S_+ = 0$, $S_- = \infty$

$\gamma = 20$, equals 39, 24, 13 for values of $\alpha = \infty$, 10, 0, respectively. At all values of α, the concentration decrease with increasing depth of the crystal occurs rather quickly, and if $S_+ = S_- = 0$, practically throughout the whole crystal ($-0.8 \leqq \xi \leqq -1$), the local carrier concentration is lower than the equilibrium one (Fig. 1b), but if $S_\pm = \infty$, in the same region $f(\xi) \equiv 1$ due to intensive carrier generation on the surface $\xi = 1$ (Fig. 1a). But if the carriers are extracted to the face with great surface recombination velocity ($S_+ = 0$, $S_- = \infty$), the s patialdistribution $f(\xi)$ practically does not depend on the parameters α, β_R (see Fig. 1c).

In the same figure (at the top) the field dependences of the total number of carriers $F(\gamma)$ in the crystal is presented. According to $F(\gamma) = \int_{-1}^{1} f(\xi)\, d\xi$, the total number of carriers in the crystal, expressed in dimensionless units, equals $F(0) = 2$. Under the conditions of Auger recombination the effect has an interesting feature: a non-monotonous function $F(\gamma)$ in crystals with asymmetrical treatment of the faces (Fig. 1a). The maximum on the curve $F(\gamma)$ occurs in all cases if $\alpha \neq \infty$. And the larger the Auger recombination contribution, the lower the values of F_{max}, and the earlier the function $F(\gamma)$ begins to decrease on the γ scale. The function $F(\gamma)$ has a sharp maximum (if $\alpha \neq \infty$) and the value of the field γ corresponds to the condition

$$\gamma_{max} = \sqrt{\beta_R^2 + \beta_A^2}\, \sqrt{\tfrac{3}{2}(\alpha + 3) + 2\sqrt{6(1 + \alpha)}}\ . \tag{11}$$

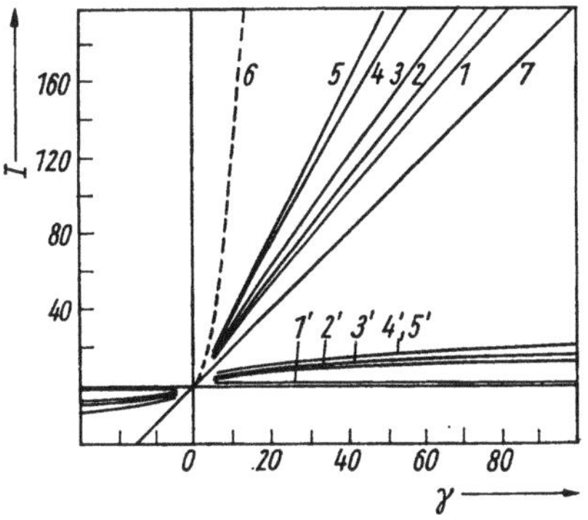

Fig. 2

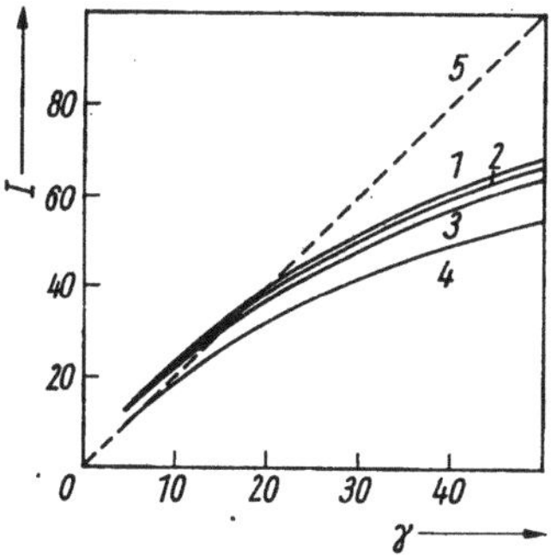

Fig. 3

Fig. 2. The current–voltage characteristics $I(\gamma) = \gamma \int_{-1}^{1} f(\xi)\, d\xi$ at different values of the surface recombination velocities and $\beta_R = 1$ (curves 2 to 5, 2′ to 5′), $\beta_A = 1$ (curves 1, 1′). (1 to 6) $S_+ = \infty$, $S_- = 0$; (1′ to 5′) $S_+ = S_- = 0$. (1, 1′) $\alpha = 0$, Auger recombination, (2, 2′) 3, (3, 3′) 10, (4, 4′) 100, (5, 5′) ∞ — quadratic recombination, (6) linear recombination, (7) ohmic characteristic, $F = 2$

Fig. 3. The same as in Fig. 2 at $\alpha = 0.06$, $\beta_A = 1$, $S_+ = 50$. (1) $S_- = 0$, (2) 0.5, (3) 1, (4) 5, (5) ohmic characteristic, $F = 2$

The decrease of the total number of carriers with the field γ is connected with the Auger recombination. In the range $\gamma \to \infty$ the total number of carriers tends to its equilibrium value $F(0) = 2$.

If $S_+ = S_- = 0$, the total number of carriers decreases with γ growth the faster, the smaller α is (result of non-linearity of recombination). When $S_+ = 0, S_- = \infty$ the function $F(\xi)$ practically does not depend on α and decreases sharply with increase of γ.

Knowing the dependence $F(\gamma)$, it is easy to observe a behaviour of CVC according to (9). The analysis of (9) shows that with assymmetrically treated surfaces ($S_+ = \infty$, $S_- = 0$), the sections with linear and sublinear dependence of the current I on the field γ can be distinguished on the CVC curves. The linear section of the dependence $I(\gamma)$ is replaced by the sublinear one the faster, the smaller α is, and the value of I at a fixed value of γ is the lower, the smaller α is (see Fig. 2). Thus, curve 1 describes the enrichment regime at Auger recombination, and curve 5 at quadratic recombination, curve 6 (dashed line) describing the CVC at linear recombination, is given for comparison. In the other direction of the field γ (exhaustion regime), the CVC are saturated and the value of the saturation current practically does not depend on α (it is not shown in the figure). When surfaces are symmetrically treated $S_+ = S_- = 0$, in the general case the CVC (curves 2 to 5, Fig. 2) have two sublinear sections (if $\alpha \neq 0$) in weak fields $I \sim \sqrt{\gamma}$ and further, $I \sim \sqrt[3]{\gamma}$. The transition to the dependence $I \sim \sqrt[3]{\gamma}$ occurs the faster, the greater the contribution of the Auger recombination is (the less α is). In the hypothetical case, when $\alpha = 0$ (heat generation of carriers is absent), the CVC are saturated and the saturation current has the value $I = \sqrt{3}$ (curve 1). Naturally, in this case the CVC form does not depend on the sign of γ.

When $S_+ = S_- = \infty$, the CVC remain ohmic at any values of α and γ (curve 7).

At real treatment of the surface $S_\pm \neq 0, \infty$, the asymptotic expressions for CVC and $F(\gamma)$, given above, remain true, if the following conditions are satisfied: $\gamma \ll S_+$, $S_- \ll \beta_R$ at $\alpha > 0$ and $\gamma \ll S_+$, $S_- \ll \sqrt[3]{\gamma \beta_A^2}$ at $\alpha = 0$.

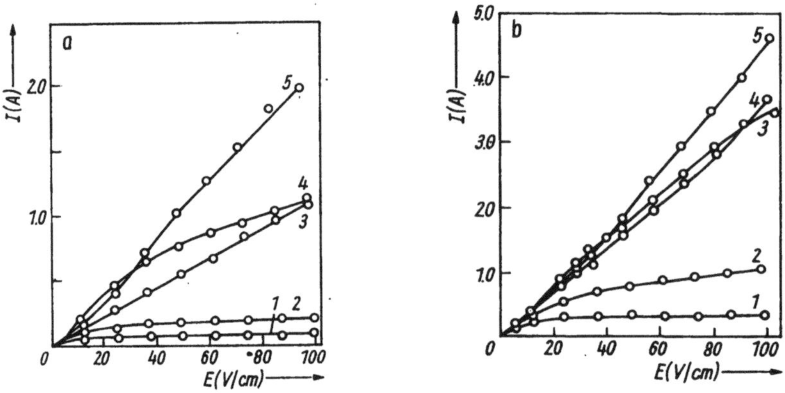

Fig. 4. The current–voltage characteristics of p-InSb in magnetic field at a) $T = 240$ and b) $T = 300$ K for $S_+ \gg S_-$, (1) and (2) exhaustion regime, $E < 0$; (4) and (5) enrichment regime, $E > 0$; (1) and (4) $H = 6$, (2) and (5) 2, (3) 0 kOe

In Fig. 3 there are CVC for the practically interesting case of a real treatment of the faces $S_- \neq S_+ \neq 0, \infty$. The value $\alpha = 0.06$ occurs in InSb at room temperature.

Let us discuss the experimental results. The CVC in the magnetic field are given in Fig. 4 for two temperatures (the faces are treated asymmetrically). It is obvious that their form is in a qualitative agreement with the influence of the parameter α, considered above: the degree of crystal enrichment is the lower, the higher the temperature is (intensive Auger recombination). In the exhaustion regime it is easy to observe a CVC tendency to saturation. Fig. 5 presents the CVC for sepcimens with symmetrical treatment of the faces, obtained under similar conditions. In this case as it was expected, the tendency to enrichment is absent, the value of the current in magnetic field is always lower than without it, the degree of CVC sublinearity increases with temperature growth. Such a behaviour of curves can be explained by intensive non-linear recombination in the volume and the absence of surface generation of carriers.

The comparison of the CVC, obtained experimentally, with the calculated ones makes it possible to estimate the velocity of surface recombination on the etched

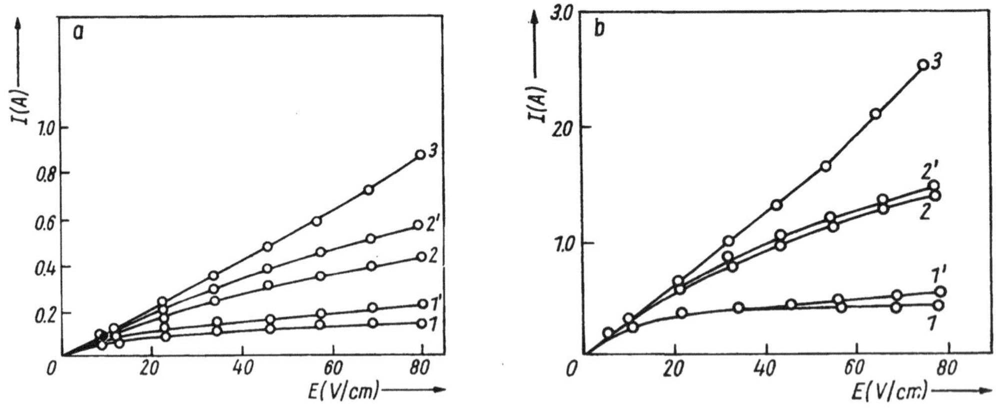

Fig. 5. The same as in Fig. 4 for $S_+ \approx S_-$. (1) and (1′) $H = \pm 6$, (2) and (2′) ± 2, (3) 0 kOe

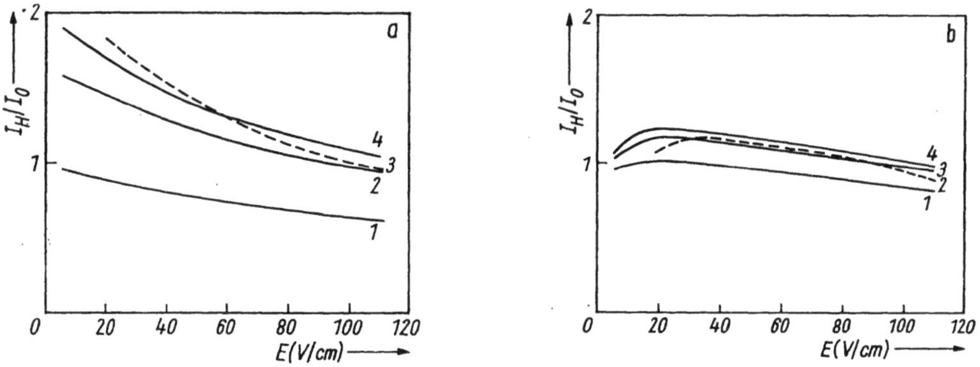

Fig. 6. The field dependence of the magnetosensitivity of p-InSb at a) $T = 240$ and b) $T = 300$ K in the enrichment regime a) (1, 2, 4) calculation at $s_+ = 10^6$ cm/s, $H = 2$ kOe, (3) experiment; b) (1, 2, 4) calculation at $s = 10^6$ cm/s, $H = 6$ kOe, (3) experiment; (1) $s_- = 5 \times 10^4$, (2) 10^4, (3) 5×10^3 cm/s in both cases

surface of the sample. For this purpose, let us compare the field dependences of the I_H/I_0 ratio, where I_H is the current value in the enrichment regime, I_0 is the current value at $H = 0$. The temperature dependence of the parameter α is taken from [12], the values $U_n(T)$ and $U_p(T)$ from [13]. As it is shown in [4], the degree of surface enrichment at MCE in InSb is not so great, that neglect of degeneration could considerably influence the results of numerical treatment (when $T = 300$ K, the Fermi quasi-level for electrons is situated above the bottom of the conduction band by kT at strongest enrichment).

The results of comparison are presented in Fig. 6. As one can see, the surface recombination velocity on the etched face varies in the range $5 \times 10^3 < s < 10^4$ cm/s and does not depend much on the temperature. The same value s at $T = 300$ K has been recently obtained in [14].

In conclusion we shall note the practical aspect of the obtained results. It is known [15, 16] that MCE can be used for the creation of highly sensitive sensors of magnetic field (the total number of carriers depends on the value of H). It is interesting, that in spite of the intensive Auger recombination in InSb at room temperature, this material can be used successfully for magnetic sensors. In this case the measured magnetic field should be oriented in such a way that it could ensure the exhaustion regime in the crystal (see Fig. 1, straight line 7 and curve 5). The measurements show that the current sensitivity γ_i at $E = 10^2$ V/cm (impulse regime) reaches the value 10^{-3} A/Oe in fields up to $H \leqq 10^3$ Oe. The time constant of such sensors is not less than 10^{-8} s.

References

[1] H. Welker, Z. Naturf. 6a, 184 (1954).
[2] I. I. Boiko and V. A. Romanov, Fiz. Tekh. Poluprov. **11**, 817 (1977).
[3] V. K. Malyutenko, S. S. Bolgov, and Yu. M. Malozovskii, phys. stat. sol. (a) **50**, 723 (1978)
[4] L. A. Almazov, A. I. Liptuga, and V. K. Malyutenko, Fiz. Tekh. Poluprov. **13**, 52 (1979)
[5] I. I. Boiko and V. K. Malyutenko, Ukr. fiz. Zh. **14**, 1449 (1969).
[6] V. V. Vladimirov and V. M. Chernousenko, Fiz. Tekh. Poluprov. **5**, 685 (1971).
[7] A. Chovet, phys. stat. sol. (a) **28**, 633 (1975).
[8] H. Fujisada, J. appl. Phys. **45**, 3530 (1974).
[9] Y. P. Varshni, Radiative Recombination in Semiconductors, Nauka, Moscow 1972.

[10] D. BLAKEMORE, Electron Statistic in Semiconductors, Mir, Moscow 1964.
[11] D. BLAKEMORE, Proc. Internat. Conf. Phys. Semicond., Prague 1960 (p. 981).
[12] V. K. MALYUTENKO, S. S. BOLGOV, V. I. PYPA, and V. I. CHAIKIN, Fiz. Tekh. Poluprov. **14**, 781 (1980).
[13] J. E. HOLLIS, S. C. CHOO, and E. L. HEASELL, J. appl. Phys. **38**, 1626 (1967).
[14] W. HANUS and M. OSZWAŁDOWSKI, phys. stat. sol. (a) **36**, 445 (1976).
[15] M. MORVIC and K. MĚŘÍNSKÝ, phys. stat. sol. (a) **5**, K23 (1971).
[16] I. S. LEVITAS, YU. K. POZHELA, and A. P. SASHUK, Liet. fiz. Rink. **5**, 825 (1972).

(Received September 25, 1980)

phys. stat. sol. (a) **65**, 141 (1981)

Subject classification: 12.1; 21

Kharkov State University[1])

Anisotropy of Yttrium Microhardness

By

I. I. PAPIROV, V. V. VOROBEV, A. I. PIKALOV, and N. I. MOREVA

Polar and azimuthal diagrams of the microhardness of yttrium single crystals are constructed. The polar diagram of the microhardness in the $(11\bar{2}x)$ planes with the c axis is close in its form to an ellipsoid of rotation around the [0001] direction with axes 90 and 40 kg/mm². The azimuthal anisotropy of microhardness in the (0001) and $(10\bar{1}0)$ planes of single crystals with anisotropy coefficients of the second kind of 1.1 and 2.45 is found. The first kind anisotropy coefficients have a maximum of ≈ 1.6 in the $(10\bar{1}0)$ and $(11\bar{2}0)$ planes.

Построены полярные и азимутальные диаграммы микротвердости монокристаллов иттрия. Полярная диаграмма микротвердости в плоскостях $(11\bar{2}x)$, содержащих ось c, по форме близка к эллипсоиду вращения вокруг направления [0001] с осями 90 и 40 kg/mm². Обнаружена азимутальная анизотрорпия микротвердости в плоскостях (0001) и $(10\bar{1}0)$ монокристаллов с коэффициентами анизотропии второго рода равными соответственно 1,1 и 2,45. Коэффициенты анизотропии первого рода максимальны (≈ 1,6) в плоскостях $(10\bar{1}0)$ и $(11\bar{2}0)$.

1. Introduction

The anisotropy of the microhardness of metals has not yet been completely studied [1, 2]. However, information about the hardness changes in crystals in different directions is of great importance for the complete description of their elastoplastic and strength characteristics. Among metals of hexagonal structure microhardness diagrams in a wide range of orientations have been constructed only for beryllium [3 to 5] and rhenium [6]. According to published data these diagrams, in their form, are close to ellipsoids with rotation around the [0001] hexagonal axis and axes ratio $H_{(0001)}/H_{(10\bar{1}0)}$ of 1.6 and 2.1 for beryllium and rhenium, respectively. More detailed analyses have shown that there is a weak anisotropy of hardness in the base plane and planes lying in the [0001] axis zone. Thus in the base plane of beryllium, the hardness measured by the scratching method may vary from 115 to 133 kg/mm², while in planes parallel to the [0001] axis from 225 to 208 kg/mm² [3, 5]. For other metals with hexagonal structure there is information only about correlation of microhardness along the main axes c and a [1 to 2, 7]. Thus the ratio $H_{(0001)}/H_{(10\bar{1}0)}$ equals 3, 2, and 0.5 for titanium, yttrium, and zinc, respectively. For zinc and cadmium (metals with axes ratio $c/a > (c/a)_{ideal}$, where $(c/a)_{ideal} = 1.633$) the hardness measured on the (0001) plane is less than on the $(10\bar{1}0)$ plane. For all other metals with hexagonal structure which were investigated having $c/a < (c/a)_{ideal}$ there is $H_{(0001)} > H_{(10\bar{1}0)}$. The general information on microhardness for yttrium is limited only to the paper of Terekhova and Savitsky [7] where by the mean values of hardness on the (0001) and $(10\bar{1}0)$ planes reported to be 112 and 61 kg/mm², respectively. It is also noted that the indenter imprints on the base plane have a square form, the diagonal of the imprint being practically independent of the rotation angle of the indenter with respect to its own axis. On the $(10\bar{1}0)$ plane the anisotropy, of microhardness depends on the rotation angle of the indenter, the imprints having on irregular form.

[1]) 4, Dzerzhinskii pl., 310077 Kharkov, USSR.

The purpose of the present work was to study the anisotropy of the microhardness for yttrium single crystals in a wide range of orientations together with the construction of microhardness diagrams. Yttrium as a object of study was chosen for two reasons: firstly, information about hardness, plasticity, and strength characteristics of yttrium single crystals is extremely limited, secondly, yttrium is a metal with a minimum ratio of lattice periods c/a (the c/a value of yttrium equals 1.571). Together with beryllium and some other rare earth metals (Tb, Ho, Er, Tm) yttrium closes the row of metals with hexagonal structure built in the order of decreasing c/a ratio.

2. Material and Method of Measurement

As the object of study yttrium single crystals, grown by the method of recrystallization annealing in vacuum (1×10^{-8} Torr) and at temperatures 50 to 100 K lower than the temperature of polymorphous transition [7], were used. The general purity of the metal was 99.99%.

Large crystals cut from ingots by electroerosion cutting were oriented using the method of precisional orientation [8]. The accuracy of the crystal cutting of a given orientation was about 0.5°. The crystals were oriented in the (0001) and $(10\bar{1}0)$ planes, then cross-sections parallel to these planes were made, and cylinders of 6 mm diameter were cut, their axes being perpendicular to the base and prism planes.

After thin grinding and polishing in 5% solution of nitric acid in alcohol, the cylindrical single crystals were mounted in a holder of a special device for measuring the microhardness anisotropy which allows rotation of the sample at any angle around the cylinder axis on the objective table of the microhardness measuring instrument. The device was provided with a limb for registering the rotation angle. All measurements were performed by using the microhardness meter PMT-3 with a standard pyramidal indenter. The load, while using the microhardness indentation method, was 50 g.

For studying of the anisotropy of microhardness of yttrium in (0001) and $(10\bar{1}0)$ planes the method of indentation of the indenter with a square basis proved to be unsuitable. For this reason the scratching method, whereby the sample is shifted under a diamond with a speed of approximately 1 mm/min, was used. The load was 10 and 20 g. At lower loads the groove width was very small and the width-measurement error increased.

As a rule all tests were repeated on three different crystals or they covered the necessary angles interval by a factor of 2 to 3. The number of tests at each point was from 5 to 15. In the diagrams given below, either mean microhardness values calculated as a result of 5 to 15 tests (circles) or scattering of measured values (from minimum to maximum) for different samples or different angles intervals (error bars) are shown.

3. Results and Discussion

A polar diagram of microhardness of yttrium single crystals in the $(11\bar{2}x)$ planes enclosed between $(11\bar{2}0)$ and (0001) is plotted in Fig. 1. For comparison, data of [3] for beryllium are shown by the dashed line. When measuring microhardness in the $(11\bar{2}0)$ plane and planes inclined to it at 40°, a very strong anisotropy of hardness of the first kind[2]) was observed: imprints had a rhombic form with $K_I \approx 1.3$ to 1.8.

[2]) For a quantitative characteristic of anisotropy in the present work we used anisitropy coefficients of the first and second kind [9]: $K_I = (d_{max}/d_{min})^2$ and $K_{II} = H_{max}/H_{min}$ where d_{max} and d_{min}, H_{max} and H_{min} are corresponding maximum and minimum values of diagonal lengths of imprints and microhardness numbers.

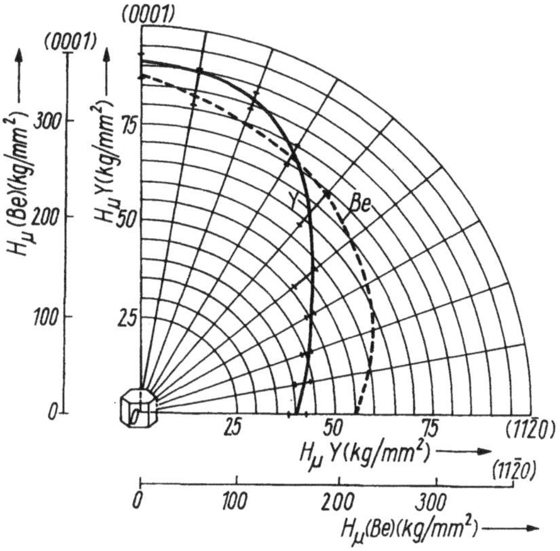

Fig. 1. Polar diagrams of microhardness of yttrium and beryllium single crystals in planes containing the [0001] axis

$K_I \to 1$ when approaching the base plane. Therefore the value of microhardness was determined from the values of a mean geometrical length of the diagonal $d = \sqrt{d_{max} d_{min}}$.

From Fig. 1 it follows that the microhardness anisotropy of yttrium is a little bit higher than for beryllium.

Fig. 2 shows angle dependences of the anisotropy coefficient $K = H_{(0001)}/H_{(11\bar{2}x)}$. The value $K_{II} = H_{(0001)}/H_{(11\bar{2}0)}$ for yttrium is 2.25 to 2.5, i.e. it is close to the value K_{II} given in a previous paper [7]. As to absolute values of microhardness, they are somehow lower. This is probably due to the higher purity of crystals used in this study. The strong dependence of yttrium microhardness on orientation allows an accurate determination of the orientation of separate grains of the polycrystal by the microhardness using a polar diagram (Fig. 1).

Fig. 3 and 4 show azimuthal diagrams of the microhardness of yttrium in the (0001) and (10$\bar{1}$0) planes measured by the scratching method. In a base plane the microhard-

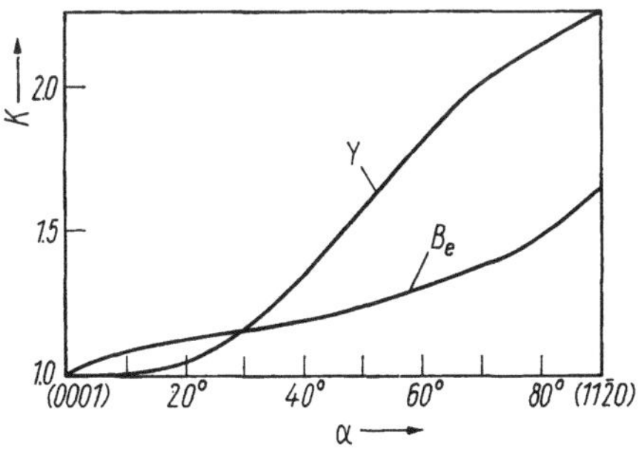

Fig. 2

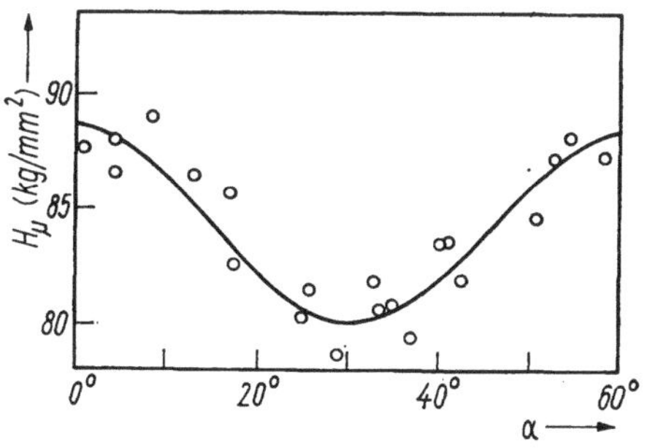

Fig. 3

Fig. 2. Angular dependences of coefficients of microhardness $K = H_{(0001)}/H_{(11\bar{2}x)}$ for yttrium and beryllium

Fig. 3. Azimuthal diagram of microhardness of yttrium in the base plane (0001)

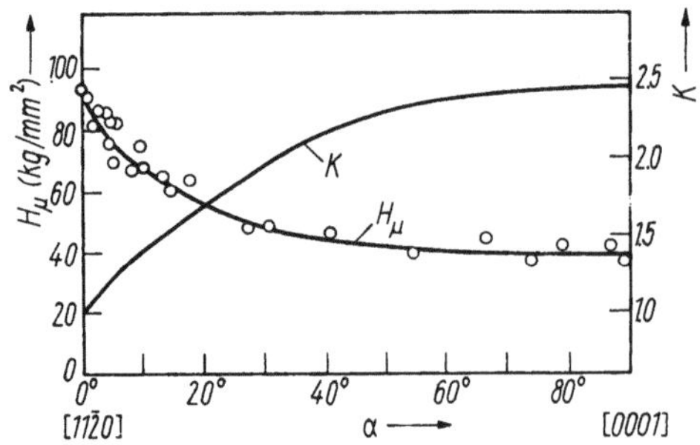

Fig. 4. Azimuthal diagram and coefficient of anisotropy of yttrium in the prism plane (10$\bar{1}$0)

ness varies periodically (with 60° period) from 80 to 89 kg/mm², i.e. depending on orientation, microhardness deviations of $\approx$ 10% are possible.

The azimuthal diagram of the microhardness for beryllium and the amount of deviations are similar to those of yttrium. The change of microhardness is more pronounced in the (10$\bar{1}$0) plane for yttrium (Fig. 4). Here it changes from $\approx$ 95 kg/mm² for directions parallel to the [11$\bar{2}$0] axis to $\approx$ 40 kg/mm² for directions parallel to the [0001] axis.

Fig. 4 also shows values of the coefficients of microhardness anisotropy $K = H_{[11\bar{2}0]}/H_{[11\bar{2}x]}$. The large microhardness anisotropy for yttrium, while transmitting from the (0001) plane to the (11$\bar{2}$0) plane and from the [11$\bar{2}$0] direction in the (10$\bar{1}$0) plane to the [0001] direction, is a result of peculiarities of the bonding forces and plastic flow of single crystals. It is known that the main sliding system for yttrium is that of {10$\bar{1}$0} ⟨11$\bar{2}$0⟩ and the twinning system is {10$\bar{1}$2} ⟨10$\bar{1}\bar{1}$⟩ [7, 10]. The slide is less active on the base plane {0001} ⟨11$\bar{2}$0⟩ and the twinning on the planes {11$\bar{2}$1} ⟨11$\bar{2}$6⟩ and {11$\bar{2}$2} ⟨11$\bar{2}$3⟩. Thus, when the indenting into the (0001) plane or scratching the (10$\bar{1}$0) plane along the [11$\bar{2}$0] direction, a plastic deformation is difficult because of the absence of stress components in the main sliding systems {10$\bar{1}$0} ⟨11$\bar{2}$0⟩ and {0001} ⟨11$\bar{2}$0⟩, but a specific crystallographical shear in the main twinning system {10$\bar{1}$2} ⟨10$\bar{1}\bar{1}$⟩ is positive. The latter means that twinning occurs not during compression but on extension along the c axis. On the contrary, indentation of the indenter into the (11$\bar{2}$0) plane or the (10$\bar{1}$0) plane and scratching of the (10$\bar{1}$0) plane along the [0001] direction deformation is done easily because of sliding in the {10$\bar{1}$0} ⟨11$\bar{2}$0⟩ system and different kinds of twinning. Thus the hardness change of yttrium in different planes and directions reflects its character of resistance to deformations and is determined by the anisotropy of the shear stress and of the strengthening in the indicated planes and directions. In this respect hardness measurements may prove to be a very effective method of analysis of change connected with the presence and the redistribution of impurities while hardening, ageing and other material treatment methods.

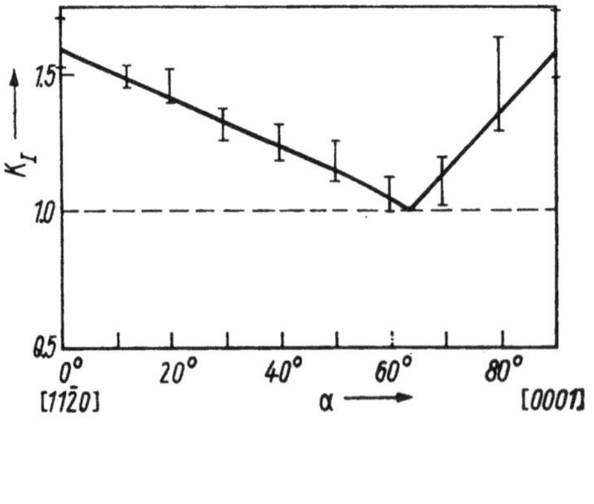

Fig. 5. Change of the anisotropy coefficient of first kind $K_I = (d_{max}/d_{min})^2$ versus direction in the (10$\bar{1}$0) plane of an yttrium single crystal

Existence of a strong anisotropy of the microhardness of yttrium leads to substantial deviations of the form of the indenter imprint from a regular square. The anisotropy of first kind is observed almost for all orientations of yttrium single crystals excluding planes close to (0001). Fig. 5 shows the dependence of the anisotropy coefficient of first kind in the direction of the $(10\bar{1}0)$ plane. It is seen that in a prism plane the anisotropy of first kind is maximum when imprint diagonals coincide with a and c axes. It is absent only on rotating the indenter so that one of its diagonals forms an angle ≈ 20 to $30°$ to the [0001] axis, however, even in this case, the imprint projection differs from a square.

In conclusion, it should be noted that in spite of a large azimuthal anisotropy on scratching, the hardness during indentation, determined by the mean geometrical length of the imprint diagonal, is practically independent of the indenter rotation around its own axis.

4. Conclusions

1. Orientational dependences of microhardness of yttrium single crystals have been plotted. The polar diagram of the microhardness in planes having the c axis is in its form close to an ellipsoid of rotation around the c axis having axes 90 and 40 kg/mm^2.

2. The azimuthal anisotropy of microhardness in the (0001) and $(10\bar{1}0)$ planes of single crystals with anisotropy coefficients of the second kind of 1.1 and 2.45, respectively, has been determined.

3. The anisotropy coefficients of first kind are maximum in the $(10\bar{1}0)$ and $(11\bar{2}0)$ planes where they reach values of ≈ 1.6. The anisotropy of first kind is absent in the base plane (0001).

4. The considerable anisotropy of the microhardness of yttrium can be used for an approximate determination of grain orientation in polycrystals by measuring their hardness.

5. The obtained results are explained on the basis of data on the plastic deformation nature of yttrium single crystals.

Acknowledgements

The authors are extremely grateful to A. S. Kapcherin for help in orientation of yttrium single crystals.

References

[1] B. W. Mott, Micro-Indentation Hardness Testing, Butterworths Scientific Publications, London 1956.
[2] V. M. Glazov and V. N. Vigdorovich, Microhardness of Metals, Metallurgizdat, Moscow 1969 (in Russian).
[3] R. I. Garber, R. I. Zalivadni, and F. S. Gorokhovatsky, Fiz. Metallov i Metallovedenie 9, 274 (1960).
[4] N. A. Hill and J. W. S. Jones, J. nuclear Mater. **3**, 138 (1961).
[5] V. N. Lange and T. I. Lange, Fiz. Metallov i Metallovedenie **16**, 638 (1963).
[6] V. M. Azhazha, G. P. Kovtun, V. A. Elenskii, and V. S. Belovol, Fiz. Metallov i Metallovedenie **39**, 202 (1975).
[7] V. F. Terekhova and E. M. Savitsky, Yttrium, Nauka, Moscow 1967 (in Russian).
[8] A. S. Kapcherin, I. I. Papirov, and S. S. Avotin, Method of Precisional Orientation and Cutting Single Crystals, Preprint, Phys. Techn. Inst., Kharkov 1974 (in Russian).
[9] V. K. Grigorovich, in: Methods of Microhardness Testing, Nauka, Moscow 1965 (p. 223) (in Russian).
[10] E. J. Rapperport and C. S. Hartley, Trans. MS AIME **215**, 1071 (1959).

(Received October 17, 1980)

phys. stat. sol. (a) 65, 147 (1981)

Subject classification: 10.2 and 20.3; 22.8.1

Department of Physics, Indian Institute of Technology, Kharagpur[1])

On the Thermoluminescence of Sb-Doped Calcium Fluorophosphate

By

R. Jayaprakash and V. V. Ratnam

The thermoluminescence (TL), thermoluminescence spectra, and monochromatic thermoluminescence (MTL) of X-irradiated, Sb-doped calcium fluorophosphate (FAP) are reported. The TL of these powder phosphors, synthesized in the laboratory show glow peaks at 150, 220, and 365 °C. The TL emission spectra have peaks at 310, 400, 563, and 620 nm for low Sb-doping (0.1%) and at 300, 450, 572, and 625 nm for high Sb-doping (10%). The emissions around 563 nm at low Sb concentration and around 572 nm at high Sb concentration are attributed to the presence of manganese at Ca (II) and Ca (I) sites, respectively. From MTL measurements it is observed that for FAP:Sb (0.1%) the main contribution to the 150 °C TL glow peak is from the 570 nm emission, while both 400 and 570 nm emissions contribute to the 365 °C TL glow peak. The effect of Sb concentration on TL and TL spectra is discussed.

Es wird über Thermolumineszenz (TL), Thermolumineszenzspektren und monochromatische Thermolumineszenz (MTL) von röntgenbestrahltem, Sb-dotiertem Kalziumfluorophosphat (FAP) berichtet. Die TL dieser im Laboratorium synthetisierten Pulverphosphore zeigt Glowmaxima bei 150, 220 und 365 °C. Die TL-Emissionsspektren haben Maxima bei 310, 400, 563 und 620 nm für niedrige Sb-Dotierung (0,1%) und bei 300, 450, 572 und 625 nm für hohe Sb-Dotierung (10%). Die Emissionen bei etwa 563 nm bei niedrigen Sb-Konzentrationen und etwa 572 nm bei hohen Sb-Konzentrationen werden der Anwesenheit von Mangan auf Ca (II) bzw. Ca (I)-Plätzen zugeschrieben. Bei den MTL-Messungen wird beobachtet, daß für FAP: Sb (0,1%) der Hauptbeitrag zu dem 150 °C-Peak aus der 570 nm-Emission herrührt, während sowohl die 400 als auch die 570 nm Emissionen zum 365 °C-TL Glowmaximum beitragen. Der Einfluß der Sb-Konzentration auf die TL und die TL-Spektren wird diskutiert.

1. Introduction

In our earlier paper [1] we have reported results on thermoluminescence (TL) and TL spectra of X-ray induced defect centres in commerical grade calcium fluorophosphate (FAP) doped with varying concentrations of Mn. In the commercial grade starting materials used in the preparation of FAP, manganese and antimony are present as inherent impurities. However, the exact role of antimony in the defect centre formation and in the luminescence process can be understood better when studies are made on samples containing antimony in sufficient and widely varying concentrations. A model for a new Sb-associated centre, called E-II (Sb) centre was proposed by Warren et al. [2]. In FAP there are two inequivalent calcium sites; viz., Ca(I) and Ca(II); the former has C_3 point group symmetry and the latter has C_{1h} point group symmetry [3]. Generally Sb^{3+} substitutes Ca^{2+} in the Ca(II) site and the charge compensation is accomplished by a nearby O^{2-} ion substituting for a F^- ion [4]. The fluorescence of Sb^{3+} occurs around 490 nm. Its excitation spectra extend from 180 to 300 nm [4]. TL, TL spectra, and monochromatic thermoluminescence (MTL) of Sb-doped FAP irradiated by X-rays are reported in this paper.

[1]) Kharagpur 721302, India.

2. Experimental Methods

The method of preparation of the sample and the apparatus used for optical, TL, and TL spectra measurements have been described already [1]. The TL spectra given in the figures are corrected for photomultiplier (PM) tube response and monochromator response. Antimony incorporation was accomplished by mixing Sb_2O_3 to the starting materials in desired concentrations before firing. The same arrangement used for recording TL spectra was used for recording MTL except that the monochromator, set at the desired wavelength, was arranged between the vacuum chamber containing the sample and the PM tube. All the MTL glow curves are corrected for the PM tube pick-up of the thermal glow at all wavelengths.

3. Results

3.1 Diffuse reflectivity

The diffuse reflectivity measurements of Sb-doped FAP taken in the 300 to 700 nm region are shown in Fig. 1. Absorption bands induced by X-irradiation occur around 375 and 450 nm. Compared to the unactivated FAP, the absorption of FAP:Sb above 350 nm has decreased. With increase in Sb concentration, there is a small increase in the induced absorption, particularly above 400 nm.

3.2 Thermoluminescence

The TL glow curves of X-irradiated FAP:Sb exhibit three glow peaks around 150, 220, and 365 °C (Fig. 2), the glow peaks occurring approximately in the same region as in unactivated FAP; only the relative strengths of the different peaks vary. At low Sb concentration (i.e., Sb/Ca $= 10^{-3}$) the high temperature peak at 365 °C is stronger than the peak around 220 °C. As the Sb concentration is increased, the 365 °C peak is suppressed and the first two peaks almost merge together. At 0.1% Sb concentration the area of the glow curve falls to less than half of that of unactivated FAP, while at 10% Sb concentration it falls to about 1/6.

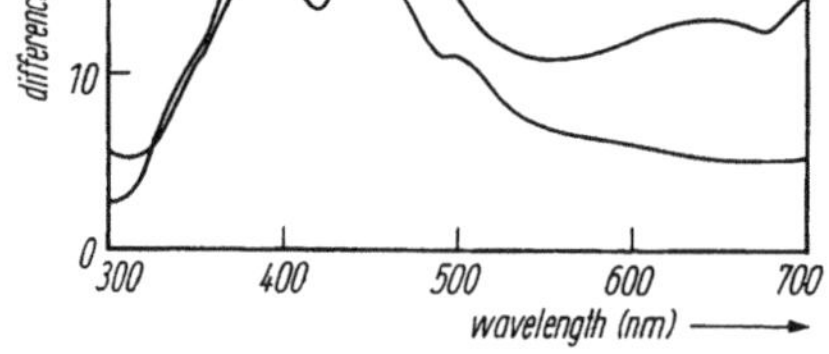

Fig. 1

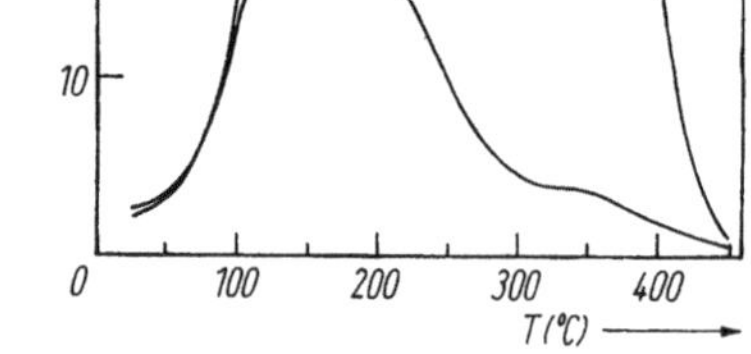

Fig. 2

Fig. 1. The difference in diffuse reflectivity before and after X-irradiation (30 kV, 10 mA) plotted against wavelength, of (1) FAP:Sb (0.1%) and (2) FAP:Sb (10%)

Fig. 2. The TL glow curves of X-irradiated (1) FAP:Sb (0.1%) and (2) FAP:Sb (10%)

3.3 Thermoluminescence spectra

At low Sb concentration (0.1%), the TL spectra show peaks around 310, 400, 563, and 620 nm (Fig. 3a). At 150 and 220 °C, the 563 nm peaks is most prominent. As the temperature is increased, the peak shifts to lower wavelengths. At 365 °C, the 560 nm emission is reduced and the emission around 400 and 620 nm becomes dominant. The 310 nm emission is prominent at 150 °C. The spectra of the sample containing 10% Sb show a marked shift in the peak positions. The peak positions occur around 300, 450, 572, and 625 nm (Fig. 3b). At this concentration also the emission around 572 nm is the most prominent at 150 and 220 °C, whereas at 365 °C the emission around 572 nm is almost totally quenched. The new emission peak around 450 nm is present only at high Sb concentration. Except the emission around 450 nm, the emission in all other regions is similar to that of the undoped FAP, though slight variations in relative intensities and spectral peak positions occur.

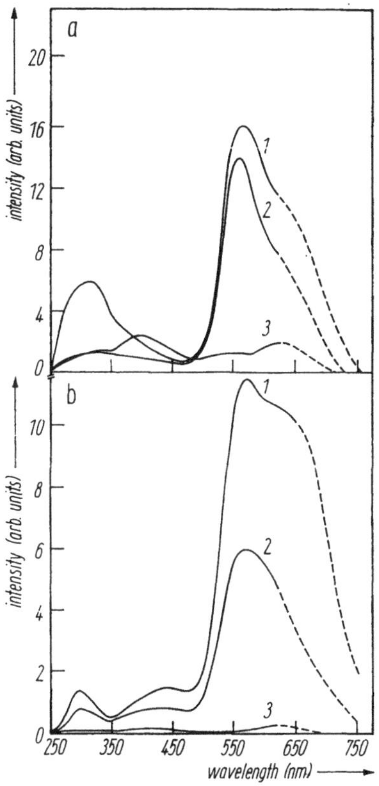

Fig. 3

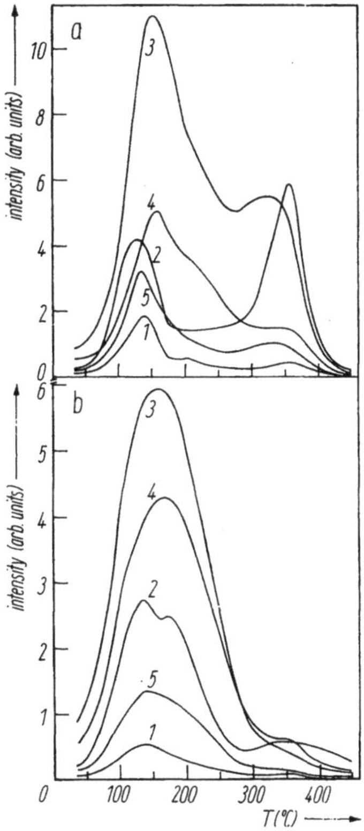

Fig. 4

Fig. 3. The TL spectra of a) X-irradiated FAP:Sb (0.1%) and b) X-irradiated FAP:Sb (10%). (1) $T = 150$, (2) 220, and (3) 365 °C

Fig. 4. The MTL of a) X-irradiated FAP:Sb (0.1%) at (1) 310, (2) 400, (3) 570, (4) 585, and (5) 610 nm and b) X-irradiated FAP:Sb (10%) at (1) 310, (2) 450, (3) 570, (4) 590, and (5) 620 nm

3.4 Monochromatic thermoluminescence

This gives information about the relative intensities of the different emissions at any temperature. The MTL of FAP:Sb (0.1%) is recorded setting the monochromator at 310, 400, 570, 585, and 610 nm (Fig. 4a). From Fig. 4a it can be seen that preferential recombination is taking place at different temperatures. The 150 °C glow peak is contributed by mainly two competing emissions around 400 and 570 nm. For the sample having 10% Sb, MTL was recorded at 310, 450, 570, 590, and 620 nm (Fig. 4b). Fig. 4b shows that the glow peak around 150 °C is a composite one and contains at least two peaks.

4. Discussion

In our earlier paper on the TL of X-irradiated FAP:Mn, we have shown that the glow peaks occur around 160, 265, and 368 °C. At all concentrations of manganese the 368 °C glow peak was more prominent than the other two peaks. In the case of FAP:Sb the 365 °C peak is suppressed and the two low temperature peaks around 150 and 220 °C become prominent, though, however, the TL glow peak positions in both cases (FAP:Mn and FAP:Sb) are nearly the same as that of unactivated FAP. It is apparent that manganese stabilises deep traps whereas antimony stabilises relatively shallow traps in the FAP lattice. Results further show that in all the above systems the nature of traps and trapping levels is the same except that they differ in their relative concentrations.

As regards to TL emission spectra of Mn-doped FAP, manganese emission peaks appear at 563 nm for low (up to 1% Mn) and at 574 nm for high (around 10% Mn) manganese concentrations and they have been attributed [1] to the presence of Mn^{2+} at Ca(II) and Ca(I) sites, respectively. It was also observed that at low concentrations of manganese, the oxygen–vacancy (O–V) complexes in the FAP lattice attract manganese ions and as a result manganese occupies Ca(II) sites [5, 6]. In the FAP:Sb samples at low antimony concentration, sufficient intrinsic defects, viz., O–V complexes are expected to be present, as a result of which Mn^{2+} ions, present inherently in small concentrations in the starting materials, occupy the Ca(II) sites and therefore the emission peak is observed at 563 nm. At high concentrations of antimony, Sb^{3+} ions preferentially occupy Ca(II) sites [4] which are nearer to halogen ions chains, and the charge compensation is supposedly accomplished by the replacement of a nearby F^- ion with the O^{2-} ion. As a result, the manganese present as inherent impurity is pushed to Ca(I) sites and hence the emission peak shifts to 572 nm.

In the TL emission spectra of FAP:Sb in addition to manganese emission which is the strongest, a new emission band occurs around 450 nm for high concentrations of antimony (at 10% Sb). This is not present in the TL spectra of undoped and Mn-doped FAP [1, 7]. Hence the 450 nm band occurring only when antimony is present in sufficient concentration should be attributed to the presence of antimony. The emission around 400 nm at 365 °C is stronger in FAP:Sb (0.1%) than in unactivated FAP. In our study of the effect of quenching of samples during preparation on their TL performance we have observed that when the sample is quenched from 1200 °C to RT, the TL emission in the 400 nm region shows an increase compared to that of the sample which is cooled slowly to RT. In the case of the former sample, the 400 nm emission is strong only at the high temperature peak. The high temperature TL spectrum of FAP:Sb (10%) does not show strong emission in this region. From these results we infer that the 400 nm emission does not occur at the Sb centres. It occurs at the sites of some (intrinsic) defects inherently present in the FAP lattice. Also the traps involved in this emission process are deep and are only destroyed at temperatures above 300 °C. This is quite evident in the MTL of FAP:Sb (0.1%) at 400 nm.

In addition to these, another salient feature in the TL of FAP:Sb is the comparative increase in intensity above 600 nm. In unactivated FAP some emission occurs in this region. Above 620 nm, due to the diminished sensitivity of the PM tube, the magnitude of the correction factor for the emission intensities is large, and so the error in intensities in this region may be large. The O^{2-} ions replacing F^- ions are reported to have absorption in the UV region around 220 nm [4] with its fluorescence occurring around 680 nm. Because of the charge compensation requirement the number of O^{2-} ions replacing the lattice F^- ions will be more at high Sb concentration. Hence, the increase in the emission intensity above 600 nm might be due to the increase in the number of O^{2-} ions.

The MTL shows that at different temperatures preferential recombination occurs. In FAP:Sb (0.1%) manganese emission around 570 nm mainly contributes to the 150 °C TL glow peak. The contributions of other emissions to this glow peak, though present, are very small, whereas manganese emission and 400 nm emission contribute equally to the 365 °C peak. Small shifts are observed in the glow peak positions of MTL at different wavelengths because the overlapping TL glow peaks vary in their relative strengths for different emissions. Though no spectral peak is present in the 585 nm region, MTL was recorded setting the monochromator at 585 nm. The identical nature of the two MTL curves at 570 and 585 nm shows that only a single peak is present.

The MTL of FAP:Sb (10%) sample at different wavelengths shows a special feature. In the place of the 150 °C TL glow peak two partially resolved overlapping TL peaks appear. This is evident in the case of MTL at 450 nm. Unlike the undoped FAP in which there is clearly a single glow peak around 150 °C, which is possibly due to some O–V complex, the FAP:Sb at high Sb concentration shows two glow peaks in this region. This closeness of the glow peaks can be interpreted as being due to different trapping centres or the same type of trapping centres in different environments. Since at high Sb concentration, the O–V complexes with separate identities (i.e., unassociated to Sb or other impurities) decreases in number this appearance of the two glow peaks can be attributed to antimony occupying Ca(I) and Ca(II) sites.

5. Conclusions

1. Basically the same type of colour centres are formed by X-irradiation in FAP:Sb as in unactivated FAP, but only the equilibrium between the different colour centres formed is changed by Sb doping. 2. Compared to FAP and FAP:Mn, colour centre formation in FAP:Sb is less. 3. TL results show that Sb stabilises the shallower traps. 4. The 450 nm emission in the TL spectra of FAP:Sb is attributed to the presence of Sb, whereas the 400 nm emission is due to some basic lattice defects (O–V complexes). The emission above 600 nm is due to the presence of oxygen. 5. In FAP:Sb, the Mn emission at low Sb concentration (0.1%) peaks around 563 nm, whereas at high Sb concentration (10%) it peaks around 572 nm, because at low Sb concentrations Mn occupies Ca(II) sites and at high Sb concentrations they are pushed into Ca(I) sites. 6. MTL shows that preferential recombination takes place at all temperatures. In the TL of FAP:Sb, mainly 570 nm emission contributes to the 150 °C TL glow peak, whereas both 570 and 400 nm emissions contribute to the 365 °C TL glow peak. 7. Increase in Sb concentration reduces the TL emission which is possibly due to reduction in the number of oxygen–vacancy complexes with increase in Sb concentration.

References

[1] R. Jayaprakash and V. V. Ratnam, phys. stat. sol. (a) **61**, 259 (1980).

[2] R. W. Warren, F. M. Ryan, R. H. Hopkins, and J. van Broekhoven, J. Electrochem. Soc. **122**, 752 (1975).

[3] P. D. JOHNSON, Luminescence of Organic and Inorganic Materials, Wiley, 1962.
[4] J. S. PRENER and W. W. PIPER, Res. Rep. No. 68-C-396 of GEC, Schenectady, New York.
[5] F. M. RYAN, R. C. OHLMAN, J. Murphy, R. MAZELSKY, G. R. WAGNER, and R. W. WARREN, Phys. Rev. B 2, 2341 (1970).
[6] R. W. WARREN and R. MAZELSKY, Phys. Rev. B 10, 19 (1974).
[7] V. V. RATNAM, R. JAYAPRAKASH, and N. P. DAW, J. Lum., to be published.

(Received January 8, 1981)

phys. stat. sol. (a) 65, 153 (1981)

Subject classification: 9 and 10.1; 21

Sektion Mathematik/Physik der Pädagogischen Hochschule Halle (Saale)[1])

On the n-Dimensional Diffusion-Controlled Growth of Precipitates in a Finite Matrix

By

O. SIMMICH and H. LÖFFLER

The diffusion-controlled n-dimensional growth ($n = 1; 2; 3$) in a finite matrix is considered solving the diffusion equation (DE) in the approximation of the stationary solution and having regard to the mass balance of the alloyed atoms. The results are compared with those obtained for the infinite matrix as well as for the non-stationary solutions of the DE. The derived results show that the method applied in this work is a useful approach to gaining information on the diffusion-controlled growth of precipitates.

Das diffusionskontrollierte n-dimensionale Wachstum ($n = 1;, 2; 3$) von Ausscheidungen in einer begrenzten Matrix wird untersucht in der Näherung der stationären Lösung der Diffusionsgleichung (DE) und des Massengleichgewichts der zulegierten Atome. Die Resultate werden mit denjenigen verglichen, die sowohl für eine unendliche Matrix als auch aus den nichtstationären Lösungen der DE abgeleitet werden. Die gewonnenen Ergebnisse zeigen, daß die in dieser Arbeit verwendete Methode ein zweckmäßiger Schritt für die Gewinnung von Aussagen über das diffusionskontrollierte Wachstum von Ausscheidungen ist.

1. Introduction

The influence of a finite matrix on the decomposition kinetics had been extensively theoretically investigated by Ham [1] on the basis of the time-dependent (non-stationary) diffusion equation (DE), taking the diffusion coefficient D to be independent of the concentration, i.e. putting $D(c) = D_0$, and neglecting the movement of the interface between the growing particle and the matrix. Zener [2] treated the n-dimensional growth of precipitates in an infinite matrix solving the DE presuming that $D =$ const. A solution of this problem with spherical particles was also given by Wert and Zener [3] in the frame of stationary conditions and a constant diffusion coefficient. The well known papers of Wagner [4] and Lifshits and Slyozov [5] concerning the phenomenon of coalescence start with expressions for the diffusion current, which directly follow from these stationary solutions of DE. Comparing the results obtained in the frame of the different approximations applied in [1 to 3] it became transparent that the particle growth follows one and the same growth law, namely $R = \beta(D_0 t)^{1/2}$, with differences only in the coefficient β. The influence of the concentration dependence of the diffusion coefficient $D = D(c)$ on the particle growth in an infinite matrix has recently been considered by the authors of this paper [6, 7] in the frame of the stationary solution of the DE.

It is the purpose of this paper to investigate the possibilities to extend the stationary solution of the DE on the one-, two-, and three-dimensional particle growth in a finite matrix taking simultaneously into account the concentration dependence of the diffusion constant D. These investigations seem to be useful, since on the basis of the stationary solution of the DE relatively easily overlookable results of the growth law of the precipitates can be obtained, which is important to comparisons with experimentally

[1]) Kröllwitzer Str. 44, DDR-4020 Halle (Saale), DDR.

found results. Furthermore, in the frame of this approximation the movement of the interface between the particle and the matrix as well as the influence of the concentration dependence of D can be taken into account. The latter aspect seems particularly worth mentioning as in many real systems the diffusion constant D essentially depends on the concentration as shown for the Al–Zn systems for example by Krupkowski [8] and the authors of this paper [9].

2. General Outline of the Model Applied

Taking into account the concentration dependence of the diffusion coefficient D the DE has to be written

$$\frac{\partial c}{\partial t} = \frac{\partial D}{\partial c}\left(\frac{\partial c}{\partial r}\right)^2 + D\left[\frac{\partial^2 c}{\partial r^2} + \frac{(n-1)}{r}\frac{\partial c}{\partial r}\right]. \tag{1}$$

Applying the stationary approximation of (1) the left-hand side has to be taken as equal to zero, i.e. $\partial c/\partial t = 0$. As shown by the authors in [6, 7] the dependence of the concentration gradient on the distance r can be written in the frame of the stationary solution,

$$\frac{\mathrm{d}c(r, R)}{\mathrm{d}r} = -\frac{a_n(R, R_\mathrm{E})}{D(c)\, r^{n-1}}, \tag{2}$$

where the exponent n refers to the dimension of the growing particle, and R to its size. $n = 3$ belongs to the growth of a sphere of radius R, $n = 2$ to a cylinder of infinite length, and radius R, and $n = 1$ to an infinite plate of the thickness R. Analogously to the model of Rogalla and Schmalzried [10] each particle should grow at the expense of the alloyed elements available in a finite drainage area of radius $R_\mathrm{E} > R$, i.e. the total flux of the solved atoms has to vanish at R_E. Equation (2) fulfils (1) also considering a finite matrix. The only difference to the growth problem in an infinite lattice is that in the finite one the coefficient a_n may depend on both R and R_E.

From (2) we easily obtain for three different dimensions

$$\int_{c_\mathrm{OM}}^{c} D(c)\,\mathrm{d}c = -a_n(R, R_\mathrm{E}) \int_R^{R_\mathrm{E} \geqq r > R} r^{1-n}\,\mathrm{d}r = \begin{cases} -a_1(R, R_\mathrm{E})\,(r - R) & \text{(3a)} \\ -a_2(R, R_\mathrm{E})\,\ln(r/R) & \text{(3b)} \\ -a_3(R, R_\mathrm{E})\,(1/R - 1/r)\,, & \text{(3c)} \end{cases}$$

where c_OM means the equilibrium concentration at the particle–matrix interface, i.e. at $r = R$. From (3) $c(r, R)$ can be calculated at least numerically, if $D = D(c)$ is known. Especially for $D(c) = D_0$ we obtain

$$c_1(r, R) = -\frac{a_1(R)}{D_0}(r - R) + c_\mathrm{OM}\,, \tag{4a}$$

$$c_2(r, R) = -\frac{a_2(R)}{D_0}\ln\left(\frac{r}{R}\right) + c_\mathrm{OM}\,, \tag{4b}$$

$$c_3(r, R) = -\frac{a_3(R)}{D_0}\left(\frac{1}{R} - \frac{1}{r}\right) + c_\mathrm{OM}\,. \tag{4c}$$

Furthermore we demand

$$c_\mathrm{OM} \leqq c_n(r, R) \leqq c_0\,, \tag{5}$$

where c_0 denotes the mean concentration of the sample.

From (3) and (4) it can be easily concluded that (5) is only fulfilled in a limited range $R \leqq r \leqq Q(R)$, where $c(r = R) = c_\mathrm{OM}$, $c(r = Q) = c_0$, and $c(Q < r \leqq R_\mathrm{E}) > c_0$ may

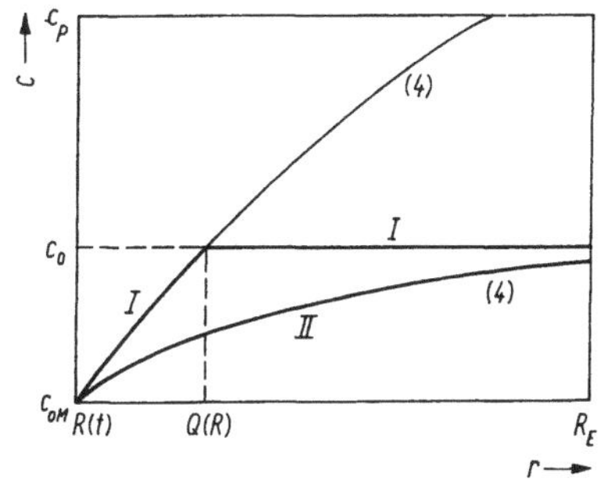

Fig. 1. Concentration profile for the ranges I and II

hold. The solutions obtained from (3) and (4) for $Q < r$ are unsuitable from the physical point of view. We substitute them by $c(Q < r \leqq R_E) = c_0$. For better understanding we demonstrate the dependence $c = c(r)$ for the two cases of actual interest, namely $Q(R) < R_E$ (range I) and $Q(R) > R_E$ (range II) in Fig. 1 taking the simple case $D(c) = D_0$ (see (4)). (One may object that the cut-off of the "normal" course of $c = c(r)$ is more or less "artificial". But solving this problem approximatively in the frame of the stationary solution one has to accept this fact as a consequence of the approximation. The decision whether this approximation is a proper one for solving growth problems in a finite matrix should be drawn from the results derived by means of this method and their comparison with the results of other authors obtained in the frame of other models. That will, of course, be done in the following.) From these considerations it can be concluded that the coefficients a_n are generally different for the ranges I and II, but have to pass continuously from range I to range II.

Taking into account the "mass balance" of the alloyed atoms it follows for the two ranges considered

$$\text{range I:} \quad \frac{1}{n}(Q^n c_0 - R^n c_P) = \int\limits_R^{Q(R)} c(r)\, r^{n-1}\, dr \tag{6a}$$

and

$$\text{range II:} \quad \frac{1}{n}(R_E^n c_0 - R^n c_P) = \int\limits_R^{R_E} c(r)\, r^{n-1}\, dr\,, \tag{6b}$$

respectively, where c_P denotes the concentration within the precipitate. $c(r)$ has to be grasped from (2), that means (6) holds generally for $D = D(c)$. But the cases $D = D(c)$ require the knowledge of $D = D(c)$ of the special binary system considered. A few solutions of $c = c(r)$ have been delivered in a former paper of the authors [7] in the frame of the stationary solution of the DE and assuming an infinite matrix, i.e. $R_E \to \infty$, but the integrals (6) can mostly be solved only numerically after substitution of these results. Therefore, we will restrict our further efforts to the simple cases $D(c) = D_0$, i.e. applying (4) instead of (3). Doing this $Q(R)$ can easily be determined from (4) putting $c(r = Q) = c_0$.

3. One-Dimensional Growth of Infinite Plates

At first we consider the growth law in range I. From (4a) we obtain

$$Q(R) = R + \frac{D_0(c_{0M} - c_0)}{a_1(R, R_E)}.$$

Inserting this into (6a) one finds

$$a_1(R, R_E) = -\frac{D_0(c_0 - c_{0M})^2}{2(c_P - c_0)\, R} = a_1(R)\,. \tag{7}$$

From (7) it can be concluded that in range I i.e. during the early growth, a_1 does not depend on R_E, that means the precipitates are not aware of growing in a finite matrix. Consequently for the range I from (2) for the concentration gradient at the particle–matrix interface $(\mathrm{d}c/\mathrm{d}r)_{r=R} \equiv g_1(R)$ follows

$$g_1(R) = \frac{(c_0 - c_{0M})^2}{2R(c_P - c_{0M})}\,. \tag{8}$$

From (2) the growth law of the precipitates can easily be derived. According to (2) the net flux

$$J(r) = -D(c)\,\frac{\mathrm{d}c}{\mathrm{d}r} = \frac{a_n(R;\, R_E)}{r^{n-1}}\,,$$

that means the integral flux through each area placed symmetrically to the precipitate (an area parallel to the plate in case $n = 1$, a cylinder ($n = 2$), or a sphere ($n = 3$) having their middle in the middle of the particle) is independent of r, $J(r)\, r^{n-1} = J(R)\, R^{n-1}$. Furthermore taking into account $(c_P - c_{0M})\,\mathrm{d}R = J(R)\,\mathrm{d}t$ we obtain from (8) the well-known parabolic growth law

$$R = \beta_1(D_0 t)^{1/2} \tag{9a}$$

with

$$\beta_1 = \frac{c_0 - c_{0M}}{(c_P - c_0)^{1/2}\,(c_P - c_{0M})^{1/2}} \tag{9b}$$

obtained also by Zener in the approximation of a linear composition profile.

Equation (9a) approximately agrees with the result also obtained by Zener [2] by means of the non-stationary solution of the DE. Only the growth factor β_1 in (9a) has to be replaced by $\beta_1' = K_1\beta_1$, where according to Zener [2], $1.13 \leq K_1 \leq 1.41$ in dependence on c_{0M}, c_0, and c_P. We now consider range II. From (4a) and (6b) we find

$$a_1(R, R_E) = -2D_0\,\frac{R_E(c_0 - c_{0M}) - R(c_P - c_{0M})}{(R_E - R)^2}\,, \tag{10a}$$

that means in range II the coefficient a_1 depends on both R and R_E. In accordance with the mass balance the maximum value of R is $R_m = R_E(c_0 - c_{0M})/(c_P - c_{0M})$. Substituting it into (10a) we get

$$a_1(R, R_E) = -\frac{2D_0(c_P - c_{0M})\,(R_m - R)}{(R_E - R)^2} \tag{10b}$$

and

$$g_1(R, R_E) = 2\,\frac{(c_P - c_{0M})\,(R_m - R)}{(R_E - R)^2}\,. \tag{11}$$

From (11) it can easily be concluded that the growth rate of the particle tends towards zero in cases $R \to R_m$. Furthermore, by comparison of (7) and (10a) and with the help of (4a) it can be shown that a_1 continuously passes from range I to range II. That happens at $R_k = R_E(c_0 - c_{0M})/(2c_P - c_0 - c_{0M})$. It seems worth mentioning that in range II for $R \ll R_E$ from (11) a growth law of Avrami type, namely

$$Y = \frac{R}{R_m} = 1 - \exp\left[-\frac{2D_0 t}{R_E^2}\right] \tag{12a}$$

holds. In the frame of the non-stationary DE and taking into account the so-called "soft-impingement" interaction [11] the authors obtained

$$Y = 1 - \frac{8}{\pi^2} \frac{R_E - R}{R_E} \sum_{n=0}^{\infty} \frac{1}{(2n+1)^2} \exp\left[-\frac{D(2n+1)^2 \pi^2}{4(R_E - R)^2} t\right]$$

and from this for cases $R \ll R_E$ the approximation

$$Y = 1 - \frac{8}{\pi^2} \exp\left[-\frac{2D_0 t\pi^2}{8R_E^2}\right], \tag{12b}$$

i.e. nearly the same expression as presented in (12a), as $\pi^2/8 \approx 1$. The same holds for the concentration gradient at the interface. The fact that the deviations between the solutions of the stationary and those of the non-stationary DE are fairly small may be a hint that generally the results obtained in the frame of the stationary solution of the DE are proper approximations of the growth law of precipitates even in cases $D = D(c)$, at least considering one-dimensional problems.

4. Two-Dimensional Growth of Cylindrical Precipitates

Analogously to Section 3 we again treat at first the growth of the precipitate in range I. Applying (4b) one finds

$$Q(R) = R \exp\left[-\frac{D_0(c_0 - c_{0M})}{a_2}\right] \tag{13a}$$

and from this with the help of (6a)

$$a_2(R, Q) = \frac{2D_0(c_P - c_{0M})\, R^2}{R^2 - Q^2} = a_2 \, . \tag{13b}$$

From (13 a, b) it can be concluded that a_2 is (contrary to a_1) independent of both Q and R within range I and one obtains the condition

$$\frac{2D_0(c_P - c_{0M})}{a_2} = 1 - \exp\left[-\frac{2D_0(c_0 - c_{0M})}{a_2}\right], \tag{14}$$

i.e. a transcendental equation for determining a_2, which can only be solved approximately. As a_2 is independent of R we obtain from (4b) for the concentration gradient at the interface

$$g_2(R) = \left(\frac{dc}{dr}\right)_{r=R} = \frac{a_2}{D_0} \frac{1}{R} \sim \frac{1}{R} \tag{15a}$$

which means a result similar to that obtained in Section 3 for g_1. We approximate a_2 by developing the right-hand side of (14) into a series and break it off after the quadratic term. With this approximation we get

$$a_2 = \frac{D_0(c_0 - c_{0M})^2}{c_P - c_0} \tag{15b}$$

and

$$g_2(R) = \frac{(c_0 - c_{0M})^2}{(c_P - c_0)\, R}. \tag{15c}$$

Comparing (15c) with $g_1(R)$ delivered in (8) one finds $g_2(R) = 2g_1(R)$, i.e. when assuming in both cases the same R-values the concentration gradient at the infinite cylinder is in the frame of the approximations applied twice that of the infinite plate.

Considering range II we obtain

$$a_2(R, R_E) = -\frac{2D_0(c_P - c_{OM})(R_m^2 - R^2)}{2R_E^2 \ln(R_E/R) - R_E^2 + R^2}, \tag{16}$$

$$g_2(R, R) = \frac{2(c_P - c_{OM})(R_m^2 - R^2)}{R[2R_E^2 \ln(R_E/R) - R_E^2 + R^2]} \tag{17}$$

and for the maximum value of R one finds $R_m^2 = R_E^2(c_0 - c_{OM})/(c_P - c_{OM})$. In this case (16) and (17) also describe the saturation behaviour of the growth as for $R \to R_m$ follows $g_2(R) \to 0$, which causes $(\mathrm{d}R/\mathrm{d}t) \to 0$. Furthermore, by means of (13b) and (16) it can be shown that the values of a_2 obtained for the ranges I and II, respectively, pass continuously from one range to the other. This happens at $R = R_k$ and $g_2 = g_{2k}$, respectively, where for the critical value of R

$$\ln\left(\frac{R_E}{R_k}\right) = \frac{1}{2}\left(\frac{R_E^2}{R_k^2} - 1\right)\frac{c_0 - c_{OM}}{c_P - c_{OM}}$$

and

$$g_{2k} = \frac{2(c_P - c_{OM})}{R_k(R_E^2/R_k^2 - 1)}$$

holds.

5. Three-Dimensional Growth of Spherical Precipitates

Considering the growth of a single precipitate in an infinite matrix the stationary solution of the DE had already been delivered by Wert and Zener [3]. Of course, putting $D(c) = D_0$ and $R_E \to \infty$, we obtain the same solution as Zener by means of (2), namely

$$c_3(r, R) = -\frac{a_3(R)}{D_0}\left(\frac{1}{R} - \frac{1}{r}\right) + c_{OM}\,. \tag{18}$$

Equation (18) fulfils the boundary condition $c(r \to \infty) = c_0$ with $a_3 = RD(c_{OM} - c_0)$ but does not fulfil the mass balance as necessary in cases of a finite matrix (finite R_E). Thus we treat this problem as performed in Sections 3 and 4. In a manner analogous to these sections at first we turn to the range I. Applying (4c) one finds

$$Q(R) = \frac{Ra_3}{a_3 + RD_0(c_0 - c_{OM})}\,. \tag{19}$$

Taking into account the mass balance one obtains by the help of (18) and (6a)

$$Q^3 c_0 - R^3 c_P = 3 \int_R^Q \left[c_{OM} - \frac{a_3}{D_0}\left(\frac{1}{R} - \frac{1}{r}\right)\right] r^2 \,\mathrm{d}r\,.$$

From this equation follows

$$R^3(c_P - c_{OM}) - Q^3(c_0 - c_{OM}) = -\frac{a_3}{2RD_0}(2Q^3 + R^3 - 3Q^2R)\,. \tag{20}$$

From (19) we get

$$a_3 = \frac{RQ(c_0 - c_{OM})\,D_0}{R - Q}$$

and inserting this into (20)

$$R^3(c_P - c_{OM}) - Q^3(c_0 - c_{OM}) = \frac{Q(c_0 - c_{OM})}{2(R - Q)} (2Q^3 + R^3 - 3Q^2R)$$

holds, from which follows, using the abbreviations $Q/R = x$ and

$$2(c_P - c_{OM})/(c_0 - c_{OM}) = q\,, \qquad x^2 + x - q = 0$$

with the solution (as $x > 0$)

$$x_1 = \frac{Q}{R} = \frac{1}{2}\left(\sqrt{1 + 8\,\frac{c_P - c_{OM}}{c_0 - c_{OM}}} - 1\right). \tag{21}$$

Substituting (21) into (19) we find

$$a_3(R) = -\,\frac{x_1 R(c_0 - c_{OM})\,D_0}{x_1 - 1} \tag{22}$$

and

$$g_3(R) = \frac{x_1(c_0 - c_{OM})}{(x_1 - 1)\,R}. \tag{23}$$

From $\mathrm{d}R(c_P - c_{OM}) = \mathrm{d}t\,D_0 g_3(R)$ we obtain the growth law

$$R(t) = \varkappa_3\sqrt{D_0 t} \tag{24}$$

with

$$\varkappa_3 = \sqrt{\frac{2x_1}{x_1 - 1}\,\frac{c_0 - c_{OM}}{c_P - c_{OM}}}\,. \tag{25a}$$

Now we would like to compare the growth coefficient obtained in the frame of our model for the early stage of aging (range I) with those found by other authors.

In the frame of the stationary solution and considering the growth in an infinite matrix Wert and Zener [3] got the growth coefficient

$$\beta_3 = \sqrt{2\,\frac{c_0 - c_{OM}}{c_P - c_{OM}}} \tag{25b}$$

but, as already mentioned above, this growth coefficient does not fulfil the mass balance, but our $\varkappa_3$ do. Furthermore, in the frame of the non-stationary solution of the DE and assuming an infinite matrix Zener [2] obtained the growth coefficient as

$$\alpha_3 = \alpha_3\left(\frac{c_0 - c_{OM}}{c_P - c_0}\right). \tag{25c}$$

α_3 in dependence on c_0, c_{OM} and c_P can be grasped from Fig. 4 in Zener's work [2]. In Fig. 2 these three growth coefficients are shown in dependence on the initial supersaturation $C_{is} = c_0 - c_{OM}$ and assuming $c_P = 1$ and $c_{OM} = 0.01$. From this figure it can easily be concluded that for small supersaturations ($C_{is} \ll 1$) the deviations between the three growth coefficients may be neglected. The differences between these growth coefficients increase with the rise of C_{is}. Additionally from Fig. 2 it can be learned that the growth coefficient derived in the frame of our approximation, namely $\varkappa_3$, fits fairly well this one obtained by Zener [2] in the frame of the non-stationary solution and assuming the growth in an infinite matrix. From this result we conclude that the stationary solution of the DE combined with the accomplishment of the mass balance yields a proper method to approximately solve growth problems of precipitates in a finite matrix at

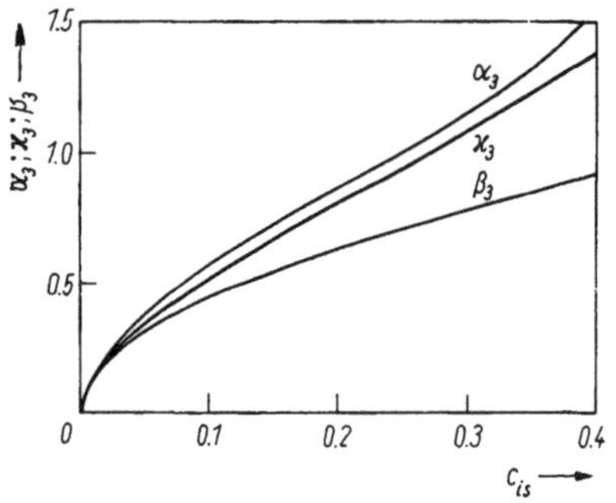

Fig. 2. Growth coefficients in dependence on the initial supersaturation for three-dimensional growth

least in cases where the initial supersaturation of the alloy is not too high (less than about 0.3).

Considering the growth in range II and taking into account the finite drainage area of radius R_E we obtain by means of (18) and (6b)

$$a_3(R) = \frac{2D_0(c_P - c_{0M})\,(R^3 - R_m^3)}{2(R_E^3/R) - 3R_E^2 + R^2} \tag{26a}$$

and thereupon the concentration gradient at the particle–matrix interface

$$g_3(R) = \frac{2(c_P - c_{0M})\,(R_m^3 - R^3)}{2R_E^3R - 3R_E^2R^2 + R^4}. \tag{26b}$$

From (26b) once more follows that in the range II the growth rate becomes zero for $R \to R_m$.

6. Discussion

Even in the frame of the stationary solution of the classical DE one obtains a growth law of the type $t^{1/2}$ in the early stage of aging, where the model of an infinite matrix is still a good approximation. This is likewise true for spherical, cylindrical, and plate-like precipitates. In each of the three dimensions considered the growth rate tends to become zero for prolonged times (saturation of the growth) as reasonable for the growth of precipitates in a finite matrix. (These results are close to those of Ham [1]. The authors intend to show these relations in another paper in detail.)

Furthermore it seems profitable to consider the fulfilment of some initial and boundary conditions which have to be realized from the physical point of view.

C1) At $t = 0$, i.e. $R = 0$, we have to demand $c(R = 0, r) = c_0$. This condition is fulfilled for each dimension considered.

C2) After attaining the maximal radius R_m of the precipitate one has to claim $c(R = R_m, r) = c_{0M}$ for $R_m < r \leqq R_E$, which is (see (4)) also realized in each case, since $a_n(R \to R_m) \to 0$.

C3) For the concentration at the precipitate–matrix interface, i.e. at $r = R$ we demand $c(r = R) = c_{0M}$ which is also fulfilled according to (4).

But unfortunately in the frame of our approximation we cannot possibly fulfil the condition that the concentration gradient (disregarding the particle–matrix interface) should be continuous with respect to r, because "jumps" of dc/dr happen at $r = Q$ as well as at $r = R_E$ (see Fig. 1).

We would like to discuss at first the concentration gradient at the boundary at the finite drainage area ($r = R_E$). Taking into account a finite drainage area we have to demand that the flux of the solute atoms must vanish at $r = R_E$, which can be easily fulfilled by claiming $(dc/dr)_{r=R_E} = 0$. But this condition cannot possibly be fulfilled in the frame of the stationary solution. One is only able to realize the condition C4:

C4) $(dc/dr)_{r=R_E-0} = -(dc/dr)_{r=R_E+0}$, from which the vanishing of the total flux through the boundary area at $r = R_D$ follows.

From the discontinuity of dc/dr at $r = Q$ that of the flux of the solute atoms, namely

$$J(r = Q - 0) = D_0(dc/dr)_{r=Q-0} \neq 0 ,$$

results, but

$$J(r = Q + 0) = D_0(dc/dr)_{r=Q+0} = 0 .$$

This discontinuity is the weak point of this method, which can hardly be overcome.

One may object that these two weak points mentioned in C4 turn the results delivered before to be more or less doubtful. Our point of view is the following. Of course the stationary solution of the diffusion equation is an approximation. The main advantage of this method is that it allows to solve the diffusion equation including the variation of the diffusion coefficient with the composition of the material. The results obtained by the help of this method, as far as solutions of other authors are available, fit fairly well those results (see, e.g., Fig. 2). Furthermore one is able to fulfil the mass balance and the conditions C1 till C3 without any difficulty strictly. Of course, the condition C4 is a compromise not to circumvent in the frame of the stationary solution. Therefore, we would finally like to state that the method of the stationary solution of the classical DE seems to be a proper tool for gaining information on the diffusion-controlled growth of precipitates in infinite as well as finite matrices. The results delivered in this work outline the possibilities of applying this method but show, on the other hand, the limits of this method.

References

[1] F. S. Ham, J. Phys. Chem. Solids **6**, 335 (1958).
[2] C. Zener, J. appl. Phys. **20**, 950 (1949).
[3] C. Wert and C. Zener, J. appl. Phys. **21**, 5 (1950).
[4] C. Wagner, Z. Elektrochem., Ber. Bunsenges. phys. Chem. **65**, 581 (1961).
[5] I. M. Lifshits and V. V. Slyozov, J. Phys. Chem. Solids **19**, 35 (1961).
[6] O. Simmich and H. Löffler, phys. stat. sol. (a) **60**, K43 (1980).
[7] O. Simmich and H. Löffler, Exper. Tech. Phys., in the press.
[8] A. Krupkowski, Rev. Métall. **59**, 34 (1962); Bull. Acad. Polon. Sci., Ser. Sci. Tech. **17**, 87 (1969).
[9] H. Löffler and O. Simmich, phys. stat. sol. (a) **30**, 495 (1975).
[10] W. Rogalla and H. S. Schmalzried, Ber. Bunsenges. phys. Chem. **72**, 615 (1968).
[11] O. Simmich and H. Löffler, phys. stat. sol. (a) **57**, 577 (1980).

(Received January 23, 1981)

phys. stat. sol. (a) **65**, 163 (1981)

Subject classification: 2 and 20; 22.6.1

Department of Physical Sciences, University of Turku[1])

Anomalous Soft X-Ray Reflectivity of Borosilicate Glass by Successive Grazing Reflections

By

L. KAIHOLA

Finite throughput will result in successive reflections of X-rays from a curved surface when the real and imaginary parts of the complex refractive index $n = 1 - \delta - i\beta$ are related to each other by $\beta/\delta < 1$ and the angle of incidence is less than the critical one. Strong attenuation of the transmitted intensity will be observed when the critical angle is exceeded. The feasibility of the phenomenon for X-ray dispersion studies is investigated with multiple reflections by borosilicate glass in the L resonance region of silicon. A soft X-ray pass band is found beyond the edges.

Eine errechenbare Ausbeute wird in sukzessiven Reflexionen von Röntgenstrahlen aus einer gekrümmten Oberfläche geliefert, wenn die Real- und Imaginärteile des komplexen Brechungsindex $n = 1 - \delta - i\beta$ im Verhältnis $\beta/\delta < 1$ zueinander stehen und der Einfallswinkel kleiner als der Grenzwinkel der Totalreflexion ist. Eine starke Dämpfung der transmittierten Intensität wird gefunden, wenn der Grenzwinkel überschritten wird. Die Brauchbarkeit des Phänomens für die Untersuchung der Dispersion der Röntgenstrahlung wird mit Vielfach-Reflexionen von Borsilikatglas im Gebiet der L-Resonanz des Siliziums untersucht. Ein Übertragungsbereich für weiche Röntgenstrahlen wird außerhalb der Absorptionskanten gefunden.

1. Introduction

At X-ray frequencies all real materials have their refractive indices slightly less than unity apart from the resonances. A total external reflection is then possible at grazing angles of incidence θ less than the critical one $\theta_c = \sqrt{2\delta}$. In the absence of currents or charges of external origin at the reflecting plane boundary and with the lack of magnetic effects at X-ray frequencies in normal conditions, the Fresnel coefficient representing a single reflection can be expressed as

$$R = 1 - \sqrt{2}\,\beta\delta^{-3/2}\theta = 1 - 2\,\frac{\beta}{\delta}\,\frac{\theta}{\theta_c}, \tag{1}$$

when $\beta/\delta < 1$ and $\theta \ll \theta_c$ [1]. At fixed θ fluctuations will be caused for the reflectivity in the energy regions where δ and β vary, e.g. in the vicinity of the characteristic frequencies of the reflecting material [2]. The optical "constants" δ and β are related to each other by the dispersion relations and their resonance behavior is thus exposed in the reflectivity. Very low reflectivity will be found when $\theta > \theta_c$. This is the case just on the low-energy side of the absorption edges in the fixed θ mode.

When X-rays are reflected repeatedly in a plane of incidence perpendicular to the axis of a circular cylinder, a deflection 2θ of the beam takes place each time. A limit is found for the multiple reflectivity R^N, when the grazing angle is decreased and the conditions of (1) are in effect

$$R^N = \left(1 - \sqrt{2}\,\beta\delta^{-3/2}\,\frac{\varphi}{2N}\right)^N \approx \exp\left(-\frac{\varphi}{\sqrt{2}}\,\beta\delta^{-3/2}\right), \tag{2}$$

[1]) SF-20500 Turku 50, Finland.

where φ is the total deflection after N encounters [1]. The expected reflectivity is of the order 10^{-3} at soft X-ray frequencies for carbon, polystyrene, silver, and titanium in $\varphi = 2\pi$ deflection. A soft X-ray pass band was verified on the low-energy side of C K edge in reflections from a cylindrical polystyrene surface [3].

Glass was chosen as the reflector material in this work. Glass tubing is known to be capable of propagating soft X-rays through curved paths above 17 nm [4]. Equation (2) applies also for a straight tubing with $\varphi = 2N\,|\theta|$ and $N = l\theta/d$, where l and d are tube length and diameter, respectively. The main component of glass, SiO_2 shows the $SiL_{2,3}$ structure both in absorption and reflectivity spectra beyond 9 nm [5 to 11] and from the measured values of the optical constants one would predict $R^N > 10^{-6}$ in 2π deflection by successive reflections.

2. Experimental

A soft X-ray spectrometer with an electron excited tungsten bremsstrahlung source and a 2 m gold coated glass grating with 600 lines/mm was used in the experiment to produce monochromatized radiation of 50 pm bandwidth. Higher-order interferences were prevented by a spherical polystyrene coated mirror placed at an angle 8.0° which sets the high-energy cut-off at 8.5 nm.

An 80° section of a straight circular cylinder of radius 13.8 mm acted as the soft X-ray propagator, the inside of which was polished borosilicate glass [12]. The grazing angle of incidence was 5.3° corresponding to seven repeated reflections from the mirror. The reflected radiation was measured with a flow proportional counter which had 0.4 μm polypropylene window and Ar–CH_4 filling in the atmospheric pressure.

3. Results and Discussion

Fig. 1 shows the multiple reflectivity of the borosilicate glass in which the percentages of SiO_2, B_2O_3, Na_2O, and Al_2O_3 are roughly 80, 10, 5, and 2, respectively. The resonance structure of silicon is clearly indicated which was found earlier by photoabsorption [5 to 7], reflectivity [5, 8 to 10], and X-ray photoemission [11] of pure quartz and silicate glasses. Analyses of the absorption spectra of the tetrahedral SiO_4^{4-} complexes through the construction of molecular orbitals are given in several works [7, 11, 13].

The reflectivity spectrum shows the resonance peaks shifted to $\lambda = 8.9$, 10.6, 11.3, and 11.6 nm from their respective positions $\lambda = 9.5$, 10.8, 11.4, and 11.7 nm in the absorption spectra [5]. The shift is understood through the asymmetry of δ with respect to β. The unit decrement $\delta(\lambda)$ resembles the positive derivative of $\beta(\lambda)$ via the Kramers-Kronig dispersion relation linking together the optical constants [14]. The

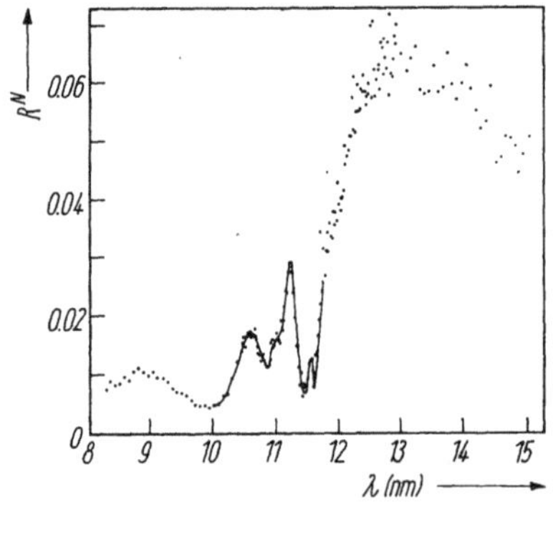

Fig. 1. Multiple reflectivity of soft X-rays by an 80° section of a cylindrical borosilicate glass mirror of radius 13.8 mm. The X-ray beam was deflected by seven successive reflections at the grazing angle of incidence of 5.3°.

reflectivity is not very sensitive to β changes at the high δ values now under discussion, viz. 0.01 to 0.06 in the range $\lambda = 10$ to 20 nm. The reflectivity variation is almost solely determined by δ fluctuation. The reflectivity peak is therefore observed close to the high-energy inflection point of the respective photoabsorption peak. A very sharp minimum is found at the low-energy inflection point of the 11.4 nm absorption peak which falls close to the $L_{2,3}$ threshold of silicon in SiO_2, $\lambda = 11.6$ nm [11] and the subsequent logarithmic resonance minimum of the continuum part of δ [2]. Also the reflectance peak arising from a transition from the Si 2p core state to highly localized s-type molecular orbitals is resolved in this experiment at 11.6 nm. Delayed onset of the transition from Si 2p states to the d-type continuum states is found at 9 nm in addition to the localized transitions at $\lambda = 10.6$ and 11.3 nm due to a potential barrier.

The multiple reflectivity of (2) depends exponentially on $\beta\delta^{-3/2}$ while a linear relationship is present in the single reflectivity formula. When $\theta > \theta_c$, e.g. just beyond the absorption edge, strong attenuation of the propagated beam results from the repeated reflections. An improved contrast over the single reflection experiment is therefore obtained in the current case.

Absorption data is not usually available through the entire frequency range as would be required for the Kramers-Kronig analysis to derive δ from β and further R^N. A limited interval of known photoabsorption is yet useful for such an analysis when approximations are made as to the course of absorption far from the edge region [2]. Photo-yield [11] and absorption data [6] were used for the analysis which produced multiple reflectivity well in accord with the observed shape of the reflectivity spectrum but roughly double in intensity. Rising multiple reflectivity towards longer wavelengths is expected theoretically beyond 13 nm. The decline of the observed reflectivity together with the loss of intensity is understood on the basis of roughness induced scattering which reduces R^N by $N\delta/2$ from the values given by equation (2) [1]. The prominent soft X-ray pass band in the borosilicate glass reflectivity spectrum above the Si $L_{2,3}$ threshold follows from the fast growth of δ from the logarithmic minimum at the threshold and low, although rising β in this region.

4. Conclusion

Enhancement of structures present in X-ray reflection spectra can be brought about by multiple reflections, which effectively raise the single reflectivity to the power indicated by the number of reflections.

Moderate intensities are found to be propagated along a cylindrical borosilicate glass surface in an 80° deflection of the soft X-rays. Larger deflection would still increase the contrast at the expense of intensity.

The curvature of the interface does not affect the multiple reflectivity. The curvature can therefore be very strong as illustrated by this experiment.

Glass tubing has been found to propagate soft X-rays above 17 nm in an earlier experiment [4]. Now it is shown spectroscopically that in addition to the resonance structure a broad pass band exists beyond the $L_{2,3}$ edge region of silicon. In this energy range glass possesses a quality of a good X-ray waveguide, a small ratio β/δ simultaneously with high δ and the critical angle of incidence. The conditions mean high single reflectivity at angles of incidence less than the critical one. The same requirements are also fulfilled beyond the C K threshold of low density carboneous materials.

Acknowledgements

The author is obliged to J. F. McGee of Saint Louis University, St. Louis for supplying the cylindrical mirror. The work was supported by the Academy of Finland.

References

[1] J. BREMER and L. KAIHOLA, Appl. Phys. Letters **37**, 360 (1980).
[2] J. BREMER, L. KAIHOLA, and R. KESKI-KUHA, J. Phys. C **13**, 2225 (1980).
[3] L. KAIHOLA, Appl. Optics **20**, 8 (1981).
[4] D. MOSHER and S. J. STEPHANAKIS, Appl. Phys. Letters **29**, 105 (1976).
[5] O. A. ERSHOV, D. A. GOGANOV, and A. P. LUKIRSKII, Fiz. tverd. Tela **7**, 2355 (1965); Soviet Phys. – Solid State **7**, 1903 (1966).
[6] F. C. BROWN, R. Z. BACHRACH, and M. SKIBOWSKI, Phys. Rev. B **15**, 4781 (1977).
[7] I. A. BRYTOV and YU. N. ROMASHCHENKO, Fiz. tverd. Tela **20**, 664 (1978); Soviet Phys. – Solid State **20**, 384 (1978).
[8] A. P. LUKIRSKII, E. P. SAVINOV, O. A. ERSHOV, I. I. ZHUKOVA, and V. A. FOMICHEV, Optika i Spektroskopiya **19**, 425 (1965); Soviet Phys. – Optics and Spectroscopy **19**, 237 (1965).
[9] O. A. ERSHOV and S. I. CHERNOVA, Optika i Spektroskopiya **26**, 597 (1969); Soviet Phys. – Optics and Spectroscopy **26**, 327 (1969).
[10] J. RIFE and J. OSANTOWSKI, J. nuclear Instrum. Methods **172**, 297 (1980).
[11] Y. IGUCHI, Sci. Light (Japan) **26**, 161 (1977).
[12] L. KAIHOLA and J. F. MCGEE, Adv. X-Ray Analysis **16**, 150 (1973).
[13] J. A. TOSSELL, J. Amer. Chem. Soc. **97**, 4840 (1975).
[14] F. STERN, Solid State Phys. **15**, 300 (1965).

(Received January 5, 1981)

phys. stat. sol. (a) 65, 167 (1981)

Subject classification: 13.4 and 14.3

Department of Physics, University of Hull[1])

A New Method for the Analysis of Thermally Stimulated Current Data

By

D. SHAW and H. R. MOGHADDAM

A general iterative procedure is described, with specific reference to TSC, for evaluating trap parameters in thermally excited transients when the heating rate varies during the transient. The method is not restricted to half-width parameters and gives the trap depth to an accuracy of $< 0.1\%$ after the second iteration step. An empirical method of analysis with an estimated accuracy in the trap depth of $<1\%$ is also described.

Es wird ein allgemeines Iterationsverfahren, insbesondere für TSC, zur Berechnung der Trap-Parameter bei thermisch angeregten Übergängen beschrieben, wenn sich die Heizgeschwindigkeit während des Übergangs ändert. Die Methode ist nicht auf die Halbwertsbreitenparameter beschränkt und liefert die Traptiefe mit einer Genauigkeit von $< 0{,}1\%$ nach dem zweiten Iterationsschritt. Eine empirische Analysenmethode mit einer berechneten Genauigkeit der Traptiefe von $< 1\%$ wird ebenfalls beschrieben.

1. Introduction

Wide use of thermally stimulated electronic transients such as thermally stimulated luminescence (TSL), current (TSC), and capacitance (TSCAP), has been made in the study of defects in phosphors and semiconductors. The behaviour of the transient is governed primarily by the emission rate, α, of a trapped carrier from a defect, which is expressed by

$$\alpha = BT^s \exp(-E/kT) , \tag{1}$$

where T is the absolute temperature, k Boltzmann's constant, B a temperature-independent characteristic of the defect, and $-2 \leqq s \leqq 2$ [1]. The purpose of thermally stimulated measurements is to obtain values for B and E assuming a value for s. The literature abounds with methods of evaluating these two parameters from experimental data and all are based on various approximate solutions to the rate equations describing the transient. The error in the approximation varies considerably between the different methods and Chen and Winer [2]were the first to make a systematic study of the accuracy to which the trap parameters could be determined. They did this by considering hypothetical cases (i.e. specific values of B, s, and E) and calculating exactly the form of the transient by numerical methods. The transient was then analysed by one of the approximate methods. Attention was restricted to those methods based on the peak width, or half-width, at half-maximum amplitude. The heating rate, β, was also taken to be constant throughout the transient. More recently Kivits and Hagebeuk [1] have carried out a comprehensive assessment of the accuracies attainable by all the approximate methods in the literature. Their procedure followed that of Chen and Winer [2] and they concluded that the methods based on the effect of β (held constant during a transient) on the transient, e.g. the shift in the temperature at which the transient peaks, T_m, with β, are the most reliable

[1]) Hull HU6 7RX, Great Britain.

and can give E to an accuracy of $<1\%$ and typically of $\approx 0.1\%$. It is highly likely that in these circumstances the uncertainty in E will be limited now by experimental error. Nonetheless these assessments of approximation errors can only give estimates in a given instance and it would be preferable if the error in a specific case could be established more precisely. Balarin [3, 4] has described a general procedure whereby E can be evaluated to any desired degree of accuracy from a single transient response curve obtained at constant β. The method utilises the half-width parameters with E being obtained from a polynomial in $(T_2 - T_1)/T_m$ where T_1 and T_2 are the half-amplitude temperatures on the rise and decay of the transient, respectively. The polynomials converge rapidly so that E-values accurate to $\lesssim 0.5\%$ are easily obtained. Lushnikov et al. [5] have drawn attention to the effect of a non-constant β during a transient and have given semi-empirical expressions which enable E to be found from the half-amplitude parameters and a known variation of β. These expressions are based on $s = 0$ (equation (1)) and an accuracy of $<1\%$ in E is claimed.

This paper describes a procedure for evaluating to any desired accuracy the emission parameters of a trap from a single transient peak for any value of s and for a known variation of β during the transient. The procedure is discussed first for a TSC peak arising in a semiconductor junction depletion layer.

2. TSC in a Semiconductor Junction Depletion Layer

2.1 Basic equations and the first approximation

If N is the uniform concentration of occupied traps of a given species in a semiconductor junction depletion layer which has an area A and width L the TSC, i, parallel to L is [6]

$$i = \tfrac{1}{2} qLA\alpha N , \tag{2}$$

where q is the electron charge and α is the thermal emission rate of carriers from occupied traps given by (1). For negligible re-trapping the concentration of occupied traps decays according to

$$\frac{dN}{dt} = -\alpha N , \tag{3}$$

where t denotes time. If $N = N_0$ at $t = 0$ then integration of (3) with substitution into (2) yields

$$i = \tfrac{1}{2} qLAN_0\alpha \exp\left(-\int_0^t \alpha \, dt\right). \tag{4}$$

Setting $\theta = (T - T_m)/T_m$ and assuming that the heating rate is represented by[2])

$$\frac{dT}{dt} \equiv \beta = \beta_m(1 + \gamma\theta) , \tag{5}$$

where β_m is the heating rate at $\theta = 0$ ($T = T_m$) and γ is a constant, enables (4) to be recast as

$$i = \tfrac{1}{2} qLAN_0\alpha \exp\left[-(s + a) I(\theta_0, \theta)\right] , \tag{6}$$

where

$$I(\theta_0, \theta) = \int_{\theta_0}^{\theta} \frac{(1 + \theta')^s}{(1 + \gamma\theta')} \exp\left(\frac{a\theta'}{1 + \theta'}\right) d\theta' ,$$

[2]) It is easily shown that (5) leads to $1 + \gamma\theta = \exp\left[\gamma\beta_m(t - t_m)/T_m\right]$ in which t_m is the time when i has its maximum value i_m. This form of dependence is the same as that used in [5].

$a = E/kT_m$ and $\theta = \theta_0$ at $t = 0$. The TSC maximum occurs when $di/dt = 0$. From (1) to (3) this condition is readily shown to lead to

$$\frac{\alpha_m T_m}{\beta_m} = s + a\,, \tag{7}$$

where α_m is the emission rate at $\theta = 0$. Equation (7) has been used in the derivation of (6). Equation (1) can also be rewritten as

$$\alpha = \alpha_m(1 + \theta)^s \exp\,[a\theta/(1 + \theta)] \tag{8}$$

and with substitution into (6) leads to

$$\ln i/i_m = a\theta/(1 + \theta) + s \ln (1 + \theta) - (a + s)\, I(0, \theta)\,. \tag{9}$$

To obtain E it is necessary to find the value of a which satisfies (9) when i/i_m, θ, and γ are experimentally defined: a value for s must also be assumed. The first step is to obtain a first approximation, a_1, to a by noting that generally $-0.1 \leqq \theta \leqq 0.1$, $a \gtrsim 20 \gg s$, and if $\gamma \approx 1$ we can neglect θ, θ', $\gamma\theta'$ and s/a relative to unity in (9). The integrand of $I(0, \theta)$ simplifies to $\exp\,(a_1\theta')$ so that (9) becomes

$$\ln i/i_m = 1 + a_1\theta - \exp\,(a_1\theta)\,. \tag{10}$$

Equation (10) shows that i/i_m is a universal function of $a_1\theta$ and that, for a given value of i/i_m, there are two corresponding values of θ one positive and the other negative. Haering and Adams [7] obtained essentially the equivalent to (10) in their equation (2.7). Clearly $a \neq a_1$ and if we set $a = a_1 + \Delta$ where Δ can be regarded as a correction term we shall be concerned with determining Δ. It should be noted that a_1 is independent of both s and γ so that Δ must include the effects of both of these parameters.

The influence of γ is illustrated in Table 1 which gives some values of Δ for $a = 20$, 40, $\gamma = 0$, 1, 2, and $-0.09 \leqq \theta \leqq 0.1$. i/i_m was obtained from (9) for given a, θ, γ using $s = 2$ [8] which then enables a_1 to be found from (10) so yielding $\Delta = a - a_1$. Table 1 clearly shows that Δ is a significant correction with Δ depending on both θ and γ: increasing γ causes $|\Delta|$ to decrease. $\gamma = 1$ represents a 20% variation in β over

Table 1

Values of Δ for various a, γ, θ

θ	$a = 20$, γ			$a = 40$, γ		
	0	1	2	0	1	2
0.10	−1.270	−0.633	−0.053	—	—	—
0.08	−1.588	−0.987	−0.419	—	—	—
0.06	−1.912	−1.334	−0.780	—	—	—
0.04	−2.243	−1.694	−1.155	−1.951	−1.007	−0.410
0.03	−2.411	−1.878	−1.349	−2.117	−1.365	−0.790
0.02	−2.580	−2.063	−1.544	−2.285	−1.728	−1.177
0.01	−2.760	−2.252	−1.731	−2.621	−2.095	−1.570
−0.01	−3.117	−2.645	−2.160	−3.315	−2.856	−2.389
−0.02	−3.300	−2.850	−2.380	−3.671	−3.247	−2.810
−0.03	−3.487	−3.053	−2.600	−4.038	−3.648	−3.241
−0.05	−3.876	−3.480	−3.053	−4.805	−4.478	−4.131
−0.07	−4.286	−3.924	−3.527	−5.627	−5.353	−5.060
−0.09	−4.719	−4.389	−4.023	−6.506	−6.252	−6.029
−0.10	−4.944	−4.631	−4.279	—	—	—

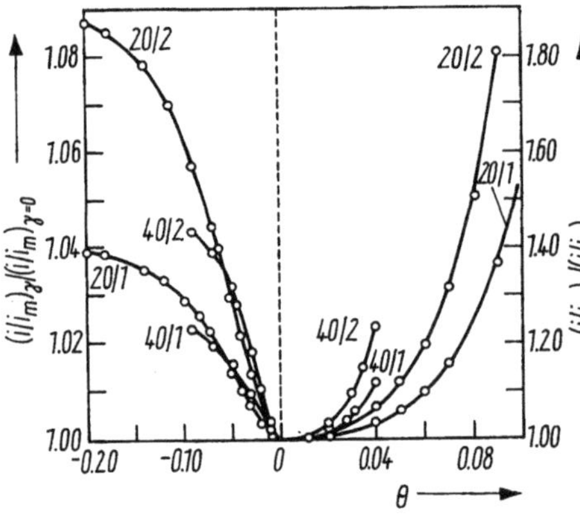

Fig. 1. The variation of $(i/i_m)_\gamma : (i/i_m)_{\gamma=0}$ with θ for $\gamma = 1, 2$, $a = 20, 40$, and $s = 2$ according to equation (9). The labels against each curve denote the values of a/γ

the interval $-0.1 \leqq \theta \leqq 0.1$. For $20 \leqq a \leqq 40$ the data in Table 1 show that neglect of γ up to $\gamma = 2$ could introduce an error in E as large as -6% depending on the value of θ. Fig. 1 shows how the ratios $(i/i_m)_{\gamma=1} : (i/i_m)_{\gamma=0}$ and $(i/i_m)_{\gamma=2} : (i/i_m)_{\gamma=0}$, computed from (9) with $s = 2$, vary with θ for $a = 20$ and 40. The curves show that if β increases during the transient the peak is broadened, relative to the $\gamma = 0$ peak, on both the rise and the decay but is more marked on the decay.

2.2 *Evaluation of Δ using Newton's iteration method*

If we denote the right-hand side of (9) by the function F then

$$\ln i/i_m = F(a) = F(a_1 + \Delta) \tag{11}$$

and by applying Newton's method of approximation

$$\ln i/i_m = F(a_1) + F'(a_1)\, \Delta_1$$

which yields

$$\Delta_1 = [\ln i/i_m - F(a_1)]/F'(a_1)\,, \tag{12}$$

where Δ_1 is the first approximation to Δ and $F'(a_1) = \partial F(a)/\partial a$ evaluated at $a = a_1$. The second approximation to a is therefore $a_2 = a_1 + \Delta_1$ and clearly the $(j + 1)$-th approximation to a is

$$a_{j+1} = a_j + \Delta_j\,, \tag{13}$$

where

$$\Delta_j = [\ln i/i_m - F(a_j)]/F'(a_i)\,.$$

It follows directly from (9) that

$$F'(a_j) = \frac{\theta}{1+\theta} - \int\limits_0^\theta \frac{(1+\theta')^{s-1}}{(1+\gamma\theta')} [1 + (1 + s + a_j)\,\theta'] \exp\left(\frac{a_j\theta'}{1+\theta'}\right) \mathrm{d}\theta'. \tag{14}$$

Finally we see from (13) that

$$a = a_1 + \sum_{j=1}^{\infty} \Delta_j \tag{15}$$

as $\Delta = \sum\limits_{j=1}^{\infty} \Delta_j$. The rapid convergence of this iterative process is illustrated in Table 2 which shows a_1, a_2, and a_3 for various values of a, i/i_m, and θ and with $s = 2$, $\gamma = 0$.

Equation (9) was used to compute i/i_m and a_1 was obtained from (10). The tabulated data show that a_2 is within 1% of the true value and that the difference between a and a_3 is considerably less than 0.1%. The iterative method described above for determining Δ is a simple and generally valid procedure which easily deals with any values of s and γ.

Table 2

a	i/i_m	θ	a_1	a_2	a_3
20	0.01082	−0.2	27.612	20.023	20.001
	0.9755	−0.01	23.117	20.166	20.002
	0.9724	0.01	22.760	20.170	20.005
	0.01924	0.09	21.428	20.094	20.000
40	0.1070	−0.07	45.627	40.040	40.001
	0.9001	0.01	42.621	40.093	40.000
	0.07282	0.04	41.618	40.060	40.000

2.3 Empirical determination of Δ

It is found that if Δ is plotted against θ for constant values of a, s, and γ the dependence can be very well approximated by two straight lines:

$$\theta > 0; \qquad \Delta = \Delta_+ + g_+\theta_+ \,, \tag{16a}$$

$$\theta < 0; \qquad \Delta = \Delta_- + g_-\theta_- \,, \tag{16b}$$

where Δ_+ and Δ_- are the intercepts and g_+ and g_- are the gradients. The near linearity of the Δ-versus-θ relation is illustrated in Fig. 2 with Δ being computed from (9) and (10). Table 3 gives the values of g_+, g_-, Δ_+ and Δ_- obtained by the least-squares fitting of (16a) and (16b) to Δ, θ data with $s = 2$ and a range of a-values. For $a = 20$, 25, 30 the fits were made over the range $-0.06 \leqq \theta \leqq +0.06$ and for $a = 40$, $-0.03 \leqq \leqq \theta \leqq +0.035$. Several important features are evident in Table 3:

(i) at constant γ, Δ_+ and Δ_- differ by less than $\approx 3\%$ for $20 \leqq a \leqq 40$;
(ii) Δ_+ and Δ_- both show a substantial and nearly linear dependence on γ;
(iii) g_+ and g_- both show strong variations with a and γ and at constant γ the variation with a is nearly linear.

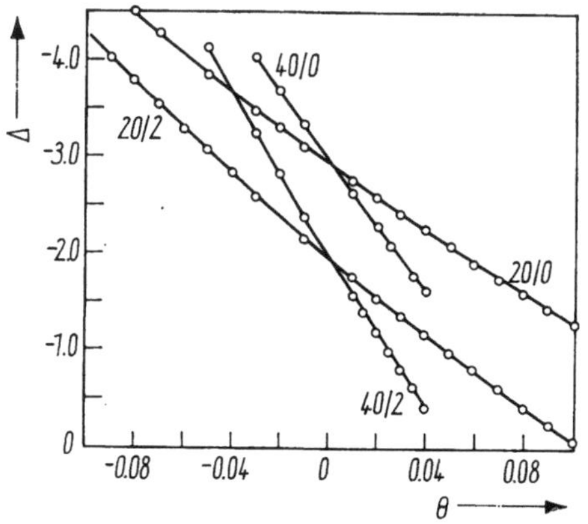

Fig. 2. The dependence of Δ on θ from equations (9) and (10) for $\gamma = 0$, 2, $a = 20$, 40, and $s = 2$. The labels against each line denote the values of a/γ

Table 3

γ	a	g_+	g_-	Δ_+	Δ_-	Δ_0
0	20	16.91	19.29	−2.922	−2.915	
	25	20.94	23.88	−2.933	−2.923	−2.935
	30	24.96	28.54	−2.938	−2.923	±0.020
	40	33.47	36.15	−2.955	−2.952	
1	20	18.35	21.08	−2.431	−2.428	
	25	22.71	26.05	−2.439	−2.434	−2.443
	30	27.12	31.05	−2.445	−2.434	±0.015
	40	36.42	39.60	−2.458	−2.458	
2	20	18.81	22.52	−1.907	−1.929	
	25	23.99	27.88	−1.938	−1.937	−1.934
	30	28.93	33.33	−1.946	−1.935	±0.027
	40	38.89	42.60	−1.957	−1.961	

Because of (i) we have introduced Δ_0 into Table 3 as the average of the Δ_+ and Δ_- values at constant γ and the error in so doing is $<1.5\%$. This enables (16a) and (16b) to be rewritten as

$$\Delta = \Delta_0 + g_\pm \theta_\pm \tag{17}$$

to within the limits of Δ_0 shown in Table 3. We then obtain

$$a = a_1 + \Delta = a_1 + \Delta_0 + g_\pm \theta_\pm \tag{18}$$

which permits a ($s = 2$) to be found from an experimental input of i/i_m, γ, and $\theta_\pm : a_1$ is obtained from (10); Δ_0 is found from Table 3 using linear interpolation and the measured γ. Both a and γ are needed for $g_\pm$ but as only γ is known the best compromise is to use $(a_1 + \Delta_0)$ as the a-value in order to obtain $g_\pm$ from Table 3 by interpolation. Some results to illustrate this procedure are given in Table 4. Equations (9) and (10) were again used to determine i/i_m, $\theta_\pm$, and a_1. The final column in this table gives the value of a obtained by means of (18). It will be seen that most of the $\theta_\pm$ values shown lie well outside the ranges used to derive the $\Delta_\pm$ and $g_\pm$ values. Even so (18) provides values for a which are within $\approx 1\%$ of the true value.

Table 4

	γ	$\theta_\pm$	i/i_m	a_1	Δ_0	$g_\pm$	a (equation (18))
$a = 40$	0	−0.09	0.04073	46.51	−2.935	38.87	40.07
$s = 2$	1	−0.09	0.04166	46.25	−2.443	42.86	39.95
	1	0.04	0.08076	41.01	−2.443	35.09	39.97
	2	−0.09	0.04249	46.03	−1.934	46.40	39.92
$a = 20$	0	−0.18	0.02086	27.01	−2.935	23.03	19.93
$s = 2$	0	0.05	0.4018	22.08	−2.935	16.22	19.95
	0	0.10	0.00518	21.27	−2.935	15.57	19.89
	2	−0.18	0.02264	26.55	−1.934	26.83	19.79
	2	−0.10	0.2196	24.28	−1.934	25.03	19.84
	2	−0.05	0.6260	23.05	−1.934	23.72	19.93
	2	0.05	0.4473	20.97	−1.934	21.48	20.11
	2	0.10	0.01200	20.05	−1.934	21.14	20.23

As it stands equation (18), in common with the procedure of Section 2.2, can provide a value for a from a single experimental value of i/i_m at a known θ. If several data points $(i/i_m, \theta)$ are available, equation (18) can be developed so that a can be obtained with an improved accuracy provided each θ_+ is matched or paired with a θ_- datum. We rewrite (18) as

$$a\theta_+ = (a_1\theta_+) + \Delta_0\theta_+ + g_+\theta_+^2 \tag{19a}$$

and

$$a\theta_- = (a_1\theta_-) + \Delta_0\theta_- + g_-\theta_-^2 \,. \tag{19b}$$

Summing separately over the θ_+ and θ_- data then gives

$$a \sum \theta_+ = \sum (a_1\theta_+) + \Delta_0 \sum \theta_+ + g_+ \sum \theta_+^2 \tag{20a}$$

and

$$a \sum \theta_- = \sum (a_1\theta_-) + \Delta_0 \sum \theta_- + g_- \sum \theta_-^2 \,, \tag{20b}$$

respectively. It will be recalled that $(a_1\theta_\pm)$ is determined from (10) using the experimental value of i/i_m. As each θ_+ datum is paired with a θ_- datum we can subtract (20b) from equation (20a) to obtain

$$a = \frac{\sum [(a_1\theta_+) - (a_1\theta_-)]}{\sum (\theta_+ - \theta_-)} + \Delta_0 + \frac{g_+ \sum \theta_+^2 - g_- \sum \theta_-^2}{\sum (\theta_+ - \theta_-)} \,. \tag{21}$$

The improvement in accuracy is due to the opposition of the two quantities $g_+ \sum \theta_+^2$ and $g_- \sum \theta_-^2$ in the third term on the right-hand side of (21). The consequence is that the accuracies of g_+ and g_- are then less important. Table 5 gives the results of the application of (21) to some of the data in Table 4 taken in single pairs and the final column shows that values for a differing by less than 0.5% from the true value can be obtained. A further advantage in practice of this pairing procedure based on (21) is the appearance of the term $(\theta_+ - \theta_-)$ which is considerably less sensitive to any experimental error in T_m than are the separate θ_+ and θ_- values. This feature has most effect on the dominant first term on the right-hand side of (21).

Table 5

a	γ	θ_-; θ_+	a (equation (21))
40	1	−0.09; 0.04	39.96
20	2	−0.05; 0.05	20.02
		−0.10; 0.10	20.04
		−0.18; 0.10	19.95
	0	−0.18; 0.10	19.92

2.4 *Half-width parameters*

The evaluation of the trap parameters using half-maximum amplitude parameters has received wide attention in the literature and it is appropriate therefore to consider the effect of γ on the half-width parameters defined for $i/i_m = \frac{1}{2}$ when $\theta_- = \theta_1$ and $\theta_+ = \theta_2$. In some respects the use of $(\theta_2 - \theta_1)$ to evaluate E can be regarded as a special case of the pairing method described in Section 2.3. Table 6 summarises values of $a(\theta_2 - \theta_1)$ and $-\theta_2/\theta_1$ computed from (9) for $s = 2$ and a representative range of a- and γ-values. The bracketed figures in the Table were obtained from Balarin's [4] polynomial expressions for $a(\theta_2 - \theta_1)$ and θ_2/θ_1 $(s = 2)$ summed as far as the cubic term in $1/a$. The data in Table 6 clearly demonstrate the relative insensitivity of $a(\theta_2 - \theta_1)$ and θ_2/θ_1 to both a and γ: $a(\theta_2 - \theta_1)$ changes by $<5\%$ as γ varies between

0 and 2 and the corresponding changes in θ_2/θ_1 are $<2\%$. The heating rate increases by 50% over the interval $-0.1 \leqq \theta \leqq +0.1$ when $\gamma = 2$. It is also apparent from the table that $(\theta_2 - \theta_1)$ increases with increasing γ, a feature noted earlier in Fig. 1. Balarin [4] pointed out that $\theta_2 \approx -0.75\theta_1$ for $\gamma = 0$ and Table 6 shows that this relationship is unaltered when $0 \leqq \gamma \leqq 2$. The effect of reducing s to zero (with $\gamma = 0$) is to cause the associated $a(\theta_2 - \theta_1)$ values to fall by $\approx 5\%$ [4]. This is of a similar magnitude and direction to the effect of γ.

Table 6

a	$a(\theta_2 - \theta_1)$ γ			$-\theta_2/\theta_1$ γ		
	0	1	2	0	1	2
20	2.102 (2.097)	2.144	2.189	0.7328 (0.7213)	0.7412	0.7488
25	2.161 (2.158)	2.197	2.234	0.7225 (0.7175)	0.7291	0.7350
30	2.203 (2.200)	2.234	2.266	0.7153 (0.7134)	0.7206	0.7255
40	2.258 (2.256)	2.282	2.307	0.7060 (0.7064)	0.7098	0.7134

2.5 Evaluation of B and N_0

Once a is known α_m can be obtained from (7), using an appropriate value for s and the experimental values for β_m and T_m. B can then be found from (1).

Setting $\theta = 0$ in (6) gives

$$i_m = \tfrac{1}{2}\, qLAN_0\alpha_m \exp\left[-(s + a)\, I(\theta_0, 0)\right] \tag{22}$$

from which N_0 can be found. Computation of $I(\theta_0, 0)$ shows that $\theta_0 = -0.3$ at $a = 20$ and $\theta_0 = -0.2$ at $a = 40$ are sufficient to give N_0 accurate to four significant figures. It is also found that $\exp[-(s + a)\, I(\theta_0, 0)]$ is quite insensitive to a, s, and γ ($20 \leqq \leqq a \leqq 40$, $0 \leqq s \leqq 2$, $0 \leqq \gamma \leqq 2$) with the exponent ≈ -1. Some representative values of $\exp\{-(s + a)\, I(\theta_0, 0)\}$ are given in Table 7.

Table 7
Some representative values of $\exp[-(s + a)\, I(\theta_0, 0)]$

a	θ_0	s	$\gamma = 0$	$\gamma = 2$
20	−0.3	0	0.4015	0.3661
		2	0.3975	0.3649
40	−0.20	0	0.3854	0.3674
		2	0.3843	0.3671

3. TSC Analysis in an Electrically Neutral Semiconductor

The thermally stimulated electrical conductivity, σ, in a homogeneous electrically neutral semiconductor sample is expressed by

$$\sigma = q\mu n\,, \tag{23}$$

where μ is the free-carrier mobility and n the thermally stimulated free-carrier concentration. If re-trapping is negligible then for mono-molecular recombination kinetics

$$n = -\tau \frac{\mathrm{d}N}{\mathrm{d}t} = -\alpha N\,, \tag{24}$$

where τ is the free-carrier lifetime [9]. Integration of (24) followed by substitution into (23) yields

$$\sigma = q\mu\tau N_0\alpha \exp\left(-\int_0^t \alpha\,\mathrm{d}t\right). \tag{25}$$

This is very similar in form to (4) except that there are now additional temperature dependences from μ and τ. μ varies as $T^{3/2}$, $T^{-3/2}$, and $\approx T^{-2}$ for ionised impurity, acoustic mode, and combined acoustic/optical mode scattering, respectively [10]. Assuming that τ can be described by a similar power law in T the product

$$\mu\tau = \mu_0\tau_0 T^z\,, \tag{26}$$

where z represents both the μ and τ temperature variation. Substituting (26) into (25) and introducing θ gives

$$\sigma = q\mu_0\tau_0N_0\alpha_\mathrm{m}(1+\theta)^{s+z} \exp\left[\frac{a\theta}{1+\theta} - (a+s+z)\,I(\theta_0, \theta)\right] \tag{27}$$

from which

$$\sigma_\mathrm{m} = q\mu_0\tau_0N_0\alpha_\mathrm{m} \exp\left[-(a+s+z)\,I(\theta_0, 0)\right] \tag{28}$$

and therefore

$$\ln\frac{\sigma}{\sigma_\mathrm{m}} = \ln\frac{i}{i_\mathrm{m}} = \frac{a\theta}{1+\theta} + (s+z)\ln(1+\theta) - (a+s+z)\,I(0, \theta)\,. \tag{29}$$

The condition for the peak maximum now becomes

$$\frac{\alpha_\mathrm{m}T_\mathrm{m}}{B_\mathrm{m}} = a + s + z\,. \tag{30}$$

The procedure described in Section 2.2 is still applicable for the determination of a. As before a_1 is found from (10) but in order to calculate the correction terms Δ_j in (13) we must set $F(a_j)$ equal to the right-hand side of (29) and so obtain

$$F'(a_j) = \frac{\theta}{1+\theta} - \int_0^\theta \frac{1+(1+a+s+z)\,\theta'}{1+\gamma\theta'}\,(1+\theta')^s \exp\left(\frac{a\theta'}{1+\theta'}\right)\mathrm{d}\theta'\,. \tag{31}$$

Table 8

z	θ	i/i_m	a_1	a_2	a_3
3/2	0.07	0.02419	26.980	25.242	25.016
	0.01	0.9564	28.436	25.279	25.012
	−0.01	0.9626	28.950	25.246	25.004
	−0.15	0.01832	33.286	28.232	24.991
−2	0.07	0.03795	25.776	25.049	25.002
	0.01	0.9615	26.760	25.070	25.004
	−0.01	0.9669	27.094	25.060	25.000
	−0.15	0.02869	30.271	24.984	25.000

The effect of z and the use of the iterative procedure of Section 2.2 is shown in Table 8 for $a = 25$, $s = 2$, $\gamma = 0$. The i/i_m $(= \sigma/\sigma_m)$ values were computed from (29). The tabled data confirm the effectiveness and rapidity of the iteration procedure for determining a and indicate that z has a role similar in magnitude to that of γ. Once a result for a has been obtained, α_m can be got from (30) and hence B. To obtain N_0 from (28) $\mu_0\tau_0$ as well as z must be known.

4. Conclusions

A general iterative method has been described for evaluating B and E (equation (1)) from a single TSC peak and requires a minimum of one datum point $(i/i_m, \theta)$ taken from any part of the TSC peak. The accuracy in E/T_m is $<0.1\%$ after the second iteration (a_3). The method will deal with a time-dependent heating rate and also any value of s (equation (1)) or z (equation (26)). An empirical method is also described for $s = 2$ and a variable heating rate based on paired θ_+ and θ_- data points, which is conservatively estimated to give E/T_m to an accuracy of $<1\%$ and is particularly suited to cases where there is a large experimental uncertainty in T_m.

The approach described in Sections 2.1, 2.2, and 2.3 could be applied directly to the analysis of glow curves (TSL) where re-trapping is negligible and mono-molecular kinetics obtain [11] as well as to other thermal transients obeying an equation of the same form as (4) or (25).

Acknowledgement

We wish to thank T. L. Shaw for considerable computational assistance.

References

[1] P. KIVITS and H. J. L. HAGEBEUK, J. Lum. **15**, 1 (1977).
[2] R. CHEN and S. WINER, J. appl. Phys. **41**, 5227 (1970).
[3] M. BALARIN, J. thermal Analysis **17**, 319 (1979).
[4] M. BALARIN, Solid State Commun. **34**, 419 (1980).
[5] N. A. LUSHNIKOV, G. A. ZHDAN, and V. B. SANDOMIRSKII, Soviet Phys. – Semicond. **13**, 247 (1979).
[6] C. T. SAH, L. FORBES, L. L. ROSIER, and A. F. TASCH, JR., Solid State Electronics **13**, 759 (1970).
[7] R. R. HAERING and E. N. ADAMS, Phys. Rev. **117**, 451 (1960).
[8] C. T. SAH, L. FORBES, L. L. ROSIER, A. F. TASCH, JR., and A. B. TOLE, Appl. Phys. Letters **15**, 145 (1969).
[9] K. H. NICHOLAS and J. WOODS, Brit. J. appl. Phys. **15**, 783 (1964).
[10] R. J. ELLIOT and A. F. GIBSON, An Introduction to Solid State Physics, MacMillan Press, Ltd., London 1974 (p. 303).
[11] W. HOOGENSTRAATEN, Philips Res. Rep. **13**, 515 (1958).

(Received December 8, 1980)

phys. stat. sol. (a) 65, 177 (1981)

Subject classification: 14.4; 14.3.3; 22

Institute of Semiconductor Physics, Academy of Sciences of the USSR, Siberian Branch, Novosibirsk[1])

Study of Thin Silicon Nitride Films by Isothermal Depolarization of MNOS Structures

By

V. M. EFIMOV

Electrophysical properties of thin films of silicon nitride obtained by various techniques are investigated by the method of isothermal depolarization of MNOS structures. For Si_3N_4 grown in a hydrogen flow the obtained estimate of trap depth amounts to 1.6 to 2 eV, for Si_3N_4 obtained in a low-pressure reactor about 1.2 eV. It is shown that the difference in electron spectra of these samples may be due to the strong influence of hydrogen on the trap spectrum in Si_3N_4. A direct comparison of the method of isothermal depolarization and the conduction method is made to estimate the parameters of electrical conductance of silicon nitride layers.

Методом изотермической деполяризации МНОП-структур исследованы электрофизические свойства тонких пленок нитрида кремния полученного разными методами. Для Si_3N_4, выращенного в потоке водорода, полученная оценка для глубины залегания ловушек составляет 1,6 до 2 eV. Для нитрида кремния, полученного в реакторе пониженного давления — около 1,2 eV. Показано, что разница в электронном спектре для этих образцов может быть связана с сильным влиянием водорода на спектр ловушек в Si_3N_4. Проводится прямое сравнение ИТД-методики и методики сквозной проводимости для оценки параметров электропроводности слоев нитрида кремния.

1. Introduction

In memory cells based on the metal–silicon nitride–silica–silicon (MNOS) structures Si_3N_4 serves as a functional element. The dielectric volume is used here to record an information that has been stored for a long time in the form of trapped charge [1]. The storage time is mostly determined by the depth of trapping centres and may be estimated only by using the results of accelerated tests. These cases require a knowledge of the band structure of the dielectric and, particularly, the principal characteristics of trapping centres: depth, trapping cross-section, and concentration. The cross-section and concentration of deep traps in silicon nitride are 10^{-12} cm^2 and 10^{-19} cm^{-3}, respectively [15, 16]. The values of the depth of trapping centres presented by various authors [2 to 14] range from 0.5 to 2.5 eV (Fig. 1). [2 to 8] give an interpretation of experimental data on the basis of a one-level model, [9 to 11] on the basis of a set of levels, in [12, 13] a quasi-continuous spectrum of states is assumed to exist in the forbidden band of silicon nitride. On the one hand, the divergence is due to the sensitivity of Si_3N_4 properties to the processing technology. It has been shown in [4, 6] that electrophysical properties of silicon nitride films are strongly dependent on the SiH_4/NH_3 ratio at which the films are grown. However, as can be seen in Fig. 1, this ratio does not determine unambiguously the trapping spectrum of Si_3N_4. A more essential cause of the divergence of the above data is likely to be the complication of the subject of research and, consequently, the complication of interpretation of experimental data received by conventional methods. In this concern, the use of

[1]) Prospekt Nauki 13, 630090 Novosibirsk, USSR.

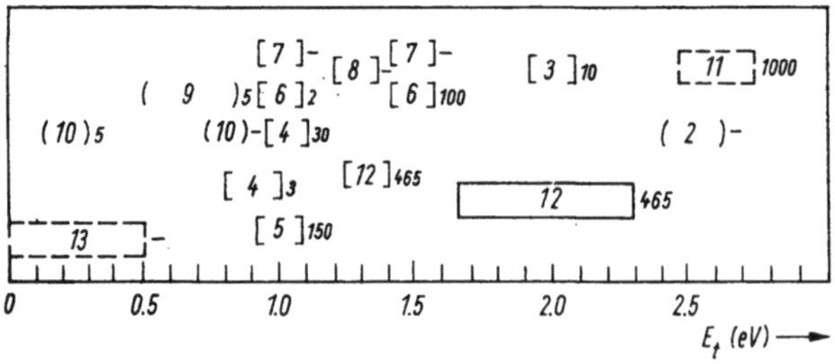

Fig. 1. Trap depth in silicon nitride according to various authors. In the brackets is the number of reference; to the right of the brackets is the ratio NH_3/SiH_4 (for [8, 13], $NH_3/SiCl_4$) at which the film has been grown. Brackets dc data, dashed rectangle optical data, parantheses TSC measurements, rectangle ITD measurements

new techniques to study a spectrum of traps in Si_3N_4 is of great importance. [12] proposed a new experimental technique to determine a trap spectrum by using isothermal depolarization (ITD) of MNOS structures with a thin SiO_2 sublayer. However, our studies show that Lundkvist's model well describes the experimental results only in a few cases. The errors arising in using this technique are due, as we think, to the influence of SiO_2–Si_3N_4 interface states and of the finite SiO_2 sublayer width on the charge transfer from nitride to silicon. In [14] we proposed a technique for determining the parameters of deep trapping centres (≈ 1 eV) that excludes the influence of these factors. That paper presented the preliminary results of investigation of Si_3N_4 films by this technique.

This paper presents an analysis of the method of isothermal depolarization of MNOS structures with a relatively thick sublayer of SiO_2 (≈ 100 Å) to determine the parameters of deep trapping centres in Si_3N_4 and experimental results on the investigation of silicon nitride films obtained by various techniques. Along with the investigations by ITD, we made measurements of current–voltage characteristics in reference structures with a thin tunnel SiO_2 sublayer to compare directly the two techniques.

Results are presented for measurements on samples that have been annealed at increased temperatures in vacuum and hydrogen atmosphere.

2. The Model of Discharge

As the simplest model of discharge we shall consider the process of thermal escape of electrons with concentration N_t, trapping cross-section S_t, and depth E_t. Assume that at the initial moment electrons are captured by traps in the form $n_0(x) = n_0 \exp(-x/L)$. This condition is realized in MNOS structures at small quantities of the charge trapped in the volume of a dielectric [15]. In the same case one can achieve the fulfilment of conditions if the concentration of electrons on traps, n_t, is several times smaller than the total concentration of trapping centres, i.e. the following condition is met:

$$n_t \ll N_t \, .$$

Thus, in the first assumption we shall consider that the form of the function of spatial carrier distribution is not varied with time and after thermal escape from traps electrons are intercepted after a typical time $\tau_t = (S_t v N_t)^{-1}$ or leave the dielectric volume after a typical time $\tau_d = (d_N/\mu \mathcal{E})$ (Fig. 2), where v and μ are the thermal velocity and mobility of the electrons, respectively. Then the equations describing

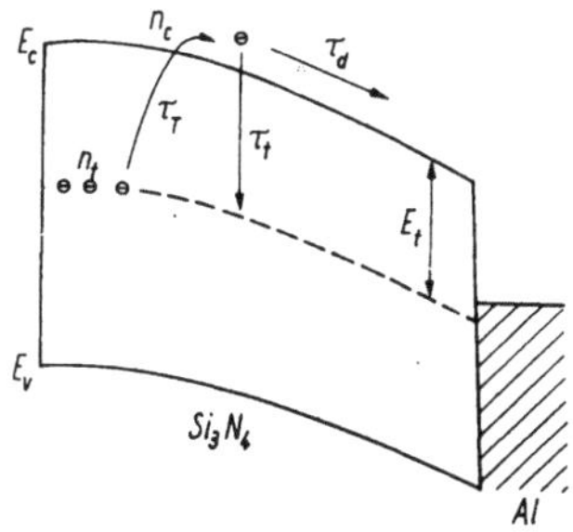

Fig. 2. The emptying of the traps in the volume of silicon nitride

the process of discharge are written as

$$\frac{dn_t}{dt} = \frac{n_c}{\tau_t}\left(1 - \frac{n_t}{N_t}\right) - \frac{n_t}{\tau_T}, \tag{1}$$

$$\frac{dn_c}{dt} = -\frac{dn_t}{dt} - \frac{n_c}{\tau_d}, \tag{2}$$

where n_c is the electron concentration in the conduction band, $\tau_T = \tau_1 \exp(E_t/kT)$ is the typical time of thermal excitation from the trapping centres, τ_1 is a constant independent of temperature and field. Usually this system of equations is used to describe the process of devastation of traps in studying thermally stimulated currents (TSC). However, in the case of isothermal discharge solution of the set of equations (1) and (2) is considerably simplified. Differentiating (1) over time, neglecting the term n_t/N_t in (1), and using (2) we obtain for n_t the simple second-order equation

$$n_t'' - \frac{n_t}{\tau^*} - \frac{n_t}{\tau_d \tau_t} = 0\,, \tag{3}$$

where $\tau^* = (\tau_t^{-1} + \tau_d^{-1} + \tau_T^{-1})^{-1}$. The general solution of this equation is

$$n_t = A \exp\left(-\frac{t}{T_1}\right) + B \exp\left(-\frac{t}{\tau}\right). \tag{4}$$

When fulfilling the condition

$$\tau_t, \tau_d \ll \tau_T \tag{5}$$

for T_1 and τ we have

$$T_1 \approx \frac{\tau^*}{\alpha}, \qquad \tau \approx \tau^*(1 - \alpha) \approx \tau^*\,,$$

where $\alpha = \tau^{*2}/\tau_d\tau_T \ll 1$, i.e. the relation $\tau \ll T_1$ is fulfilled. At the initial conditions $n_t(0) = n_0$ and $n_c(0) = 0$ from (1) and (2) we have $|B| \ll |A|$ and $A \approx n_0$, i.e. the final solution

$$n_t \approx n_0 \exp\left(-\frac{t\tau^*}{\tau_d \tau_T}\right). \tag{6}$$

The result (6) agrees with the solution of the set (1), (2) in meeting the so-called condition of quasi-stationarity, when

$$\frac{dn_c}{dt} \ll \frac{dn_t}{dt}.$$

Relation (5) is well fulfilled at devastation of deep trapping centres as at $E_t \gtrsim 1$ eV the typical time of thermal excitation amounts to tens of seconds and more, while at typical values of S_t and N_t the times τ_d and τ_t range from 10^{-8} to 10^{-13} s.

If the shift of flat-band voltage in MNOS structures U_{FB} is proportional to the quantity of trapped charge, then with $z = \Delta U_{FB}(t)/\Delta U_{FB}(0)$ expression (6) may be rewritten in the form

$$\ln \frac{t_z}{\tau_0} = \frac{E_t}{kT} + y(z) \ , \tag{7}$$

where $y(z) = \ln \ln (z^{-1})$ is a quantity approaching zero, when z is close to $\frac{1}{3}$, t_z is the time for which the given level of depolarization, z, is achieved and

$$\tau_0 = \tau_1 \left(1 + \frac{\tau_d}{\tau_t}\right). \tag{8}$$

As can be seen from (7), with $z =$ const (i.e. $U_{FB} =$ const) the slope of the dependence $\lg t(1/T)$ determines the quantity E_t. If the charge is transported through Coulomb centres, E_t in (6) should be substituted by $E_t - \beta\sqrt{\mathcal{E}_g}$, where $\beta = (e^3/\pi\varepsilon_N\varepsilon_0)^{1/2}$ is the Frenkel constant, e the electron charge, $\mathcal{E}_g$ the external electric field in Si_3N_4, ε_N the relative permittivity for silicon nitride, ε_0 the vacuum permittivity. A decrease of the barrier for thermal ejection $\beta\sqrt{\mathcal{E}}$ depends in general, on the trapped charge field, but with charge flowing to the metal it is easy in experiment to achieve the condition that this field may be neglected in comparison with the external one. This requires that the SiO_2 thickness and $\Delta U_{FB}(0)$ would be several times smaller than the Si_3N_4 film d_N and the external applied voltage U_g. In this case the Frenkel constant can be found from the slope of the dependence of $\lg t(\sqrt{\mathcal{E}_g})$ at $\Delta U_{FB} =$ const.

With fulfilling (5) two relations between τ_d and τ_t are possible. If $\tau_t \ll \tau_d$, one can assume that a quasi-equilibrium exists between traps and the conduction band. Then $\tau_1 = (S_t v N_t)^{-1} = \tau_t N_t/N_c$ and from (8) we have

$$\tau_0 \approx \frac{N_t}{N_c} \frac{d_N^{\prime}}{\mu\mathcal{E}}, \tag{9}$$

where N_c is an effective density of states in the silicon nitride conduction band. In fulfilling the inverse relation between τ_t and τ_d, $\tau_0 \approx \tau_1 \sim \nu^{-1}$, where ν is a frequency factor. Note that (7) describes usually a relaxation dependence with a relaxation time constant. In this case the fall of the dependence of $z(\lg t)$ from $z = 0.9$ to $z = 0.1$ should occur in the interval $\Delta(\lg t)$ that is given by

$$D_0 = \lg t_1(z_1 = 0.1) - \lg t_2(z_2 = 0.9) = \lg \frac{\ln z_1}{\ln z_2} \approx 1.3 \ .$$

3. Samples and Measurement Technique

We used MNOS structures obtained by conventional methods of planar technology. Preliminarily cleaned silicon plates were oxidized in the flow of dry oxygen at temperatures 900 to 1000 °C up to the thickness of 80 to 100 Å. Then 600 to 800 Å silicon nitride was deposited by one of the three following techniques: "chloride" Si_3N_4 was grown by ammonolysis of silicon tetrachloride and in the presence of hydrogen flow at the temperature 920 °C (X-samples); "silane" Si_3N_4 was obtained by silane ammonolysis in the hydrogen flow at 900 °C and the ratio $SiH_4/NH_3 < 1/40$ (C-samples); "vacuum" nitride was also obtained by silane ammonolysis but at an increased pressure of reagents (≈ 1 Torr) and with no hydrogen flow (V-samples).

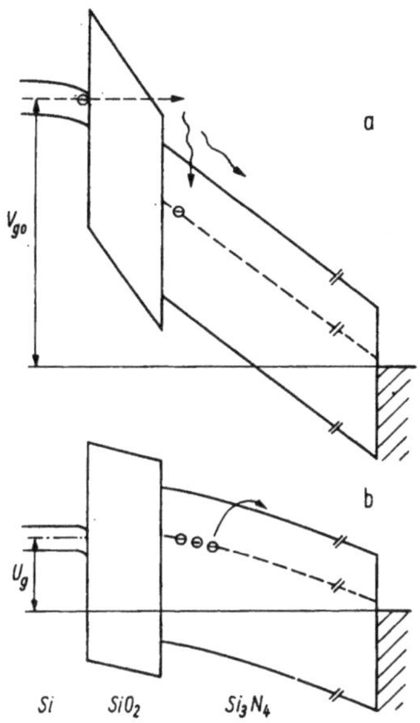

Fig. 3. The process of a) polarization and b) depolarization of a MNOS structure

The traps in the silicon nitride bulk were filled at room temperature by applying a voltage pulse of duration 10^{-1} to 10^{-3} s and amplitude $U_{go} = 40$ to 60 V to the MNOS structure. After this the sample was placed on a massive table heated up to a certain temperature T, so the time of heating the sample up to this temperature amounted to several seconds. Then the direct pulling voltage U_g from 0 to 20 V (Fig. 3) was applied to a metal electrode and the dependences $\Delta U_{FB}(\lg t)$ were measured. The measurements were carried out in 5×10^{-3} cm² capacitors. In order to avoid errors that were due to a redistribution of the trapped charge, each curve was recorded in a new capacitor.

4. Experimental Results

4.1 Temperature and field dependences

A typical time dependence of the shift of flat-band voltage at a constant temperature is given in Fig. 4 (X_1-sample). Parameter is the temperature. As had been shown in [14], in the case with no preliminary polarization of the MNOS structure under the same conditions there is no appreciable variation of the flat-band voltage, which is indicative of the absence of carrier injection from metal or from silicon. Thus, the observed variation is due to a discharge of the traps in the silicon nitride only. Dependences similar to those given in Fig. 4 have been obtained for silicon nitride grown in the other technological processes. For instance, Fig. 5 shows the temperature dependence of the discharge curves for the S_1-sample. It is also seen from this figure that, according to this mechanism, almost the whole trapped charge leaves the nitride bulk. Unlike this, after recording the TSC curves a considerable part of the initially trapped charge can remain in the dielectric bulk [10, 17].

Using the families of discharging curves at various temperatures that are similar to those given in Fig. 4 and 5, one can plot the dependence (7) $\lg t(1/T)$ at $z = \text{const}$ the slope of which determines an activation energy of the process of discharging in

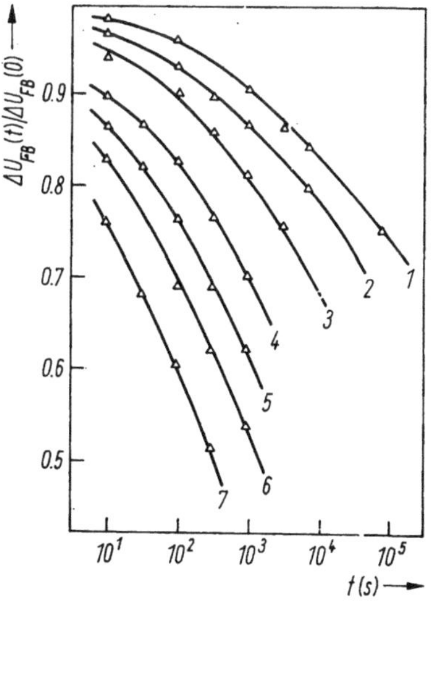

Fig. 4. Typical set of curves of depolarization at various temperatures: (1) $T = 427$, (2) 445, (3) 470, (4) 492, (5) 510, (6) 530, (7) 545 K; $\mathcal{E}_g = 1.5 \times 10^6$ V/cm, sample X_1, $d_{ox} = 90$ Å, $d_N = 820$ Å, $\Delta U_{FB}(0) = 10$ V

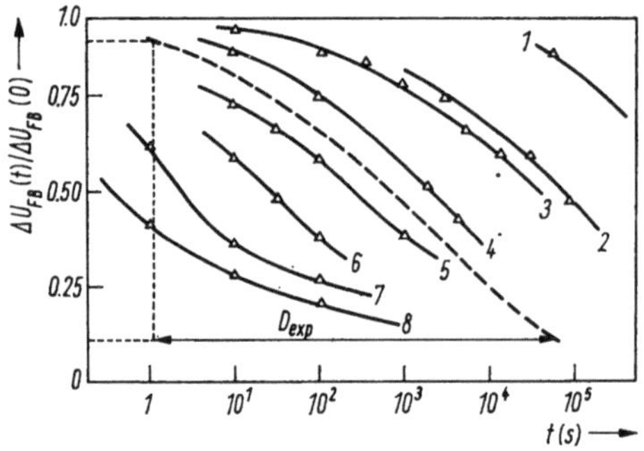

Fig. 5. The dependences z vs. $\lg t$ at various temperatures for sample S_1. (1) $T = 378$, (2) 420, (3) 409, (4) 448, (5) 477, (6) 513, (7) 530, (8) 567 K; $\mathscr{E}_g = 10^6$ V/cm, $d_{ox} = 90$ Å, $d_N = 590$ Å, $\Delta U_{FB}(0) = 3$ V

the given field. Fig. 6 shows the dependence $\lg t(10^3/T)$ for various samples at $\mathscr{E}_g = 10^6$ V/cm. Noteworthy is the following fact: (i) the dependence is little dependent on the technological process of obtaining the films if there is hydrogen in the reactor and if the ratios of SiH_4/NH_3 and $SiCl_4/NH_3$ are less than 1/40 (Fig. 6, dependence 2); (ii) the Si_3N_4 films obtained in the low-pressure reactor (with no hydrogen flow) have a conductivity several orders higher, the activation energy decreasing by two times (Fig. 6, straight line 1). In the first case it is 1.6 to 1.8 eV, in the second about 1 eV for $\mathscr{E}_g = 10^6$ cm. With allowing for the barrier reduction due to the Frenkel effect we have for the trap depth 2.0 and 1.4 eV, respectively. The activation energy determined from the ITD curves, is strongly dependent on the field both for the samples grown in a hydrogen flow and for the Y-samples.

Fig. 7 shows the dependences $\lg t(1/T)$ obtained at the level $z = 0.66$ for the V_1-samples. The parameter is the external electric field $\mathscr{E}_g = U_g/d_N^*; d_N^* = d_N + d_{ox}(\varepsilon_N/\varepsilon_{ox})$. It is seen that the slope of the dependences changes by more than a factor two. In fields of about 10^6 V/cm this change is stronger than that expected from the relation $\Phi = E_t - \beta \sqrt{\mathscr{E}_g}$ with $\beta = 0.38 \times 10^{-3}$ eV.

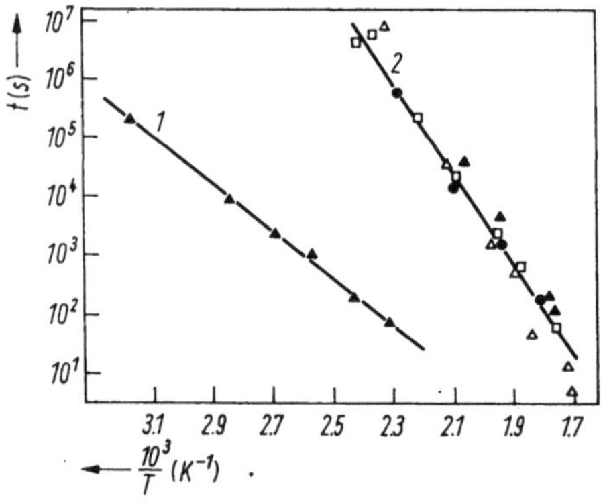

Fig. 6

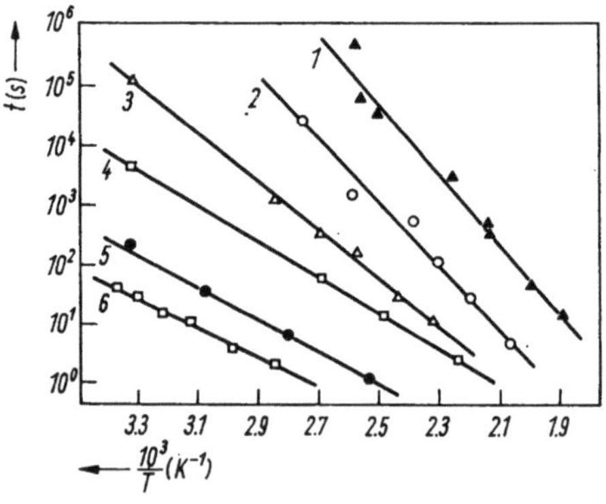

Fig. 7

Fig. 6. The dependences $\lg t$ vs. $10^3/T$ obtained at the level of depolarization $z = \frac{2}{3}$, $\mathscr{E}_g = 10^6$ V/cm, sample number: ▲ U_1, △ S_1, ● S_2, □ S_3

Fig. 7. The influence of the electric field on the activation dependences $\lg t$ vs. $10^3/T$. Sample V_1, (2) $\mathscr{E}_g = 0.6 \times 10^6$, (3) 1.0×10^6, (4) 1.5×10^6, (5) 2.0×10^6 V/cm; (1) presents the discharge in the field of the trapped charge $\mathscr{E}_g \approx 10^5$ V/cm

Knowing the value of the activation energy one can find the value of τ_0 from the dependence lg $t(1/T)$. For Si_3N_4 structures with low conduction (S- and X-samples) the value of τ_0 approaches 10^{-13} to 10^{-14} s. For the films obtained at a decreased pressure τ_0 strongly differs from the value that can be obtained from the estimate of a frequency factor $\tau_0 \sim \nu^{-1} \sim h/E_t$ and amounts to $\approx 10^{-10}$ s. Measurements on films of various thicknesses (from 400 to 3000 Å) indicated that the discharge time is not increased with increasing d_N. Thus, the value of τ_0 must have been determined by the value of the frequency factor. However, presently we cannot indicate the physical parameters to determine the value of τ_0 that ranges from 10^{-8} to 10^{-15} s for different samples.

Along with the structures investigated by the ITD method, control structures with a thin tunnel SiO_2 sublayer were produced in depositing a nitride film. Current–voltage characteristics were measured in these samples. Fig. 8 shows the dependences of lg $J(10^3/T)$ for three different samples. It is seen that as well as in depolarization experiments the structure conduction in V_2-samples is much higher than that in X_2- and S_2-samples, the activation energy of conduction exceeding that for the latter by more than a factor two (in the temperature range of about 500 K). In addition, in this temperature range the variation of the conduction activation energy with field for X- and S-samples is very slight, e.g. curves 2 and 4.

In this case a strict fulfilment of the dependence lg $J \sim \beta \sqrt{\mathcal{E}_g}$ on the current–voltage characteristics is observed with a slope close to the calculated one.

The slopes of the dependence lg $t(\sqrt{\mathcal{E}_g})$ obtained from the ITD measurements are close to those of J–U characteristics of lg $J(\sqrt{\mathcal{E}_g})$. Fig. 9 shows these dependences reduced to the same temperature $T = 400$ K. The dependences are arbitrarily displaced along the vertical axis, we are interested in the slope of these characteristics only. The dashed line marks the calculated slope equal to β/kT, where the value of the high-frequency permittivity $\varepsilon = n_N^2$ (n_N is the refractive index of Si_3N_4) is used for $\beta = (e^3/\pi\varepsilon\varepsilon_0)^{1/2}$. The spread in β for the ITD measurements is more considerable, and the dependence of lg $t \sim \sqrt{\mathcal{E}_g}$ is observed not so strictly as for the curves lg $J(\sqrt{\mathcal{E}_g})$. Nonetheless the value of the slope of these straight lines is in all cases close to the theoretical one and to the experimental value of the slope for the $J - U$ characteristics. The small value of the conduction activation energy in strong fields and its weak dependence on this parameter can be explained by the fact that at temperatures of 300 to 500 K and fields $\gtrsim 4 \times 10^6$ V/cm a pure activation mechanism of charge transport in Si_3N_4 does yet not manifest itself (at least, for silicon nitride layers with low conduction). It can be seen from Fig. 8 that at higher temperatures of about 60 K the curves S for X-samples turn upwards. The activation energy is approximately the same as that expected for centres with $E_t \approx 2$ eV. The experimentally observed

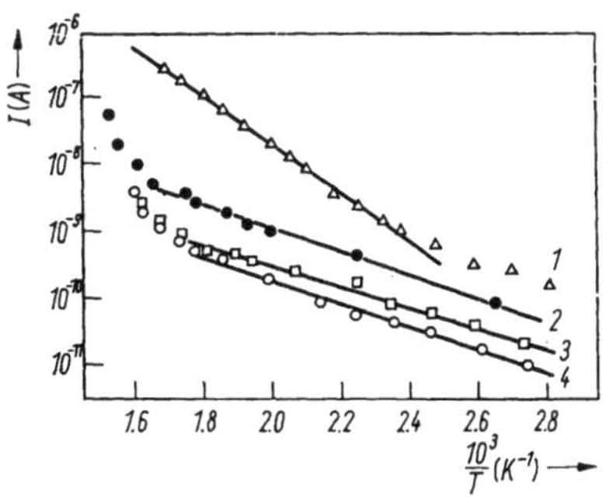

Fig. 8. The dependences of the current on the temperature: (1) sample V_1, $U_g = 24$ V, $\Phi = 0.74$ eV, $d_N = 595$ Å; (2) S_1, 30 V, 0.35 eV, 592 Å; (3) X_1, 32.2 V, 0.35 eV, 794 Å; (4) S_1, 24 V, 0.35 eV, 592 Å; (1 to 4) $d_{ox} \approx 18$ Å

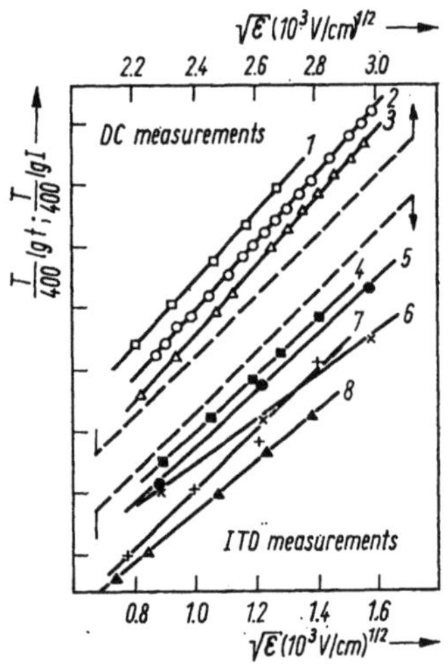

Fig. 9. The dependences lg t vs. $\sqrt{\mathscr{E}}$ and lg I vs. $\sqrt{\mathscr{E}}$ for $T = 400$ K. (1) Sample S_1, T (of measurement) = 520 K; (2) X_1, 434 K; (3) V_1, 393 K; (4) X_1, 433 K; (5) X_1, 454 K; (6) X_2, 500 K; (7) V_1, 433 K; (8) V_1, 387 K

strict fulfilment of the root dependence $\lg J \sim \sqrt{\mathscr{E}_g}$ in the temperature range from 300 to 500 K appears to be only an imitation of the field-assisted thermo-activation Poole-Frenkel mechanism, though it is beyond doubt that the reduction of the barrier by $\beta \sqrt{\mathscr{E}_g}$ has some effect on the ionization probability of trapping centres. However, we think, the estimation of the trap depth from the slope of the "activation" dependence of $\lg J(1/T)$ in this temperature range is doubtful.

4. 2. Influence of hydrogen on the spectrum of deep levels

As has been noted in most papers, the designation of silicon nitride obtained by the method of planar technology with Si_3N_4 is conventional. The quantities of N–H and Si–H bonds strongly differ in silicon nitride obtained in different technological processes [18, 19]. Both optical and electrophysical properties of silicon nitride are to a great extent, determined by the SiH_4/NH_3 ratio at which film growth occurs [19 to 21]. So the difference in ITD spectra for S-, X-, and V-samples is not surprising. Of interest is the connection between the properties of these groups of samples that arises at high-temperature annealings. Fig. 10 shows the dependence lg $t\,(10^3/T)$ for

(i) the initial sample with silicon nitride obtained in the flow of hydrogen (S_3-sample);

(ii) the sample annealed in hydrogen at 950 K for 60 min;

(iii) the sample annealed in vacuum at 950 K for 60 min.

For the operations (2) and (3) sample 1 was split. The dashed line in Fig. 10 marks the dependence of lg $t\,(10^3/T)$ for a V_1-sample in which silicon nitride has been obtained in a low-pressure reactor. As can be seen from this figure, the film properties for electrical conduction are strongly worsened by annealing. In addition, these properties become close to those of films obtained in the synthesis of silicon nitride without

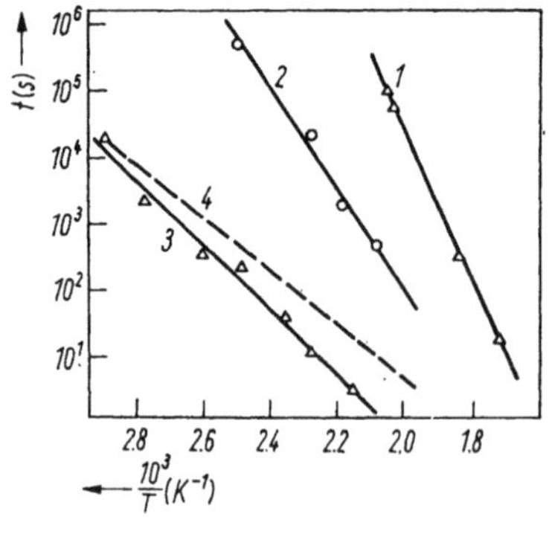

Fig. 10. The influence of the annealing in (2) H_2 and (3) vacuum on the dependences lg t vs. $10^3/T$. (1) Initial dependence, (4) dependence lg t vs. $10^3/T$ for sample V; $Z = \frac{2}{3}$, $\mathscr{E}_g = 10^6$ V/cm

hydrogen. If the annealing occurs in a hydrogen atmosphere, the worsening of the properties is not so rapid. These data indicate the strong influence of the presence of hydrogen in a silicon nitride film on its trapping spectrum.

4.3 The form of the spectrum of deep levels in silicon nitride

A very important result of ITD measurements is the "logarithmic" character of discharging curves, i.e. the large value D of their "extension" in the logarithmic time scale. As it is seen from Fig. 5 (dotted line), it amounts to four decades at $T = 475$ K. Discharging curves that are more extended in time are known for MNOS structures with a thin SiO_2 sublayer [12, 21]. Its logarithmic character in the low-temperature range is well explained by the tunnel mechanism of discharging of carriers distributed deep in the silicon nitride bulk [21]. However, in our case this factor appears not to work. The question can be only the consideration of the energy distribution of traps introduced to explain the logarithmic character of spreading of charge (in MNOS structures with a thin SiO_2 sublayer) at increased temperatures. The distribution of carriers along the coordinate gives a small correction to the magnitude D that we shall estimate from the following simple model. Assume that at the initial moment of time carriers are distributed according to the law $n_1(x) = n_0 \exp(-L/x)$, where L is a quantity close to that for the trapped charge centroid observed in experiment [15]. Also assume that the discharging occurs rapidly (in doing so we shall obtain, apparently, an upper estimate for the correction to D) that moves with time according to

$$t(x) = \nu^{-1} \exp\left(-\frac{E_t - \beta\sqrt{\mathcal{E}_N(x)}}{kT}\right)\frac{d_N - x}{l}. \qquad (10)$$

Here $\mathcal{E}_N(x)$ is the electric field in the region discharge front–metal electrode (insert in Fig. 11). For the distribution of trapped electrons $n(x) = n_0 \exp(-x/L)$ the expression for $\mathcal{E}_N(x)$ is obtained by elementary integration

$$\mathcal{E}_N(x) = \frac{1}{d_N^*}\left\{U_g + \frac{\Delta U_{FB}(0)}{(d_N - L)}\left[\left(L + d_{ox}\frac{\varepsilon_N}{\varepsilon_{ox}}\right)(1 - p(x)) - xp(x)\right]\right\}, \qquad (11)$$

where $p(x) = \exp(-x/L)$, ε_{ox} is the permittivity of oxide and

$$\Delta U_{FB}(0) = \frac{e}{\varepsilon_N\varepsilon_0}\int_0^{d_N} n(x)\frac{(d_N - x)}{d_N^*}\,dx\,.$$

The term $(d_N - x)/l$ in (10) allows the carriers distributed over the Si_3N_4 bulk to be several times recaptured when moving towards the aluminium electrode. Here l is the average distance an electron travels before being recaptured. For l we used the value 40 Å [15]; $d_N - x$ is the distance from the discharge front to the electrode.

Thus, the present model qualitatively allows for both distribution of carriers along the coordinate and the influence of the intrinsic field of the trapped charge on the process of discharge. Fig. 11 shows the variation of the form of ITD curves at different relations between external voltage and initial value of $\Delta U_{FB}(0)$. The curves were calculated in accordance with (10) and (11) with the values of parameters $\beta = 0.38 \times 10^{-3}$ eV/(V/cm)$^{1/2}$ and $\varepsilon_N/\varepsilon_{ox} = 2$. The dashed line marks the dependence of $Z = \exp(-t/\tau_0)$ at $\tau_0 = $ const. Curve 3 for which $U_g = 3\,\Delta U_{FB}(0)$ is in agreement with the experimental conditions, Fig. 5. For this curve $D = 2$ decades. The value D_0 must be considered as correction to D as we assume that the electrons escape, the

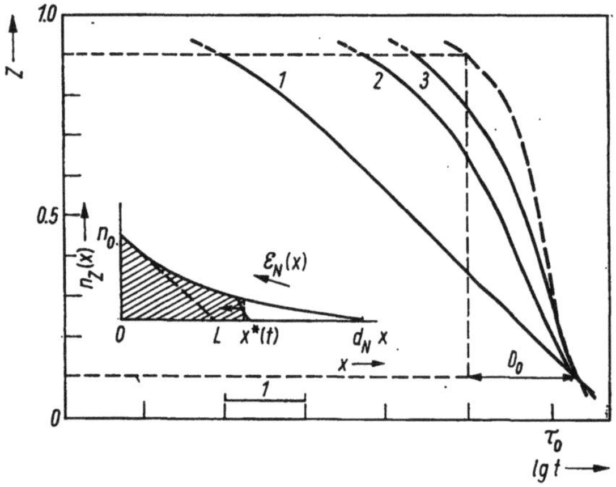

Fig. 11. The influence of the spatial distribution of the trapped charge on the shape of the ID curve. The calculation was made according to (10), (11) with the values of parameters $d_{ox} = 100$ Å, $d_N = 700$ Å, $L = 200$ Å, $l = 40$ Å, $\Delta U_{FB}(0) = 3$ V; (1) $U_g = 0$, (2) 3, (3) 9 V; ——— ordinary relaxational dependence $Z = \exp(-t/\tau)$, $\tau = \text{const}$

traps immediately at a definite moment $t(x)$. Then, if the large experimental value of D (from Fig. 5 $D_{exp} = 4$) is explained by the difference ΔE_t of the depth of trapping centres [12], we obtain for ΔE_t

$$\Delta E_t \lesssim 2.3kT(D_{exp} - D - D_0) \approx 0.1 \text{ eV},$$

$T = 475$ K. This value is smaller than ΔE_t received in [12].

5. Conclusions

The results of investigations discussed in this paper may by summarized as follows:

1. Data on the ITD of MNOS structures with a thick sublayer of SiO_2 indicate the presence of deep electron trapping levels of $E_t \approx 2$ eV in silicon nitride obtained by ammonolysis of monosilane or silicon tetrachloride with the ratio of SiH_4/NH_3 and $SiCl_4/NH_3 \lesssim 1/40$ and in the presence of hydrogen flow. For the silicon nitride obtained in the absence of a hydrogen flow (LPD-nitride) the ground level depth for electrons decreases down to ≈ 1.4 eV. Data on the ITD of MNOS structures with a thick sublayer of SiO_2 indicate that the spectrum width of deep levels in the silicon nitride is less than 0.1 eV.

2. A comparison of the ITD data with the results of measurements of the direct current passing through MNOS structures with a tunnel sublayer (silicon nitride in both cases was produced in one process) indicates that a determination of the depth of deep trapping levels for electrons from the activation dependence of current within the temperature range 300 to 400 K can give a considerably lower value of E_t.

3. Vacuum annealing of the structures manufactured by depositing a nitride film in the presence of a hydrogen flow results in electrophysical properties of silicon nitride similar to the properties of nitride obtained in a low-pressure reactor. Simultaneously annealing in hydrogen changes the silicon nitride properties only partially. This is indicative of the strong influence of hydrogen in the film on a spectrum of deep levels in silicon nitride.

Acknowledgements

The author would like to thank Dr. S. P. Sinitsa for valuable discussion of the results and Dr. A. S. Ginovker for preparing the samples.

References

[1] J. J. Chang, Proc. IEEE **64**, 1039 (1976).
[2] T. Katsube, Y. Adashi, and I. Ikoma, Solid State Electronics **19**, 11 (1976).
[3] J. R. Yeargan and H. L. Taylor, J. appl. Phys. **39**, 5600 (1968).
[4] Y. Misava and H. Yogi, Japan. J. appl. Phys. **15**, 1045 (1976).
[5] P. K. Choudhari, J. M. Franz, and C. P. Asker, Solid State Sci. Technol. 491 (1973).
[6] G. A. Brown, W. C. Robinett, and H. G. Carlson, J. Electrochem. Soc. **115**, 948 (1968).
[7] J. E. Yeargan and H. L. Taylor, Solid State Sci. **115**, 2273 (1968).
[8] S. M. Sze, J. appl. Phys. **38**, 2951 (1967).
[9] E. J. M. Kendall, Canad. J. Phys. **16**, 2509 (1968).
[10] E. J. M. Kendall, phys. stat. sol. (a) **32**, 763 (1969).
[11] E. P. Jacobs and G. Dorda, Solid State Electronics **20**, 61 (1977).
[12] L. Lundkvist, C. Svensson, and B. Hansson, Solid State Electronics **19**, 221 (1976).
[13] J. Bouer, phys. stat. sol. (a) **39**, 411 (1977).
[14] V. M. Efimov, V. A. Kolosanov, and S. P. Sinitsa, phys. stat. sol. (a) **49**, 217 (1978).
[15] P. C. Arnett and B. H. Yun, Appl. Phys. Letters **26**, 94 (1975).
[16] N. Endo, Solid State Electronics **21**, 1153 (1978).
[17] H. A. Mar and J. G. Simmons, IEEE Trans. Electron Devices **24**, 540 (1977).
[18] V. I. Belyi, F. A. Kuznetsov, T. D. Smirnova, L. V. Cramova, and L. K. Kravchenko, Thin Solid Films **37**, 39 (1976).
[19] C. Stubnya, I. C. Szep, G. Hoffman, Zs. Horvath, and F. Tütö, Rev. Phys. appl. **13**, 579 (1978).
[20] V. V. Voskoboinikov, V. A. Gritsenko, N. D. Dikovskaya, and S. P. Sinitsa, Mikroelektronika **5**, 369 (1976).
[21] L. Lundkvist, J. Lundström, and C. Svensson, Solid State Electronics **16**, 811 (1973).

(Received September 18, 1980)

phys. stat. sol. (a) **65**, 189 (1981)

Subject classification: 12.2 and 14.4.1; 1.2; 22.8

Section of Physics, Martin-Luther-University, Halle/Wittenberg, Halle (Saale)[1]) *(a) and Institute of Crystallography, Academy of Sciences of the USSR, Moscow (b)*

Elastic, Dilatational, and Electrostrictive Properties of TSCC Crystals

By

G. SORGE (a), U. STRAUBE (a), and N. R. IVANOV (b)

The temperature dependence of some elastic compliance and stiffness coefficients, linear thermal expansion coefficients, and coefficients of electrostriction from tris-sarcosine calcium chloride (TSCC) are presented. The results are discussed using a simple thermodynamical model. Obviously at the transition deformations take place only in the (001)-plane.

Die Temperaturabhängigkeiten einiger elastischer Nachgiebigkeiten, elastischer Steifigkeiten, linearer thermischer Ausdehnungskoeffizienten und Elektrostriktionskoeffizienten von Tris-Sarcosin-Calcium-Chlorid (TSCC) werden vorgestellt. Die Ergebnisse werden anhand eines einfachen thermodynamischen Modells diskutiert. Offenbar treten bei der Phasenumwandlung Verzerrungen nur in der (001)-Ebene auf.

1. Introduction

Tris-sarcosine-calcium chloride (TSCC) exhibits a ferroelectric phase transition from a high-temperature paraelectric orthorhombic non-polar phase into a polar orthorhombic phase ($D_{2h}^{16} \rightarrow C_{2v}^{9}$) [1]. It is an uniaxial ferroelectric with a second-order phase transition of the order–disorder type at about 130 K [1 to 4]. Analysing an overdamped soft mode Prokhorova et al. [5] suggested a partially displacive phase transition. The crystals which are grown from aqueous solution show a nearly hexagonal symmetry and are always twinned at room temperature. Sawada et al. [6] found ferroelastic properties of TSCC within the room temperature paraelastic phase. The ferroelastic domain walls (twin boundaries) may be moved by an external mechanical stress. The hypothetical hexagonal paraelelastic high-temperature phase [6] cannot be reached because of decomposition of the crystals at about 230 °C. A first review of dielectric, structural, elastic, optical, and thermal properties of TSCC crystals was given by Windsch and Sorge [7].

The aim of this paper is to present some new results on dilatation, ultrasonic, and electromechanical investigations.

2. Experimental

In order to investigate the elastic properties of TSCC crystals it was necessary to use ferroelastic monodomain samples in each case. These samples could be obtained by the procedure described by Sawada et al. [6], if the specimens are not too large, or by cutting monodomain samples from grown crystals. The ferroelastic monodomain state of the specimens was controlled by means of a polarizing microscope.

For thermal expansion measurements an automatic quartz dilatometer of Strelkov type was used [8]. The samples had an approximate size of $2 \times 2 \times 8\ mm^3$. Details on experimental arrangements for ultrasound investigations are described in [9, 10], for resonance investigations in [11, 12], and for the measurements of the electrostrictive

---

[1]) Friedemann-Bach-Platz 6, DDR-4020 Halle (Saale), GDR.

coefficients by a quasistatic method in [12]. Using the latter method due to the ferroelasticity of TSCC the resulting strains perpendicular to the pseudohexagonal axis were not reproducible because of the point stress influencing the sample.

The orientation of the samples was chosen in the following way according to Makita [1]: The orthorhombic phase was used as prototypic phase because the hypothetical hexagonal phase is not reached. The pseudohexagonal axis is the a-axis (cf. Fig. 2) and the optical plane lies perpendicular to the c-axis. The Cartesian coordinates x, y, z correspond to the crystallographic directions of a, b, c, respectively.

Ultrasonic measurements were carried out at 20 MHz on samples with propagations lengths from 5 to 10 mm.

Samples for the resonance measurements were bars with a geometry described in [11]. In ordering to measure the elastic compliances s_{11} and s_{33} the alternating field was applied in y-direction of samples oscillating in x- or z-direction. Because of a small restpiezoelectricity it was possible to excite the sample for the measurement of s_{11} piezoelectrically some K above the phase transition. A small electrostrictive excitation of this sample some K above the transition was possible, too. But the error in determining the values of the electrostrictive coefficient Q_{221} was rather high. The coefficient s_{33} could be estimated only 2 K around the maximum of the dielectric constant because of a very small restpiezoelectricity. A weak electrostrictive excitation could be found in this temperature region, but a measurement of Q_{223} was impossible.

3. Experimental Results

The temperature dependence of the elastic stiffness coefficients c_{11} c_{22}, c_{33}, and c_{44} is shown in Fig. 1 a and b. Results on c_{22} and c_{11} were already publihsed earlier [9].

While the temperature dependence of c_{11}, c_{22}, and c_{33} shows a step-like change at the transition temperature, this is not the case for c_{44}. Such an elastic behaviour of c_{11}, c_{22}, and c_{33} can be explained and understood with the assumption of an electrostrictive coupling between strain and order parameter (see Section 5). From observed Brillouin shift Prokhorova et al. [5] calculated the elastic stiffness coefficients c_{11}, c_{22}, c_{12}, c_{46}, and $c_{44} + c_{55}$. Their values for c_{11} and c_{22} are $\approx 15\%$ higher than our values. It is possible that the reason for these differences is the frequency dependence of the elastic coefficients.

Below the transition temperature the mean contribution of the strong temperature dependence of c_{11}, c_{22}, and c_{33} can be explained by the fact that the transition of second order is close to a critical point. However, it is not yet possible for us to explain the real deviation from the Landau behaviour above and also below T_c, which can be seen in a rather wide temperature region (below T_c about 3 K, above T_c about 15 K).

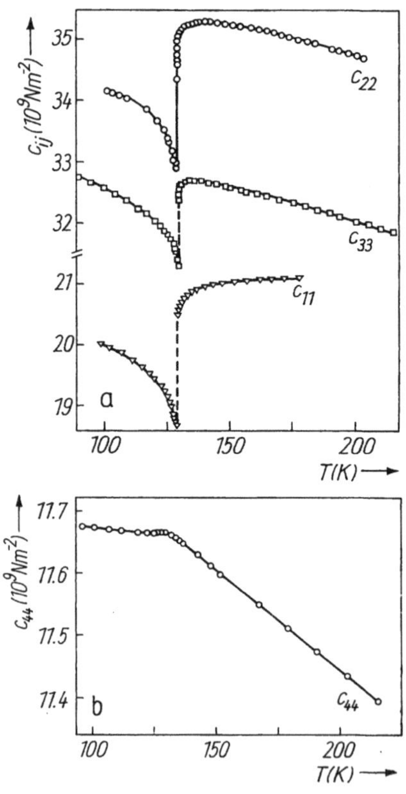

Fig. 1. Elastic stiffness coefficients a) c_{11}, c_{22}, c_{33} and b) c_{44} of TSCC versus temperature ($f = 20$ MHz)

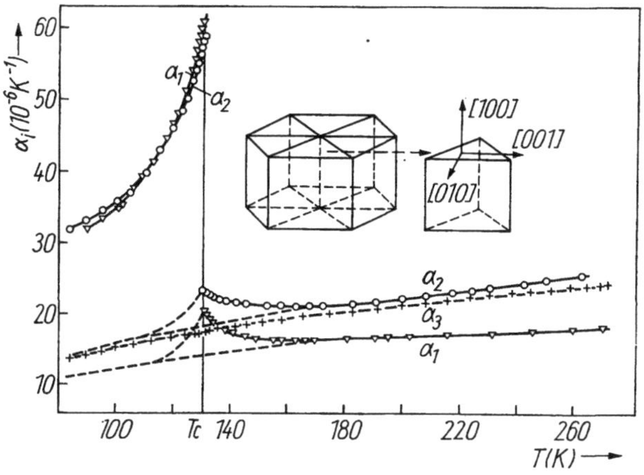

Fig. 2. Temperature dependence of dilatational coefficients α_1, α_2, and α_3 of TSCC and the orientation of specimens

Results on the temperature dependence of dilatational and thermal expansion coefficients α_1, α_2, and α_3 are shown in Fig. 2. The absolute values of c_{22} and c_{33} as well as α_2 and α_3 differ only by 5% from each other in the paraelectric phase. This is surely due to the pseudohexagonal structure. Fig. 2 shows a very large anomaly in the temperature dependence of α_2 and α_1. The jump has a magnitude of about 200% relative to the paraelectric value. The temperature dependence of α_3 shows no anomaly. These facts indicate that the deformation change at T_c takes place obviously only in the plane (001) perpendicular to the z-axis [3].

Elastic stiffness coefficients c_{ij} and thermal expansion coefficients α_i are not directly comparable. Therefore, the elastic compliances s_{11} and s_{33} were measured. The result of the measurement of s_{11} is shown in Fig. 3. No jump of the s_{33}-coefficient could be detected in the vicinity of the phase transition. It is possible that the jump of s_{33} is more than three orders of magnitude smaller than that of s_{11}. For the value of s_{33} at the phase transition we obtained constant values of 3.86×10^{-11} m^2 N^{-1}.

Using the quasistatic method it could be measured $Q_{221} = (15 \pm 5)$ m^4 C^{-2} at room temperature [13]. With the electrostrictive resonator we obtained Q_{221} values between 50 m^4 C^{-2} ≈ 5 K above T_c up to 150 m^4 C^{-2} ≈ 0.3 K above T_c. These measurements showed also a high rate of error. Q_{223} could not be measured but only roughly estimated. It is smaller than 0.5 m^4 C^{-2}. If we applied an electric dc field of 4 kV cm^{-1} to the sample ≈ 1.5 K below T_c (linearized electrostriction) we could get $Q_{221} = 220$ m^4 C^{-2} and $Q_{223} \leqq 4$ m^4 C^{-2}.

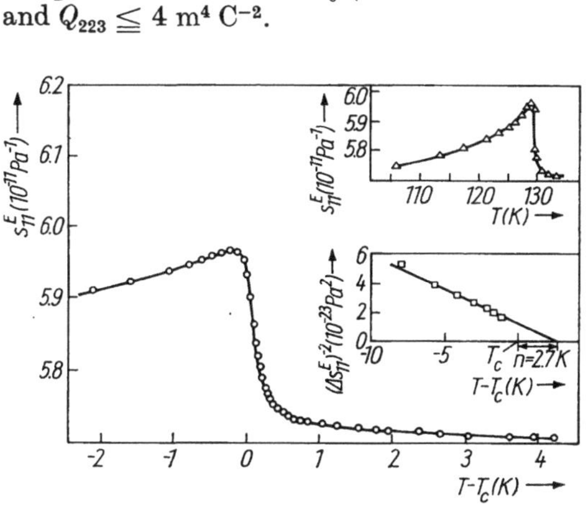

Fig. 3. The elastic compliance coefficient s_{11}^{E} and the relation $(1/\Delta s_{11}^{\mathrm{E}})^2$ as a function of temperature

4. Discussion

4.1 Potential expansion

For the explanation of the temperature dependence of the elastic coefficients we use the Helmholtz free-energy function expanded by Makita [1]. We assume that TSCC is an uniaxial proper ferroelectric with the spontaneous polarization P_2 parallel to the y-axis. The term $\frac{1}{6}CP_2^6$ was taken into account because the transition is of second order but near to a tricritical point [14]. $F = F_0 + \frac{1}{2}c_{ij}^P S_i S_j + q_{22i} S_i P_2^2 + \frac{1}{2}a' P_2^2 + \frac{1}{4}BP_2^4 + \frac{1}{6}CP_2^6$, with $A' = A(T - T_c)$, where the q_{mni} are electrostrictive coefficients and the S_i, S_j are components of strain. The electrostrictive term may be written in the longer form [1]: $(q_{221}S_1 + q_{222}S_2 + q_{223}S_3)\, P_2^2$. From this formula the conclusion may be drawn that c_{11}, c_{22}, and c_{33} have jumps and c_{44} has no jump [15, 16]. Besides the values of c_{44} are much smaller than those of c_{11}, c_{22}, and c_{33}. This can be understood taking into account the ferroelastic nature of TSCC, namely that the c_{44} shear coefficient [17] is significant for the hypothetical transition from the orthorhombic to the hexagonal phase. Perhaps c_{44} cannot be very high because of a possible soft mode.

The free elastic enthalpy is more instructive to explain the results of thermal expansion and mechanical resonance. The potential has the same form as the free-energy expansion,

$$G_1 = G_{10} - \tfrac{1}{2}\, s_{ij}^P T_i T_j - Q_{22i} T_i P_2^2 + \tfrac{1}{2}\, A' P_2^2 + \tfrac{1}{4}\, BP_2^4 + \tfrac{1}{6}\, CP_2^6\,,$$

where Q_{mmi} are electrostrictive coefficients and the T_i, T_j are mechanical stress components.

4.2 Linear thermal expansion and the pressure dependence of the phase transition

Using the Ehrenfest relation [18] the stress-induced shift of the Curie temperature may be calculated from the jump of α_i

$$\frac{dT_c}{dT_i} = -\frac{T_c\, \Delta\alpha_i}{\varrho\, \Delta c}\,,$$

where T_c is the Curie temperature, T_i are the external mechanical stress components, $\Delta\alpha_i$ the jumps of the expansion coefficients α_i at T_c, Δc the jump of the specific heat, and ϱ the density. With the values

$$T_c = 130\ \text{K}\,, \qquad \varrho = 1533\ \text{kg m}^{-3}\ [1]\,,$$

$$\Delta c = 16.5\ \text{J K}^{-1}\,\text{mol}^{-1} = 43.7\ \text{J K}^{-1}\,\text{kg}^{-1}\ [3]\,,$$

$$\Delta\alpha_1 = 40.4 \times 10^{-6}\ \text{K}^{-1}\,, \qquad \Delta\alpha_2 = 35.8 \times 10^{-6}\ \text{K}^{-1}\,, \qquad \Delta\alpha_3 = 0$$

we obtained the results

$$\frac{dT_c}{dT_1} = -7.9 \times 10^{-8}\ \text{KPa}^{-1}\,, \qquad \frac{dT_c}{dT_2} = -7.0 \times 10^{-8}\ \text{K Pa}^{-1}\,, \qquad \frac{dT_c}{dT_3} = 0$$

and for the hydrostatic pressure p

$$\frac{dT_c}{dp} = -\left(\frac{dT_c}{dT_1} + \frac{dT_c}{dT_2} + \frac{dT_c}{dT_3}\right) = 14.9 \times 10^{-8}\ \text{KPa}^{-1}\,.$$

This value agrees very well with that found by Hegenbarth and Schmidt [18],

$$\frac{dT_c}{dp} = 13.2 \times 10^{-8}\ \text{KPa}^{-1}\,.$$

4.3 Linear expansion and the elastic compliances

There are three possibilities to determine the jump of the elastic compliances at the transition point. At first from the linear expansion coefficients by the Ehrenfest-Garland relation [18]

$$\Delta s_{ij}^{E} = -\frac{dT_c}{dT_i}\,\Delta\alpha_i\,.$$

We get

$$\Delta s_{11}^{E} = 0.32 \times 10^{-11}\ m^2\,N^{-1}\,, \qquad \Delta s_{22}^{E} = 0.25 \times 10^{-11}\ m^2\,N^{-1}\,,$$

and

$$\Delta s_{33}^{E} = 0\,.$$

Direct measurements by the piezoelectric resonance method yield

$$\Delta s_{11}^{E} = (0.24 \pm 0.02) \times 10^{-11}\ m^2\,N^{-1}\,, \qquad \Delta s_{33}^{E} < 0.0002 \times 10^{-11}\ m^2\,N^{-1}\,.$$

A very simplified calculation from ultrasound results gives, if we neglect the tensorial character of the elastic coefficients,

$$\Delta s_{11} \approx 0.37 \times 10^{-11}\ m^2\,N^{-1}\,, \qquad \Delta s_{22} \approx 0.14 \times 10^{-11}\ m^2\,N^{-1}\,,$$

$$\Delta s_{33} \approx 0.08 \times 10^{-11}\ m^2\,N^{-1}\,.$$

4.4 Electrostrictive coefficients

The electrostrictive coefficients Q_{mni} may be estimated both from the jump of the elastic compliances and of the thermal expansion coefficients. The term $\frac{1}{6}CP_2^6$ in the potential expansion is negligible near the phase transition below T_c. The compliances have a jump

$$\Delta s_{ij}^{E} = \frac{2Q_{mni}Q_{mnj}}{B}\,,$$

where B is the expansion coefficient of the P_2^4 term and has the value

$$B = 5.2 \times 10^{15}\ V\,m^5\,A^{-3}\,s^{-3}$$

not much different from the value given in [4]. We calculated Q_{mni} values of the order of magnitude $Q_{221} \approx Q_{222} \approx 80\ m^4\,C^{-2}$ in this way. Q_{223} equals zero, of course. The other possibility to calculate the Q_{mni} values is based on the relation

$$Q_{mni} = \frac{S_i^*}{P_m^* P_n^*} = \frac{\Delta\alpha_i}{d(P_m^* P_n^*)/dT}\,,$$

where the asterisk means the spontaneous values. For Q_{221} and Q_{222} we obtain from this formula values from 77 to 93 $m^4\,C^{-2}$ and $Q_{223} = 0$. All the measured and estimated Q_{mni} values are two to three orders of magnitude greater than in well-known proper ferroelectrics ($BaTiO_3$ [15], TGS [19]).

4.5 The influence of the term $\frac{1}{6}CP_2^6$ on the temperature behaviour of the elastic coefficients of TSCC below T_C

Taking into account the term $\frac{1}{6}CP_2^6$ in the free elastic enthalpy function the spontaneous polarization squared becomes

$$P_2^2 = -\frac{B}{2C} + \sqrt{\frac{A}{C}(T_n - T)}\,, \quad \text{with } T_n = T_c + n,\ n = \frac{B^2}{4AC}\,,$$

and $T_c = T_0 + \dfrac{2Q_{22i}T_i}{A}$.

The following linear equation should be fulfilled:

$$\left(\frac{\partial P_2^2}{\partial T}\right)^{-2} = \frac{4C}{A}(T_n - T)\,.$$

The term $(2Q_{22i}/A)\,T_i$ in the relation for T_c describes the shift of the transition temperature as a function of the stress. For the linear thermal expansion

$$(\Delta\alpha_i)^{-2} = \frac{1}{Q_{221}^2}\frac{4C}{A}(T_n - T)$$

should be valid. The results for the thermal expansion coefficients and the spontaneous polarization are plotted in Fig. 4. Matsuo et al. [3] showed the linear dependence $(\Delta c/T)^{-2}$ on temperature.

To obtain the jumps (the differences) $\Delta s_{ij}^{\mathrm{E}}$ it is necessary to use the relation $\Delta s_{ij}^{\mathrm{E}} = (\partial S_i/\partial P_2)\,(\mathrm{d}P_2/\mathrm{d}T_j)_{\mathrm{E}}$. The calculated $\Delta s_{ii}^{\mathrm{E}}$ have the form

$$\Delta s_{ii}^{\mathrm{E}} = \frac{Q_{22i}^2}{\sqrt{AC(T_n - T)}}\,.$$

Using $\Delta c_{ij}^{\mathrm{E}} = (\partial T_i/\partial P_2)\,(\mathrm{d}P_2/\mathrm{d}S_j)_{\mathrm{E}}$, $\Delta c_{ii}^{\mathrm{E}} = q_{22i}^2/\sqrt{AC(T_n - T)}$ may be calculated, where in this case $T_c = T_0 - 2q_{22i}S_i/A$. Thus for the functions $(\Delta s_{ii}^{\mathrm{E}})^{-2}$ and $(\Delta c_{ii})^{-2}$ plotted against the temperature linear relations should be valid. In Fig. 3 and 5 a rather good agreement between theory and experimental result is shown. The value $n = B^2/4AC$ indicates that the phase transition is near a critical one. For TSCC $\sqrt{n/T_c} = 0.31$; for TGS 0.8. We get $n \approx 1.8$ K if we take the potential coefficients from [4], and $n \approx 14.7$ using the coefficients from [20]. The difference is mainly due

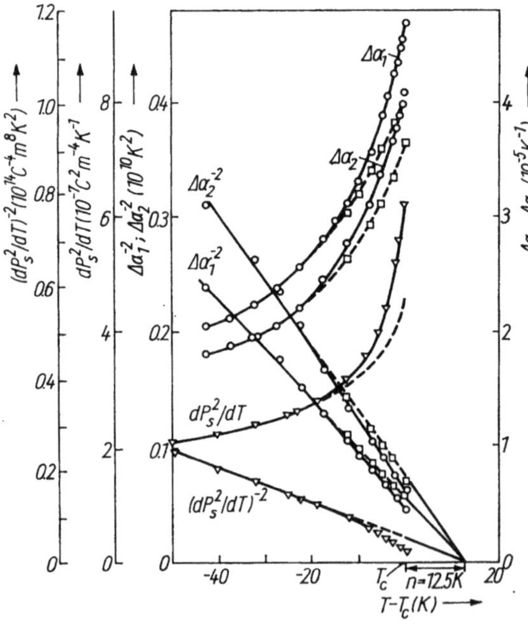

Fig. 4. Temperature dependence of the critical parts $\Delta\alpha_1$ and $\Delta\alpha_2$ of the dilatational coefficients α_1 and α_2, the functions $1/\Delta\alpha_1^2$ and $1/\Delta\alpha_2^2$, and the relations $\mathrm{d}P_s^2/\mathrm{d}T$ and $(\mathrm{d}P_s^2/\mathrm{d}T)^{-2}$ of TSCC

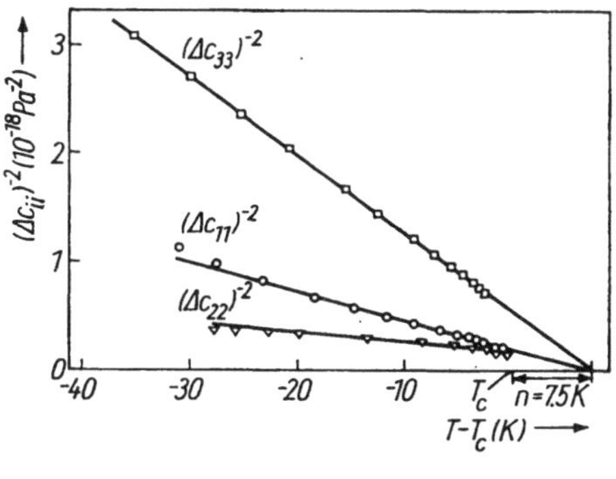

Fig. 5. Plot of the relations $(\Delta c_{11})^{-2}$, $(\Delta c_{22})^{-2}$, and $(\Delta c_{33})^{-2}$ of TSCC versus temperature

to the deviations of the Curie constants from each other that have a great influence on the values of the expansion coefficients B and C.

5. Conclusions

1. An explanation of the anomaly of the elastic stiffness coefficients c_{11}, c_{22}, c_{33} below the transition temperature has been found. The main part of the anomalies can be explained on the basis of the Landau theory. But we observe some deviations from normal Landau behaviour. We think that these deviations are not caused by fluctuations, because the temperature region is rather wide. Empirically we have found an exponential behaviour (like in [9]) but the reason for this is not yet clear.

2. We have found some results on the anisotropy properties of TSCC ($\Delta\alpha_3 = 0$, $\Delta s_{33} \approx 0$, $Q_{223} \approx 0$). Perhaps this behaviour can be interpreted assuming that deformations at the transition only take place perpendicular to the [001] direction.

3. The abnormally large electrostriction and large anisotropy found may possibly be understood by the assumption that TSCC has a pseudoproper ferroelectric phase transition and the main part of dilatation comes from an unknown order parameter and not from the polarization, which is linearly coupled with the order parameter.

References

[1] Y. Makita, J. Phys. Soc. Japan **20**, 2073 (1965).
[2] W. Windsch, Ferroelectrics **12**, 63 (1976).
[3] T. Matsuo, M. Mansson, and S. Sunner, Acta chem. Scand. **A33**, 781 (1979).
[4] G. Sorge and U. Straube, phys. stat. sol. (a) **51**, 117 (1979).
[5] S. D. Prokhorova, G. A. Smolensky, I. G. Siny, E. G. Kuzminov, V. D. Mikvabia, and H. Arndt, Ferroelectrics **25**, 629 (1980).
[6] A. Sawada, Y. Makita, and Y. Takagi, J. Phys. Soc. Japan **42**, 1918 (1977).
[7] W. Windsch and G. Sorge, KTB Martin-Luther-Universität, Halle/Wittenberg, Wissenschaftliche Beiträge **49**, 53 (1980)
[8] N. R. Ivanov, Izv. Akad. Nauk SSSR, Ser. fiz. **43**, 1707 (1979).
[9] G. Sorge and U. Straube, Ferroelectrics **21**, 533 (1978).
[10] G. Sorge, U. Straube, B. Thamke, and L. A. Shuvalov, phys. stat. sol. (a) **59**, K183 (1980).
[11] H. Beige and G. Schmidt, Wiss. Z. Univ. Halle XXVI-M **2**, 47 (1977).
[12] G. Schmidt, G. Sorge, H. Beige, Ch. Birkholz, and E. Ciesla, Wiss. Z. Univ. Halle, XXII-M **2**, 87 (1973).
[13] D. Grützmann, Univ. Halle, private communication.
[14] J. Helwig, J. Peterson, and E. Schneider, Z. Phys. **B28**, 87 (1977).

[15] A. S. SONIN and B. A. STRUKOV, Einführung in die Ferroelektrizität, Akademie-Verlag, Berlin 1974.
[16] W. REHWALD, Adv. Phys. **22**, 721 (1973).
[17] R. A. COWLEY, Phys. Rev. B **13**, 11 (1976).
[18] E. HEGENBARTH and B. SCHMIDT, Ferroelectrics **25**, 359 (1980).
[19] G. SCHMIDT and P. PFANNSCHMIDT, phys. stat. sol. **3**, 2215 (1963).
[20] N. R. IVANOV and H. ARNDT, KTB Martin-Luther-Universität, Halle/Wittenberg **39**, 159 (1978).

(Received January 5, 1981)

phys. stat. sol. (a) 65, 197 (1981)

Subject classification: 1.3; 21; 21.1

Sektion Mathematik/Physik der Pädagogischen Hochschule Halle (Saale)[1]) *(a)*
and Institute of Physics, Czechoslovak Academy of Sciences, Prague[2]) *(b)*

On the Mode of Decomposition of Al–Zn Alloys

By

H. LÖFFLER (a), V. SYNEČEK (b), M. SIMERSKÁ (b), G. WENDROCK (a),
P. BARTUŠKA (b), and R. KROGGEL (a)

The aim of this paper is to find reliable criterions whether the fast homogeneous decompositions of supersaturated Al–Zn alloys are governed by the "normal" nucleation and growth process or by the spinodal mode. Arguments are delivered that the position of the aluminium rich side of both, the chemical and the coherent spinodal calculated by means of thermodynamical methods is fairly certain. Thus indications of the start of the rapid increase of decomposition above the chemical spinodal can impossibly be related to the spinodal mode, but to the normal nucleation and growth process, which may initiate a "fast autocatalytic nucleation of zones". This statement is supported by SAXS investigations.

Ziel dieser Arbeit ist es, zuverlässige Kriterien zu finden, ob schnelle homogene Entmischungen übersättigter AlZn-Legierungen über den „normalen" Keimbildungs- und Wachstumsprozeß oder über spinodale Entmischung ablaufen. Es werden Argumente gefunden, daß die Lage der thermodynamisch berechneten chemischen und kohärenten Spinodalen, zumindest für den linken Ast, ziemlich sicher ist. Deshalb können Anzeichen für den Beginn einer schnellen homogenen Entmischung, die oberhalb der chemischen Spinodalen liegen, unmöglich mit spinodaler Entmischung verknüpft sein, sondern nur mit dem normalen Keimbildungs- und Wachstumsprozeß, der eine „schnelle autokatalytische Bildung von Zonen" initiieren kann. Diese Behauptung wird durch RKWS-Untersuchungen gestützt.

1. Introduction

The Al–Zn system is most suitable for studying the mechanisms and kinetics of phase transformations in supersaturated solid solutions. The solid solubility of Zn in Al reaches 66.5 at% at 382 °C. Due to these and other favourable facts the decomposition processes in the Al–Zn system have been studied by many authors. Thus many details of the decomposition behaviour of this system in dependence on concentration and thermo-mechanical treatment are known.

The mechanisms of the homogeneous, heterogeneous, and discontinuous structure transformation have been established in a wide range of temperature and concentration of the Al-rich Al–Zn alloys [1]. The most important continuous mode of structure changes may proceed by the phase separation of the supersaturated α-matrix comprising the full sequence of metastable phases, i.e. spherical Guinier-Preston (GP) zones, ellipsoidal GP zones, rhombohedral α'_R-phase, cubic α'_m-phase, before reaching the equilibrium state with platelets of β(Zn)-phase in epitaxial orientation to the Zn depleted α-matrix [2]. Much emphasis associated with the improvement of the mechanical properties was given to the first stage of decomposition of the supersaturated α-matrix leading to the formation of GP zones [3] fully coherent with the α-matrix. Numerous investigations have been performed delivering, e.g. data on the decomposition of GP zones given by the boundary of the corresponding metastable miscibility gap (e.g. [4 to 6]), the quasi-periodical spatial arrangement of the zones

[1]) Kröllwitzer Str. 44, DDR-4020 Halle (Saale), DDR.
[2]) Na Slovance 2, 18040 Prague 8, Czechoslovakia.

[7], the temperature of their easy formation [8], the changes of their internal structure, shape, and coherence with the α-matrix as well as the establishment of the later metastable and stable equilibria [2, 9 to 15].

Theoretical calculations have also been performed giving the position of the GP zones [5] and α'_m-phase boundaries [16] and also the respective coherent [17, 18] and chemical [18, 19] spinodal curves.

Many problems have yet to be solved to achieve a more complete understanding of the structure transformations in Al–Zn alloys. A considerable confusion and disagreement in the literature is apparently associated with the establishment of the mechanisms governing the initial decomposition of the supersaturated α-matrix in dependence on Zn content, temperature, and treatment of the alloy. The differences encountered are due to the controversial interpretation of experiments either in favour of the "normal" nucleation and growth process or of the spinodal mode of decomposition. The solution of this problem is tightly bound with the establishment of the correct position of the coherent spinodal curve in the Al–Zn system.

The aim of this paper is to contribute to a better understanding of processes governing the initial stage of decomposition of supersaturated solid solutions of Zn in Al.

2. Reliable Criteria for Spinodally Decomposing Al–Zn Alloys

In the past several efforts were done to obtain information on spinodally decomposing Al–Zn alloys. In their excellent review Gerold and Kostorz [20] divided these works into two groups, based on quantitative and qualitative experiments. In the first group of investigations the samples are aged far below the position of the spinodal curve and the results are mostly correlated with the linear-approximation theory of spinodal decomposition first developed by Cahn [21]. The existence of a cross-over point in the SAXS spectra, the ratio between the critical and the main wavelengths of the concentration fluctuations, and the time and temperature dependence of the distribution of the wavelength of the concentration fluctuations are taken into account. In this way Rundman and Hilliard [22] investigated an Al–22 at% Zn alloy at 65 °C and Acuña and Bonfiglioli [23] an Al–15 at% Zn alloy at −45 °C. They found indications that these alloys start to decompose at these temperatures by the spinodal mode, although the Al–15 at% Zn alloy yielded a somewhat larger ratio of the main and the critical wavelengths as compared to that predicted by Cahn's theory. From the results obtained with the Al–22 at% Zn alloy Rundman and Hilliard [22] estimated the position of the coherent spinodal at about 240 °C and Bartel and Rundman [24] at about 228 °C (see ■ in Fig. 1).

Such quantitative experiments were not applied to investigate Al–Zn alloys in the vicinity of the coherent spinodal till 1977. The main reason could be that at elevated temperatures the decomposition processes are remarkably accelerated and that in

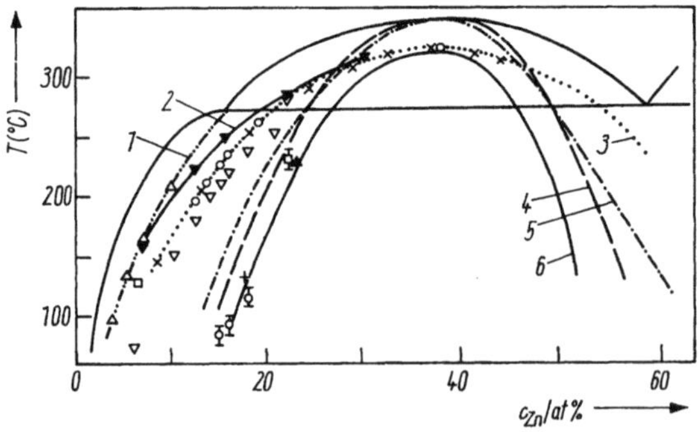

Fig. 1. Part of the phase diagram of the Al–Zn system (■ [22, 24], ● [25, 26], ▲ [27, 28], + [30], □ [31 to 34], ▽ [35], × [36], ▼ [37], ○ [38], △ [39, 40], see text); (1) α_m [16]; (2) α_R solution line [38]; (3) line of easy formation of GP zones and α'_R [36, 37]; (4, 5, 6) chemical and coherent spinodal [18, 19]

the vicinity of (but below) the spinodal curve the spinodal mode is overlapped by the normal nucleation and growth processes. In spite of these difficulties Kroggel et al. [25, 26], Ciach et al. [27], and Dutkiewicz et al. [28] tried to find information on the position of the coherent spinodal by means of SAXS [25 to 28] and TEM investigations [27, 28]. Their procedure is based on the idea that the time evolution of the spinodal structure comprising GP zones in their later stage of coarsening may provide reliable information on the mechanism initiating the decomposition of the alloy. Both experiments led to the same result indicating that the square of the reciprocal value of the main wavelength of concentration fluctuations, determined after an ageing time of about 10 min, plotted versus ageing temperature yielded straight lines in a certain temperature range. This result found for alloy compositions from 15 to 22 at% Zn agreed well with Cahn's predictions for alloys decomposing spinodally [29]. We may refer to this procedure as to a semi-quantitative one as compared to the scheme adopted by Gerold and Kostorz [20].

From these straight lines the position of the coherent spinodal can be easily extrapolated. The results of the SAXS measurements are shown in Fig. 1 by full circles (Al–15, 16, 18 at% Zn) and those of the SAXS and TEM investigations (Al–22 at% Zn) by the full triangle (▲). It seems worth mentioning that a re-evaluation of the SAXS data of the Al–18 at% Zn alloy, based on a better approximation [30] than that of Cahn [29], yielded the result that the temperature of the coherent spinodal (T_{cs}) is situated at $T_{cs} = (130 \pm 3)$ °C (+ in Fig. 1).

Qualitative methods for obtaining information on the position of the spinodal curve were applied in the past by Murakami et al. [31], Allain et al. [32], Naudon et al. [33], Delafond et al. [34], and most recently by Ungar et al. [35]. In [31 to 34] an Al–6.8 at% Zn alloy was investigated. These authors assume that the position of the coherent spinodal should be at about 127 °C for this alloy (□ in Fig. 1). This conclusion is mainly based on the assumption [32 to 34] that only in the case of a spinodally decomposing alloy a "ring" in the SAXS intensity curves and a maximum in the resistivity curve appear, but not in the case of the decomposition by the normal nucleation and growth process. Ungar et al. [36] took the steep increase of the total SAXS intensity during slow continuous cooling from the range of homogeneity as a proper test for finding the position of the coherent spinodal curve (▽ in Fig. 1).

It should be mentioned that the onset of easy decomposition of the supersaturated Al–Zn alloys was first found already in 1964 by Lašek [36] from the resistivity changes during cooling experiments with the Al–Zn alloys with zinc contents covering the entire range of GP zone miscibility gap. He associated this line with the temperature of easy formation of smaller GP zones (dotted line in Fig. 1). Later Simerská et al. [37] found by means of X-ray diffraction, TEM, and DTA investigations of alloys having zinc contents from about 10 to 40 at% the line of easy formation of α'_R-precipitates (▼ in Fig. 1) and that this line fits closely to that found by Lašek [36]. They explained the onset of easy homogeneous decomposition in terms of the classical nucleation theory, as the possibility of easy nucleation of GP zones and α'_R at temperatures near to the nose of the C-shaped T–T–T curve corresponding to the very first appearance of the α'_R-phase [37]. Most recently Wendrock et al. [38] determined the start of rapid homogeneous decomposition by the onset of the SAXS intensities near the primary beam (scattering angle 0.12°; CuK_α radiation) during slow continuous cooling from the range of homogeneity (see ●, ○, ▽ in Fig. 1 and the text delivered to Fig. 3). These results agree with the conclusions drawn in [36, 37].

From Fig. 1 it can be concluded that the information obtained by the qualitative experiments (▽, □) gives the apparent position of the coherent spinodal curve even above the calculated chemical spinodal curves [18, 19]. The position of the coherent spinodal obtained by the quantitative and semi-quantitative experiments (□, ●, ▲) is situated close to the calculated coherent spinodal curve [17, 18] (dotted line in Fig. 1) and below both calculated chemical spinodals [18, 19] as shown by the dashed and dash-dotted lines in Fig. 1.

From our point of view the controversy whether the results obtained by the qualitative experiments or those of the quantitative and semi-quantitative ones are more reliable, depends also on the question, if the calculated spinodals can be taken as reliable enough or not. Therefore, we would like to go deeper into this task. Of course, the spinodal curves are calculated on the base of an "ideal" binary system, that means neglecting vacancies, dislocations, stacking faults, and so on. But contributions of these lattice defects are at least partially taken into account in calculating the spinodals, since the thermodynamic data are taken from the upper miscibility gap boundary the position of which is also affected by these lattice defects. At least the position of the left-hand side of the chemical spinodal should be fairly sure. The boundary of the upper miscibility gap of the Al–Zn system (with the expense of the near vicinity of the top of this gap [39] not indicated in Fig. 1) is well known and therefore the lines calculated by means of these values should be fairly reliable. This statement can be supported by the calculated boundary of the metastable α_m-phase (Krupkowski [16], see the dash-double-dot line in Fig. 1) and the confirmation of the reliability of this line by experimental investigations (some are indicated in Fig. 1 by triangles) [40, 41]. From these facts it can be concluded that also the position of the chemical spinodal calculated by Krupkowski et al. [19] (dash-dotted line in Fig. 1) on the base of these data should be fairly reliable. This statement is confirmed by the position of the chemical spinodal calculated by other authors applying a somewhat different procedure [17, 18].

Of course, the position of the coherent spinodal shown in Fig. 1 is not as certain as that of the chemical spinodal, since the contributions of the coherency stresses to the free enthalpy are not so well known, but many authors [8, 18, 41, 42] agree in that the position of the coherent spinodal in the vicinity of the top of the miscibility gap should be situated about 30 K lower than that of the chemical one. In our opinion, therefore, the results obtained by the quantitative and semi-quantitative methods are with a high probability situated close to the coherent spinodal. The indications of changes in the decomposition mechanism [32 to 34] or the start of the rapid increase of decomposition [5, 35 to 38] are situated not far from the formation line of the GP zones or α'_R-phase [5, 36, 37] and above the chemical spinodal curves (Fig. 1). Therefore we doubt that the critical temperature of (127 ± 2) °C in the Al–6.8 at% Zn alloy corresponds to the spinodal curve. We completely agree with the opinion of Gerold and Kostorz [20] that qualitative indications such as the appearance or not of a ring in the SAXS curves and/or of a maximum of resistivity or a certain incubation time cannot be taken as a reliable test whether the decomposition is starting by the spinodal mode or by the normal nucleation and growth process.

3. On the Mode of Decomposition above the Temperature of the Coherent Spinodal

Ungar et al. [35] interpreted the steep increase of the total SAXS intensity during slow cooling as the onset of the spinodal decomposition, but the triangles (▽ in Fig. 1) indicating this steep rise are also clearly situated above the calculated chemical spinodal curves and, at least for the alloys with lower zinc content, not far from the curve of the start of rapid homogeneous decomposition (in the following shortly called curve of s.r.h.d.). Therefore, from our point of view, these triangles should not reflect the start of the decomposition by the spinodal mode.

The open question remains to elucidate what mechanism causes the fairly rapid start of homogeneous decomposition during the slow continuous cooling after passing this critical temperature. We will go deeper into this problem. In our opinion the mode of decomposition at temperatures between the curve of s.r.h.d. and the coherent spinodal during the cooling experiments as well as during isothermal ageing can be well understood in the following way. If one cools the Al–Zn alloys from the one-phase region by a proper rate, one can undercool it to temperatures close to the nose of the C-shaped curve for the formation of GP zones or α'_R-phase in the T–T–T diagram. In our opinion the curve of s.r.h.d. is close to this nose, at least approximately.

In order to obtain more information on the structure changes in the close vicinity of this curve we measured isothermally the SAXS curves of an Al–15 at% Zn alloy after cooling from 400 °C with a cooling rate of 1 K/min to the ageing temperature (T_a). The cooling was stopped at T_a and the SAXS intensities were measured at T_a. Some results are shown in Fig. 2.

It can be concluded from these measurements that the decomposition rate is very slow at 230 °C and starts to get rapid at 220 °C. $T = 220$ °C fits very well the curve of s.r.h.d., i.e. it corresponds to the temperature of the onset of decomposition within this alloy found by Simerská et al. [37] by means of DTA measurements applying a cooling rate of 2 K/min. From the course of the SAXS intensity curves the Guinier radii can be evaluated. The results are shown in Table 1. It can be concluded that the mean size of the precipitates is not essentially changed between 225 and 210 °C, in spite of the steep jump of the decomposition rate taking place during slow continuous cooling within this temperature range (see Fig. 3). This means most probably that above as well as below the curve of s.r.h.d. the same type of precipitates with approximately the same size has been formed.

The reason of the steep increase of the transformation rate (see, e.g., Fig. 3) can be visualized with reference to Fig. 4. By cooling the alloy at a proper rate to a temperature close below the GP zone or α'_R formation line and annealing it at this temperature the critical sizes and mutual distances of homogeneously nucleated precipitates

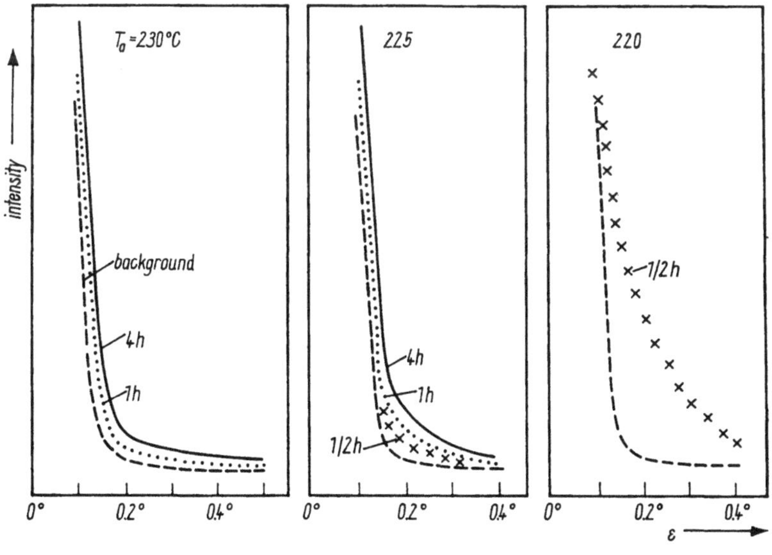

Fig. 2. Developement of the SAXS intensities during isothermal ageing at T_a of an Al–15 at% Zn alloy, cooled with 1 K/min from 400 °C to T_a

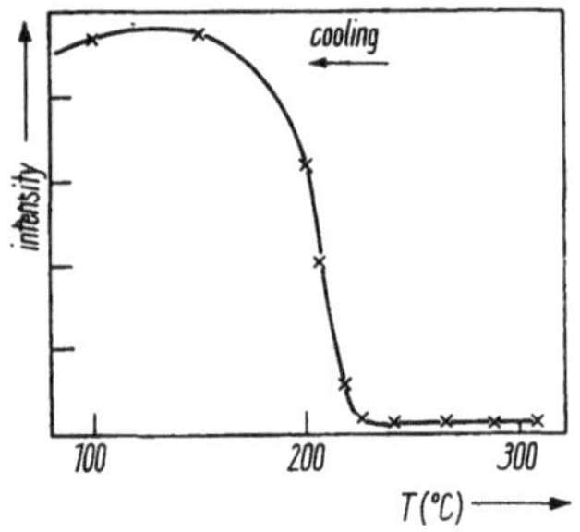

Fig. 3

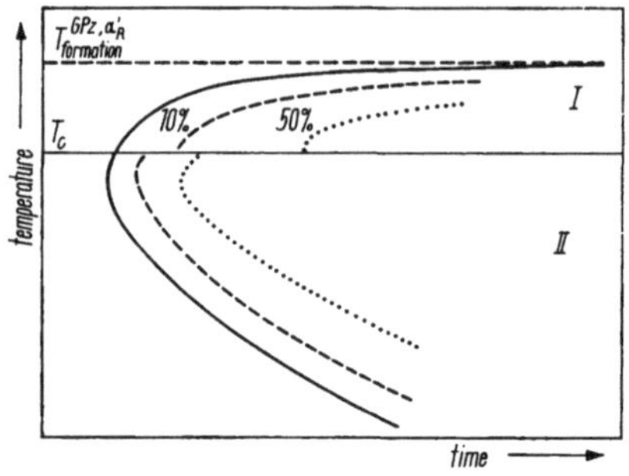

Fig. 4

Fig. 3. Course of the SAXS intensity near the primary beam ($\varepsilon = 0.12°$, CuK_α radiation) during continuous cooling (1 K/min) from 400 °C of an Al–15 at% Zn alloy

Fig. 4. T–T–T diagram (schematically) of the formation of GP zones and/or α_R'-phase above and below the critical temperature (T_c) of secondary autocatalytic nucleation of GP zones and/or α_R'-phase. I normal nucleation and growth; II normal nucleation and growth followed by secondary autocatalytic nucleation of GP zones

Table 1

Guinier radii of precipitates developed during isothermal ageing at various ageing temperature T_a (the experimental error of radii is about ±10%)

ageing time (min)	Guinier radii (nm)		
	$T_a = 225$ °C	$T_a = 220$ °C	$T_a = 210$ °C
10	10.3	10.4	11.5
30	10.8	10.4	10.7
60	10.6	10.4	12.0
120	10.6	11.5	11.3
180	10.2	11.5	12.3
240	10.0	11.5	12.3

are large. The attainment of metastable equilibrium proceeds by a nucleation of precipitates almost independent of each other and by their growth by volume diffusion from the surrounding matrix. By undercooling the alloy to a certain critical temperature (T_c) the sizes of regions containing the critical nucleus with its adjoining Zn-depleted surrounding become sufficiently small to initiate the autocatalytic nucleation of precipitates within the matrix adjoining such regions. Such secondary nucleation promoted by the primary nucleation process allows the alloy to attain much faster the metastable state as compared to processes above T_c. Such a cooperative nucleation process involves local long-range electronic and stress-field interactions. The cooperative nucleation leads to a fairly good quasi-periodic alignment of secondary nucleated zones around the primary nucleus. The fast mode of decomposition is completed by impingement of regions with internally aligned zones. This explains the ring in SAXS since the intensity profile reflects the interparticle interference effect and is the superposition of intensities from regions comprising local zones with fairly good quasi-periodic arrangement as already proved quantitatively by Syneček [43] and Syneček and Sebo [7]. One cannot get otherwise a strong interparticle effect in such polydisperse system like GP zones as already mentioned by Syneček in 1962 [43].

We would like to mention that this is also supported by Lorimer [44] who explained the step-like alignment of GP zones in Al–Cu, reported by Rioja and Laughlin [45], by an autocatalytic effect. "The strain field associated with one GP zone could trigger the nucleation of a second zone." [44]. This mode is further supported by analysis of the time dependence of the course of the electrical resistivity during isothermal ageing of various Al–6 at% Zn, Al–(3 to 6) at% Zn–(0.1 to 2) at% Mg alloys [46]. The *R–t* curves can be fitted from the start of decomposition up to the time at which the maximum of resistivity appears by means of the theory developed recently by Radomsky and Löffler [47, 48] assuming that only the amount of GP zones increases, but the size of the zones remains constant. The decomposition by the autocatalytic mode could cause such a behaviour at the start of decomposition.

The distinction between the spinodal mode and the normal nucleation process followed by the fast autocatalytic reaction is that by the operation of the latter mode the alloy is made not unstable against concentration fluctuation as a whole (as it is assumed in the spinodal theory) but strictly local from the moment when, e.g. the local coherency stresses can be partially releaved by a cooperative mechanism of secondary nucleation of a quasi-periodic array of zones. This can start in the case where the critical size of zones is sufficiently small and where a reasonable amount of zones can be created, i.e. fairly below the solvus of GP zones but sufficiently above the spinodal temperature.

From this work we conclude that the confusion in the literature concerning the initial decomposition in the Al–Zn alloys in the temperature range above the spinodal is just due to the fact that the possibility of cooperative autocatalytic nucleation was not taken into consideration.

Acknowledgement

We would like to express our thanks to Prof. R. Ciach from Cracow for his valuable hints.

References

[1] M. Simerská, P. Bartuška, and V. Syneček, Acta cryst. **A34**, 304 (1978).
[2] M. Simerská and V. Syneček, Acta metall. **15**, 223 (1967).
[3] A. Guinier, Mètaux, Corrosion, Usure **18**, 209 (1943).
[4] V. Gerold and W. Schweizer, Z. Metallk. **52**, 76 (1961).
[5] J. Lašek, Czech. J. Phys. **B15**, 848 (1965).
[6] M. Simerská, V. Syneček, and V. Šima, Czech. J. Phys. **B24**, 543 (1974).
[7] V. Syneček and P. Sebo, Czech. J. Phys. **B14**, 622 (1964).
[8] J. Lašek, J. Inst. Metals **95**, 320 (1967).
[9] V. Syneček and M. Simerská, Proc. Annu. Pittsburgh Diffraction Conf., Pittsburgh 1964 (p. 28).
[10] R. D. Garwood, A. L. Davies, and G. L. Richards, J. Inst. Metals 88, 375 (1960).
[11] M. Simerská and P. Bartuška, Czech. J. Phys. **B27**, 655 (1977).
[12] M. Simerská and V. Syneček, Proc. II. Internat. Conf. Age-Hardenable Aluminium Alloys, Visegrad (Hungary) 1979 (p. CSS2).
[13] R. Ciach, Arch. Hutnictwa **19**, 22 (1974).
[14] W. Merz and V. Gerold, Z. Metallk. **57**, 607 (1966).
[15] H. Löffler, O. Kabisch, and R. Kroggel, phys. stat. sol. (a) **49**, 299 (1978).
[16] A. Krupkowski, Rev. Métall. **59**, 34 (1962).
[17] H. Löffler, H. Bönning, and O. Simmich, Bull. Acad. Sci., Ser. sci. Tech. **22**, 7 (1974).
[18] H. Löffler and O. Simmich, phys. stat. sol. (a) **30**, 495 (1975).
[19] A. Krupkowski, R. Ciach, and J. Krol, Arch. Hutnictwa **13**, 17 (1968).
[20] V. Gerold and A. Kostorz, J. appl. Cryst. **11**, 376 (1978).
[21] J. W. Cahn, Acta metall. **10**, 907 (1962).

[22] K. B. RUNDMAN and J. E. HILLIARD, Acta metall. **15**, 1025 (1967).
[23] R. ACUÑA and A. BONFIGLIOLI, Acta metall. **22**, 399 (1974).
[24] T. L. BARTEL and K. B. RUNDMAN, Metallurg. Trans. A **6**, 1887 (1975).
[25] R. KROGGEL, H. LÖFFLER, and O. KABISCH, Kristall und Technik **12**, 737 (1977).
[26] R. KROGGEL, H. LÖFFLER, and O. KABISCH, phys. stat. sol. (a) **40**, K163 (1977).
[27] R. CIACH, H. LÖFFLER, J. SALAWA, and G. WENDROCK, phys. stat. sol. (a) **53**, 441 (1979).
[28] J. DUTKIEWICZ, L. LYTINSKA, J. SALAWA, G. WENDROCK, and R. KROGGEL, see [12] (p. CSS3).
[29] J. W. CAHN, Acta metall. **14**, 1685 (1965).
[30] O. SIMMICH and H. LÖFFLER, phys. stat. sol. (a) **59**, K207 (1980).
[31] M. MURAKAMI, O. KAWANO, and Y. MURAKAMI, Acta metall. **17**, 1517 (1969).
[32] J. ALLAIN, A. NAUDON, J. DELAFOND, A. JUNQUA, and J. MIMAULT, Scripta metall. **8**, 831 (1974).
[33] A. NAUDON, J. ALLAIN, J. DELAFOND, A. JUNQUA, and J. MIMAULT, Scripta metall. **8**, 1105 (1974).
[34] J. DELAFOND, A. JUNQUA, J. MIMAULT, and J. R. RIVIERE, Acta metall. **23**, 405 (1975).
[35] T. UNGAR, J. LENDVAI, and I. KOVACS, privat communication.
[36] J. LAŠEK, phys. stat. sol. **5**, K117 (1964).
[37] M. SIMERSKÁ, V. ŠIMA, and P. BARTUŠKA, Czech. J. Phys. **B24**, 654 (1974).
[38] G. WENDROCK, K. SIEBERT, D. SELKA, and K. SKALITZ, Wiss. Z. PH Halle **1**, 23 (1981).
[39] A. A. PRESNYAKOV, A. Y. GORBAN, and V. V. CHERVYAKOVA, Zh. fiz. Khim. **35**, 1289 (1961).
[40] C. J. G. CARPENTER and R. D. GARWOOD, J. Inst. Metals **96**, 301 (1966).
[41] J. E. HILLIARD, in: Phase Transformation, Chap. 12, Amer. Soc. Metals, Cleveland (Ohio) 1970 (p. 536).
[42] J. W. CAHN, J. chem. Phys. **42**, 93 (1965).
[43] V. SYNEČEK, J. Phys. Radium **23**, 828 (1962).
[44] G. W. LORIMER, Phase Transformation Spring Residential Conf., London 1979, Vol. 2 (p. 69).
[45] R. J. RIOJA and D. E. LAUGHLIN, Metallurg. Trans. A **8**, 1257 (1977).
[46] W. OETTEL, C. RADOMSKY, and H. LÖFFLER, Kristall und Technik **15**, 713 (1980).
[47] C. RADOMSKY and H. LÖFFLER, phys. stat. sol. (a) **54**, 615 (1979).
[48] C. RADOMSKY and H. LÖFFLER, Exper. Tech. Phys. **28**, 211 (1980).

(Received January 5, 1981)

phys. stat. sol. (a) **65**, 205 (1981)

Subject classification: 14.3; 22.8.2

Laboratoire de Recherches sur la Réactivité des Solides, associé au C.N.R.S., Faculté des Sciences Mirande, Dijon[1]) *(a)*
and Laboratoire de Chimie des Matériaux Inorganiques, Université Paul Sabatier, Toulouse III[2]) *(b)*

Electrical Conductivity of Zinc–Iron Ferrites in Vacuum and in the Presence of Oxygen

By

B. Gillot (a), R. M. Benloucif (a), and A. Rousset (b)

The electrical conductivity of finely-divided spinels of composition $(Zn_x^{2+}Fe_{1-x}^{2+}Fe_2^{3+})O_4^{2-}$ is studied in vacuum as a function of temperature. From this study it is concluded that fast electron exchange takes place amongst Fe^{3+} and Fe^{2+} ions situated on B-sites in the spinel lattice. For some compositions a break of the conductivity–temperature curves is found near the Curie point. The profile of the $\sigma = f(t)$ curves during oxidation of those n-type spinels in defect phases γ-$x(ZnFe_2O_4)$, $(3/2)(1 - x)Fe_2O_3$ in the temperature range 190 to 300 °C is related to the presence of vacancies on B-sites which upset the electron exchange Fe^{3+}–Fe^{2+}. When the temperature is raised ($t > 400$ °C) the $\log \sigma = f(1/T)$ curves show discontinuities attributed to the precipitation of hematite.

La conductivité électrique des spinelles finement divisés $(Zn_x^{2+}Fe_{1-x}^{2+}Fe_2^{3+})O_4^{2-}$ est étudiée sous vide en fonction de la température. A partir de cette étude il est conclu que le rapide échange d'électrons se fait entre les ions Fe^{3+} et les ions Fe^{2+} sur les sites B du réseau spinelle. Pour plusieurs compositions une cassure dans les courbes conductivité–température a été trouvée dans le domaine du point de Curie. L'évolution des courbes $\sigma = f(t)$ lors de l'oxydation de ces spinelles de type n en phase lacunaire γ-$x(ZnFe_2O_4)$, $(3/2)(1 - x)Fe_2O_3$ dans le domaine de température 190 à 300 °C est reliée à l'introduction de lacunes dans les sites B qui perturbent l'échange électronique Fe^{2+}–Fe^{3+}. Lors d'une élévation de température ($t > 400$ °C) les courbes $\log \sigma = f(t)$ présentent une discontinuité attribuée à la précipitation d'hématite.

1. Introduction

In previous papers [1 to 3] the electrical conductivity of fine-grained magnetites substituted by trivalent ions $(Fe^{2+}Fe_{2-x}^{3+}M_x^{3+})\,O_4^{2-}$ $(0 < x < 2,\ M^{3+} = Al^{3+}, Cr^{3+})$ and iron-II chromi-aluminates $(Fe^{2+}Al_{2-x}^{3+}Cr_x^{3+})\,O_4^{2-}$ $(0 < x < 2)$ has been investigated as well as the evolution of conductivity in the presence of oxygen during the oxidation of these spinels at low temperature ($T < 300$ °C) to γ defect phases of same spinel structure.

Magnetites highly substituted by trivalent ions ($x > 1.60$) and iron-II chromialuminates which are direct spinels initially show a conductivity increase due to the formation of Fe^{3+} ions on tetrahedral sites (A sites) by oxidation, which favour the "hopping" of electrons on these sites. Maximum conductivity occurs approximately when the amuont of Fe^{2+} and Fe^{3+} ions on A-sites is equivalent (Fig. 1a). Then, when the amount of Fe^{3+} ions becomes large, the electronic exchange is upset by vacancies and conductivity decreases considerably. On the contrary, for inverse

[1]) B.P. 138, 21004 Dijon Cédex, France.
[2]) 118, route de Narbonne, 31000 Toulouse, France.

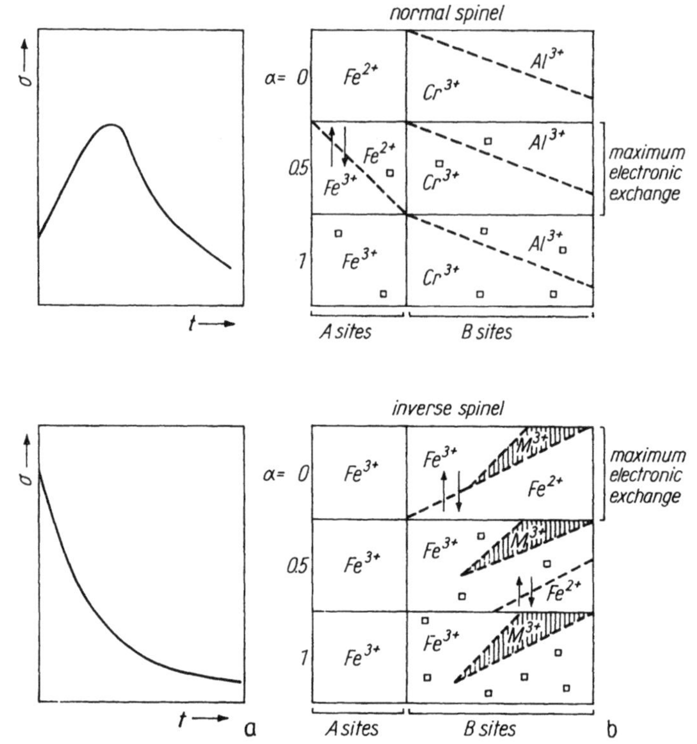

Fig. 1. a) Scheme of the variation of conductivity σ vs. time t for normal and inverse spinels. b) Scheme of the various Fe^{2+}–Fe^{3+} "hopping" possibilities for normal and inverse spinels. ↑↓ "hopping" possibilities, □ vacancies

spinels such as slightly aluminium- or chromium-substituted magnetites ($x < 0.80$) a conductivity decrease is always observed in the presence of oxygen (Fig. 1a) during oxidation to the γ defect phase due to the introduction of vacancies on octahedral sites, which upset the electronic exchange between Fe^{2+} and Fe^{3+} ions both present on equivalent sites, here octahedral sites (B-sites). Fig. 1b shows the various possibilities of exchange depending on the nature of the spinel and oxidation extent α. Only direct spinels and spinels totally oxidized into the γ-phase no longer possess this possibility of exchange and hence have a low conductivity.

A similar case is that of divalent-ion-substituted magnetites. The electrical properties of ferrous ferrites of type $(Fe_2^{3+}M_x^{2+}Fe_{1-x}^{2+})\,O_4^{2-}$ ($0 < x < 1$ and $M^{2+} = Zn^{2+}$, Co^{2+}, Ni^{2+}, ...) [4 to 10] or oxyfluorized with spinel structure $(Fe_{2-x}^{3+}N_x^{2+}Fe^{2+})\,O_{4-x}^{2-}F_x^-$ ($N = Fe^{2+}$, Ni^{2+}) [11] in fact depend basically on the possibility of exchange between cations of different valency on equivalent sites and the low conductivity of pure ferrites $(M^{2+}Fe_2^{3+})\,O_4^{2-}$ is explained by the absence of exchange [12, 13]. The oxidation of ferrous ferrites usually leads directly to the precipitation of the rhombohedral phase α-Fe_2O_3 with the large size of the crystallites from the initial phase [14 to 16].

Following the studies of Rousset and co-workers [17, 18] who prepared magnetites substituted by divalent ions sufficiently reactive to oxidize to the γ defect phase, the present paper reports the results of electrical conductivity for zinc–iron ferrites and for these phases either during oxidation to the γ defect phase or during the precipitation of α-Fe_2O_3 from γ-phases by temperature rise.

2. Experimental

2.1 Samples

In principle, the preparation conditions of magnetites substituted by divalent ion do not differ from those already used to obtain magnetites substituted by trivalent ions [19]. The thermal decomposition of mixed iron and zinc oxalate of type $(Fe_{1-x}Zn_x)C_2O_4$, $2\,H_2O$ $(0 < x < \frac{1}{3})$ leads, through adequate low-temperature treatments (<500 °C), to solid solutions $Fe_3O_{4(1-x)}ZnFe_2O_{4x}$ $(0 < x < 1)$ which are extremely fine grained solids and thus highly reactive.

The techniques used to characterize these samples, namely, chemical analysis of Fe^{2+}, X-ray analysis, SEM, and DTA, have also been already reported [19]. Crystallite size (determined from specific area and by electron microscopy) is around 600 Å. Synthesized materials are mixed spinels (Fe^{2+} ions are on octahedral sites and Zn^{2+} ions on tetrahedral sites) and the lattice parameter varies linearly with x rising from 8.397 Å for magnetite to 8.443 Å for zinc ferrite (Table 1). The formula of these compounds may be written as $(Zn_x^{2+}Fe_{1-x}^{3+})_A\,(Fe_{1-x}^{2+}Fe_{1+x}^{3+})_BO_4^{2-}$ $(0 < x < 1)$. $CoFe_2O_4$ was also prepared by an identical process [20] for a comparison with $ZnFe_2O_4$; the Co^{2+} ions have a stronger preference for the octahedral sites and the cation distribution is $(Fe^{3+})_A\,(Co^{2+}Fe^{3+})_BO_4^{2-}$.

Table 1

Physical properties of spinels $Zn_xFe_{3-x}O_4$ and of $CoFe_2O_4$

samples x	lattice parameter (Å)	θ (μV K^{-1})	E_A (eV) (temperature range 350—150 °C)	σ at 300 °C (Ω^{-1} cm^{-1})	T_C (°C)
0	8.397	− 55	—	3×10^{-1}	576
0.15	8.402	− 71	0.041	1.8×10^{-1}	527
0.27	8.409	− 77	0.055	1.4×10^{-1}	480
0.57	8.420	−153	0.059	1.1×10^{-1}	398
0.76	8.427	−270	0.107	5×10^{-2}	10
1	8.442	−530	1.310	10^{-6}	—
$CoFe_2O_4$	8.391	+232	0.723	4×10^{-5}	487

Defect spinel-type cubic sesquioxides were obtained by low-temperature oxidation (<300 °C) of the previous solid solutions according to the reaction

$$(Zn_x^{2+}Fe_{1-x}^{2+}Fe_2^{3+})O_4^{2-} + \tfrac{1}{4}(1-x)\,O_2 \rightarrow \gamma x(Zn^{2+}Fe_2^{3+}O_4^{2-})\,,\quad \tfrac{3}{2}(1-x)\;Fe_2^{3+}O_3^{2-}$$

The crystallographic and magnetic studies [21] showed that the vacancies were on octahedral sites, resulting in the structure

$$(Fe_{3(1-x)}^{3+}Zn_{3x}^{2+})_A\,(Fe_{5+x}^{3+}\,\square_{\,1-x})_B\;O_{12}^{2-} \qquad (0 < x < 1)\,.$$

Nevertheless, the lattice parameter varies linearly with x (Table 1). Finally the morphology of these defect spinels is similar to that of the initial phases; especially the crystallite sizes are almost the same. The study is not only restricted to the stability range of the γ-phases but extended to higher temperatures (>450 °C) where these phases are converted to α-phases according to the reaction

$$(Zn_{3x}^{2+}Fe_{8-2x}^{3+}\,\square_{\,1-x})\;O_{12}^{2-} \rightarrow 3xZn^{2+}Fe_2^{3+}O_4^{2-} + 4(1-x)\;Fe_2^{3+}O_3^{2-}\,.$$

2.2 *Measurements*

The dc conductivity of zinc–iron ferrites is measured between 700 and −40 °C with a method already reported [1, 3]. Before each measurement the sample is degassed at 700 °C for 3 h, the residual pressure being 10^{-8} atm. It is mounted in the measuring apparatus as cylindrical pellets of about 0.8 cm diameter and 0.2 cm thickness. The Seebeck coefficient measurements are made at temperature intervals of $\Delta T \approx 40$ K (against platinum). Oxidation the γ defect phase is conducted at 300 °C with 2×10^{-2} atm pure oxygen (N 48) containing less than 3 ppm H_2O. The conductivity evolution of these γ-phase is then determined without this oxygen pressure for higher temperatures.

3. Results and Discussion

Fig. 2 shows the overall behaviour of the electrical conductivity of both zinc-substituted magnetites of largely varied composition, in vacuum and in the presence of oxygen. Whereas the conductivity varies little in vacuum over a large temperature range (curves 1 and 3), it is observed, however, that in the presence of oxygen (curves 2 and 4), from a given temperature (point B), the behaviour is largely different, similar to that already observed for aluminum or chromium slightly substituted magnetites [1]. Part BC is attributed to defect ferrite; part DC is specific to this defect spinel, and DE corresponds, from X-ray analysis, (just before D and after E) to the precipitation of the rhombohedral phase α-Fe_2O_3.

Thus, conductivity in vacuum, the formation kinetics of the γ defect phase, and the conversion to the α-Fe_2O_3 phase were investigated.

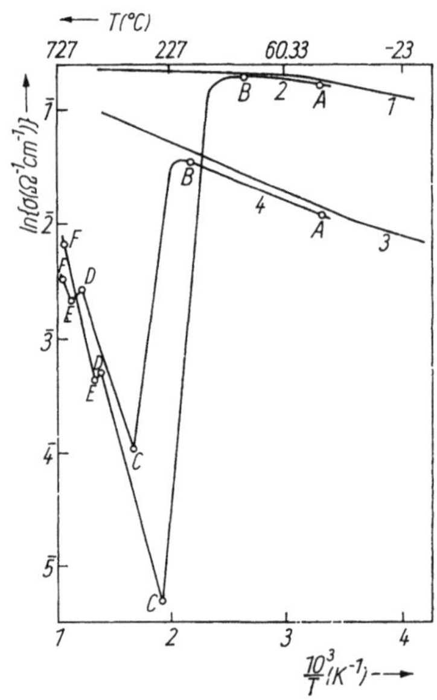

Fig. 2

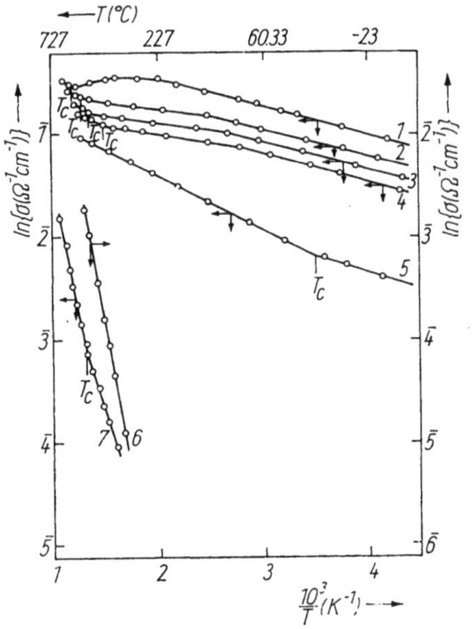

Fig. 3

Fig. 2. Influence of vacuum and oxygen pressure on the conductivity of spinels $(Zn_x^{2+}Fe_{1-x}^{2+}F_2^{2+}) \cdot O_4^{2-}$. (1) $x = 0.15$ and (3) $x = 0.76$: spinel in vacuum, (2) $x = 0.15$ and (4) $x = 0.76$: oxygen pressure 2×10^{-2} atm

Fig. 3. Conductivity of spinels $(Zn_x^{2+}Fe_{1-x}^{2+}Fe_2^{3+})O_4^{2-}$ plotted on a logarithmic scale against reciprocal temperature. (1) $x = 0$, (2) 0.15, (3) 0.27, (4) 0.57, (5) 0.76, (6) 1, (7) $CoFe_2O_4$

3.1 *Zinc-substituted magnetites* $(Zn_x^{2+}Fe_{1-x}^{3+})_A\ (Fe_{1-x}^{2+}Fe_{1-x}^{3+})_B O_4^{2-}$

The variation of conductivity with temperature over the range −44 to 700 °C for all ferrites is shown in Fig. 3. A $\ln \sigma$ versus $10^3/T$ plot yields straight lines. Three types of straight lines could be distinguished. Composition $x = 1$ reveals a straight line over the entire temperature range. Those of $x = 0.15$ and $x = 0.57$ yield straight lines with two breaks, whereas for $x = 0.76$ and for $CoFe_2O_4$ only one break could be detected. All breaks (noted T_C in Fig. 3) are in the vicinity of the Curie points of the respective spinels [22, 23] and it can be seen that as the zinc concentration is decreasing T_C is moving towards higher temperatures (Table 1). Magnetite is an exception among ferrites in that it behaves as a semiconductor above the Curie point of approximately 578 °C, exhibits an anomalous drop in resistance with the appearance of magnetic order, and has a positive temperature coefficient of resistivity characteristic of metallic conduction in the range 120 to 500 °C [24] (Fig. 3).

The activation energies (E_A) of all ferrites along with conductivity (σ) at room temperature and the Seebeck coefficient (θ) are plotted in Table 1 against the composition x. The increase in E_A (the activation energy increases as the conductivity decreases) for $x > 0.76$ is likely to be a consequence of the fact that in this composition range the number of Fe^{2+} ions on the B-sites of the spinel lattice becomes very small (whereas the B-site Fe^{3+} concentration remains high). In terms of a model of electron hopping between B-site Fe^{3+} and Fe^{2+} ions, this means that the electron exchange is suppressed to a large extent and a second conduction mechanism may be effective. An analogous observation of an increase in E_A associated with a pronounced drop in the concentration of B-site Fe^{2+} ions had been observed for magnetites substituted by trivalent ions [1] and for several ferrite spinels with a variable content of divalent iron ions [25 to 27]. For an analysis of E_A in terms of a detailed conduction model it is important that a plot of E_A versus composition refers to the same physical state, for instance to a ferrimagnetic or paramagnetic state. Therefore, the results presented in Table 1 are slightly altered since E_A is related to ferrimagnetic samples for $x = 0.15$, 0.57 and paramagnetic ones for $x = 0.76$. It might be that in the various spinel series mentioned above the difference in E_A below and above the Curie points are at least to a certain extent responsible for the change observed in E_A with composition.

The Seebeck coefficient measurements showed that zinc-substituted magnetites are n-type semiconductors (Table 1) and in the region $0.1 < x < 1$ (for $T > 300$ K) θ was decreasing slowly with temperature in disagreement with the results of Niclau et al. [5] who reported that θ was temperature-independent. $CoFe_2O_4$ has a positive Seebeck coefficient (Table 1) and the plus sign can be ascribed to electron jumps between the Co^{2+} and Co^{3+} ions

3.2 *Conductivity in the presence of oxygen*

3.2.1 *Study of oxidation versus time*

The samples were oxidized in the conductivity cell under 2×10^{-2} atm oxygen and the dependence of σ versus time is shown in Fig. 4. It is evident that for an increase in time the electrical conductivity decreases in all the investigated samples The electrical conductivity of the defect γ-phase was several orders of magnitude lower than that of unoxidized zinc-substituted magnetites The higher the vacancy concentration (the Fe^{2+} ion concentration decreases) the lower the conductivity and this until the γ-compound is obtained. The conductivity decrease is due to the presence of vacancies on octahedral sites acting as "foreign" ions which upset the B-sublattice and a second conduction mechanism may be effective. These partially oxidized zinc-substituted magnetites can be structurally formulated as

$$(Zn_x^{2+}Fe_{1-x}^{3+})_A\ (Fe_{(1-x)(1-3\delta)}^{2+}Fe_{x+1+2\delta(1-x)}^{3+}\ \square_{\delta(1-x)})_B O_4^{2-}\ ; \qquad 0 < \delta < \tfrac{1}{3}\,.$$

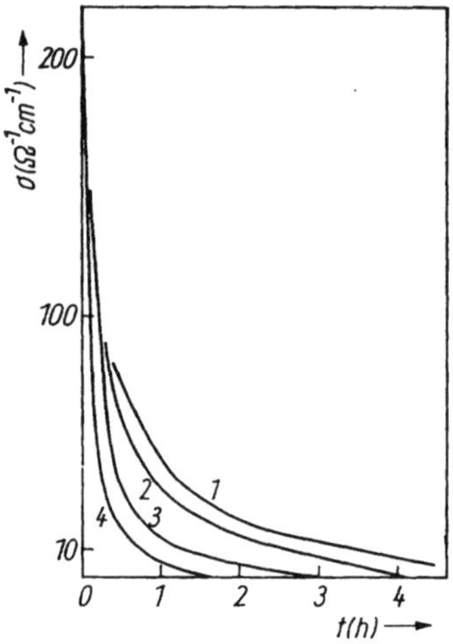

Fig. 4. Variation of conductivity σ vs. time t for different oxidation temperatures of spinels $(Zn_x^{2+}Fe_{1-x}^{2+}Fe_2^{3+})O_4^{2-}$.(1) $x = 0.76$ ($T = 290$ °C), (2) 0.57 (235 °C), (3) 0.27 (185 °C), (4) 0.15 (160 °C): oxygen pressure 2×10^{-2} atm

This behaviour in the presence of oxygen is similar to that already observed for aluminium of chromium slightly substituted magnetites [2], but is different from that reported for solid solutions $(Fe^{2+}Al_{2-x}^{3+}Cr_x^{3+})O_4^{2-}$ where conductivity increases initially with time [3].

From the above results we have plotted the $\alpha = f(t)$ curve, α being the transformation degree defined by $\alpha = (\sigma_0 - \sigma_t)/(\sigma_0 - \sigma_f)$, where σ_0 is the conductivity in vacuum at the given temperature, σ_t conductivity at time t, σ_f the the conductivity of the γ-phase (totally oxidized product). Such curves are reported in Fig. 5. They are similar to the $\alpha = f(t)$ curves obtained in thermogravimetry by recording the weight gain corresponding to oxygen fixation [28] (α is determined by the ratio of partial oxidation weight gain to completed oxidation weight gain). In particular, they can practically be superimposed (Fig. 6) whatever the composition; thus, we can deduce that the weight change recorded in thermobalance is equivalent to the conductivity change resulting from the introduction of vacancies on octahedral sites.

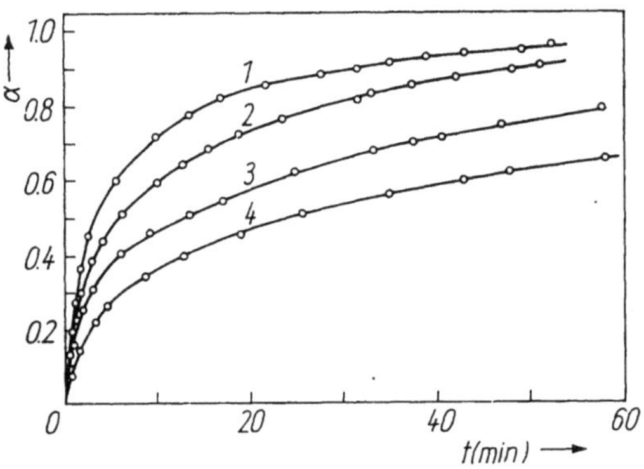

Fig. 5

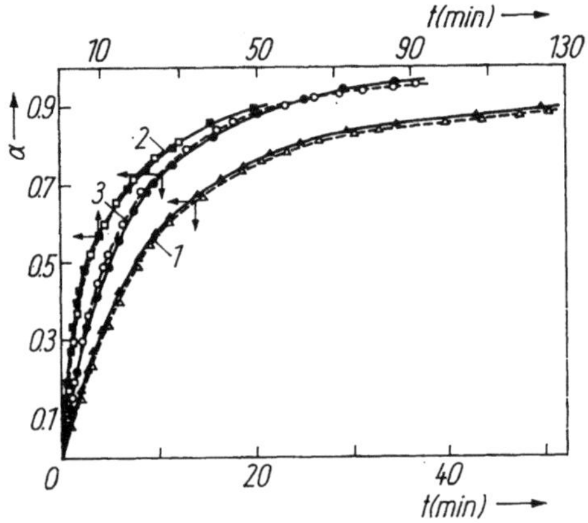

Fig. 6

Fig. 5. $\alpha = f(t)$ curves. (1) $T = 225$, (2) 212, (3) 203, (4) 193 °C, oxygen pressure 2×10^{-2} atm, $x = 0.57$

Fig. 6. Analogy between the $\alpha = f(t)$ curves obtained by electrical conductivity and by thermogravimetry. ----- conductivity (1) $x = 0.15$ ($T = 190$ °C), (2) 0.57 (212 °C), (3) 0.76 (283 °C); ——— thermogravimetry (1) $x = 0.15$ ($T = 181$ °C), (2) 0.57 (200 °C), (3) 0.76 (268 °C)

3.2.2. Precipitation of α-Fe_2O_3 from γ defect phases

When the γ-phases previously obtained over the temperature range 190 to 300 °C are submitted to successive temperature steps (interval from 10 to 100 °C) the $\ln \sigma = f(t)$ curves show discontinuities, which are represented in dashed lines (Fig. 7). The straight line part relating to temperatures lower than that of the discontinuity (line CD) corresponds to the defect phase φ and the straight line relating to temperatures above and below the discontinuity (line EFG) corresponds to the precipitation of α-Fe_2O_3, these phases being identified by X-ray analysis on either part of the discontinuities. Fig. 8 reports the temperatures T_D and T_E of the beginning and the end of each discontinuity versus composition x as well as the temperature T_M related to $(T_D + T_E)/2$. This temperature T_M is close to the transformation temperature of the γ- to the α-phase obtained by DTA [21] and for a same crystallite size it is the higher (Fig. 8), the greater the zinc-substitution degree is.

An increase in the temperature of precipitation as caused by the decrease in α-Fe_2O_3 concentration might be closely associated with the cation vacancy in the spinel lattice since homogeneous oxidation of ferrous ferrites generates cation vacancies whose concentration is proportional to the α-Fe_2O_3 content. Such an association is also indicated by the fact that the stability of solid solutions Fe_3O_4–γ-Fe_2O_3 increases with decreasing α-Fe_2O_3 content in the solid solution Fe_3O_4–γ-Fe_2O_3 [29] because the precipitation of hematite decreases the concentration of cation vacancies in the host phase.

Table 2 also includes the activation energies of the $\gamma(E_\gamma)$ and $\alpha(E_\alpha)$ phases determined from segments CD and EFG, the activation energy of the γ-phase being always higher than that of the corresponding α-phase. However, this energy is little significant

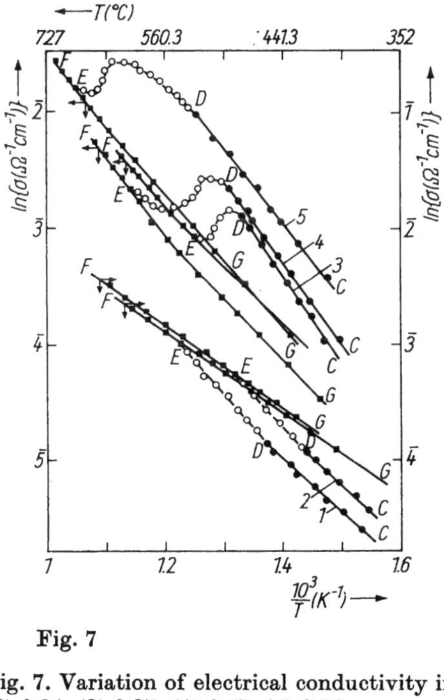

Fig. 7

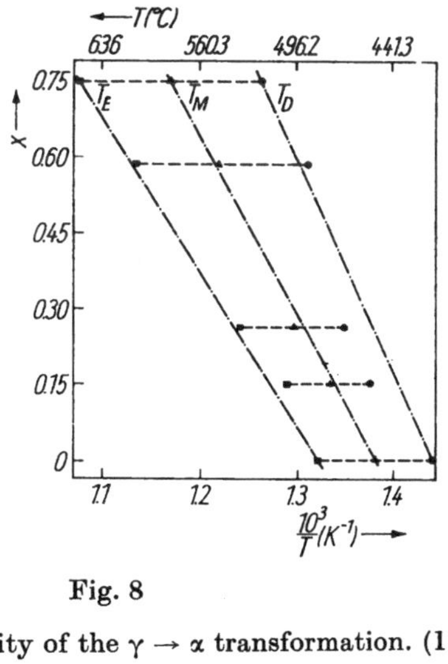

Fig. 8

Fig. 7. Variation of electrical conductivity in the vicinity of the $\gamma \rightarrow \alpha$ transformation. (1) $x = 0$, (2) 0.14, (3) 0.27, (4) 0.57, (5) 0.76

Fig. 8. Variation of T_D (●), T_E (■), and T_M (▲) with composition x

in the case of α-phases resulting from γ-phases substituted by divalent ions of during precipitation two phases occur in variable proportion depending on the substitution degree in divalent ions.

Table 2
Physical properties of the defect phase γ and of the γ → α transformation

samples x	lattice parameter (Å)	E_γ (eV)	θ (μV/K^{-1})	E_α for $T < 560$ °C (eV)	T_M (°C)
0	8.346	1.32	−180	1.16	423
0.15	8.360	1.34	−150	0.96	480
0.27	8.372	1.38	−195	1.15	513
0.57	8.401	1.57	−200	1.20	557
0.76	8.417	1.60	−305	1.36	590

Finally, for a given composition, the activation energy increases when the unoxidized spinel is converted to the totally oxidized spinel and the sign of the Seebeck coefficient involves the presence of n-type semiconduction (Table 2).

4. Conclusion

The results of the investigation of the electrical properties of the $(Zn_x^{2+}Fe_{1-x}^{2+}Fe_2^{3+})O_4^{2-}$ system in the composition range $0 < x < 0.76$ can be explained by a hopping mechanism of conduction with a low activation energy, but for $x > 0.76$ the Fe^{2+} ion concentration on B-sites becomes vanishingly small and the activation energy increases appreciably. It may be noted that in this system the appearance of magnetic order is related to a change in the activation energy of semiconduction and to a displacement in the Curie point T_C; this change becomes progressively more pronounced with increasing Fe^{2+} concentration and the Curie point is moving towards higher temperatures as the zinc content is decreasing.

These inverse spinels, which are n-type semiconductors, undergo a large conductivity decrease during oxidation in the defect phase γ, related to the generation of Fe^{3+} and vacancies on B-sites upsetting the electron exchange between the Fe^{2+} and Fe^{3+} ions on B-sites. The activation energy of the defect phase γ is within the range of 1.22 and 1.62 eV for $x = 0$ and $x = 0.7$, respectively. In addition, we showed, referring to the temperatures of the beginning and the end of the discontinuity in the $\ln \sigma = f(1/T)$ curves that the transformation temperature of the γ-phases to α-Fe_2O_3 increases with the zinc substitution degree.

References

[1] B. Gillot, J. F. Ferriot, and A. Rousset, J. Phys. Chem. Solids **37**, 857 (1976).
[2] B. Gillot, J. Phys. Chem. Solids **38**, 751 (1977).
[3] B. Gillot, F. Bouton, F. Chassagnaux, and A. Rousset, phys. stat. sol. (a) **50**, 109 (1978).
[4] Clinton F. Jefferson and C. Kenneth Baker, IEEE Trans. Magnetics **4**, 460 (1968).
[5] P. Nicolau, I. Bunget, M. Rosenberg, and I. Belciu, IBM J. Res. Developm. **14**, 248 (1970).
[6] E. Schmidbauer, Z. Geoph. **37**, 425 (1971).
[7] A. A. Samokhvalov and A. G. Rustamov, Soviet Phys. — Solid State **7**, 961 (1965).
[8] V. R. K. Murthy and J. Sobhanadri, phys. stat. sol. (a) **38**, 647 (1976).
[9] N. M. Botrous el Badramany, E. F. Mina, H. D. Merchant, S. Arafa, and R. P. Poplawsky, J. Amer. Ceram. Soc. **62**, 113 (1979).

[10] O. S. JOSYULU and J. SOBHANADRI, phys. stat. sol. (a) **59**, 323 (1980).
[11] A. CASALOT, J. CLAVERIE, and P. HAGENMULLER, J. Phys. Chem. Solids **34**, 347 (1973).
[12] E. J. W. VERWEY, in: Semi-Conducting Materials Butterworths, London 1951 (p. 15).
[13] A. ROSENCWAIG, Canad. J. Phys. **47**, 2309 (1969).
[14] C. OKAZAKI, J. Phys. Soc. Japan **15**, 2013 (1960).
[15] R. J. MISHRA and G. THOMAS, J. Amer. Ceram. Soc. **62**, 293 (1979).
[16] P. VESSIERE, J. RABIER, H. GAREM, and J. GRILLE, Phil. Mag. A **40**, 553 (1979).
[17] A. ROUSSET, P. MOLLARD, and A. GIRAUD, C.R. Acad. Sci. (France) **275**, 709 (1972).
[18] P. MOLLARD, A. COLLOMB, J. DEVENYI, A. ROUSSET, and J. PÂRIS, IEEE Trans. Magnetics **11**, 894 (1975).
[19] A. ROUSSET, J. PÂRIS, and P. MOLLARD, Ann. Chim. **7**, 119 (1972).
[20] P. MOLLARD, J. PÂRIS, and A. ROUSSET, Brevet ANVAR n° 7214215, Paris 1972.
[21] G. DUPRE, A. ROUSSET, and P. MOLLARD, Mater. Res. Bull. **11**, 473 (1976).
[22] R. PARKER and C. J. TINSLEY, In: Proc. Internat. Conf. of Ferrites Japan 1970 (p. 591).
[23] N. REZLESCU, C.R. Acad. Sci. (France) **268**, 136 (1969).
[24] D. O. SMITH, Progr. Rep. Lab. Insulator Res. M.T.I. **11**, 53 (1952).
[25] S. K. BANERJEE and W. O'REILLY, IEEE Trans. Magnetics **2**, 463 (1966).
[26] Z. SIMSA and N. ANDREJEV, Czech. J. Phys. **B 19**, 1389 (1969).
[27] Z. FUNATOGAWA, N. MIYATA, and S. USAMI, J. Phys. Soc. Japan **14**, 854 (1959).
[28] B. GILLOT, F. CHASSAGNEUX, and A. ROUSSET, J. Solid State Chem. **38**, 451 (1981).
[29] B. GILLOT, Mater. Res. Bull. **13**, 783 (1978).

(Received January 30, 1981)

phys. stat. sol. (a) **65**, 215 (1981)

Subject classification: 10.1 and 14.3; 1.2; 22.8

Department of Physics, Aristoteles University of Tessaloniki[1])

On the Disordered Nature of Single Crystalline $TlBiTe_2$ and Its Influence on Its Electrical Properties

By

O. VALASSIADES, E. K. POLYCHRONIADIS, J. STOEMENOS, and N. A. ECONOMOU

$TlBiTe_2$ single crystals are examined by TEM and it is found that their structure consists of coherent (100) and (110) twins and semicoherent type twins. The strain during the transformation from the high temperature cubic phase to the low temperature rhombohedral phase is relieved by the formation of lenticular dislocations and low angle boundaries. Electrical measurements indicate a transformation from a highly anisotropic to an isotropic structure at 672 K with a sudden jump in the number of carriers. The number of carriers released agrees well with the estimated number of the dangling bond atoms which disappear during the transformation.

$TlBiTe_2$-Einkristalle werden mittels TEM untersucht, und es wird gefunden, daß ihre Struktur aus kohärenten (100)- und (110)-Zwillingen und halbkohärenten Twins besteht. Die Spannung, die während des Übergangs von der kubischen Hochtemperaturphase zur rhomboedrischen Tieftemperaturphase auftritt, wird durch die Bildung von linsenförmigen Versetzungen und Kleinwinkelgrenzen beseitigt. Elektrische Messungen zeigen einen Übergang von einer stark anisotropen zu einer isotropen Struktur bei 672 K mit einem plötzlichen Sprung der Ladungsträgeranzahl. Die Zahl der beseitigten Träger stimmt gut mit der berechneten Zahl der „dangling-bond"-Atome überein, die während des Übergangs verschwinden.

1. Introduction

The compound $TlBiTe_2$ may be regarded as belonging to the IV–VI group of materials since the elements Tl and Bi precede and follow Pb in the periodic table. Hein and Swiggard [1] reported a superconducting behaviour of this material, while it presents a high thermoelectric figure of merit which makes it suitable for thermoelectric applications. Recently it has been proposed as a material for acoustoelectric applicaton [2]. Nevertheless the work on this compound is meager due to the fact that single-phase material has not been available.

$TlBiTe_2$ was first synthesized by Bither [3], while Hockings and White [4] determined its melting point to be 535 °C and found that its structure at room temperature has a rhombohedral distortion from the NaCl lattice with an elongation along the [111] direction. The pertinent crystallographic data are as follows; space group $R\bar{3}m–D_{3d}^5$, $a_0 = 8.137$ Å, and $\alpha = 32°18'$. Polychroniadis et al. [5] have suggested a structure with a non-primitive cell of 16 Tl, 16 Bi, and 32 Te atoms, a lattice parameter $a_0 = 12.988$ Å and a rhombohedral angle $\alpha = 88°21'$. In this structure the atoms are not placed randomly at the cation sublattice but rather form layers normal to the three-fold axis in the sequence–Tl–Te–Bi–Te– (Fig. 1).

On the basis of the behaviour of the isostructural compound $AgBiSe_2$ [6] which has a rhombohedral phase belonging to the space group D_{3d} and transforms to the disordered NaCl state at 287 °C, a transformation of the type $O_h \rightarrow D_{3d}$ is also ex-

[1]) Tessaloniki, Greece.

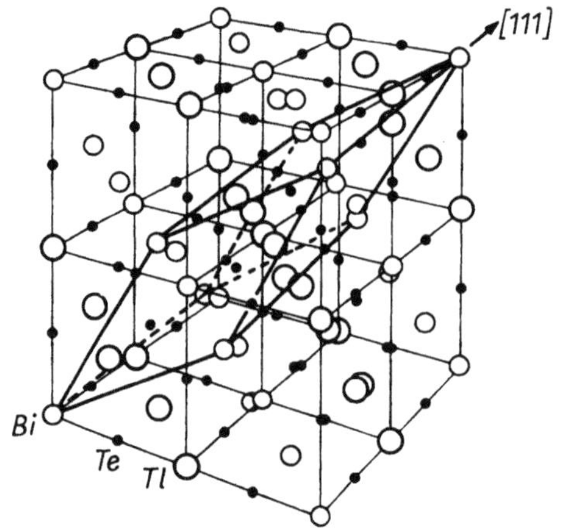

Fig. 1. The structure of ordered $TlBiTe_2$. The heavy lines represent the primitive rhombohedral unit cell

pected at some elevated temperature for $TlBiTe_2$. On the basis of this assumption and using the group theoretical treatment advanced by van Tendeloo and Amelinckx [7] the kinds of defects that are to be expected in the low-temperature phase of $TlBiTe_2$ may be obtained.

The electrical properties of $TlBiTe_2$ are not well defined. Jensen et al. [8] dealt with the differences in properties of $TlBiTe_2$ and TlTe, while Allgaier et al. [9] reported that it is a highly degenerate n-type semiconductor. The number of carriers is estimated to be of the order of $8.2 \times 10^{19}\ cm^{-3}$ [10]. On the basis of resistance discontinuities they predicted a phase transformation to occur at 170 K. Nevertheless they considered their results as not completely satisfactory due to the lack of single-phase material.

From the above considerations the study of $TlBiTe_2$ on well defined crystals was appropriate.

2. Crystal Growth

The $TlBiTe_2$ samples were prepared by the Bridgman technique from commercially available 99.999% pure elements. The elements were placed in exact stoichiometric proportions in an evacuated quartz ampoule. The lower end of the ampoule was tapered in order to favour crystal growth at a single point. The ampoule was placed in a vertical furnace at the temperature of 650 °C for more than 48 h to secure complete reaction and homogeneous melt. Afterwards it was lowered mechanically through a vertical temperature gradient with a speed of 1 K/h. The material obtained was single phase and consisted of single crystals with dimensions about 1 cm in diameter and 2 cm in length.

X-ray powder data and the m.p. agreed with those of Hockings and White [4]. The DTA measurements performed to establish the m.p. indicated also an exothermic peak at 699 K which lead to a speculation of a phase change at this temperature.

3. Structural Considerations

The transformation from the high-temperature cubic phase to the rhombohedral low-temperature that may be advanced as an hypothesis based on the behaviour of the isosturctural compounds $AgBiSe_2$, $AgBiTe_2$ [6] is strengthened by the DTA findings that establish the existence of a transformation occurring at 699 K and the electrical properties which indicate (see below) that indeed the material undergoes a phase change at this temperature from a structure highly anisotropic in resistance to an

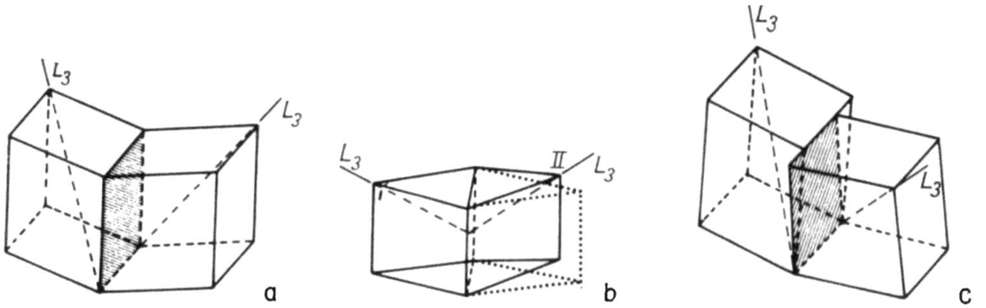

Fig. 2. a) Coherent {100}-type twin, b) Coherent {110}-type twin, c) semicoherent {100}-type twin. L_3 represents the three-fold axis

isotropic one. With this in mind using the group theoretical approach proposed by van Tendeloo and Amelinckx [7] the extended defects that are expected to occur in the low-temperature phase are easily predicted. Since this treatment has been applied previously for $TlBiTe_2$ [5] it is omitted here, but the main findings are that the ex-

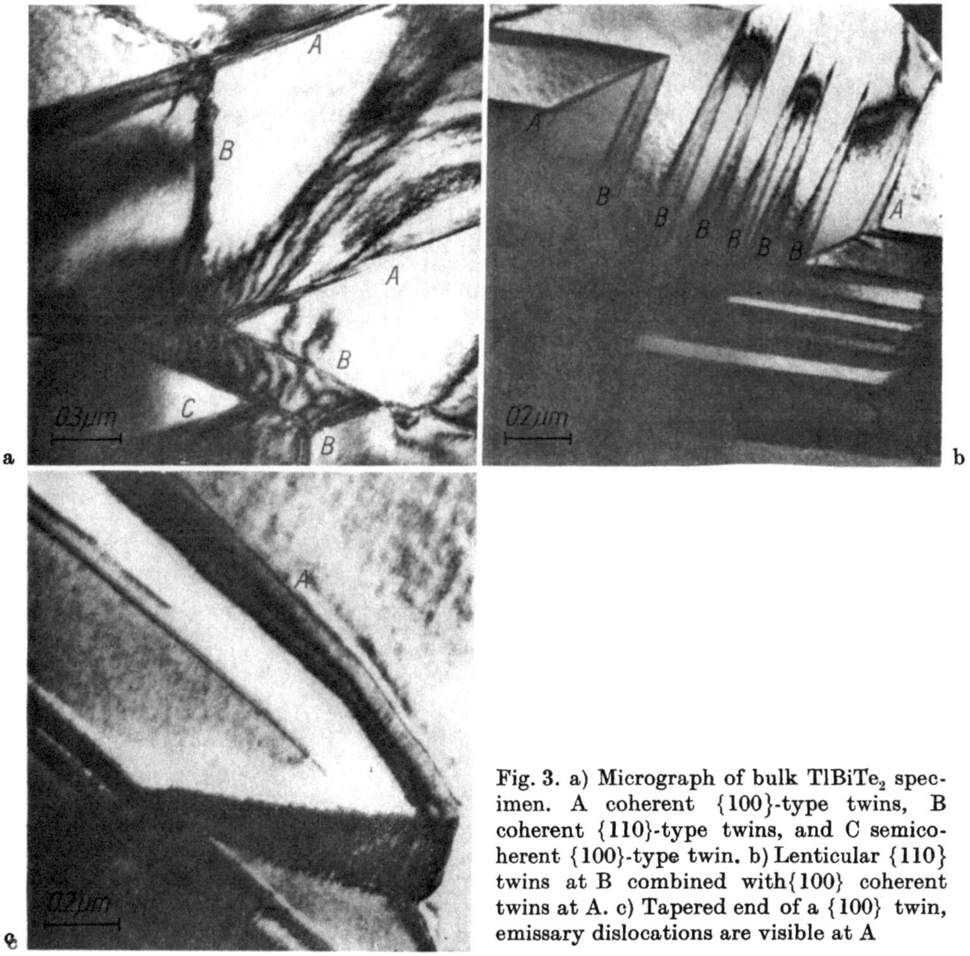

Fig. 3. a) Micrograph of bulk $TlBiTe_2$ specimen. A coherent {100}-type twins, B coherent {110}-type twins, and C semicoherent {100}-type twin. b) Lenticular {110} twins at B combined with {100} coherent twins at A. c) Tapered end of a {100} twin, emissary dislocations are visible at A

tended defects that are predicted by this approach consist of coherent {100} type twins, coherent {110} type twins, and semicoherent {100} type twins (Fig. 2a, b, c). Actually during the transformation $O_h \rightarrow D_{3d}$ the low-temperature phase of the lower symmetry develops along a given direction of the higher-symmetry parent structure; therefore the lower symmetry phase develops with equal probability in directions that are equivalent by a symmetry operation of the higher-symmetry structure which though is not a symmetry operation of the lower phase.

The electron microscopical findings confirmed the existence of the predicted extended defects as it is evident in Fig. 3a, b, c. The twin geometry was determined by calculating the twin vectors $\Delta \boldsymbol{g}$ from the diffraction patterns combined with image contrast tilting experiments. Thus in Fig. 3a twins labeled with the letter A were identified as coherent {100}-type twins, twins labeled with B were identified as coherent {110}-type twins, while C labels a semicoherent {100} twin. When two different coherent twins intersect, a low-angle boundary may be formed which separates the two regions of the crystal differing in orientation by 3°18′. Sets of dislocations are distributed uniformly in the low-angle boundary (L.A.B.) and are clearly visible (Fig. 4b). These dislocations relieve part of the strain due to the transformation from the high-temperature cubic phase to the low-temperature rhombohedral one.

Since the L.A.B. is perpendicular to the coherent {100} twin (Fig. 4a) the dislocations should be of the edge type, all parallel to the common false cube axis of the two grains. These dislocations have [100] Burgers vectors with $b = 12.9$ Å, which is the distance for the structural identity along this direction. The angle of the two grains is $\varphi = 3°18'$ so the space D between these dislocations is given by

$$\frac{b}{D} = 2 \sin \varphi/2 \approx \varphi\,. \tag{1}$$

Numerical calculations from equation (1) result in $D = 215$ Å. This number is in agreement with the spacing of the dislocations in the L.A.B., observed for instance in Fig. 4b, which is equal to $D = 240$ Å. Thus the L.A.B. introduce a linear density of [100] edge dislocations of the order $4 \times 10^5\ \text{cm}^{-1}$ along the boundary.

Evidence for the progressive nature of the twin formation and for the mechanism for microstructural development is provided by the frequent observation of twins

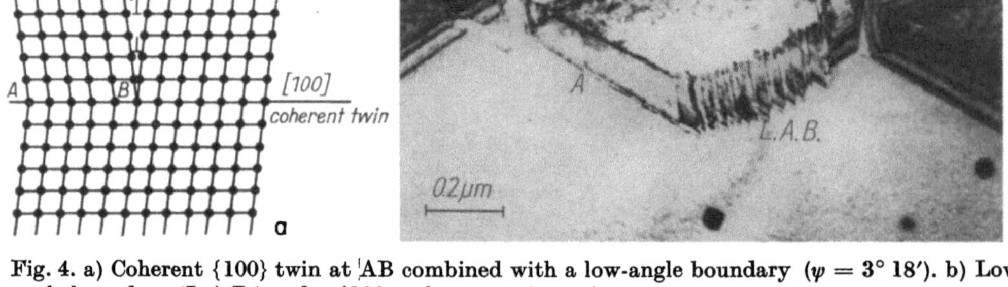

Fig. 4. a) Coherent {100} twin at AB combined with a low-angle boundary ($\psi = 3°\ 18'$). b) Low-angle boundary (L.A.B.) and a {100} coherent twin at A

with lenticular cross-sections which terminate within the crystal (Fig. 3b). These tapered cross-sections are typical for deformation twins, which arise from the atempt to retain during growth the twin interface parallel to the coherent twin boundary and reduce the shear strain at the twin periphery. Such a twin lamella consists of twinning dislocations (Fig. 3c). Since the ends of the boundaries are non-coherent they reduce the energy by creating emissary dislocations.

4. Electrical Properties

The electrical properties were measured using the van der Pauw configuration. In the experimental investigation of $TlBiTe_2$ the rhombohedral distortion should be taken into consideration, since this introduces an anisotropy. Therefore, the results from the measurements should be analysed according to the method proposed by Wascher [11]. The measurements were performed on circular discs cut in various crystallographic sections. Pressure contacts of Pt were used.

At first, room-temperature measurements were performed to establish the existing anisotropy. It was established that we were dealing with a resistance ellipsoid. We use here the cubic equivalent descriptions, defining as [111] dierction the c-axis of the rhombohedral structure.

In Fig. 5 a polar plot of the dispersion of the resistivity values on the two principal sections $(10\bar{1})$ and (111) is presented. From these results it is obvious that the small axis of the resistance ellipsoid lies in the c-axis direction, while the large axis lies in the $[1\bar{2}1]$ direction, perpendicular to the c-axis. Note that on the $(10\bar{1})$ section the resistivity can be described by the equation

$$\varrho = \varrho_{\parallel} + (\varrho_{\perp} - \varrho_{\parallel}) \cos^2 \omega \,, \tag{2}$$

where $\varrho_{\perp}$ and $\varrho_{\parallel}$ are the resistivities along the c-axis and perpendicular to it in the $[1\bar{2}1]$ direction. The values found for these resistivities were of the order 1×10^{-3} and 1.2×10^{-3} Ω cm, leading to an anisotropy value of $\lambda = 1.2 = \varrho_{\perp}/\varrho_{\parallel}$. The Hall coefficient was found isotropic with a room-temperature value equal to $R_H \approx -5 \times$ $\times\ 10^{-2}$ cm^{-3} C^{-1}, which leads to a nominal concentration of electrons

$$n^* = \frac{1}{R\,|e|} \approx 1.2 \times 10^{20}\ \mathrm{cm}^{-3}\,. \tag{3}$$

The temperature dependence of the resistivity was investigated in the temperature interval 13 to 740 K. The results are given in Fig. 6.

We may think of two separate behaviours, the low-temperature one where there are two different branches of resistivities to $\varrho_{\perp}$ and $\varrho_{\parallel}$ and the high-temperature behaviour from the temperature 672 K and above, where the two branches coincide to one. At 672 K an abrupt and substantial drop in the resistivity is observed. The

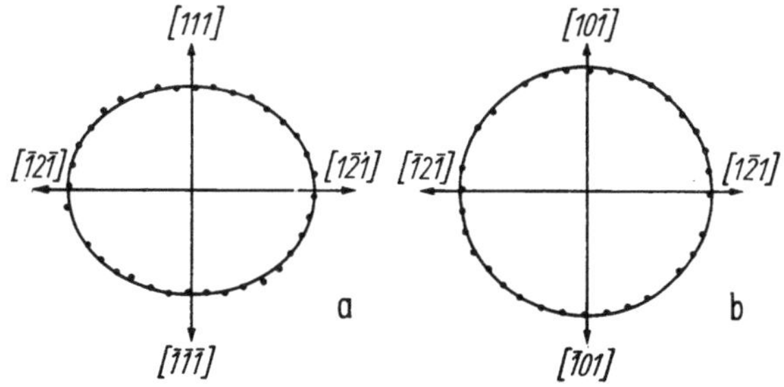

Fig. 5. A polar plot of the resistivity of $TlBiTe_2$ on two sections a) $(10\bar{1})$ and b) (111) showing the resistance anisotropy. The small axis of the resistance ellipsoid lies in the [111] direction, while the large axis is in the $[1\bar{2}1]$ direction

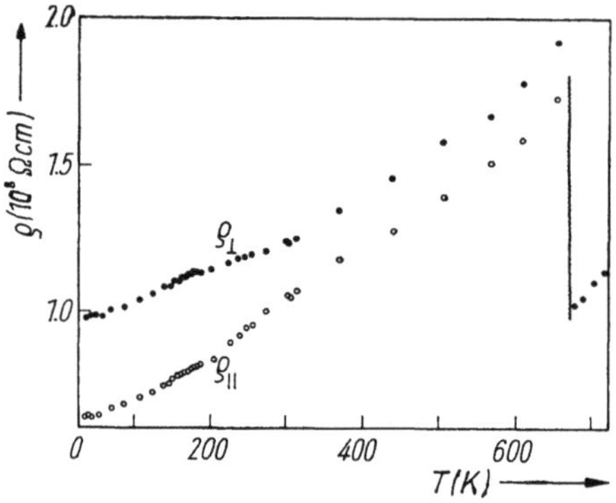

Fig. 6

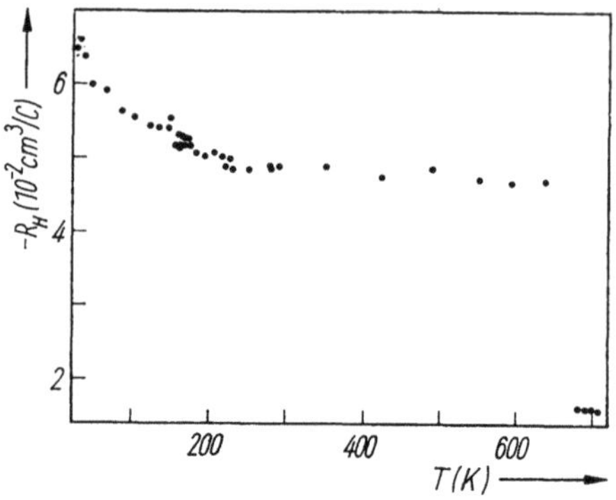

Fig. 7

Fig. 6. The dependence of the two principal resistivities upon temperature

Fig. 7. The dependence of the Hall coefficient upon temperature

behaviour at this temperature, where the disappearance of the resistance anisotropy is observed and the material at temperatures above behaves isotropically in resistance, combined with the exothermic peak observed at 699 K in the DTA measurements, suggests a phase transformation of the type $D_{3D} \rightarrow O_H$, as it is expected.

In the low-temperature region, as mentioned above, two distinctive resistivity branches are observed corresponding to $\varrho_\perp$ and $\varrho_\parallel$, respectively, with an almost parallel behaviour. We were careful in the measurements especially around 170 K where a resistance anomaly was observed previously [9]. The only noticeable feature in this region is a change in slope in both branches around 200 K. Nevertheless it is worth emphasizing that the curve reported by Allgaier et al. [9] is very close to the $\varrho_\perp$ data presented here. Incidentally there is a common feature showing a change in a lower slope above about 170 K.

Therefore, we may summarise the overall resistivity behaviour by saying that the resistivity increases with temperature in the whole range of temperatures studied, with an abrupt discontinuity occurring at 672 K, while the same trend with temperature continues also above this temperature. The temperature dependence of the Hall coefficient is presented in Fig. 7. It is found to be isotropic in all temperatures, with an abrupt change occurring again at 672 K. Here also we may notice a partial agreement with [9] having a very similar character in the $R(T)$ behaviour except that the previous results are 1.4 and $\approx 2\times$ bigger than ours. From these data we may calculate the carrier concentration. Since we have no information on the band structure of this material and there are no material constants available from the literature we may define a nominal concentration of the electrons by relation (3). A semilogarithmic plot of n^* versus $1/T$ leads to an activation energy of the order 1 meV (Fig. 8) below 200 K, while above there is a slight deviation from this value. This behaviour is reminiscent of highly defective semiconductors, especially GeTe.

With these results we may proceed to calculate the mobility, using the simple relation $\mu = \sigma R_H$. In Fig. 9 we present on a log–log basis the plot of μ_H versus T. It is obvious that the behaviour can be represented by a relation

$$\mu \sim T^{-a} \tag{4}$$

with the constant a assuming different values in different temperature ranges. Thus below 200 K both mobility branches have almost the same slope ≈ 0.3 with $\mu_\parallel > \mu_\perp$. Around 200 K the slope in both branches changes to 0.6 for $\mu_\parallel$ and 0.4 for $\mu_\perp$, respec-

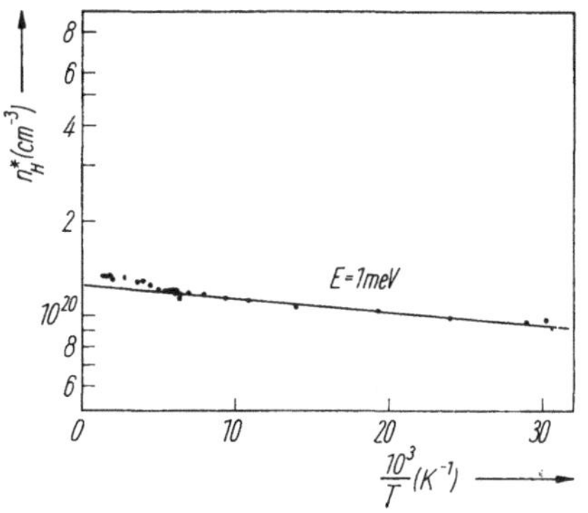

Fig. 8

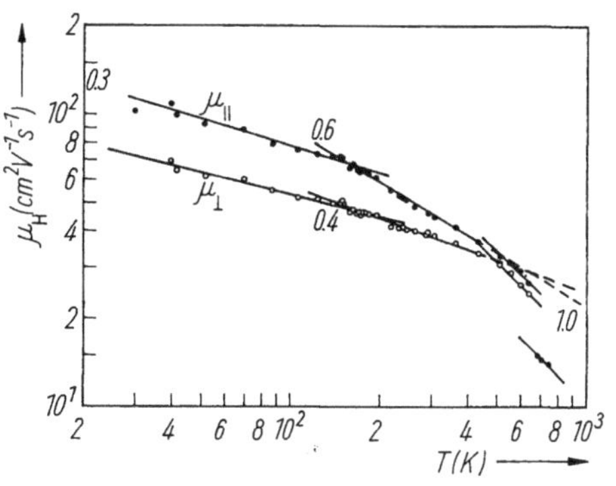

Fig. 9

Fig. 8. The dependence of the nominal concentration of electrons on the reciprocal temperature

Fig. 9. The dependence of the mobilities on temperature. A power law is observed with varying slopes

tively. At the temperature 500 K the slope changes again to a value of 1 for both branches, continuing with the same slope above the transformation temperature.

To assign a definite scattering mechanism will be premature, since our measurements were performed on samples of one composition and one should wait for an examination relating to compositional dependence. Nevertheless piezoelectric scattering should not be ruled out, modified by ionized defect scattering.

In conclusion we may say that the electrical measurements demonstrate the semiconducting behaviour of $TlBiTe_2$. The shallow temperature dependence of the carrier concentration on temperature is characteristic of highly disordered extrinsic semiconductor material, a view which is supported by the highly defective structure of this material. The present results support a phase change of the type $O_H \rightarrow D_{3D}$ occurring at 672 K, while at 200 K the change should not be attributed to a structural transition.

5. Discussion

It is an established fact that reorganization of the lattice accompanied by changes in the phonon frequency plays an important role in the phase transition of IV–VI compounds, the phase transition being aided by the reduction of the phonon frequencies [12]. Actually by keeping the ground state of the high-temperature phase, the equations of motion result in imaginary frequencies for the low-temperature phase. These vibrations are active in mixing the quasi-degenerate electron levels of the active centres that suffer the relative displacement due to a cooperative Jahn-Teller-like effect.

The findings that have been presented in the previous paragraphs introduce, apart from the relative distortion of the low-temperature phase, the notion of extended defects, semicoherent twin boundaries, and low-angle boundaries, that are formed as a requirement of the reduction in the lattice strain due to the transformation. The defects are actually connected with dangling bonds. The ordering of the DB atoms lead to a local reduction of the symmetry, the percursions of which are strongly anharmonic vibrations. This leads to an "equilibrium" deformation of the off-positioned atoms that form the semicoherent or low-angle boundaries and lead to states that may be considered as excited states with respect to their true equilibrium states on the normal sites. Thus a cup exists in the potential at the centre of these interstitices, the potential away it from being flat, since the off-centre symmetry is an equilibrium position as well as the normal site. Creation energy of the defects is gained by a summary stabilization of the lattice, introducing the notion of non-intimate valence alternation pairs [13]. The spatial configuration of the DB atoms has a drastic influence on the electronic states. Thus localized states are introduced which are associated with the strong electron–phonon coupling confined around the dangling bonds. The corresponding modification of the electronic structure can be characterized by the change of the local density of states on atoms with an atypical chemical composition of the nearest-neighbour sphere.

Dangling bonds act as recombination centres besides acting as acceptors or donors which alter the electrical properties of this material. Especially at the transformation from the low-temperature phase to the high-temperature phase where these interfaces are expected to disappear, a substantial change in the electrical properties is expected.

At this point it is worthwhile to compare the number of atoms with DB with the results obtained from the electrical measurements. At the phase transformation of $TlBiTe_2$ apart from the rhombohedral distortion a partial disorder is set due to the off-site atoms connected with the DB. An exact estimation of the size of the twins is not possible and inevitably it would be an approximation since it will depend on two-dimensional observations and on the statistical distribution of the size of the developed twins. Nevertheless a rough estimate can be made resulting in the number of twins per unit length equal to $N_L = 10^4$ twins cm^{-1} and a mean diameter of the twin area 1 μm^2. Thus the mean volume enclosed by the twin boundaries is 1 μm^3 or $N_V = 10^{12}$ grains cm^{-3}, while the total area of the twin boundaries is $S_V = 3 \times 10^4$ cm^2. The number of atoms per unit volume in $TlBiTe_2$ is 2.9×10^{22} cm^{-3}. Therefore, the number of atoms on the twin boundaries per unit volume should be 5.7×10^{19} cm^{-3}. From these atoms only those that belong to the boundaries of semicoherent-type twins suffer a shear displacement on the twin boundary (Fig. 2c), while the atoms on the coherent twin boundary are located on ideal sites and only second-neighbour interactions exist. In this calculation we did not take into account the atoms connected with the dislocations at low-angle boundaries.

Estimating that 40% of the twins are of the non-coherent type and taking into account the low-angle boundaries and the emissary dislocations one may assume that nearly half of the atoms that resulted from the crude estimate above for the misfit present dangling bonds. Therefore, a better estimate will increase this number to the order of 10^{20} which is compatible with the increase in the nominal concentration of electrons at the phase transformation, which was found to be 1.56×10^{20} cm^{-3}. It is therefore reasonable to assume that the sudden release of electrons in going to the cubic phase at the phase transformation is due to the disappearance of the dangling bonds, since these atoms are forced to occupy normal sites. The disorder-induced localization of band-edge states could explain the $R(T)$ behaviour and also it will produce an $\varrho(T)$ which grows at low temperatures, a trend which may be noticed in our results.

The model presented above based on the partial disorder that prevails in the rhombohedral phase of $TlBiTe_2$ introduces a new dimension on the nature of the phase transformation of the IV–VI compounds, since these effects are common in GeTe, SnTe as well as in their ternary compounds. An investigation along these lines on these materials is in progress.

References

[1] R. A. Hein and E. M. Swiggard, Phys. Rev. Letters 24, 53 (1970).

[2] R. S. Feigelson and R. K. Route, Acousto-Optic Bulk-Wave Devices, Society of Photo-Optical Instrumentation Engineers, Conference of Ternary Compounds, Tokyo 1980.

[3] T. A. BITHER, US Patent 2893891, July 7, 1959.
[4] E. F. HOCKINGS and J. G. WHITE, Acta cryst. **14**, 328 (1961).
[5] E. K. POLYCHRONIADIS, G. L. BLERIS, and J. STOEMENOS, Thin Solid Films **37**, 407 (1976).
[6] S. GELLER and J. H. WERNICK, Acta cryst. **12**, 46 (1959).
[7] G. VAN TENDELOO and S. AMELINCKX, Acta cryst. **A30**, 431 (1974).
[8] J. P. JENSEN, J. R. BURKE, D. W. ERNST, and R. S. ALLGAIER, Phys. Rev. B **6**, 319 (1972).
[9] R. S. ALLGAIER, R. PERL, and E. M. SWIGGARD, in: Physics of Semimetals and Narrow Gap Semiconductors, Ed. D. L. CARTER and R. T. BUTE, Pergamon Press, New York 1971 (p. 437).
[10] R. S. ALLGAIER, Conf. Physics of Semimetals and Narrow Gap Semiconductors, Nice-Cardiff 1973.
[11] J. D. WASCHER, Philips Res. Rep. Suppl. 8 (1969).
O. VALASSIADES and N. A. ECONOMOU, Proc. Faculty Math. Phys. Sc., University of Tessaloniki **17**, 323 (1977).
[12] N. N. KRISTOFEL, Soviet Phys. — Solid State **21**, 523 (1979).
[13] M. KASTNER, D. ADLER, and H. FRITZSCHE, Phys. Rev. Letters **37**, 1504 (1976).

(Received February 9, 1981)

phys. stat. sol. (a) 65, 225 (1981)

Subject classification: 11 and 20.1; 22.1.2

Sektion Physik der Friedrich-Schiller-Universität Jena[1])

Correlation between Structural Defects and Optical Properties in Ion-Implanted Silicon

By

W. WESCH, E. GLASER, G. GÖTZ, H. KARGE, and R. PRAGER

Measurements of optical transmission and reflection spectra and Rutherford backscattering experiments (RBS) are performed to investigate the influence of radiation damage on optical properties of silicon crystals implanted with Ar^+-, P^+-, and B^+-ions at different temperatures ($T_i = 77$; 300; 500 K). At sufficiently low temperatures the absorption coefficient and refractive index increase with defect concentration and reach saturation values after formation of completely amorphous layers. No change of the refractive index is found for higher implantation temperatures, where even at high doses no amorphization may be reached. However for these conditions the absorption coefficient increases significantly. Obviously the change of refractive index is mainly caused by heavily damaged or amorphous regions, whereas the absorption coefficient is influenced by point defects also.

Zur Untersuchung des Einflusses von Strahlenschäden auf die optischen Eigenschaften von Silizium-Einkristallen, die bei verschiedenen Temperaturen ($T_i = 77$; 300; 500 K) mit Ar^+-, P^+- und B^+-Ionen implantiert worden waren, werden Transmissions- und Reflexionsmessungen sowie Rutherford-Weitwinkelstreuungsmessungen (RBS) durchgeführt. Für ausreichend niedrige Temperaturen wachsen Absorptionskoeffizient und Brechungsindex mit der Defektkonzentration an und erreichen Sättigungswerte nach der Bildung vollständig amorpher Schichten. Für höhere Implantationstemperaturen, bei denen auch bei großen Implantationsdosiswerten keine Amorphisierung erreicht werden kann, wird keine merkliche Brechzahländerung festgestellt. Dagegen wächst unter diesen Bedingungen der Absorptionskoeffizient stark an. Offensichtlich wird die Änderung des Brechungsindex hauptsächlich durch stark zerstörte und amorphe Bereiche hervorgerufen, während der Absorptionskoeffizient auch durch Punktdefekte beeinflußt wird.

1. Introduction

In recent years the interest of electronic and optical device technology in ion implantation as a method to influence the electrical and optical properties of solids in a definite manner has vehemently grown. In this connection extensive investigations were carried out concerning the physical processes taking place during production and annealing of implantation-induced radiation damage and their influence on electrical and optical properties.

The generation of radiation defects in silicon causes discrete energy levels in the forbidden gap. With increasing defect concentration an increasing band tail extending into the gap and a broadening of the bands are observed. This results in a shift of the absorption edge towards smaller photon energies and an increased near-edge absorption [1, 2]. The defect-induced near-edge absorption is connected with an increase of the refractive index of the damaged layer in the near-infrared spectral region, which also influences the reflection and yields the appearance of interferences in the transmission and reflection spectra at sufficiently high damage concentrations [1 to 4]. In the case of point defects additionally discrete absorption bands appear in the near-

[1]) Max-Wien-Platz 1, DDR-6900 Jena, DDR.

infrared region. An absorption band at a wavelength $\lambda = 1.8\ \mu m$ caused by the divacancy was also found in ion-implanted silicon [1, 2].

All these previous investigations show that for a given ion species and implantation temperature both refractive index n and absorption coefficient K increase with ion dose and reach saturation values after formation of completely amorphous layers [5].

In the present paper we give a summary of our results concerning the influence of different ion masses, implantation and annealing temperatures on the changes of refractive index n and absorption coefficient K near the fundamental absorption edge of silicon.

2. Experimental

Silicon wafers of $\langle 111 \rangle$ orientation (thickness 200 μm, specific resistivity $\varrho = (16$ to $18)\ \Omega$cm, mechanically and chemically polished at both surfaces) have been implanted with 150 keV boron ions and with 200 keV argon and phosphorus ions. The ion fluences varied between 10^{13} and $10^{16}\ cm^{-2}$, ions were implanted at temperatures $T_i = 77$, 200, 300, and 500 K, respectively.

Informations about depth distributions of defect concentration $N'(t)$ and integrated defect concentration $\tilde{N}' = \int_0^{\tilde{t}} N'(t)\ dt$ ($\tilde{t}$ = thickness of damaged layer) were received by Rutherford backscattering (RBS) measurements with 1.4 MeV helium ions.

The optical investigations were carried out with a single-beam spectrometer which allows measurements of transmission $T = I_d/I_0$ and reflection $R' = I_r/I_0$ in the wavelength region $0.9\ \mu m \leqq \lambda \leqq 10\ \mu m$ at nearly normal incidence.

An average absorption coefficient $K = 4\pi k/\lambda$ (k absorption index, λ wavelength in vacuum) was calculated from transmission spectra using the formula for the transmission of plane-parallel samples [6] and for samples with a constant change of n and K throughout the depth of the implanted layer [7] in a first approximation averaging over the inflexion points of the interference curves. The refractive index was determined by evaluating the position of interference extrema in transmission and reflection spectra taking the thickness of the implanted layers from RBS results.

In this work we discuss the change of the absorption coefficient ΔK at $\lambda = 1.2\ \mu m$ as a measure for the near-edge absorption and the average relative refractive index change $\Delta n/n_{cr}$ (n_{cr} bulk material value) in the wavelength region $\lambda = (1$ to $3)\ \mu m$. The absorption exponent ΔKt and the optical path length nt are chosen for the discussion in the case of annealed samples. Further informations about the profiles of the optical constants were obtained by means of a computer program calculating transmission and reflection spectra of multi-layer systems with different values of n and k in each layer [8].

3. Results

3.1 Influence of the ion mass on the change of n and K

Fig. 1 shows the wavelength dependence of the transmission of silicon samples implanted with boron ions at a temperature $T_i = 77$ K. The near-edge absorption due to the band tailing extends to wavelengths of about $\lambda = 2.5\ \mu m$ and grows up with increasing ion dose. The amplitudes of the interferences observed for dose values $N_i \geqq 10^{14}\ cm^{-2}$ increase with ion dose. The positions of the interference extrema are shifted towards greater wavelengths. Besides interferences with small amplitudes for the dose $N_i = 1 \times 10^{14}\ cm^{-2}$ the divacancy band at $\lambda = 1.8\ \mu m$ is observed disappearing for doses $N_i \geqq 2.5 \times 10^{14}\ cm^{-2}$. Fig. 2 shows the dose dependences of

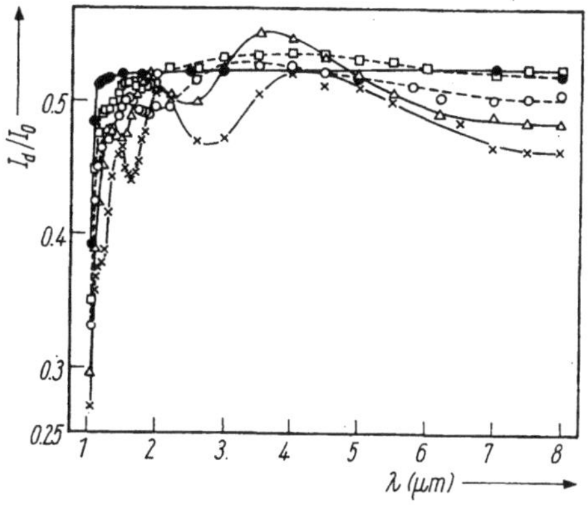

Fig. 1

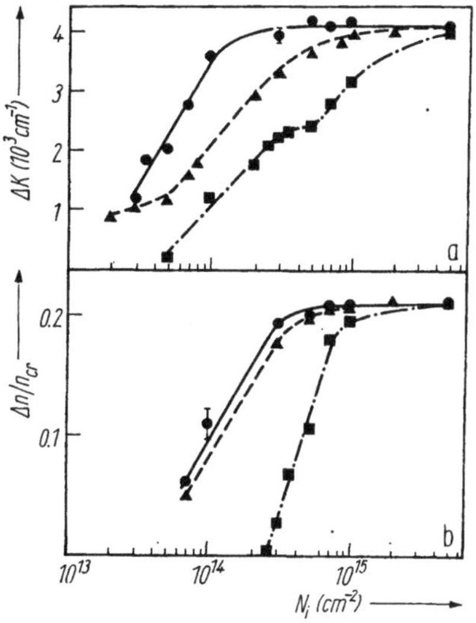

Fig. 2

Fig. 1. Optical transmission spectra for boron-implanted silicon samples (150 keV) (implantation temperature $T_i = 77$ K). ● unimplanted, × $N_i = 5 \times 10^{15}$, △ 7×10^{14}, ○ 3×10^{14}, □ $1 \times$ $\times\ 10^{14}$ cm^{-2}

Fig. 2. Dose dependences of the change of near-edge absorption coefficient ΔK and of relative refractive index $\Delta n/n_{cr}$ for argon-, phosphorus-, and boron-implanted silicon samples ($T_i = 77$ K). ● Si: Ar^+ (200 keV), ▲ Si: P^+ (200 keV), ■ Si: B^+ (150 keV)

ΔK and $\Delta n/n_{cr}$ in the case of B^+, P^+, and Ar^+ implantation into silicon at $T_i = 77$ K. Both $\Delta n/n_{cr}$ and ΔK increase with ion dose, i.e. with increasing nuclear deposited energy density [9, 10], reaching saturation for doses at which amorphous layers are formed. With decreasing ion mass the curves for ΔK are shifted towards greater ion fluences and the slopes of the curves become smaller (Fig. 2a). Whereas the slopes of $\Delta n/n_{cr}$ curves are the same for P^+ and Ar^+ implantation, for boron-implanted samples the refractive index change increases steeper with the ion dose (Fig. 2b).

3.2 Influence of the implantation temperature on the change of n and K

The influence of implantation temperature on the optical transmission is illustrated in Fig. 3 for implantation of 5×10^{15} P^+ cm^{-2}. As in the case of

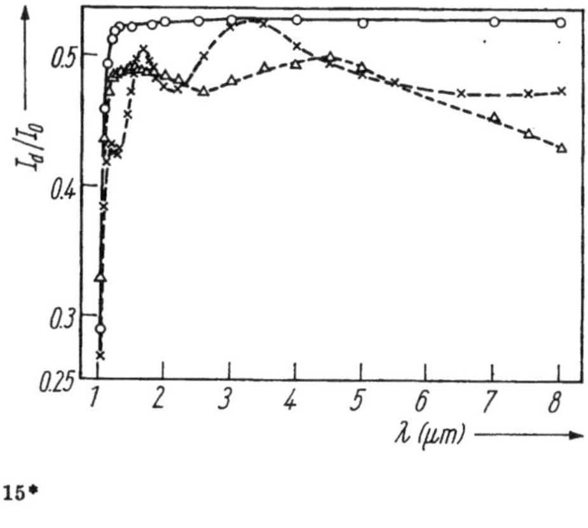

Fig. 3. Optical transmission spectra for phosphorus-implanted silicon (200 keV) samples. ○ unimplanted, × $N_i = 5 \times 10^{15}$ cm^{-2} ($T_i = 77$ K), △ $N_i = 5 \times 10^{15}$ cm^{-2} ($T_i = 500$ K)

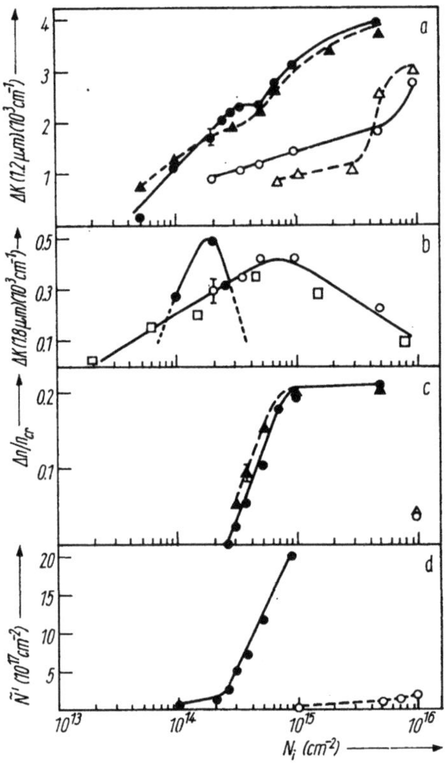

Fig. 4. Dose dependences of a) the change of near-edge absorption coefficient ΔK ($\lambda = 1.2$ µm), b) the divacancy absorption band ΔK ($\lambda = 1.8$ µm), c) the relative refractive index change $\Delta n/n_{cr}$ (averaged for $\lambda =$ (1 to 3) µm), and d) the defect concentration $\tilde{N}'$ (from RBS measurements) for phosphorus- and boron-implanted silicon samples. Si: B^+ (150 keV): ● $T_i = 77$ K, ○ 300 K; Si: P^+ (200 keV): ▲ $T_i = 300$ K, △ 500 K; □ Si: B^+ (80 keV) (Baranova et al. [5])

boron implantation Fig. 1 it can be seen that for low implantation temperatures ($T_i = 77$ K) a sharp increase of the near edge absorption and interferences with high amplitudes are found. An increase of irradiation temperature up to $T_i = 500$ K leads to a high near-edge absorption, whereas there are only weak interferences.

The dose dependence of the absorption coefficient ΔK at $\lambda = 1.2$ µm and 1.8 µm (divacancy absorption), of the refractive index change $\Delta n/n_{cr}$, and of the defect concentration $\tilde{N}'$ is illustrated in Fig. 4 for boron and phosphorus implantation into silicon at different temperatures. In the case of boron implantation at $T_i = 77$ K the near-edge absorption increases with rising ion dose and reaches the saturation value $\Delta K = 4000$ cm^{-1} after formation of a completely amorphous layer at doses between 3×10^{15} and 5×10^{15} cm^{-2} [10, 11]. A sharp increase of refractive index change and defect concentration $\tilde{N}'$ is observed in the same dose region, where the slope of the dose dependence of ΔK decreases (3×10^{14} cm$^{-2} \lessapprox N_i \lessapprox 5 \times 10^{14}$ cm^{-2}). The increase of $\Delta n/n_{cr}$ and $\tilde{N}'$ is accompanied by a decrease of the divacancy absorption for doses $N_i > 2 \times 10^{14}$ cm^{-2}. A noticeable divacancy absorption is not detected for $N \gtrapprox 5 \times 10^{14}$ cm^{-2}.

At $T_i = 300$ K we have a slight increase of ΔK for boron implantation only for doses $N_i > 5 \times 10^{15}$ cm^{-2}. At doses $N_i < 3 \times 10^{15}$ cm^{-2} ΔK reaches a value of about 35% of that of amorphous layers. The refractive index change has a value of about 3% even at $N_i = 1 \times 10^{16}$ cm^{-2}. The defect concentration $\tilde{N}'$ increases only unessentially up to this dose (Fig. 4c, d). The divacancy absorption is observed up to the dose $N_i \approx 1 \times 10^{16}$ cm^{-2}, its amount decreases for $N_i > 8 \times 10^{14}$ cm^{-2}.

Phosphorus implantation at $T_i = 300$ K results in a similar behaviour of $\Delta n/n_{cr}$ and ΔK with ion dose as in the case of low-temperature boron implantation (Fig. 4a, c). At $T_i = 500$ K the refractive index change of phosphorus-implanted samples for the dose $N_i = 1 \times 10^{16}$ cm^{-2} is in the order of 3%, like in the case of boron implantation at $T_i = 300$ K.

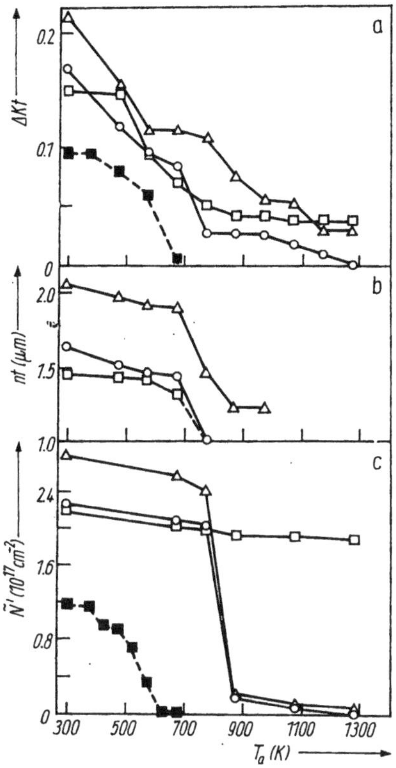

Fig. 5. a) Absorption exponent ΔKt, b) optical path length nt and c) defect concentration $\tilde{N}'$ in dependence on the annealing temperature for argon-, phosphorus-, and boron-implanted silicon samples ($T_i = 77$ K). $N_i = 5 \times 10^{15}$ cm^{-2}: □ Si: Ar^+ (200 keV), ○ Si: P^+ (200 keV). △ Si: B^+ (150 keV); $N_i = 1 \times 10^{14}$ cm^{-2}, ■ Si: Ar^+ (200 keV)

3.3 Annealing of radiation defects

Defects created during ion bombardment can be annealed by thermal treatment after implantation. For annealing of continuous amorphous layers temperatures higher than 800 K are necessary, isolated amorphous regions and nonamorphous clusters anneal already below 800 K [12]. Different point defects anneal in characteristic temperature regions [6].

Amorphous layers which extend to the crystal surface are created in silicon at $T_i = 77$ K by implantation of 5×10^{15} Ar^+, P^+, and B^+ ions cm^{-2} with energies of 200 and 150 keV, respectively. In Fig. 5 absorption exponent ΔKt, optical path length nt, and integrated defect concentraion N' are illustrated in dependence on annealing temperature T_a for samples implanted under these conditions.

In the case of P^+ and B^+ implantations nt is nearly constant up to $T_a = 700$ K (Fig. 5b) and shows an annealing step in the temperature region between 700 and 900 K, where the defect concentration $\tilde{N}'$ also strongly decreases (Fig. 5c). Different to that ΔKt already decreases for temperatures up to 600 K. The absorption exponent of amorphous layers created by Ar^+ implantation continuously decreases for temperatures $T_a > 500$ K. For these samples nt decreases towards the crystalline value which is reached at $T_a \approx 800$ K. On the contrary, a noticeable reduction of the defect concentration $\tilde{N}'$ could not be found up to $T_a = 1300$ K. If the implanted layers are not yet amorphous, like in the case of implantation of 10^{14} Ar^+ cm^{-2} at $T_i = 77$ K, total defect annealing up to $T_a = 700$ K is observed (Fig. 5a, c).

4. Discussion

Ion implantation into silicon at $T_i = 77$ K leads to an increase of the defect concentration into dependence on the ion dose. In critical dose regions which depend on the ion mass, amorphous layers are formed [9, 12, 13]. Because of the high part of energy deposited in nuclear processes, heavier ions produce heavily damaged zones around the ion tracks [11]. In the case of medium-mass-ion implantation (Ar^+, P^+) ΔK and $\Delta n/n_{cr}$ increase in the same dose region, where a sharp increase of the defect concentration $\tilde{N}'$ is also detected (Fig. 3) [12]. The damage clusters created under these conditions obviously influence both absorption coefficient and refractive index. The stronger increase of ΔK in the case of argon implantation in respect to the phosphorus implantation may be caused by differences in the structure of the defect clusters due to the different amount of nuclear deposited energy and to the different solubility of phosphorus and argon atoms in silicon. In the case of light-ion implantation (B^+) defect clusters with high point defect concentrations are produced around the ion tracks [14] because of the smaller part of energy deposited into nuclear processes. Due to the formation of amorphous regions by overlap of the defect clusters with proceeding implantation dose, the defect concentration $\tilde{N}'$ as well as the refractive index change $\Delta n/n_{cr}$ strongly increase in a narrow dose region between 2×10^{14} and 5×10^{14} cm^{-2} (Fig. 2, 4). Investigations of the kinetics of the formation of amorphous regions during ion implantation at low temperatures by means of EPR have shown that there exists a good agreement between the dose dependences of the intensity of the EPR line with Landé factor $g = 2.0055$, which is attributed to amorphous material, and the defect concentration $\tilde{N}'$ [15]. A connection between the existence of amorphous regions and refractive index change can be concluded according to the simultaneous increase of refractive index change and defect concentration. A comparison of the dose dependence of $\tilde{N}'$ and $\Delta n/n_{cr}$ in the framework of the overlap damage model [16] shows that the initially created damage clusters are not completely amorphous even in the case of heavier-ion implantation, i.e. an overlap of clusters is necessary to create amorphous regions [3, 17]. For boron implantation the absorption coefficient already increases in the dose region between 4×10^{13} and 2×10^{14} cm^{-2}, where a significant increase of $\tilde{N}'$ and $\Delta n/n_{cr}$ is not yet observed (Fig. 4). For these doses the number of overlaps necessary to form amorphous areas is not yet reached. Evidently the near-edge absorption under these conditions is mainly caused by point defects. This assumption is supported by the rising absorption due to divacancies up to a dose of about 2×10^{14} cm^{-2}, indicating the existence of point defects.

The decrease of the divacancy absorption for doses $N_i > 2 \times 10^{14}$ cm^{-2} as well as the modification of the slope of the dose dependence of the near-edge absorption between 3×10^{14} and 5×10^{14} cm^{-2} are a consequence of the reduction of the number of point defects with proceeding implantation. Besides this, the sheet resistivity of highly doped silicon decreases for doses $N_i > 7 \times 10^{14}$ cm^{-2} [12], due to the reduction of the number of compensating point defects during the formation of extended amorphous areas. On the other hand, the near edge absorption also seems to be influenced by heavily damaged zones and amorphous regions, which results in the observed increase of ΔK for dose values $N_i > 5 \times 10^{14}$ cm^{-2} (Fig. 4).

The results discussed above are in good agreement with computer simulations of transmission and reflection spectra of boron implanted silicon by means of a program for multi-layer systems [8]. After the formation of amorphous regions, i.e. for dose values $N_i \gtrsim 3 \times 10^{14}$ cm^{-2}, the profiles of both the optical parameters agree well with the profiles of the defect concentration measured by RBS. On the other hand, for $N_i < 3 \times 10^{14}$ cm^{-2} the ratio between the change of absorption coefficient and the saturation value for amorphous layers in the ΔK profile at a given dose is larger than the corresponding values for refractive index and defect concentration. This means that point defects existing in this dose region mainly influence the absorption coefficient. A correlation between the existence of heavily damaged or amorphous regions and the change of refractive index can also be found by annealing experiments. The steep decrease of the defect concentration $\tilde{N}'$ in the temperature region $800 \text{ K} \leq T_A \leq 900 \text{ K}$ for amorphous layers, created by P^+ and B^+ implantation is accompanied by a decrease of the optical path length nt (Fig. 5b, c). For these conditions we could not find a significant decrease of the defect concentration $\tilde{N}'$ for amorphous layers created by high-fluence Ar^+ implantation, although the optical path length decreases in the same temperature region as in the case of phosphorus- and boron-implanted samples. Electron microscopic investigations of these samples have shown that due to the small solubility of argon in the silicon lattice thermal annealing results in the formation of bubbles preventing the recrystallization and so the annealed layers have a polycrystalline structure [12] giving similar RBS yields as for amorphous layers. Thus, the observed decrease of the optical path length between 700 and 900 K is oviously caused by a recrystallization process yielding to polycrystalline layers.

Whereas the recrystallization of amorphous layers in a narrow temperature region is connected with a decrease of refractive index in the mentioned temperature region, the absorption exponent ΔKt already decreases at lower temperatures. Due to the depth dependence of nuclear deposited energy [9], for the discussed implantation parameters ion implantation creates a transition region behind the continuous amorphous layer which is not completely amorphous. In this region heavily damaged zones as well as point defects and simple clusters exist at the same time. Because no reduction of the thickness of the amorphous layers could be found below 800 K the strong decrease of ΔKt up to 600 K is correlated with annealing of defects in this transition region. A correlation between annealing of simple defect clusters, point defects, and near-edge absorption is also indicated by the simultaneous reduction of defect concentration and absorption exponent for samples with defect concentrations below the amorphization level (Fig. 5a, c) [10]. At higher implantation temperatures ionization-stimulated annealing effects during implantation lead to a decay of the initially produced damage clusters [13]. For ions with intermediate masses (Ar^+, P^+) the out-diffusion of mobile point defects from the damage clusters entails the formation of heavily damaged cores with environments of point defects [17]. As a consequence the defect concentration as well as the values of the optical constants are reduced. Thus the defects produced by phosphorus ions at $Ti = 300$ K obviously influence the optical constants in a similar manner as boron implantation at $T_i = 77$ K, as can be seen from the nearly identical dose dependences of ΔK and $\Delta n/n_{cr}$ in Fig. 4. By a further increase of the temperature the formation of amorphous regions becomes more unlikely for medium mass ion implantation. The damage clusters completely dissociate at $T_i = 500$ K, and an amorphous layer cannot be formed at this temperature [13]. A similar behaviour can be assumed for light-ion implantation (B^+) at $T_i = 300$ K [17]. According to this we only found a relatively small value of the refractive index change (about 3%) and of the defect concentration (about 15% of the amorphous concentration) even for a dose of $N_i = 1 \times 10^{16}$ cm^{-2}. The comparatively high values of the near-edge absorption show, on the other hand, that a high amount of point defects and simple defect clusters is present. As an example divancies are observed up to a dose $N_i = 1 \times 10^{16}$ cm^{-2} in the case of boron implantation at room temperature (Fig. 4).

5. Summary

The generation of amorphous layers in silicon implanted with ions of intermediate (Ar^+, P^+) and small (B^+) masses at sufficiently low temperatures ($T_i = 77$ K, 300 K) is connected with an increase of absorption coefficient and refractive index in the near-edge region. After formation of completely amorphous layers saturation values are reached. The nearly identical dose dependence of defect concentration, refractive index change, and change of absorption coefficient for Ar^+- and P^+-implanted samples ($T_i = 77$ K) indicate that defect clusters influence both refractive index and absorption coefficient. For B^+-implanted samples defect concentration and refractive index change only increase in a narrow dose region, in which heavily damaged and amorphous clusters are formed due to manyfold overlap of the initially produced defect clusters. However, a different absorption behaviour is found for smaller doses.

At elevated temperatures (for Ar^+, P^+ $T_i = 500$ K; for B^+ $T_i = 300$ K) annealing processes during implantation prevent the formation of amorphous areas. Only a very small refractive index change of about 3% is caused in these cases even at high doses.

From our results we conclude that the change of refractive index is strongly correlated with the existence of amorphous and heavily damaged regions. However, the absorption coefficient is determined by the defects existing in the damaged layers, i.e. by point defects, damage clusters, and amorphous zones.

Acknowledgements

The authors wish to thank N. A. Sobolev, D. Baither, and H. Bartsch for helpful discussions, C. Dierse and G. Lenk for the careful preparation of the samples, and U. Barth and W. Ahnefeld for technical assistance during RBS measurements.

References

[1] H. J. STEIN, F. L. VOOK, D. K. BRICE, J. A. BORDERS, and S. T. PICRAUX, Radiat. Eff. **6**, 19 (1970).
[2] E. K. BARANOVA, V. M. GUSEV, and L. N. STRELTSOV, Fiz. Tekh. Poluprov. 7, 1851 (1973).
[3] W. WESCH and G. GÖTZ, Radiat. Eff. **49**, 137 (1980).
[4] G. K. HUBLER, C. N. WADDELL, W. G. SPITZER, J. E. FREDRICKSON, S. PRUSSIN, and R. G. WILSON, J. appl. Phys. **50**, 3294 (1979).
[5] E. K. BARANOVA, V. M. GUSEV, J. V. MARTYNENKO, C. V. STARININ, and I. B. HAIBULLIN, Radiat. Eff. **25**, 157 (1975).
[6] H. J. STEIN, F. L. VOOK, and J. A. BORDERS, Appl. Phys. Letters **14**, 328 (1969).
[7] M. H. BRODSKY, R. S. TITLE, K. WEISER, and G. D. PETTIT, Phys. Rev. B **1**, 2632 (1970).
[8] K. HEHL and W. WESCH, phys. stat. sol. (a) **58**, 181 (1980).
[9] D. K. BRICE, Radiat. Eff. **6**, 77 (1970).
[10] E. GLASER, G. GÖTZ, W. WESCH, and H. FREY, Radiat. Eff. **48**, 19 (1980).
[11] V. M. GUSEV, J. B. MARTYNENKO, and K. V. STARININ, Kristallografiya **14**, 1050 (1969).
[12] E. GLASER, G. GÖTZ, W. WESCH, and N. A. SOBOLEV, to be published.
[13] F. F. MOREHEAD and B. L. CROWDER, Radiat. Eff. **6**, 27 (1970).
[14] L. T. CHADDERTON, Radiat. Eff. **8**, 77 (1971).
[15] N. A. SOBOLEV, G. GÖTZ, W. KARTHE, and B. SCHNABL, Radiat. Eff. **42**, 23 (1979).
[16] J. F. GIBBONS, Proc. IEEE **90**, 1062 (1972).
[17] G. GÖTZ, E. GLASER, W. WESCH, and N. A. SOBOLEV, Proc. Internat. Conf. Radiation Physics of Semiconductors and Related Materials, Tbilissi 1979.

(Received December 17, 1980)

phys. stat. sol. (a) 65, 233 (1981)

Subject classification: 13.4 and 20.3; 22.2.2

Sektion Physik der Humboldt-Universität zu Berlin[1]), *Bereich Halbleiteroptik (a) and Sektion Physik der Karl-Marx-Universität Leipzig, Arbeitsgruppe Halbleiter (b)*

Minority Carrier Lifetime in VPE-GaP:N, Te at Various Excitation Levels

By

K. WANDEL (a), G. WURBS (a), R. BINDEMANN (b), A. KIRSCH (b), and R. SCHWABE (b)

The photoluminescence decay at room temperature in GaP:N, Te grown by vapor phase epitaxy exhibits a nonexponential behaviour. It is pronounced at excitation densities giving rise to a minority carrier density about equal to the equilibrium majority carrier density. The experimental results are in agreement with the dominating nonradiative recombination on saturable acceptor-like centres, described by a Shockley-Read-model. From fitting the calculated decay curves in the nonexponential region the centre concentration $N_t = 7 \times 10^{14}$ cm^{-3} and the capture cross-sections $\sigma_{pt} = 7 \times 10^{-15}$ cm^2 and $\sigma_{nt} = 1 \times 10^{-18}$ cm^2 are estimated. A nonsaturable parallel recombination channel should exist with an effective capture probability of about 5×10^6 to 1×10^7 s^{-1}.

Im Abklingen der Photolumineszenz in GaP:N, Te, das durch Züchtung aus der Gasphase hergestellt wurde, wird ein nichtexponentielles Verhalten beobachtet. Es tritt besonders bei solchen Anregungsdichten auf, die eine Minoritätsträger-Überschußkonzentration in der Größenordnung der Majoritätsträger-Gleichgewichtskonzentration erzeugen. Dieses Ergebnis ist zu erwarten, wenn für die strahlungslose Rekombination ein sättigbares, akzeptorartiges Zentrum verantwortlich ist, das durch ein Shockley-Read-Modell beschrieben werden kann. Durch numerische Anpassung des Abklingverlaufes können die Zentrenkonzentration $N_t = 7 \times 10^{14}$ cm^{-3} sowie die Einfangsquerschnitte $\sigma_{pt} = 7 \times 10^{-15}$ cm^2 und $\sigma_{nt} = 1 \times 10^{-18}$ cm^2 abgeschätzt werden. Außerdem existiert ein nichtsättigbarer paralleler Rekombinationskanal mit einer effektiven Einfangwahrscheinlichkeit von $\approx 5 \times 10^6$ bis 1×10^7 s^{-1}.

1. Introduction

The investigation of minority carrier lifetime by means of the near band edge photoluminescence (PL) decay is frequently used to estimate the quality of optoelectronic materials (e.g. [1 to 4]). The value of minority carrier lifetime gives a general impression of the dominant recombination processes and is directly related to the luminescenc efficiency. Usually, however, one cannot distinguish between different recombination channels and cannot get detailed informations concerning the parameters of special recombination channels, e.g. point defect centres, extended defects, or interfaces.

Contrary to lifetime measurements such methods as DLTS and photocapacitance spectroscopy are able to measure the energy level E_t, the concentration N_t, and the capture cross-sections σ_{nt} and σ_{pt}, respectively, of deep centres, the role of which in the total recombination process must be determined by another way, however.

Because of the great practical and theoretical interest in the identification and physical understanding of the efficiency limiting nonradiative recombination processes (e.g. [5, 6]), we have investigated the excitation density dependence of the recombination at point defect centres in order to determine the character and some parameters of nonradiative recombination channels in VPE-GaP:N, Te in more detail.

---

[1]) Neue Schönhauser Str. 20, DDR-1020 Berlin, DDR.

Although the dependence of minority carrier lifetime on the excess carrier density in this material seems to be a well established fact, the discussion of the underlying recombination mechanisms are rather qualitative [2, 7] and sometimes controversial [8]. Our results on this topic agree well with a recombination model, which includes the radiative channel related to nitrogen and nonradiative recombination at point defect centres. These results cannot be explained qualitatively with the nonradiative diffusion controlled recombination at extended defects (dislocations or interfaces) only

2. Experimental

The GaP:N, Te wafers were grown and doped by vapour phase epitaxy on 5° misoriented ⟨100⟩-LEC substrates. After a first layer doped only with tellurium a second one doped with tellurium and nitrogen was grown. The thickness of the second layer was greater than 20 μm.

The concentration of optically active nitrogen atoms has been determined from the absorption spectrum at 77 K. Before measuring the PL decay the samples were not specially prepared. The PL was excited by dye-laser pulses at $\lambda = 480$ nm. This wavelength secures an acceptable penetration depth of excitation, which therefore can be assumed as nearly homogeneous. Further, in this case the influence of surface recombination is small. On the other hand, it is possible to separate the luminescence emission from excitation with the aid of an undoped GaP wafer even at low intensities. The pulse halfwidth (< 2 ns) is small with respect to the average minority carrier lifetime. Therefore, the excitation can be approximated by a δ-function. Usually the measurements were performed with a repetition rate of 3 to 5 Hz.

A maximum pulse power of 1 kW was achieved with 400 kW pump power of a nitrogen laser. The laser beam was focussed to a 600 to 800 μm diameter spot on the sample.

With respect to the measurements at low excitation densities the luminescence was not spectrally resolved. Because the luminescence emitted through the substrate side of the sample was observed, the spectral range was limited by the absorption edge of GaP and by the spectral sensitivity of the photomultiplier. The latter was a fast one with a SbKCs-cathode and a fall time of 2 ns. The signals were sampled and averaged by a boxcar integrator with a 1 ns gate so that the time resolution of the whole apparature was about 3.5 ns.

The decay process could be logarithmically recorded over two decades in luminescence intensity.

All measurements were performed at room temperature.

3. Experimental Results

The decay of the green emission in VPE-GaP:N, Te after dye-laser pulse excitation was investigated in an excitation intensity range from 2 to 10^4 kW cm^{-2} at room temperature.

Above 2×10^2 kW cm^{-2} nonlinear recombination effects as a consequence of the high density of excited carriers are observed which are not of interest in this paper.

For intensities lower than 2×10^2 kW cm^{-2} (the actual value depends on the net doping concentration n_0 as described later on) two groups of samples can be discerned by their decay behaviour.

That of the first and most frequently observed type is demonstrated in Fig. 1 by the PL decay curves of a sample with a net doping concentration $n_0 = 2.6 \times 10^{17}$ cm^{-3} and a nitrogen concentration $[N] = 1 \times 10^{18}$ cm^{-3}. Parameter in this picture is the excitation intensity, which was varied from 190 kW cm^{-2} down to 2 kW cm^{-2}. To discuss the experimental results the density of excited carriers must be known.

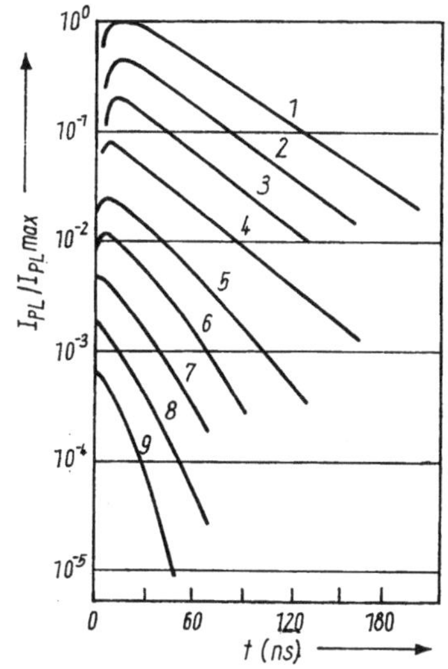

Fig. 1. PL decay (arb. units) in n-type GaP: Te, N after dye-laser impulse excitation with varying excitation densities: (1) 190 ($\delta p(0) = 5.7 \times 10^{17}$), (2) 127 ($3.8 \times 10^{17}$), (3) 70 ($2.1 \times 10^{17}$), (4) 48 ($1.4 \times 10^{17}$), (5) 27 ($0.8 \times 10^{17}$), (6) 18 ($0.5 \times 10^{17}$), (7) 10 ($0.3 \times 10^{17}$), (8) 4 ($0.1 \times 10^{17}$),(9) 2 kW cm^{-2} ($0.07 \times 10^{17}$ cm^{-3}); ($\lambda_{exc} = 480$ nm, [N] = $= 10^{18}$ cm^{-3}, [Te] $= 2.6 \times 10^{17}$ cm^{-3}, $d = 18$ μm)

Neglecting the inhomogeneity of excitation (the penetration depth of excitation $1/\alpha$ is much greater than the diffusion length L_p) we use the initial value $\delta p(0)$ immediately after the excitation pulse ($t = 0$) on the surface. This is justified because the main fraction of luminescence intensity comes from a layer with the thickness of the penetration depth $1/\alpha$. $\delta p(0)$ then corresponds to the total concentration of generated carriers divided by $1/\alpha$ ($1/\alpha \approx 8$ μm for $\lambda_{exc} = 480$ nm). With an excitation power of 190 kW cm^{-2} $\delta p(0)$ lies at about 5.7×10^{17} cm^{-3}.

If 1.4×10^{17} cm^{-3} $< \delta p(0) <$ 5.7×10^{17} cm^{-3} we observe an exponential decay with slightly varying time constants τ_{PL} between 39 and 48 ns on this sample. With concentrations $\delta p(0)$ below 1.4×10^{17} cm^{-3} the decay becomes nonexponentially and the decay rate rises with decreasing excitation intensity, i.e. minority carrier concentration. The distances between the curves in Fig. 1 reflect the actual decrease of PL intensity with decreasing excitation intensity.

With respect to the discussion it is worth to note that the initially excited carrier concentration $\delta p(0)$ at the beginning of the nonexponential decay behaviour is about 1.4×10^{17} cm^{-3}, i.e. slightly smaller than the equilibrium carrier concentration in this sample.

Some samples exhibit additionally to the accelerated decay at small excited carrier concentrations a fast initial part at moderate excited carrier concentrations ($\delta p < n_0$; Fig. 2). This part is followed by a slower decay region, so that the complete decay curves show a characteristic s-like shape. There are no intentional differences in

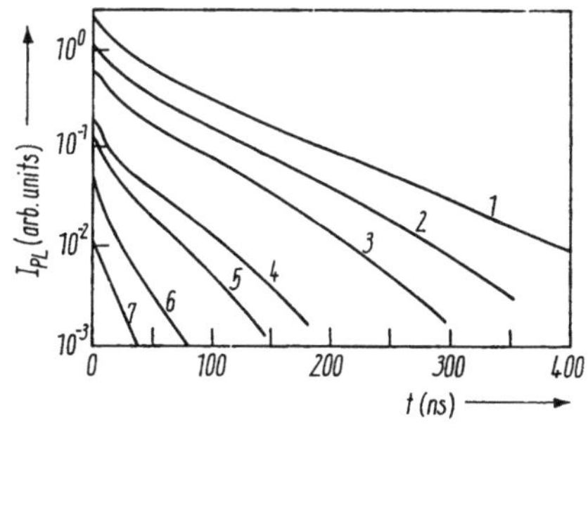

Fig. 2. PL decay with a fast initial part. (1) $\delta p =$ $= 2 \times 10^{16}$, (2) 1×10^{16}, (3) 5.6×10^{15}, (4) 2×10^{15}, (5) 1.2×10^{15}, (6) 4×10^{14}, (7) 1×10^{14} cm^{-3}

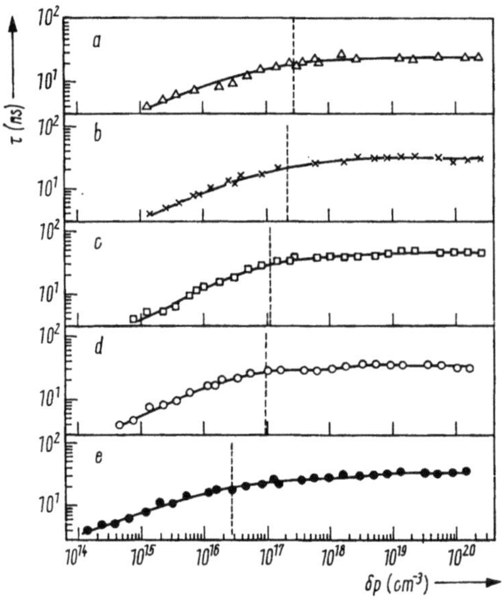

Fig. 3. Influence of the net doping concentration n_0 on the excitation dependence of the PL decay in VPE-GaP:Te, N; a) $n_0 = 2.9 \times 10^{17}$, b) 2.3×10^{17}, c) 1.17×10^{17}, d) 9.4×10^{16}, e) 2.7×10^{16} cm^{-3}

doping concentrations and growth technology between these and the samples mentioned above.

The influence of doping with tellurium and nitrogen, respectively, is shown in Fig. 3 and 4. To compare the intensity-dependent decay behaviour of different samples in the whole intensity range, the nonexponential curves are characterized by an average decay time $\bar{\tau}_{PL}$, resulting from the slope at a fixed time t_1 after the pulse maximum. These values are plotted against $\delta p(t_1)$ for each value of the net doping concentration n_0 in Fig. 3. It is seen, that the differences between the "high" and "low" intensity values τ_{PL}^{∞} and τ_{PL}^{0}, respectively, are nearly the same in all samples. The range of excited carrier density, in which the decay times strongly depend on δp, is shifted to lower δp with lower doping concentration.

An analogous picture for varying nitrogen concentrations (4×10^{17} cm^{-3} to $3 \times$ $\times$ 10^{18} cm^{-3}) is shown in Fig. 4. The net donor concentration is constant in these samples ($n_0 = 1 \times 10^{17}$ cm^{-3}). No significant shift in the excited carrier concentration necessary for substantial alteration of the PL decay time is observed.

The values of τ_{PL}^{0} at low excitation densities seems to be nearly the same in different samples. At high excitation densities the values of τ_{PL}^{∞} are markedly different. Other mesaurements have shown a great influence of dislocations on the PL decay in this region of excitation intensity.

4. Discussion

The kinetics of green emission in nitrogen-doped GaP is usually described by the shunt-path model ([9], see Fig. 5). According to this model the relative luminescence

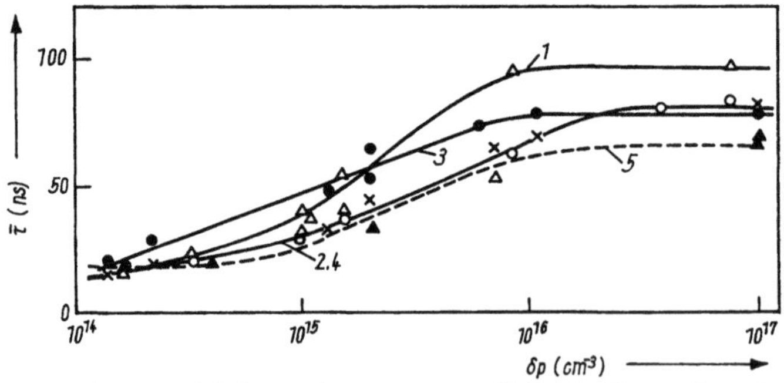

Fig. 4. Average PL decay time in n-type GaP:Te, N as a function of initial excess carrier concentration with varying nitrogen concentration; (1) [N] = 4×10^{17}, (2) 5×10^{17}, (3) 2×10^{18}, (4) 3×10^{18}, (5) 3×10^{18} cm^{-3} ($n_0 = 1 \times 10^{17}$ cm^{-3})

intensity in n-type GaP at room temperature for any time (greater 0.6 ns, [2]) after the excitation pulse is given by

$$I_{PL}(t) = (n_0 + \delta n(t))\, \delta p(t)\, N_N \tag{1}$$

(n_0 equilibrium majority carrier concentration; N_N concentration of nitrogen traps; δn, δp excess carrier concentrations).

In the total recombination rate this radiative channel is neglected compared with the effective non-radiative channel. The influence of the excitation level on the PL kinetics can be discussed in this model, if it is assumed additionally, that single level point defect centres play an essential role in the recombination process. In this case, according to the Shockley-Read model, a dependence of the minority carrier lifetime on the nonequilibrium carrier concentration is expected. We shall show that the above-mentioned two types of decay curves can be explained in this manner with different concentrations N_t of saturable point defect centres only.

The concentrations $\delta n(t)$, $\delta p(t)$ in (1) then must be evaluated from the rate equations

$$\frac{d\delta p}{dt} = - c_{pt}\, \delta p n_t, \tag{2a}$$

$$\frac{d\delta n_t}{dt} = - c_{pt} \delta p n_t + c_{nt}(\delta n + n_0)\,(N_t - n_t) - c_{nt} n_1 n_t\,, \tag{2b}$$

$$\delta n = \delta p + N_t - n_t\,. \tag{2c}$$

(For thermal equilibrium conditions $p_0 = 0$ in n-type GaP and $n_t = N_t$; N_t concentration of the dominant point defect centres, n_t electron occupied point defect centres $c_{nt} = \sigma_{nt} v_{th}^{n}$; $c_{pt} = \sigma_{pt} v_{th}^{p}$ recombination coefficients for electrons and holes, respectively, and $n_1 = N_c \exp\left(-(E_c - E_t)/kT\right)$. It is assumed, that the centres create a level in the upper half of the gap and therefore the thermal release of holes can be neglected.

Further parallel recombination channels (e.g. dislocations or radiative processes) can be taken into account approximately by a constant effective capture probability $1/\tau_{||}$. This leads to an additional term $-\delta p/\tau_{||}$ in (2a). It is used for the quantitative discussion later on but it should not alter the qualitative decay behaviour.

δ-pulse excitation leads to an excited carrier concentration $\delta n(0) = \delta p(0)$ at $t = 0$. A variation of excitation density enters the rate equations by this initial condition. Generally, equations (2) must be solved numerically. Such calculations, performed with the initial condition $\delta p = \delta n = \delta p(0)$, $n_t = N_t$ at $t = 0$ and varying $\delta p(0)$ agree qualitatively with the experimental curves of Fig. 2.

The decay process according to (2) can be understood as follows. At $t = 0$ we have

$$\delta \dot{p} = - c_{pt} n_t\, \delta p = - c_{pt} N_t\, \delta p = - \frac{\delta p}{\tau_{p0}}$$

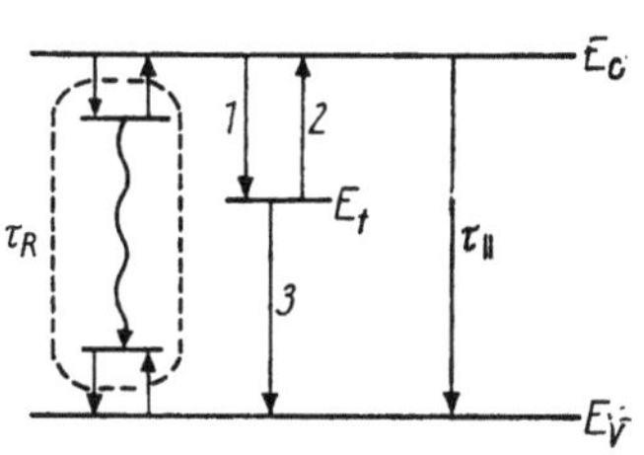

Fig. 5. Shunt-path model of recombination in GaP:N. (1) Capture rate for electrons $c_{nt} n (N_t - n_t)$, (2) emission rate for electrons $e_{nt} n_t = c_{nt} n_1 n_t$, (3) capture rate for holes $c_{pt} p n_t$

because all centres are filled with electrons and are not in a quasiequilibrium with the bands.

With elapsed time n_t becomes smaller because of hole capture and therefore $|\delta\dot{p}|$ and hence the slope of the PL decay curve decrease. This change of occupation is more pronounced for smaller ratios of capture cross-sections σ_{nt}/σ_{pt} and at moderate excited carrier concentrations ($n_0 > \delta p > N_t$).

The hole filling of the centres causes an increased electron capture, and the further decay is characterized by equal capture rates for electrons and holes as long as enough holes are present (nearly exponential part). At lower hole concentrations the recombination centres are refilled with electrons. In this range the simultaneous change of centre occupation n_t and hole concentration δp gives rise to a nonexponential decay behaviour. Finally, all centres are occupied by electrons and δp decays exponentially with τ_{p0}. All these parts are observable in the PL decay, if it starts at $\delta p(0) \approx N_t$ and can be measured over a large intensity range or by stepwise decreasing of $\delta p(0)$ by the excitation intensity as it has been done in our measurements.

At high concentrations of excited carriers only a small part of them is needed to empty the recombination centres. Therefore, a fast initial part is not obtained. At low excess carrier concentrations the occupation of the deep centres with electrons is nearly unaffected by excitation and only an exponential decay with the time constant τ_{p0} should exist.

In Fig. 6 numerical results are shown together with experimental decay curves from Fig. 2. The calculations were performed with the following parameter values: $N_t = 7 \times 10^{14}$ cm^{-3}, $\sigma_{pt} = 6.8 \times 10^{-15}$ cm^2, $\sigma_{nt} = 1.2 \times 10^{-18}$ cm^2, and $\tau_{||} = 100$ ns. These values give a rough estimation to characterize the recombination centres. An important influence of nonsaturable recombination processes is demonstrated by a value of $\tau_{||}$, comparable with that of τ_{n0}. Probably it is related to extended defects (surface or dislocations).

There are, however, some other mechanisms which can lead to a fast initial decay. Effects of high carrier concentrations (e.g. Auger effects) could be excluded, because δp was smaller than 10^{16} cm^{-3} and the initial part decreases at higher excitation densities. Surfaces recombination [11] should influence the PL decay markedly at higher concentrations of excited carriers because of the higher value of bulk lifetime. This explains the deviations of the calculated curves from the measured ones in Fig. 6 in the region of higher excitation.

At small concentrations of recombination centres N_t the part of excited holes necessary for filling the centres with holes is negligible and a fast initial part should not be observed. This is the case in the most obtained PL decay characteristics (see Fig. 1). Then the minority carrier decay can be calculated analytically with the

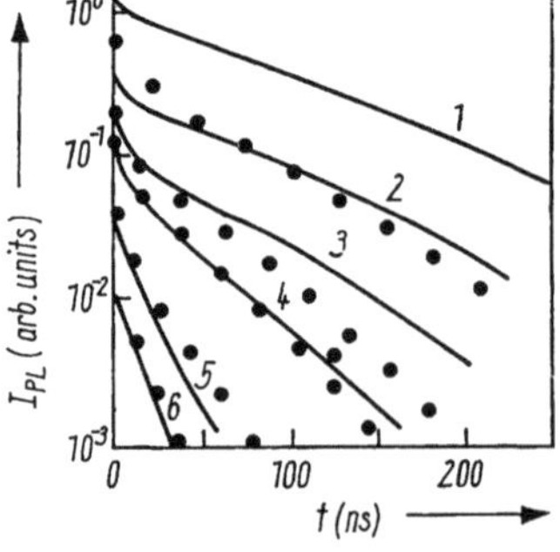

Fig. 6. Logarithmic plot of I_{PL} versus t as calculated from (2a), (2b), and (1) with $\sigma_{pt} = 6.8 \times 10^{-15}$ cm^2, $\sigma_{nt} = 1.2 \times 10^{-18}$ cm^2, $N_t = 7 \times 10^{14}$ cm^{-3}, $\tau_{||} = 100$ ns, $n_1 = 0$ for different $\delta p(0)$ (curves 1 to 6 correspond to curves 2 to 7 of Fig. 2)

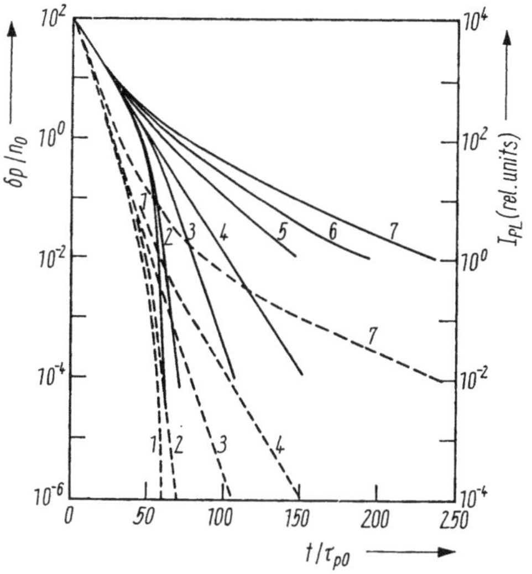

Fig. 7. Logarithmic plot of $\delta p/n_0$ (solid lines) and I_{PL} (broken lines, arbitrary units) versus t/τ_{p0}, respectively, as calculated from (3) and (1) with $\tau_{p0}/\tau_{n0} = 0.1$ and n_1/n_0 as parameter: (1) 10^{-5}, (2) 10^{-3}, (3) 5, (4) 10, (5) 20, (6) 30, (7) 50

condition $N_t/n_0 \ll c_n/c_p$ according to [10] as

$$T = \frac{1 - \gamma b}{\gamma} \ln\left(\frac{y_0 + 1}{y + 1}\right) + (1 + b) \ln \frac{y_0}{y} \tag{3}$$

with

$$T = \frac{t}{\tau_{p0}}, \quad \gamma = \frac{\tau_{p0}}{\tau_{n0}}, \quad b = \frac{n_1}{n_0}, \quad y = \frac{\delta p}{n_0}, \quad \tau_{p0} = \frac{1}{c_{pt}N_t}, \quad \text{and} \quad \tau_{n0} = \frac{1}{c_{nt}N_t}.$$

y_0 is the initial value at $t = 0$ and responsible for the excitation density dependence. $y(T)$ as calculated from (3) is shown in Fig. 7 (solid lines, left scale) with $\gamma = 0.1$ and b as parameter. The physical picture is the same as discussed above.

The decay (in the figure shown from $y_0 = 100$) consists of two exponential parts in the limiting cases

(i) $\delta p \gg n_0$ with the time constant

$$\tau_p^\infty = \tau_{n0} + \tau_{p0} \tag{4}$$

and (ii) $\delta p \ll n_0$ (but $\delta p \gg N_t$) with

$$\tau_p^0 = \tau_{p0}\left(1 + \frac{n_1}{n_0}\right). \tag{5}$$

If

$$\frac{\tau_{p0}}{\tau_{n0}} \frac{n_1}{n_0} < 1\,,$$

from (4) and (5) follows that $\tau_p^\infty > \tau_p^0$ as observed experimentally. It is possible to describe the nonexponential decay by a momentary minority carrier lifetime which is defined as $\tau_p = \delta p(t)/\delta\dot{p}(t)$ and therefore from (3) we get

$$\tau_p = \tau_{p0} \frac{(1 + b)\,\gamma + y(1 + \gamma)}{1 + y}. \tag{6}$$

To evaluate the PL decay according to (1) the values $\delta p(t)$ from (3) are used with $\delta p(t) = \delta n(t)$ if $\delta n(t) > n_0$ because of the low concentration N_t. This leads to the broken lines in Fig. 7.

At low densities the exponential part follows the hole concentration, but at high densities, according to (1), the influence of excited electrons leads to a lower time constant of PL with respect to the minority carrier decay. As in this case $\tau_n = \tau_p$ holds with (1) and (4) the time constant of luminescence decay must be

$$\tau_{PL}^{\infty} = \frac{\tau_{n0} + \tau_{p0}}{2}. \tag{7}$$

For a constant minority carrier lifetime τ_p one would expect an exponential decay with the time constant $\tau_p/2$ at high and τ_p at low densities, i.e. qualitatively different from the behaviour observed experimentally.

The decay curves for different excitation intensities, i.e. excited carrier densities, correspond to different values y_0 in (3) and can be obtained for $y_0 < 100$ from the curves in Fig. 7 by an appropriate shift of the origin to the requested value y_0. This was used for fitting the calculated curves at low excited carrier concentrations ($\delta p < < n_0$). The results are shown in Fig. 8 with $\tau_{p0} = 9.3$ ns, $\gamma = 0.07$, and $b = 1$.

A value $\gamma = \tau_{p0}/\tau_{n0} < 1$ is necessary for the explanation of the great difference between the high and low density lifetimes (see Fig. 1). This is a typical property of acceptor-like centres having a greater cross-section for holes than for electrons because of the Coulomb field of the firstly trapped electrons.

In this case, however, a greater difference in the capture cross-sections is expected ($\gamma < 10^{-2}$). It must be noted, therefore, that there are two causes at least, which can lead to an overestimation of γ. Firstly, for a shallow recombination centre (greater value of $b = n_1/n_0$) actually a smaller value of γ is necessary to explain the variation of $\tau(\delta p)$ (cf. (5) and Fig. 7).

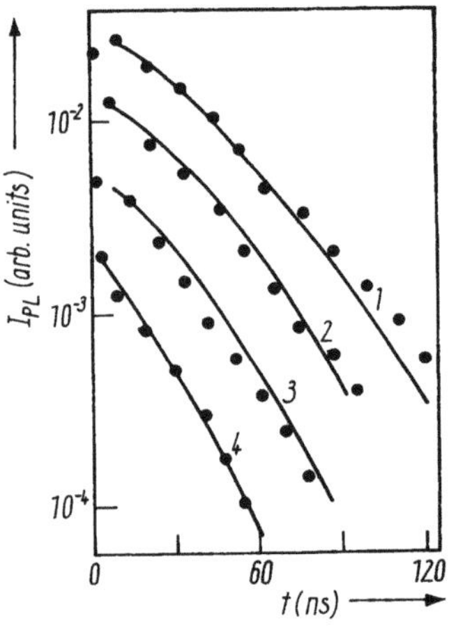

Fig. 8

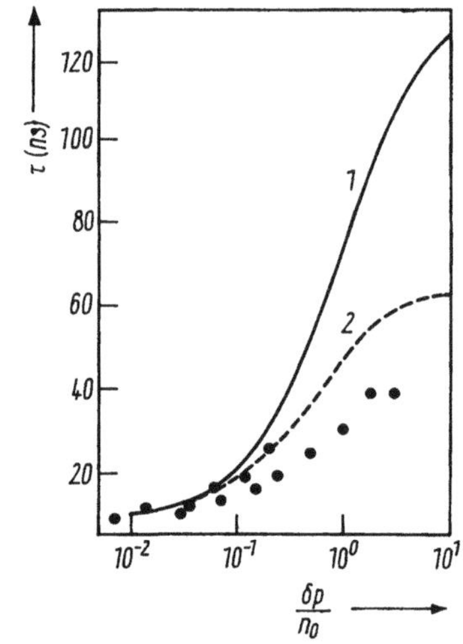

Fig. 9

Fig. 8. Fit of nonexponential PL decay curves of GaP:N, Te for low level excitation intensities according equation (3) for $\tau_{p0} = 9.3$ ns, $\gamma = 0.07$, $b = 1$ (points mean experimental decay, lines calculated curves); $P_{max} = 27$ kW cm^{-2}, $\lambda_{exc} = 480$ nm, $T = 300$ K (curves 1 to 4 correspond to curves 5 to 8 of Fig. 1

Fig. 9. Dependence of the momentary PL-decay time on the injection level $\delta p/n_0$ for GaP:N, Te at 300 K; (1) minority carrier lifetime as calculated on the basis of τ_{p0} and γ, determined by fitting (cf. Fig. 5), (2) PL decay time as calculated on the basis of curve 1, ● experimental values

This would lead, however, to a pronounced temperature dependence of $\tau_{PL}^{\infty}/\tau_{PL}^{0}$ which could not be observed experimentally. From this it is concluded that $n_1/n_0 \leqq 1$. Secondly, the influence of parallel recombination channels gives rise to a lower value τ_{PL}^{∞} and therefore a greater experimental value of γ.

A parallel channel seems to play an essential role also in these samples at higher concentrations of excited carriers. This is demonstrated by the fact that the calculated value

$$\tau_{PL}^{\infty} = \tau_{p0}\left(1 + \frac{1}{\gamma}\right)\Big/2$$

(τ_{p0} and γ from fitting at low δp) is higher than the value observed experimentally. Fig. 9 shows the agreement of experimental and calculated decay times as functions of the excited carrier concentration at low levels but a disagreement at high levels. This discrepancy can be removed as discussed above by a nonsaturable recombination channel with $\tau_{||} = 248$ ns for the sample used in Fig. 8.

If equation (3) holds (point defect centres of small concentration), then it must be possible to compose the experimental decay curves for different initial excess carrier concentrations to an unique function over the whole range by an appropriate shift of the origin (cf. Fig. 7). This was done in Fig. 10 for the sample of Fig. 1. The achieved agreement is a further confirmation that in a wide range of excess carrier density a point defect centre is a dominant one.

In the same model the influence of the donor concentration can be discussed. It is a general conclusion from the reaction kinetics in the case of one type of point defect centres of small concentration that it is not of importance whether the injection level $\delta p/n_0$ is produced by constant n_0 and varying δp or by constant δp and varying n_0 (e.g. equation (6)).

The experimental results of Fig. 3 confirm therefore the role of point defects in the excitation dependence of minority carrier lifetime: With decreasing net doping concentration n_0 the values of δp are decreased, for which τ_{PL} depends markedly on excitation intensity ($\delta p \lessapprox n_0$).

In Fig. 11 the average PL decay time for each sample of Fig. 3 is plotted against $\delta p/n_0$ instead of δp.

The very good agreement between the curves of different n_0 confirms quantitatively the validity of the point defect recombination mechanism in the intensity region $\delta p(0) < n_0$. The variation of nitrogen concentration at constant donor concentration (Fig. 4) essentially influences the decay only at high excitation intensities.

This may be explained by nonradiative parallel channels of different efficiencies. Because n_0 was nearly constant in these samples, no significant shifts of the excitation-dependent part are observed.

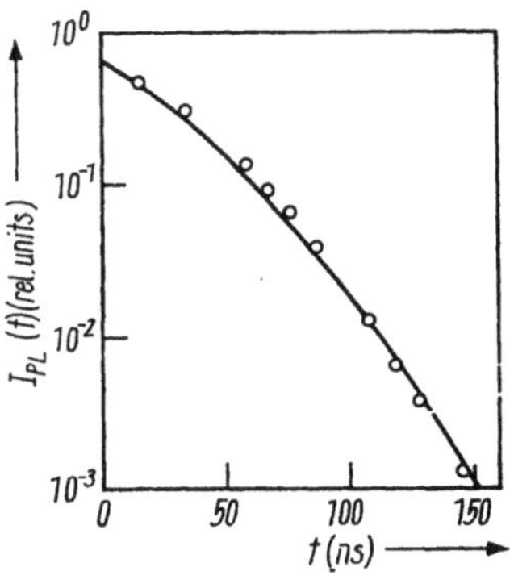

Fig. 10. PL decay from an excitation level $\delta p/n_0 = 0.4$, composed from the decay curves in Fig. 8 (solid line) in comparison with the calculated curve ($\tau_{p0} = 9.3$ ns, $\gamma = 0.07$, $b = 1$,dots) for GaP:N, Te at 300 K

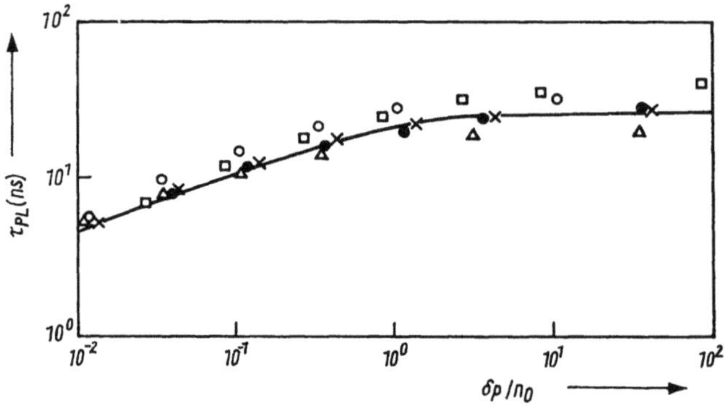

Fig. 11. Average PL decay time as a function of the injection level $\delta p/n_0$ for different net doping concentration n_0 (replotted from Fig. 3): $n_0 = 2.9 \times 10^{17}$ (△), 2.3×10^{17} (×), 1.17×10^{17} (□), 9.4×10^{16} (○), $2.7 \times 10^{16}\ cm^{-3}$ (●)

5. Conclusions

From the discussed nonexponential PL decay we conclude that point defect centres, but not dislocations or interfaces are responsible for the fast nonradiative recombination in VPE-GaP:N, Te, at least for low excitation densities with $\delta p < n_0$. Some parameters of the dominant point defect centres can be determined from the nonexponential decay. This makes it possible to characterize the nonradiative recombination by the centre parameters rather than by a minority carrier lifetime, measured at an arbitrary excitation intensity.

The experimental results are in agreement with observations of other authors [7] and seem to be characteristic of VPE-GaP. In LPE-material the excitation dependence of PL decay is less pronounced according to our results on this material (e.g. $\tau_{PL}^{\infty} = 87$ ns, $\tau_{PL}^{0} = 56$ ns for LPE and $\tau_{PL}^{\infty} = 60$ ns, $\tau_{PL}^{0} = 7$ ns for VPE-material in the same excitation range). According to [14] in LPE-material dislocations and interfaces at dislocation densities below $2 \times 10^5\ cm^{-2}$, respectively, are responsible for the luminescence efficiency. Therefore, the reported behaviour of the minority carrier lifetime in VPE-GaP:N, Te seems to be typical for kinetic controlled recombination on point defects. Using the excitation dependence of minority carrier lifetime it is possible, therefore, to separate the recombination mechanisms at extended and point defects, respectively.

An excitation-dependent minority carrier lifetime gives rise to a variation of the bulk efficiency with the injection level. With respect to the lux–ampere characteristics of LED it must be noted that for a current density of 20 A cm^{-2} (typical for most green LED) the excited hole concentration on the n-side varies from about $10^{16}\ cm^{-3}$ at the junction down to the equilibrium concentration ($\delta p \approx 0$) far from the junction. Because typical values for donor concentration are $10^{17}\ cm^{-3}$, in this region according to our results (see e.g. Fig. 3) a local variation of minority carrier lifetime and diffusion length, respectively, must be included in the numerical analysis. This was done, e.g. in [12], but a ratio $\gamma = \tau_{p0}/\tau_{n0} > 1$ was needed to explain the experimental lux–ampere characteristics.

Such a value of γ would manifests itself also in an excitation-dependent luminescence decay but with an increasing decay time at lower excitation levels. This was not observed in the PL decay in bulk VPE-material. The observation of an increasing efficiency with higher injection levels should be connected also with the increasing minority carrier lifetime.

Concerning the evaluation of bulk luminescence efficiency from PL decay measurements in the case of low concentrations of deep centres the momentary value of lifetime τ_p can be used, but at higher concentrations N_t this value does not agree with the stationary lifetime [13].

Acknowledgements

The authors are grateful to Dr. J. Voigt and Prof. E. Gutsche for helpful discussions and continuous interest. We also wish to thank Dipl.-Phys. Maege for the Hall measurements.

The department of research and development of the "Werk für Fernsehelektronik Berlin" is gratefully acknowledged for supplying the samples and for stimulating discussions.

References

[1] C. J. Hwang, J. appl. Phys. **42**, 4408 (1971).
[2] R. Z. Bachrach and O. G. Lorimor, J. appl. Phys. **43**, 500 (1972).
[3] J. A. W. van der Does de Bye and A. T. Vink, J. Lum. **5**, 108 (1972).
[4] G. B. Stringfellow and D. Kerps, Solid State Electronics **18**, 1019 (1975).
[5] B. Hamilton, A. R. Peaker, and D. R. Wight, J. appl. Phys. **50**, 6373 (1979).
[6] A. R. Peaker and B. Hamilton, Microelectronics **7**, 5 (1976).
[7] F. P. J. Kuijpers, L. Blok, and A. T. Vink, J. Crystal Growth **31**, 165 (1975).
[8] B. W. Wessels, J. appl. Phys. **46**, 2143 (1975).
[9] P. D. Dapkus, W. H. Hackett, O. G. Lorimor, and R. Z. Bachrach, J. appl. Phys. **45**, 4920 (1974).
[10] K. C. Nomura and J. S. Blakemore, Phys. Rev. **112**, 1607 (1958).
[11] M. Boulou and D. Bois, J. appl. Phys. **48**, 4713 (1977).
[12] C. Batailler, J. Bernard, and P. Garcia, Rev. Phys. appl. **14**, 587 (1979).
[13] D. J. Sandiford, Phys. Rev. **105**, 524 (1957).
[14] A. T. Vink, C. Werkhoven, and C. van Opdorp, Inst. Phys. Conf. Ser. **43**, 259 (1979).

(Received March 19, 1981)

phys. stat. sol. (a) 65, 245 (1981)

Subject classification: 1.4; 10.2; 22.1.2

Zentralinstitut für Elektronenphysik der Akademie der Wissenschaften der DDR, Berlin[1])

Simulated X-Ray Contrast of Striations for Plane Waves in the Laue Case

By

B. JENICHEN and R. KÖHLER

The X-ray contrast of a crystal containing relatively thin layers of rather steep gradients in the concentration profile is mainly due to the partial reflection of wavefields at these layers. A simple explanation of the contrast on the basis of the plane wave theory is confirmed by contrast simulations for the case of peak shaped or sawtooth shaped concentration profiles. For carbon striations in silicon, the 444-reflection, MoKα radiation and the Laue case the detection limit of X-ray plane wave topography is about 10^{15} carbon atoms/cm^3 (amplitude of the fluctuations in the carbon content).

Der Röntgenkontrast eines Kristalls, der relativ dünne Schichten mit sehr steilem Gradienten im Konzentrationsprofil enthält, rührt vor allem von der partiellen Reflexion von Wellenfeldern an diesen Schichten her. Eine einfache Erklärung des Kontrastes auf der Basis der Planwellentheorie wird durch Kontrastsimulationen für den Fall eines peak- bzw. eines sägezahnförmigen Konzentrationsprofils bestätigt. Für Kohlenstoff-Wachstumsstreifen in Silizium, den 444-Reflex, MoKα-Strahlung und den Lauefall beträgt die Nachweisgrenze der Röntgen-Planwellen-Topographie ungefähr 10^{15} Kohlenstoffatome/cm^3 (Amplitude der Fluktuation im Kohlenstoffgehalt).

1. Introduction

The high sensitivity to minute strains makes double crystal X-ray topography a useful tool for the investigation of inhomogeneous impurity incorporation during the growth process. Provided the lattice parameter change caused by a given and dominating impurity is known the local deviation of the impurity content from the mean value can be calculated from the strain. The relation between strain and X-ray contrast is not a simple one. We will show here that the contrast in the symmetrical Laue case of X-ray diffraction is essentially due to partial reflection of wave fields if the impurity profile has narrow regions of a rather steep concentration gradient. This holds e.g. for the boundaries of growth sectors in quartz and for growth striations caused by rather fast changes in the growth velocity, especially by remelt. Under the presumption that the width of these regions of steepest concentration gradient is much smaller than a pendellösung distance the contrast formation over so small a distance cannot be explained without taking into account the formation of new wave fields. In the limits of a simple model we will explain this as a partial reflection of wave fields.

Contrast simulations on the basis of a numerical solution of Takagi's equations [1] show that the simple model describes the main features of the contrast rather well. In order to allow for a comparison with experiment these simulations were done for the Si 444 reflection and MoKα_1 radiation assuming periodically arranged growth striations. The latter assumption sufficiently approximates the real situation in silicon grown by the Czochralski or the floating zone methods.

[1]) Mohrenstr. 40/41, DDR-1080 Berlin, DDR.

These simulations show that a direct correspondence between contrast profile and concentration profile exists only up to a crystal thickness of about one pendellösung distance.

All the simulations presented here were done for plane waves. In a forthcoming paper [2] we will compare experimental double crystal topographs with simulations adapted more closely to the experimental situation in that the real divergence of the collimated beam is taken into account.

2. Theoretical Interpretation of the Contrast

For the following considerations and the contrast simulations in Section 3 a sample cut parallel to the growth direction is approximated by a crystal plate with plane striations exactly perpendicular to the surface and parallel to the reflecting net planes.

From simulation we know that the contrast is mainly due to regions of largest gradient for profiles like those sketched in Fig. 4a, b. If the thicknesses of the corresponding layers are essentially smaller than a pendellösung length only the differences between the crystal lattices on both sides of this layer play a role. The layer itself can be replaced by a plane. We will call this plane the defect plane because it induces the difference between both adjacent crystal parts. So the contrasts of profiles according to Fig. 4a and b can be qualitatively explained by a defect plane inducing a difference in the Bragg condition (like a misfit boundary [3, 4]) and a phase difference (like a stacking fault [4 to 7]), respectively. The crystal parts on both sides of a defect plane are assumed to be perfect.

The contrast behaviour of the defect plane can be deduced by plane wave theory. The dispersion surfaces with the tie points of the wave fields and their directions of propagation are schematically given in Fig. 1. The dispersion surface of the left

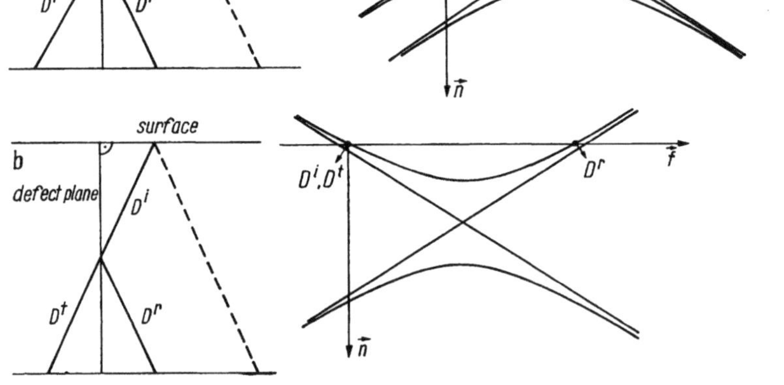

Fig. 1. Ray paths and dispersion surfaces for crystals with defect planes caused by a) a step or b) a peak in the concentration profile. Ray paths and tie points correspond to the wave fields starting at the surface right of the defect plane (s_h-side of the defect plane)

crystal part is shifted relative to that of the right one in the case of a difference in the Bragg condition (Fig. 1a). In both parts one wave field is propagating towards the defect plane. Because of the additional boundary conditions at the defect plane additional wave fields may arise from these two. The tie points for the new wave fields are the additional intersections of the normal $\boldsymbol{f}$ of the defect plane drawn through the already existing tie points of the simple Laue case. The two wave fields which propagate towards the defect plane are partly reflected.

For a peak-shaped profile and the symmetrical case considered here we get the relation

$$\frac{D_h^r}{D_0^i} = \frac{D_0^r}{D_h^i} \tag{1}$$

for the plane wave components of the wave fields (absorption neglected, indices i and r for incident and partly reflected wave field, respectively). The absolute values of the ratio (1) are the same for the wave fields running towards the defect plane from the s_0- and from the s_h-side. On account of (1) and the relatively high intensity of the wave field coming from the s_h-side the reflection to this side contributes more than the reflection to the other side to the reflected beam that forms after the decoupling of the wave fields at the exit surface. The pattern due to partial reflection is therefore more pronounced on the s_h-side of the defect plane in topographs taken in the reflected beam as usual.

For the sawtooth profile the reflection of the two wave fields incident on the defect plane is of different strength. Nevertheless the fringe system at the s_h-side dominates as a rule.

The interference of three wave fields on each side of the defect plane results in a new fringe system parallel to the defect plane in addition to the usual pendellösung fringes parallel to the crystal surface.

The intensity amplitude of the additional fringe system for the case of a peak-shaped profile depends on the phase difference induced by this peak and on the angle of incidence θ of the plane wave. The situation is quite identical to that of a stacking fault. For a sawtooth profile the intensity amplitude depends on the step in the Bragg condition $\Delta\alpha$ at the defect plane and on the angle of incidence as is shown in Fig. 2. There exists a certain range of total reflection (the flat tops of the curves in Fig. 2) at the defect plane due to the fact that for certain ranges of $\Delta\theta = \theta - \theta_B$ and of $\Delta\alpha$ there is no intersection with the dispersion surface in the crystal behind the defect plane. This is roughly equivalent to the usual Bragg case.

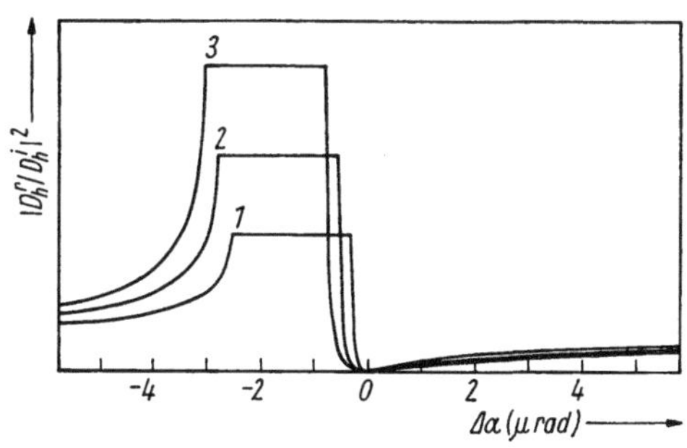

Fig. 2. Reflection coefficient for a wave field coming from the s_h-side of a defect plane caused by a step in the concentration profile (absorption neglected). $\Delta\alpha$ is the difference in the Bragg conditions between the s_0- and the s_h-sides of the defect plane. $\Delta\theta$ is the deviation from the Bragg condition at the s_h-side. (1) $\Delta\theta = 1.8$, (2) 2.6, (3) 3.2 μrad

3. Contrast Simulation

For the contrast simulation we assumed perfect periodicity of the impurity content. Because of that periodicity the integration of the Takagi equations is performed on the basis of the half-step algorithm [8] along the s_0-direction in the width of one period with a periodic boundary condition (i.e. the last and first elements of a line are used to calculate the first element of the next line). This corresponds in the limits of the algorithm exactly to an incident plane wave. Two periods are plotted in Fig. 3, 4. The intensities inside the crystal are given by a pseudoperspective plot.[2]) In the calculation the following values were used: period of the striations 230 μm, crystal thickness 390 μm, for the Si 444 reflection and $MoK\alpha_1$ $\theta_B = 0.46948$ rad, $\chi_{0r} = 0$, $\chi_{0i} = -1.38 \times 10^{-8}$, $\chi_{hr} = \chi_{\bar{h}r} = -9.65 \times 10^{-7}$, $\chi_{hi} = \chi_{\bar{h}i} = -1.10 \times 10^{-8}$.

For different impurity profiles the corresponding deformations were calculated on the basis of the linear elasticity theory of an isotropic medium. The lattice deformation by a periodic striation system [9] is illustrated in Fig. 3a for a sawtooth profile (Fig. 4b). The lattice curvature is neglible in the bulk and there is only a periodic modulation of the lattice constant parallel to the growth axis. Near the surface the part of the deviation from the Bragg condition due to stress relaxation at the surface overcomes the part that is due to the deformation in the bulk. But this surface relaxation declines very fast with the depth in the crystal. This holds especially for the higher harmonics of the concentration profile. Fig. 3b and Fig. 4d show contrast calculations for the same reflection conditions taking into account and neglecting stress relaxation at the surface, respectively. Because the effect of the surface relaxation is so small in all subsequent calculations the deformation in the bulk was used throughout the whole crystal.

Fig. 4 shows several contrast simulations for different deviations from the Bragg condition and for two different concentration profiles. The deformation was calculated on the basis of the $\Delta d/d$ values of carbon in silicon (-6.5×10^{-24} cm³/atom [10]). The interference pattern calculated for the flanks of the reflection curve (Fig. 4c, d, g, h) corresponds well to the model explained in Section 2.

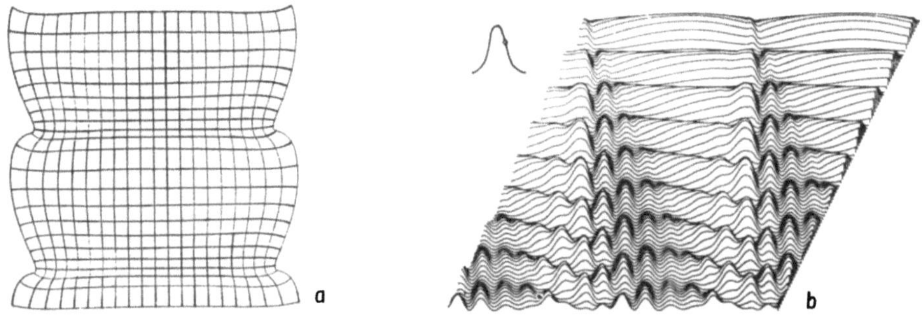

Fig. 3. b) Plane wave contrast simulation calculated for Si 444 reflection, $MoK\alpha_1$ radiation, $\Delta\theta = 1.2$ μrad, and a sawtooth-shaped carbon concentration profile (cf. Fig. 4b) taking into account stress relaxation at the surface a). The deformation in a) is magnified by a factor of 3×10^6, $A = 13.5$

[2]) For several depths in the crystal the intensity times a factor B plus the depth is given as a function of the lateral coordinate. $B = A \times$ thickness/incident intensity (A see figures).

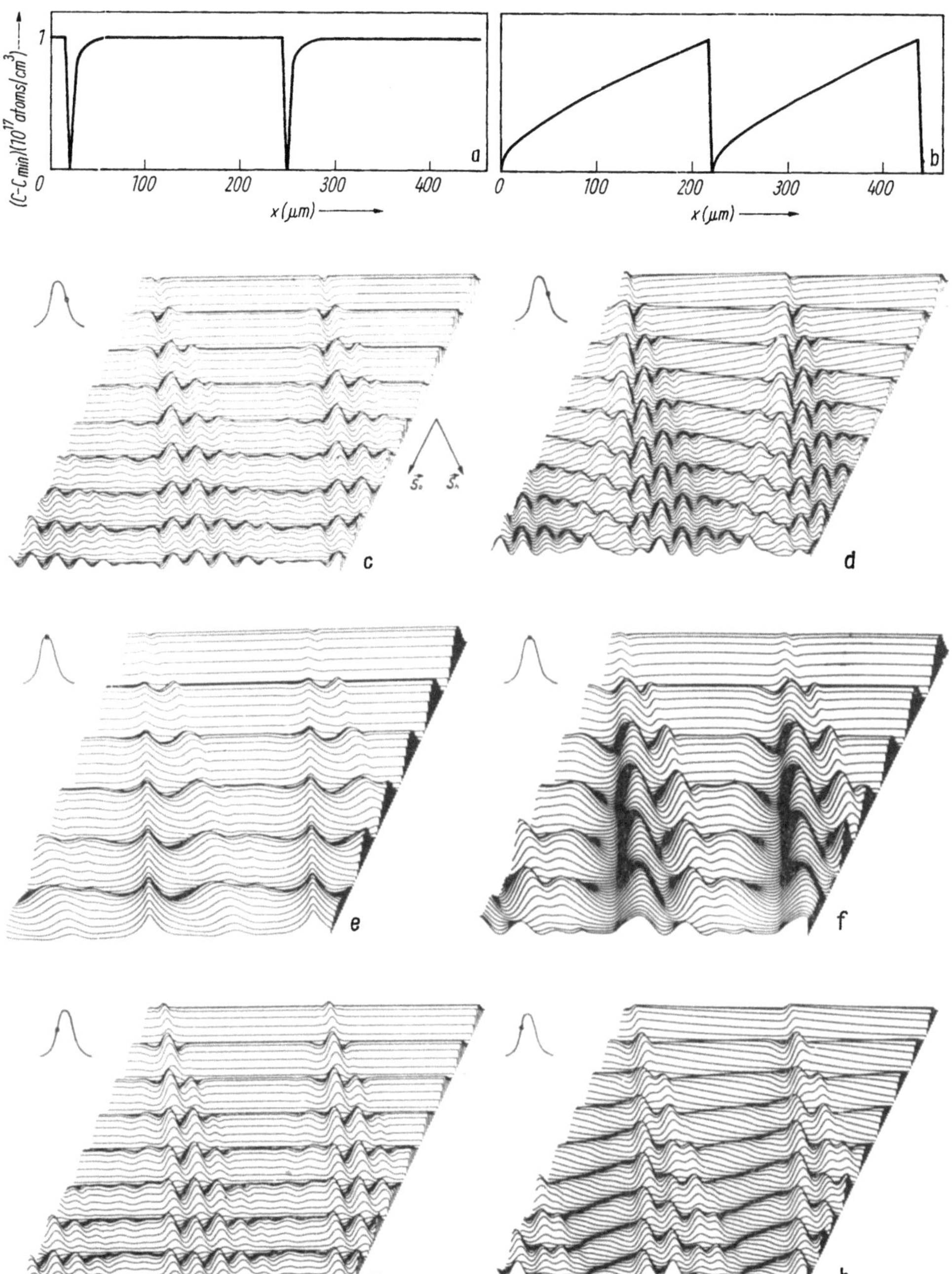

Fig. 4. Plane wave contrast simulations calculated for Si 444 reflection, $MoK\alpha_1$ radiation and mean deviations from the Bragg condition: c) and d) $\Delta\theta = 1.2\ \mu rad$, e) and f) $\Delta\theta = 0\ \mu rad$, and g) and h) $\Delta\theta = -1.2\ \mu rad$. c), e), g) and d), f), h) correspond to the carbon concentration profiles in a) and b), respectively. $A = 13.5$

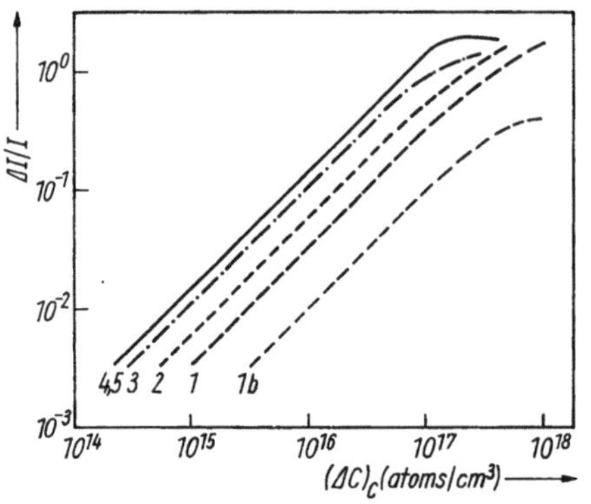

Fig. 5. Dependence of the X-ray contrast on the amplitude of carbon concentration fluctuations calculate for Si 444 reflection, MoKα_1 radiation, $\Delta\theta = 1$ µrad, and two crystal thicknesses: 75 µm (second pendellösung maximum, curves 1 to 5) and 25 µm (first pendellösung maximum, curve 1b). Curves 1 to 4 correspond to peak-shaped profiles with different halfwidths: (1) 2 (2) 3.7, (3) 7.3, and (4) 15.2 µm and curve 5 to a sawtooth profile (cf. Fig. 4b)

Changing the flanks of the reflection curve has the same result as changing the sign of the concentration peak. The latter means a change of the sign of the phase shift introduced by the defect plane.

Therefore, the contrast for the peak-shaped profile (Fig. 4a) is almost reversed going from one flank to the other (cf. Fig. 4c and g). The contrast at the top of the reflection curve (Fig. 4e) cannot be understood with the help of the ray theory or the considerations of Section 2 because the ray directions should be parallel to the defect plane outside the narrow region of the peak. But because of diffraction effects (similar to the diffraction of light at small obstacles) the wave fields are affected by the deformation also outside the deformed region.

The patterns for the sawtooth profile are quite similar to those calculated for the peak-shaped profile. At the flanks of the reflection curve the reason of that similarity is quite obvious from the considerations in Section 2, the contrast is determined by the partial reflection of wave fields for both profiles. At the top of the reflection curve the Bragg condition is fulfilled approximately in the middle of the step of the sawtooth profile. The directions of the wave fields on both sides of the step are slightly inclined to the defect plane. Therefore, the contrast is mainly determined by the strong reflection of the wave field running towards the region of the step from the right.

For a thickness of half a pendellösung distance the contrast at the flank follows closely the concentration profile. At the next pendellösung fringe the contrast approaches its maximum value but still we can guess the shape of the concentration profile. Later on the partly reflected wave fields extend so far from the peak or from the step that the structure due to this reflection overlaps even with the rough details of the concentration profile. Nevertheless there are characteristic differences in some details of the pattern and in the intensities for the different profiles, e.g. the usual pendellösung fringes are sensitive to the slope of the slowly increasing flanks of the sawtooth.

It should be noted that the shape of the contrast is not very sensitive to the amplitude of the concentration profile. Over a wide range the amplitude of the contrast changes proportionally to the amplitude of the concentration profile as is shown in Fig. 5.

4. Discussion

Assuming a sinusoidal concentration profile the dynamical contrast can be calculated analytically on the basis of the ray theory [11]. The contrast is then mainly due to the focussing effect [12]. But there is experimental evidence (e.g. [13 to 15]) that the concentration profile usually contains regions of rather steep gradients. These regions can be explained by remelt [16, 17]. Then the contrast formation might be due to partial reflection of wave fields as indicated by additional contrast fringes in a section topograph [13]. This assumption is verified by our contrast simulations. It is shown that we get quite similar additional interference fringes in a plane wave topograph as well for peak-shaped as for sawtooth concentration profiles. The stress relaxation at the surface has no essential influence on the contrast.

At the first pendellösung maximum the contrast follows the shape of the concentration profile quite closely. The contrast amplitude can be calculated from the shape of the reflection curve for this thickness taking into account the oscillations in the reflection curve. This is true in the flanks of the reflection curve but for the peak-shaped profile only above a certain halfwidth of the peak (about 5 μm under our conditions). Below that width the peak-shaped profile behaves exactly like a stacking fault with a Burgers vector given by the integral lattice shift introduced by the peak. The same holds at the second pendellösung maximum. Therefore, curves 1 to 4 in Fig. 5 will coincide if the contrast amplitude is not related to the concentration amplitude as is done here but to the integral lattice shift, i.e. to a value proportional to the area of the peak of the profile in Fig. 4a. At this thickness (second pendellösung maximum) the partial reflection of wave fields begins to play a dominating role and the contrast already approaches its maximum value.

Contrast differences down to about 1 to 2% are visible in topographs. Fig. 5 gives then a lower limit for the detection of fluctuations in the carbon content in silicon of about 10^{15} atoms/cm^3 for the 444 reflection and MoKα_1. The contrast according to Fig. 5 with our conditions is higher by a factor of 5 compared to the value estimated on the basis of the slope of an averaged reflection curve. The averaged reflection curve means an approximation neglecting the oscillations in the reflection curve. This is an appropriate approximation of the real topographical experiment, because the divergence of the collimated beam in double crystal or so-called plane wave topography is not small enough to resolve these oscillations. This means on the other hand, that we cannot compare experimental topographs directly to the contrast simulations given in this paper, because these simulations presume ideal plane wave conditions. Taking into account the real divergence of the collimated beam does not eliminate the additional fringe system but changes the contrast pattern considerably [2]. This is verified also by experiments in a forthcoming paper [2].

Acknowledgements

The authors are indebted to A. Authier, V. L. Indenbom, W. Möhling, and H. Peibst for valuable discussions.

References

[1] S. Takagi, J. Phys. Soc. Japan **26**, 1239 (1969).
[2] B. Jenichen and R. Köhler, phys. stat. sol. (a) **65**, No. 2 (1981).
[3] G. N. Dubnova, V. L. Indenbom, S. A. Pikin, and F. N. Chukhovskii, Kristallografiya **16**, 18 (1971).
[4] N. Kato, in: X-Ray Diffraction, Ed. L. V. Azaroff, McGraw-Hill Publ. Co., 1974 (p. 380).
[5] A. Authier, phys. stat. sol. **27**, 77 (1968).
[6] V. L. Indenbom and F. N. Chukhovskii, Kristallografiya **19**, 35 (1974).
[7] Y. Epelboin, J. appl. Phys. **50**, 1312 (1979).
[8] A. Authier, C. Malgrange, and M. Tournarie, Acta cryst. **A24**, 126 (1968).
[9] V. L. Indenbom and F. Kroupa, Czech. J. Phys. **B29**, 338 (1979).
[10] J. A. Baker, T. N. Tucker, N. E. Moyer, and T. C. Buschert, J. appl. Phys. **39**, 4365 (1968).

[11] B. Dietrich and J. Härtwig, Kristall und Technik **15**, 1377 (1980).
[12] V. L. Indenbom and F. N. Chukhovskii, Kristallografiya **16**, 1101 (1971).
[13] T. Abe, Y. Abe, and J. Chikawa, in: Semiconductor Silicon 1973, Ed. H. R. Huff and R. R. Burgess, Electrochem. Soc., Princeton (N.J.) 1973 (p. 95).
[14] U. Bonse, Z. Phys. **184**, 71 (1965).
[15] A. Murgai, H. C. Gatos, and W. A. Westdorp, J. Electrochem. Soc. **126**, 2240 (1979).
[16] L. O. Wilson, J. Crystal Growth **48**, 451 (1980).
[17] A. M. J. G. van Run, J. Crystal Growth **47**, 680 (1979).

(Received January 2, 1981)

phys. stat. sol. (a) **65**, 253 (1981)

Subject classification: 14.3.3; 13.3; 22.1.2

Institute of Electron Technology, Warsaw[1]) *(a) and Warsaw Technical University*[2]) *(b)*

A Simple Technique of Work Function Difference Determination in MOS Structures

By

H. M. Przewłocki (a), S. Krawczyk (b), and A. Jakubowski (b)

A new measurement technique of work function differences in MOS structures is described. Results obtained for the Al–SiO_2–Si system are presented and discussed in comparison with the results obtained by other authors.

Описан новый метод измерения контактной разницы потенциалов в структурах МОП. Представлены результаты измерений, полученные для структур Al–SiO_2–Si. Эти результаты обсуждаются в сравнении с результатами других авторов.

1. Introduction

The threshold voltage (U_T) which is a fundamental parameter of MOS devices depends directly on the work function difference (Φ_{MS}) between the gate and substrate materials [1, 2]. In designing modern MOS circuits with low $|U_T|$, it is particularly important to know the exact values of Φ_{MS}.

In spite of this Φ_{MS}-values found in the literature [3 to 6, 13] differ considerably.[3]) The values of work function difference found in [4], for the Al–SiO_2–Si (n-type) system are by more than 0.4 V higher than the values obtained in [5], and according to [4], Φ_{MS} is positive in this case, while other authors claim that it is negative [3, 5].

All the above-mentioned results (with the exception of the results of Deal et al. [3]) were obtained using the well-known technique of flat-band voltage determination for MOS capacitors with varying thickness of the dielectric layer [1, 4 to 6, 13].

In an attempt to explain the differences in Φ_{MS}-values found in the literature, it seemed appropriate to find a new (possibly simpler, and more accurate) technique of Φ_{MS}-determination. This paper presents such a new technique, together with some experimental results, which support its validity.

2. Theory

The energy band diagram of the MOS system is shown in Fig. 1, in which the various potentials used in this work are also defined. The balance of potentials in this system is expressed by the equation

$$U_G + \Phi_M = \Phi_I + \Phi_S + X - \frac{E_G}{2q} + \Phi_B \,. \tag{1}$$

The work function difference Φ_{MS} is defined here as

$$\Phi_{MS} = X - \Phi_M - \frac{E_G}{2q} + \Phi_B \,, \tag{2}$$

---

[1]) Al. Lotników 32/46, 02-668 Warsaw, Poland.

[2]) Koszykowa 75, 00-662 Warsaw, Poland.

[3]) These differences are clearly visible in Fig. 4 where previously published results are shown for comparison with the results obtained in this work.

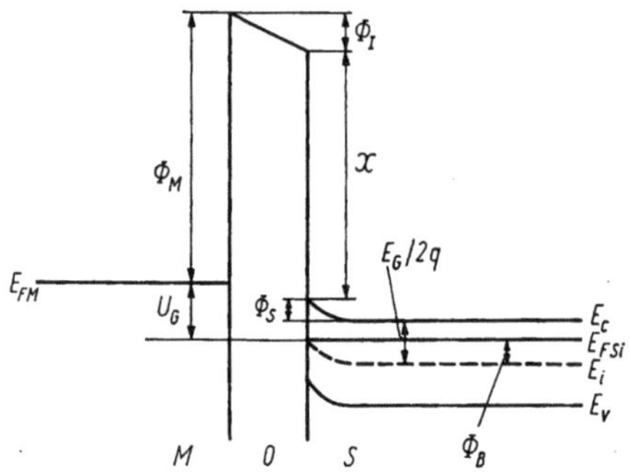

Fig. 1. The energy band diagram of the MOS system. The potentials used are indicated

where Φ_M- and X-values are assumed to be negative, in accordance with the sign convention used in the energy band structure theory.[4]) Combining (1) and (2) results in a general expression for the gate voltage

$$U_G = \Phi_I + \Phi_S + \Phi_{MS} \tag{3}$$

which is used as a basis of the measurement technique proposed.

Let us consider now two situations illustrated in Fig. 2, which may prevail in a MOS system. The first situation (Fig. 2a) is the well-known flat-band situation. In this case $\Phi_S = 0$, and the gate voltage, U_{FB}, is given by

$$U_G = U_{FB} = \Phi_I(\Phi_S = 0) + \Phi_{MS} . \tag{4}$$

Assuming that the fixed oxide charge Q_f[5]) is the only charge present in the dielectric, and at the dielectric–semiconductor interface, which usually is the case for MOS systems used in practice, the familiar expression for U_{FB} is obtained

$$U_{FB} = \Phi_{MS} - \frac{Q_f}{C_0}, \tag{5}$$

where C_0 is the dielectric capacitance.

The second situation (Fig. 2b) takes place when the voltage drop in the dielectric Φ_I is equal to zero. The gate voltage, called now U_{G0}, is given by

$$U_G = U_{G0} = \Phi_S(\Phi_I = 0) + \Phi_{MS} . \tag{6}$$

In this case

$$Q_f = -Q_S , \tag{7}$$

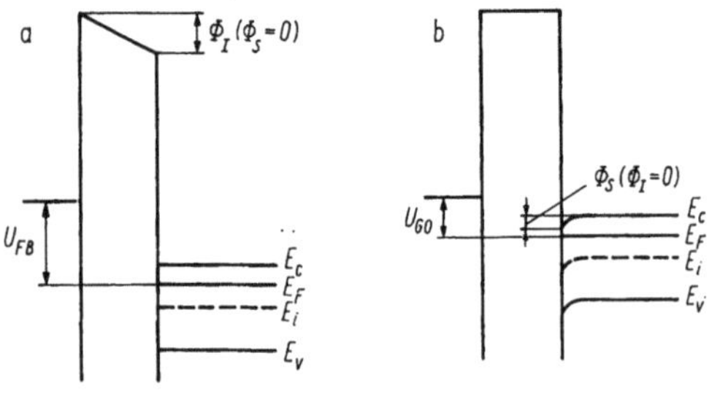

Fig. 2. The energy band diagrams of the MOS system at two different gate voltages. a) $U_G = U_{FB}$, the bands in the semiconductor are flat, i.e. $\Phi_S = 0$; b) $U_G = U_{G0}$, the bands in the dielectric are flat, i.e. $\Phi_I = 0$

[4]) This sign convention is different from the one which is commonly used [1, 2], but simplifies the interpretation of band diagrams representing situations encountered in practice.

[5]) The standardized terminology for oxide charges [7] is used here.

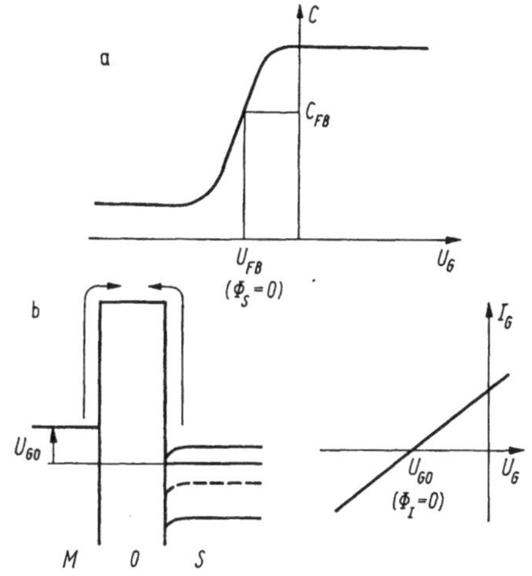

Fig. 3. Illustration of the methods used to measure a) the gate voltage U_{FB} (at which $\Phi_S = 0$) and b) the gate voltage U_{G0} (at which $\Phi_I = 0$)

where Q_S is the charge accumulated in the semiconductor, which is a well-known function of the surface potential Φ_S given by the Kingston formula [8, 9]. This means that

$$\Phi_S(\Phi_I = 0) = F(Q_f) . \tag{8}$$

Subtracting (5) from (6) yields

$$U_{G0} - U_{FB} = \Phi_S(\Phi_I = 0) + \frac{Q_f}{C_0} . \tag{9}$$

The value of $U_{G0} - U_{FB}$ depends on Q_f, as given by (9) and (8), but does not depend on Φ_{MS}. Using (9) and (8) Q_f can be calculated numerically, if the values of U_{G0} and U_{FB} are measured. Once the value of Q_f is known, Φ_{MS} can be found using (5) or (6).[6])

The values of U_{FB} and U_{G0} can be measured as shown in Fig. 3. The simplest way to measure U_{FB} is the standard method of $C(U)$ characteristics, although there are other methods which sometimes are more advisable [10, 11]. U_{G0} can be measured by a photoelectric method [12] of $I_G(U_G)$ current–voltage characteristics of an illuminated MIS system. If the illumination is such that electrons from the semiconductor and the gate are excited into the conduction band of the dielectric, a current may flow in the external circuit. This current changes sign (goes through zero) for $\Phi_I = 0$, which happens for $U_G = U_{G0}$.

The measured values of U_{FB} and U_{G0} are used in numerical calculations of Q_f, and Φ_{MS}, as described above.

3. Results and Discussion

This technique has been applied to measure Φ_{MS}-values in a series of differently processed Al–SiO_2–Si capacitors. Some of the results are shown in Fig. 4 against the background of the values obtained by other authors. Circles indicate average values, while vertical bars indicate deviations of results obtained for a group of identically processed capacitors. Usually five capacitors of one group were measured.

Two features of these results are immediately apparent:

(i) The dispersion of Φ_{MS}-values obtained for one group of identically processed MOS capacitors is small, the deviations being usually less than 0.035 V.

(ii) There are appreciable differences between results obtained for differently processed groups of otherwise identical capacitors. Trying to explain these results, an experiment was made in which two groups of Al–SiO_2–Si (p, ⟨100⟩) capacitors were produced, using two different silicon oxidation processes, but with no other differences between the groups in processing, equipment, and materials used. The first group was made using "dry O_2" wafer oxidation, the other on wafers oxidized in "wet O_2" atmosphere. Measurement results obtained for these two groups are indicated in

[6]) Simple programs have been developed, which allow an accurate calculation of Φ_S ($\Phi_I = 0$), Q_f, and Φ_{MS} using hand held programmable calculators.

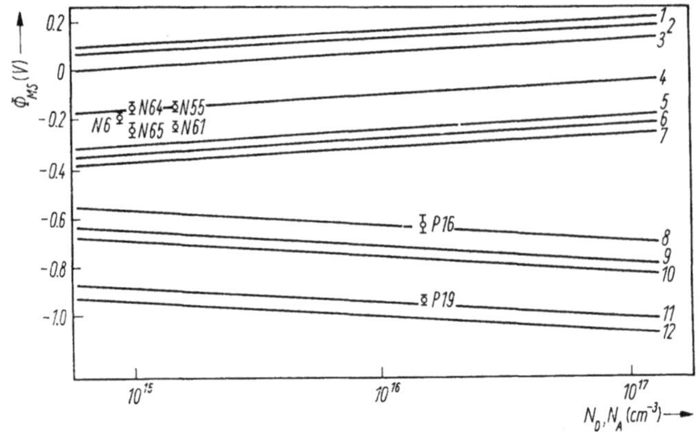

Fig. 4. Results of Φ_{MS}-measurements obtained in this work for differently processed Al–SiO_2–Si capacitors shown in comparison with the results obtained by other authors. The silicon wafers used for the P16 samples were oxidized in dry O_2 atmosphere; wafers for the P19 samples were oxidized in the wet atmosphere. Circles indicate average values, while vertical bars indicate deviations obtained for one group of identically processed capacitors. Results obtained by other authors are shown as Φ_{MS} vs. substrate doping (N_D, N_A) lines. (1) reference [6], substrate orientation ⟨100⟩, type n-Si; (2) [4], ⟨100⟩, n-Si; (3) [13], ⟨100⟩, n-Si; (4) [13], ⟨111⟩, n-Si; (5) [3], ⟨100⟩, ⟨111⟩, n-Si; (6) [6], ⟨111⟩, n-Si; (7) [5], ⟨100⟩, n-Si; (8) [4], ⟨100⟩, p-Si; (9) [13], ⟨100⟩, p-Si; (10) [13], ⟨111⟩, p-Si; (11) [3], ⟨100⟩, ⟨111⟩, p-Si; (12) [5], ⟨100⟩, p-Si

Fig. 4, showing that Φ_{MS}-values obtained for "dry O_2" MOS capacitors (P16) are close to the results of Werner [4] (who also used "dry oxides"), while those for "wet O_2" capacitors (P19) are close to the results obtained by Kar [5] (who also used "wet oxides"). This behaviour strongly suggests that the Φ_{MS}-value in an Al–SiO_2–Si system depends on the presence of hydrogen in the silicon oxidation atmosphere. Such a dependence was already mentioned in [13], and is very likely in the light of the well-known influence of hydrogen on the work function measured on clean silicon surfaces [14]. Further work is necessary to explain the observed changes in Φ_{MS}-values.

The results presented above have shown the applicability of the measurement technique proposed. This technique is much simpler than the previously used one, allowing determination of Φ_{MS} using only one MIS capacitor as a test sample. This facilitates the observation of the influence of various processing parameters on the work function difference in a MIS system. Also, a series of measurements can be easily made allowing a statistical interpretation of the results. Moreover, the technique proposed is believed to be more accurate than the previously used one.

The main limitations of this technique are:

(i) The necessity of transparent gate electrodes of the measured MIS capacitors.

(ii) Fulfilment of the assumption that Q_f is the only charge present in the dielectric and at the dielectric–semiconductor interface.

4. Conclusions

A new technique of work function difference measurement in MOS systems has been proposed. This technique is simpler, and is believed to be more accurate than the techniques used previously. Results obtained for the Al–SiO_2–Si system show that the work function difference depends not only on the type and resistivity of the substrate, but also on other parameters, e.g. on the processing history of the system.

References

[1] A. S. Grove, Physics and Technology of Semiconductor Devices, J. Wiley, New York 1967.
[2] S. M. Sze, Physics of Semiconductor Devices, J. Wiley, New York 1969.
[3] B. E. Deal, E. H. Snow, and C. A. Mead, J. Phys. Chem. Solids 27, 1873 (1966).
[4] W. M. Werner, Solid State Electronics 17, 769 (1974).
[5] S. Kar, Solid State Electronics 18, 169 (1975).
[6] K. Haberle and E. Fröschle, J. Electrochem. Soc. 126, 878 (1979).
[7] B. E. Deal, J. Electrochem. Soc. 127, 979 (1980).
[8] R. H. Kingston and S. F. Neustadter, J. appl. Phys. 26, 718 (1955).
[9] A. S. Grove, B. E. Deal, E. H. Snow, and C. T. Sah, Solid State Electronics 8, 145 (1965).
[10] B. H. Yun, Appl. Phys. Letters 21, 194 (1972).
[11] A. Jakubowski and S. Krawczyk, Electron Technology 11, 23 (1978).
[12] C. R. Viswanathan and S. Ogura, J. appl. Phys. 42, 5865 (1971).
[13] L. A. Kasprzak and A. K. Gaind, Proc. Internat. Topical Conf. Physics of SiO_2 and Its Interfaces, Ed. S. T. Pantelides, Pergamon Press, New York 1978 (p. 438).
[14] T. Sakurai and H. D. Hagstrum, Phys. Rev. B 12, 5349 (1975).

(Received January 19, 1981)

phys. stat. sol. (a) 65, 259 (1981)

Subject classification: 14.3.2 and 16; 22.2.1

Institute of Semiconductor Physics, Academy of Sciences of the USSR, Siberian Branch, Novosibirsk[1])

Effect of Electron Heating by Surface Electric Fields on Oscillating Photoresponse Spectra of Gallium Arsenide–Metal Structures

By

V. L. ALPEROVICH, A. F. KRAVCHENKO, N. A. PAKHANOV, and A. S. TEREKHOV

The effect of surface electric fields on the oscillating photoresponse spectra of gallium arsenide–metal structures is established. A shift of the oscillation extrema to the high- and low-energy range is observed. The observed shifts of the oscillation extrema are explained as additional heating of the electrons by the external electric field and by the Dember field caused by illumination of the structures.

Обнаружено влияние приповерхностных электрических полей на осциллирующие спектры фотоответа структур GaAs–металл. Наблюдались смещения экстремумов осцилляций как в длинноволновую, так и в коротковолновую сторону. Эти смещения объясняются дополнительным (по отношению к свету) разогревом электронов внешним электрическим полем и полем Дембера, возникающим при освещении структуры.

1. Introduction

In [1] we reported the observation of oscillations in the photoresponse spectra of n-GaAs–metal structures. The energy positions of the oscillation extrema point to the fact that they are due to the thermalization of photoexcited electrons via emission of optical phonons. The effect of the surface electric fields on the shape of the oscillating photoresponse spectra was discovered in this work. Surface electric fields were changed by external bias and additional illumination (AI). By surface electric fields we mean the field of the space charge region (SCR) and the Dember field, which appear near the surface under illumination.

2. Experimental

The samples to be measured were prepared by sputtering of a semitransparent layer of aluminium on the surface of epitaxial n-GaAs, which was grown on an n^+ substrate. The electron concentration of the studied GaAs at 300 K was $n_0 \lessapprox 10^{13}\ \text{cm}^{-3}$. The film surface was not treated before sputtering in order not to remove the natural oxide layer. The high nonideality coefficients $K \approx 6$ for the investigated structures indicate the presence of such layers. K was measured from the direct branch of the I–U characteristics at $T = 300$ K. The alternating photovoltage $U_{ph}^{\sim}$, which appears on the structure upon illumination by monochromatic radiation, modulated with the frequency 500 Hz, was registered by a lock-in amplifier. The dc component of the photovoltage, U^{-}, was also controlled during recording the spectra, as it was done in [2]. The additional modulated illumination (AI) was made by a filament lamp.

[1]) Prospekt Nauki 13, 630090 Novosibirsk–90, USSR.

The part of the lamp spectrum with photon energy $\hbar\omega$, exceeding the width of the forbidden gap, E_g, was separated with the help of filters. The measurements were carried out at 4.2 K.

3. Results and Discussion

Fig. 1 shows the spectra $U_{ph}^{\sim}(\hbar\omega)$ of one of the structures (I-240) at different reverse bias U. The features near the fundamental absorption edge are of exciton nature. The oscillation due to emission of longitudinal optical phonons by photoexcited electrons is seen in the spectra at photon energies above the band gap, $\hbar\omega > E_g$. (The energy of the LO phonon, $\hbar\omega_{LO}$, is equal to 36.7 meV for GaAs.) As the voltage U increases, the absolute value of the photoresponse increases, the relative amplitude of the oscillations decreases, and at $U \gtrapprox 6$ V they disappear completely. Simultaneously, a long-wavelength shift of the oscillation extrema is observed. Such a shift was observed in a few structures, the biggest being at the given one (I-240).

It is well known that under usual conditions the photovoltage arising upon Schottky barrier illumination is due to the existence of an electric field in SCR. In this case the photovoltage is determined by the minority carrier flux to the surface.

Hence, we connected the oscillations in photoresponse spectra of GaAs–metal structures with the oscillation of the lifetime of the minority carriers, i.e. holes [1]. However, further investigations [2] showed that an oscillating photoresponse was observed at flat bands as well. The flat-band condition was indicated by a constant positive voltage $U^{=} \approx 1$ V, which appeared on the structures at illumination. In some structures this band condition was retained, when AI was switched off [3].

It is assumed in [2] that the Dember effect on hot electrons is the most probable cause of the oscillating photoresponse under these conditions. The essence of such a mechanism of photoresponse is as follows: At helium temperature and low excitation level the thermalization rate of photoexcited electrons with initial energy $E_0 \lessapprox \hbar\Omega_{LO}$ (passive band) is determined by emission of acoustic phonons, and the momentum loss (mobility) is determined by the scattering on charge impurities. The time of thermalization on acoustic phonons can be comparable with the overall lifetime. Thus the hot, nonthermalized electrons can give a considerable contribution to the Dember effect. The mobility and lifetime of nonthermalized electrons increase with E_0, i.e. with increase of light frequency ω. As a result an increase of Dember photovoltage is observed. At $E_0 > \hbar\Omega_{LO}$ a fast thermalization of an electron to the passive band via emission of optical phonos occurs. This process leads to a periodic dependence of the electron distribution function and hence the Dember photovoltage on photon energy $\hbar\omega$.

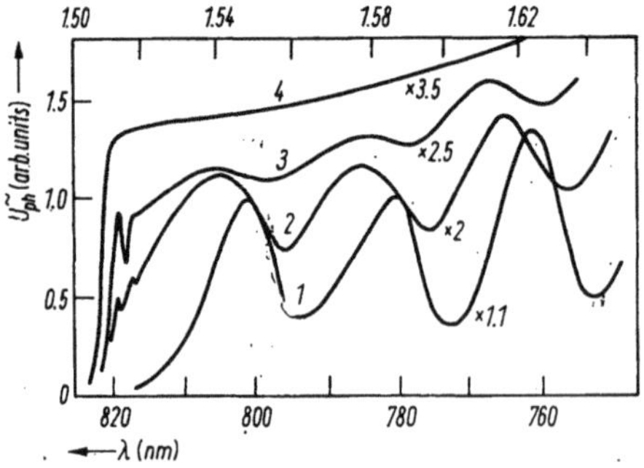

Fig. 1. The spectra of the structure I-240 at different reverse bias. (1) Initial spectra at $U^{=} = +1$ V, (2) reverse bias $U = -2$, (3) -4, (4) -6 V

The influence of surface electric fields being absent, the energy positions of the minima in oscillating spectra must be described by the well-known relation [5]

$$\hbar\omega_n = E_g + n\hbar\Omega_{\text{LO}}\left(1 + \frac{m_e^*}{m_h^*}\right),$$

where $\hbar\omega_n$ is the energy position of the n-th minimum, m_e^*/m_h^* the ratio of effective masses of electron and heavy hole.

The shift of the extrema to the long-wave side, when applying an inverse bias to the structure, shows that the electron thermalization at the bottom of the conduction band becomes possible at lower energies of exciting photons. Following [5] we connected the dependence of the positions of the extrema on voltage, $\hbar\omega_n(U)$, with heating of the electrons diffusing from the surface by the external electric field. This means that the energy

$$\Delta E \approx E_g + n\hbar\Omega_{\text{LO}}\left(1 + \frac{m_e^*}{m_h^*}\right) - \hbar\omega_n(U)$$

is gained by an electron by the electric field, a LO phonon being emitted. With increasing voltage the electron "heating" increases and, hence, extremum shifts also increase.

The difference in the magnitude of the displacement of the extrema in the photoresponse spectra of different structures is probably connected with the different spatial distribution of the electric field. In the structure with the maximum shift of extrema (I-240) the field was apparently most uniform and penetrated into the whole film depth. In this case the energy gain of photoexcited electrons weakly depends on the place of its generation. In the structures where the field is nonuniform, the addition to electron energy due to the electric fields depends on the point of excitation. The effect of field on electron thermalization in such structures manifests itself mainly as oscillation smoothing. The conclusion on the degree of field uniformity was made through the analysis of C–U characteristics. Thus, the capacity of the I-240 structure did not depend on inverse voltage at 300 K and its magnitude was close to the magnitude of the geometric capacity. This indicates that the field in this structure at reverse bias is close to a uniform one. At the same time the capacities of other structures were dependent on the reverse bias.

Changes in $U_{\text{ph}}^{\sim}(\hbar\omega)$ under reverse bias were observed in all structures and were qualitatively clear. At the same time the lineshape of $U_{\text{ph}}^{\sim}(\hbar\omega)$ as a function of constant potential $U^{=}$, created by additional illumination, was more complex and was not quite clear. The band flattening phenomenon in a real Schottky barrier is apparently dependent on subtle processes of accumulation, capture, and transition of carriers through a thin oxide layer [4]. As a result the connection between the shape of spectra and the constant photovoltage $U^{=}$ can differ from sample to sample and can be dependent on the initial charge of the subsurface region and oxide layer. For example, the capture of holes in the interface layer or their accumulation in a narrow layer near the surface can actually flatten the bands in the main region of electron–hole pair generation, but $U^{=}$ does not appear. In due course, the holes can penetrate into the metal and so create a constant positive potential, but the shape of $U_{\text{ph}}^{\sim}(\hbar\omega)$ in this case would not depend on its magnitude [2]. This process, to our mind, needs further investigation.

To check up the proposed mechanism of photoresponse formation, we investigated the effect of additional illumination on the spectral shape. The photoresponse spectra at different intensities of AI are shown in Fig. 2. In this case the relative amplitude of oscillations increases, but the signal magnitude decreases by one or two orders.

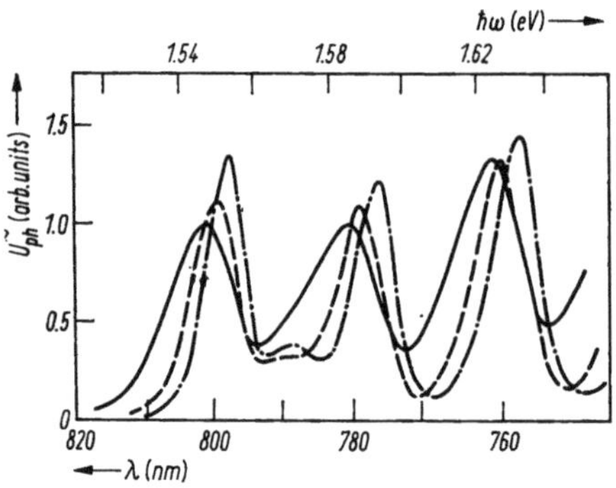

Fig. 2. Photoresponse spectra of the structure I-240 at different intensities of additional illumination I_{AI}: ——— $I_{AI} = 0$, - - - - $I_{AI} \approx 4 \times 10^{15}$ photons/ cm²s, —·— $I_{AI} \approx 8 \times 10^{15}$ photons/ cm²s¹. Spectra are plotted in different scales

The intensity of additional illumination I_{AI} exceeds sufficiently the intensity of alternating monochromatic radiation. It is seen that the extrema shift towards the short-wave side with increasing I_{AI}. In our opinion, the cause of such a displacement is the constant Dember photovoltage $U_D^=$ appearing near the surface during additional illumination. This field is oriented in opposite direction to the external field, so it "cools" the electrons diffusing from the surface. In other words, the hot electrons with an excess of energy above the bottom of the conduction band, $E_0 \gtrsim \hbar\Omega_{LO}$, that are responsible for the appearance of the alternating signal $U_{ph}^{\sim}(\hbar\omega)$, move some distance from the point of generation into the bulk of the semiconductor during the time of LO-phonon emission. Since due to the presence of a constant Dember photovoltage $U_D^=$ bands are raised in the bulk of the semiconductor (see Fig. 3), the excess of electron energy above the bottom of the conduction band can become smaller than $\hbar\Omega_{LO}$. In this case the emission of optical phonons is impossible. For the electron to emit an LO phonon, the initial energy E_0 must be larger than the energy of the

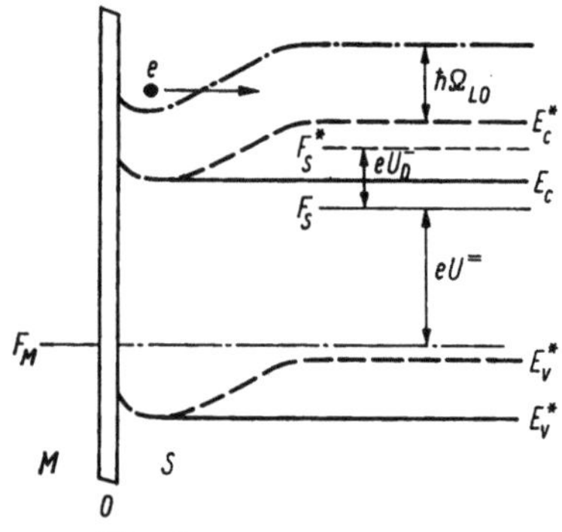

Fig. 3

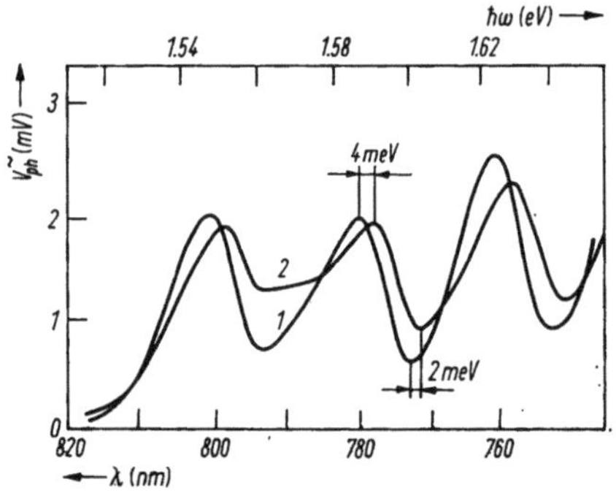

Fig. 4

Fig. 3. The illustration of the "cooling" of electrons by the Dember field which causes the shift of the extrema to the short-wavelength side. M metal, O oxide, S semiconductor. ● is the electron excited by the monochromatic modulated radiation. E_c and E_v are the edges of conduction and valence bands, F_M and F_S are the Fermi levels in the metal and in the semiconductor bulk, respectively. Solid lines correspond to almost flat-band conditions without AI, dashed lines — with AI. * indicate energy level positions at AI switched on

Fig. 4. Photoresponse spectra of the structure I-240 at different intensities of modulated monochromatic radiation. (1) corresponds to the intensity $\approx 10^{12}$ photons/cm²s, (2) to the intensity $\approx 10^{14}$ photons/cm²s. The scale on the ordinate is plotted for (2). The values of the signal for curve 1 are approximately two orders of magnitude smaller

optical phonon by $eU_D^{\sim}$ (e is the electron charge). This is the cause of the shift of the extrema to the short-wave range.

The magnitude of the displacement of the extrema at increasing intensity of alternating monochromatic radiation (at switched-off AI) is an important quantitative confirmation of such an interpretation of the shift of the extrema. The photoresponse spectra at two different intensities are shown in Fig. 4. In this case the displacement of extrema is due to the increase of alternating voltage $U_{ph}^{\sim}$, which also cools the electrons. At large intensities the magnitude of $U_{ph}^{\sim}$ can become sufficient for a noticeable shift of the extrema. Since the displacement depends on the magnitude of the signal $U_{ph}^{\sim}$, the shift of the maxima must be larger than that of the minima and approximately equal to the signal value at high intensity multiplied by the electron charge. It is seen from Fig. 4 that the quantitative correspondence between the absolute value of the photoresponse magnitude and the shift magnitude is really rather good.

In conclusion it is necessary to pay attention to the fact that the external voltage shifts the extrema not as effectively as the Dember effect. It is evident from a comparison of Fig. 1 and Fig. 4 that the external voltage of a few volts leads to nearly the same shift of the extrema as the Dember voltage of a few millivolts. The reasons of this difference are not clear at present. It is possible that the essential part of external voltage falls off on the dielectric layer during illumination of the structure.

The main results of the paper are as follows: The dependence of the shape of oscillations and positions of extrema in low-temperature photoresponse spectra of GaAs–metal structures with an interfacial oxide layer on external bias and additional illumination is investigated. This dependence is due to "heating" and "cooling" of the electrons by the external electric field and the Dember field, respectively. The different effect of these fields on the process of electron thermalization is explained by their different directions with respect to the diffusive flux of electrons responsible for the appearance of photoresponse.

References

[1] V. L. Alperovich, A. F. Kravchenko, N. A. Pakhanov, and A. S. Terekhov, Zh. eksper. teor. Fiz., Pisma **28**, 551 (1978).

[2] V. L. Alperovich, A. F. Kravchenko, N. A. Pakhanov, and A. S. Terekhov, Fiz. Tekh. Poluprov. **14**, 1768 (1980).

[3] N. L. Dmitruk, V. I. Lyashenko, O. Yu. Borkovskaya, and O. I. Maeva, Fiz. Tekh. Poluprov. **9**, 2261 (1975).

[4] M. I. Abaev, A. L. Gerasimov, A. A. Gudkin, and V. E. Sedov, Fiz. Tekh. Poluprov. **13**, 746 (1979).

[5] H. J. Stoker and Kaplan, Phys. Rev. **150**, 150 (1966).

(Received December 15, 1980)

phys. stat. sol. (a) 65, 265 (1981)

Subject classification: 1.3; 21.1

Institute of Solid State Physics, M. V. Lomonossov State University, Moscow (a) and Institute of Solid State Physics, Eötvös Loránd University, Budapest[1] *(b)*

Thermal Decomposition of the B.C.C. β-Solid Solution of Titanium Alloy Containing 6.7 at% Mo, 3 at% Zr, and 1.8 at% Sn

I. Changes in Crystal and Electron Structure

By

M. I. Zakharova (a), A. K. Khundzhua (a), L. Kertész (b), and A. Szász (b)

Changes in the crystal structure of the titanium alloy, containing 6.7 at% Mo, 3 at% Zr, and 1.8 at% Sn, during thermal decomposition are followed by means of X-ray and electron diffraction methods. Parallel to these tests the alteration in the electron structure and chemical bonds of the alloy are investigated with the help of the soft-X-ray emission (SXES) method. Attention is focussed on the at room temperature not equilibrated b.c.c. β-solid solution, on the metastable transition phase ω, and on the equilibrium phase α.

Änderungen in der Kristallstruktur einer Titanlegierung mit 6,7 At% Mo, 3 At% Zr and 1,8 At% Sn werden während des thermischen Zerfalls mittels Röntgen- und Elektronenbeugungsmethoden untersucht. Parallel zu diesen Untersuchungen wird die Änderung der Elektronenstruktur und der chemischen Bindungen der Legierung mit Hilfe der Emission weicher Röntgenstrahlen (SXES) verfolgt. Insbesondere werden die bei Zimmertemperatur nicht im Gleichgewicht befindliche k.r.z.-β-Legierung, die metastabile ω-Phase und die Gleichgewichtsphase α untersucht.

1. Introduction

In recent years a great number of papers and monographs dealing with the investigation of titanium alloys have been published [1 to 4]. Attention was focussed on the phase composition of the alloys and on the morphology of the components making up the various phases, that is, on the changes in morphology under mechanical stress. The Ti–Mo–Zr–Sn alloys and their metastable phases represent an important group of new structural materials.

The dependence of the properties of the Ti–Mo–Zr–Sn alloys upon the concentration of the alloying components has already been investigated [5].

We shall report in this paper on the structure of single crystals of titanium alloys of composition $Ti_{88.5}Mo_{6.7}Zr_3Sn_{1.8}$ by menas of X-ray and electron diffraction methods.

Parallel to the investigation of the crystal structure, the electron structure and the chemical bonds in the b.c.c. β solution being not in equilibrium at room tempeıature, in the metastable transition phase ω, and in the equilibrium phase α of the alloy were studied with the help of the soft-X-ray emission (SXES) method.

2. Experimental

The alloy was prepared from high purity metals (Ti 99.9%, Mo 99.9%, Zr 99.9%, Sn 99.9%) in the light arc furnace and remelted six times.

The electropolished single crystals were small cylinders with diameters ranging

[1]) Múzeum krt 6-8, Budapest VIII, Hungary.

from 0.5 to 0.8 mm and were used in this form for X-ray diffraction and SXES tests. The thin foils for electron diffraction tests were electrolytically produced.

The b.c.c. β-solid solution was decomposed by isothermal treatment at 623, 723, and 823 K under a pressure of 8×10^{-3} Pa.

Since during decomposition of this solid solution the phases α and ω appear in highly dispersed small quantities, we used for their identification not Debye's method, but the far more sensitive X-ray diffraction test in which the [110] and [001] axes of the single crystals were set parallel to the X-ray beam. The diffraction pattern was recorded for both orientations after every ageing phase. The pictures were compared with the theoretically calculated reflections of the phases α and ω for both orientations of the β-crystal. The method of calculation can be found in one of our earlier publications [6]. For the X-ray diffraction tests the K_α radiation of molybdenum was used.

The SXES tests were performed for the M_{II}–M_{III} transition of titanium, for the M_{IV}–M_V transition of zirconium, and for the N_I level of tin. By means of these soft X-ray levels it was possible to study the fine structure and the ageing processes.

The method and apparatus were described in one of the earlier communications [7].

3. Results

The diffraction patterns recorded immediately after hardening at 1373 K contain only the β-matrix reflections of the b.c.c. β-solid solution.

The electron diffraction pictures show in addition to the β-matrix reflections certain diffuse effects in the form of quasi-circular arcs.

The deviations of the electron configurations on the SXES curves recorded in this state (β) indicate quite clearly the presence of the alloying components. (For the sake of comparison the curve of pure titanium was also recorded, see Fig. 1.) The characteristic peaks of the alloying components (A, Zr; B, Sn) changed in a manner which is typical of the chemical bonds in the alloy (Fig. 2).

After heat treatment at 623 K for 30 min beside the reflections of the b.c.c. β-matrix the reflections characteristic of the ω-phase also appear in the X-ray and electron diffraction pictures (Fig. 3 and Table 1). When the period of heat treatment at 623 K is prolonged to 2, 6, 15 and 40 h, respectively, the intensity of the reflections of the ω-phase in the X-ray diffraction pictures will increase, and in the electron diffraction pictures an f.c.c. phase with a lattice constant of 0.44 nm will appear beside the reflections of the b.c.c. β-matrix.

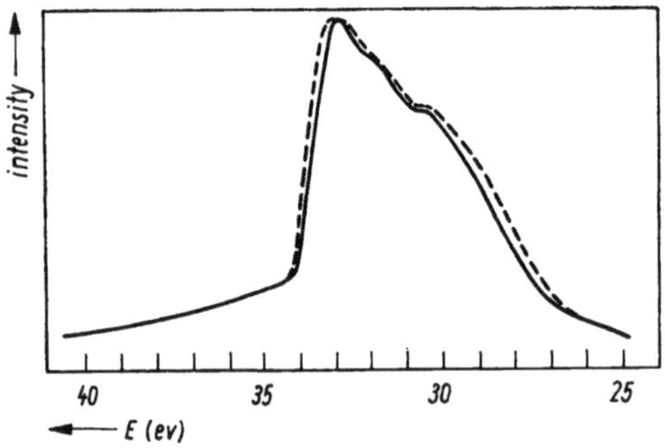

Fig. 1

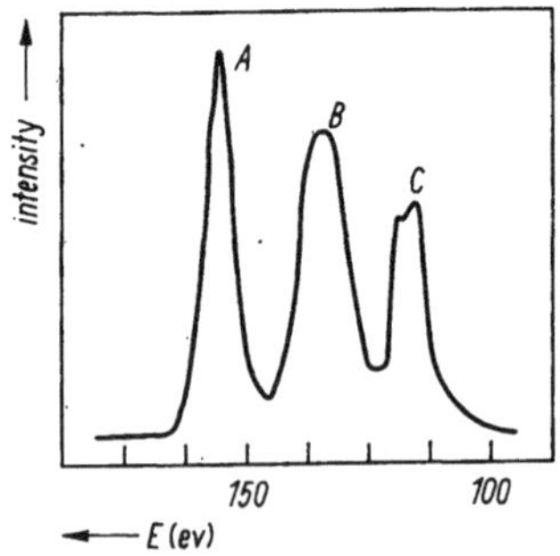

Fig. 2

Fig. 1. The SXES curve of the β-phase of Ti–Mo–Zr–Sn alloy (– – –) compared to the Ti $M_{II\text{-}III}$ level. As reference we also plotted the pure Ti curve (———)

Fig. 2. The SXES peaks of alloying element in the β-phase of the alloy. (The reference Ti curve as base line is subtracted!)

Table 1

Summary of the measured phase compositions during different heat treatments of the Ti–Mo–Zr–Sn alloy

ageing time (h)							
temperature (K)	0	0.5	2	6	15	40	method
623	β + D.E.	β + ω	β + ω	β + ω	β + ω	β + ω	X.D.
623	β + D.E.	β + ω	β + f.c.c.	β + f.c.c.	β + f.c.c.	β + f.c.c.	E.M.
723	β + D.E.	β + α	β + α	β + α	β + α	β + α	X.D.
723	β + D.E.	β + α_1	β + α_1 + α_2	β + α_2	β + α_2	β + α_2	E.M.
823	β + D.E.			β + α		β + α	X.D.
823	β + D.E.			β + α_2		β + α_2	E.M.

X.D. X-ray diffraction, E.M. electronmicroscopy, D.E. diffuse effects

Heat treatment at 623 K for 40 h will result beside the reflections of the β- and ω-phases in the appearance of reflections characteristic of the α-phase.

The SXES records of the β + ω phases show a fundamental change, namely a definite increase of the interval containing the characteristic peaks of the alloying components (Table 2).

Isothermal treatment of the alloy at 723 K showed that this temperature rise had greatly accelerated the decomposition of the b.c.c. β-solid solution. At this temperature the α-phase appeared within 30 min. Prolonged heat treatment up to 40 h caused no essential change in the β + ω phase composition (Table 1), only the quantity of the α-phase increased parallel with the prolongation of heat treatment, as indicated by the higher intensities of the characteristic reflections.

On the electron diffraction pictures the situation after 30 min heat treatment corresponds to the Burger orientation of the α-phase (Fig. 4). After heat treatment for 2 h the β-matrix, Burger's α-phase, and all reflection lines characteristic of the $\{10\bar{1}2\}$ $\langle 10\bar{1}1\rangle$ double (α_2) phase appear together in the electron microscopic picture. When thermal treatment is continued, only the characteristic reflections of the β- and α-phases will be visible on the electron microscopic picture (Fig. 5).

On the foils the α-phase is doubled only by electrolytic polishing [8], no similar effect can be observed on single crystals.

Heat treatment at 823 K also results in the α-phase. Experiments on foils showed the repeated doubling of the α-phase.

After the appearance of the α-phase (at 823 K, after 5 h) the changes in the electron structure and chemical bonds continued in the direction started at the development of the ω-phase. The equilibrium conditions stabilized and corresponding to the equilibrium the new bonds produced a higher intensity. The C peak with characteristic fine structure appearing in the equilibrated β + α phase is shown in Fig. 6.

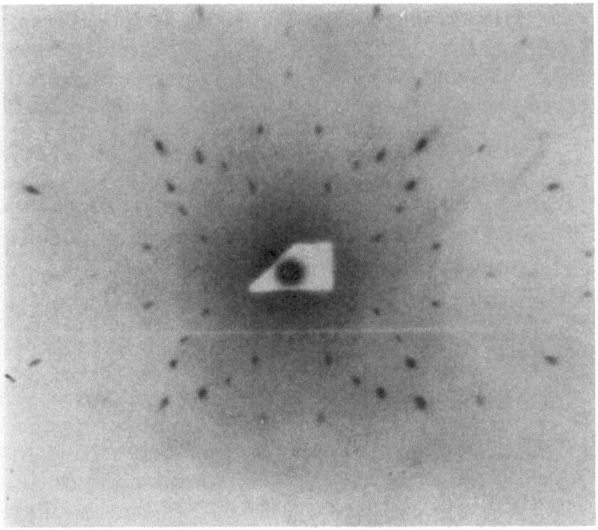

Fig. 3. The X-ray diffraction pattern of a Ti–Mo–Zr–Sn alloy monocrystal after 6 h heat treatment at 623 K. The reflections of β- and ω-phases are well seen

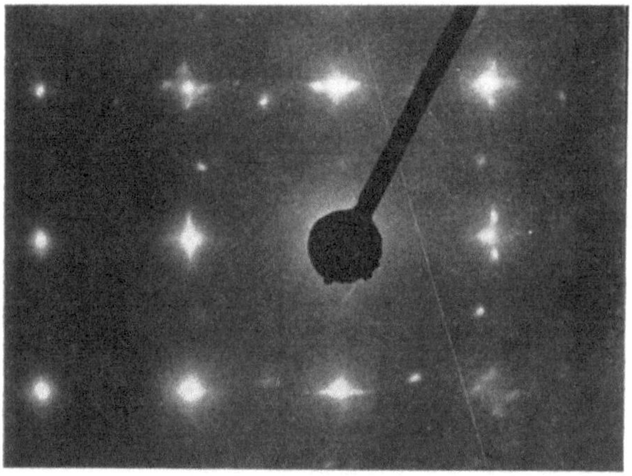

Fig. 4. Electron diffraction pattern of Ti–Mo–Zr–Sn alloy after 30 min heat treatment at 723 K. The reflections of β- and α_1-phases are well seen

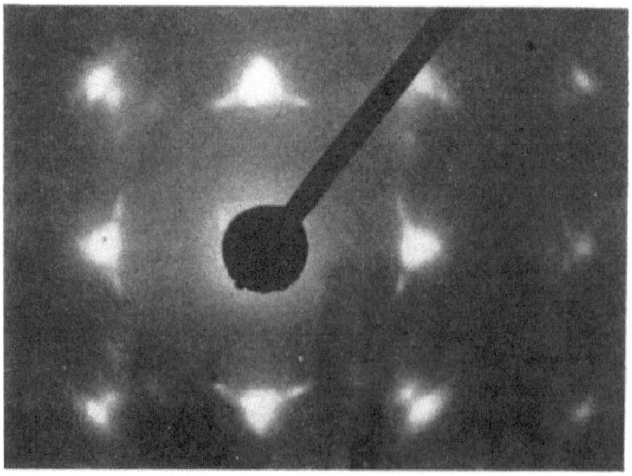

Fig. 5. Electron diffraction pattern of Ti–Mo–Zr–Sn alloy after 40 h heat treatment at 723 K, with the reflections characteristic of β- and α_2-phases

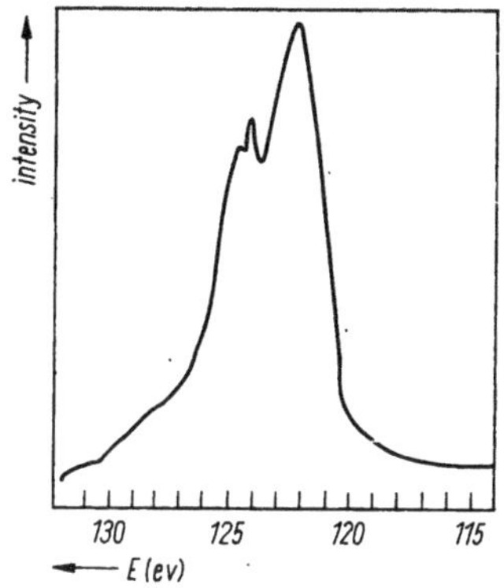

Fig. 6. The peak C in the SXES record of the α-phase of the alloy

Table 2

Summary of the measured Ti M_{II-III} peaks of different phases of the Ti–Mo–Sn alloy

phase	peak energy (eV) (±0.2 eV)	asymmetry at half intensity	half-width (eV)
Ti reference	32.9	0.2	11.0
Ti–Mo–Zr–Sn β	33.1	0.2	11.7
Ti–Mo–Zr–Sn β + ω	33.0	0.2	12.4
Ti–Mo–Zr–Sn β + α	33.0	0.3	12.1

Table 3

Relative intensity of the C peak of SXES records in different phases of the Ti–Mo–Zr–Sn alloy

phase	C (%)
Ti	—
β	20
β + ω	25
β + α	100

4. Discussion

The diffuse effects appearing beside the reflections of the b.c.c. β-matrix are the results of not-correlated linear crystal defects parallel to the direction of the crystal axis [9]. Increase in the half-width of the SXES curve plotted under the same conditions and the rather more markedly blurred fine structure indicate a dissolved state within the alloy [10] (Fig. 1). The half-width of the typical peaks of the alloying components which were plotted at the same time also increased and a new peak (peak C, Fig. 2) due probably to mixed bonds developed. (The properties of this mixed bond will be studied later, and we intend to report on an acceptable explanation in the next part of our series of communications. The work is now in progress.)

Effects typical of a new phase appear in all three methods by which the decomposition process is followed, though in electron configuration only quantitative changes were observed (increase of peak C). The appearance of an f.c.c. phase with a lattice constant of 0.44 nm on the electron diffraction pictures is due to the re-arrangement of the elastic stresses arising during heat treatment, during polishing [11]. The metastable transition phase which is formed during of heat treatment is responsible for the appearance of elastic stresses. The re-arrangement of stresses by way of a martensitic mechanism gives rise to a f.c.c. phase in certain places of the foil.

Appearance of the stable α-phase leads to a further considerable increase (see Table 3) of peak C, as testified by the study of the electron structure, which suggests that the development of the α-phase is accompanied by significant complex bond formation.

References

[1] T. Headley, and H. Rack, Metallurg. Trans. **10A**, 909 (1979).
[2] C. G. Rhodes and N. E. Paton, Metallurg. Trans. **8A**, 1749 (1977).
[3] C. Yolton, F. Froes, and R. Malone, Metallurg. Trans. **10A**, 132 (1979).
[4] U. Zwicker, Titan und Titanlegierungen, Springer-Verlag, Berlin/Heidelberg/New York 1974.
[5] S. G. Fedotov and E. M. Sidonova, in: Kristallicheskaya struktura i svoistva metallicheskikh splavov, Izd. Nauka, Moskva 1978 (p. 221).
[6] N. A. Khatanova, S. A. Kirov, and M. I. Zakharova, Zav. Labor. **43**, 1086 (1977).
[7] A. P. Lukirskii, I. A. Britov, and N. I. Komyak, in: Apparatura i metody rentgenovskogo analiza, Vol. 2, Leningrad 1967 (p. 4).
[8] M. I. Zakharova, S. A. Kirov, and A. G. Khundzhua, phys. stat. sol. (a) **49**, 803 (1978).
[9] I. C. Williams, D. de Fontaine, and N. E. Paton, Metallurg. Trans. **4**, 2701 (1973).
[10] L. Kertész, J. Kojnok, and A. Szász, Acta phys. Hungar., in the press.
[11] M. I. Zakharova and A. G. Khudzhua, Izv. Akad. Nauk SSSR, Ser. fiz. **41**, 1463 (1977).

(Received February 5, 1981)

phys. stat. sol. (a) **65**, 271 (1981)

Subject classification: 2 and 4; 10.2; 22.6

Sektion Physik der Wilhelm-Pieck-Universität Rostock[1])

Chemical Bond and Related Properties of SiO_2

VIII. Experimental and Theoretical Investigation of the Structure of SiO_x[2])

By

R. ENGELKE, TH. ROY, H.-G. NEUMANN, and K. HÜBNER

With the help of a high-intensity X-ray generator with rotating anode diffraction measurements are performed on commercial SiO powder and vapour deposited SiO_x samples obtained from the commercial SiO. From an extensive analysis of corresponding radial distribution functions the atomic distances within the short-range order, the average chemical composition x, and the bonding statistics of SiO_x are determined. Different statistical and fitted models for the chemical disorder of SiO_x are tested. It is shown that a statistical model in which network defects are simulated by O–O bond sleads to an excellent reproduction of the experimental data. The random-bonding character of vapour deposited SiO_x and other structural and chemical differences between differently prepared types of SiO_x are shown.

Mit Hilfe eines Hochleistungsröntgengenerators mit rotierender Anode werden Beugungsmessungen an kommerziellem SiO-Pulver sowie an SiO_x-Proben, die durch Verdampfung des kommerziellen SiO erhalten wurden, durchgeführt. Aus einer ausführlichen Analyse entsprechender radialer Verteilungsfunktionen werden der Atomabstand in der Nahordnung, die mittlere chemische Zusammensetzung x und die Bindungsstatistik von SiO_x bestimmt. Unterschiedliche statistische und angepaßte Modelle für die chemische Unordnung des SiO_x werden getestet. Es wird gezeigt, daß ein statistisches Modell, in dem Netzwerkdefekte durch O–O-Bindungen simuliert werden, eine ausgezeichnete Reproduktion der experimentellen Daten ermöglicht. Der random-bonding-Charakter des aus der Dampfphase niedergeschlagenen SiO_x und andere strukturelle und chemische Unterschiede zwischen unterschiedlich präparierten SiO_x-Typen werden nachgewiesen.

1. Introduction

In Part VII of this work [1] we have discussed the relation between differently prepared SiO_x forms and their bonding statistics and demonstrated the important influence of the distribution of Si–Si and Si–O bonds on the average electronic properties of SiO_x. In the present paper we represent an extensive experimental and theoretical investigation of the structure of two different types of SiO_x. We use for the average chemical composition x of our SiO_x samples $x \approx 1$, because this composition has the greatest "distance" from the limiting cases with chemically ordered states i.e. Si and SiO_2. Some preliminary results of this work were already published in [2] and [3].

Since a corresponding investigation by Potter [4] it is known that SiO (silicon monoxide) can be obtained from the evaporation of an 1:1 mixture of Si and SiO_2. A similar material can be prepared by the evaporation of SiO_2 under reducing conditions or by evaporating Si in the presence of oxygen. The designation SiO_x $(0 < x < 2)$ has been adopted because of the strong dependence of its oxygen content on the mixing ratio and the physical conditions during preparation. The physico-chemical properties of SiO_x were investigated by means of different methods [5]. Some authors tried to

---

[1]) Universitätsplatz 3, DDR-2500 Rostock, DDR.

[2]) Parts I to VII see phys. stat. sol. (a) **40**, **133**, 487 (1977); **42**, 501 (1977); **46**, 451 (1978); **48**, 147 (1978); **52**, 541 (1979), and **61**, 665 (1980), respectively.

establish a crystalline structure of SiO_1 (see [5, 6], and references therein), but these structures resulted from carbon contaminated during the preparation procedure or from the partial crystallization of samples showing a decomposition into Si and SiO_2.

The stability of neutral SiO molecules was proved by spectroscopical methods for $T > 1400$ K [7]. Astrophysical investigations demonstrated the existence of such molecules in the solar atmosphere and within the interstellar space. Metastable solid SiO_x forms were found to be at room temperature in an amorphous state with a tendency to decomposition into amorphous silicon (a-Si) and silica (a-SiO_2). This decomposition occurs at $T \approx 700$ K and is complete at $T \approx 1200$ K.

Recent structural investigations of SiO_x were particularly induced by the existence of SiO_x regions at microelectronic Si–SiO_2 interfaces [1]. They were favoured, for example, by advances of the experimental diffraction techniques. Wide-angle diffraction methods can provide now structural information on the atomic scale by the investigation of the radial distribution function (RDF) [8, 9]. Brady [10] concluded from the RDF extracted from the X-ray diffraction of SiO powder that this material is a macroscopic mixture of equal amounts of a-Si and a-SiO_2. Results published by Coleman and Thomas [11] and by Lin and Yoshi [12] seemed to be in agreement with this assumption of a mixture model (MM). However, the resolution of the RDF curves of these investigations was relatively poor because of experimental conditions (employed range of scattering angle and errors at high scattering angles). Therefore, the validity of this macroscopic MM for SiO_x was questioned by the results of electron diffraction experiments on thin SiO_x films [13]. Yasaitis and Kaplow [14] demonstrated in the case of SiO_1 powder that there are significant differences between scattering curves obtained from X-ray diffraction and calculated scattering curves basing on the MM. These differences reveal a certain degree of chemical "disorder" in SiO_x, which exists additionally to the geometrical disorder expected from a simple mixture of a-Si and a-SiO_2. There are essentially two models for this chemical disorder, which have to be verified experimentally (see also [1] and references therein). (1) Referring to the data of [14] Temkin [15] proposed the assumption of a mixing on an atomic scale, which is in fact a modified MM. He found his "microscopic" MM to be in agreement with the experimental RDF data. George and d'Antonio [16] have confirmed this result by electron diffraction measurements. However, Temkin's calculation of the model RDF contains some errors as will be shown in the present paper. (2) On the basis of optical measurements Philipp [17] proposed a random-bonding model (RBM) for an SiO_x network without O–O or broken Si–O bonds, assuming a completely random distribution of oxygen atoms among the Si bonds. Implications of the RBM and different physical properties of SiO_x forms with random-bonding character were extensively investigated in [1] and related papers.

The main purpose of the present paper is the discussion of our high-resolution X-ray diffraction measurements in relation to these and other models and a new consideration of the problem of the chemical disorder in SiO_x.

2. Experimental Procedure

First, we performed systematic measurements on a-Si and a-SiO_2 to test our experimental equipments and computational methods. The corresponding results are in good agreement with known structural data of these materials [18, 19]. The diffraction measurements were carried out with a horizontal counter goniometer HZG-3 and a high-intensity X-ray generator "Rotaflex" with rotating anode from Rigaku, Ltd. Typical power data of the generator were 50 kV acceleration voltage and 200 mA anode current. Both AgK_α and MoK_α radiations were used; monochromatization was obtained by a focusing graphite monochromator.

Two different types of SiO_x samples were prepared in capillaries of quartz glass or Lindemann glass with an inner diameter of 1.0 mm and a wall thickness of 0.1 mm. (1) SiO_1 powder samples, commercially available from Chemapol, Prague (com-SiO_x), and (2) vapour deposited SiO_x samples basing on the evaporation of the com-SiO_x under high-vacuum conditions (ev-SiO_x). The background intensity of each empty capillary was measured with reduced accuracy (1×10^5 pulses per data point) and was subtracted from the sample scattering. The registration of 2×10^5 pulses per sample data point was possible by multiple measurements. Absorption corrections for the capillaries used were shown to be negligible [20]. The experimental data output was realized by a paper-tape-punch device for direct data treatment by the computer system HP-9825A. The stability of the X-ray generator during the time of measurement was checked with the help of reference measurements on an a-SiO_2 sample at the beginning and the end of the experiment. In the case of weak instabilities a corresponding correction procedure was applied. For further details of the experimental techniques used see [21].

3. Calculation of the Radial Distribution Function

The reduced intensity $i(s)$ is derived for the computation of RDF curves from the experimental scattering data by the relation

$$i(s) = \left[I_{\text{eu}}(s)/N - \sum_{\text{uc}} f_i^2(s)\right] \Big/ f_e^2(s)\,, \tag{1}$$

where $I_{\text{eu}}(s)$ is the corrected and scaled experimental intensity in electron units. The calculation of the independent coherent scattering $\sum_{\text{uc}} f_i^2(s)$ has to take into account all atomic constituents in a corresponding basic unit of composition (uc). In the mean scattering factor [9]

$$f_e(s) = \sum_{\text{uc}} f_i(s) \Big/ \sum_{\text{uc}} Z_i \tag{2}$$

the atomic scattering factors $f_i(s)$ were calculated with the help of an analytical approximation [22] (Z_i nuclear charge number). A similar approximation allows also the estimation of the incoherent intensities [23]. The experimental intensity functions were scaled using the independent scattering at high values of the scattering variable $s = 4\pi \sin(\theta/2)/\lambda$, where θ is the scattering angle and λ the wavelength of the X-ray radiation. The RDF $J(r)$ follows by Fourier inversion of the reduced intensity function,

$$J(r) = 4\pi r^2 \varrho(r) = 4\pi r^2 \varrho_0 + \frac{2r}{\pi} \int_{s_{\text{min}}}^{s_{\text{max}}} s\, i(s) \sin(rs)\, ds\,, \tag{3}$$

where $\varrho(r)$ describes the radial electron density distribution of the sample (ϱ_0 is the average electron density, which can be derived from macroscopic sample data and chemical information). The part of the RDF sensitive to structural details is given by $D(r) = J(r) - 4\pi r^2 \varrho_0$. Artificial damping factors for reducing any termination errors were not used for the RDF curves. For further details see [21, 24].

4. Experimental Results

The reduced intensity function $i(s)$ is shown in Fig. 1 for both the com-SiO_x and the ev-SiO_x. The different preparation methods result in different scattering behaviour. In Fig. 2 the structure-sensitive part $D(r)$ of the RDF's derived from these scattering functions are shown. Scattering function and RDF of the com-SiO_x powder are very similar to the corresponding curves given in [14]. The estimation of the actual average

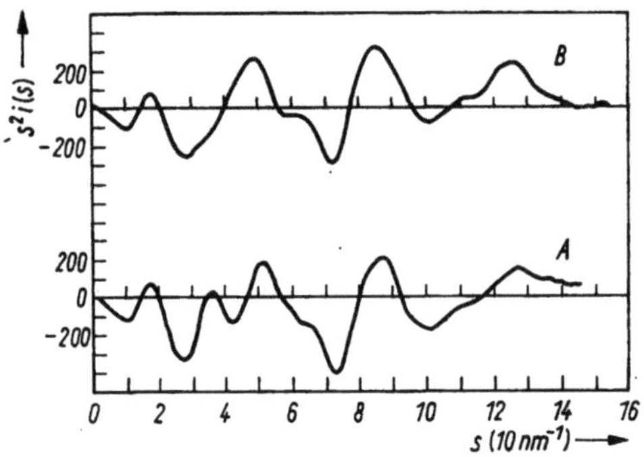

Fig. 1

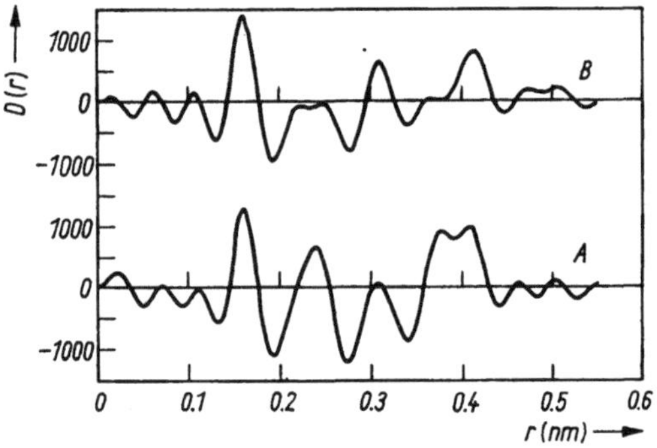

Fig. 2

Fig. 1. Reduced intensity function $i(s)$ (equation (1)) of commercial (A) and vapour deposited SiO_x (B)

Fig. 2. Structure dependent part of the radial distribution function (3) of commercial (A) and vapour deposited SiO_x (B)

chemical composition x of this SiO_x form from the number $N_O = 1.97$ of nearest O neighbours around an Si atom, which determines the area of the first RDF peak at the Si–O distance 0.162 nm, leads to $x = 0.98$. In contrast to [14] we find in all cases the Si–O distance to be 0.162 nm, in agreement with the value of this bond length in a-SiO_2. The second RDF peak is located at 0.244 nm and can be assumed to be a superposition of contributions from the Si–Si distance (0.235 nm in a-Si) and the O–O distance (0.265 nm in a-SiO_2). The third peak at 0.312 nm is in good agreement with the distances between the Si atoms of two neighbouring Si–O_4 tetrahedra of a-SiO_2 and with the Si–O edge lengths of an Si–Si_yO_{4-y} ($y = 1, 2, 3$) tetrahedron. If the possible range of angle values of the scattering function is used completely, the fourth maximum splits into a double peak at 0.38 and 0.412 nm. The 0.38 nm peak can be linked with the edge length of an Si–Si_4 tetrahedron assuming an Si–Si bond length of 0.235 nm. Also RDF's of a-SiO_2 show a peak at about 0.38 nm, but with a very different maximum height. Therefore, the interpretation of the 0.38 nm peak of SiO_x as a contribution from a-Si seems to be reasonable.

The first peak of the RDF of ev-SiO_x is also at 0.162 nm. Compared with the powder material (com-SiO_x) the number of nearest O neighbours, however, is increased ($N_O = 2.27$). Therefore, an increased value of $x \approx 1.15$ can be estimated for the vapour deposited material. This increase results from the oxygen implantation during the preparation procedure using liquid-N_2 cooled cryosurfaces. The peak at 0.38 nm is drastically decreased in its height, but without significant change in its position. This behaviour can be understood as a result of a reduction of the number of pure Si–Si_4 tetrahedra due to the evaporation procedure. Furthermore, we observe that the second RDF peak splits into two peaks at 0.227 and 0.251 nm after vacuum deposition. We interpret this fact with a separation of the Si–Si and the O–O distances due to the decreased Si–Si contributions with regard to the ratio of the atomic scattering factors for Si and O. The differences with respect to the expected locations at 0.235 and 0.265 nm, respectively, are due to ripples from neighbouring peaks.

5. Structural Models

Three factors determine the local chemical structure [25], the atomic coordination, the statistics of bond distribution, and the "molecular" structure of the network-forming groups. From a comparison of the RDF of $SiO_{x\approx 1}$ with the RDF's of a-Si

Table 1

Correspondence of atomic correlation pairs to the first three peaks i at positions r_i of the radial distribution function of commercial $SiO_{0.98}$ (com-SiO_x) and vapour deposited $SiO_{1.15}$ (ev-SiO_x)

i	1	2	3
r_i (nm) com-SiO_x	0.162	0.244	0.312
pair	Si–O	Si–Si O–Si–O	Si–O–Si Si–Si–O
r_i (nm) ev-SiO_x	0.162	0.227/0.251	0.312
pair	Si–O	Si–Si/O–Si–O	Si–O–Si Si–Si–O

and a-SiO_2 follows that the network-forming "molecules" have tetrahedral structure (cf. Section 4). Table 1 shows the correspondence of atomic correlation pairs to the first three peaks of the RDF, which is in agreement with the physical coexistence of Si–Si and Si–O bonds discussed in [1]. The total number $(2 + x)\,N_{Si}$ of chemical bonds in SiO_x follows from the fact that each O atom has two and each Si atom has four nearest neighbours (N_{Si} being the total number of Si atoms of the sample). Concerning the statistics of their distribution, two extreme cases can be considered, namely a random covalent network (RCN) and a chemically ordered network (CON) [25]. The RCN model neglects relative bond energies. Therefore, the distribution of different bonds is completely statistical and its application to SiO_x would include Si–Si, Si–O, *and* O–O bonds. The CON model, which takes into account bond energies, should give a more realistic distribution of bonds. It assumes a maximum amount of heteropolar bonds and excludes therefore the existence of O–O bonds in SiO_x. A more general description of the bond statistics, which includes the RCN and the CON models as boundary cases, is proposed now. Starting from a CON distribution we may use the relative number n_{O-O} of O–O bonds as a parameter for the variation of the bond distribution between the RCN and the CON case. Then, the relations

$$n_{Si-Si} = \frac{2-x}{2+x} + n_{O-O}\,, \qquad n_{Si-O} = \frac{2x}{2+x} - 2n_{O-O} \tag{4}$$

describe the relative bond numbers in this covalent intermediately ordered network (CION). The maximum of n_{O-O} should be equal to the RCN value of n_{O-O}, namely $x^2/(2 + x)^2$. Applied to SiO_x the RCN model is only of theoretical interest and very unrealistic. The real situation in SiO_x should be similar to a CON. The "classical" SiO_x models, i.e. the RBM and the MM (cf. Section 1 and [1]), are in fact CON models. However, SiO_x is a system in which two types of directed bonds with very different bond lengths (Si–Si 0.235 nm, Si–O 0.162 nm), bond energies, and other bond parameters coexist [1]. Owing to the directional character of its covalent-ionic bonds it forms a network structure. From the differences in bond lengths and other parameters there follows, however, an increased tendency to the formation of network defects in comparison with a-Si and a-SiO_2. Although the density of typical intrinsic defects such as valence alternation pairs of threefold coordinated and non-bridging (single coordinated) oxygen atoms is very high (of the order of 10^{19} cm^{-3}) even in SiO_2, where only one type of bonds exists [26], this defect type does not change the average bond distribution and has no significant influence on our analysis. Owing to the higher degree of dilution of oxygen and the increased tendency to form broken Si–O bonds in $SiO_{x<2}$ [1], however, the probability for finding threefold coordinated partners to form valence alternation pairs should be reduced in comparison with SiO_2. Another defect type, which could have in an RDF analysis a similar effect as non-bridging O atoms, is the OH group [27]. Additionally, Revesz [27] considered an Si–O–O–Si peroxide complex. All these defects have the effect that the bond distribution tends from a CON to an RCN distribution, since two non-bonding oxygen atoms or one Si–O–O–Si peroxide complex can be assumed to represent one virtual O–O bond.

The atomic coordination is the main factor for the local chemical structure. It influences the RDF and is connected with the bond statistics and the "molecular" structure of the network-forming groups. For the calculation of coordination numbers we have introduced in [2] the method of transition probabilities in random sequences. The probability for finding a j-type atom bonded to an atom of type i under the condition that a k-type atom is already bonded to the atom i was defined as $p_{ij}^{(k)}(m)$. The variable m gives the coordination shell of the atom j within the RDF. The initial distribution

$$\{p_i(0)\} = \begin{cases} \dfrac{1}{1+x}, & i = \mathrm{Si}, \\ \dfrac{x}{1+x}, & i = \mathrm{O} \end{cases} \tag{5}$$

of the random sequence is determined only by the chemical composition of SiO_x. The matrix of transition probability for the first step of the random "process", which corresponds to the first coordination shell in the RDF, is determined by the bond distribution in the following way:

$$\left.\begin{aligned} p_{\mathrm{Si-Si}}(1) &= (1 + x/2)\, n_{\mathrm{Si-Si}}\,, \\ p_{\mathrm{Si-O}}(1) &= (1 + x/2)\, n_{\mathrm{Si-O}}/2\,, \\ p_{\mathrm{O-Si}}(1) &= 1 - (1 + 2/x)\, n_{\mathrm{O-O}}\,, \\ p_{\mathrm{O-O}}(1) &= (1 + 2/x)\, n_{\mathrm{O-O}}\,. \end{aligned}\right\} \tag{6}$$

If $n_{\mathrm{O-O}} = 0$, these relations lead to the known transition probabilities for the RBM and the MM

$$\left.\begin{aligned} p_{\mathrm{Si-Si}}(1) &= (1 - x/2)\,, \quad & p_{\mathrm{Si-O}}(1) &= x/2\,, \\ p_{\mathrm{O-Si}}(1) &= 1\,, & p_{\mathrm{O-O}}(1) &= 0\,. \end{aligned}\right\} \tag{7}$$

Any tendency to a decomposition into Si and SiO_2 can be described beyond the first coordination sphere through the randomness within the sequence. In the RBM the O atoms are randomly distributed between the Si atoms. Together with the CON statistics follows that

$$p_{ij}^{\mathrm{RBM}}(m > 1) = p_{ij}^{\mathrm{RBM}}(1)\,. \tag{8}$$

In the MM only $Si–Si_4$ and $Si–O_4$ tetrahedra are considered. Therefore, the transition probabilities are

$$\left.\begin{aligned} p^{(\mathrm{Si})}_{\mathrm{Si-Si}}(m) &= 1\,, \quad & p^{(\mathrm{Si})}_{\mathrm{Si-O}}(m) &= 0\,, \\ p^{(\mathrm{O})}_{\mathrm{Si-Si}}(m) &= 0\,, & p^{(\mathrm{O})}_{\mathrm{Si-O}}(m) &= 1\,, \\ p_{\mathrm{O-Si}}(m) &= 1\,, & p_{\mathrm{O-O}}(m) &= 0, \quad m > 1\,. \end{aligned}\right\} \tag{9}$$

For the quantitative application of the statistical analysis introduced above, our experience on the local chemical structure of SiO_x can be summarized as follows. (1) The bond lengths of Si–Si and Si–O bonds have essentially the same values as in a-Si

(0.235 nm) and a-SiO_2 (0.162 nm), respectively. (2) Each Si atom is tetrahedrally coordinated with nearest neighbouring Si and O atoms. (3) The SiO_x network is formed by Si–Si_yO_{4-y} tetrahedra ($y = 0, \dots, 4$) and particularly cracked due to the existence of non-bridging O atoms and other defects.

6. Coordination Numbers

Temkin [15] has shown that even a chemical decomposition process, which does not change the average bond distribution, may be visible in the RDF of SiO_x. Therefore, an RDF analysis can be used to investigate the local chemical structure of this complicated material. The RDF (3) is related to coordination numbers $N(\underline{i}- \dots -\underline{j})$ through the peak areas

$$A(r) = \sum_{i,j} K_{ij}(x)\, p_i(0)\, N(\underline{i}- \dots -\underline{j}) \tag{10}$$

of the RDF maxima at positions r ($K_{ij}(x)$ includes the atomic X-ray, electron, or neutron form factors of the atomic species i and j) [15].

The coordination number can be calculated for a model structure in two different ways. (1) There is a relation between the mathematical description of the chemical disorder introduced in Section 5 and the coordination number. This number for a special configuration $\underline{i}$–k–l– ... –$\underline{j}$ of m atoms "seen" by the atom i is given by the relation

$$N(\underline{i}- \dots -\underline{j}) = z_i(z_k - 1)\,(z_l - 1) \dots (z_{j-1} - 1)\, p_{ik}(1)\, p_{kl}^{(i)}(2) \dots p_{(j-1)j}^{(j-2)}(m)\,, \tag{11}$$

where z_i is the number of bonds per atom i. This relation is correct if rings within the m-th coordination shell around the atom i are neglected. This assumption is valid for an Si structure with six-membered rings up to the second coordination sphere and for SiO_2 with rings of six Si–O_4 tetrahedra or twelve atoms (6 Si and 6 O atoms) up to the fifth shell. Thus, equation (11) is sufficient for the aim of this paper. (2) The second possibility follows from the statistics for the random-bonding tetrahedra. The probability $\alpha_y(x)$ for finding in SiO_x an Si–Si_yO_{4-y} tetrahedron with $y = 0, 1, 2, 3$, or 4 follows from purely statistical considerations (see equation (1) and Fig. 2 of [1]). The correspondence between these random-bonding probabilities and the coordination numbers is given in Table 2 for the atomic correlation pairs of Table 1. It has to be compared with corresponding relations deduced by Temkin [15]. Applied to the limiting RBM and MM, Temkin's relations are not consistent with equation (11), whereas the second column of Table 2 leads to the same RBM and MM coordination numbers as equation (11) as demonstrated, for example, for $N(\underline{Si}-\underline{O})$, $N(\underline{Si}-\underline{Si})$, and $N(\underline{O}-Si-\underline{O})$ in

Table 2

Relations between the distribution probabilities $\alpha_y(x)$ of the bonding tetrahedra Si–Si_y–O_{4-y} ($y = 0, \dots, 4$) and the coordination numbers $N(\underline{i} - \dots - \underline{j})$ for a chemically ordered network of SiO_x

$N(\underline{i} - \dots - \underline{j})$	this work	Temkin [15]
$N(\underline{Si}-\underline{O})$	$4\alpha_0 + 3\alpha_1 + 2\alpha_2 + \alpha_3$	$4\alpha_0 + 3\alpha_1 + 2\alpha_2 + \alpha_3$
$N(\underline{O}-\underline{Si})$	2	2
$N(\underline{Si}-\underline{Si})$	$\alpha_1 + 2\alpha_2 + 3\alpha_3 + 4\alpha_4$	$\alpha_1 + 2\alpha_2 + 3\alpha_3 + 4\alpha_4$
$N(\underline{O}-Si-\underline{O})$	$(2/x)\,(6\alpha_0 + 3\alpha_1 + \alpha_2)$	$2(3\alpha_0 + 2\alpha_1 + \alpha_2)$
$N(\underline{Si}-O-\underline{Si})$	$4\alpha_0 + 3\alpha_1 + 2\alpha_2 + \alpha_3$	$4\alpha_0 + 3\alpha_1 + 2\alpha_2 + \alpha_3$
$N(\underline{O}-Si-\underline{Si})$	$(1/x)\,(3\alpha_1 + 4\alpha_2 + 3\alpha_3)$	$2(\alpha_1 + 2\alpha_2 + 3\alpha_3)$
$N(\underline{Si}-Si-\underline{O})$	$3\alpha_1 + 4\alpha_2 + 3\alpha_3$	$4(3\alpha_1 + 2\alpha_2 + \alpha_3)$

the following. If $\alpha_y(x)$ is the probability that an Si atom is centre of an $Si\text{–}Si_yO_{4-y}$ tetrahedron, then the number of nearest O neighbours around this Si atom is $4-y$ and the number of nearest Si neighbours is y. Therefore, we find

$$N(\underline{Si}\text{–}\underline{O}) = \sum_{y=0}^{4} (4-y)\,\alpha_y(x)\,, \qquad N(\underline{Si}\text{–}\underline{Si}) = \sum_{y=0}^{4} y\alpha_y(x)\,. \tag{12}$$

Concerning $N(\underline{O}\text{–Si–}\underline{O})$ we have to ask for the probability $\beta_y(x)$ for finding an O atom as a member of an $Si\text{–}Si_yO_{4-y}$ tetrahedron. According to its nature this quantity is a conditional probability. The condition determining the value of $\beta_y(x)$ is the fact that an O atom is bonded with certainty to the centre of an $Si\text{–}Si_yO_{4-y}$ tetrahedron. If this O atom is bonded to two $Si\text{–}O_4$ tetrahedra it has six nearest O neighbours; if it connects two $Si\text{–}SiO_3$ tetrahedra the number of its O neighbours is four, and so on. The resulting relation for the average number of $\underline{O}\text{–Si–}\underline{O}$ pairs around an O atom is

$$N(\underline{O}\text{–Si–}\underline{O}) = 6\beta_0(x) + 4\beta_1(x) + 2\beta_2(x) = (2/x)\,[6\alpha_0(x) + 3\alpha_1(x) + \alpha_2(x)]\,. \tag{13}$$

From Table 2 it can be seen that the correspondence between the statistics for the bonding tetrahedra and the coordination numbers is more complex than deduced by Temkin [15], who used in fact only $\alpha_4(x)$ as a parameter. Therefore, the random-sequence formalism and equation (11) are applied in the following section to the interpretation of our experimental RDF's of SiO_x.

7. Results

The calculated peak areas of model RDF's corresponding to the RBM and the MM are given in Table 3 in comparison with the experimental values and the peak areas of other model RDF's. In FIT 2 and FIT 3 we assume a CON without dangling bonds. Therefore, $n_{O\text{-}O} = 0$ in these models. From the experimental value of the peak area $A(0.244\text{ nm})$ of com-SiO_x (as an example) we obtain

$$p_{Si\text{-}O}^{(0)}(2) = \frac{A(0.244\,\text{nm}) - 4K_{Si\text{-}Si}p_{Si}(0)\,p_{Si\text{-}Si}(1)}{6K_{O\text{-}O}p_{O}(0)\,p_{O\text{-}Si}(1)} \tag{14}$$

Table 3

Peak areas (in numbers of atoms) of experimental and theoretical radial distribution functions of commercial (com) and vapour deposited (ev) SiO_x with $x = 0.98$ and 1.15, respectively. For further details see text

	com-SiO_x						
peak position (nm)	RBM	FIT 2	FIT 3	mMM	MM	CION	EXP
0.162	1.70	1.70	1.70	1.70	1.70	1.71	1.70
0.244	2.49	2.63	2.83	2.84	3.12	2.62	2.63
0.312	4.42	3.86	3.04	2.94	1.81	3.05	3.03
	ev-SiO_x						
peak position (nm)	RBM	FIT 2	FIT 3		MM	CION	EXP
0.162	1.96	1.96	1.96		1.96	1.95	1.95
0.227/0.251	2.34	2.26	2.37		2.93	2.25	2.25
0.312	4.56	4.91	4.44		2.11	4.49	4.46

for the probability that an O atom is bonded to an Si atom if already another O atom is linked with this Si atom (FIT 2). FIT 3 means that this probability is fitted to the area $A(0.312\text{ nm})$ of the third peak of the experimental RDF,

$$p^{(0)}_{\text{Si-O}}(2) = 1 - \frac{A(0.312\text{ nm}) - 4K_{\text{Si-Si}}p_{\text{Si}}(0)\ p_{\text{Si-O}}(1)\ p_{\text{O-Si}}(2)}{12K_{\text{Si-O}}p_{\text{O}}(0)\ p_{\text{O-Si}}(1)}. \tag{15}$$

Also Temkin's microscopic MM (mMM) was applied to the com-SiO_x. In this model it is assumed that $(N_{Si}/3)$ Si atoms are centres of Si–Si_4 tetrahedra, $(N_{Si}/3)$ are centres of Si–O_4 tetrahedra, and $(N_{Si}/3)$ act as centres of Si–Si_yO_{4-y} units with $y = 1, 2, 3$. The corresponding value of $p^{(0)}_{\text{Si-O}}(2)$, which describes the tendency to decomposition of SiO_x, follows to be $7/9 \approx 0.78$ (see Table 4, where these magnitudes of the different models are compared). Finally, it should be mentioned that the value x of the average chemical composition was determined from the area of the first RDF peak.

Taking into account also a variation of the bond distribution as discussed in Section 5, we get the best agreement with the experimental results and a quantitative estimation of the relative number $n_{\text{O-O}}$ of effective O–O bonds within the SiO_x network. The peak areas of this covalent intermediately ordered network (CION), which has an intermediate position between the CON and the RCN models, are

$$\left.\begin{aligned} A(0.162\text{ nm}) &= 8K_{\text{Si-O}}p_{\text{Si}}(0)\ p_{\text{Si-O}}(1)\,,\\ A(0.244\text{ nm}) &= 4K_{\text{Si-Si}}p_{\text{Si}}(0)\ p_{\text{Si-Si}}(1) + 6K_{\text{O-O}}p_{\text{O}}(0)\ p_{\text{O-Si}}(1)\ p^{(0)}_{\text{Si-O}}(2)\,,\\ A(0.312\text{ nm}) &= 4K_{\text{Si-Si}}p_{\text{Si}}(0)\ p_{\text{Si-O}}(1)\ p_{\text{O-Si}}(2) +\\ &\quad + 12K_{\text{Si-O}}p_{\text{O}}(0)\ p_{\text{O-Si}}(1)\ p^{(0)}_{\text{Si-Si}}(2)\,, \end{aligned}\right\} \tag{16}$$

if com-SiO_x is taken again as an example. It can be seen from Tables 3 and 4, that a network structure in which the dominating defects are simulated by O–O bonds leads to an excellent reproduction of the experimental RDF peak areas and a high amount of O–O bonds in both the com-SiO_x and the ev-SiO_x. Going from the com-SiO_x to the ev-SiO_x the x-values deduced within the CION model have the same trend as within the other approximations. Whereas FIT 2 and FIT 3 seem to be other useful models for the reproduction of the experimental RDF's of both the com-SiO_x and the ev-SiO_x, only the structure of com-SiO_x can be approximated by the microscopic MM,

Table 4

Average chemical composition x, relative number $n_{\text{O-O}}$ of O–O bonds and decomposition parameter $p^{(0)}_{\text{Si-O}}(2)$ of the two SiO_x forms characterized in Tables 1 and 3

	com-SiO_x					
	RBM	FIT 2	FIT 3	mMM	MM	CION
x	0.98	0.98	0.98	0.98	0.98	1.22
$n_{\text{O-O}}$	—	—	—	—	—	0.06
$p^{(0)}_{\text{Si-O}}(2)$	0.49	0.60	0.76	0.78	1.00	0.70
	ev-SiO_x					
	RBM	FIT 2	FIT 3		MM	CION
x	1.15	1.15	1.15		1.15	1.30
$n_{\text{O-O}}$	—	—	—		—	0.04
$p^{(0)}_{\text{Si-O}}(2)$	0.58	0.52	0.60		1.00	0.56

which should, however, not be identified with a simple mixture of a-Si and a-SiO_2. As can be seen from Tables 3 and 4 a model basing on the RBM and corrected with respect to the clustering of oxygen (FIT 3, CION) is more useful for com-SiO_x than the MM.

The strong influence of the preparation method on the structure of SiO_x, already discussed in [1], can be deduced from the comparison of com-SiO_x and ev-SiO_x. The ev-SiO_x, obtained from electron evaporation of the com-SiO_x and subsequent deposition, reveals a typical random-bonding character. Furthermore, the results of the RBM are between FIT 2 and FIT 3 for ev-SiO_x. The reduced tendency of ev-SiO_x to decomposition is revealed by the CION values of the decomposition parameter $p^{(O)}_{Si-O}(2)$. Thus, it is shown for the first time that the RBM, which follows from purely statistical considerations, is consistent with the RDF data of a real SiO_x form. The usefulness of this model for the investigation of the dependence of various physical properties of SiO_x on the average chemical composition x has already been shown in previous papers (see [1] and corresponding references therein). The slight increase of x going from com-SiO_x to ev-SiO_x comes from the preparation method and a corresponding oxygen redistribution. Finally, it should be mentioned that the deduced relative numbers of effective O–O bonds (Table 4) are consistent with the high density of network defects in SiO_x [1, 25 to 27].

References

[1] K. HÜBNER, phys. stat. sol. (a) **61**, 665 (1980).
[2] K. HÜBNER and R. ENGELKE, phys. stat. sol. (a) **53**, K79 (1979); in: SiO_2 – Herstellung, Struktur und Eigenschaften, Vol. II, Ed. K. HÜBNER and G. ZUTHER, Wilhelm-Pieck-Universität Rostock, 1979 (p. 7).
[3] H.-G. NEUMANN and TH. ROY, phys. stat. sol. (a) **54**, K107 (1979); see [2], Vol. I (p. 129).
[4] H. N. POTTER, Trans. Amer. Electrochem. Soc. **12**, 191 (1907).
[5] L. BREWER and R. K. EDWARDS, J. phys. Chem. **58**, 351 (1954).
[6] N. FUSCHILLO, A. D. MCMASTER, and M. L. GIMPL, J. appl. Phys. **36**, 574 (1965).
[7] F. A. COLBOURN, J. M. DYKE, E. P. F. LEE, A. MORRIS, and I. R. TRICKLE, Mol. Phys. **35**, 873 (1978).
[8] H. P. KLUG and L. E. ALEXANDER, X-Ray Diffraction Procedures for Polycrystalline and Amorphous Materials, Wiley & Sons, New York 1974 (p. 791).
[9] B. E. WARREN, X-Ray Diffraction, Addison & Wesley, Reading (Mass.) 1969 (p. 105).
[10] G. W. BRADY, J. phys. Chem. **63**, 1119 (1959).
[11] M. V. COLEMAN and D. J. D. THOMAS, phys. stat. sol. **22**, 593 (1967).
[12] S. C. H. LIN and L. YOSHI, J. Electrochem. Soc. **116**, 1740 (1969).
[13] B. J. MOLNAR and D. B. DOVE, Bull. Amer. Ceram. Soc. **51**, 374 (1972).
[14] J. A. YASAITIS and R. KAPLOW, J. appl. Phys. **43**, 995 (1972).
[15] R. J. TEMKIN, J. non-crystall. Solids **17**, 215 (1975).
[16] C. F. GEORGE and P. D'ANTONIO, J. non-crystall. Solids **34**, 323 (1979).
[17] H. R. PHILIPP, J. non-crystall. Solids **8–10**, 627 (1972).
[18] R. L. MOZZI and B. E. WARREN, J. appl. Cryst. **2**, 172 (1969).
[19] J. F. GRACZYK, in: Physics of Structurally Disordered Solids, Ed. S. S. MITRA, Plenum Press, New York 1976 (p. 7).
[20] D. KITTA, Diploma Work, Wilhelm-Pieck-Universität Rostock, 1978.
[21] H. STEIL, Thesis, Wilhelm-Pieck-Universität Rostock, 1980.
[22] F. HAJDU, Acta cryst. **A28**, 250 (1972).
[23] G. PALINKAS and T. RADNAI, Acta cryst. **A32**, 666 (1976).
[24] TH. ROY, Thesis, Wilhelm-Pieck-Universität Rostock, 1980.
[25] G. A. N. CONNELL and G. LUCOVSKY, J. non-crystall. Solids **31**, 123 (1978).
[26] G. LUCOVSKY, Phil. Mag. **B41**, 457 (1980).
[27] A. G. REVESZ, phys. stat. sol. (a) **58**, 107 (1980).

(Received January 5, 1981)

phys. stat. sol. (a) 65, 281 (1981)

Subject classification: 14.3.2 and 18; 14.3.3; 22.1.1; 22.1.2

Laboratoire Physique des Composants à Semiconducteurs, Institut National Polytechnique de Grenoble, Equipe de Recherche associée au C.N.R.S., n° 659, E.N.S.E.R., Grenoble[1])

Magnetic Field and Surface Influences on Double Injection Phenomena in Semiconductors

II. The Magnetodiode Effect Theory for the Diffusion Regimes, Magnetodiode Experiments and Applications[2])

By

S. Cristoloveanu

The magnetodiode effect occurs when a carrier Lorentzian deflection is superimposed to the double injection phenomena. The regimes dominated by the carrier longitudinal diffusion are analysed with a general two-dimensional theory, for wide ranges of injection rates and magnetic fields. It is found that the operating of p^+nn^+ and p^+n magnetodiodes is similar. For low inductions, the current–voltage characteristics strongly depend on the surface quality and induction direction, while, for high inductions, the magnetodiode effect is relatively masked by the magnetoresistance. In this part also experimental results are reported on Ge and integrated SOS magnetodiodes, which are found to be in very good agreement with the whole proposed theory. The measured sensitivities (300 VT^{-1} or 30 AT^{-1} in Ge and 10 VT^{-1} or 0.15 AT^{-1} in SOS) are largely exceeding those of conventional semiconductor sensors and allow a wide field of applications.

L'effet magnétodiode se manifeste lorsqu' une déflection Lorentzienne de porteurs est superposée aux phénomènes de double injection. Une théorie bidimensionnelle générale nous permet d'analyser les régimes dominés par la diffusion longitudinale de porteurs, pour des gammes étendues d'injection et d'inductions magnétiques. A faible induction, les caractéristiques courant-tension dépendent principalement de l'état des surfaces et du sens de l'induction, alors que pour de fortes inductions l'effet magnétodiode est relativement masqué par la magnétorésistance. Dans cette partie nous présentons également des résultats expérimentaux sur des magnétodiodes en Ge ou intégrées sur SSI, qui sont en très bon accord avec l'ensemble de la théorie proposée. Les sensibilités mesurées (300 VT^{-1} ou 30 AT^{-1} dans le Ge et 10 VT^{-1} ou 0,15 AT^{-1} dans le SSI) dépassent largement les performances des capteurs conventionnels à semiconducteurs et permettent d'envisager un vaste champ d'applications.

1. Introduction

In a double injecting structure provided with finite surface recombination rates a transverse carrier gradient can be produced when applying a magnetic field perpendicularly to the current. This is the magnetodiode (MD) effect which involves considerable perturbations of the typical current–voltage laws of double-injection phenomena. In Part I of this work [1] we have established the basic formalism of the MD effect; the particular relationships corresponding to the semiconductor and insulator regimes in p^+nn^+ structures have been derived by neglecting the carrier longitudinal diffusion. The purpose of Sections 2 and 3 is to relate, for p^+nn^+ and for p^+n diodes, the double injection regimes due to the diffusion dominated regions near

[1]) 23, Avenue des Martyrs, 38031 Grenoble Cédex, France.

[2]) Part I see phys. stat. sol. (a) 64, 683 (1981).

the contacts growing everywhere in the sample; such a situation is especially encountered in relatively short structures or under very high injection levels. The experimental results on Ge p^+nn^+ and p^+n magnetodiodes and on SOS micromagnetodiodes are described in Section 4 in connection with various applications in semiconductor physics or as ultrasensitive devices.

2. The Longitudinal Diffusion Regime in p^+nn^+ Magnetodiodes

2.1 Special considerations

In the MD effect, when the roles of the doping or of the space charge become negligible, the carrier repartition $n(x, y)$ is derived from [1]

$$D^* \frac{\partial^2 n}{\partial y^2} + \frac{D^*}{l} \frac{\partial n}{\partial y} - R_v = -S_3 , \tag{2.1}$$

where

$$S_3 = D^* \frac{\partial^2 n}{\partial x^2} - \frac{D^*}{2\beta} (|\mu_{H_n}| + \mu_{H_p}) \, BE_y \frac{\partial n}{\partial x} .$$

Here, a distinction must be made between the longitudinal diffusion, which is the original reason of the carrier distribution, and the transverse diffusion which gives evidence for surface and magnetic field influences (*magnetoconcentration* (MC) effect).

The evaluation of the source term S_3 (2.1) shows that the part containing E_y can be neglected. In the absence of all transverse influences and for linear recombinations (2.1) gives the longitudinal carrier distribution

$$\delta n_1(x) = \left(\delta n_1(0) \operatorname{sh} \frac{L - x}{L^*} + \delta n_1(L) \operatorname{sh} \frac{x}{L^*} \right) \Big/ \operatorname{sh} \frac{L}{L^*} , \tag{2.2}$$

where $\delta n_1(0)$, $\delta n_1(L)$ are the carrier excesses at the boundaries and L^* is the ambipolar diffusion length.

The MC effect transversely disturbs this diffusion dominated distribution; in other words, $\delta n_1(x)$ acts as a suitable source term,

$$S_3 = D^* \frac{\partial^2 n_1}{\partial x^2} . \tag{2.3}$$

Then the carrier two-dimensional distribution is given by [1]

$$\delta n(x,y) = \delta n_k(y) + \delta n_1(x) \left(1 + \frac{\delta n_k(y)}{n_0} + \frac{\delta n_s(y)}{n_0} \right) . \tag{2.4}$$

With the same notations as in Part I, the current equation becomes

$$J = \alpha_1 e(|\mu_n| + \mu_p) (1 + g_k + g_s) \delta n_1(x) E_x - eBD^*(|\mu_{H_n}| + \mu_{H_p}) (f_k + f_s) \delta n_1(x) - \\ - e\beta\alpha_2(\mu_n + \mu_p) (1 + g_k + g_s) \, \partial n_1/\partial x . \tag{2.5}$$

The forward characteristic $J(U)$ is found by integrating (2.5),

$$\frac{J}{\alpha_1 e(|\mu_n| + \mu_p) (1 + g_k + g_s)} Y = (1 + a_6) \, U + U_{LD} , \tag{2.6}$$

where U_{LD} is the potential drop due to the longitudinal diffusion,

$$U_{\mathrm{LD}} = -\frac{\beta\alpha_2(\mu_n + \mu_p)}{\alpha_1(|\mu_n| + \mu_p)} \log \frac{\delta n_1(L)}{\delta n_1(0)} \tag{2.7}$$

and for the case of practical interest $L \gtrapprox 2$ to $3L^*$,

$$Y = \int_0^L \frac{dx}{\delta n_1(x)} \approx \frac{\pi L^* \exp(L/2L^*)}{2\sqrt{\delta n_1(0)\, \delta n_1(L)}}. \tag{2.8}$$

2.2 Boundary conditions

The boundary excess $\delta n_1(0)$ (or $\delta n_1(L)$) is derived from the current continuity of minority carrier across the p^+–n (or n–n^+) junction [2]. In the heavily doped region p^+ the electron current is essentially dominated by the longitudinal diffusion, while in the n region the electron current is constituted of diffusion and drift contributions (influenced by the MD effect). Then $\delta n_1(0)$ is given by two asymptotic expressions corresponding to relatively "moderate" and "very high" injection levels,

$$\delta n_1(0) = \alpha_4 J \quad \text{if} \quad \delta n_1(0) \ll \alpha_3 p_p\,, \tag{2.9}$$

$$\delta n_1(0) = \alpha_5 \sqrt{J} \quad \text{if} \quad p_p \gg \delta n_1(0) \gg \alpha_3 p_p\,, \tag{2.10}$$

where p_p is the doping of the p^+ region and α_i are magnetoresistive coefficients.

An injection rate greater than in (2.10) $(\delta n_1(0) \gtrapprox p_p)$ would be responsible for the semiconductor junction effacement. We suppose $\delta n_1(L)$, which is expressed by similar relations, of the same order as $\delta n_1(0)$; (else, if $\delta n_1(0) \gg \delta n_1(L)$ the low injection junction n–n^+ can be better approached by an ohmic contact (cf. Section 3)). Then U_{LD} is negligible in (2.6).

2.3 Moderate injection

This case corresponds to the approximation of zero minority currents across the junctions. Using (2.9) the master equation (2.6) becomes independent of J. This means that the voltage drop across the base is constant, i.e. the current increases for a fixed potential U_0.

In such an abrupt characteristic the MC effect modulates the threshold voltage U_0, which would exist without the *influence of surfaces* and in the absence of the magnetic field,

$$U = \frac{\alpha_5}{\alpha_3(1 + a_6)(1 + g_k + g_s)} U_0 \gtrless U_0\,, \text{ with } U_0 = \frac{\pi D^*}{2\sqrt{|\mu_n|\,\mu_p}} \exp\left\{\frac{L}{2L^*}\right\}. \tag{2.11}$$

At high magnetic field the magnetoresistive decrease of L^* leads to an exponential enhancement of U_0 and the modifications due to the MD effect have a relatively small influence; nevertheless the transverse diffusion always acts (by means of the coefficient $a_6 \gg 1$) in order to attenuate the magnetoresistance effect. On the contrary at small and moderate inductions (where $a_6 \ll 1$) the voltage across the base is largely perturbed by the transverse conditions as it is illustrated in Fig. 1.

The current for which the transition between insulator and diffusion regimes takes place is obtained by inserting (2.11) in the insulator $J(U)$ characteristic [1].

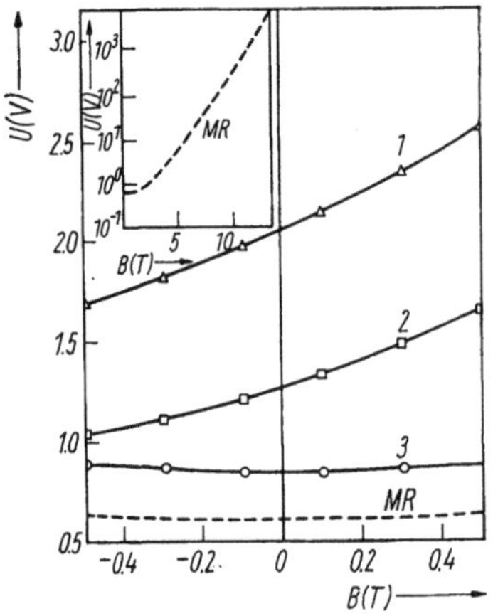

Fig. 1

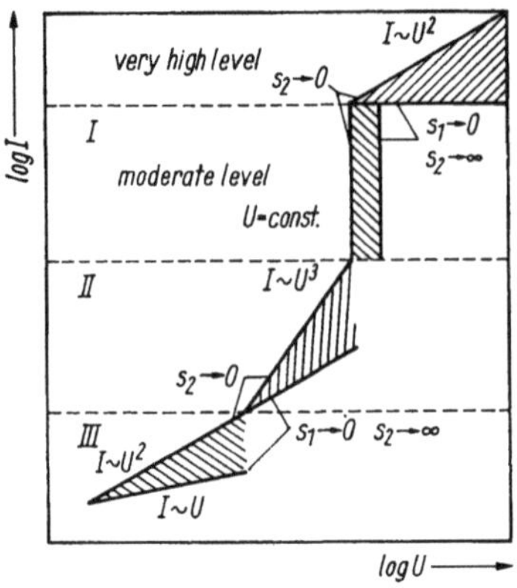

Fig.2

Fig. 1. Voltage drop U across the base of a Ge p^+nn^+ structure against the (small) magnetic induction, in the "abrupt" diffusion regime (moderate level); surface recombination velocities ($s_{1,2}$) are parameters. In the pure magnetoresistive curve (MR) the surface influence is not included (no MC effect); the case of high magnetic inductions is given in the insert. ($U = 20$ V, length $L = 4.5$ mm, thickness $b = 1$ mm, bulk lifetime $\tau_v = 100$ μs, $T = 300$ K, doping $n_0 = 4 \times \times 10^{19}$ m^{-3}, carrier mobilities (at $B = 0$) $|\mu_n^0| = 0.38$ m^2 V^{-1} s^{-1}, $\mu_p^0 = 0.19$ m^2 V^{-1} s^{-1}); (1) (s_1; s_2) = (2; 200), (2) (0.2; 20), (3) (2; 2) (ms^{-1})

Fig. 2. Schematic representation of the double injection regimes in p^+nn^+ structures provided with small densities of deep centers. The hatched region shows the possible evolution of characteristics due to magnetic field ($B > 0$) and surface influences. The extreme cases correspond to a carrier deflection towards very low or very high recombining surfaces. I diffusion regime, II insulator regime, III semiconductor regime

2.4 Very high injection

Equations (2.6) and (2.10) lead to a "quadratic" law

$$J = A(1 + g_k + g_s)^2 (1 + a_6)^2 U^2 \tag{2.12}$$

with $A \sim \exp(-L/L^*)$.

It is clear that the semiconductor regime (defined by bulk recombinations) and the very high diffusion regime (imposed mainly by the contacts) cannot be confused, in spite of the similarity of their current–voltage dependences ($J \sim U^2$); the higher action of MC in (2.12) than in the semiconductor law [1] is attributed to the increase of the carrier number submitted to the Lorentzian deflection.

So when carriers are deflected towards a low recombining surface ($s_2 \ll l/\tau_v$) a maximum conductivity is obtained,

$$J = A(1 + a_6)^2 U^2 . \tag{2.13}$$

Moreover, when $a_6 \ll 1$, this expression corresponds to that derived by Kleinman [3].

In the opposite situation ($s_2 \gg l/\tau_v \gg s_1$), the current *saturates* to a value essentially governed by B and s_2,

$$J = A\left(\frac{2\beta L(1 + a_6)}{s_2\tau_v(|\mu_{H_n}| + \mu_{H_p})\, B}\right)^2 . \tag{2.14}$$

This treatment may be easily extended to the case of short ($L \sim L^*$) or very short ($L \ll L^*$) magnetodiodes, by modifying only the boundary conditions and the Y expression [4].

A summary of the successive injection regimes in structures provided with two injecting contacts is shown in Fig. 2. Nevertheless one or more regimes can be missing in a particular sample, owing to contact, length, or power dissipation limitations.

3. Magnetodiode with One Injecting Contact

3.1 General considerations

This section is devoted to "long diodes" ($L \gtrapprox 3L^*$) provided with one injecting contact $(\delta n_1(0) \gg n_0)$, the other contact being ohmic $(\delta n_1(L) = 0)$. The carrier longitudinal gradient is then very important and regimes similar to the previous semiconductor or insulator are masked by the diffusion dominated regime. Moreover, in the proximity of the ohmic contact the doping is no longer negligible and the hypothesis $n = p$ fails there. The variation of the ambipolar diffusion coefficient causes the non-linearity of the MD surface equations [1]; to avoid this we shall consider further a near-intrinsic base[3]) (p^+i structure: $n_0 = p_0 = n_i$).

Then, the carrier distribution is given again by (2.2) and (2.4). Rigorously, the ohmic contact is responsible of a more important variation of E_x along the sample than in p^+nn^+ structures; near $x = L$ a higher Lorentz force acts on a small carriers density and in the proximity of $x = 0$, numerous carriers are deflected by a lower force. Then the assumption of a mean MC effect remains convenient for the analytical description of p^+i magnetodiodes.

The current equation characterizing the problem is

$$J = \alpha_1 e(|\mu_n| + \mu_p) \langle n \rangle_y E_x - eBD^*(|\mu_{H_n}| + \mu_{H_p}) \left\langle \frac{\partial n}{\partial y} \right\rangle_y - \\ - \alpha_2 e\beta(\mu_n + \mu_p) \left\langle \frac{\partial n}{\partial x} \right\rangle_y . \tag{3.1}$$

The total voltage (U) is the sum of basis (U_B) and junction (U_J) potential drops; U_J is usually derived from the ratio between the boundary density and the minority carrier density, which is here a function of the MC effect,

$$U_J = \beta \log \frac{(1 + g_k + g_s)\, \delta n_1(0)}{(1 + g_k)\, n_i} . \tag{3.2}$$

Integrating (3.1) we obtain

$$U_B = \frac{J}{\alpha_1 e(|\mu_n| + \mu_p)} Y^* - U_{TD} - U_{LD} . \tag{3.3}$$

U_{TD} and U_{LD} are the contributions of transverse and longitudinal diffusion; Y^* and the boundary excess $\delta n_1(0)$ are given by equations similar to (2.8) to (2.10). We shall assume that $L/L^* \ll U_J/\beta$; this hypothesis is not restrictive, longer structures being governed by highly resistive bases.

[3]) However, the doping level slightly modifies only the surface conditions [1]; as the bulk equation (2.1) is not affected, the results deduced for a p^+i diode can be generalized for a p^+n diode.

3.2 Moderate injection

The forward characteristic can be expressed by algebraic manipulations of (2.9), (3.2), and (3.3) in a conventional exponential form

$$J = J_s \exp \frac{U}{c\beta}. \tag{3.4}$$

The coefficient c accounts for the potential drop division between the junction and the base,

$$c = 1 + \frac{1}{(1 + a_6)(|\mu_n| + \mu_p)} \left(\frac{2\alpha_7 \mu_p \operatorname{ch}\left(\frac{L}{L^*}\right)}{|1 + g_k + g_s} + \frac{\alpha_2}{\alpha_1}(|\mu_n| - \mu_p) \right). \tag{3.5}$$

When all transverse influence is eliminated ($B = 0$, $s_1 = s_2 = 0$), the coefficient c takes a particular form c_0, illustrating the case studied by Stafeev [5],

$$c_0 = 2 \frac{\mu_p \operatorname{ch}\left(\frac{L}{L^*}\right) + |\mu_n|}{|\mu_n| + \mu_p}. \tag{3.6}$$

The magnetosensitivity of "long diodes" has been previously ascribed only to the magnetoresistive modification of c_0: when B increases, $1/L^*$ and c_0 increase, too; the current becomes smaller like the base conductivity and the total voltage is redistributed in order to increase U_B; so U_J and the current are again reduced, etc. [6]. It is clear that such a description cannot include surfaces, field sense, or thickness influences, like in the general equation (3.5); only at *high magnetic inductions*, the coefficient c is mainly dominated by the magnetoresistance effect (exponential dependence, so that the MC effect has a relatively weak influence).

It has been shown that the current variation with c is important for structures with optimum lengths; for example in Ge: $L \approx 3$ to $7L^*$ [7]. Then U_{LD} and U_J become negligible and the expressions of c and J_s are reproduced as

$$c = \frac{\alpha_7 \mu_p \exp\left(\frac{L}{L^*}\right)}{(1 + a_6)(1 + g_k + g_s)(|\mu_n| + \mu_p)}, \quad J_s = \frac{\alpha_1 \alpha_7 (1 + g_k)\, e\beta \mu_p n_i \exp\left(\frac{L}{L^*}\right)}{2(1 + g_k + g_s) L^*}. \tag{3.7}$$

It is shown in Fig. 3 that at low inductions ($a_6 \ll 1$) the coefficient c is always exceeding c_0 during the MD effect. An interesting consequence is that the slope of the ($\log J \sim U$) curve is no more constant, even at fixed B. The current increase can be limited and its saturation becomes plausible when $s_1 \ll l/\tau_v \ll s_2$ [1],

$$J \to \frac{e\beta \mu_p n_i \exp\left(\frac{L}{L^*}\right)}{L^*} \exp \frac{2L}{s_2 \tau_v \mu_p B \exp\left(\frac{L}{L^*}\right)}. \tag{3.8}$$

The magnetosensitivity around $B = 0$ is given by

$$\log J \sim (1 + g_k + g_s)\, U \sim BU^2\,. \tag{3.9}$$

However, this apparently large sensitivity is balanced by the important values of c. For this reason p^+i diodes are less interesting than p^+nn^+ structures.

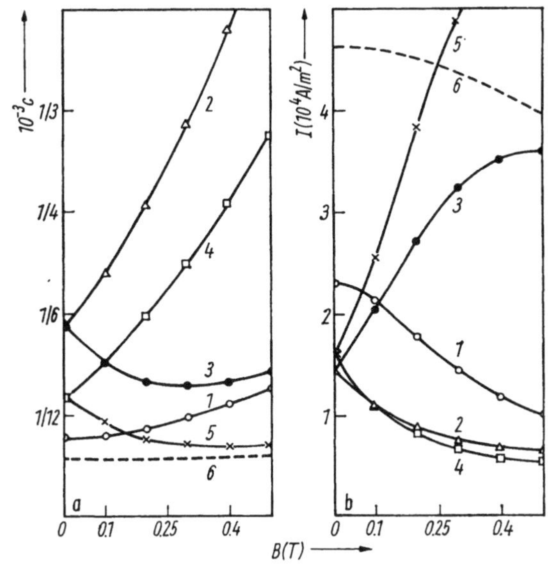

Fig. 3. a) Coefficient c and b) total current against induction B, in Ge p^+i MD (exponential regime); $U = 5$ V, $L = 4$ mm, other parameters like in Fig. 1. (1) $(s_1; s_2) = (2; 2)$, (2) (2; 200), (3) (200; 2), (4) (0.2; 20), (5) (20; 0,2), (6) pure MR effect

3.3 Very high injection

This case corresponding to the boundary condition (2.10) seems to have been neglected in the previous studies concerning "long diodes"; nevertheless its physical reality is associated with that of the quadratic diffusion law of Section 2.4.

In the practical case where $L = 3$ to $7L^*$, (2.10), (3.2), (3.3) yield an unusual characteristic,

$$\frac{U}{U_0} = \sqrt{\frac{J}{J_0}} \log \frac{J}{J_0} . \tag{3.10}$$

Both U_0 and J_0 are functions of the MC effect,

$$U_0 = \frac{\beta L^* n_i \operatorname{sh}(L/L^*) \left(\operatorname{ch}(L/2L^*)\right)^2}{2\alpha_8 L_n p_p} \frac{1 + g_k}{(1 + a_6)(1 + g_k + g_s)^2}, \tag{3.11}$$

$$J_0 = \frac{\alpha_1 e \beta n_i^2 (|\mu_n| + \mu_p) \left(\operatorname{ch}(L/2L^*)\right)^4}{a_8 L_n p_p} \left(\frac{1 + g_k}{1 + g_k + g_s}\right)^2 . \tag{3.12}$$

At this point the similitude of p^+in^+ and p^+i magnetodiodes is to be underlined. In p^+in^+ structures the diffusion regime is successively depicted by abrupt and quadratic characteristics, while in p^+i structures similar laws are obtained: exponential and underquadratic. It follows that the absence of a second injecting contact penalizes, in each case, the current law by a factor proportional to $(\log J)$.

The comparison of (2.11) and (3.4) or (2.12) and (3.10) giving the action of the MC effect, also denotes the similarity of p^+in^+ and p^+i magnetodiodes. The distinction between these two situations, made up to now, seems then to be not quite justified.

4. Experiments

4.1 Germanium p^+nn^+ structures

Samples are cut from an n-type germanium ingot (30 to 50 Ω cm) in slab geometry ($L = 2$ to 10 mm, $b = 0.2$ to 2 mm, width $= 1$ to 4 mm). The p^+–n or n–n^+ junctions are prepared by In or Sn diffusion at about 550 °C. Low recombining surfaces are obtained by mechanical and chemical treatments; strong sandblasting confers important recombination rates.

A typical $I(B)$ curve of a structure provided with very different surfaces is plotted in Fig. 4. Notice the agreement with the semiconductor regime theory [1]:

— The current increases when carriers are deflected towards the lower recombining surface ($B > 0$); the maximum (for $B \approx 0.1$ T) corresponds to the limitation of the carrier effective lifetime ($\tau_{eff} \leqq \tau_v$) [1]. For $B > 0.1$ T the current is decreasing because the recombination velocity of the surface where the carriers are accumulated cannot be zero.

— Around $B = 0$, $I(B)$ is linear and the slope varies as U^3.

— There are two regions (for $B \approx 0$ and $B \approx 0.2$ T) where the sensitivity is very high.

— The current–voltage law is quadratic ($|B| \lessapprox 0.1$ T) or "ohmic" ($|B| \geqq 0.25$ T).

When the two surfaces are chemically etched, the characteristic becomes symmetrical (Fig. 5) and the maximum sensitivity $\Delta I/\Delta B$ is displaced to $|B| \approx 0.1$ to 0.15 T. The magnetosensitivity is quite suppressed in diodes with two highly recombining surfaces; this proves the prevailing role of the MC effect.

Looking for an increase of the sensitivity we have slightly modified the conventional geometry, by placing the p^+–n junction on a surface. So not only the effective length is reduced but also the space charge layer of this lateral junction is directly influenced by the carrier accumulation or depletion (as in the case of Schottky mag-

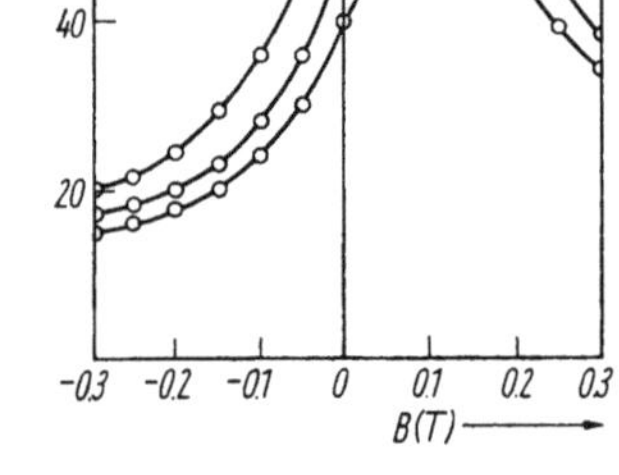

Fig. 4

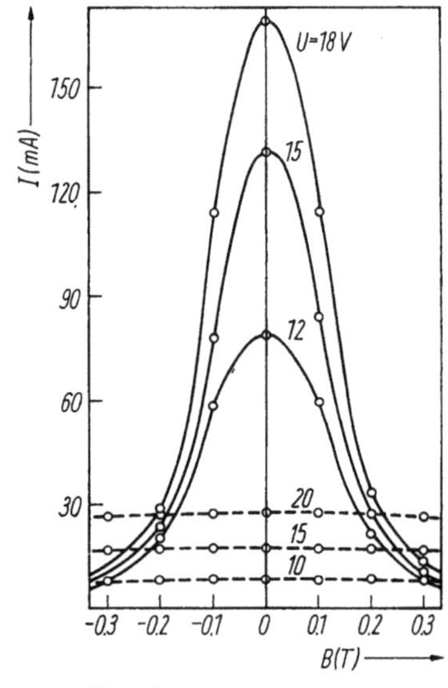

Fig. 5

Fig. 4. Experimental current variation against induction B in a Ge p^+nn^+ structure provided with very different surface recombination rates (semiconductor regime). Length $L \approx 7$ mm, thickness $b \approx 0.25$ mm, width $w \approx 1$ mm, base conductivity $\sigma \approx 2.5\ \Omega^{-1}\ m^{-1}$, $T = 295$ K

Fig. 5. Experimental current against induction in two Ge p^+nn^+ magnetodiodes provided with *two low* (———) or *two high* (– – –) recombining surfaces ($L \approx 5$ mm, $b \approx 0.5$ mm, $w = 2$ mm, $T = 295$ K). When the surfaces are abraded the carrier injection is compensated by recombination: the current is then reduced and the semiconductor regime is not yet established

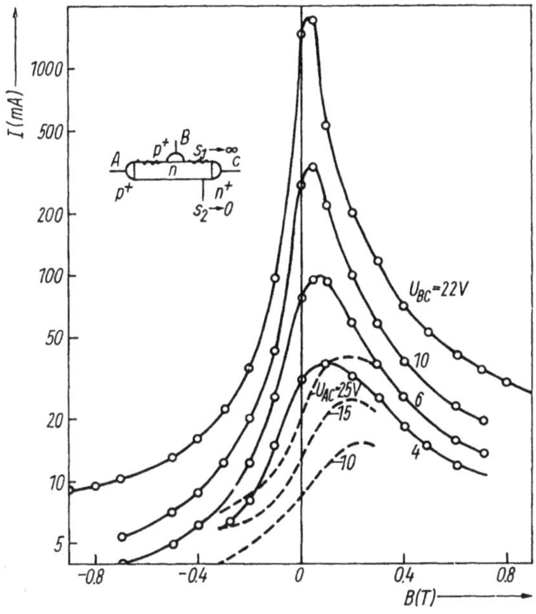

Fig. 6. Experimental current versus induction for ultrasensitive Ge p^+nn^+ structures (semiconductor regime). Full lines indicate characteristics of the "special" magnetodiode (BC); note the current decrease around $B = 0$. Dashed lines represent characteristics of the "usual" magnetodiode (AC). ($b \approx 1$ mm, $w \approx 2$ mm, $L_{AC} \approx 10$ mm, $\sigma \approx 2.5\ \Omega^{-1}\ m^{-1}$)

netic sensors [8]); this leads to a better coupling between double injection and MC effects. Indeed, Fig. 6 denotes an excellent magnetosensitivity: $(\Delta I/\Delta B) \approx 30$ A/T or $(\Delta U/\Delta B) \approx 300$ V/T. Sensitivities exceeding 100 V/T are available even for very low currents in short and thin diodes (Fig. 7).

Characteristics corresponding to insulator or diffusion regimes have not been observed; this is due to heating limitations and to the difficulty to realize very short structures.

4.2 SOS p^+nn^+ structures

The silicon on sapphire technology brings naturally an agreement with the magnetodiode effect requirements: the thickness of Si films deposited epitaxially on sapphire corresponds to the diffusion length L^* (≈ 1 μm) and the recombination rates at Si/Al_2O_3 and Si/SiO_2 interfaces are very different. Typical characteristics of SOS p^+nn^+ micromagnetodiodes, conceived in our laboratory and LETI-CEA Grenoble are reproduced in Fig. 8. The measured sensitivity (10 V/T or 0.15 A/T) [9] is two orders higher than in integrated Hall sensors. A more detailed analysis and experiments on noise, temperature, or high magnetic field are given elsewhere [10].

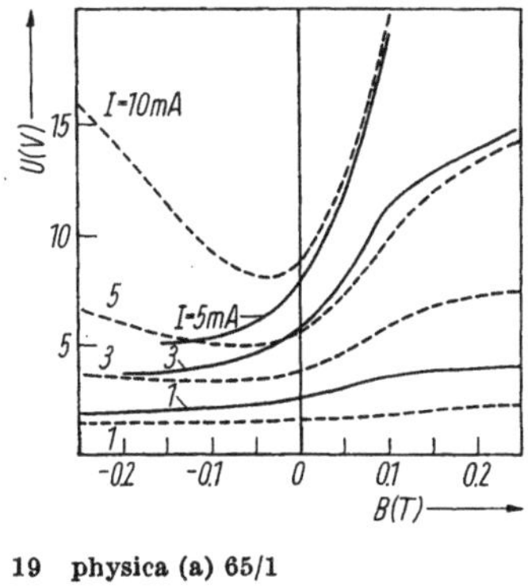

Fig. 7. Experimental voltage versus magnetic induction, for low values of the current, in Ge p^+nn^+ structures ($L \approx 3$ mm, $b \approx 0.2$ mm, $w \approx 1$ mm; ——— $s_2 \gg s_1$, – – – $s_2 \gtrsim s_1$)

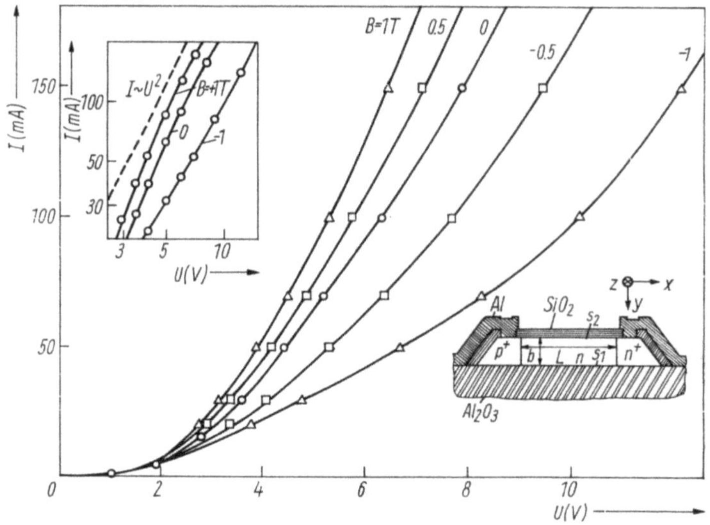

Fig. 8. Experimental current–voltage characteristics of a SOS integrated structure at room temperature ($L = 18\ \mu m$, $b = 7\ \mu m$, $w = 250\ \mu m$, $n_0 \approx 2 \times 10^{21}\ m^{-3}$). The insert indicates the excellent fit of experiments with the semiconductor quadratic law

The interest in SOS technology is due not only to the VLS integration of magnetodiodes and associated amplifier units but also to the high frequency performances ($\tau_v \approx 10$ ns, then a range of 10^8 Hz is available).

4.3 Long diodes

Several p$^+$i or p$^+$n structures of different lengths or thicknesses have been realized on germanium. Fig. 9 shows the influence of the length of the structure and of surface treatments on the exponential current increase: whatever may be the magnetic field direction for $|B| \geqq 0.2$ T the slope is reduced, in comparison with the $B = 0$ case. The sensitivity is here one order of magnitude lower (10 V/T) than in p$^+$nn$^+$ structures fabricated from the same ingot. Silicon magnetodiodes with one injecting contact have been largely described by Karakushan et al. [6, 7].

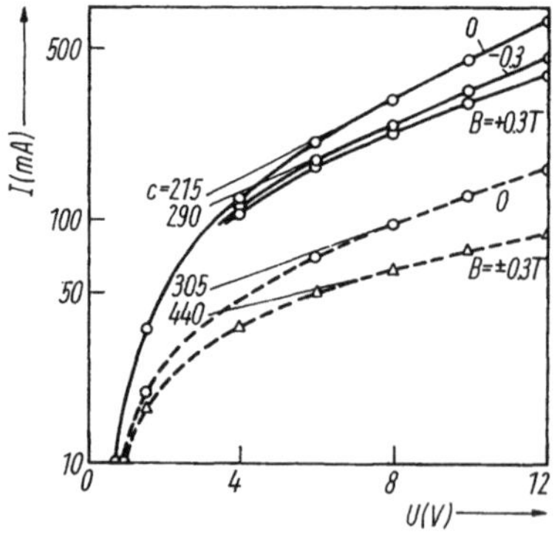

Fig. 9. Experimental current–voltage characteristics of two Ge p$^+$n magnetodiodes provided with *different* (——— $s_1 > s_2$, $L \approx 4$ mm) or *two etched* (– – – $s_{1,2} \to 0$; $L \approx 5$ mm) surfaces ($\sigma \approx 2.5\ \Omega^{-1}m^{-1}$, $b \approx 1$ mm, $w \approx 3$ mm). Note that (i) the exponential law is well verified for $U > 6$ V, and (ii) the sensitivity $\Delta U/\Delta B$ of the longer structure is better than in the shorter one, although the current is less important

4.4 Applications

The two basic effects (injection and magnetoconcentration) are dominated by the recombination process; conversely the fitting of theoretical and experimental magnetodiode results leads to a high accuracy in the determination of recombination parameters. The MD effect is the only one able to give simultaneously τ_v, s_1, and s_2 in a non-intrinsic semiconductor. It is moreover possible to analyse non-linear recombination mechanisms or inhomogeneous thin films (SOS) [11].

The applications of the MD effect beyond the field of physical characterization are based on the high sensitivities available: $\Delta U/\Delta B > 300$ V/T, $\Delta I/\Delta B > 30$ A/T, or $\Delta R/\Delta B > 10$ kΩ/T. These values have been experimentally found around moderate biases ($I \approx 10$ to 100 mA, $U \lessgtr 20$ V) so that relative sensitivities ($\Delta U/U \, \Delta B$, $\Delta U/I \, \Delta B$, etc.) are simultaneously important. For comparison, performances of conventional Hall or magnetoresistance sensors are 1 to 4 orders of magnitude lower.

Experiments show a minimum measurable induction of 10^{-9} T, for a noise to signal unit ratio, in the range of 0.1 Hz to 100 kHz; improvements are possible by using field concentrators or by sample cooling. The optimization of the structure to favour the current flow (relative high width to length ratio, highly injecting contacts, etc.) could cause an increase of sensitivity by a factor 10 to 100; so a power slope of 1 kW/T is plausible.

The MD effect offers a large set of applications based on the measurement of small magnetic fields or small variations around a fixed magnetic induction (with appropriate sample surface treatment). In a general point of view the system magnetic source–magnetodiode would detect all phenomena able to perturb the source.

5. Conclusion

This study of surface and magnetic field influence on the double injection phenomena demonstrates very similar behaviours for structures apparently quite different (with one or two injecting contacts) in a wide range of magnetic and electric fields.

Our theory gives evidence of the essential role of surface recombination: the single surface influence is responsible for the carrier density reduction; the current magnitude is modified, but the current–voltage dependence remains unchanged.

On the contrary, when a magnetic field is applied the classical laws (which are depicting only the maximum conductivity for each injection level) can switch to unusual characteristics; very useful magnetosensitive devices are thus obtained. For the diffusion regimes dominated by boundaries, it is no more possible to define an effective carrier lifetime, illustrating in a condensed form (as for the semiconductor and insulator regimes) the MD effect.

We also noted that the transverse Dember currents always act in order to enhance the total current; in other words the transverse diffusion attenuates the magnetoresistive decrease of the sample conductivity.

The analytical expressions given in this paper as well as the discussion of the interesting particular cases simplify the meaning of the magnetoconcentration–injection interaction and allow the rational optimization of sensitive structures.

Acknowledgements

I wish to thank Dr. A. Chovet and Dr. G. Kamarinos for their critical and detailed reading of this manuscript.

References

[1] S. Cristoloveanu, phys. stat. sol. (a) **64**, 683 (1981).
[2] R. Baron and J. W. Mayer, in: Semiconductor and Semimetals, Vol. 6, Ed. R. K. Willardson and A. C. Beer, Academic Press, New York 1970 (p. 201).
[3] D. A. Kleinman, Bell Syst. tech. J. **35**, 685 (1956).
[4] S. Cristoloveanu, Thesis, INP Grenoble, 1981.
[5] V. I. Stafeev, Soviet Phys. — Tech. Phys. **3**, 1502 (1958).
[6] E. I. Karakushan and V. I. Stafeev, Soviet Phys. — Solid State **3**, 493 (1961).
[7] E. I. Karakushan, V. Ya. Kovarskii, K. F. Komarovskikh, E. I. Gamolin, and V. I. Stafeev, Soviet Phys. — Semicond. **3**, 1453 (1970).
[8] S. Cristoloveanu, A. Mohaghegh, and J. de Pontcharra, J. Physique **41**, L235 (1980).
[9] G. Kamarinos, P. Viktorovitch, S. Cristoloveanu, J. Borel, and R. Staderini, IEDM, Technical Digest, Washington 1977 (p. 114A).
[10] A. Mohaghegh, S. Cristoloveanu, and J. de Pontcharra, IEEE Trans. Electron Devices 28, 237 (1981).
[11] S. Cristoloveanu, A. Chovet, and G. Kamarinos, Solid State Electronics **21**, 1563 (1978).

(Received June 18, 1980)

phys. stat. sol. (a) **65**, 293 (1981)

Subject classification: 10.2; 12.2; 22.6

Departamento de Física Fundamental, Facultad de Ciencias, Universidad de Zaragoza[1])

The Effect of Lithium Impurity on Flow Stress in MgO

By

V. M. Orera

The presence of Li_2O precipitates contributes to the hardening of MgO crystals at room temperature. The enhanced flow stress value is unaffected by thermal treatment, below annealing temperatures of 1100 K. However, above $T \approx 1100$ K point defect hardening begins to occur and the room temperature flow stress becomes a strong function of the thermal history of the crystal and the atmosphere in which it is incubated. In an oxidizing atmosphere, the enhancement of the flow stress begins to occur at $T \approx 1100$ K and is attended by the appearance of the $[Li]^\circ$ optical absorption band at 1.8 eV. In an inert or reducing atmosphere, the flow stress increase occurs at $T \approx 1450$ K, coinciding with the appearance of an optical band at 5.3 eV. The flow stress depends on the square root of the point defect concentration both, for $[Li]^\circ$ defects and for those responsible of the 5.3 eV band, where the hardening is the same for both defects between experimental error. The 5.3 eV band is assigned to some substitutional configuration of lithium ions.

La presence des precipités de Li_2O contribute au durcissement dans les cristaux de MgO a la temperature ambiante. Au dessous de $T \approx 1100$ K la valeur accrue de la limite élastique ne change pas avec le traitement thermique. Cependant, au dessus de $T \approx 1100$ K commence le durcissement pour les défauts ponctuels et la limite élastique à température ambiante devient fortement dépendante de l'histoire termique des cristaux et de l'atmosphere de recuit. Dans une atmosphère oxidante, l'acroissement de la limite élastique commence à $T \approx 1100$ K et est relié à l'apparition de la bande d'absorption optique du $[Li]^\circ$ à 1,8 eV. Dans une atmosphère inerte ou réductrice, l'accroissement de la limite élastique commence à $T \approx 1450$ K coincidant avec l'apparition d'une bande d'absorption optique à 5,3 eV. La limite élastique croit linéairement avec la racine carrée du nombre de centres $[Li]^\circ$ et de ceux qui produisent la bande à 5,3 eV, le durcissement est pareil pour tous les deux defauts, dans les limite d'erreur expérimentale. Nous assignons la bande à 5,3 eV à une configuration substitutionelle des ions de lithium.

1. Introduction

Defects (vacancies, interstitials, impurities and their aggregates) are known to contribute to the hardening of a material, thereby affecting the resolved flow stress of a solid [1, 2]. Also whether the defects are present in the matrix in isolated or aggregated forms they will also have a strong influence on the flow stress. The dependence of the flow stress on the defect concentration can yield information pertaining to the nature of the defects. Single-crystal refractory oxides can serve as ideal systems for such investigations, since isolated defects, and sometimes even aggregates, can often be characterized by powerful techniques such as optical absorption and magnetic resonance. In the past decade, deformation and flow stress investigations in MgO have utilized these techniques in order to determine the types and concentrations of the defects present [3 to 9]. The presence of impurities has been demonstrated to affect certain properties of MgO [10, 11]. Lithium, in particular, affects the electrical conductivity [12] and there is evidence that it suppresses radiation damage [13]. The present work concerns lithium impurities and how they contribute to the flow stress in MgO.

The MgO crystals in this study were doped [14] with Li to ≈ 0.03 to 0.05 at%. In the as-grown (virgin) state, the lithium impurities are present in MgO primarily as

[1]) Ciudad Universitaria, Zaragoza 6, España.

Li_2O precipitates. The size of these precipitates is typically in the 10^2 Å range [15]and their concentration $\approx 10^{13}$ cm^{-3}. In an oxidizing atmosphere at elevated temperatures, lithium ions diffuse out from the precipitates, these ions are optically active and paramagnetic, due to the holes which must be trapped in order for the defects to remain electrically neutral [16]. The dissociation of molecular oxygen at the surface into oxygen ions provides the required source of holes [17]. Each of the lithium ions thereby becomes a $[Li]^0$ defect [18] with linear configuration of Mg^{2+}–O^{2-}–Li^+–O^-–Mg^{2+}. EPR measurements have shown that the lithium impurities are present in predominantly single isolated $[Li]^0$ defects. At RT the hole hops between the O^{2+} ions adjacent to the Li^+ ion.

In an atmosphere of oxygen, the threshold temperature for production of the $[Li]^0$ defect is $T \approx 1100$ K [19]. The production of these defects is strongly dependent on the annealing temperature as well as the ambient atmosphere. Therefore, it is possible to determine the different contributions of the precipitates and the solute ions to the flow stress by varying the annealing temperature.

2. Experimental Procedure

The crystals used for this work were grown at Oak Ridge National Laboratory by the arc-fusion method technique using high-purity MgO powder from the Kanto Chemical Co. (Tokyo) as starting material by Chen et al. [14]. The total concentration of lithium in the resulting single crystals was determined by chemical analyses. Analyses for other impurities yielded results similar to those published earlier [13].

Thermal treatments were performed in a quartz tube, inserted in the horizontal axial hole of a Sentry Model 7AV furnace. For treatment at temperatures higher than 1600 K, an alumina tube was used. Different controlled atmospheres, such as air and O_2, N_2, H_2, or Ar have been used during the heat treatments. After each thermal anneal, the samples were quickly pulled out of the furnace portion of the tube to the room temperature portion of the tube (remaining in their respective atmospheres) in order to quench in the defects generated by the heat treatments.

Optical absorption measurements were carried out on a Cary 17 spectrophotometer and electron paramagnetic resonance (EPR) was used to determine the absolute concentrations of the paramagnetic defects.

Compression tests were performed on rectangular sample platelets, cleaved along (100) planes. The length-to-width ratio was maintained at >3 in order to achieve reproducible data [9]. The samples were compressed along their longest $\langle 100 \rangle$ directions in an Instron testing machine with a crosshead speed of 0.05 cm/min. Wherever possible, both optical and mechanical measurements were made on the same sample. A minimum of six samples was used in order to obtain each experimental point on the flow stress curves. A standard deviation less than 10% of the mean was observed in all cases. The samples were measured at room temperature following the thermal treatments.

For the series of experiments on isochronal and isothermal annealing, samples cleaved from the same large crystal were used in order to minimize differences due to crystal non-homogeneity.

3. Experimental Results

3.1 Optical bands and deformation

Plastic deformation of nominally pure MgO crystals produces a broad optical absorption band, which peaks at 5.7 eV [3, 5, 7, 9]. Although the defect responsible of that band has not been unambiguously identified it is most probably associated with multiple vacancies [7]. Deformation of virgin MgO:Li crystals also produces the

5.7 eV band, shown in Fig. 1a. The dependence of the absorption coefficient of that band with deformation is linear and similar to that obtained in pure MgO [20]. Hence the concentration of the defect absorbing at 5.7 eV appears to be the same for Li-doped and undoped MgO, the presence of the Li_2O precipitates does not affect the production of the defects responsible for the 5.7 eV band. Since this band is the result of deformation and also is not associated with the lithium impurity, we need not consider it further.

An MgO:Li crystal, heated in an oxidizing atmosphere at $T > 1100$ K, exhibits a broad absorption band at 1.8 eV (blue color), which has been identified as due to [Li], defects [18, 21]. This band is not produced if the heat treatment is performed in an oxygen-free atmosphere. At $T > 1450$ K, a broad band at 5.3 eV emerges, regardless of the ambient atmosphere. The defect responsible for this band has not been identified, but is most probably associated with lithium since the band is not observed in crystals which do not contain lithium. Hence, a MgO:Li crystal heated in an oxidizing atmosphere at $T > 1450$ K exhibits both the 1.8 and 5.3 eV bands such as shown in Fig. 1b. Deformation of such a crystal lowers the intensity for the 1.8 eV band, but produces the 5.7 eV deformation band, which overlaps the 5.3 eV band.

The decrease of the 1.8 eV band occurs along the slip bands [20] and has been associated to hole transfer from the $[Li]^0$ to the defects produced by the deformation.

3.2 Effect of thermally generated defects on flow stress

Fig. 2 illustrates the resolved flow stress for three samples: "pure" MgO, virgin MgO:Li, and oxidized MgO:Li crystals. The "pure" crystal was the optically clear type. The average resolved flow stress was determined to be 1.8 kp/mm², consistent with values reported by previous investigators [4, 9]. The average values for the virgin and oxidized MgO:Li crystals were 4.4 and 9.0 kp/mm², respectively. It is clear that the presence of lithium impurities increases the flow stress of MgO.

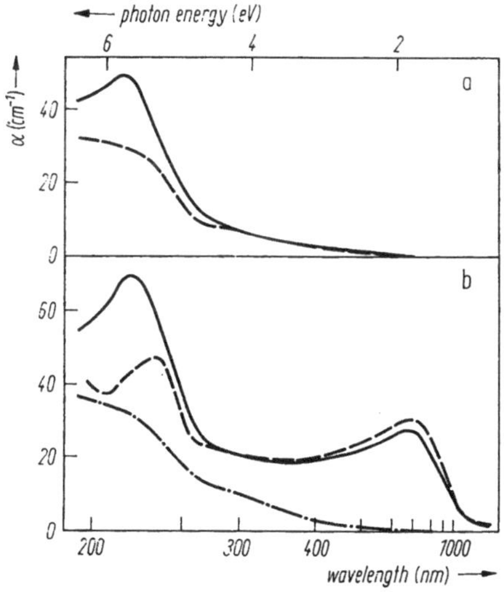

Fig. 1

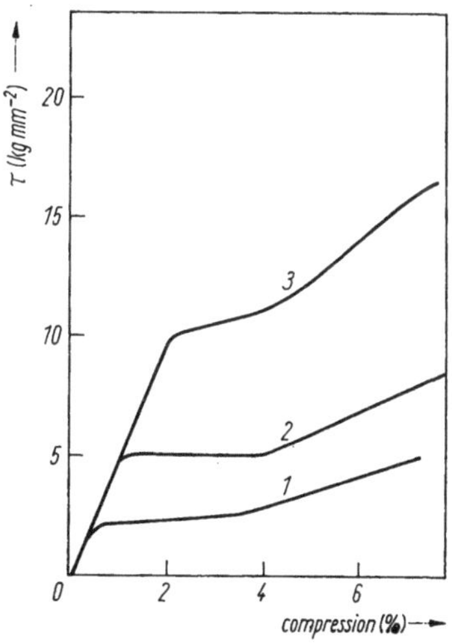

Fig. 2

Fig. 1. Optical absorption spectra of a a) ----- "virgin" MgO:Li crystal, ——— after 1.7% of plastic deformation; b) –·–·– "virgin" MgO:Li crystal, -------- after 10 min heating in air at 1500 K, ——— after 2% of plastic deformation

Fig. 2. Stress vs. strain curves for three samples: (1) a "pure" transparent MgO crystal, (2) a "virgin" MgO:Li crystal, (3) a MgO:Li crystal previously heated at 1825 K in O_2 for 25 min

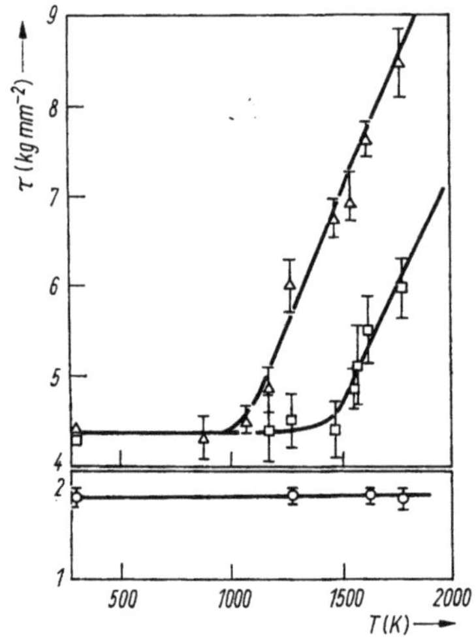

Fig. 3. Flow stress vs. incubating temperature for MgO:Li crystals. △ heated in oxygen, □ heated in nitrogen, ○ "pure" transparent MgO crystals. All the crystals were heated during 5 min

In order to study the annealing temperature dependence of MgO:Li on room-temperature flow stress, 5 min treatments at different temperatures up to 1800 K were carried in flowing O_2 and N_2 atmospheres. The results are plotted in Fig. 3. At temperatures below $T \approx 1100$ K the flow stress values remained unchanged, regardless of the atmospheres used and neither the 1.8 nor the 5.3 eV band was produced. For the oxidized crystals the 1.8 eV band appeared after the 1100 K anneal, and the 5.3 eV band after annealing at ≈ 1450 K. The flow stress of these specimens grew monotonically with annealing temperature beginning at $T \approx 1100$ K. On the other hand, samples heated in nitrogen atmosphere exhibited no 1.8 eV band, but the 5.3 eV band appeared at $T \approx 1450$ K and increased with rising annealing temperature. The corresponding flow stress curve likewise began to rise at 1450 K. Crystals heated in hydrogen or argon gave the same flow stress values as those heated in nitrogen.

To demonstrate that the temperature dependence of the flow stress in these crystals was indeed due to lithium impurities, the same procedure was followed with pure crystals grown by the same technique using similar high-purity MgO powder. The flow stress remained unchanged with thermal treatment in oxygen or hydrogen as shown in the bottom of Fig. 3.

3.3 Dependence of flow stress and optical-band generation on annealing time

The effect of the isothermal generation of the two optical bands on the flow stress was determined for different temperatures. Fig. 4a illustrates the evolution of the 1.8 and 5.3 eV bands at 1550 K in air (top), and the corresponding flow stress as measured for the same time intervals (bottom). Both bands grew during the early stage and reached saturation which persisted for long times. The measured flow stress responded in a similar manner.

Above 1700 K, a different isothermal behavior of the optical bands was observed. Fig. 4b (top) illustrated the absorption coefficient of the two bands at different time intervals when the crystals were heated at 1725 K. In contrast to the 1550 K isotherms in Fig. 4a, the 5.3 eV band initially had a larger absorption coefficient than the 1.8 eV band but after $t \approx 1$ h, a reversal in their relative intensities was observed.

3.4 Dependence of flow stress and optical-band annihilation on treatment time

Thermally generated lithium defects can also be thermally annihilated [19]. If the temperature is held slightly below the threshold temperature for formation, the 1.8 and the 5.3 eV bands can be readily annealed out. Specimens previously heat-treated at 1550 K in air to form the bands were annealed at 1075 K for different time intervals.

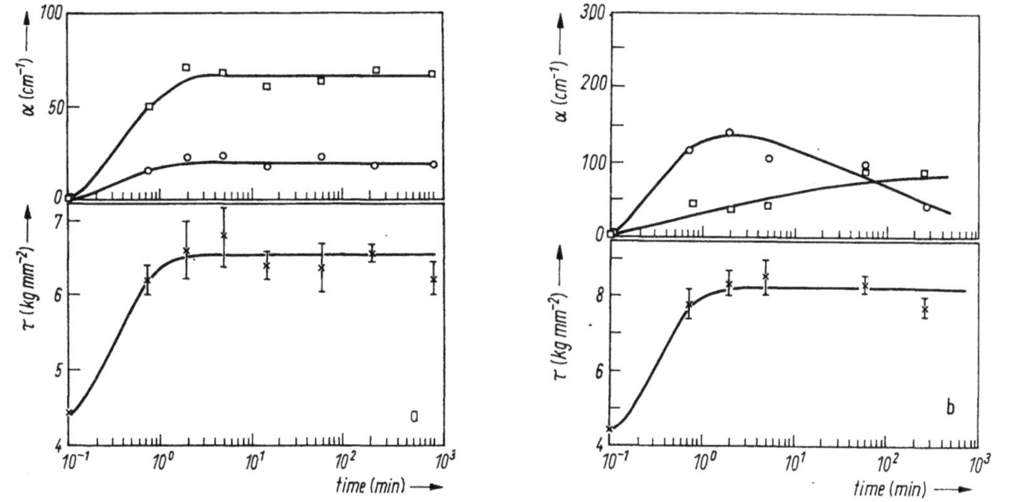

Fig. 4. Time dependence of the optical absorption coefficient on □ the 1.8 eV band, ○ the 5.3 eV band, and × the flow stress. Measurements resulting from annealing at a) 1550 and b) 1725 K in air

The results are shown in Fig. 5. In the top of the figure, the 5.3 eV band was observed to be more stable than the 1.8 eV band. When both optical bands were annealed out, at $\approx 10^2$ min, the flow stress resumed the virgin state value.

4. Discussion and Conclusions

In Fig. 2 we see that the presence of Li impurities in MgO single crystals modifies the stress–strain curves giving rise to changes in the mechanical properties of these crystals, particularly in the flow stress τ. The flow stress, on the other hand, is strongly dependent on the thermal treatment of the samples and on the atmosphere at which the crystals are heated (see Fig. 3).

Two stages of the hardening as a function of the temperature of treatment can be distinguished: Below 1100 or 1450 K in samples heated in an oxidizing or inert

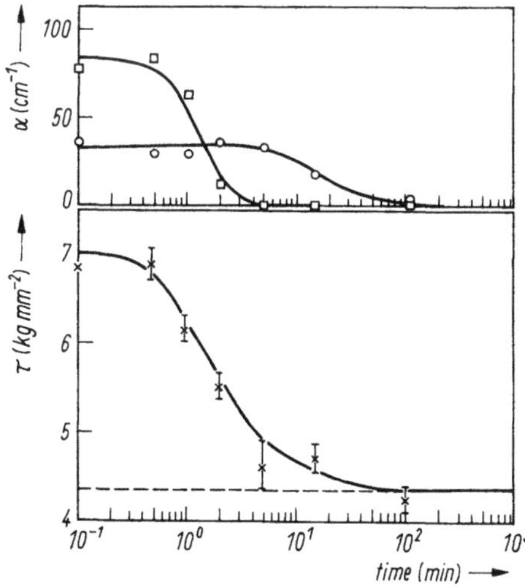

Fig. 5. Time dependence of the optical absorption coefficient on □ the 1.8 eV band, ○ the 5.3 eV band, and × the flow stress. Measurements due to isothermal annealing at 1075 K in air of a MgO:Li crystal previously heated at 1575 K in air

atmosphere, the resolved flow stress does not change from the value obtained for the as-grown (virgin) crystals. The increased hardening of the virgin MgO:Li crystals with respect to pure MgO is attributed to the Li_2O precipitates. ESR and optical studies suggest that the lithium from the precipitates does not dissolve in the MgO matrix to form Li defects at that stage. Above 1100 K in oxidized samples and above 1450 K for samples heated in an inert atmosphere a rise in the flow stress is detected. The fact that these temperatures coincide with the threshold temperatures for the formation of the 1.8 and the 5.3 eV bands, respectively, indicates that the defects associated with these bands are responsible for the enhanced hardening. A further verification of the association between the enhanced flow stress and the optical bands can be seen in the time dependence curves for defect formation (Fig. 4 a, b) and annihilation (Fig. 5).

The hardening produced by the presence of Li_2O precipitates can be explained if one assumes that the flow stress associated with the particles can be expressed as [22]

$$\tau \approx \frac{kGb}{\bar{l}}, \tag{1}$$

where k is a dimensionless coefficient which defines the strength of the interaction between the dislocation and the obstacle, G the shear modulus $\approx 1.2 \times 10^4$ kp mm^{-2} for MgO, b the magnitude of Burgers vector, and $\bar{l}$ the mean obstacle spacing.

The electron microscopy studies carried out on virgin MgO:Li [15] show that the Li_2O precipitates are coherent cubes in the matrix with average sides $\bar{a} \approx 220$ Å with a number density $n_p \approx 2.5 \times 10^{13}$ cm^{-3}. In MgO the active slip system is (110) ⟨110⟩ and $b \approx 3 \times 10^{-8}$ cm. In order to give a rough estimation of the mean interparticle distance $\bar{l}$ we may assume that the obstacles are all of the same size and distributed in a regular square array, in such a case

$$\bar{l} = \tfrac{1}{2} (1 + \sqrt{2})\, N_g^{1/2}, \tag{2}$$

where

$$N_g \approx n_p (\sqrt{2})\, \bar{a}. \tag{3}$$

Substituing (2) into (1) and if on takes $k \approx 1$ the calculated flow stress due to the Li_2O precipitates can be estimated to be ≈ 2.6 kp mm^{-2}. This result confirms our previous hypothesis that the increased hardening observed either in the as-grown or in the MgO:Li crystals heated at temperatures below the point defect formation threshold temperature can be associated to the presence of Li_2O precipitates.

The second stage of hardening which corresponds to samples heated in oxidizing atmospheres above 1100 K or in inert atmospheres above 1450 K is attributed to the presence of $[Li]^0$ defects for oxidized samples between 1100 and 1450 K, to the defect responsible for the 5.3 eV band for samples heated in an inert atmosphere above 1450 K, and to the combined effect of both defects for samples heated in oxidizing atmospheres above 1450 K.

In the study of the hardening produced by point defects an expression for the flow stress as given for

$$\tau = Kc^{1/2} \tag{4}$$

is generally obtained [23]. In (4) c is the defect mole concentration and K a constant related to the interaction of the dislocation with the defect.

In order to apply (4) we have found that the $[Li]^0$ defects, the defects responsible for the 5.3 eV band, and the Li_2O precipitates are present in different concentrations and thus contribute simultaneously to the hardening. In fact electron microscopy studies suggest that depletion of the precipitates occurs in air-heated samples at ≈ 1700 K (in that case we found both 1.8 and 5.3 eV bands) or above our temperature limit (≈ 1825 K) in nitrogen-heated samples.

We have compared the influence of the hardening of the $[Li]^0$ defects and of the defects responsible for the 5.3 eV band by using (4).

The concentration of $[Li]^0$ defects was measured by electron paramagnetic resonance (EPR) [17] using several standards [24]. An oscillator strength $f(1.8) \approx 0.1$ for the 1.8 eV band was obtained. In Fig. 6 the experimental flow stress is plotted versus the square root of the mole fraction for crystals which exhibit primarily the 1.8 eV band, the experimental points fall on a line represented by

$$\tau = A + Bc^{1/2} \tag{5}$$

being $A \approx 4.3$ and $B \approx 2.1 \times 10^2$ kp mm^{-2}.

The defect which produces the 5.3 eV band, on the other hand, is not paramagnetic and therefore the oscillator strength of the absorption band cannot be directly deduced from EPR measurements. We have to rely on indirect methods to estimate the oscillator strength of the optical absorption band.

Firstly, we note in Fig. 4 and 5 that changes in either the 5.3 or the 1.8 eV bands appear to affect the flow stress equally. If the oscillator strengths were drastically different from one another, we would likely have observed a preferential response. Secondly, during short isochronal anneals in O_2 atmospheres, the 1.8 eV band peaks after annealing at 1600 K and decreases for anneals above this temperature. However, the 5.3 eV band continues to increase up to annealing temperatures of approximately 1700 K. As shown in Fig. 4b, at 1725 K the 5.3 eV band grows rapidly initially, and then decreases. Similar behavior has been found at 1825 K which is the upper limit of our present apparatus. These results are compatible with our previous conclusion from TEM, EPR, and optical studies that the precipitates are gone at $T \approx 1700$ K, indicating that all or nearly all the lithium ions are in solution. Consider the isothermal generation of the two optical bands at 1725 K in air, such as shown in the top portion of Fig. 4b. The decrease in the 5.3 eV band is attended by the increase of the 1.8 eV band. If we make the assumption that this is due to one lithium ion annihilating one defect and creating another, we can obtain a ratio of the oscillator strengths for the two bands. Two sets of isothermal annealing data corresponding to 1725 K in Fig. 4b and 1825 K (not shown) were used. Using the measured half-widths of 1.4 and 0.9 eV for the 1.8 and 5.3 eV bands, respectively, we obtain the ratio of $f(5.3)/f(1.8) = 1 \pm 0.3$. Hence we shall use $f = 0.1$ to calculate the concentration of the 5.3 eV defects.

We have also plotted in Fig. 6 the flow stress versus the square root of the mole fraction for those crystals which were annealed in an inert or reducing atmosphere at $T > 1450$ K and hence do not contain the 1.8 eV band (squares). The points appear to lie on the same straight line as that corresponding to the $[Li]^0$ defects. In Fig. 6 the symbols △ represent those crystals which exhibit both bands where the total mole concentration is taken to be the sum of the mole fractions for each band. These points also lie on the same straight line.

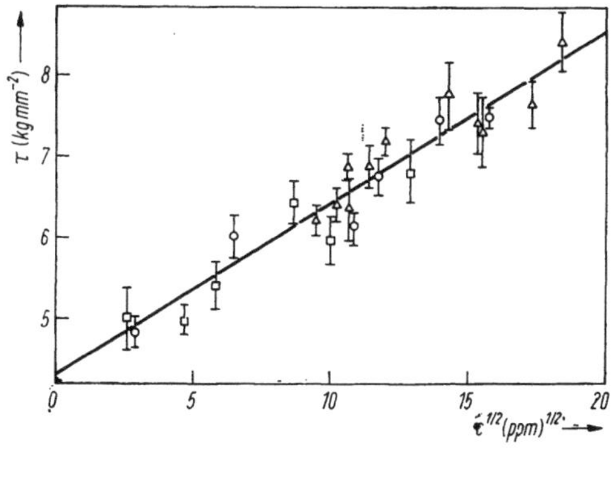

Fig. 6. Flow stress vs. the square root of mole fraction of defects in ○ crystals with 1.8 eV band, □ 5.3 eV band, and △ both bands

The extrapolated strength for MgO:Li with zero concentration of point defects is given by $A \approx 4.3 \times 10^2$ kp mm^{-2} which coincides with the experimental value obtained for virgin MgO:Li. The value of B that we have obtained by applying (5) is $\approx 2.1 \times \times 10^2$ kp mm^{-2}. This value has to be taken as tentative since we have not accounted for the depletion of the Li_2O precipitates. It can be compared to that of $\approx 4.9 \times \times 10^2$ kp mm^{-2} found by Ahlquist [23] in MgO containing Al^{3+} and Cr^{3+}, similar value can be found for Fe^{3+} but it is much smaller than the one observed for interstitials in electron irradiated pure MgO ($\approx 2 \times 10^3$ kp mm^{-2}) [4].

The solution hardening produced in MgO by the presence of the Al^{3+}–, Cr^{3+}–, and Fe^{3+}–magnesium vacancy complexes is due to the interaction of dislocations with elastic dipoles [25]. The same mechanism has been proposed for the radiation produced hardening [4]. If we assume that this mechanism is responsible for the hardening that we have observed in MgO:Li when point defects associated to the Li impurity are produced, we conclude that the defect which produces the 5.3 eV band is associated to some substitutional configuration of lithium ions rather than to an interstitial configuration.

Acknowledgements

I am indebted to the Solid State Division of ORNL where I was visiting in 1978 when this work was initiated, to Dr. Y. Chen for suggesting that work, and to Dr. Y. Chen and Dr. M. M. Abraham for helpful discussions.

References

[1] C. D. CLARK and J. H. CRAWFORD, Adv. Phys. **22**, 117 (1973).
[2] R. W. WHITWORTH, Adv. Phys. **24**, 203 (1975).
[3] W. A. SIBLEY, J. L. KOLOPUS, and W. C. MALLARD, phys. stat. sol. **31**, 223 (1969).
[4] W. C. McGOWAN and W. A. SIBLEY, Phil. Mag. **19**, 967 (1969).
[5] T. J. TURNER, N. N. ISENHOWER, and P. K. TSE, Solid State Commun. **7**, 1661 (1969).
[6] T. J. TURNER and T. E. SCHULTHEISS, J. Physique **34**, C9–231 (1973).
[7] Y. CHEN, M. M. ABRAHAM, T. J. TURNER, and C. M. NELSON, Phil. Mag. **32**, 99 (1975).
[8] Y. CHEN, M. M. ABRAHAM, L. C. TEMPLETON, and E. SONDER, Solid State Commun. **18**, 61 (1976).
[9] R. D. NEWTON and W. A. SIBLEY, phys. stat. sol. (a) **41**, 569 (1977).
[10] R. W. DAVIDGE, J. Mater. Sci. **2**, 339 (1967).
[11] M. SRINIVASAN and T. G. STOEBE, J. appl. Phys. **41**, 3726 (1970).
[12] Y. CHEN, M. M. ABRAHAM, J. L. BOLDÚ, and V. M. ORERA, J. Physique **41**, C6–398 (1980).
[13] Y. CHEN and M. M. ABRAHAM, J. Amer. Ceram. Soc. **59**, 101 (1976).
[14] M. M. ABRAHAM, C. T. BUTLER and Y. CHEN, J. chem. Phys. **55**, 3752 (1971).
[15] J. NARAYAN, M. M. ABRAHAM, Y. CHEN, and H. T. TOHVER, Phil. Mag. A **37**, 909 (1978).
[16] Y. CHEN, H. T. TOHVER, J. NARAYAN, and M. M. ABRAHAM, Phys. Rev. B **16**, 5535 (1977).
[17] J. L. BOLDÚ, M. M. ABRAHAM, and Y. CHEN, Phys. Rev. B **19**, 4421 (1979).
[18] M. M. ABRAHAM, Y. CHEN, L. A. BOATNER, and R. W. REYNOLDS, Phys. Rev. Letters **37**, 849 (1976).
[19] J. B. LACY, M. M. ABRAHAM, J. L. BOLDÚ, Y. CHEN, J. NARAYAN, and H. T. TOHVER, Phys. Rev. B **18**, 4136 (1978).
[20] V. M. ORERA, Y. CHEN, and M. M. ABRAHAM, Phil. Mag. **41**, 431 (1980).
[21] F. A. MODINE, Solid State Commun. **20**, 1097 (1976).
[22] J. FRIEDEL, Dislocations, Pergamon Press, Oxford, 1964.
[23] C. N. AHLQUIST, J. appl. Phys. **46**, 14 (1975).
[24] Y. CHEN, M. M. ABRAHAM, L. C. TEMPLETON, and W. P. UNRUH, Phys. Rev. B **11**, 881 (1975).
[25] T. E. MITCHELL and A. H. HEUER, Mater. Sci. Engng. **28**, 81 (1977).

(Received January 2, 1981)

phys. stat. sol. (a) 65, 301 (1981)

Subject classification: 10.1; 21.1.1

Ecole des Mines, Saint-Etienne[1])

Cyclic Plastic Deformation of B.C.C. Fe–26Cr Single Crystals at Room Temperature

By

T. MAGNIN, A. FOURDEUX, and J. H. DRIVER

The low cycle fatigue behaviour of single crystals of a high purity ferritic stainless steel (Fe–26Cr–1Mo) is studied at room temperature using both thermal activation theory and transmission electron microscopy. A correlation is established between the mechanical results and the evolution of the dislocation microstructure as a function of the number of cycles, the strain amplitude, and the strain rate $\dot{\varepsilon}_T$. Three different deformation regimes are defined for this alloy, (1) a low temperature regime when $\dot{\varepsilon}_T > 2 \times 10^{-3}\ s^{-1}$, for which plastic deformation is controlled by the movement of rigid screw dislocations, (2) a high temperature regime, when $\dot{\varepsilon}_T < 2 \times 10^{-5}\ s^{-1}$, for which curved and mixed dislocations are observed, and (3) an intermediate regime between the high and low temperature regimes. The consequences of the single crystal behaviour on the fatigue properties of polycrystals are discussed.

Le comportement en fatigue oligocyclique de monocristaux d'un acier inoxydable ferritique industriel de haute pureté (Fe–26Cr–1Mo) est étudié, à température ambiante, en utilisant d'une part la théorie de l'activation thermique, d'autre part l'analyse en microscopie electronique en transmission. Les résultats mécaniques sont interprétés par l'étude de l'évolution de la microstructure des dislocations en fonction du nombre de cycles, de l'amplitude de la déformation et de la vitesse de déformation $\dot{\varepsilon}_T$. Trois régimes de déformation sont définis: (1) un régime "basse température", lorsque $\dot{\varepsilon}_T > 2 \times 10^{-3}\ s^{-1}$, pour lequel la déformation est contrôlée par des dislocations vis rigides, (2) un régime "haute température", lorsque $\dot{\varepsilon}_T < 2 \times 10^{-5}\ s^{-1}$, pour lequel on observe la présence de dislocations mixtes et courbes, et (3) un régime intermédiaire entre les deux précédents. Enfin, les conséquences du comportement des monocristaux sur celui des polycristaux sont discutées.

1. Introduction

As a consequence of the special core structure of screw dislocations in the b.c.c. lattice, dislocation glide in b.c.c. metals often occurs on different slip systems in tension and compression [1 to 3]. Because of the thermally activated motion of screw dislocations, this asymmetry of slip is strongly temperature dependent [2], and is much more pronounced at temperatures below a critical temperature T_0, ($T_0 \approx 0.1$ to $0.2T_m$, T_m melting temperature) than at higher temperatures.

In recent papers [4 to 9], it has been shown how this slip asymmetry can lead to stress asymmetries and/or shape changes of b.c.c. crystals cyclically deformed in symmetric tension–compression. As suggested by Mughrabi [7], these crystal shape changes are of great importance for the mechanisms of fatigue crack initiation in b.c.c. polycrystals. This was clearly demonstrated in an earlier paper [8] which describes the influence of strain rate on the fatigue behaviour of single crystals of two ferritic alloys (Fe–3Si and Fe–26Cr–1Mo). Both the stress asymmetries and shape changes are very sensitive to strain rate, being very large at high strain rates ($\approx 10^{-2}\ s^{-1}$) but substantially reduced at low strain rates ($\approx 10^{-5}\ s^{-1}$). The aim of the present work is to correlate the mechanical behaviour of Fe–26Cr–1Mo single

[1]) 158, cours Fauriel, 42023 Saint Etienne Cédex, France.

crystals at room temperature with the change in the dislocation microstructure as a function of orientation, number of cycles, strain amplitude, and strain rate, using both thermal activation theory and transmission electron microscopy (TEM).

2. Experimental Procedure

The ferritic alloy examined in the present study is a stainless steel with (25.5 to 26.1) wt% Cr and (0.94 to 1.15) wt% Mo, containing about 15 ppm C and 25 ppm N. Single crystals of various orientations ([001], [011], [013], [$\bar{1}$49], [$\bar{1}$12], [$\bar{1}$11], and [$\bar{2}$34]) were grown by controlled solidification in horizontal furnaces. The fatigue specimens with a square cross-section (4×4 mm^2) and a gauge length of 10 mm were obtained by spark-cutting. They were first annealed at 1200 °C for 16 h in a vacuum of 10^{-5} Torr and then water-quenched after a second treatment at 950 °C for 30 min in argon. The crystals were brazed into threaded shoulders and electropolished before testing.

Symmetrical tension–compression tests were performed under total strain control in a servohydraulic machine, at constant strain rate $\dot{\varepsilon}_t$ (in the range 2×10^{-5} to $2 \times$ $\times 10^{-1}$ s^{-1}). The techniques of specimen testing are described in further detail in [8, 10].

The mechanical behaviour of single crystals was analysed using both (1) thermal activation theory and (2) TEM observations:

1) The internal (σ_i) and effective (σ^*) components of the flow stress σ ($\sigma = \sigma^* + \sigma_i$ [11]) and the activation volume V^* were determined by the cyclic strain rate change method due to Michalak [12] and Li [13] as indicated elsewhere [8]. For a nominal strain rate $\dot{\varepsilon}_1$, the values of σ^*, σ_i, and V^* may be expressed as

$$\sigma_1^* = \frac{\Delta\sigma}{\left(\frac{\dot{\varepsilon}_2}{\dot{\varepsilon}_1}\right)^{1/m^*} - 1}, \qquad \sigma_{1i} = \sigma_1 - \sigma_1^*,$$

$$V_1^* = kT \frac{\ln\left(\frac{\dot{\varepsilon}_2}{\dot{\varepsilon}_1}\right)}{\Delta\sigma} \frac{1}{S},$$

where $\Delta\sigma$ is the stress increment corresponding to the strain rate change, m^* the strain rate sensitivity factor (≈ 15 for all the orientations [10]), and S the Schmid factor.

2) TEM observations were carried out in a JEM 100C electron microscope operating at 100 kV. Slices oriented parallel to the slip plane and about 0.2 mm thick were sectioned from the specimen gauge length using an electrolytic saw, and were then mechanically thinned to 0.07 mm. Disks of 3 mm diameter were cut out by chemical polishing and final thinning was performed using a jet method with an electrolyte containing 90% CH_3COOH:10% $HClO_4$ at a temperature of about 10 °C.

3. Mechanical Results

3.1 Cyclic hardening

3.1.1 Cyclic hardening curves at constant strain rate $\dot{\varepsilon}_t = 2 \times 10^{-3}$ s^{-1}

Fig. 1 shows the variation of the flow stress σ with the cumulative plastic strain $\varepsilon_{p\,cum}$ ($\varepsilon_{p\,cum} = 2\Delta\varepsilon_p N$, where $\Delta\varepsilon_p/2$ is the plastic strain and N the number of cycles) for three different orientations as a function of strain amplitude. From this set of curves, we deduce that

1. some orientations exhibit a stress asymmetry Δ between tension (t) and compression (c). Δ, which is expressed by the ratio $2(\sigma_c - \sigma_t)/\Delta\sigma$ can be either positive (e.g. [001] crystals) or negative ([011] crystals);

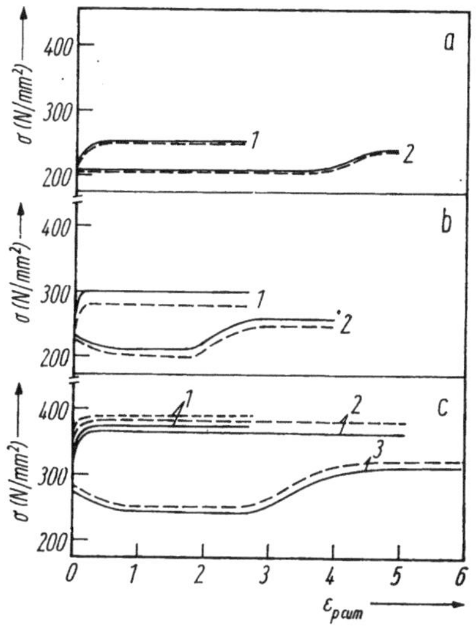

Fig. 1

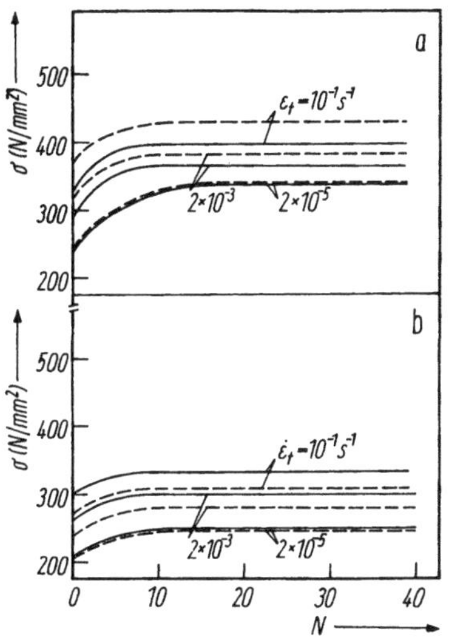

Fig. 2

Fig. 1. Cyclic hardening curves, $\sigma = f(\varepsilon_{p\,cum})$, of Fe–26Cr–1Mo crystals a) [$\bar{1}$49]; (1) $\Delta\varepsilon_t = 1.2 \times 10^{-2}$, $\Delta\varepsilon_{ps} = 9.6 \times 10^{-3}$; (2) $\Delta\varepsilon_t = 4 \times 10^{-3}$, $\Delta\varepsilon_{ps} = 2 \times 10^{-3}$, b) [001]; (1) $\Delta\varepsilon_t/2 = 6 \times 10^{-3}$, $\Delta\varepsilon_{ps}/2 = 3.8 \times 10^{-3}$; (2) $\Delta\varepsilon_t/2 = 2 \times 10^{-3}$, $\Delta\varepsilon_{ps}/2 = 8 \times 10^{-4}$. c) [011]; (1) $\Delta\varepsilon_t = 2 \times 10^{-2}$, $\Delta\varepsilon_{ps} = 1.5 \times 10^{-2}$; (2) $\Delta\varepsilon_t = 1.2 \times 10^{-2}$, $\Delta\varepsilon_{ps} = 7.6 \times 10^{-3}$; (3) $\Delta\varepsilon_t = 4 \times 10^{-3}$, $\Delta\varepsilon_{ps} = 1.7 \times 10^{-3}$. ——— compression, — — — tension

Fig. 2. Influence of strain rate on the cyclic hardening of Fe–26Cr–1Mo crystals cycled at $\Delta\varepsilon_t = 1.2 \times 10^{-2}$ for a) [011] and b) [001]. ——— compression, — — — tension

2. the cyclic hardening rate increases with strain amplitude $\Delta\varepsilon_t/2$ and is less pronounced for single crystals oriented for single slip ([$\bar{1}$49]) than for the others;

3. at low strain amplitudes ($\Delta\varepsilon_t/2 = 2 \times 10^{-3}$), a cyclic softening followed by a plateau is observed before the hardening occurs. In this case, the saturation regime is obtained after a cumulative plastic strain much greater than in the case of higher strain amplitudes.

3.1.2 Influence of strain rate on cyclic hardening

The influence of strain rate is shown in Fig. 2. In all cases the flow stress levels are sensitive to the strain rate. Moreover the cyclic hardening increases when $\dot{\varepsilon}_t$ decreases: for [011] crystals the hardening increases from ≈ 20 to 40% when $\dot{\varepsilon}_t$ changes from 10^{-1} to 2×10^{-5} s^{-1}.

It can also be noticed that the stress asymmetry decreases with the strain rate. Finally the strain rate also influences the cyclic softening behaviour at low $\Delta\dot{\varepsilon}_t$. Thus both the degree of cyclic softening and the length of the plateau between softening and hardening, are greater at high than at low strain rates.

3.2 Cyclic stress–strain curves

The influence of the crystallographic orientation on the saturation stress levels, σ_s, is presented in Fig. 3 for a range of total strain amplitudes of 2×10^{-3} to 10^{-2} and at a constant strain rate, (2×10^{-3} s^{-1}). These cyclic stress–strain curves were obtained using one specimen for each value of the imposed strain amplitude. It can be noted

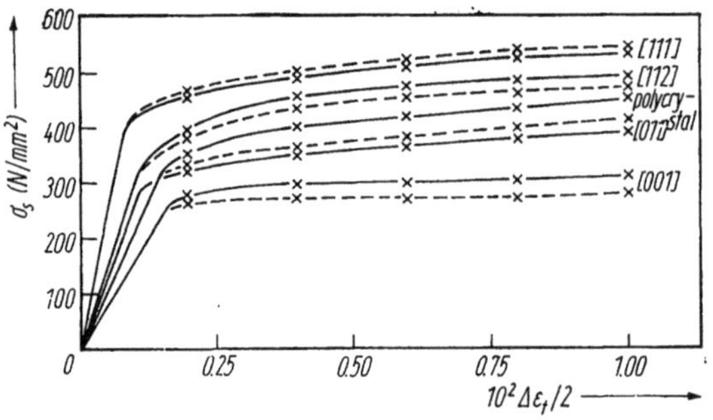

Fig. 3. Influence of orientation on the cyclic stress–strain curves at $\dot{\varepsilon}_t = 2 \times 10^{-3}\ s^{-1}$

that for a given value of $\Delta\varepsilon_t$, σ_s for the [111] crystals is about twice as high as σ_s for the [001] crystals. The influence of strain rate on these curves will be discussed later (Section 4.2).

3.3 Stress asymmetry and shape change

These two phenomena have been studied and discussed in an earlier paper [8] as a function of orientation, strain amplitude, and strain rate. They are both consequences of the special core structure of screw dislocations in the b.c.c. lattice. Stress asymmetry is observed in the following cases:

1. Slip occurs on the same systems in tension and compression but which are not crystallographically equivalent in both shear senses, e.q. {112} ⟨111⟩ and {123} ⟨111⟩.

2. Slip systems in tension are different from those in compression.

Shape changes due to cyclic deformation arise from the accumulated effect during cycling of changes in the slip systems and or the slip amplitudes between the tensiles and compressive parts of each cycle. The cross-sectional changes of the specimens during cycling have been studied in detail for [011] single crystals oriented such that the initially square cross-section become diamond-shaped [8]. The most important features of this behaviour are [8, 10]:

1. Stress asymmetry and shape change are very pronounced at high strain rates ($\approx 10^{-1}\ s^{-1}$) but much reduced at low strain rates ($\approx 10^{-5}\ s^{-1}$).

2. At a given strain rate the stress asymmetry and the rate of shape change decrease with the plastic strain amplitude.

3. Stress asymmetry is invariant with respect to the number of cycles for a given strain amplitude and strain rate. However, the rate of shape change (which persists during saturation) decreases at large cumulative plastic strains.

4. Thermal Activation Analysis

4.1 Cyclic hardening at $\dot{\varepsilon}_t = 2 \times 10^{-3}\ s^{-1}$

The variation of the internal (σ_i) and effective (σ^*) components of the flow stress σ was determined as a function of the number of cycles for several orientations. In fact the different orientations were found to behave very similarly and, since we are essentially concerned with interpreting the mechanical properties described in Section 3, we have chosen to only present the results obtained on the [011] crystals, (Fig. 4), which exhibit both stress asymmetry and shape change. It is obvious that hardening is due to an increase of σ_i and that σ^* remains nearly constant with N. This result emphasizes the role of long-range internal stresses during cycling.

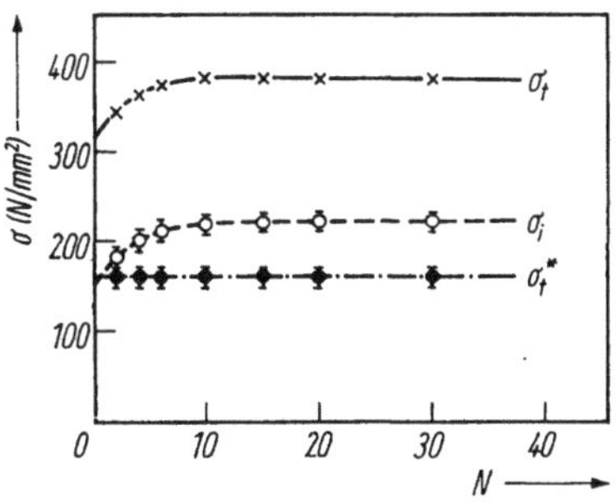

Fig. 4. Dependence of σ_t (the cyclic flow stress in tension), σ_i and σ_t^* on the number of cycles for [011] crystals cycled at $\Delta\varepsilon_t = 1.2 \times 10^{-2}$ and $\dot{\varepsilon}_t = 2 \times 10^{-3}\ s^{-1}$

4.2 Influence of strain rate

The values, at saturation, of the internal and effective stresses of [011] crystals are plotted against the strain amplitude $\Delta\varepsilon_t$ for two different strain rates $\dot{\varepsilon}_t = 2 \times 10^{-2}\ s^{-1}$ (Fig. 5a) and $\dot{\varepsilon}_t = 2 \times 10^{-5}\ s^{-1}$ (Fig. 5b). From these curves, the following remarks can be made:

1. At high strain rate ($2 \times 10^{-2}\ s^{-1}$) σ^* represents the major part of the flow stress.
2. At low strain rate ($2 \times 10^{-5}\ s^{-1}$), σ_i is the predominant component of the flow stress.

These results are in good agreement with those obtained on pure iron single crystals oriented for single slip [14, 15]. There is a complete reversal of the ratio σ^*/σ_i between high and low strain rates, which seems to indicate an important change in the deformation regime when $\dot{\varepsilon}_t$ varies from 10^{-1} to $10^{-5}\ s^{-1}$. This fact is clearly emphasized by the variation of the activation volume V^* with $\dot{\varepsilon}_t$ (Fig. 6). In this figure, it can be seen that V^*, which is constant with strain amplitude, decreases when $\dot{\varepsilon}_t$ increases ($V^* \approx \approx 120b^3$ at $\dot{\varepsilon}_t = 2 \times 10^{-5}\ s^{-1}$ and $40b^3$ when $\dot{\varepsilon}_t = 2 \times 10^{-2}\ s^{-1}$, $\boldsymbol{b}$ is the Burgers vector). Finally, it should be pointed out that the influence of orientation on the values of V^* is not very significant: for instance, at $\dot{\varepsilon}_t = 2 \times 10^{-3}\ s^{-1}$, $V^* \approx 55b^3$ for [011] crystals, $\approx 50b^3$ for [$\bar{1}$49] crystals, and $65b^3$ for [013] crystals.

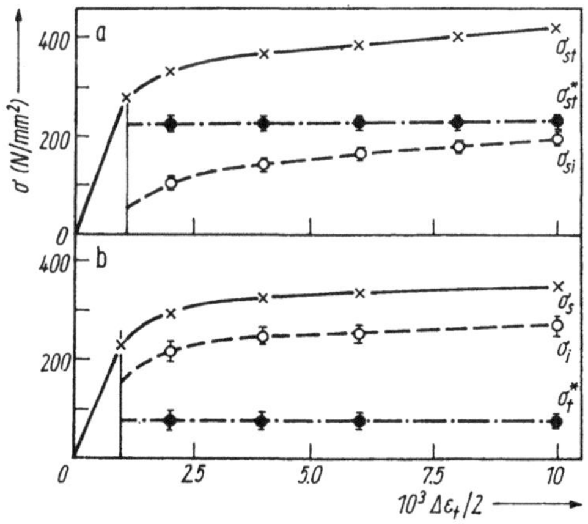

Fig. 5

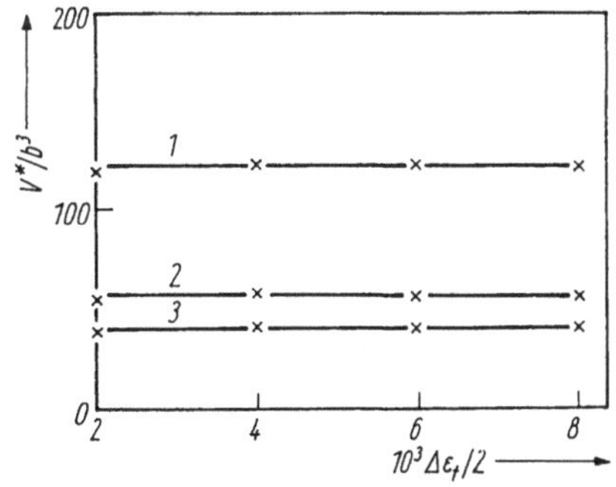

Fig. 6

Fig. 5. Dependence of σ, σ_i, σ^* at saturation on $\Delta\varepsilon_t/2$ for [011] single crystals cycled at a) $\dot{\varepsilon}_t = 2 \times 10^{-2}\ s^{-1}$ and b) $\dot{\varepsilon}_t = 2 \times 10^{-5}\ s^{-1}$

Fig. 6. Influence of strain rate on the activation volume of [011] single crystals. (1) $\dot{\varepsilon}_t = 2 \times 10^{-5}$, (2) 2×10^{-3}, (3) $2 \times 10^{-2}\ s^{-1}$

5. Microstructural Observations

The dislocation substructure was studied by TEM on two different orientations: [$\bar{1}$49] oriented for single slip on ($\bar{1}$01) planes, and [011] crystals oriented for double slip on ($\bar{2}$11) and (211) planes.

5.1 Dislocation arrangements at $\dot{\varepsilon}_t = 2 \times 10^{-3}\,s^{-1}$

Fig. 7 shows a general view of the dislocation arrangement on the slip plane of a [$\bar{1}$49] specimen cycled at $\Delta\varepsilon_t/2 = 6 \times 10^{-3}$ for 20 cycles. This stage corresponds to the end of hardening and just precedes saturation so that the micrograph reveals the different substructures leading to saturation. The dominant features are:

1. Elliptical dislocation loops with long jogged screw components, as detailed in Fig. 8a, and short edge dislocation dipoles.

2. Dislocation bundles composed of tangled edge dipoles, termed multipoles (Fig. 8b), and a large number of small prismatic loops which form debris (Fig. 8c).

The general view of Fig. 7 also indicates the beginning of cell formation (still very large and open) and that the dislocations between the bundles are predominantly of screw character. Similar configurations have been observed on the ($\bar{2}$11) slip planes of [011] crystals cycled at $\Delta\varepsilon_t/2 = 6 \times 10^{-3}$ and $\dot{\varepsilon}_t = 2 \times 10^{-3}$ for 15 cycles.

Fig. 9 illustrates the characteristic configuration obtained at saturation on the ($\bar{2}$11) slip plane of [011] crystals deformed at $\Delta\varepsilon_t/2 = 6 \times 10^{-3}$ for 500 cycles ($\varepsilon_{p\,cum} = 7.6$). A well-developed cell structure is observed with highly elongated cells and a smallest cell spacing of $\approx 0.5\,\mu m$. The cell walls are roughly perpendicular to the Burgers vector. A relatively low dislocation density can be seen between the cell walls. Finally it should be noted that most cells are completely closed in contrast with the more open structure observed in early saturation.

At small strain amplitudes, the density of mobile dislocations appears to be relatively low as indicated in Fig. 10a for [011] crystals cycled at $\Delta\varepsilon_t/2 = 2 \times 10^{-3}$ for 400 cycles. However, the dislocation density in the walls is so high that individual loops are not discernible. Finally it can be noted that dislocations between the walls are emitted from cell walls (Fig. 10b); their elongated shapes emphasize the predominance of screw over edge dislocations in the dislocation-poor regions.

5.2 Influence of strain rate on the dislocation configurations

The dislocation arrangements were examined on the ($\bar{2}$11) slip plane of [011] crystals cycled at $\Delta\varepsilon_t/2 = 6 \times 10^{-3}$ for 10 cycles at two different strain rates (2×10^{-2} and $2 \times 10^{-5}\,s^{-1}$). The following conclusions can be drawn:

(1) At $\dot{\varepsilon}_t = 2 \times 10^{-2}\,s^{-1}$ (Fig. 11a), the dislocation arrangement consists of a substantial density of long straight screw dislocations and of small dislocation clusters, the latter being distributed without any well-defined order.

(2) At $\dot{\varepsilon}_t = 2 \times 10^{-5}\,s^{-1}$ (Fig. 11b) the dislocation arrangement is quite different. The total dislocation density is higher than at $\dot{\varepsilon}_t = 2 \times 10^{-2}\,s^{-1}$, and long screw dislocations are not observed. Dislocation bundles composed of edge multipoles are well defined and oriented perpendicular to the Burgers vector. The dislocations between the bundles are very curved and of mixed character. This difference in dislocation arrangements at high and low-cyclic strain rates is in very good agreement with the results obtained by Mughrabi et al. [14] on pure iron single crystals oriented for single slip.

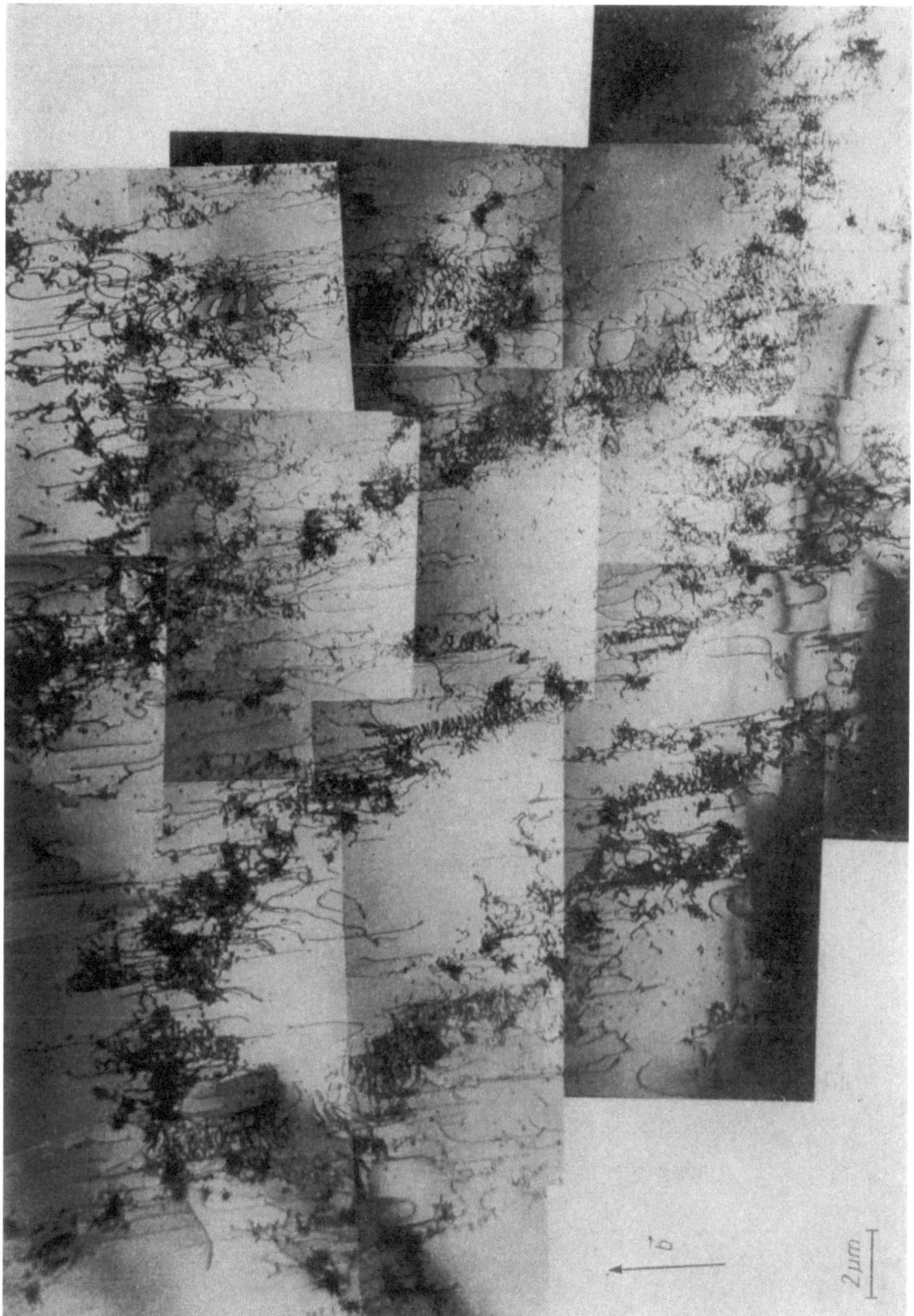

Fig. 7. Dislocation arrangement on the $(\bar{1}01)$ plane of a $[\bar{1}49]$ crystal after 20 cycles at $\Delta\varepsilon_t = 1.2 \times 10^{-2}$ ($\varepsilon_{p\,cum} \approx 0.4$) and $\dot{\varepsilon}_t = 2 \times 10^{-3}\ s^{-1}$

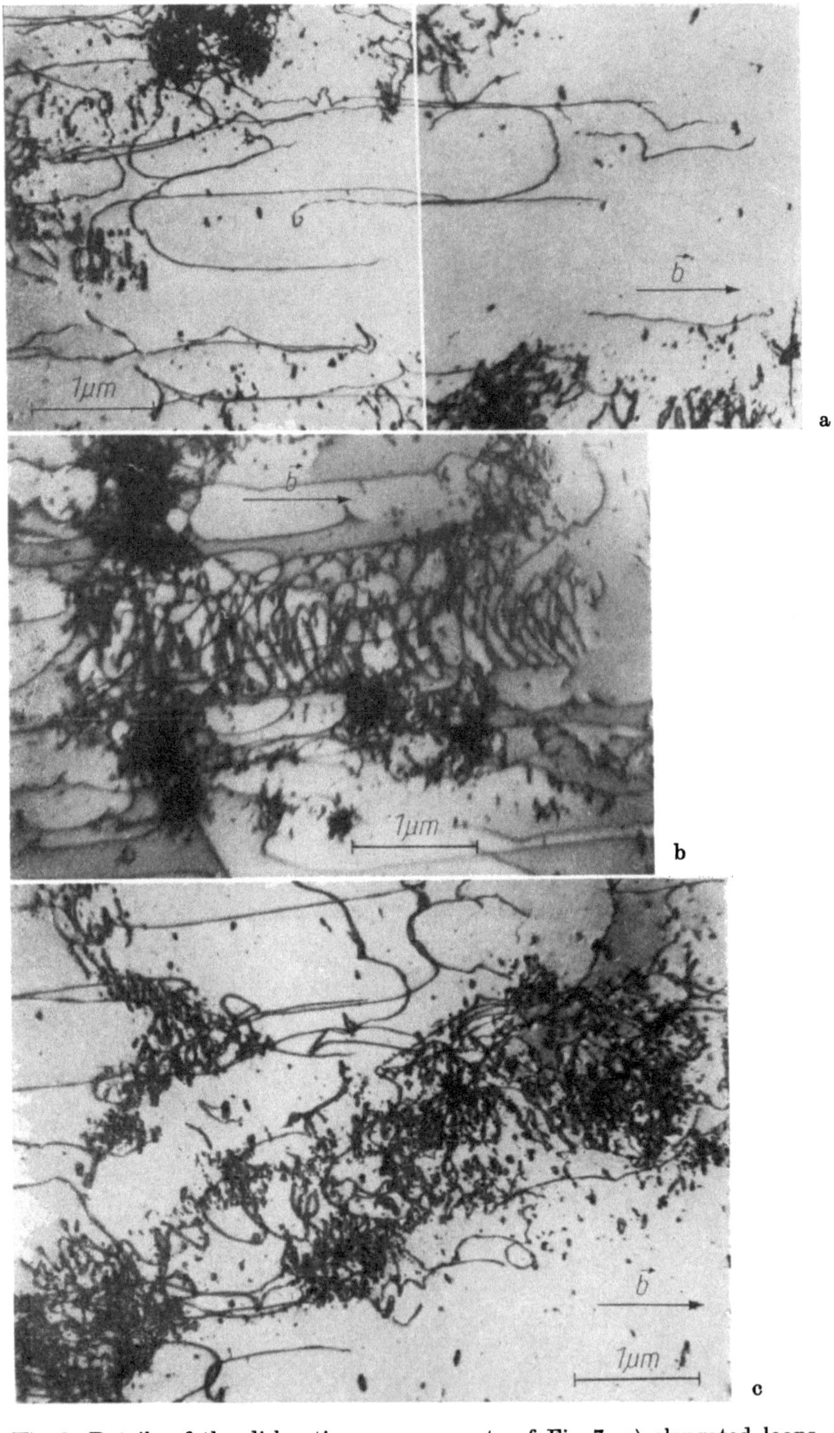

Fig. 8. Details of the dislocation arrangements of Fig. 7. a) elongated loops. b) multipoles, c) debris

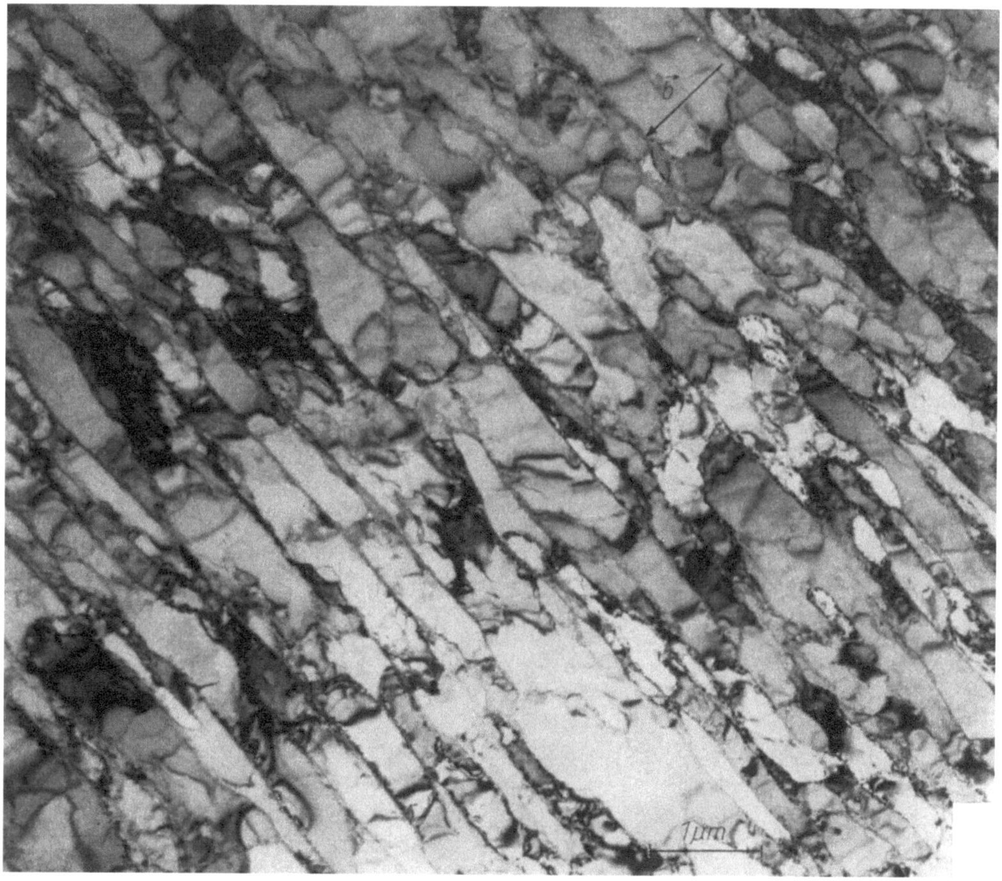

Fig. 9. Dislocation cells on the $(\bar{2}11)$ slip plane of a [011] crystal after 500 cycles at $\Delta\varepsilon_t = 1.2 \times 10^{-2}$ ($\varepsilon_{p\,cum} \approx 7.6$) and $\dot{\varepsilon}_t = 2 \times 10^{-3}\ s^{-1}$

6. Discussion

6.1 Fatigue behaviour of Fe-26Cr-1Mo crystals at $\dot{\varepsilon}_t = 2 \times 20^{-3}\ s^{-1}$

6.1.1 The deformation regime and consequences

Results obtained from thermal activation analysis have shown that, at $\dot{\varepsilon}_t = 2 \times 10^{-3}\ s^{-1}$, the long-range internal stress σ_i and the short-range effective stress σ^* of iron-chromium crystals are of the same order of magnitude and that the activation volume V^*, independent of strain, is about $55b^3$. From these results, the deformation regime is expected to correspond to an intermediate regime between the high and low temperature regimes. The friction stresses remain, however, relatively elevated. Electron microscopy confirms this idea. The presence of elliptical dislocation loops with large screw components indicates that the mobility of screw dislocations is lower than the mobility of screw dislocations. The deformation regime of Fe–26Cr at $\dot{\varepsilon}_t = 2 \times 10^{-3}\ s^{-1}$ is an intermediate temperature regime [16] closer however to the low than to the high temperature regime.

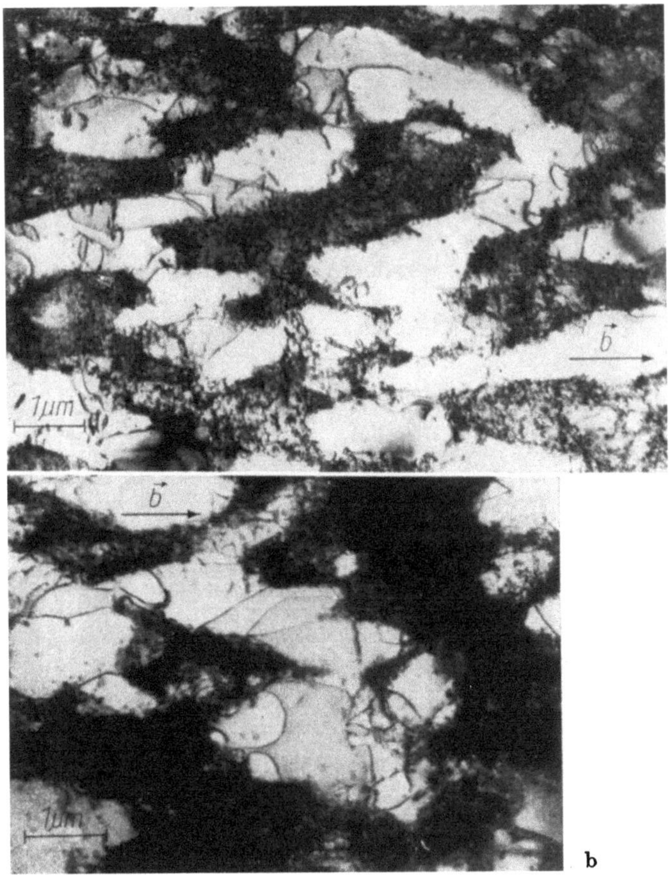

a

b

Fig. 10. Dislocation bundles in [011] crystals after 400 cycles at a relatively low strain amplitude $\Delta\varepsilon_t = 4 \times 10^{-3}$ ($\dot{\varepsilon}_t = 2 \times 10^{-3}\ s^{-1}$)

In this regime the deformation is largely, but not entirely, controlled by the thermally activated motion of screw dislocations (the double kink mechanism). Because of the special core structure of screw dislocations in the b.c.c. lattice a slip asymmetry occurs, often associated with a stress asymmetry, and leading to shape changes of cyclically deformed b.c.c. crystals. Since the slip asymmetry depends on the relative contribution of screw dislocations to glide, both shape change and stress asymmetry will decrease with the plastic strain amplitude because of the difference in the mobility of edge and screw dislocations.

Finally, it has been shown experimentally [5, 9, 20] that while crystals continue to change shape during saturation, the rate of shape change decreases at high values of $\varepsilon_{p\,cum}$. This behaviour cannot be attributed to a decreasing contribution of the screw dislocations to the deformation when $\varepsilon_{p\,cum}$ increases for two reasons: (i) electron microscopy has clearly shown that deformation at saturation is largely controlled by screw dislocations, and (ii) the stress asymmetry which depends on the screw dislocation contribution remains constant with $\varepsilon_{p\,cum}$. The rate of decrease of shape change with $\varepsilon_{p\,cum}$ can be explained in the following way: when $\varepsilon_{p\,cum}$ increases the internal stress increases, dislocation walls become more dense, dislocation cells are finer but remain

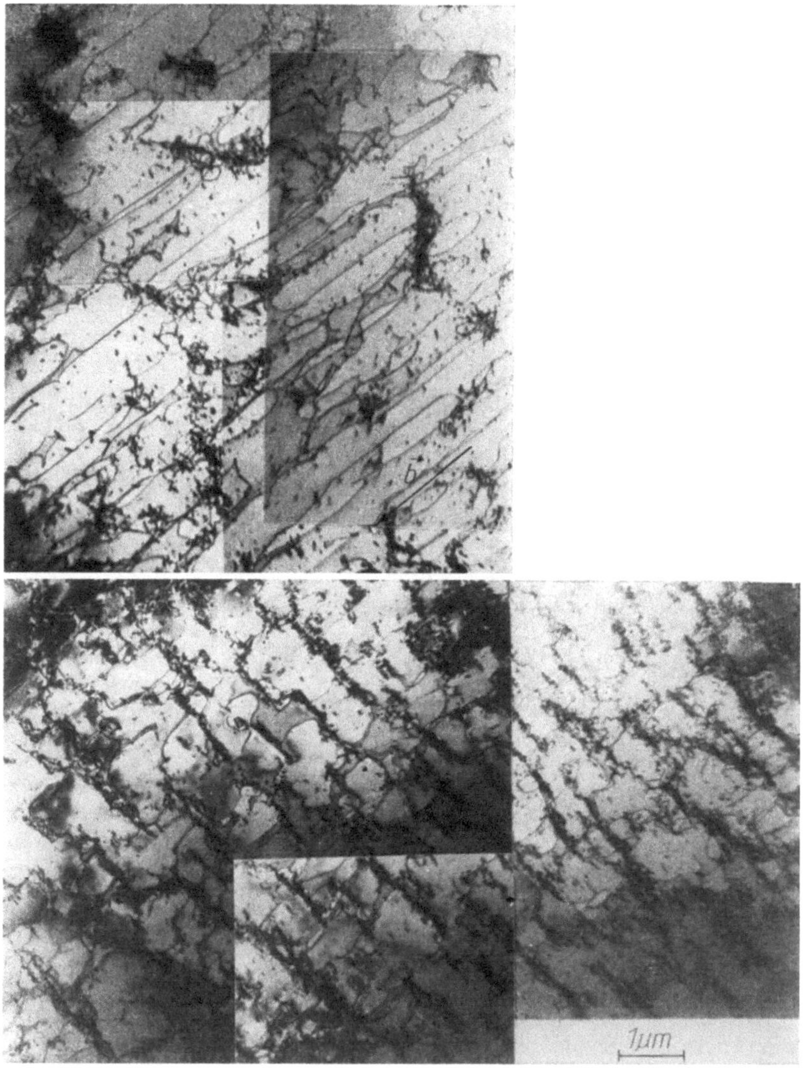

b

Fig. 11. Dislocation arrangements on the $(\bar{2}11)$ slip plane of [011] crystals cycled at $\Delta\varepsilon_t = 1.2 \times 10^{-2}$ for 10 cycles at a) $\dot{\varepsilon}_t = 2 \times 10^{-2}\ s^{-1}$ and b) $\dot{\varepsilon}_t = 2 \times 10^{-5}\ s^{-1}$. (In Fig. 11b the arrow ***b*** has the same direction as in Fig. 11a)

open in certain directions as clearly shown in TEM. Thus the inhomogeneous stress fields, induced by the open dislocation cells on screw dislocations moving between the cell walls, increasingly affect the screw dislocation motion when $\varepsilon_{p\,cum}$ increases. The change in slip systems between tension and compression becomes less favourable due to the inhomogeneity of the induced stress field. The crystal shape changes, which depend on slip asymmetry will therefore decrease at large cumulative plastic strains.

6.1.2 Cyclic hardening

Cyclic hardening results when a proportion of the mobile dislocations produced by cyclic strain interact in the crystal, forming new obstacles which impede the movement of dislocations. Thus the internal stress increases and dislocation multiplication is necessary to accommodate the deformation. It has been clearly shown that σ_i is responsible for the increase of the cyclic flow stress before saturation. The dislocation arrangement leading to cyclic hardening can be summarized as follows: during the first cycles, dislocations, mainly screws, interact, the dislocation density increases, jogs are formed on screw dislocations, and an edge dipolar structure may be developped. Dislocation clusters, built up of edge multipoles and debris, appear. On further cycling screw dislocations interact with the clusters, and are obliged to cross slip to overcome the clusters. The debris density increases leading to the growth of clusters [17]. Bundles are then formed from dislocation clusters by the same mechanism and become increasingly sharper. The subsequent stage is the formation of walls and cells which lead to the saturation state.

Cyclic softening, which precedes hardening when iron–chromium crystals are cycled at low strain amplitudes ($\Delta\varepsilon_t/2 = 2 \times 10^{-3}$) is probably related to the inter action between edge dislocations (at low strains the screw dislocation contribution is reduced) and interstitials (C and N) which anchor edge dislocations at the beginning of cycling. Softening can take place when a large number of edge dislocations is liberated. The deformation is then accomodated by the to-and-fro movement of the free dislocations (during the plateau) until the creation of new dislocations becomes necessary and leads to the hardening phase. At low strain rates the dislocation density is higher and the length of the plateau will be reduced, as experimentally observed.

6.1.3 Saturation stage

At large cumulative plastic strains saturation is characterized by a well-defined cell structure at high strain amplitudes ($\Delta\varepsilon_t/2 = 6 \times 10^{-3}$) and at low strain amplitudes by a high volume fraction of dislocation bundles. It is clear, however, that although the stress remains essentially constant during saturation, significant changes in the dislocation substructure can take place; the cell size and shape and the overall dislocation density change during saturation. It would appear therefore that saturation is not related to a particular dislocation arrangement but probably corresponds to an equilibrium between the interaction of dislocations with the cell walls and the liberation or creation of dislocations from the walls as shown by Fig. 10.

6.2 Influence of strain rate

Results obtained with thermal activation theory are in good agreement with the observations of the dislocation arrangements. Thus it has been shown that for [011] crystals cycled at $\Delta\varepsilon_t/2 = 6 \times 10^{-3}$, the ratio σ^*/σ_i varies from 0.3 to 1.4 when $\dot{\varepsilon}_t$ changes from 2×10^{-5} to 2×10^{-2} s^{-1}. Similarly the activation volume V^* is divided by a factor of 3 for the same variation in the strain rate. In addition TEM observations indicate that straight screw dislocations control the deformation at high strain rates and that mixed and curved dislocations are observed at low strain rates. From these results three different regions can be defined.

1. A high strain rate region ($\dot{\varepsilon}_t > 2 \times 10^{-3}$ s^{-1}), where cyclic deformation is controlled by rigid screw dislocations, the mobility of which is much lower than the mobility of edge dislocations. This dislocation configuration, the large values of σ^* and the low values of V^* suggest that the friction stress dominates. Thus, at room temperature and high strain rate, the deformation regime corresponds to the "low temperature regime". This regime is characterized by an asymmetric behaviour of screw dislocations between tension and compression leading to the mechanical properties of stress asymmetry and shape change. The sign of the stress asymmetry depends on crystal

orientation (it may be either positive or negative) and can be understood from slip system analysis [8]. Shape change, which is due to the irreversibility of plastic deformation between tension and compression, depends on the crystal orientation, and increases with strain amplitude and strain rate.

2. A low strain rate region ($\dot{\varepsilon}_t < 2 \times 10^{-5}\ s^{-1}$), where the thermal activation is important and the mobilities of screw and edge dislocations are of the same order. Curved dislocations accomodate the deformation and the internal stress component σ_i represents the major part of the flow stress. Thus, the regime of deformation corresponds to a "high temperature regime". The asymmetric behaviour between tension and compression is suppressed, shape change and stress asymmetry are negligible. The dislocation density is higher than in the high strain rate region, which leads to a more pronounced cyclic hardening. Bundle formation is more rapid than at high strain rates because of the ease of cross-slip which favours debris formation. This general behaviour can be compared with that of f.c.c. metals [18].

3. An intermediate strain rate region ($2 \times 10^{-5}\ s^{-1} < \dot{\varepsilon}_t < 2 \times 10^{-3}\ s^{-1}$), where the deformation is controlled by screw dislocations of increasing curvature, the difference in mobility between screw and edge dislocations decreasing as $\dot{\varepsilon}_t$ decreases. Internal and effective stresses are of the same order of magnitude and stress asymmetry and shape change are less pronounced than at high strain rates.

If we consider that the cyclic plastic deformation of Fe–Cr alloys is dependent on the thermally activated motion of screw dislocations (double kink mechanism), two critical temperatures (T_b and T_0) can be defined:

T_b below which the friction stress represents the major part of the flow stress.

T_0, ($T_0 > T_b$), such that between T_b and T_0 the friction stress decreases and that for $T > T_0$, the dislocation behaviour is nearly athermal. Thus, when the test temperature is kept constant, changing the strain rate leads to a variation of the values of T_b and T_0. If the flow stress can be described by an Arrhenius law, the evolution of T_0 with $\dot{\varepsilon}$ can be calculated

$$\Delta G(\sigma^*) = kT(\ln \dot{\varepsilon}_0 - \ln \dot{\varepsilon}) ,$$

where $\Delta G(\sigma^*)$ is the activation energy for double kink formation ($\Delta G(\sigma^*) \approx 0.75$ eV at $T = T_0$ for iron [19]) and $\dot{\varepsilon}_0$ the pre-exponential term. T_0 can be written as

$$T_0 = \frac{0.75}{k(\ln \dot{\varepsilon}_0 - \ln \dot{\varepsilon})} .$$

From TEM experiments, we can conclude that T_0 is about 300 K when $\dot{\varepsilon} \approx 10^{-4}\ s^{-1}$. Thus

$$T_0\,(\mathrm{K}) = \frac{8720}{19.8 - \ln \dot{\varepsilon}} .$$

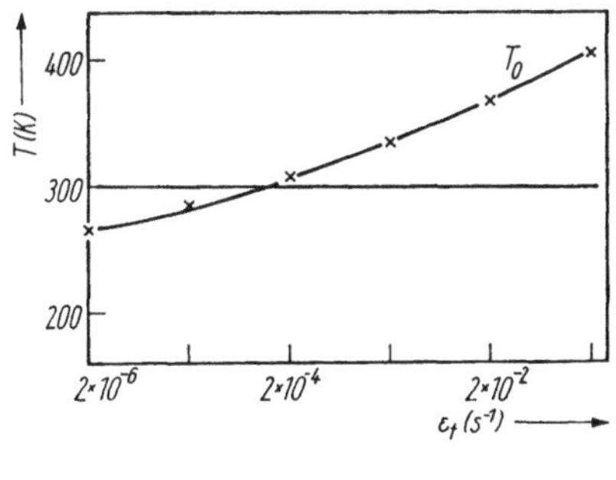

Fig. 12. Dependence of the athermal temperature T_0 on strain rate for Fe–26Cr–1Mo crystals

Fig. 12 gives the curve $T_0 = f(\dot{\varepsilon})$. It will be noticed that T_0 is about 100 K above room temperature when $\dot{\varepsilon}_t \approx 10^{-1}\ s^{-1}$ (low temperature regime at 300 K) and is below room temperature when $\dot{\varepsilon}_t \approx 10^{-5}\ s^{-1}$ (high temperature regime at 300 K).

6.3 Final remarks

It is evident that the pronounced effect of the cyclic strain rate on iron–chromium crystals will be of great importance on the fatigue behaviour of polycrystals. The strain rate effect on fatigue lifetimes and fatigue crack initiation mechanisms in polycrystals has been largely described in [8, 10]: crack initiation is intergranular at high strain rates (because of the shape changes of surface grains in the low temperature regime) and transgranular at low strain rates (no shape changes in the high temperature regime). Thus, an interpretation of polycrystal behaviour can be provided by the present single crystal study. The good agreement between TEM observations and thermal activation analysis suggests a correlation between the values of the ratio σ^*/σ_i (which can be easily calculated from a single simple mechanical test) and the regime of deformation. When $\sigma^*/\sigma_i < 0.8$, the deformation regime corresponds to the high temperature regime and when $\sigma^*/\sigma_i > 1.2$ to the low temperature regime. From this correlation, it is possible to propose a kind of criterion for the prediction of fatigue crack initiation in polycrystals, using the values of the ratio σ^*/σ_i .The validity of such a criterion has been successfully examined [20] on three high purity ferritic alloys (Fe–26Cr–1 Mo, Fe–3 Si, Fe–30 ppm C) as a function of strain rate, strain amplitude, grain size, and type of alloying elements.

Acknowledgements

The authors would like to thank Dr. H. Mughrabi and Dr. L. P. Kubin for several useful discussions and comments.

References

[1] J. W. CHRISTIAN, Proc. ICSMA2, Asimolar (Calif.), ASM 1970 (p. 31).
[2] V. VITEK, Crystal Lattice Defects **5**, 1 (1974).
[3] B. SESTAK, Mater. Sci. Engng. **25**, 171 (1976).
[4] R. NEUMANN, Z. Metallk. **66**, 26 (1975).
[5] H. MUGHRABI and CH. WUTHRICH, Phil. Mag. **33**, 963 (1976).
[6] T. MAGNIN, J. DRIVER, and J. RIEU, Mém. sci. Rev. Métall. **75**, 293 (1978).
[7] H. MUGHRABI, Z. Metallk. **66**, 719 (1975).
[8] T. MAGNIN and J. H. DRIVER, Mater. Sci. Engng. **39**, 175 (1979).
[9] M. ANGLADA, and F. GUIU, Proc. ICSMA5, Aachen, Vol. 2, Pergamon Press, Oxford/Toronto 1979 (p. 1237).
[10] T. MAGNIN, Thèse Doctorat d'Etat, University of Grenoble, 1980.
[11] A. SEEGER, Phil. Mag., Ser. 7 **45**, 771 (1954).
[12] J. T. MICHALAK, Acta metall. **13**, 213 (1965).
[13] J. C. LI, Canad. J. Phys. **45**, 493 (1967).
[14] H. MUGHRABI, K. HERZ, and X. STARK, Acta metall. **24**, 659 (1976).
[15] H. MUGHRABI, K. HERZ, and X. STARK, Proc. ICSMA4, Nancy, Vol. 3, 1976 (p. 1244).
[16] F. LOUCHET, L. P. KUBIN, and D. VESELY, Phil. Mag. **A39**, 433 (1979).
[17] H. MORI, M. TOKUWANE, and T. MIYAZAKI, Phil. Mag. **A40**, 409 (1979).
[18] H. MUGHRABI, F. ACKERMANN, and K. HERZ, Fatigue Mechanisms, ASTM-STP, n° 675, 1979 (pp. 675, 69).
[19] J. P. COTTU, J. P. PEYRADE, P. CHOMEL, and P. GROH, Acta metall. **26**, 1179 (1978).
[20] T. MAGNIN and J. H. DRIVER, Proc. ICF5, Cannes 1981, Vol. 1 (p. 307).

(Received November 20, 1980)

phys. stat. sol. (a) 65, 315 (1981)

Subject classification: 2; 21

Institute for Solid State Physics, The University of Tokyo[1])

An Interpretation of the Pair Distribution Function of a Metallic Amorphous Structure

By

S. TAKEUCHI and S. KOBAYASHI

The characteristic shape of the pair distribution function (PDF) of metallic glasses is interpreted in terms of the tetrahedral close packing. Average distances and coordination numbers of atomic shells are estimated by a geometrical consideration. The PDF obtained after the Gaussian broadening agrees well with the experimental one both in positions and relative intensities of the peaks.

Die charakteristische Form der Paar-Verteilungsfunktion (PVF) der metallischen Gläser werden mit einer dichten Tetraederpackung erklärt. Die mittleren Abstände und Koordinationszahlen der Atomschalen werden durch eine geometrische Betrachtung berechnet. Die PVF, die nach Gauss-Verbreiterung erhalten wird, stimmt gut mit den experimentellen Werten der Lagen und relativen Intensitäten der Maxima überein.

1. Introduction

It has been shown that the pair distribution function (PDF) of metallic glasses, especially of monatomic metallic glasses, exhibits a common characteristic feature [1 to 3]. Various modellings have been tried to simulate the amorphous structure from the point of the dense random packing of hard spheres (DRPHS) by many authors [4 to 9]. In recent years, computer modellings through the relaxation of a random aggregate of atoms using pair interatomic potentials have succeeded in reproducing the experimental PDF's fairly well [10 to 16]; the calculated PDF's being rather insensitive to the potentials employed. Geometrically, some interpretations of the appearance of the second and third peaks (often referred to as the first and second components of the split second peak) have been given [7, 17], but no satisfactory explanation seems to have been offered to further peaks. The purpose of the present paper is to interpret the overall shape of the PDF based on a geometrical consideration using the concept of the tetrahedral close packing (TCP), a term used by Dixmier and Sadoc [18]. Here, the term "tetrahedral" means the smallest unit of clusters composed of four atoms, and is different from the tetrahedral bonding characteristics in semiconductors.

Briant and Burton [19] proposed from their computer simulation that the structure of metallic glasses can be regarded as an aggregate of 13-atom icosahedra. Hoare and Pal [20] have shown that the icosahedral arrangement is the lowest-energy configuration of 13-atom clusters. An icosahedron can be further decomposed into 20 tetrahedral units, which are close to the regular tetrahedron. If we consider every atom, the central atom or the surrounding atom of an icosahedron, as statistically equivalent, then the amorphous structure may be regarded as an aggregate of tetrahedral units. Incidentally, it has been verified by Pinsker [21] that the three-dimensional space can be filled up with trigonal bipyramids composed of two tetrahedra.

2. Modelling

Let us take the origin at an atom in an amorphous structure. We consider a sequence of tetrahedral clusters connected in the radial direction, sharing each triangular face.

[1]) Roppongi, Minato-ku, Tokyo 106, Japan.

The shape of each tetrahedral unit will deviate considerably from the regular tetrahedron, but if we consider a large ensemble of such sequences, the average positions of atoms will constitute a sequence of regular tetrahedra with the edge length equal to the average nearest-neighbour distance r_1. The sequence of the first nine regular tetrahedra is shown in Fig. 1, which is constructed under the principle that any new

Table 1

Radial distances of peaks in PDF for monatomic amorphous metals [2] and those of shells in the present model in units of r_1, and the coordination numbers estimated

shell No.		1	2	3	4	5	6
radial distance	exp.	1.00	1.67 ± 0.02	1.93 ± 0.03	2.52 ± 0.04	(≈3.00)	3.40 ± 0.05
	model	1.00	1.65	1.99	2.49	2.97	3.38
coordination number of model		13.4	30.8	13.4	≈91.7	≈13.4	≈182.0

tetrahedron is put on one of three faces of the previous tetrahedron making contact face to face, so that the vertex of the new tetrahedron comes to the farthest distance from the origin. Such a "spiral-polytetrahedral" cluster has already been considered by Bernal [22] as "pseudonucleus" in the DRPHS model of liquids. The distances from the origin to vertices of tetrahedra are also given in Fig. 1. Suppose the atoms around an atom are divided into shells, then the average distance to the atoms belonging to each shell will be assumed to correspond to each distance given in Fig. 1. Regarding the close distances, OB_1 and OB_2, OD_1 and OD_2, and OF_1 and OF_2 as belonging to the same shell, respectively, the radii of the first, second, third, fourth, fifth, and sixth shells are 1, 1.65, 1.99, 2.49, 2.97, 3.38, respectively, in units of r_1. In Table 1, these values are compared with the average peak positions of experimental PDF's for many deposited monatomic amorphous metals compiled by Leung and Wright [2]. The fifth peak has not been assigned so far but is recognizable in some cases [3]. It is found that the agreement of the peak positions is excellent.

In the next step, we must consider the way of filling the space with tetrahedra to estimate the coordination numbers for each shell. The typical first shell may form the well-known 13-atom icosahedra with twelve five-ray vertices and 20 faces, and thus a typical coordina-

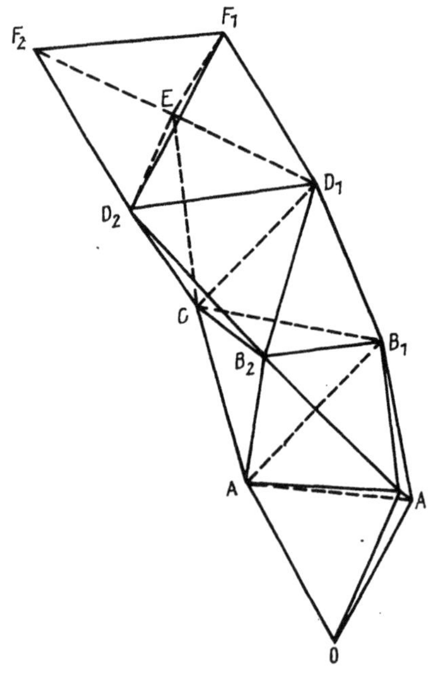

Fig. 1. Average radial sequence of tetrahedral units in an amorphous structure. $OA \equiv r_1$, $OB_1 = 1.6330r_1$, $OB_2 = 1.6667r_1$, $OC = 1.9907r_1$, $OD_1 = 2.4369r_1$, $OD_2 = 2.5402r_1$, $OE = 2.9650r_1$, $OF_1 = 3.2902r_1$, $OF_2 = 3.4797r_1$

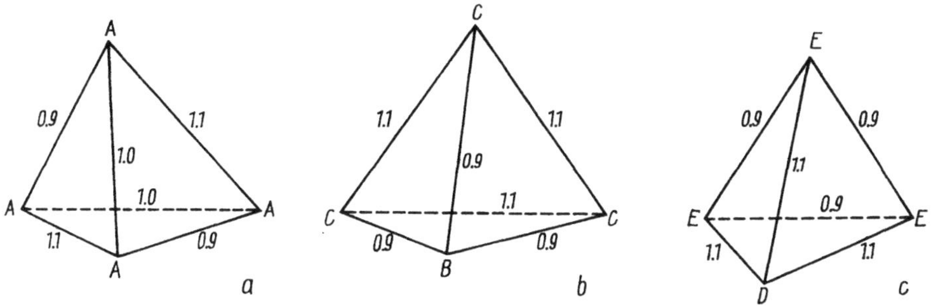

Fig. 2. Three variations of a tetrahedron. The lengths of the edges are shown in units of r_1. Solid angles of apices are: $\Omega_A = 0.5201$, $\Omega_B = 0.9607$, $\Omega_C = 0.4434$, $\Omega_D = 0.3379$, and $\Omega_E = 0.6362$

tion number of the first shell may be twelve. However, the value of 20 times the solid angle Ω_0 ($= 0.5513$) of a vertex of a regular tetrahedron is 11.03 which is considerably smaller than 4π. At this stage, we should take the distribution of tetrahedral shapes in the amorphous structure into account to estimate the average solid angle of vertices. Taking account of the typical distribution $r_1 \pm 0.1r_1$ of the nearest neighbour distance, we consider three typical variations of distorted tetrahedra, as shown in Fig. 2. The average solid angles of four vertices for a), b), and c) in the figure are calculated to be 0.5201, 0.5727, and 0.5616, respectively. The further average of these three values is 0.5515, which is very close to Ω_0. Hence, it seems reasonable to assume that the average number of tetrahedra constituting the first shell polyhedron is $4\pi/\Omega_0 = 22.8$. The average coordination number of the first shell n_1 or the average number of vertices of the first shell polyhedron is then obtained based on Euler's theorem, i.e.

$$n_1 - \tfrac{1}{2}(3 \times 22.8) + 22.8 = 2$$

which yields $n_1 = 13.4$.

The second shell may be composed of atoms at apices of tetrahedra added on the first-shell polyhedron, face to face in contact with those in the first shell, and the third shell is formed by filling in between with further tetrahedra. Taking account of the requirement of the statistical equivalence between the central atom and atoms in the first shell, the polyhedron constituted by the first-, the second-, and the third-shell atoms is considered to be constructed by interpenetrating new first-shell polyhedra centred at every corner of the original first-shell polyhedron. Fig. 3 shows such

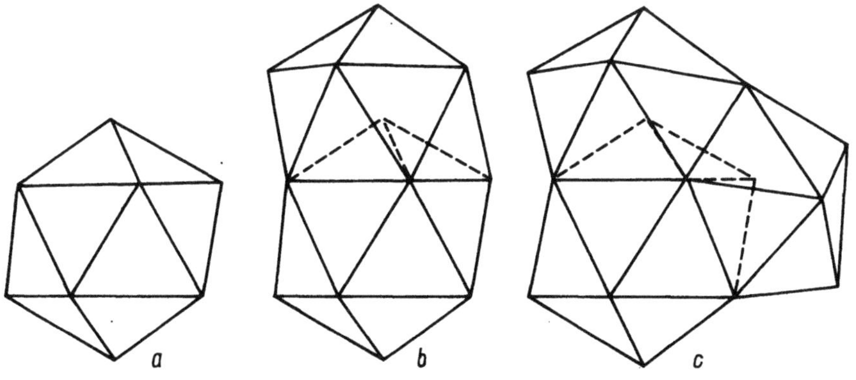

Fig. 3. The process of interpenetration of icosahedra centred at the corners of the original one

a process of interpenetration of polyhedra which are drawn in this case by icosahedra for clarity. The value of n_2 can be estimated in the following way: Let us call edges of tetrahedra as bonds and denote the number of bonds of an atom in the i-th shell that connect with the atoms in the j-th shell as n_i^j. Recalling that the average number of triangular faces of the first-shell polyhedron is 22.8 and that of vertices 13.4, the average value of n_1^1 is obtained as $n_1^1 = 22.8 \times 3/13.4 = 5.10$. Due to almost collinear configuration of the central atom, the first-shell atom and the third-shell atom (see, Fig. 1), n_1^3 value is most probably unity. Thus, owing to the statistical equivalence of the central atom and the first-shell atom, the n_1^2 value should be

$$n_1^2 = 13.4 - (n_1^1 + n_1^0 + n_1^3) = 6.3\,,$$

where the superscript 0 means the central atom. This requires the average total number of n_1^2 to be $6.3 \times 13.4 = 84.4$. Now, the n_2^1 value for each of the second-shell atoms at the apex of tetrahedra which share a face with the first-shell polyhedron is three. Hence, the total n_2^1 value of these second-shell atoms amounts to $22.8 \times 3 = 68.4$, which is less than the required value of 84.4 by 16.0. Thus, extra second-shell atoms must be squeezed in the above 22.8 second shell-atoms. In contrast to the above 22.8 "capping" second-shell atoms, the n_2^1 value of each "squeezed-in" atoms is two, as illustrated in Fig. 4, where I and II denote capping atoms and squeezed-in atoms, respectively. Therefore, to fulfil the equivalence of the central atom and the first-shell atoms, eight more second-shell atoms must be added, resulting in $n_2 = 22.8 + 8 = 30.8$.

The value of n_3 may be assumed to be equal to n_1 due to the collinear arrangement, as already mentioned. It is hard to construct bigger polyhedra composed of larger shells, but coordination numbers of further shells may be estimated from the condition that the spatial density should be equal everywhere. The density estimated from the region within the third-shell distance is about 1.37. The density here is defined by the number of atoms in a cube with the dimension of r_1. We assume further that coordination numbers of odd shells, n_5 and n_7, are equal to n_1 for the same reason as n_3. Thus, n_4 and n_6 are estimated as presented in Table 1 to give the homogeneous density of 1.37.

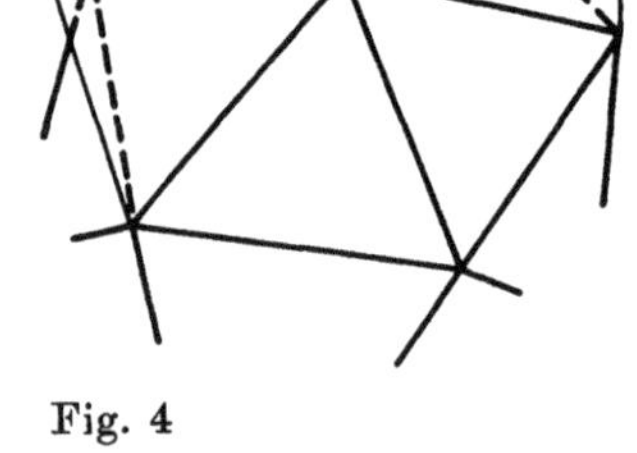

Fig. 4

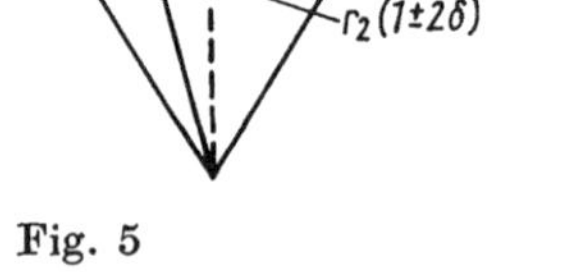

Fig. 5

Fig. 4. Schematic representation of two kinds of second-shell atoms: capping atoms (I) and a sequeezed-in atom (II)

Fig. 5. A bipyramidal unit showing that a variation of the nearest-neighbour-distance by $r_1(1 \pm \delta)$ results in the variation of the diagonal distance by $r_2(1 \pm 2\delta)$

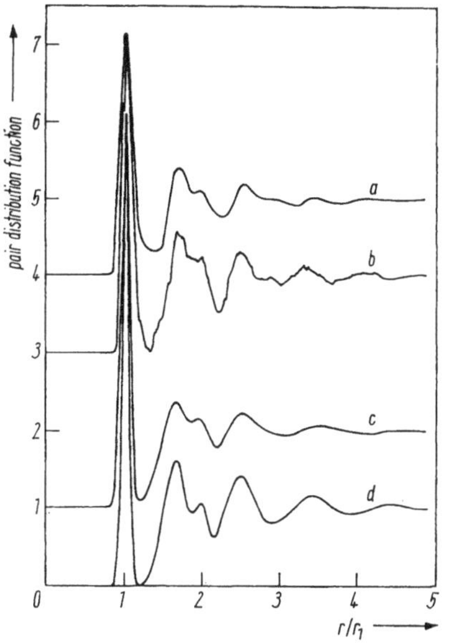

Fig. 6. (a) shows experimental PDF of amorphous iron [3]; (b) the PDF of model amorphous iron simulated by a computer [15]; (c) the PDF obtained in the present model for distribution widths $\Delta r_{\text{odd}}/r_{\text{odd}} = 0.08$ and $\Delta r_{\text{even}}/r_{\text{even}} = 0.16$; (d) for $\Delta r_{\text{odd}}/r_{\text{odd}} = 0.065$ and $\Delta r_{\text{even}}/r_{\text{even}} = 0.13$, where r_{odd} and r_{even} are the radial distances of the odd and even shells, respectively, and $\Delta r_{\text{odd (even)}}$ means the standard deviation of the Gaussian distribution

The scatter of atomic positions around the average distance of a shell will become larger as the radial distance of the shell increases; the standard deviation of the distribution is set to be a linear function of the distance. However, an important difference in the width of the distribution should be noticed between the shells of collinear atomic configuration and other shells, if the broadening arises mainly from the variation of the nearest-neighbour distances, which seems to be the case in metallic glasses. In a trigonal bipyramid as shown in Fig. 5, it can readily be calculated that the variation of the nearest-neighbour distance by $r_1(1 \pm \delta)$ results in the variation of the second-nearest-neighbour distance by $r_2(1 \pm 2\delta)$. Thus, the relative distribution width, $\Delta r_i/r_i$, of the second peak must be twice as large as that of the first peak. In the following Gaussian broadening procedure of the PDF, the $\Delta r_i/r_i$ values of all even shells are set double that of all odd shells of nearly collinear coordination. Fig. 6 (c) and (d) are PDF's thus obtained for two sets of $\Delta r_i/r_i$ values. For comparison an experimental result of iron [3] and an example of computer modelling [15] are shown in Fig. 6 (a) and (b), respectively. It is found that the present modelling is in overall agreement with experiment and computer modelling.

3. Discussion

The present trial seems to be the first one to reproduce successfully the total PDF of a metallic amorphous structure based purely on geometrical considerations, without relying on computer simulation. Though the present derivation depends to some extent on intuitive reasoning, good agreement of the result with experiment may indicate the validity of the concept of the TCP employed in the present modelling.

If an amorphous alloy is a random mixture of two species of atoms with a small difference in size, as is thought to be the case for metal–metal amorphous alloys, then the structure can be regarded as composed of more distorted tetrahedra than in the case of a monatomic amorphous metal. As a result, the total PDF of such alloy will show a curve similar to the present one but with broadened peaks. On the other hand, if an alloy comprises some chemical order as is thought to be the case in metal–metalloid alloys, then the form of the PDF will deviate considerably from the present result [23].

An important point in the present derivation of the realistic PDF is the difference in distribution widths between collinear and other coordination shells. At high temperature or in the liquid state, the ratio of them will become close to unity due to the equal contribution of large thermal broadening. It is interesting to note that as the

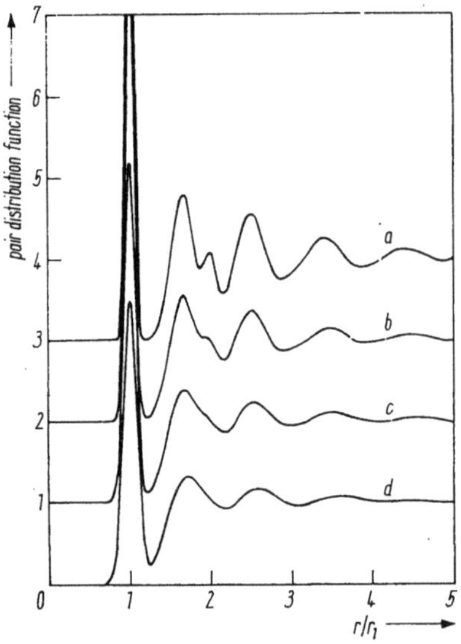

Fig. 7. PDF's for different distribution widths for odd and even shells. $\Delta r_{odd}/r_{odd} = 0.06 + \alpha$, $\Delta r_{even}/r_{even} = 0.12 + \alpha$; (a) $\alpha = 0$, (b) 0.02, (c) 0.04, (d) 0.06

ratio approaches unity the second and third peaks of the calculated PDF merge into a single peak as illustrated in Fig. 7; i.e. the so-called second peak splitting disappears, as the experimental PDF's in liquid metals show [24].

References

[1] L. B. DAVIES and P. J. GRUNDY, phys. stat. sol. (a) **8**, 189 (1971).
[2] P. K. LEUNG and J. G. WRIGHT, Phil. Mag. **30**, 185, 995 (1974).
[3] T. ICHIKAWA, phys. stat. sol. (a) **19**, 707 (1973).
[4] J. D. BERNAL, Nature (London) **183**, 141 (1959).
[5] G. D. SCOTT, Nature (London) **194**, 956 (1962).
[6] D. J. ADAMS and A. J. MATHESON, J. chem. Phys. **56**, 1989 (1972).
[7] C. H. BENNETT, J. appl. Phys. **43**, 2727 (1972).
[8] J. F. SADOC, J. DIXMIER, and A. GUNIER, J. non-crystall. Solids **12**, **46** (1973).
[9] J. L. FINNEY, Mater. Sci. Engng. **23**, 199 (1976).
[10] J. A. BARKER, M. R. HOARE, and J. L. FINNEY, Nature (London) **257**, 120 (1975).
[11] L. v. HEIMENDAHL, J. Phys. F **5**, L141 (1975).
[12] D. C. KOSKENMAKI, Mater. Sci. Engng. **23**, 207 (1976).
[13] W. D. KRISTENSEN, J. non-crystall. Solids **21**, 303 (1976).
[14] R. YAMAMOTO, H. MATSUOKA, and M. DOYAMA, phys. stat. sol. (a) **45**, 305 (1978).
[15] K. MAEDA and S. TAKEUCHI, J. Phys. F **12**, L283 (1978).
[16] H. KOIZUMI and T. NINOMIYA, J. phys. Soc. Japan **49**, 1022 (1980).
[17] J. L. FINNEY, Proc. Soc. **A319**, 479 (1970).
[18] J. DIXMIER and J. F. SADOX, in: Metallic Glasses, Amer. Soc. Metals, Ohio 1978 (p. 97).
[19] C. L. BRIANT and J. J. BURTON, phys. stat. sol. (b) **85**, 393 (1978).
[20] M. R. HOARE and P. PAL, Adv. Phys. **20**, 161 (1971).
[21] G. Z. PINSKER, Soviet Phys. – Doklady **22**, 391 (1977).
[22] J. D. BERNAL, Proc. Roy. Soc. **A280**, 299 (1964).
[23] Y. WASEDA and S. Z. TAMAKI, Z. Phys. **B23**, 315 (1976).
[24] Y. WASEDA, in: Liquid Metals, Ed. R. EVANS and D. A. GREENWOOD, The Institute of Metals, Bristol 1977 (p. 230).

(Received January 12, 1981)

phys. stat. sol. (a) **65**, 321 (1981)

Subject classification: 1.2; 14.4.1; 22.8.1

Institute of Solid State Physics, Latvian State University, Riga[1])

X-Ray Temperature Study of Crystallographic Characteristics of Barium Titanate

By

L. A. SHEBANOV

A precise experimental diffractometric study of unit cell parameters of polycrystalline $BaTiO_3$ in the temperature range from −170 to 220 °C is made, using an X-ray diffraction line refinement computer method. The values of unit cell volume jump, the widths of phase coexistence regions, and the hysteresis characteristics at phase transitions are determined. Analytical expressions for unit cell parameters as functions of the temperature are given. The concept of crystallographic parameters at phase transition points is discussed.

Es wird die Gitterkonstantenänderung von polykristallinem $BaTiO_3$ im Temperaturbereich von −170 bis 220 °C röntgendiffraktometrisch untersucht, wobei die Zerlegung der zusammengesetzten Beugungslinienprofile mit Hilfe der EDV erfolgt. Die numerischen Werte der Einheitszellenvolumensprünge, Breiten der Phasenkoexistenzbereiche, Charakteristiken der Hysteresiserscheinungen bei Phasenübergängen werden bestimmt. Die Temperaturabhängigkeiten der Zellenparameter werden in analytischer Form gegeben. Der Begriff der kristallographischen Parameter an Phasenumwandlungspunkten wird diskutiert.

1. Introduction

The crystal structure of barium titanate $BaTiO_3$ is concerned in a wide spectrum of papers rather different in defining the problem as well as in experimental methods, including X-ray diffraction.

The basic crystallographic properties of perovskites ABO_3 have been first investigated in barium titanate by Megaw [1]. Soon after this, complete and detailed (considering the experimental possibilities of that time) X-ray studies of the thermal behaviour of the unit cell parameters of $BaTiO_3$ in the range $T = -120$ to 150 °C, covering all the observable phase transitions of barium titanate, were reported by Kay and Vousden [2, 3]. A number of attempts have later been made to increase the accuracy of cell parameters, the temperature range of the experiments, however, being narrower. (Rhodes [4] has made his experiment in the temperature range $T = -160$ to 20 °C and Bland [5] at $T = 350$ to 1050 °C.) The values of the cell parameters were given by differences exceeding the claimed accuracies due to systematic errors (the higest accuracy of these measurements is $\pm 5 \times 10^{-4}$ Å). This has stimulated Joho [6] to repeat the Kay-Vousden experiments, using a precision powder method. His studies were restricted to the tetragonal–cubic transition in $BaTiO_3$ (temperature range of the experiment $T = 16.8$ to 206 °C, phase transition temperature $T_c = 122.6$ °C at heating). The accuracy of the experiment, including its temperature constancy, allows to regard these results [6] as the most reliable X-ray data at present time on the crystallographic characteristics of the higher-temperature phase transition in barium titanate.

---

[1]) Kengaraga 8, 226063 Riga, USSR.

However, a number of problems concerning the crystallographic aspects of $BaTiO_3$ phase transitions require a further refinement:

1. The photographic method of registration used in [6], even though much more accurate as compared to the earlier studies, still does not solve the problem of determining the structure parameters near the phase transition region, where the accuracy of cell parameters drops down to 10^{-3} Å, which is obviously less than that required to evaluate with any assurance the difference between c and a, describing the structure rearrangement at the phase transition.

2. It is desirable to have a refinement of the Kay-Vousden results also for the lower-temperature phase transitions in $BaTiO_3$. Data provided by different authors concerning the structure parameters of these phase transitions are rather inconsistent. According to Kay and Vousden the phase transitions in $BaTiO_3$ single crystals occurring at −90 °C and +5 °C are characterized by positive jumps of unit cell volume (at heating), in contrast to a negative volume change at the tetragonal–cubic phase transition at 120 °C. Later dilatometric studies [7] to polycrystalline $BaTiO_3$ have shown that a positive change of volume (at heating) is observed only at the rhombohedral–orthorhombic phase transition, while according to the most recent precise dilatometric measurements [8] the volume jumps at all the $BaTiO_3$ phase transitions have only a different value (positive values are absent).

3. The fundamental characteristics of the structure rearrangement within the range of coexistence of adjacent phases for the sequence of phase transitions and the quantitative characteristics of the hysteresis are still to be considered in more detail. Since the rearrangement of the crystal structure of $BaTiO_3$ at phase transitions occurs over a range of temperatures, it is necessary to define more precisely some crystallographic characteristics (the unit cell volume in particular) at the very "phase transition point".

2. Experimental

With account for the above-mentioned problems we have studied barium titanate ceramics in the temperature range −170 to 220 °C by the method described in [9] and based on the analysis of complex X-ray diffraction lines. The results of thermal measurements are represented according to the following rules: To determine the errors due to the production technology, preliminary measurements of elementary cell parameters were made at room temperature for a group of five to seven samples, each sample from a different technological batch. The samples were selected from a considerably large number to comply with strong requirements to their quality, the latter being checked by surface microphotographs as well as by electron microscope and X-ray microanalyzer. The functions of the type $Y(T)$ represented hereafter are obtained for a particular sample, the cell parameters of which at room temperature correspond to the average of the group of samples.

All the experiments have been made on the X-ray diffractometer "DRON-2,0" (FeK_α radiation, Mnβ-filter) according to the conventional procedure considered, e.g., in [10]. The high-temperature measurements were performed by means of an originally designed attachment. In the temperature range −180 to 20 °C the commercial low-temperature attachment "URNT-180" was used. Over the whole temperature range the thermal stability has been not worse than ±0.1 K. The accuracy of the absolute temperature value determined by the thermocouple calibration accuracy is estimated to ±0.5 K. Diffraction profiles were processed on an electronic computer by special software [9].

The lattice parameters were determined from the following reflections: In the case of the tetragonal phase from the group of lines 200, 002; in the case of monoclinic distortion (orthorhombic cell) from the lines 222, 22$\bar{2}$ and 20$\bar{2}$, 202, 220; in the case of rhombohedral symmetry the calculations were made on the basis of the group of lines 222, 22$\bar{2}$.

3. Results and Discussion

The unit cell parameter measurements are represented in Fig. 1. The analytical expressions of cell parameters as a function of temperature obtained by the least-squares method from processed experimental data are given in Table 1. It is worth noting that angular parameters (monoclinic angle β and rhombohedral angle α) are little changed in the range of corresponding phase stability ($\beta = 90.14°$ to $90.15°$ and $\alpha = 89.83°$ to $89.80°$ for the temperature intervals given in Table 1). The angular distortion increases as the temperature goes down.

The results of the present study (see Fig. 1) as compared to those obtained in [3] are different as far as they concern the sign of the volume jump at the lower-tem-

Table 1

Unit cell parameters (in Å) of $BaTiO_3$ as a function of temperature (in °C)

temperature interval (°C)	crystal symmetry	analytical expressions for thermal behaviour
−170 to −68.9	rhombohedral (R)	$a_R = 4.00758 + 0.1881 \times 10^{-4}T$
−98.5 to 8.2	monoclinic (M)	$a' = 4.01566 - 0.2569 \times 10^{-4}T$
		$c' = 3.99021 + 0.1069 \times 10^{-3}T$
−5.9 to 121.4	tetragonal (T)	$a = 3.99153 + 0.6284 \times 10^{-4}T - 0.3443 \times 10^{-7}T^2 + 0.3495 \times 10^{-8}T^3$
		$c = 4.03641 - 0.8647 \times 10^{-4}T + 0.9716 \times 10^{-6}T^2 - 0.9753 \times 10^{-8}T^3$
118.6 to 220	cubic (C)	$a = 4.00692 + 0.1537 \times 10^{-4}T + 0.8277 \times 10^{-7}T^2$

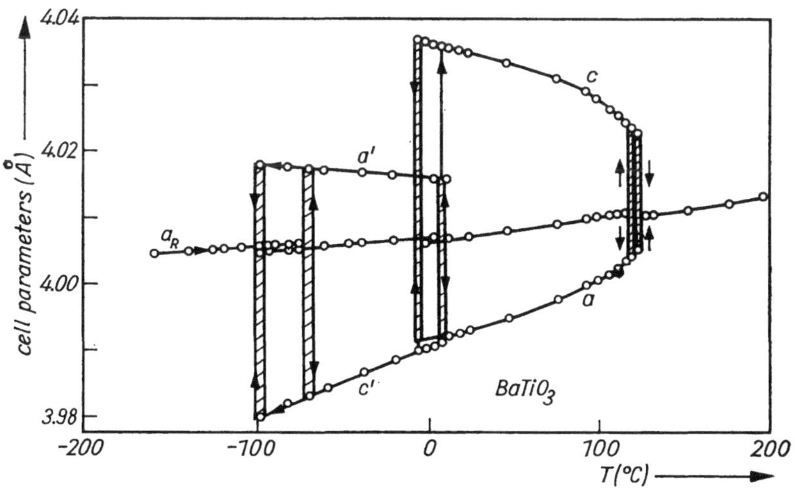

Fig. 1. The sequence of phase transitions and thermal behaviour of structure characteristics of $BaTiO_3$ in cubic, tetragonal, orthorhombic, and rhombohedral phases (The shaded areas correspond to the intervals of adjacent-phase coexistence determined from X-ray data)

perature phase transitions. We find a negative value of the unit cell volume change to be characteristic of all the phase transitions in $BaTiO_3$ at heating. The values are $\Delta V = 0.021$, 0.038, 0.047 Å^3 for R → M, M → T, T → C transitions, respectively. Our results are consistent with data of [6] for the tetragonal–cubic phase transition within the accuracy of experimental measurements. Essential attention has to be paid to the given values of volume jump ΔV at the phase transition point.

According to the latest views on phase transitions in perovskite structures [12 to 16], the classical concepts of parameter (entropy S, volume V, polarization P_s) jumps at the first-order phase transitions are applicable only to so-called elementary regions being thermodynamic subsystems of the whole macroscopic crystal. The existence of the elementary regions bas heen confirmed in Känzig's experiments [11], his views being later developed in [12 to 17]. The main results of a theoretical consideration are as follows:

1. The phase transition of a macroscopic crystal is essentially "diffused" in the sense that there is no transition "point" at which a jump of crystal symmetry occurs causing a relevant change of macroscopic parameters of the whole crystal. An interval of the phase transition is observed instead, wherein the phase transition in the elementary region is statistically completed.

2. The relative amount n_p of the polar phase and n_n of the non-polar phase in the phase transition interval is described by the so-called order function [13]

$$\frac{n_p}{n_p + n_n} = L_p(T) = \left[1 + \exp\left(\frac{T - T_c}{\xi}\right)\right]^{-1} \tag{1}$$

where ξ, describing the diffusion of the phase transition, is determined by macroscopic and microscopic parameters of the substance,

$$\xi = \frac{2kT_cC_W^*}{V_{er}P_{sc}^2}, \tag{2}$$

where k is Boltzmann's constant, C_W^* the Curie-Weiss constant, P_{sc} the spontaneous polarization jump at T_c, T_c the phase transition temperature corresponding to the bending point of the function $L(T)$, V_{er} the elementary region volume.

Analysis of (1) shows that $L_p(T)$ approaches unity asymptotically at temperatures lower than T_c (complete ordering of the lower-temperature phase). At $T = T_c$ $L_p(T) = 1/2$, which corresponds to equal quantities of the two phases. Obviously physical characteristics at the phase transition point must be related to T_c. Since there is little information on $L_p(T)$ in most cases of macroscopic experiments, the determination of T_c is practically somewhat difficult. To reduce dilatometric measurements to the phase transition point, in [8] a rather cumbersome method of nonlinear extrapolation has been used. The X-ray method and the complex diffraction line refinement techniques [9] provide reliable values of T_c at which the quantities of adjacent phases are equal. Really, in the region of coexistence of adjacent phases an overlap of diffraction lines of phases of different symmetry is observed. The relative amount of adjacent phases is determined from the integrated intensities I^T and I^C of the two different polymorphic modifications. Thus, e.g., in the case of the T → C transition the concentration of the tetragonal phase is equal to the ratio

$$\frac{I_{200}^T + I_{002}^T}{I_{200}^T + I_{002}^T + I_{200}^C} = C_T . \tag{3}$$

The values of ΔV given for the sequence of phase transition in $BaTiO_3$ are determined for temperatures at which $c_1 = c_2 = 0.5 \pm 0.05$. This problem has not been considered in [6], though an overlap of diffraction lines of the adjacent phases is reported over an interval of about 2 K. The regions of coexisting adjacent phases in the sequence of phase transitions in $BaTiO_3$ are shown in Fig. 1 by the shadded areas. The noted limits correspond to the minor phase content 0.1 ± 0.05. The values are $\Delta T\downarrow = 7.3, 4.7, 2.5$ K and $\Delta T\uparrow = 6.3, 3.7, 2.2$ K at cooling and heating, respectively, the order in both being from lower to higher temperature of phase transition.

We have as well studied the hysteresis at structural transformation in $BaTiO_3$. The thermal hysteresis follows from the basic assumption of the first-order phase transition thermodynamics. An application of the general theory to the relaxation model of structure change in the region of the T → C phase transition has been considered in [13]. The observed hysteresis of ≈ 2 K at the T → C phase transition (2.8 K according to the present experiment) is mainly related to formation and annihilation of domain structure, while the relaxation of polar and non-polar regions is of minor importance. According to our data the phase transitions R → M → T → C in $BaTiO_3$ occur at temperatures (determined in the way as discussed above) $T = -98.5, -5.9,$ 118.6 °C at cooling and $T = -68.9, 8.2,$ 121.4 °C at heating. Thus, the width of hysteresis is growing towards lower temperatures, its values being 2.8, 14.0, 29.6 K for the C → T, T → M, and M → R phase transitions, respectively.

There is a good agreement between the thermal behaviour of elementary cell parameters of $BaTiO_3$ obtained in the present experiment and the ones calculated by Devonshire [18].

Acknowledgement

The author is thankful to Prof. V. Fritsberg for revising the manuscript and for valuable remarks.

References

[1] H. D. Megaw, Proc. Roy. Soc. **A189**, 261 (1947).
[2] H. F. Kay, Acta cryst. **1**, 229 (1948).
[3] H. F. Kay and P. Vousden, Phil. Mag. **40**, 1019 (1949).
[4] G. Rhodes, Acta cryst. **4**, 105 (1951).
[5] J. A. Bland, Canad. J. Phys. **37**, 417 (1959).
[6] P. Joho, Z. Krist. **120**, 329 (1964).
[7] G. Shirane and A. Takeda, J. Phys. Soc. Japan **7**, 1 (1952).
[8] V. Ya. Fritsberg and K. Ya. Borman, Uchen. Zap. Latv. Un. **189**, 99 (1974).
[9] L. A. Shebanov, Uchen. Zap. Latv. Un. **189**, 150 (1974).
[10] D. M. Kheiker and L. S. Zevin, Rentgenovskaya difraktometriya, Fizmatgiz, Moskva 1963.
[11] W. Känzig, Helv. phys. Acta **24**, 175 (1951).
[12] V. Ya. Fritsberg, Fiz. tverd. Tela **10**, 385 (1968).
[13] V. Ya. Fritsberg, Uchen. Zap. Latv. Un. **235**, 5 (1975).
[14] V. Ya. Fritsberg, Uchen. Zap. Latv. Un. **250**, 5 (1976).
[15] B. N. Rolov, Razmitie fazovie perekhodi, Izd. Zinatne, Riga 1972.
[16] V. Ya. Fritsberg, Izv. Akad. Nauk LatvSSR **2**, 6 (1977).
[17] T. B. Romanovskis and B. N. Rolov, in: Fazovie perekhodi v segnetoelektrikakh, Ed. V. Ya. Fritsberg Izd. Zinatne, Riga 1971 (p. 67).
[18] A. F. Devonshire, Phil. Mag. **42**, 1040 (1949).

(Received December 15, 1980)

phys. stat. sol. (a) 65, 327 (1981)

Subject classification: 14.1; 21.1

Laboratoire d'Electronique, Université de Nancy[1])

Thickness Dependence of the Temperature Coefficient of Resistivity of Polycrystalline Films in a Three-Dimensional Conduction Model

By

C. R. PICHARD, C. R. TELLIER[2]), and A. J. TOSSER

The temperature coefficient of the total film resistivity of thin polycrystalline films is calculated from the three-dimensional model of grain-boundaries. This study reveals that the analytical equations so derived are convenient tools to describe both, the grain-boundary and the thickness dependence of the t.c.r. of fine-grained films. The fit of the data obtained for polycrystalline zinc films is successful; values of the transmission coefficient t and specularity parameter p determined from measurements on annealed zinc films are reasonable.

Der Temperaturkoeffizient des Gesamtschichtwiderstands dünner polykristalliner Schichten wird mit dem dreidimensionalen Modell der Korngrenzen berechnet. Die Untersuchung zeigt, daß die so abgeleiteten analytischen Gleichungen geeignete Mittel sind, um sowohl die Korngrenze als auch die Dickenabhängigkeit der t.c.r. von feingekörnten Schichten zu beschreiben. Die Anpassung der für polykristalline Zinkschichten erhaltenen Werte ist erfolgreich; die Werte für den Transmissionskoeffizienten t und des Reemissionsparameters p, die aus Messungen an getemperten Zinkschichten bestimmt werden, sind vernünftig.

1. Introduction

The theoretical expressions for electronic transport properties of polycrystalline metal films have been established by some authors [1 to 3] who started from the direct calculation of the simultaneous effects of three electronic scatterings: background scattering, scattering at grain boundaries and at external film surfaces.

The background scattering is generally represented by a relaxation time related to the bulk material [1 to 4]; in most cases, the scattering at external film surfaces is represented by a specular reflection coefficient p independent of the scattering angle [1, 6]. Several alternative models have been proposed [5, 6] but no further development has been proposed except in a recent paper [7] derived from Soffer's model [8]. The scattering at grain boundaries has been treated from two different points of view. Mayadas and Shatzkes [1] assumed that the grain boundaries could be represented by arrays of planar scatterers whose effects are described by a Dirac potential; they assumed that the scatterers parallel to the electric field have no influence on the electronic conductivity and that the geometrical repartition of scatterers perpendicular to the electric field obeys a Gaussian law, so that the average scatterer spacing can be identified with the average grain diameter. A quantum mechanical formalism is then used for calculating the transition rate of conduction electrons in the wave vector space, in a way similar to that used for expressing the perturbation of the Hamiltonian of the free electron due to the presence of dislocation lines in a crystal [9]. It is clear that this calculation procedure gives a one-dimensional model which is not suitable

[1]) Case Officielle 140, 54037 Nancy Cedex, France.

[2]) Laboratoire de Chronomètrie et Piezoélectricité, E.N.S.M.M., Route de Gray, 25030 Besancon, France.

for analyzing the transport phenomena occurring in a three-dimensional space, such as the Hall effect.

An alternative model of grain boundaries has been first presented by Warkusz [10, 11] but the mathematical framework was questionable [12]; further studies have shown that this model could correspond the cylindrical grain boundaries [13], i.e. could describe the effect of grain boundaries in monocrystalline films. In a similar way, a statistical representation of grain boundaries based on an electronic transmission coefficient t through grain boundary was proposed [14]. Comparisons with experiments suggest that a tree-dimensional array of scatterers could conveniently describe the effects of grain boundaries in polycrystalline films [15].

Further introducing the Cottey model [16] for representing the electronic scattering at external film surfaces allows the calculation of the total polycrystalline film conductivity, in good agreement with experimental data [2, 3]. From this three-dimensional model theoretical linearized accurate expressions for the conductivity [3] and theoretical expressions for the Hall coefficient [17, 18] and for the strain coefficient of resistivity [19] have been successively derived. A good fit with experiments was obtained for linearized conductivity equations [3]. In a recent paper [20] a two-dimensional array of scatterers has been proposed for expressing monocrystalline film conductivity: general expressions for thermoelectric power [21] have been derived from this model and give an adequate description of experimental data [22]; moreover the identification of imperfections contributing to monocrystalline film conductivity have been recently reported [23]. These features sustain the validity of the two-dimensional model [20] and suggest that the statistical representation of a grain boundary is also suitable for polycrystalline films. Therefore, the purpose of this paper is to examine the physical validity of the theoretical expressions of the temperature coefficient of resistivity of polycrystalline films as an acceptable physical basis for further developments of the three-dimensional conduction model. Further extensions are probable since the thermoelectric power can be expressed in term of t.c.r. [24, 25], when the effects of thermal strains [26, 27] are neglected; it has also been recently demonstrated that it is also the case for the Hall coefficient [17] and probably for strain coefficients of resistivity [28].

Moreover in the framework of the three-dimensional model it can be theoretically established that the arrays of scatterers parallel to the electric field give a contribution to the film resistivity [29]; this feature markedly differs from the MS theory in which these contributions are neglected. Therefore a further requirement exists for making a comparison of theoretical expressions of the t.c.r. of polycrystals in the three-dimensional model with experimental data.

2. Theoretical

In the three-dimensional model [13], three arrays of scatterers represent the grain boundaries; the average effect of a grain boundary is represented by a specular transmission coefficient of electrons, t, which gives the fraction of electrons whose velocity in the electric field direction is not altered by the grain boundary, whereas the remainder electrons are diffusely scattered and do not contribute to the current [14, 15]. Assuming that the probability for an electron to travel a given distance without being diffusely scattered is given by an exponential law [14], a mean free path can be ascribed to the three-dimensional array of scatterers whose spacing is identified with the average grain diameter a_g; the total conductivity can be calculated [2, 3, 29] under the assumption that the reciprocal total relaxation time due to all the sources of scattering is the sum of the reciprocal relaxation times related to any scattering, as generally assumed for such calculations

[4]. For convenience a grain parameter, ν, is introduced; ν is defined from the background mean free path l_0, the average grain size a_g, and the coefficient t of "specular" transmission by [15]

$$\nu = a_g l_0^{-1} \left[\ln \frac{1}{t}\right]^{-1}. \tag{1}$$

The ratio of infinitely thick polycrystalline film conductivity to that in a bulk material was found to be [15, 29]

$$\sigma_g/\sigma_0 = \varrho_0/\varrho_g = \frac{3}{2} \frac{\nu}{1-c} \left\{\gamma - \frac{1}{2} + (1-\gamma^2) \ln(1+\gamma^{-1})\right\}, \tag{2}$$

i.e.

$$\sigma_g/\sigma_0 = \frac{3}{2} \frac{\nu}{1-c} U(\gamma) \tag{3}$$

with

$$\gamma = (\nu + c^2)(1-c)^{-1}; \qquad c = 4/\pi. \tag{4}$$

In the more general case of partially diffuse scattering of electrons on the external surfaces, the external surface effect can be analyzed in terms of Cottey's model [16], which states that a mean free path can be associated with scattering; an "external surface" parameter can be introduced to decribe the size effect [12], i.e.

$$\mu = a l_0^{-1} \left[\ln \frac{1}{p}\right]^{-1}, \tag{5}$$

where a and p refer to film thickness and specularity parameter [5], respectively. Consequently, in thin polcrystalline films in which three types of electron scattering (background, grain boundary, and external surface scattering) are simultaneously operative, the electrical conductivity σ_{F_p} takes the simple analytical form [3]

$$\sigma_{F_p}/\sigma_0 = \frac{3}{2} \frac{1}{b} \left\{\alpha - \frac{1}{2} + (1-\alpha^2) \ln(1+\alpha^{-1})\right\} = b^{-1} U(\alpha) \tag{6}$$

with

$$b = \mu^{-1} + \nu^{-1}(1-c), \tag{7}$$

$$\alpha = (1 + c^2 \nu^{-1}) b^{-1}. \tag{8}$$

Data on film t.c.r. are usually interpreted in terms of size effect theories by considering the following usual assumptions [4]:

1) the rigid band model of metals is valid,

2) the number of conduction electrons per unit volume is temperature independent in the experimental range,

3) the thermal expansions of the grain and the film thickness are negligible with respect to that of the mean free path.

We also retain these assumptions and further neglect the effect of the thermal expansion of the substrate since it only gives a correcting term [27].

2.1 The grain boundary t.c.r. β_g

Logarithmic differentiation of equation (3) gives

$$\frac{d\sigma_g}{\sigma_g} - \frac{d\sigma_0}{\sigma_0} = \frac{d\nu}{\nu} + \frac{dU(\gamma)}{U(\gamma)} = \frac{d\nu}{\nu} \left\{1 + \frac{\nu}{U(\gamma)} \frac{dU(\gamma)}{d\nu}\right\}, \tag{9}$$

where with (4)

$$\frac{\mathrm{d}U(\gamma)}{\mathrm{d}\nu} = \frac{\mathrm{d}U(\gamma)}{\mathrm{d}\gamma} \frac{1}{1-c}, \tag{10}$$

whereas from equation (1) and according to the above assumptions

$$\frac{\mathrm{d}\nu}{\nu} = -\frac{\mathrm{d}l_0}{l_0}. \tag{11}$$

Combining equations (9), (10), and (11) gives

$$\frac{\mathrm{d}\sigma_\mathrm{g}}{\sigma_\mathrm{g}} - \frac{\mathrm{d}\sigma_0}{\sigma_0} = -\frac{\mathrm{d}l_0}{l_0}\left\{1 + \frac{\nu}{1-c}\frac{1}{U(\gamma)}\frac{\mathrm{d}U(\gamma)}{\mathrm{d}\gamma}\right\}. \tag{12}$$

If the temperature variation of the background m.f.p. l_0 and bulk conductivity σ_0 are the same, as assumed above, we have

$$\beta_0 = -\frac{1}{l_0}\frac{\mathrm{d}l_0}{\mathrm{d}T}, \tag{13}$$

then the general expression for the temperature coefficient of film resistivity (t.c.r.) defined as

$$\beta_\mathrm{g} = -\frac{\mathrm{d}\ln\sigma_\mathrm{g}}{\mathrm{d}T} = \frac{\mathrm{d}\ln\varrho_\mathrm{g}}{\mathrm{d}T} \tag{14}$$

may be derived from equation (12), i.e.

$$\beta_\mathrm{g}/\beta_0 = H(\nu) = \frac{\nu}{1-c}\frac{\gamma^{-1} - 2 + 2\gamma\ln(1+\gamma^{-1})}{\gamma - \frac{1}{2} + (1-\gamma^2)\ln(1+\gamma^{-1})}. \tag{15}$$

Values of the t.c.r. β_g/β_0 as given by equation (15) are reported in Table 1.

Table 1

T.c.r. ratio β_g/β_0 as given by equation (15) for different values of the ν parameter

ν	β_g/β_0	ν	β_g/β_0
0.1	0.061994	4	0.725025
0.2	0.116728	10	0.868202
0.4	0.208991	20	0.929360
1	0.397599	40	0.963538
2	0.568807	100	0.981958

It is easy to see that equation (15) satisfies the physical requirement which states that the infinitely thick film t.c.r. β_g reduces to the bulk t.c.r. β_0 when the grain boundary effect vanishes as also predicted by the limiting form of equation (15), i.e.

$$\beta_\mathrm{g}/\beta_0|_{\nu\gg1} \approx 1 + \frac{1}{\nu}\left\{\frac{3}{8}(1-c) - c^2\right\}. \tag{16}$$

2.2 The total film t.c.r. β_{F_p}

Taking into account that according to the assumptions we have

$$\frac{db}{b} = \frac{dl_0}{l_0}, \tag{17}$$

$$\frac{d\alpha}{\alpha} = -\frac{dl_0}{l_0}\left[1 + \frac{c^2}{\nu}\right]^{-1}, \tag{18}$$

the logarithmic differentiation of equation (6) leads to

$$\frac{d\sigma_{F_p}}{\sigma_{F_p}} = \frac{d\sigma_0}{\sigma_0} - \frac{dl_0}{l_0} + \frac{\alpha}{U(\alpha)}\frac{dU(\alpha)}{d\alpha}\frac{1}{1 + c^2\nu^{-1}}\left(-\frac{dl_0}{l_0}\right). \tag{19}$$

After some calculations it yields

$$\beta_{F_p}/\beta_0 = \frac{1}{b}\frac{\alpha^{-1} - 2 + 2\alpha \ln(1 + \alpha^{-1})}{\alpha - \frac{1}{2} + (1 - \alpha^2)\ln(1 + \alpha^{-1})}, \tag{20}$$

where the film t.c.r. β_F is commonly expressed as [4]

$$\beta_{F_p} = \frac{d \ln \varrho_{F_p}}{dT} = \frac{-d \ln \sigma_{F_p}}{dT}. \tag{21}$$

It must be noted that formula (20) satisfies some qualitative physical requirements. Indeed the forms of equations (20) and (15) are quite similar except that in the limiting case $p = 1$, b is replaced by ν^{-1}; moreover the film t.c.r. β_{F_p} reduces to the t.c.r. β_g as far as the electron scattering on external surfaces becomes negligible.

From equations (7) and (8) it can be easily seen that when the grain boundary scattering does no longer contribute to the electrical resistivity (i.e. for $\nu \to \infty$), the β_{F_p}/β_0 ratio coincides with the Cottey t.c.r. ratio β_F^*/β_0 [30].

Fig. 1a shows the thickness variation of the t.c.r. ratio β_{F_p}/β_0 for different values of ν and for $p = 0.75$ whereas Fig. 1b shows the results for $\nu = 1$ and for different values of the specularity parameter p. We observe that, as expected from the above discussion

1) the t.c.r. ratio markedly depends on the grain parameter ν and approaches the grain boundary t.c.r. ratio β_g/β_0 in the limit of large k (Fig. 1a),

2) for a given value of the reduced thickness k the t.c.r. ratio increases with increasing value of the specularity parameter p (Fig. 1b) and grain boundary parameter ν (Fig. 1a).

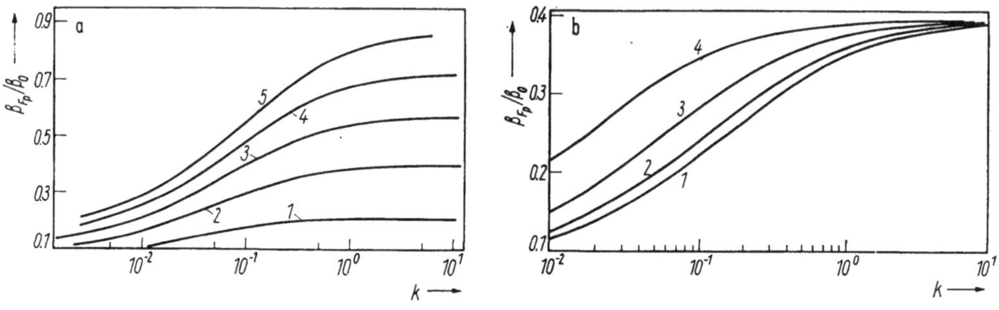

Fig. 1. T.c.r. ratio β_F/β_0 versus the reduced thickness $k = al_0^{-1}$ a) for different values of the grain parameter ν and for $p = 0.75$. (1) $\nu = 0.4$, (2) 1, (3) 2, (4) 4, (5) 10, b) for different values of the specularity parameter p and for $\nu = 1$. (1) $p = 0.4$, (2) 0.5, (3) 0.7, (4) 0.9

3. Experimental Fits

Experiments on the t.c.r. of thin r.f. sputtered zinc films [31] allow us to compare the observed size effects with the present model. Experimental details have been given elsewhere [31]. Fig. 2 shows the variation of the film t.c.r. β with thickness a for annealed zinc films of electrical resistance R_s. The thickness dependence of β may be understood in terms of the two-layer model which has been previously developed [32].

In this model the r.f. sputtered films consist of two layers of respective resistance R_{1s} and R_{2s} after aging; the second layer is homogeneous and polycrystalline with a constant grain diameter [32]. Since the total conductivity is the sum of reciprocal resistance R_{1s} and R_{2s}, the total t.c.r. β is [33]

$$\beta_2 = \beta + (\beta - \beta_1) \frac{R_{2s}}{R_{1s}}$$

where β_2 and β_1 refer to the second layer of thickness a_2 and to the first layer of critical thickness e_{c_1} ($e_{c_1} \approx 150$ Å [31]), respectively.

The experimental values of β_1 related to e_{c_1} is given on Fig. 2, i.e.

$$\beta_1 \approx 1 \times 10^{-4}\ \text{K}^{-1}\,.$$

In the light of experimental works [4, 31, 34 to 41] on the thickness dependence of the t.c.r. it is reasonable to expect the plot of $a_2\beta_2^{-1}$ versus a_2 or of $\beta_2 a_2$ versus a_2 (Fig. 3) should yield straight lines of slope β_g^{-1} or β_g when $a_2 > l_0$ ($a_2 > 120$ Å in the case of zinc films [42, 31]).

It can be seen (Fig. 3) that these predictions agree well with the experimental data related to zinc films: the values of β_g calculated from the two slopes are nearly the same, i.e.

$$\beta_g \approx 1.923 \times 10^{-3}\ \text{K}^{-1} \quad \text{and} \quad \beta_g \approx 1.93 \times 10^{-3}\ \text{K}^{-1}\,.$$

Then we may determine from (1) and (15) the experimental values of the grain parameter ν and transmission coefficient t assuming that the mean grain diameter is equal to the critical thickness e_{c_1} and taking $\beta_0 \approx 4.19 \times 10^{-3}\ \text{K}^{-1}$ [20] and $\beta_{\text{g average}} \approx 1.926 \times 10^{-3}\ \text{K}^{-1}$; we obtain

$$\nu = 1.29\,, \qquad t = 0.379\,,$$

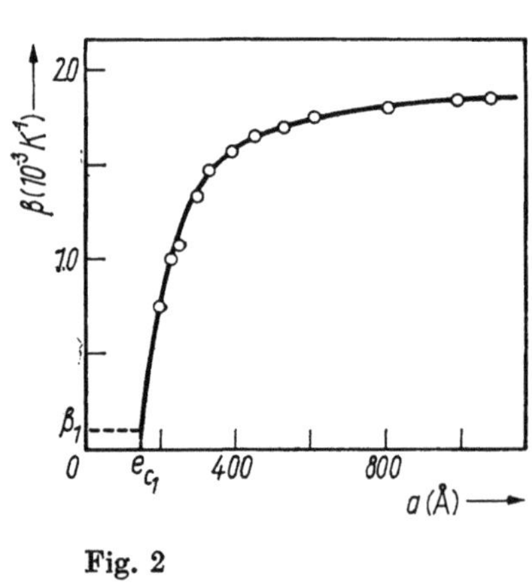

Fig. 2

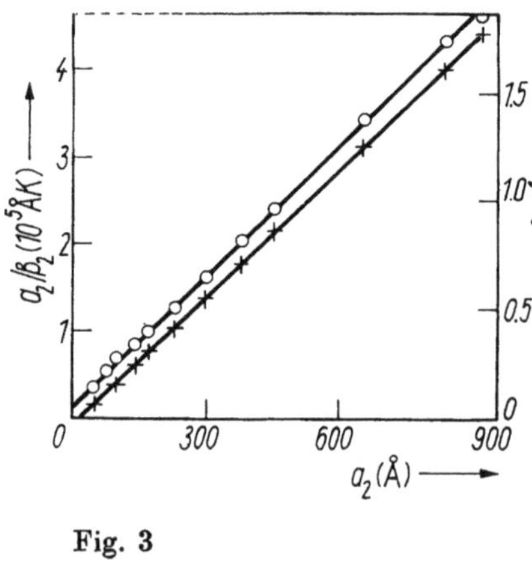

Fig. 3

Fig. 2. Thickness dependence of the t.c.r. β of annealed zinc films

Fig. 3. Plots of a_2/β_2 (○) and $\beta_2 a_2$ (+) versus the thickness a_2 of the second layer

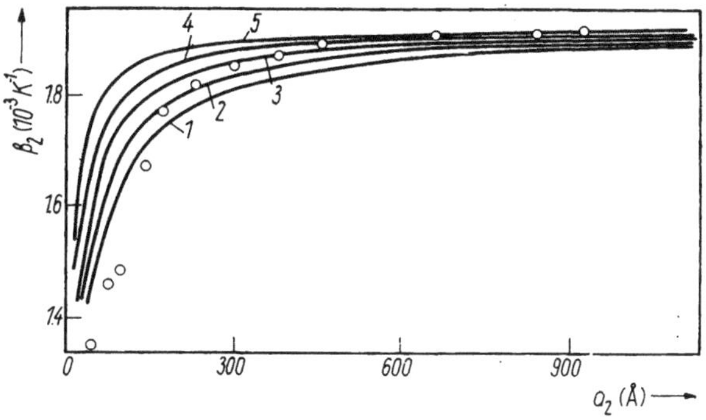

Fig. 4. Theoretical thickness dependence of the t.c.r. β_2 from equation (20) for $\nu = 1.29$ and the experimental data. (1) $p = 0.4$, (2) 0.5, (3) 0.6, (4) 0.7, (5) 0.8

these values are physically reasonable and agree with the values previously deduced from the data related to the zinc film resistivity, i.e. $\nu = 1.43$ and $t = 0.417$ [2].

Experimental values of ν and β_g as well as the usual bulk mean free path value were used to determine the theoretical variations (broken curves) of β_2 with thickness a_2 for different values of the specularity parameter p (Fig. 4). Fig. 4 shows that a reasonable fit of the thickness dependence of the t.c.r. is observed down to $a_2 \approx 150$ Å.

4. Conclusion

The simple analytical expression of the thin polycrystalline film t.c.r. presented here could be quite useful for research in transport properties of polycrystalline metal films. From this study it appears that

1. the mathematical expressions (15) and (20) agree with some essential physical requirements and allow an easy evaluation of the t.c.r. dependence on thickness and grain parameter,
2. fundamental physical parameters such as transmission coefficient t and specularity parameter p can be easily extracted from data by using the present model,
3. a good fit of the experimental data on the t.c.r. of thin r.f. sputtered zinc films can be made and reasonable values of t and p can be obtained by applying the proposed model.

References

[1] A. F. Mayadas and M. Shatzkes, Phys. Rev. B **1**, 1382 (1970).
[2] C. R. Tellier, C. R. Pichard, and A. J. Tosser, Le Vide **196 S**, 93 (1979).
[3] C. R. Pichard, C. R. Tellier, and A. J. Tosser, J. Mater. Sci. **15**, 2236 (1980).
[4] K. L. Chopra, Thin Film Phenomena, McGraw-Hill Publ. Co., New York 1969.
[5] E. H. Sondheimer, Adv. Phys. **1**, 1 (1952).
[6] T. J. Coutts, Thin Solid Films **7**, 77 (1971).
[7] J. R. Sambler and K. C. Elson, J. Phys. F, in the press.
[8] S. B. Soffer, J. appl. Phys. **38**, 1710 (1967).
[9] J. M. Ziman, Electrons and Phonons, Oxford University Press, London 1960.
[10] F. Warkusz, Electrocomp. Sci. Technol. **5**, 197 (1978).
[11] F. Warkusz, Acta phys. Polon. **A54**, 31 (1978).
[12] C. R. Tellier and A. J. Tosser, Electrocomp. Sci. Technol. **6**, 37 (1979).
[13] C. R. Pichard, A. J. Tosser, and C. R. Tellier, Electrocomp. Sci. Technol. **6**, 87 (1980).
[14] C. R. Tellier, C. R. Pichard, and A. J. Tosser, Thin Solid Films **61**, 349 (1979).
[15] C. R. Pichard, C. R. Tellier, and A. J. Tosser, Thin Solid Films **62**, 189 (1979).
[16] A. A. Cottey, Thin Films Solid **4**, 297 (1968).
[17] C. R. Pichard, C. R. Tellier, and A. J. Tosser, J. Phys. F **10**, L101 (1980).
[18] C. R. Pichard, C. R. Tellier, and A. J. Tosser, Thin Solid Films **69**, 157 (1980).

[19] C. R. Pichard, C. R. Tellier, and A. J. Tosser, J. Mater. Sci. **15**, 2991 (1980).
[20] C. R. Tellier and A. J. Tosser, Thin Solid Films **70**, 225 (1980).
[21] C. R. Tellier, L. Hafid, and A. J. Tosser, Appl. Phys. **23**, 357 (1980).
[22] C. R. Tellier, C. R. Pichard, and A. J. Tosser, Thin Solid Films, to be published.
[23] C. R. Tellier, L. Hafid, and A. J. Tosser, Rev. Phys. appl. **15**, 1573 (1980).
[24] C. R. Tellier, C. R. Pichard, and A. J. Tosser, Thin Solid Films **42**, L31 (1977).
[25] J. B. Thompson, Thin Solid Films **48**, L25 (1978).
[26] P. M. Hall, Appl. Phys. Letters **12**, 212 (1968).
[27] C. R. Pichard, C. R. Tellier, and A. J. Tosser, J. Phys. D **13**, 1325 (1980).
[28] C. R. Pichard and C. R. Tellier, Rev. Phys. appl. **14**, 743 (1979).
[29] C. R. Pichard, C. R. Tellier, and A. J. Tosser, phys. stat. sol. (b) **99**, 353 (1980).
[30] See for example C. R. Tellier and A. J. Tosser, Thin Solid Films **52**, 53 (1978).
[31] C. R. Tellier, Vacuum **28**, 321 (1978).
[32] C. R. Tellier and A. J. Tosser, Thin Solid Films **37**, 207 (1976).
[33] C. R. Tellier and A. J. Tosser, Thin Solid Films **43**, 261 (1977).
[34] P. Sen and A. K. Pal, Japan. J. appl. Phys. **14**, 1473 (1975).
[35] B. Singh, C. C. Ling, and N. A. Surplice, Thin Solid Films **23**, 550 (1974).
[36] R. Suri, A. P. Thakoor, and K. L. Choppa, J. appl. Phys. **46**, 2574 (1975).
[37] A. K. Pal and S. Chaudhuri, J. Mater. Sci. **11**, 872 (1976).
[38] A. K. Pal and P. Sen, J. Mater. Sci. **12**, 1472 (1977).
[39] S. Chaudhuri and A. K. Pal, J. appl. Phys. **48**, 3455 (1977).
[40] S. K. Bandyopadhyay and A. K. Pal, J. Phys. D **12**, 953 (1979).
[41] K. Narayandas, M. Radakrishnan, and C. Balasubramian, Thin Solid Films **67**, 357 (1980).
[42] C. Reale, Indian J. pure appl. Phys. **7**, 729 (1969).

(Received January 28, 1981)

phys. stat. sol. (a) 65, 335 (1981)

Subject classification: 10.1; 12.1; 21

A. F. Ioffe Physico-Technical Institute, Academy of Sciences of the USSR, Leningrad[1])

Acoustic Technique Application in Determining Structure and Dislocation–Point Defect Force Interaction Parameters

By

V. I. Ivanov, B. K. Kardashev, and S. P. Nikanorov

It is shown that absolute values of the structure and force parameters peculiar to the system "dislocation–point defects" in crystals with low Peierls energy barrier can be estimated by the data of the acoustic study of the dislocation amplitude-dependent internal friction and yield point measurements. The results of the problem solved are illustrated by the experimental data obtained on aluminium single crystal specimens at temperatures varied from room temperature down to 4.2 K.

Показано, что по данным акустических исследований дислокационного амплитудно-зависимого внутреннего трения и измерениям предела текучести можно оценивать абсолютные значения структурных и силовых параметров, характеризующих систему ,,дислокация–точечные дефекты" в кристаллах с низким барьером Пайерлса. Ресультаты решения задачи иллюстрируются на премере экспериментальных данных, полученных на монокристаллических образцах алюминия при температурах от комнатной до 4,2 K.

1. Introduction

Recently a number of papers appeared dealing with the study of alkali halide crystals [1, 2] and dilute metal alloys [3, 4] where a correlation between the temperature dependences of the yield point and acoustical stress amplitude $\tau_0(T)$ at constant amplitude-dependent decrement (δ_h = const) was found. Those observations correspond to the thermal activation theory of the dislocation hysteresis damping developed in [5, 6]. This theory makes possible to reproduce the profile of the dislocation-pinning point interaction law. According to [5, 6] the dependence $\tau_0(T)$ at δ_h = const corresponds to the stresses responsible for the escaping of the dislocation segments of the same length at different temperatures and with the exactness of constant factors reproduces the dependence $H(F)$, $H \sim T$ being the activation energy of the dislocation depinning, while $F \sim \tau_0$ is the dislocation–point defect binding force. The derivation $\partial H/\partial F = -d(F)$ is the activation length. The inverse dependence $F(d)$ is the profile of the force law in question.

To make clear the nature of the point defects controlling the dislocation mobility in crystals it is necessary to know the absolute values of the parameters F and d, as well as the geometrical arrangement of the pinning points along the dislocation line. The estimation of the absolute values of F and d for alkali halides NaF and NaCl was made in [7] with the aid of the experimental technique of simultaneous measurements of the amplitude dependence of internal friction and the dislocation electric polarisation of the specimen arising due to the oscillating motion of charged dislocation [8, 9]. These measurements enabled us to estimate the magnitude of the average length $\bar{L}$ of dislocation segments and to perform calculations for obtaining F and d. Further analysis [1] has shown that for alkali halides in which we can study simul-

[1]) Leningrad K 21, USSR.

taneously the internal friction of the sample and its dislocation electrical polarisation [9], it is possible to estimate with sufficiently good accuracy the value of the yield point τ_s usually measured in a testing machine using the electro-acoustical experimental data only.

Electric polarisation is not observed in metals due to the high electric conductivity. In this case the determination of dislocation microscopic parameters requires the application of some additional tests alongside with internal friction results.

This paper deals with an experimental study of the yield point $\tau_s(T)$ and the amplitude dependent internal friction in Al single crystal at various fixed temperatures in the range 295 to 4.2 K. It is shown how, knowing τ_s and $\delta_h(\tau_0)$ at different T, one can obtain the absolute values of dislocation parameters F, d, and $\bar{L}$.

2. Theory

The technique used for calculations is based on Koehler's model [10], which considers a dislocation as a string pinned by immobile point defects. The theory [5] of dislocation hysteresis developed by Indenbom and Chernov was applied, too. According to [5] the amplitude-dependent decrement δ_h can be expressed with the aid of a single parameter L_h which has the physical significance of a minimum value of the dislocation segment length, the segments being released at a certain temperature T, a certain frequency ν of stress oscillations, and a certain shear stress amplitude τ_0

$$\delta_h = L_h^2 \int_{L_h}^{\infty} L N(L) \, dL \,, \tag{1}$$

where

$$L_h = \frac{F(T)}{b\tau_0} . \tag{2}$$

Here $N(L)$ is the dislocation segment length distribution function, b the magnitude of the Burgers vector. The amount of the activation energy for the depinning process is estimated in [5] in the framework of the theory of the absolute chemical reaction rates:

$$H(F) = kT \ln \frac{\nu_0}{\nu} , \tag{3}$$

where k is the Boltzmann constant, ν_0 the effective frequency of the dislocation segment oscillations. After [5] $\nu_0 \approx 10^{10}$ Hz.

From the measurements of the amplitude dependent internal friction $\delta_h(\tau_0)$ a function $H(F)$ characterizing the dislocation–pinning point force interaction, can be obtained. It is sufficient for this to plot the dependence $\tau_0(T)$ at $\delta_h = \text{const}$. The distribution function $N(L)$ shape can be determined as well [5]:

$$N(L) = \text{const} \left\{ \tau_0^3 \frac{\partial}{\partial \tau_0} \left[\tau_0^2 \delta_h(\tau_0) \right] \right\}_{|\tau_0 = F(T)/bL} . \tag{4}$$

It should be noted here that if a researcher has data on the amplitude-dependent internal friction only, the values of F on the $H(F)$ plot as well as distribution function $N(L)$ can be found with the accuracy of a constant factor. To calculate the absolute values of F, it is necessary according to (2) to know the magnitude of L_h. It is possible to solve the problem as follows.

If the distribution function $N(L)$ shape (4) is known from the $\delta_h(\tau_0)$ measurements, then limiting $N(L)$ from the side of short lengths by a certain magnitude of L_{min} so that the normalizing condition

$$\lambda = \int_{L_{\min}}^{\infty} LN(L)\,dL \tag{5}$$

is fulfilled, it is possible by (2) to obtain the value of the shear stress

$$\tau_y = \frac{F(T)}{bL_{\min}}. \tag{6}$$

The decrement δ_h at $\tau = \tau_y$ according to (1) is

$$\delta_h = L_{\min}^2 \lambda . \tag{7}$$

In (5) and (7) λ is the dislocation density in the crystal.[2])

The magnitude of τ_y has the physical significance of a stress providing the depinning of virtually all the dislocation segments and their subsequent mobility in crystals under load. It has been shown in [1] on NaF and NaCl crystals that τ_y and the yield point τ_s have almost the same values within the temperature range 295 to 4.2 K. It makes reasonable to formulate the following problem: to determine F, L, and $N(L)$ by the available values of $\delta_h(\tau_0)$ and τ_s.

If an experimentally measured decrement δ can be expressed, for instance, in the form of an empirical exponential equation

$$\delta = \delta_i + \delta_h = \delta_i + B\tau_0^n . \tag{8}$$

δ_i being an amplitude-independent background internal friction, then from (1), (2), (4), and (6) to (8) the value of $L_{\min}$ can be obtained

$$L_{\min} = \sqrt{\frac{B\tau_y^n}{\lambda}} \approx \sqrt{\frac{B\tau_s^n}{\lambda}} \tag{9}$$

for a certain dislocation density λ, which can be estimated, for example, by etch pits. Knowing $L_{\min}$ and τ_y, F can be calculated by (6). An average length $\bar{L}$ of the dislocation segments is found as follows:

$$\bar{L} = \frac{\int_{L_{\min}}^{\infty} LN(L)\,dL}{\int_{L_{\min}}^{\infty} N(L)\,dL} = \frac{n+3}{n+2} L_{\min} . \tag{10}$$

The distribution function $N(L)$ in accordance with (4) and (8) has eventually the form

$$N(L) = [(n+2)\,\lambda L_{\min}^{(n+2)}]\,L^{-(n+4)} . \tag{11}$$

The application of the calculation technique to some crystal assumes the fulfilling of the following criteria: the dependences $\tau_0(T)$ at different $\delta_h = \text{const}$ must coincide with one another (requirement of the validity of the theory [5]) as well as with the temperature dependence of the yield stress $\tau_s(T)$ by changing the scale of the stress axis.

3. Results and Discussion

The technique for calculating the absolute values of the dislocation parameters given above was tested by the authors on an Al f.c.c. single crystal.

The crystals for this study were grown from the melt of commercial purity 99.9%

[2]) The parameter $L_{\min}$ here has a meaning similar to that of Debye frequency in the theory of the heat capacity of solids [11]. $N(L)$ and λ have analogous in the heat capacity theory namely the spectrum and the quantity of atomic oscillators, respectively.

aluminium by Stepanov's method [12] which allowed to obtain any cross-section shape of a rod during its growth from the melt and to avoid the mechanical treatment of the specimen faces. The samples in question had rectangle cross-section of $3 \times \times 3$ mm². The axes of the rods corresponded to the crystallographic direction [010].

The internal friction was studied by the composite oscillator technique at longitudinal modes with frequency $\nu \approx 100$ kHz. The set-up description is given in [8, 13]. The specimens were pre-strained at room temperature by four-point plastic bending and thereby a certain amount of fresh dislocations was introduced. The yield stress τ_s was measured at the compression strain rate $\approx 10^{-4}$ s^{-1}. The value of τ_s was found by the cross point of the tangentials to the elastic and unelastic parts of the stress–strain curve.

Fig. 1 shows the amplitude dependences of the decrement for an aluminium specimen within the temperature range from 270 K and down to 4.2 K. At these temperatures after a vibrational treatment of the specimen the points measured at a successive increase and decrease of the amplitude τ_0 virtually fit the same curve.[3]) It is seen from the figure that in the amplitude-dependent range the decrement decreases monotonously with decreasing test temperature.

It has been found that the experimental data within the limits of experimental errors fit an exponential function (8) with the exponent n independent of temperature. Thus, it means that for the Al crystal in question the requirement of applicability of the theory [5] is met.

Fig. 2 shows the dependences $\tau_y(T)$ and $\tau_s(T)$. It is seen that the curves coincide well. Here the magnitude of τ_y equals exactly τ_s at 130 K. In this case for other temperatures the values of τ_y differ from those of τ_s less than by 20%.

Thus, for the Al crystal studied it appeared to be possible to estimate the absolute values of $\bar{L}$ and F. An average length of the dislocation segments was found to be 10^{-5} cm (the dislocation density λ in this calculations was taken equal to 10^8 cm^{-2} [14]). The results of the calculations for F are shown in Fig. 3, where the curves $H(F)$ and $d(F) = -\partial H/\partial F$ are plotted in accordance with the theory [5, 6]. It was found that the dislocation–pinning point interaction forces in Al at low temperatures (high external stresses) coincide by the order of magnitude with the electrostatic forces

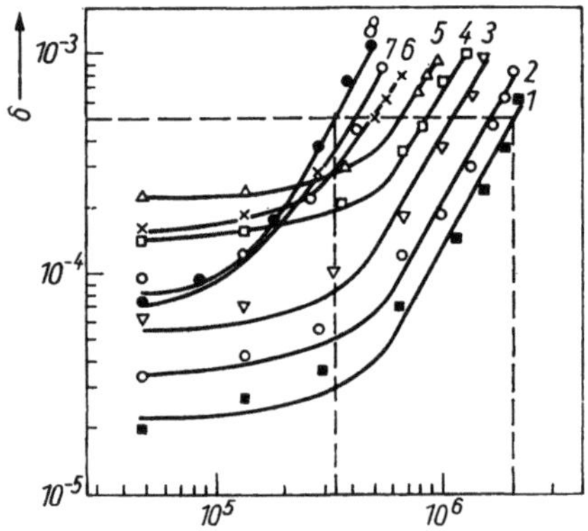

Fig. 1. The amplitude dependences of the logarithmic decrement δ for the pre-strained Al specimen at different temperatures: $T =$ (1) 4.2, (2) 17, (3) 58, (4) 97, (5) 130, (6) 180, (7) 220, (8) 270 K

[3]) At room temperature ($T = 295$ K) the curves $\delta(\tau_0)$ measured at increasing and decreasing the amplitude did not coincide. The low temperature tests were made after the large amplitude ($\tau_0 > 5 \times 10^5$ Pa) vibrational treatment at 295 K, the decrement δ exceeding the value 10^{-3}.

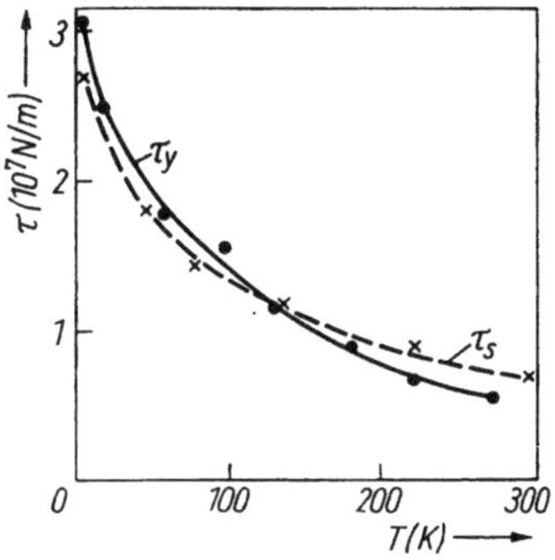

Fig. 2

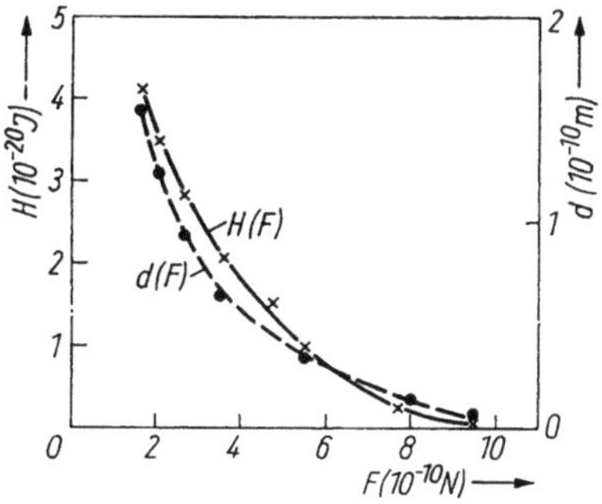

Fig. 3

Fig. 2. The temperature dependence of the yield stresses τ_s and τ_y (reduced shear stresses) determined by the data of mechanical and acoustic tests, respectively, for an Al single crystal

Fig. 3. The activation energy H and activation length d as functions of a dislocation–pinning point interaction force

which act between the ions in a perfect lattice: $F \approx e^2/a^2 \approx 10^{-9}$ N. Here e is the electron charge, a the lattice parameter. Under similar conditions (low temperatures) the characteristic values of the activation lengths are of the order of one tenth of Burgers vector.

Eventually, it should be noted that the obtained values of dislocation parameters are average magnitudes for the edges and the screws. However, it was shown that in alkali halide crystals [7 to 9], which enable us to study separately the edge dislocation behaviour, the thermal activation parameters ($H(F)$ curves) for both edge and screw components do not differ considerably.

References

[1] B. K. Kardashev, Fiz. tverd. Tela **19**, 1490 (1977).
[2] O. A. Voinova, B. K. Kardashev, and S. P. Nikanorov, Fiz. tverd. Tela **22**, 1058 (1980).
[3] R. B. Schwarz, R. D. Isaak, and A. V. Granato, Phys. Rev. Letters **38**, 554 (1977).
[4] R. D. Isaak, R. B. Schwarz, and A. V. Granato, Phys. Rev. B **18**, 4143 (1978).
[5] V. L. Indenbom and V. M. Chernov, Coll. Mechanisms of Relaxation Phenomena in Solids, Izd. Nauka, Moscow 1972; phys. stat. sol. (a) **14**, 347 (1972).
[6] R. B. Schwarz and A. V. Granato, Phys. Rev. Letters **34**, 1174 (1975).
[7] B. K. Kardashev and S. P. Nikanorov, Fiz. tverd. Tela **16**, 1073 (1974).
[8] B. K. Kardashev and S. P. Nikanorov, Fiz. tverd. Tela **13**, 160 (1971).
[9] B. K. Kardashev, S. P. Nikanorov, and O. A. Voinova, Fiz. tverd. Tela **16**, 1068 (1974); phys. stat. sol. (a) **25**, 359 (1974).
[10] J. S. Koehler, Imperfections in Nearly Perfect Crystals, J. Wiley & Sons Inc., New York 1952 (p. 197).
[11] L. D. Landau and E. M. Lifshitz, Statistical Physics, Izd. Nauka, Moscow 1964.
[12] A. V. Stepanov, Future of Metal Treatment, Leni zdat, Leningrad 1963.
[13] Yu. A. Burenkov, V. I. Ivanov, and B. K. Kardashev, Prib. i Tekh. Eksper. **5**, 210 (1979).
[14] B. Chalmers and R. King (Ed.), Progress in Metal Physics, Vol. 6, Pergamon Press, London/New York 1956.

(Received October 28, 1980)

phys. stat. sol. (a) **65**, 341 (1981)

Subject classification: 1.3; 1.4; 21; 21.1

Department of Materials Science and Technology, Kyushu University[1])

Electron Microscopic Observation of Periodic Antiphase Domain Structures in the α_2-Phase of Cu–Al Alloys

By

N. KUWANO, H. MISHIO[2]), M. TOKI[3]), and T. EGUCHI

The configuration of antiphase domains in α_2Cu–Al which has a one-dimensional long period superstructure is investigated using electron microscopy. From lattice images it is confirmed that the structure which has a non-integral antiphase period is made of two kinds of domains, each with an integral period, when the period is measured in the unit of the dimension of the basic ordered cell, $L1_2$. The fractional number of the period is regarded as an average of integral periods of the domains over the crystal. The mixing mode of the domains in the structure with an average period of a non-integer is found to be characterized as "irregular arrangement of uniform mixing" in the term of classification proposed by Fujiwara. Particularly, when the average period is nearly a half integer, the domains of different periods are mixed quite regularly. A singular configuration of antiphase domains which is ascribed to jogging of antiphase boundaries is observed: of the two kinds of domains which the antiphase structure is made of, minority domains join together in a group of a definite number, and thereby merge in the matrix.

Die Konfiguration der Antiphasendomänen in α_2Cu–Al Legierungen, die ein-dimensionale langperiodische Ordnung haben, wird mit Elektronenmikroskopie untersucht. Aus den Gitterabbildungen wird gefolgert, daß die Antiphasendomänenstruktur, die eine nicht ganzzahlige Periode in Einheiten der $L1_2$-Gitterlängen hat, aus zwei Arten Domänen jede mit einer ganzzahligen Periode zusammengesetzt. Die nichtganzzahlige Periode kann als Mittel der ganzzahligen Perioden im Kristall aufgefaßt werden. Es wird gefunden, daß die Mischungsmode der Antiphasendomänen in der Struktur mit einer nicht ganzzahligen mittleren Periode als „unregelmäßige Verteilung einheitlicher Mischung" in der Klassifikation von Fujiwara charakterisiert werden kann. Wenn die durchschnittliche Periode etwa halbzahlig ist, sind die Domänen mit verschiedenen Perioden besonders regelmäßig vermischt. Eine besondere Konfiguration der Antiphasendomänen, die auf Jogs der Antiphasengrenzen zurückgeführt ist, wird beobachtet: von den zwei Arten Domänen, aus denen die Antiphasenstruktur zusammengesetzt ist, sammeln sich die Minoritätsdomänen in einer Gruppe mit einer bestimmten Anzahl auf dem Hintergrunde der Majoritätsdomänen.

1. Introduction

In the Cu–Al system, the α_2 phase exists below 636 K around 23 at% Al [1]. Gaudig and Warlimont [2] investigated the crystal structure of the α_2 phase precipitated in the α phase by means of electron diffraction, and concluded that it has a one-dimensional long-period superstructure, the basic order lattice being an $L1_2$. The structure proposed by them is illustrated in Fig. 1, associated with a typical diffraction pattern. When a long period superstructure is characterized by the sequence of antiphase periods M, or the number of $L1_2$ lattices between adjacent antiphase boundaries, the structure shown in Fig. 1 can be described as $\{M\} = 1\bar{1}2\bar{1}1\bar{2}$, where upper bar means the domain of antiphase and $\{M\}$ a sequence of M. Alternatively, as the structure of the α_2 phase is made of domains with $M = 1$ and $M = 2$, it can be

[1]) Hakozaki, Fukuoka 812, Japan.

[2]) Now at Alps Electric Inc., Imoda, Tamayama, Iwate 028, Japan.

[3]) Now at Showa Denko, Ohmachi, Nagano 398, Japan.

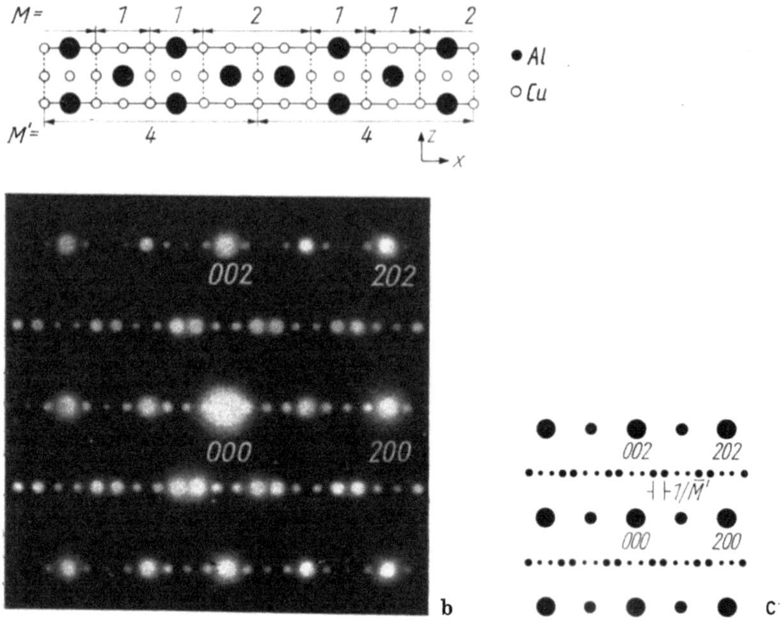

Fig. 1. a) Antiphase structure in the α_2 phase of a Cu–Al alloy ($M' = 4$). b) Typical diffraction pattern for $\overline{M}' = 4$. Spots flanking fundamental spots are due to double reflection. c) Schematic representation of pattern b)

characterized by the sequence of the distances between adjacent domains of $M = 2$. Then, $\{M'\} = 4\bar{4}$. The latter consideration, which is adopted in the present paper, corresponds to the concept that the structure of the α_2 phase can be derived from DO_{22} order lattice by inserting antiphase boundaries at every M'-th face-centred lattice. Since the distance is equal to an integral multiple of the lattice dimension of the face-centred lattice, the numercial value of M' should be integral. However, when the value is evaluated from the spacing of the splitting order-lattice reflections, the value obtained in this way being designated as $\overline{M}'$ hereafter, it is not integral in general, varying almost continuously with the Al content in the alloy [3 to 5], as in many other one-dimensional superstructural alloys. It has been confirmed for this alloy [3, 4] that the change in $\overline{M}'$ can be represented satisfactorily by the relation derived theoretically by Sato and Toth [6]. The non-integral value of $\overline{M}'$ can be interpreted as an average of different integral values of M' over the crystal. The mixing mode of the antiphase domains, however, has been clarified for very few alloys. In this work, antiphase domains in α_2Cu–Al have been observed directly by using electron microscopy in order to examine the configuration of the antiphase domains of different integral periods.

2. Experimental

The Cu–Al alloys of 22 to 25 at% Al were made from 99.99% pure copper and aluminium in vacuum by using an induction furnace. The ingots of the alloys were homogenized above 1223 K and quenched into iced brine. The quenched ingots were rolled to sheets 0.5 mm thick, and subsequently heated up to the β-phase region and

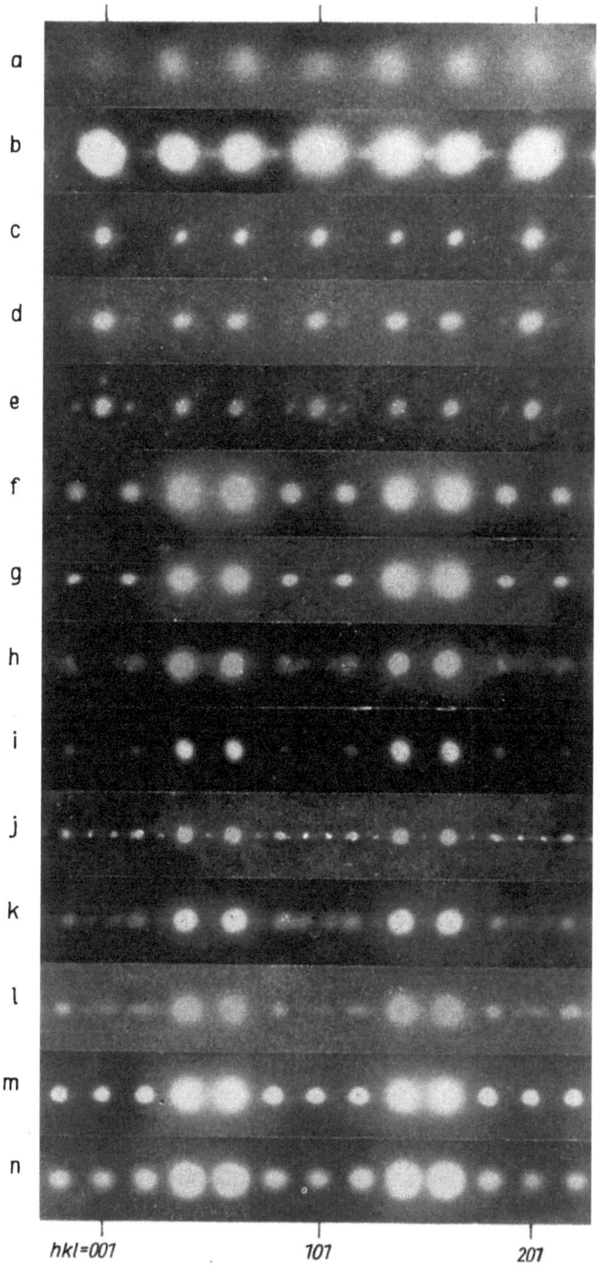

Fig. 2. A set of diffraction spots along $l = 1$ from Cu–Al alloys. The Al contents of the bulk specimens are a) to e) 19.3, f) 22.1, g) and h) 22.8, i) to k) 23.8, and l) to n), 24.7 at%, and the annealing temperatures are a) 473, b) 573, c) 546, d) 573, e) 635, and f) to n) 618 K. For a) to e) α_2 phase was formed from supersaturated α phase and for f) to n) from a martensite. The Al contents in α_2 are not equal to the composition of the bulk specimens. $\overline{M'}$ for a) to n) are: 3.2, 3.4, 3.6, 3.9, 3.9^7, 4.0, 4.1, 4.3, 4.4. 4.5, 4.6, 4.7, 4.9, 5.0, respectively

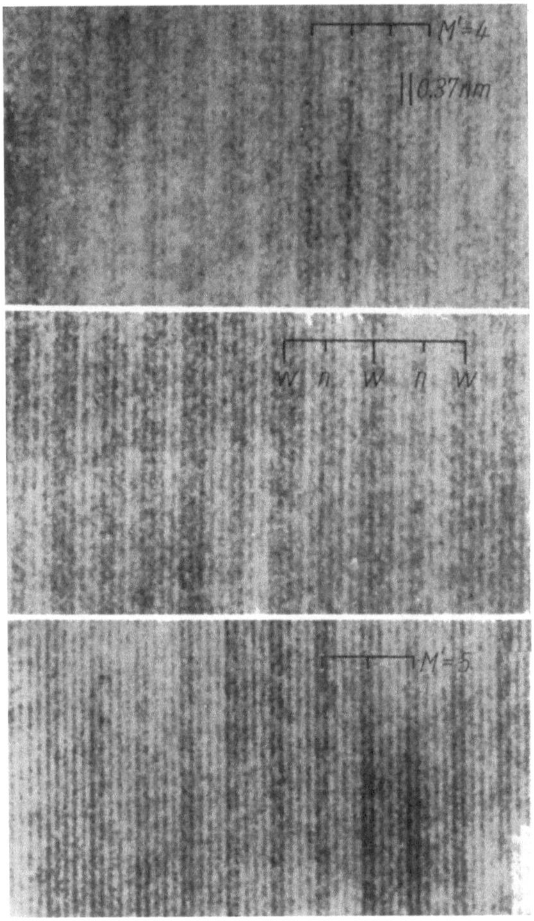

a

b

c

Fig. 3. BF images revealing lattice fringes for the specimens of a) $\overline{M'} = 4.0$, b) $\overline{M'} = 4.5$, and c) $\overline{M'} = 5.0$. In b), the modulation in fringe contrast is made of wide bands (w) and narrow ones (n)

quenched again to make martensites with no effects of the foregoing cold working. This heat treatment is necessary, because otherwise the deformed martensite decomposes rapidly into α and γ_2 phases when annealed even at a low temperature [7]. The martensites were annealed at 618 K for 30 to 63 days to promote transformation into the α_2 phase. The sheets annealed were chemically thinned in a solution of HNO_3, HCl, and H_3PO_4 (4:1:5) to foils 0.1 to 0.2 mm in thickness, from which discs of 3 mm diameter were cut out. Electropolishing was performed by a conventional jet method using an electrolyte of $HNO_3 + CH_3OH$ (1:1) at about 240 K. The alloy of 19.3 at% Al was made by the same way as for 22 to 25 at% Al. After homogenization at a high temperature the alloy was rolled to a foil 0.1 mm thick. Discs of 3 mm diameter, which were made from the foil, were quenched from 823 K and subsequently annealed at various temperatures below 632 K for prolonged times to precipitate the α_2 phase in the matrix of the α phase. Thinning was carried out by the jet method using a solution of H_3PO_4 at room temperature.

Electron microscopic observations were performed with JEM-1000 and JEM-200B of HVEM LAB, Kyushu University, equipped with side entry goniometer stages. The electron microscope JEM-1000 was operated at 500 or 750 kV in order to avoid serious radiation damage. Multibeam interference images were formed with reflections in the area within $2.5/a$ (BF image) or with order-reflections flanking a reciprocal lattice point $hkl = 101$ (DF image), where the lattice parameter a and the indices hkl are referred to Ll_2. The images were taken at rather low magnifications of 15000, 30000, or 100000× in order that the structure could be observed over a wide region of specimens.

3. Experimental Results and Discussion

3.1 Arrangement of antiphase domains

Fig. 2 shows a set of diffraction spots along $l = 1$, indicating that the spacing between the split spots, or $1/\overline{M'}$, changes with the annealing temperature or the Al content in the alloy. The reason why the value of $\overline{M'}$ changes with the annealing temperature for the alloy of 19.3 at% Al is that the Al content in the α_2 phase precipitated changes with the annealing temperature [8]. The strong spots at 001, 101, and 201 in the patterns a) to e) are attributable to α_2 precipitates of which the long-period direction is parallel to the y- or z-axis. One can see clearly that the value of $\overline{M'}$ changes almost continuously from about 3.2 to 5.0. Particularly, when $\overline{M'}$ is nearly half integral, splitting spots can be detected up to a high order. The tendency, which has been already noticed in α_2-Cu–Al [5, 8], suggests that in the structure of $\overline{M'} = 4.5$ the antiphase domains of $\overline{M'} = 4$ and 5, or the combinations of $\{M\} = 1\bar{1}2$ and $1\bar{1}1\bar{2}$, are distributed regularly, and in $\overline{M'} = 3.5$ so are those of $M' = 3$ and 4.

Fig. 3 shows typical lattice fringes from the specimens of $M' = 4.0$, 4.5, and 5.0. The spacing of the fine fringes is about 0.37 nm, corresponding to the lattice dimension of the Ll_2 cell of the alloy. On the micrograph one can see a modulation in contrast of the fringes somewhat like moiré patterns. As the modulation reveals the actual periodicity in the configuration of antiphase domains, the period of the modulation corresponds to M'. In Fig. 3a and c, the modulations have uniform periods of 4 and 5, respectively, in units of the spacing of the fine fringes. From the fact, it is clear that the structure of $\overline{M'} = 4.0$ is made of domains of $M' = 4$ uniformally and the structure of $\overline{M'} = 5.0$ of those of $M' = 5$. On the other hand, the modulation in b) is made of two kinds of dark bands, wide bands (marked by w) and narrow ones (by n). The spacing between the two adjacent wide bands is 9, revealing that the structure has a superperiod of 9. This means that the domains are mixed regularly to be $\{M'\} = 4\bar{5}4\bar{5}4\bar{5}$ with the average value of $\overline{M'} = 4.5$, as suggested by the diffraction pattern.

An attempt to observe a lattice image for the specimen of $M' = 3.5$ has been without success owing to the overlap of small precipitates of α_2 phase. However, from the appearance of splitting order-lattice spots up to a high order in the diffraction pattern (Fig. 2) and by analogy to the case of $\overline{M'} = 4.5$ it can be easily expected to have a sequence of $\{M'\} = 3\bar{4}3\bar{4}$. The structure of $\{M'\} = 3\bar{4}$ is identical to that proposed by Gaudig and Warlimont [9], who designated it as α_3. But it does not seem necessary to give a specific name to it. Because, as shown in this paper, the domains have a tendency, more or less, to distribute regularly, and no definite phase boundary between "α_2" and "α_3" has been recognized at least at the annealing temperature in the phase diagram.

Fig. 4 shows a lattice image of the specimen of $\overline{M'} = 4.37$. The spacings of the fringes correspond to M', revealing that there are only two types of domains or those of $M' = 4$ and 5. It is clearly seen that minority domains, or those of $M' = 5$ in this case, are not adjointed to each other. One can recognize a regular sequence of $\{M'\} = 4\bar{4}5\bar{4}4\bar{5}$ in one part and of $\{M'\} = 4\bar{5}4\bar{5}$ in another part. This mixing mode is characterized as

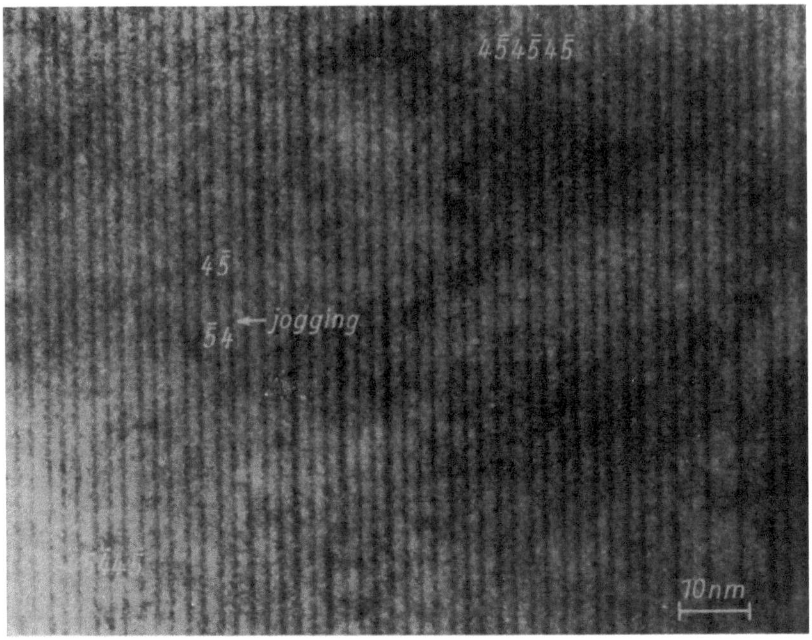

Fig. 4. DF image for the specimen of $\overline{M'} = 4.37$. The regular sequence of $\{M'\} = 4\bar{4}5\bar{4}4\bar{5}$ is observed in one part and the one of $\{M'\} = 4\bar{5}4\bar{5}4\bar{5}$ in another. The sequence of $4\bar{5}$ is interchanged to $5\bar{4}$ by jogging of an antiphase boundary at the point marked by an arrow. Note, that no domains of $M' = 5$ are adjoined to each other

"irregular arrangement of uniform mixing" in terms of the classification by Fujiwara [10].

Fig. 5 shows a lattice image of the specimen of $\overline{M'} = 4.13$. The spacings of the fringes correspond to M' and white striations to domains of $M' = 5$. It has been noticed [11] and also confirmed by a computer simulation [8, 12] that when an antiphase structure is made of two types of domains, minority domains are bright or dark in contrast compared to the majority ones under a certain observation condition. In Fig. 5 the number of domains of $M' = 4$ between adjacent domains of $M' = 5$ changes from 6 to 9, fluctuating about the average number 7. The configurations of antiphase domains shown in Fig. 3 to 5 suggest that a repulsive interaction might be present between domains of the same in period so that the minority domains tend to distribute at nearly equal intervals. The behaviour of the domains has been observed not only in this case but in the structure of $\overline{M'} = 3.5$ to 4.0 and $\overline{M'} = 4.5$ to 5.0 for α_2Cu–Al by the present authors [8], who have observed antiphase structures in α_2Cu–Al, α'Ag–Mg, and others using high resolution electron microscopy and analyzed the mixing modes statistically. In this connection, Kittel [13] suggested that such a regularity in the mixing modes is caused by the strains due to mixing of the domains of different types.

One can observe in Fig. 4 and 5 that the sequence of $\{M'\} = 5\bar{4}$ is interchanged to $4\bar{5}$ at the point indicated by an arrow, by a jog of antiphase boundary. In other words, if attention is paid to the location of the minority domain (the domain of $M' = 5$ in this case), the minority domain is considered as shifted sideways. The similar configuration to this has been observed in α'Ag–Mg [8, 14], Cu_3Sn [15], and Au_4Zn [11]. At the right-hand low corner of Fig. 5 a white striation is very broad. This is due to

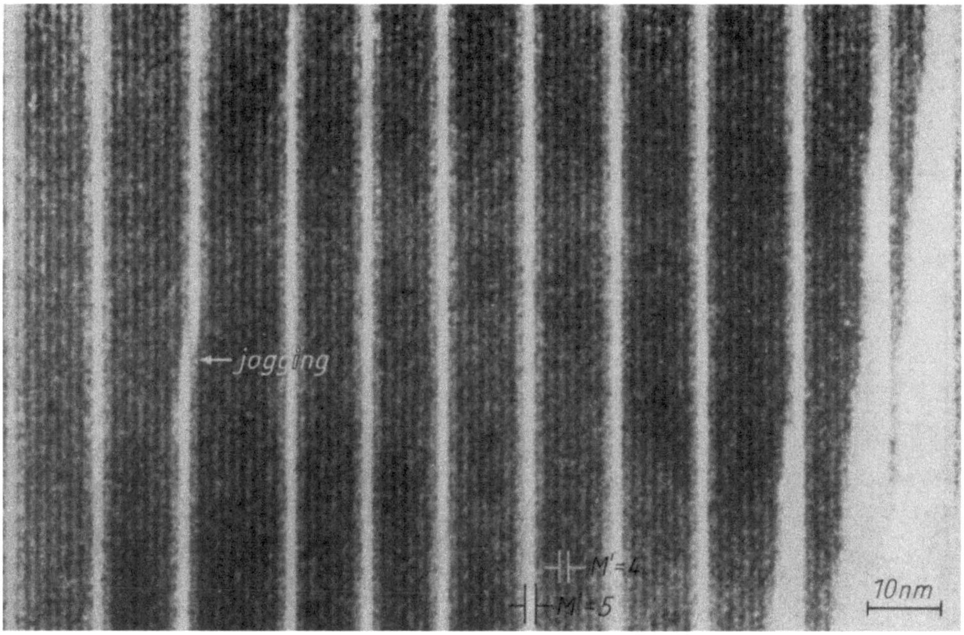

Fig. 5. DF image for the specimen of $\overline{M'} = 4.13$. White striations correspond to domains of $M' = 5$. is shifted sideways at the point marked by an arrow

a shift of domains of $M' = 5$ through jogs of the antiphase boundaries along the depth direction of the specimen. Guymont et al. [16] observed an antiphase boundary structure in AuCu II using high resolution electron microscopy, and emphasized that periodic antiphase boundaries in CuAu-II are wavy in shape, contrary to those in other alloys such as α'Ag–Mg and α''Cu–Pd; periodic antiphase boundaries in α'Ag–Mg and α''Cu–Pd are quite flat similar to those in α_2Cu–Al. The cause of the difference in electron microscopic image of antiphase boundaries between CuAu-II and other alloys can be found in the frequency of jogging of the antiphase boundary; antiphase boundaries in CuAu-II [17] jog more frequently than those in α'Ag–Mg, α''Cu–Pd, and α_2Cu–Al. Consequently, if antiphase boundaries in a somewhat thick specimen of CuAu-II are observed, the image of these should overlap and thereby look as if they were wavy.

3.2 Antiphase domain structure modified by jogging of antiphase boundaries

As shown in the previous section, when an antiphase structure is made of two kinds of domains, the minority domains can be shifted sideways through jogs of an antiphase boundary. If some minority domains are shifted towards each other, they should come into collision at last. An example of this case is shown in Fig. 6, where dark fringes correspond to antiphase boundaries and bright or dark striations to domains of $M' = 5$. It is clearly seen that domains of $M' = 5$ are joining together in a group of eight. The left junction contains nine domains of $M' = 5$, but one of these penetrates upwards (marked by p), so that the net change in the number of the domains of $M' = 5$ is still eight. Fig. 7 shows another example of the same configuration, where domains of $M' = 5$ are generated and eliminated in a complicated manner but always by eight with no exception. The model proposed for interpreting the con-

Fig. 6. A particular configuration of domains of $M' = 5$ in the matrix of $M' = 4$. Domains of $M' = 5$ are joining together in a set of eight. The left set has nine domains of $M' = 5$. One of these, however, penetrates upwards (p), so that the net change in the number of domains of $M' = 5$ is still eight

figuration is illustrated in Fig. 8. The insertion of one domain of $M' = M'_m \pm 1$ into the matrix of $M' = M'_m$ results in an advancing or receding shift of antiphase boundary by the dimension of the $L1_2$ cell. Consequently, when $2M'_m$ domains of $M'_m \pm 1$ are inserted, the total amount of the shift turns out to be just $2M'_m$, and thus the location of the antiphase boundary of the last domain of $M'_m \pm 1$ inserted is coincident with the location in the matrix. This means that when $2M'_m$ domains of $M'_m \pm 1$ get together, they can merge in the matrix.

The dark contrast at the junction of the domains of $M' = 5$ in Fig. 6 might be attributable to strains which result from the complicated jogging of the antiphase boundaries, for the contrast is changed as the specimen is tilted slightly and it is often observed as a pair of black and white lobes. The contrast of this type is frequently observed at a dislocation [18]. The darkness at the junction prevents us from observing the configuration of antiphase boundaries, but it is expected from the proposed model that the number of antiphase boundaries above the junction should be larger by two than below. In fact, the numbers between the points marked by A–A as well as A′–A′ in Fig. 6 are 32, and B–B as well as B′–B′ 30. This confirms that the present model is proper.

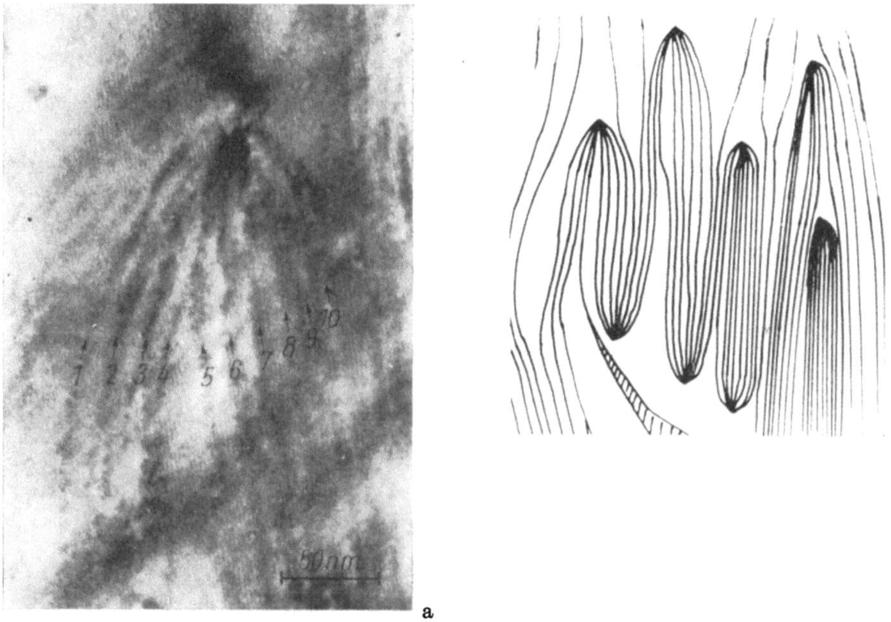

a b

Fig. 7. Complicated combination of joining of domains. Smoothly curved striations correspond to domains of $M' = 5$. Note that every junction contains eight domains of $M' = 5$

On a normally processed micrograph such as Fig. 6 one can observe hardly the configuration of antiphase boundaries in the junction of domains of $M' = 5$. But, when processed by repeating of reproduction so as to raise a contrast in the photograph, antiphase boundaries often appear even at the junction. Photographs processed in this way are shown in Fig. 9, where the configuration of antiphase boundaries near a junction of domains of $M' = 5$ can be examined. One can recognized in a) a hairpin configuration of antiphase boundary which is expected by the present model. In b), however, an antiphase boundary seems to terminate itself. Nevertheless, the difference

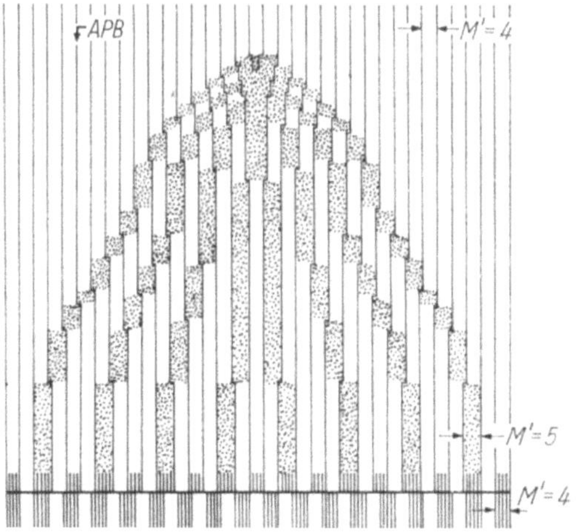

Fig. 8. Proposed model for interpreting the configuration in Fig. 6. The pointed areas are domains with $M' \neq 4$, striations are domains of in-phase region

in the number of antiphase boundaries between above and below the junction of the domains of $M' = 5$ is also two in this case. Two different interpretations for this contrast in Fig. 9b can be devised as follows.

1. As is seen in Fig. 8, antiphase boundaries are curved intensely near the junction of the domains of $M' = 5$. Since the micrograph was taken with order-lattice reflections which were split along the long-period direction, the fringe contrast perpendicular to the long-period direction is apt to be emphasized and curving parts of antiphase

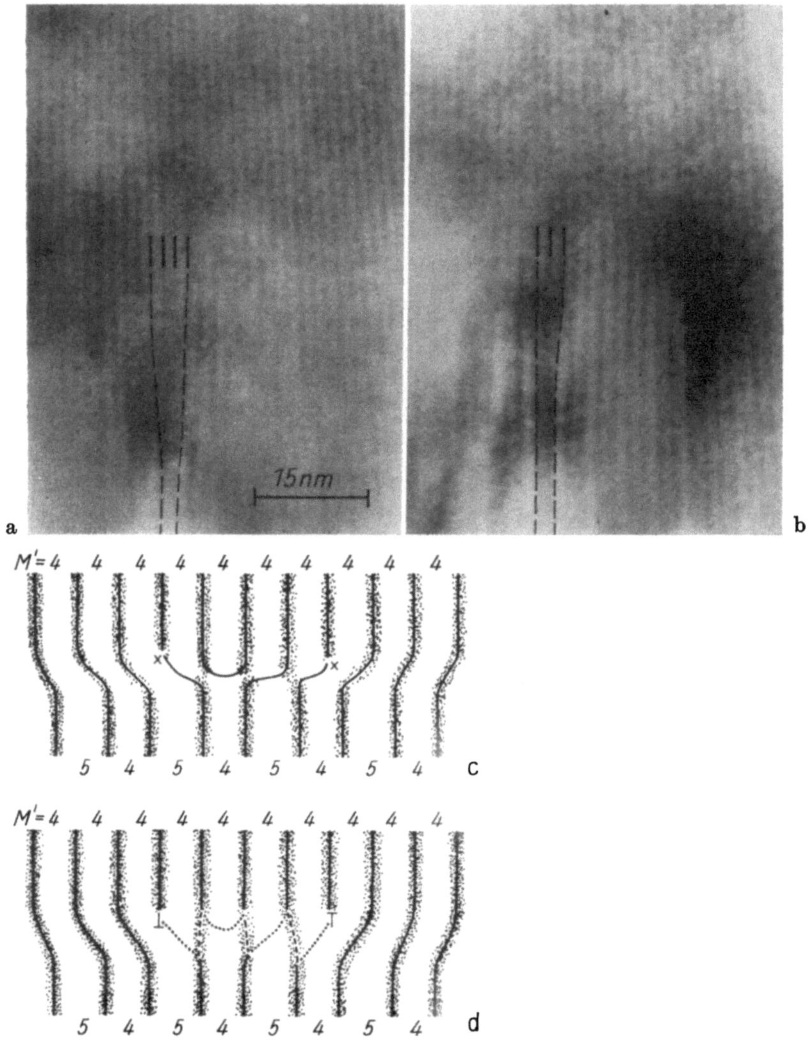

Fig. 9. Fringe contrast in the junction of domains of $M' = 5$. a) Observed fringe contrast revealing a hairpin configuration of antiphase boundary. b) Observed fringe contrast where an antiphase boundary seems to terminate itself. c) Possible interpretation with no dislocations. d) Possible interpretation with a pair of dislocations. The fringes in a) and b) are not completely parallel but somewhat radial, for these photographs were reproduced, viewed at an oblique angle so as to enlarge the contrast. | Antiphase boundary, ¦ antiphase boundary generated by dislocation, ⋮ antiphase boundary to be eliminated by dislocations, the pointed area is the fringe contrast to be observed

boundaries might be out of contrast. Then, some of the curving antiphase boundaries might seem as if they were conjoined to the adjacent antiphase boundary as illustrated in Fig. 9c, and a pair of fringes apparently terminate themselves at the points marked by × in the figure. One of the terminations is considered as having been observed in Fig. 9c.

2. When a pair of dislocations are introduced into an f.c.c. crystal, a fault plane with a shear of $\frac{1}{2}\langle 110\rangle$ is formed supplementally. If the fault plane is formed in an in-phase region of the ordered alloy, the plane is an antiphase boundary identically. But, if the plane is formed overlapping an antiphase boundary which has already existed, the antiphase boundary is eliminated. Consequently, if a pair of dislocations are introduced so that a zig-zag fault plane is formed in the manner as shown in Fig. 9d, a pair of antiphase boundaries can terminate themselves at the points where the pair of dislocations are inserted. Which of the interpretations 1. and 2. is valid for the image shown in b) is not clear in the present stage. In either case, however, the principal configuration of the antiphase boundaries in the set of domains of $M' = 5$ which are joining together is that illustrated in Fig. 8.

Since M'_m domains of $M'_m \pm 1$ can merge in the matrix in a group, domains of $M' = 4$ join together by ten in the matrix of $M' = 5$. An example of this configuration is shown in Fig. 10. The model for interpreting the configuration is illustrated in Fig. 10b, being anologous to Fig. 8.

The configurations shown in Fig. 6, 7, and 10 are apparently anologous to those, for instance, in Ni_3Mo [19] and χFe–Cr–Mo [20], where antiphase boundaries or shear planes join together in a group of definite numbers. In our case, a domain of $M'_m \pm 1$ can be thought of as made by inserting a displacement with a vector $\pm$ [100] into a domain of M'_m. As the vector does not lie on the plane formed by the displacement, the boundary is regarded as an antiphase boundary of the second kind. Thus, the structure

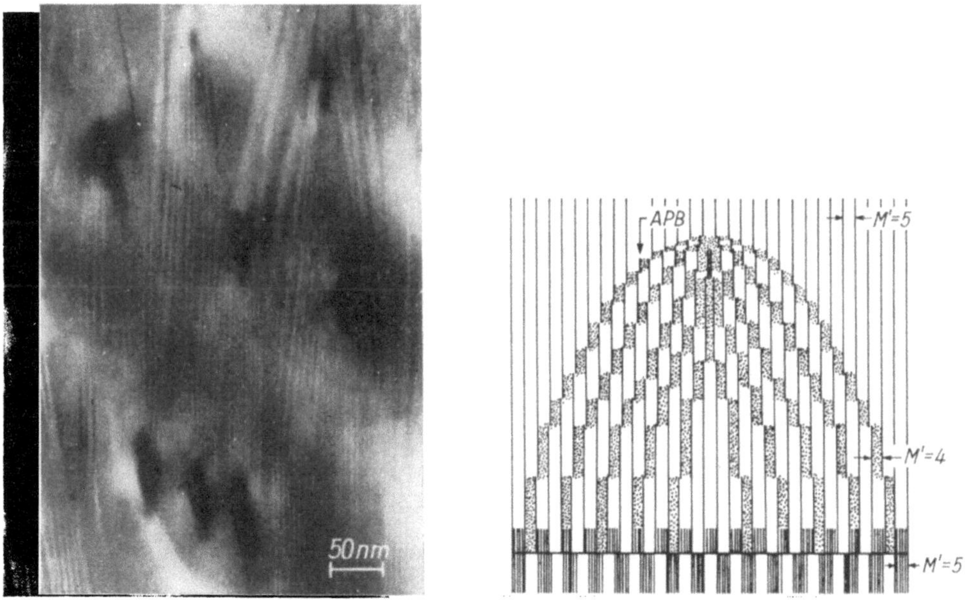

a

Fig. 10. a) A particular configuration of domains of $M' = 4$ in the matrix of $M' = 5$ and b) the schematic interpretation. Note that in the matrix of $M' = 5$ domains of $M' = 4$ join together in a set of ten

made of domains of M'_m and $M'_m + 1$ can be regarded as a fault structure which is derived from domains of M'_m, and in this sense, the group of domains of $M'_m + 1$ which join together by $2M'_m$ are identical to so-called "dissociated" antiphase boundaries in the terminology used by van Tendeloo et al. [19]. However, one can find some differences in the antiphase boundary structure between Ni_3Mo and α_2Cu–Al in the following respect. The boundary made by a displacement vector $\pm[100]$ in our case is a conservative one in the terminology used by van Tendeloo [19]. The total area of the boundaries, or the fraction of domains with $M'_m + 1$, in the crystal is unchanged by a heat treatment, because the value of $\overline{M'}$ is determined by the Al content in the alloy. On the contrary, the antiphase boundaries in Ni_3Mo are non-conservative and they are unstable thermally so that they can be eliminated by annealing the crystal.

Acknowledgements

The authors are greatly indebted to the members of HVEM LAB, Kyushu University for operation of the electron microscope and valuable discussions.

References

[1] R. P. Jewett and D. J. Mack, J. Inst. Metals **93**, 59 (1963/64).
[2] W. Gaudig and H. Warlimont, Z. Metallk. **60**, 488 (1969).
[3] E. Bernard and P. Duval, phys. stat. sol. (a) **34**, 135 (1976).
[4] N. Kuwano, T. Doi, and T. Eguchi, Trans. Japan. Inst. Metals **18**, 807 (1977).
[5] N. Kuwano, H. Mishio, and T. Eguchi, Modulated Structures-1979, AIP Conf. Proc. No. 53, 1979 (p. 273).
[6] H. Sato and R. S. Toth, Phys. Rev. **124**, 1833 (1961).
[7] N. Kuwano, T. Doi, and T. Eguchi, Trans. Japan. Inst. Metals **20**, 37 (1979).
[8] N. Kuwano and T, Eguchi, unpublished.
[9] W. Gaudig and H. Warlimont, Acta metall. **26**, 709 (1978).
[10] K. Fujiwara, J. Phys. Soc. Japan **12**, 7 (1957).
[11] G. van Tendeloo and S. Amelinckx, phys. stat. sol. (a) **43**, 553 (1977).
[12] N. Kuwano, M. Toki, N. Tanaka, and T. Eguchi, Proc. 6th Int. Conf. High Voltage Electron Microscopy, Vol. 4, Antwerp 1980 (p. 166).
[13] C. Kittel, Solid State Commun. **25**, 519 (1978); **31**, 121 (1979).
[14] R. Portier, D. Gratias, M. Guymont, and W. M. Stobbs, Acta cryst. **A36**, 190 (1980).
[15] M. van Sande, R. de Ridder, G. van Tendeloo, J. van Landuyt, and S. Amelinckx, phys. stat. sol. (a) **48**, 383 (1978).
[16] M. Guymont, D. Gratias, and R. Portier, Modulated Structures-1979, AIP Conf. Proc. No. 53, 1979 (p. 244).
[17] K. Mihama, J. Phys. Soc. Japan **31**, 1677 (1971).
[18] W. J. Tunstall, P. B. Hirsch, and J. Steeds, Phil. Mag. **9**, 99 (1964).
[19] G. van Tendeloo, J. van Landuyt, P. Delavignette, and S. Amelinckx, phys. stat. sol. (a) **25**, 697 (1974).
[20] M. Snykers, P. Delavignette, and S. Amelinckx, phys. stat. sol. (b) **48**, K1 (1971).

(Received November 26, 1980)

phys. stat. sol. (a) 65, 353 (1981)

Subject classification: 14.3.4; 2; 13.4; 22

Department of Mathematics, The University of Southampton[1]) *(a) and Institut für Theoretische Physik B der Rheinisch-Westfälischen Technischen Hochschule Aachen*[2]) *(b)*

Threshold Switching as a Generation–Recombination Induced Non-Equilibrium Phase Transition (II)[3])

By

D. J. ROBBINS (a)[4]), P. T. LANDSBERG (a), and E. SCHÖLL (b)

The model proposed in 1978 for threshold switching is explained more fully in terms of the generation–recombination induced non-equilibrium phase transition involved. This is done 1. by reference to the control parameter plane of band–trap impact ionisation coefficients in which the state of the system is represented by a trajectory on which the externally applied electric field is a parameter; 2. by a classification of various types of switching which are admitted by the model; 3. by the provision of more current–voltage curves to which the model leads; 4. by the derivation of approximate expressions for switching and delay times provided by the model, and 5. by the interpretation of these characteristic switching times in terms of phase portraits.

Das 1978 für das Schwellenschalten vorgeschlagene Modell wird mit dem Generations–Rekombinations-induzierten Nichtgleichgewichts-Phasenübergang vollständiger erklärt. Dies wird erreicht: 1. unter Bezugnahme auf die Ebene der Steuerparameter der Stoßionisationskoeffizienten, in der der Zustand des Systems durch eine Trajektorie dargestellt wird, bei der das äußere angelegte elektrische Feld ein Parameter ist; 2. durch eine Klassifizierung der verschiedenen Schalttypen, die durch das Modell geliefert werden; 3. durch den Vorschlag mehrerer Strom–Spannungs-Kurven, zu denen das Modell führt; 4. durch die Ableitung von Näherungsausdrücken für das Schalten und die Verzögerungszeiten, die das Modell angibt, und 5. durch die Erklärung dieser charakteristischen Schaltzeiten mit Phasenbildern.

1. Introduction

Threshold switching is recognised as a widely occurring phenomenon and has most recently been reviewed in the context of amorphous materials [1]. In [2] the idea that threshold switching could be viewed as a recombination–generation induced non-equilibrium phase transition was advanced. This topic has recently been discussed in [3]. In particular the semiconductor phase transition model (c_1) of [4], introduced again below, was discussed. This is the simplest of the recently found generation–recombination induced phase transition models [4 to 6] which also has the qualitative features of S-type (threshold) switching behaviour. In the notation of [2], which will be employed throughout, the reaction kinetic equations for this model depend upon seven control parameters: the five reaction constants given in Fig. 1, and the donor and trap concentrations N_D and N_t. The order parameters are the electron concentration (n) and the hole concentration (p). The rate equations are

$$\left.\begin{aligned} \dot{n} &= n[X_1 N_\text{D} - T_1^\text{S} P_\text{D} - (X_1 + T_1^\text{S})\, n - (B^\text{S} - X_1 - T_1^\text{S})\, p]\,, \\ \dot{p} &= p[X_4 P_\text{D} - T_2^\text{S} N_\text{D} - (X_4 + T_2^\text{S})\, p - (B^\text{S} - X_4 - T_2^\text{S})\, n]\,, \end{aligned}\right\} \quad (1)$$

[1]) Southampton S09 5NH, Great Brittain.

[2]) D-5100 Aachen, BRD.

[3]) The II in the title of the present paper is the result of treating [2] as Paper I.

[4]) Present address: Allen Clark Research Centre, Plessey Research (Caswell), Ltd., Caswell, Towcester, Northants NN12 8EQ, Great Brittain.

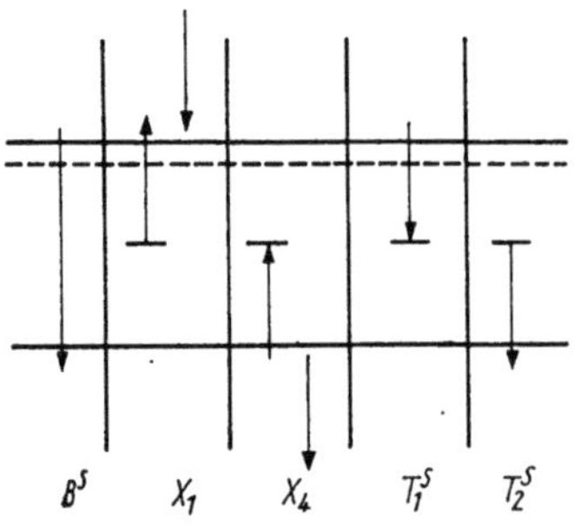

Fig. 1. Electronic processes taken into account in the model

where $P_D \equiv N_t - N_D$, and the donors are assuumed to be fully ionised. Assuming that $N_t > N_D$, there are four steady-state solutions, the stability range being given in brackets.

$$\left.\begin{array}{llll} \text{(i)} & n_1 = 0\,, & p_1 = 0 & (F_a, F_b < 0)\,, \\ \text{(ii)} & n_2 = 0\,, & p_2 = F_b/(X_4 + T_2^S) & (F_b, F_c > 0)\,, \\ \text{(iii)} & n_3 = F_a/(X_1 + T_1^S)\,, & p_3 = 0 & (F_a, F_d > 0)\,, \\ \text{(iv)} & n_4 = -F_c/D\,, & p_4 = -F_d/D & (D > 0)\,, \end{array}\right\} \tag{2}$$

where

$$\left.\begin{array}{ll} F_a \equiv N_D X_1 - T_1^S P_D\,, & F_b \equiv P_D X_4 - T_2^S N_D\,, \\ F_c \equiv B^S F_b - C\,, & F_d \equiv B^S F_a - C \end{array}\right\} \tag{3}$$

and

$$C \equiv (X_1 X_4 - T_1^S T_2^S)\, N_t\,, \qquad D = B^S(X_1 + X_4 + T_1^S + T_2^S - B^S)\,. \tag{4}$$

We shall also use $A \equiv P_D/N_D = N_t/N_D - 1$, which is a measure of the degree of compensation.

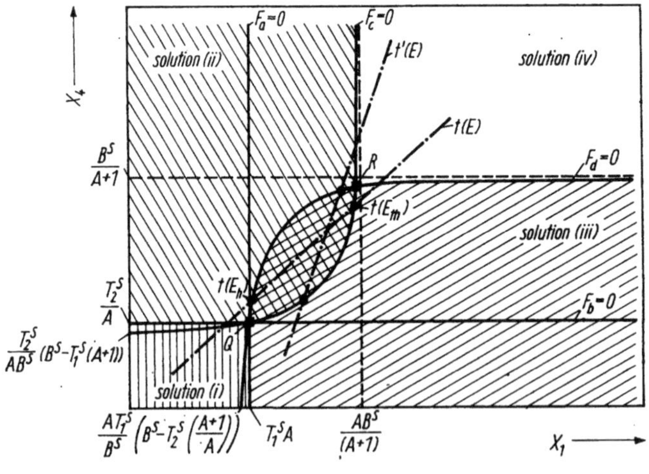

Fig. 2. A schematic plot of the stability region of the steady-state solution with positive n and p in the (X_1, X_4) control parameter plane when equations (1) and (5) hold. Trajectories $t'(E)$ and $t(E)$, and the special points Q and R, are discussed in the text

It was found in [2] that under the condition

$$\frac{B^S}{N_t} > \frac{T_1^S}{N_D} + \frac{T_2^S}{N_t - N_D} \tag{5}$$

the stability region of the steady-state solutions overlap for certain values of X_1 and X_4 as displayed in Fig. 2 below. That this overlap could be exploited to yield behaviour of the concentrations n and p as functions of the electric field (E) reminiscent of threshold switching was demonstrated by a numerical example.

2. Current–Voltage Characteristics

It is one purpose of the present paper to go beyond the discussion of electron and hole concentrations given in [2] by presenting a selection of plots of the current as a function of the electric field in Section 4, which can be obtained from this model for different values of the control parameters. These are shown in Fig. 3. The numerical values used for each curve are given in Table 1. The current $J(E)$ is plotted as the dimensionless quantity

$$\frac{J(E)}{|e|\, \mu_n N_D E} = \frac{n(E)}{N_D} + \frac{p(E)}{N_D}\, b\,, \tag{6}$$

where $b = \mu_p/\mu_n$ is the ratio of mobilities, and is assumed field-independent. In this study $N_D = 2 \times 10^{15}\ \text{cm}^{-3}$ was used. The magnitude of N_D in fact only scales the unit of current which is $|e|\, \mu_n N_D E$ in Fig. 3a to e and is therefore arbitrary. The characteristic switching times, introduced in Sections 5 and 6, vary approximately as N_D^{-1}, as is shown in Section 6. The remainder of this work is devoted to qualitative description of both the steady-state (Section 3) and the dynamic (Section 5) behaviour of our switching model.

3. Steady-States and Control Parameter Space

The qualitative features of the curves of Fig. 3 may be simply visualised from Fig. 2 in terms of a plot of the stability regions of the four steady-state solutions in the control parameter plane (X_1, X_4). The boundary curves of the stability regions are given in (2) by $F_a = 0$, $F_b = 0$, $F_c = 0$, $F_d = 0$, and $D = 0$. It is readily seen from (3) and (4) that the first four of these conditions are simultaneously fulfilled at a point Q where $X_1 = T_1^S A$ and $X_4 = T_2^S/A$. If (5) holds the curves $F_c = 0$ and $F_d = 0$ cross again at a second point R where $X_1 = AB^S/(A + 1) - T_2^S$ and $X_4 = B^S/(A + 1) - T_1^S$ and the solutions (ii) and (iii) overlap as shown in Fig. 2. From (4) it can be seen that the line $D = 0$ also passes through the point R cutting the X_1- and X_4-axes at $B^S - T_1^S - T_2^S$. However, since the condition that n_4 and p_4 are positive provides a stronger constraint, namely, $F_c < 0$, $F_d < 0$, the line $D = 0$ is omitted from Fig. 2. An overlap of other solutions is not possible in this model.

For control parameters corresponding to the point Q the two stable solutions (ii) and (iii) merge into one stationary point at the origin, while at R (ii) and (iii) merge into a stationary line. In this sense Q and R are the end points of the bistability region. Hence Q and R may be called critical points in analogy with the thermodynamic equilibrium terminology; the behaviour of the steady states near these critical points can be described by critical exponents [8].

As the electric field E is increased from low values the impact ionisation coefficients X_1 and X_4 increase and the system describes a trajectory across Fig. 2. Two examples, $t(E)$ and $t'(E)$, are shown as straight lines, though a trajectory will normally be a curve.

Table 1

Data used in Fig. 3

Fig.	X_1 (cm³ s⁻¹)	X_4 (cm³ s⁻¹)	T_1^S (cm³ s⁻¹)	T_2^S (cm³ s⁻¹)	B^S (cm³ s⁻¹)	N_t (cm⁻³)	$b = \mu_p/\mu_n$
a1, a2, a3	$3.0 \times 10^{-5}\, e^{-\frac{2 \times 10^4}{E}}$	$3 \times 10^{-5}\, e^{-\frac{2.25 \times 10^4}{E}}$	1.0×10^{-12}	1.0×10^{-14}	1.0×10^{-10}	3.0×10^{15}	10.0 (a1), 1.0 (a2), 0.1 (a3)
b1, b2	$3 \times 10^{-5}\, e^{-\frac{2.23 \times 10^4}{E}}$	$3 \times 10^{-5}\, e^{-\frac{2.0 \times 10^4}{E}}$	1.0×10^{-14}	1.0×10^{-12}	1.0×10^{-10}	6.0×10^{15}	0.2 (b1), 5.0 (b2)
c1	$3 \times 10^{-5}\, e^{-\frac{2.5 \times 10^4}{E}} + 5 \times 10^{-11}$	$3 \times 10^{-5}\, e^{-\frac{2.0 \times 10^4}{E}}$	1.0×10^{-14}	1.0×10^{-12}	1.0×10^{-10}	6.0×10^{15}	1.0
c2	$3 \times 10^{-5}\, e^{-\frac{2 \times 10^4}{E}}$	$3 \times 10^{-5}\, e^{-\frac{2.5 \times 10^4}{E}} + 6 \times 10^{-11}$	1.0×10^{-12}	1.0×10^{-14}	1.0×10^{-10}	2.5×10^{15}	1.0
d	$3 \times 10^{-5}\, e^{-\frac{2.0 \times 10^4}{E}}$	$3 \times 10^{-5}\, e^{-\frac{2.25 \times 10^4}{E}}$	1.0×10^{-17}	1.0×10^{-19}	1.0×10^{-10}	3.8×10^{-15}	0.5
e	$3 \times 10^{-5}\, e^{-\frac{2.35 \times 10^4}{E}}$	$3 \times 10^{-5}\, e^{-\frac{2.25 \times 10^4}{E}}$	1.0×10^{-12}	1.0×10^{-14}	1.0×10^{-10}	4.0×10^{15}	0.1

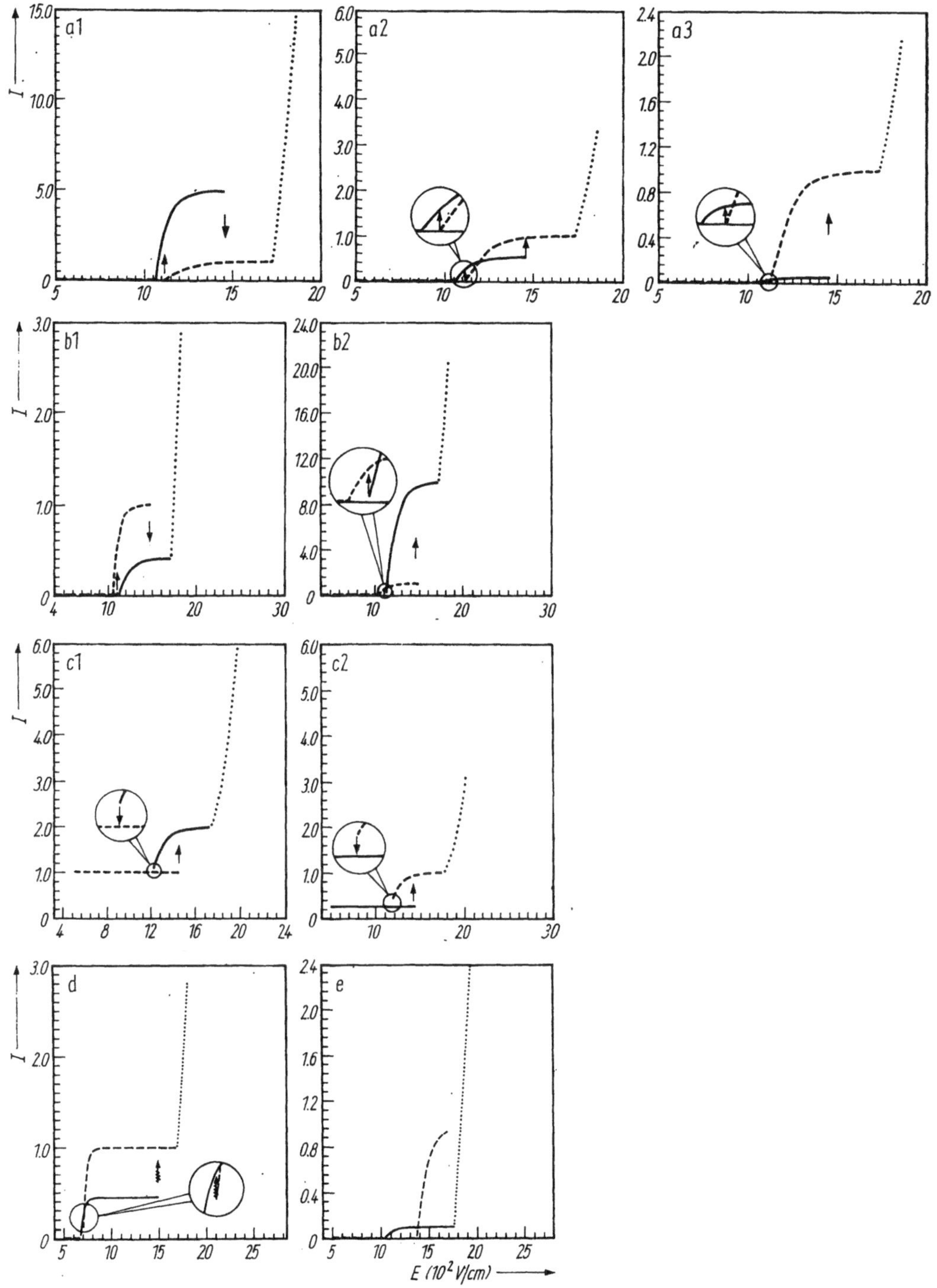

Fig. 3. Current–field characteristics for the reaction constants given in Table 1 in the four stability regions: – – – – solution (i), ——— solution (ii), - - - - - - solution (iii), ·········· solution (iv). The current is plotted in units of $|e|\ \mu_n N_D E$

If this trajectory crosses the overlap region, the necessary conditions for S-type switching may be produced. If the trajectory enters it via $F_d = 0$ and leaves via $F_c = 0$, or if it enters via $F_c = 0$ and leaves via $F_d = 0$ the usual type of switching characteristic is found. The first case corresponds to trajectory $t(E)$ in Fig. 2 with solutions (i), (ii), (iii), (iv) following each other as the field is raised. A jump occurs at $t(E_{th})$ as one passes from (ii) to (iii). The appropriate field is interpreted as the threshold field E_{th}. On reducing the field the sequence is simply reversed, but the low-field part of solution (iii) is utilised so that the jump from (iii) to (ii) occurs at a lower field, at $t(E_h)$. This field is interpreted as the holding field E_h. This situation is involved in Fig. 3a. The second case corresponds to trajectory $t'(E)$, and is involved in Fig. 3b and c1.

If T_1^S, T_2^S, or B^S were field-dependent then the structure of Fig. 2 would change with field. However, it is to be expected that the field dependence of the autocatalytic processes X_1 and X_4 which are taken into account here, will dominate at least for the overlap region. Thus one may consider the structure of Fig. 2 as representative for a range of electric fields extending at least from E_h to E_{th}.

4. Current–Voltage Characteristics: Numerical Examples

The current–field characteristics displayed in Fig. 3 show that the switching model described in the preceding sections allows a good deal of flexibility despite the relatively small variations in the control parameters employed (see Table 1), and despite the assumption that B^S, T_1^S, T_2^S and b are field-independent.

It is desirable, in view of the various alternatives, to introduce a precise nomenclature. On increasing an electric field, its value just before a jump is called a "threshold field", while the value just before a jump on decreasing the field is called a "holding field". We shall call a switch an (up, down)-switch if the jump at threshold is to a higher current, while it is to a lower current at the holding field. The normal switch is of the (up, down)-variety (cf. Fig. 3c) but our model also leads to a twisted form of S-type switching i.e. the (up, up)-variety (Fig. 3a2, 3a3, 3b2, 3d). It also leads to (down, up)-switching, which one might call reverse switching (Fig. 3a1, 3b1).

Fig. 3a is derived from the numerical example employed in [2] for three different values of the mobility ratio b, and is described qualitatively by the trajectory $t(E)$ in Fig. 2. It shows how variations in the mobility ratio b can lead from (down, up)-switching Fig. 3a1 to (up, up)-switching Fig. 3a2 and 3a3. For sufficiently small b normal (up, down)-switching would be produced. Fig. 3b shows curves corresponding to Fig. 3a when the significances of the order parameters n and p are reversed. Fig. 3c assumes a slightly modified field dependence for the ionisation coefficients (in the region of the field E_h and E_{th}). Both Fig. 3c1, which corresponds to trajectory $t'(E)$ in Fig. 2, and Fig. 3c2, in which the significances of the order parameters n and p are reversed relative to Fig. 3c1, show situations closely resembling the experimentally found (up, down) S-type switching. Fig. 3d shows the effect of very small values of the recombination coefficients T_1^S and T_2^S in reducing the holding field. In the limit $T_1^S = T_2^S = 0$ the model reduces to model (c) discussed in [4]. In this case $E_h = 0$, and solution (i) is never stable. Both solutions (ii) and (iii) can be stable from $X_1 = X_4 = 0$, and the switching behaviour is lost.

The curve shown in Fig. 3e is the result of a trajectory which crosses $F_d = 0$ on both entering and leaving the overlap region (rather than entering via $F_d = 0$, and leaving via $F_c = 0$ as for Fig. 3a). One then has the curious phenomenon that solution (iii), although overlapping solution (ii), is actually completely inaccessible by electric field manipulations alone, which provide merely the main curve shown in

Fig. 3e running continuously through solutions (i), (ii), and (iv). However, it is still possible to use this system as a switch by injecting carriers, as will be discussed in the next section.

5. Dynamics and Phase Portraits

The qualitative description given in Section 3 in terms of the control parameter plane (Fig. 2) enables discussion of the steady-state properties of switching, but it is not suitable for discussions of the dynamics of a switching event as characterised by the switching time τ_s, the delay time τ_d, and the decay time τ_{dec}. These can be described qualitatively in terms of phase portraits, which are obtained by plotting the integral curves of (1) in (n, p) phase space for fixed values of the control parameters. That given in Fig. 4 corresponds to a typical point in the overlap region of Fig. 2. Here the steady-state solutions (ii) and (iii) provide stable nodes, the solution (i) is an unstable node, and solution (iv) is a saddle (i.e. unstable) as indicated by the flow lines in Fig. 4. The region of phase space attracted towards solution (ii) is divided from the region attracted towards solution (iii) by a separatrix running through the unstable nodes and the saddle.

The dynamics of switching in our model will be discussed by supposing that the initial field $E_1 < E_{th}$ is raised abruptly to $E_2 > E_{th}$ for some time τ, and then restored to E_1, and that during this process the stable solutions are in turn (ii) and (iii), (iii), (ii) and (iii). This case corresponds to the trajectory $t(E)$ in Fig. 2. The switching dynamics along the trajectory $t'(E)$ can be discussed analogously. When (ii) and (iii) are stable, a regime such as indicated in Fig. 4 holds while another phase portrait (Fig. 5) would be applicable when only solution (iii) is stable. The numerical choices incorporated in Fig. 3a, 3c2, 3d would be appropriate for such switching. The dynamics which decide the final steady-state solution of the system are as follows. The notation prime for evaluation at field E_1 and double prime for evaluation at E_2 is adopted.

Assume that the system is initially in the stable steady state (ii) given by $n_2 = n_2' = = 0$; $p_2 = p_2'$ at the field E_1 (Fig. 4). When the field is raised to E_2, this is no longer a stable state of the reaction equations (1). Therefore, the system moves along a flow-line of Fig. 5 in a manner described by the time-dependent equations (1). If, when the field is restored to E_1, the separatrix in Fig. 4 has been crossed, the new steady-state solution (iii) will be reached along a flow-line of Fig. 4. If the separatrix in Fig. 4 has not been crossed the system will revert to solution (ii). Herein lies the origin of the delay time τ_d which, in our model, is the minimum length of time τ required for the system to cross the separatrix.

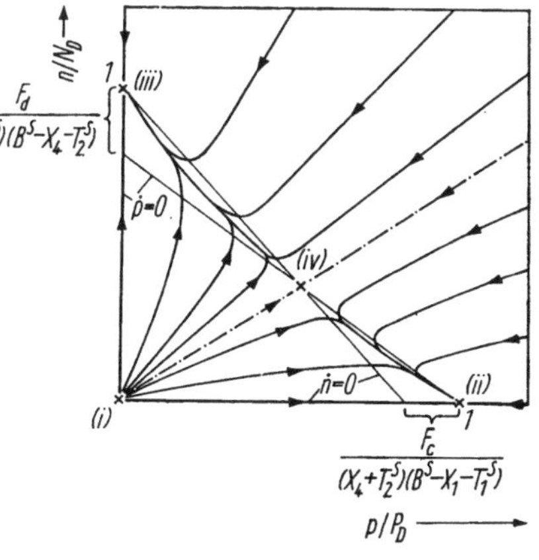

Fig. 4. A phase portrait corresponding to a typical point in the overlap region of Fig. 2. This point corresponds to the electric field E_1 in the system before and after the switching pulse

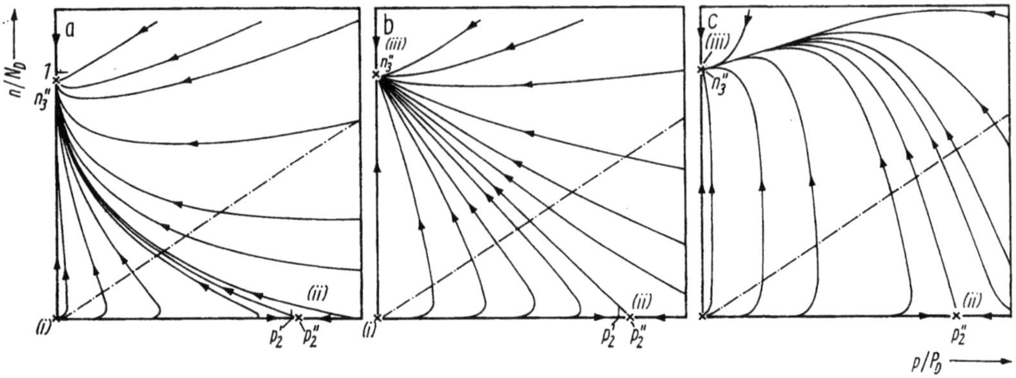

Fig. 5. Phase portraits corresponding to typical outside the overlap region of Fig. 2, but in the stability region of solution (iii). These points correspond to the electric field E_2 during the switching pulse. The chain line represents the separatrix and p_2' the steady state (ii) of Fig. 4, at field E_1. The values of T_1^S, T_2^S, B^S, and N_t are those of Fig. 3a. The values of X_1 and X_4 are (in $cm^3 s^{-1}$): a) $X_1 = 4.86 \times 10^{-11}$, $X_4 = 9.18 \times 10^{-12}$ (corresponding to $E = 1500$ V/cm); b) $X_1 = 8.31 \times \times 10^{-11}$, $X_4 = 1.67 \times 10^{-11}$ (corresponding to $E = 1563$ V/cm); c) $X_1 = 2 \times 10^{-10}$, $X_4 = 4.54 \times \times 10^{-11}$ (corresponding to $E = 1679$ V/cm, $= 15\%$ overvoltage)

Another feature of this model is that, if solution (ii) at field E_1 lies close to solution (ii) at field E_2, as it will be for small overvoltages, the initial movement of the system is very sensitive to fluctuations in the system. If there were no fluctuations, the system would follow the flow-line from p_2' to the saddle p_2'' in Fig. 5, and remain there, since this is a stationary, though unstable, state. Of course, an infinitesimally small fluctuation in concentration or in the recombination–generation coefficients will force the system away from the saddle (ii) towards the stable node (iii). Since the flow-lines are traversed more slowly the closer they are to a steady state (stable or unstable), the delay time depends upon the flow-line taken and thus upon fluctuations. A further consequence is that the system will spend most of its time "near' solutions (ii) and (iii) in Fig. 5 where $\dot{n}$ and $\dot{p}$ are small. Thus the delay time may be described as the amount of time spent by the system near state (ii), just as it arises experimentally [1]. The origin of the statistical regime for τ_d found in [1] for small overvoltages is also explained by this mechanism.

We now make three additional remarks which may be helpful.

(i) Strictly each phase portrait applies to a small region of the material, but for a uniform material in which the field is reasonably constant one can use the concept of an overvoltage as has been done above. This is $(E_2 - E_{th}) L$ where L is the length of the sample.

(ii) The switching time τ_s, in our model, is the time taken for the system to move from solution (ii) to solution (iii) at a given overvoltage minus τ_d. Numerical estimates of τ_d and τ_s were given in [2] and lay within the experimental range. These estimates stand for the present purposes and correspond here to the trajectory $t(E)$ and the characteristics of Fig. 3a.

(iii) The relation between Fig. 4 and 5 may be seen as follows. The horizontal distance marked on Fig. 4 has for small overvoltages the sign of F_c. As the field is increased beyond E_{th}, F_c changes sign but the vertical distance marked on Fig. 4 which has the sign of F_d merely decreases. This has the effect of taking solution (iv) out of the quadrant to negative values of n, and is therefore inaccessible. This yields the form of Fig. 5. If $F_c = 0$ solutions (ii) and (iv) coincide.

The decay time τ_{dec} describes the return step from solution (iii) to solution (ii) at E_h, when the field is lowered for a time τ from $E_1 < E_h$ to $E_2 < E_h$, in a completely analogous manner to τ_d.

This description of switching also holds good for a system with characteristics of Fig. 3e except that, as already noted, the high-field pulse must be accompanied or even replaced by carrier injection. The only condition is that the electric field is in the overlap region before and after the pulse. The rate equations (1) are modified during the pulse of length τ by the inclusion of additional injection terms. At the end of the pulse the rate equations (1) apply again with appropriate boundary conditions, and the excess carriers will recombine away. This lead one to solution (iii) or (ii) depending on whether the separatrix has, or has not, been crossed. If it has been crossed, electric field reduction to a value for which only solution (ii) is stable, i.e. outside the overlap region, will suffice to return the system to solution (ii). This system will therefore act as a switch if electric field manipulation is aided by injection.

6. Approximate Expressions for τ_s and τ_d

The general problem of calculating switching times from (1) according to the interpretation given in Section 5 can only be solved numerically as far as the authors are aware. However, it is possible to generate some simple expressions for τ_s and τ_d on the basis of a linearised system of equations for which time-dependent solutions for $n(t)$ and $p(t)$ may be obtained explicitly. Linearising (1) around the steady state one obtains for the deviations δn and δp from the steady state

$$\begin{pmatrix}\delta\dot{n}\\ \delta\dot{p}\end{pmatrix} = A\begin{pmatrix}\delta n\\ \delta p\end{pmatrix}, \tag{7}$$

where, near solution (ii),

$$A = A_2 = \begin{bmatrix} \dfrac{-F_c}{(X_4 + T_2^S)} & 0 \\ -p_2(B^S - X_4 - T_2^S) & -F_b \end{bmatrix}, \tag{8}$$

and, near solution (iii),

$$A = A_3 = \begin{bmatrix} -F_a & -n_3(B^S - X_1 - T_1^S) \\ 0 & \dfrac{-F_d}{(X_1 + T_1^S)} \end{bmatrix}. \tag{9}$$

This holds for all values of the electric field.

Consider a switch from solution (ii) to solution (iii) at E_{th}, as in Fig. 3a. As in the model discussed in Section 5, assume that at $t = 0$ the system is in a field E_1 in solution (ii). The eigenvalues of A_2 in (8) are

$$\lambda_1 = -\frac{F_c}{X_4 + T_2^S} \quad \text{and} \quad \lambda_2 = -F_b$$

with corresponding eigenvectors $(1, -\xi)$ and $(0, 1)$, respectively, where

$$\xi \equiv \frac{(B^S - X_4 - T_2^S)\,F_b}{(X_4 + T_2^S)\,F_b - F_c}.$$

A general solution of (7) near solution (ii) is therefore

$$\delta n(t) = a\,e^{\lambda_1 t}, \qquad \delta p(t) = -a\xi\,e^{\lambda_1 t} + b\,e^{\lambda_2 t}, \tag{10}$$

where a and b are constants determined by the initial conditions. Since both F_c' and $F_b' > 0$, λ_1' and $\lambda_2' < 0$, and (10) confirms that solution (ii) is stable at field E_1. For

$t > 0$ the field is increased to E_2. Since $F''_c < 0$ and $F''_b > 0$, $\lambda''_1 > 0$ and $\lambda''_2 < 0$, so that solution (ii) becomes a saddle.

As a first approximation it is assumed that the switching is influenced by fluctuations in the carrier concentrations of magnitude δn_0 and δp_0 arising in a negligibly small time compared to τ_d. With initial condition $\delta n(0) = \delta n_0$, and $\delta p(0) = \Delta p_2 \equiv p'_2 - p''_2 + \delta p_0$ in (10), this yields

$$n(t) = \delta n_0 \, e^{\lambda''_1 t}, \qquad p(t) = p''_2 - \delta n_0 \, \xi \, e^{\lambda''_1 t} + (\Delta p_2 + \delta n_0 \xi) \, e^{\lambda''_2 t}. \tag{11}$$

Equation (11) represents a solution which moves away from $n = \delta n_0$, $p = p'_2 + \delta p_0$ at $t = 0$. For small overvoltage F''_c is small and $|\lambda''_1| \ll |\lambda''_2|$; therefore, the term dependent upon λ''_2 in (11) can be neglected in the calculation of t_d since it represents fast decay to p''_2 at constant n. Approximately, therefore, the system moves away from the saddle (ii) along the line $n = -(p - p''_2)/\xi$. These results are assumed to hold for a large enough region extending from solution (ii) in a phase portrait appropriate to field E_2 (c.f. Fig. 5) to include the separatrix of Fig. 4 (which is appropriate to the field E_1).

The separatrix may be roughly approximated by a line $n = (F'_c/F'_d)\, p$ through the unstable node and the saddle in Fig. 4. Since the system will move very rapidly through regions of phase space with large $\dot{n}$ and $\dot{p}$ as it crosses the separatrix it is to be anticipated that τ_d will be insensitive to errors introduced by such a linear approximation. The closer E_1 is to E_{th} the better this approximation is. As E_1 approaches E_{th} the stable node (ii) and the saddle (iv) merge, as noted in Section 5. In this case the separatrix coincides with the p-axis.

The time taken for the system to reach the separatrix is therefore

$$\tau_d = \frac{1}{\lambda''_1} \ln \left| \frac{n_s}{\delta n_0} \right|, \tag{12}$$

where $n_s \equiv p''_2 F'_c/(F'_c \xi + F'_d)$.

As noted in Section 2, for a constant degree of compensation N_t/N_D, $\lambda''_1 \sim N_D$, and so approximately $\tau_d \sim N_D^{-1}$. Further, the smaller the overvoltage the closer F'' and λ''_1 lie to zero, thus leading to larger values of τ_d.

The switching time at some field E_2 may be estimated in a similar way. It is assumed that the pulse length $\tau > \tau_s + \tau_d$ in order that the switch may be completed at field E_2. (If $\tau_d + \tau_s > \tau > \tau_d$ the switch would be completed at field E_1.) The eigenvalues of A_3 in (9) at field E_2 are

$$\lambda''_m = -\frac{F''_d}{(X''_1 + T^S_1)} < 0 \quad \text{and} \quad \lambda''_M = -F''_a < 0. \tag{13}$$

The corresponding eigenvectors are $(-\eta, 1)$ and $(1, 0)$ respectively, where

$$\eta \equiv \frac{(B^S - X''_1 - T^S_1)\, F''_a}{(X''_1 + T^S_1)\, F''_a - F''_d}. \tag{14}$$

These expressions lead to a solution of the linearised rate equations (7) near solution (iii), analogous to (10) which is assumed to be adequate to describe τ_s. Since solution (iii) is a stable node, all trajectories ending at solution (iii), save two, are tangent to the eigenvector corresponding to the eigenvalue with the smaller magnitude. This observation allows three cases to be distinguished corresponding to Fig. 5a, b, and c respectively.

a) $|\lambda''_M| < |\lambda''_m|$

If $F''_d > F''_a(X''_1 + T^S_1)$ in (13), $|\lambda''_M| < |\lambda''_m|$. For $X''_1 \gg T^S_1$, as in Table 1, this corresponds to $X''_4 < (B^S - X''_1)\, N_D/N_t$ which is satisfied for small X''_4, and small overvoltage. A typical phase portrait is shown in Fig. 5a. τ_S may be approximated by the time the system takes to pass from n_S at the separatrix to within δn_∞ and δp_∞ of the final steady state (iii). The trajectory from near solution (ii) to solution (iii) becomes tangent to the n-axis near (iii), see Fig. 5a, and the system moves slowest there, so that the largest contribution to τ_S results from motion along the n-axis. That part of the solution depending upon λ''_m may therefore be neglected, then

$$\tau_S \approx \frac{1}{|\lambda''_M|} \ln \left| \frac{n''_3 - n_S}{\delta n_\infty} \right| \tag{15}$$

when n_S is defined in (12). With increasing overvoltage X''_1, and thus $|\lambda''_M|$, increase.

b) $|\lambda''_m| < |\lambda''_M|$; $\eta > 0$

With increasing overvoltage F''_d decreases and F''_a increases so that for moderate overvoltages (Fig. 5b) $X''_1 < B^S - T^S_1$ and $F''_d < F''_a(X''_1 + T^S_1)$ so that $|\lambda''_m| < |\lambda''_M|$ and $\eta > 0$. If the field dependence of X_1 and X_4 were such that at E_{th}, $X_4 > (B^S - X_1) \times \times N_D/N_t$ (assuming $X_1 \gg T^S_1$), then case b would apply to small overvoltages also.

The main contribution to τ_S is now via the slowly decaying part of the solution along the eigenvector $(-\eta, 1)$ corresponding to λ_m. In analogy with case a the λ_M solution can now be neglected leading to

$$\tau_S \approx \frac{1}{|\lambda''_m|} \ln \left| \frac{n''_3 - n_S}{\delta n_\infty} \right| . \tag{16}$$

This approximation is best for $D = 0$ when $\eta = \xi^{-1} = n''_3/p''_2$. In this case the trajectory, from (ii) to (iii), and the eigenvectors $(-\eta, 1)$ and $(1, \xi)$ are collinear.

c) $|\lambda''_m| < |\lambda''_M|$; $\eta < 0$

This holds for $X''_1 > B^S - T^S_1$, i.e. for large overvoltage (see Fig. 5c). The trajectory leaving (ii) approaches (iii) along the eigenvector $(|\eta|, 1)$. This corresponds to an overshoot in n as can be seen in Fig. 5c. The linear approximation is not very good in this case, but a crude estimate for τ_S may be obtained by replacing $(n''_3 - n_S)$ in (16) by the quantity

$$n_0 \equiv (p''_2 - n''_3 \xi)\, |\eta|/(1 + \xi\, |\eta|)$$

which is an approximation to the overshoot in n. It corresponds to the intersection of the lines defined by the eigenvectors $(|\eta|, 1)$ and $(1, -\xi)$. Now

$$\tau_S \approx \frac{1}{|\lambda''_m|} \ln \left| \frac{n_0}{\delta n_\infty} \right| \tag{17}$$

with increasing overvoltage the system moves closer to the line $F''_d = 0$ in Fig. 2, hence $|\lambda''_m|$ becomes small, resulting in an increasing "settling time" for τ_S.

The treatment of fluctuations given here is clearly inadequate; however, since τ_d is fairly insensitive to δn_0 in a reasonable range of statistical fluctuations, the above expression should still provide useful estimates. The quantity δn_∞ has a direct interpretation in experiment where, as a result of fluctuations, the theoretical steady state may never be exactly attained, and it is necessary to fix some more or less arbitrary limit such as the 10 and 90% points used in [7].

τ_s and τ_d have already been computed numerically for our model in [2], and the approximate analytical expressions (12) and (17) are in good agreement. This agreement becomes best for $\delta n_0 \approx 10^{-3}\, n_S$ and $\delta n_\infty \approx 10^{-1}\, n_0 \approx N_D/30$, at 15% overvoltage (Fig. 5c).

Analogous results for switching from solution (iii) to solution (ii) are readily obtained from (7) to (9) together with the corresponding τ_{dec}, but will not be given here.

7. Conclusions

The flexibility of the model already emphasized in [2] suggests that one need not restrict attention to the model (c_1) introduced in [4] and considered here, and one may still obtain switching behaviour in more complicated theoretical models [8]. Such models may of course furnish better agreement with experiment, if only because more processes will contribute. Nonetheless it may be useful to consider how the present model, which is the simplest of this type, may be compared with experiment.

To do this, one needs to establish experimentally likely values for the five reaction constants of Fig. 1, together with their field dependences. The model assumes B^S, T_1^S, and T_2^S to be independent of field, while dependences of X_1 and X_4 need not conform to that given in Table 1. The concentrations N_D and N_t have also to be known. If inequality (5) is then found to be violated, the model is inapplicable.

Assume the condition is satisfied. The next point to check is that for the field dependences involved the trajectory $t(E)$ generated by field manipulations crosses the overlap region in a properly constructed Fig. 2. Such a construction is straightforward using (3) and (4). If it does not cross the overlap region the model is inapplicable. If it does, equations (11) and (12) should furnish at least the order of magnitude of switching and delay times. If so, one should construct the (n, E)- and (p, E)-curves from which, using possibly field-dependent mobilities, the theoretical current–voltage characteristic should be derived and compared with experiment. One can then also solve (1) numerically for transient situations to obtain more accurate characteristic switching times as was done in [2].

If the agreement with experiment is poor, then the basic proposal [2] may still be appropriate using a more complex model.

The results obtained in this paper can be transferred to two other classes of models without changing the electronic processes involved, by simply redefining N_D. First, if in addition to N_D ionised donors and N_t traps the semiconductor contains N_A fully occupied acceptors, the charge neutrality condition is modified so that N_D is replaced by $N_D - N_A$, and P_D by $N_t + N_A - N_D$ in all of the above formulae, assuming that $N_A < N_D < N_A + N_t$. Secondly, the semiconductor may, in addition to N_D ionised donors and N_A occupied acceptors, contain N_t recombination centres with charge character (N_t^0, N_t^+), i.e. the neutral centre N_t^0 can loose an electron, in contrast to the previous models where the neutral traps N_t^0 could accept an electron: (N_t^0, N_t^-). In this case N_D has to be replaced by $N_D + N_t - N_A$ and P_D by $N_A - N_D$, assuming that $N_D < N_A < N_D + N_t$. This class of models includes the case $N_D = 0$ when all of the electrons are provided by the recombination centres.

Acknowledgements

Thanks are due to Prof. N. Klein, Haifa, for helpful comment.

References

[1] D. Adler, H. K. Henisch, and N. F. Mott, Rev. mod. Phys. **50**, 209 (1978).
D. Adler, M. S. Shur, M. Silver, and S. R. Ovshinsky, J. appl. Phys. **51**, 3289 (1980).
[2] P. T. Landsberg, D. J. Robbins, and E. Schöll, phys. stat. sol. (a) **50**, 423 (1978).
[3] P. T. Landsberg, European J. Phys. **1**, 31 (1980).
[4] E. Schöll and P. T. Landsberg, Proc. Roy. Soc. **A365**, 495 (1979).
[5] P. T. Landsberg and A. Pimpale, J. Phys. C **9**, 1243 (1976).
[6] E. Schöll, Proc. Roy. Soc. **A365**, 511 (1979).
[7] H. K. Charles, Jr. and C. Feldman, J. appl. Phys. **46**, 819 (1975).
[8] E. Schöll, to be published.

(Received March 17, 1981)

phys. stat. sol. (a) 65, 365 (1981)

Subject classification: 14.3 and 15; 22.6

Indian Association for the Cultivation of Science, Calcutta[1])

Transport Properties of Lithium and Sodium Doped Nickel Oxide

By

M. B. DUTT, R. BANERJEE, and A. K. BARUA

The electrical resistivity and thermoelectric power of ceramic samples of NiO doped with Li and Na are studied as function of temperature and dopant concentration. The resistivity ϱ is measured in the temperature range 100 to 700 K and thermoelectric power α in the range 100 to 300 K. The results obtained indicate band conduction as the mechanism for transport in NiO. The resistivity and thermoelectric power of Na doped NiO are found to be higher than the corresponding Li doped samples. This result is explained as due to the increased strain in the lattice by the introduction of Na which has a significantly higher ionic radius than nickel.

Der spezifische Widerstand und die Thermospannung von Keramikproben von mit Li und Na dotiertem NiO werden in Abhängigkeit von der Temperatur und Dotierungskonzentration untersucht. Der Widerstand ϱ wird im Temperaturbereich 100 bis 700 K und die Thermospannung α im Bereich 100 bis 300 K gemessen. Die erhaltenen Ergebnisse zeigen, daß der Transportmechanismus in NiO Bandleitung ist. Der Widerstand und die Thermospannung von Na dotiertem NiO ist höher als in entsprechenden Li-dotierten Proben. Dieses Ergebnis wird mit der erhöhten Spannung im Gitter durch die Einführung von Na erklärt, das einen beträchtlich höheren Ionenradius als Nickel hat.

1. Introduction

Owing to their interesting properties and increasing applications, oxide materials claim particular attention of physicists, chemists, and technologists. All oxides are not similar in their properties. One group of oxides with density less than that of the metal includes the transition metal oxides, such as, NiO, CoO, TiO_2, FeO. In these, the distance between the metal ions in the oxide is greater than that in the metal.

The semiconducting properties of the oxide materials arise from the presence of defects in the form of vacancies or interstitials or of foreign atoms acting as donors or acceptors. A drastic effect of deviation of stoichiometry is observed in NiO. It is well known that pure and almost stochiometric nickel oxide is an insulator with the room temperature resistivity of approximately 10^{13} Ω cm [1]. The resistivity can be lowered to about 1 Ω cm by addition of lithium. Verwey and de Boer attributed this to the creation of Li^+–Ni^{3+} acceptor centres [1]. At sufficiently high temperatures the Ni^{3+} holes can detach from the Li^+–Ni^{3+} centres becoming quasi-free. These holes can then move through the lattice by the intercharge of electrons between Ni^{3+} and Ni^{2+} ions, the activation energy of conductivity being equal to the energy necessary for detaching the hole from the Li^+ ion.

Numerous work has been done on nickel oxide, specially nickel oxide doped with lithium. However, interpretations of the results obtained by different groups of workers vary. Verwey and de Boer's model [1] was generally accepted till Morin [2] put forward

---

[1]) 700032 Calcutta, India.

the concept of hopping mechanism in NiO. According to this model, the conduction mechanism in NiO could be considered as a thermally activated diffusion process, the activation energy for the diffusion being a consequence of the trapping of electron holes by the lattice polarization. In fact, until 1965, the trapping mechanism of conduction in NiO was accepted by many investigators [3 to 6]. However, investigations carried out by other workers [7 to 11] led to results which contradicted the hopping model and pointed to the mechanism of hole transport in the narrow band.

Extensive studies on Hall and Seebeck effects and electrical conductivity in single crystals of NiO doped with Li led Austin and co-workers [12] to conclude that impurity conduction predominates upto room temperature even in samples without Li doping. There have been repeated suggestions [12 to 15] that mixed carrier conduction may occur in this oxide. On the other hand, many investigators [16 to 21] have considered NiO to be a broad-band semiconductor.

It is thus seen that there are many contradictory theories regarding the conduction mechanism in NiO. One interesting point in this connection is the fact that all workers cited above have tried to explain the conduction mechanism in NiO either by taking a pure sample or by doping with lithium. No systematic work on this oxide doped with elements of the same alkali metal as lithium has been done. Lithium has an ionic radius less than that of nickel. On the other hand, sodium, belonging to the same alkali metal group, has an ionic radius greater than that of nickel. In the present work we have studied the electrical conductivity in the temperature range 100 to 700 K and the thermoelectric power in the temperature range 100 to 300 K of Li and Na doped NiO. The experimental data have been analysed to yield information on the conduction mechanism in NiO and the effect of introduction of an impurity (Na) having a higher ionic radius than that of Ni.

2. Experimental

Grade I NiO and Li_2CO_3 were obtained from Johnson Matthey and Co, (UK) in powder form, the total impurities being less than 10 and 11 ppm, respectively. Powdered Na_2CO_3 were procured from E. Merck (FRG) and the impurity was less than 50 ppm.

To make Li and Na doped samples of NiO, the powders were taken in appropriate proportions and mixed intimately in an agate mortar. The mixture was kept in a furnace at about 1100 K for seven days. The powder was then taken out of the furnace and pellets of 14 mm diameter and thickness about 1 to 2 mm were made in a die with the help of a hydraulic press. These pellets were again sintered in air at about 1300 K for three days and then cooled in the furnace. The cooled pellets were mounted on the sample holder for the measurements of electrical resistivity and thermoelectric power.

The measurements of electrical resistivity in the temperature range 300 to 700 K were made with a sample holder of molybdenum metal. The sample was placed on a quartz disc at the base of the sample holder at the time of mounting. The electrical resistivity was measured by the in-line array four-probe technique and pressure contacts of the leads were made on the top of the sample. The sample holder was placed in an alumina tube furnace enclosed with a thick layer of quartz wool. The temperature of the furnace could be controlled within $\pm 1°$ and it was measured by a iron–constantan thermocouple. The furnace was covered by a water cooled cooling jacket made of copper. The cooling jacket was covered with a metal bell jar and the whole system was evacuated to a vacuum $\approx 10^{-5}$ Torr at the time of measurements.

The electrical resistivity and thermoelectric power in the temperature range 100 to 300 K were measured with a sample holder made of stainless steel. Suitable arrangements for maintaining a temperature difference between the two ends of the sample were made by electrical heating. The temperatures were measured by copper–constantan thermocouples. The sample holder was placed inside a cryostat which was immersed in a dewar containing liquid nitrogen.

Measurement of the resistivity of Li and Na doped NiO samples were carried out by using the standard formula of Valdes [22]. When the distances between the probes are equal, then the resistivity ϱ is given by

$$\varrho = 2\pi S(V/I)\,, \tag{1}$$

where S is the interprobe distance, V the potential between the two inner probes, and I the current through the circuit. The interprobe distances were small compared to the thickness of the sample and the outer probes were away from the edge of the sample. The voltage was measured by a Keithley model 610C electrometer. In the low temperature range (below 300 K) the current was measured by an electrometer (Electronics Corporation of India, EA815) and in the higher range by a digital multimeter. For thermoelectric power measurements the difference in voltage was measured by the Keithley 610C electrometer.

3. Results and Discussion

3.1 Resistivity

The electrical resistivity data of 0.1 and 0.5 at% Li and Na doped NiO in the temperature range 100 to 700 K are shown in Fig. 1. It is seen that the resistivity ϱ changes by factor of 10^9 in the temperature range of our study in the lithium doped samples and for sodium doped samples the change is by a factor of 10^8. The lg ϱ versus $10^3/T$ curves may be divided into several linear regions. The activation energies corresponding to the different linear regions have been evaluated and are shown in Table 1. It is seen from Table 1 that the activation energy at first increases with the increase in temperature and then decreases. This indicates not only exhaustion of majority centres but also the small temperature dependence of the mobility.

The ionic radius of Li (≈ 0.68 Å) is smaller than that of nickel (≈ 0.72 Å) so that substitution of a Ni^{2+} site by a Li^+ ion does not cause any lattice strain. NiO is a p-type semiconductor and lithium acts as an acceptor in NiO so that the resistivity decreases with increasing dopant concentration. It may be mentioned that Goodenough et al. [23] experimenting with sintered samples and Koide [24] experimenting with single crystals of NiO observed that the lattice constant decreases monotonically with the increase in lithium concentration. This result is in accordance with the lower ionic radius of lithium.

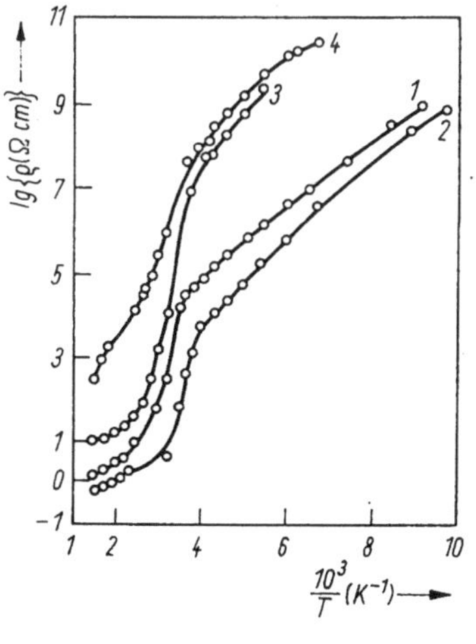

Fig. 1. Plot of lg ϱ versus $10^3/T$ of Li and Na doped NiO; (1) 0.1 at% Li doped NiO, (2) 0.5 at% Li doped NiO, (3) 0.1 at% Na doped NiO, (4) 0.5 at% Na doped NiO

Table 1

Activation energies of lithium and sodium doped NiO as obtained from the electrical resistivity data

dopant	concentration of dopant (at%)	temperature range (K)	activation energy (10^{-12} erg)
lithium	0.1	below 290	0.26
		290 to 330	1.28
		330 to 380	0.48
		380 to 700	0.16
	0.5	below 250	0.29
		250 to 310	1.17
		310 to 700	0.19
sodium	0.1	below 260	0.43
		260 to 330	2.07
		330 to 400	0.74
		400 to 700	0.18
	0.5	below 260	0.37
		260 to 330	1.06
		330 to 400	0.64
		400 to 700	0.48

On the other hand, sodium has an ionic radius (≈ 0.92Å) considerably higher than that of nickel. Thus the lattice suffers more and more strain with the increase in the sodium dopant concentration. Sodium like lithium acts as an acceptor in NiO to reduce the resistivity. But the resistivity increases with the higher dopant concentration of sodium. This result may be attributed to the increase in the lattice strain which is likely to reduce the effective carrier concentration and/or mobility so that the resistivity increases.

3.2 Thermoelectric Power

The thermoelectric power α obtained for Li and Na doped NiO has been shown in Fig. 2 in which the quantity $\alpha q/2.3k$ (q being the electronic charge and k the Boltzmann constant) has been plotted against $10^3/T$. The turn-over in $\alpha q/2.3k$ versus $10^3/T$ curves can be explained by taking into account impurity conduction as described by Bosman and Crevecoeur [16]. According to this model, the electrical conductivity is assumed to comprise two parts, viz: the free-hole conductivity σ_h and the conductivity σ_i induced by impurities. We then have

$$\sigma = \sigma_h + \sigma_i \tag{2}$$

and

$$\alpha = \frac{k}{q}\left[\frac{\sigma_h}{\sigma}\left(\frac{E_F}{kT} + A\right) + \frac{\sigma_i}{\sigma}\left(\frac{E_F - E_A}{kT}\right)\right], \tag{3}$$

where E_F is the Fermi level, E_A is the energy level of the acceptors with respect to the transport level, A is a term connected with the kinetic energy of the free holes. In the range $\sigma_i \ll \sigma$, the second term on the right-hand side of (3) is negligible. Thus as the temperature decreases, α increases. From the theory of partially compensated

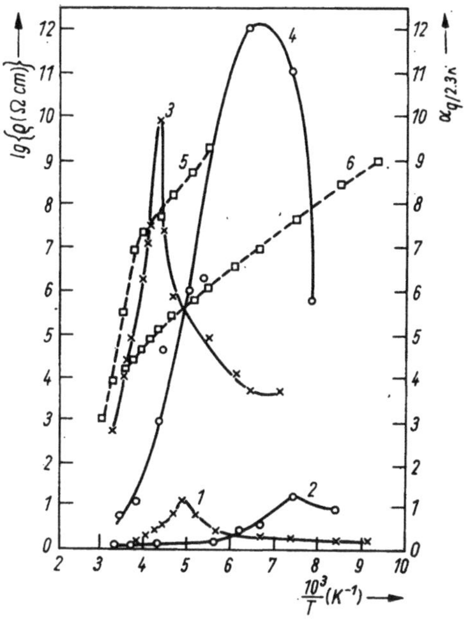

Fig. 2. Plot of lg ϱ and $\alpha q/2.3k$ versus $10^3/T$ where α is the thermoelectric power in μV/K; (1) $\alpha q/2.3k$ for 0.1 at% Li doped NiO, (2) $\alpha q/2.3k$ for 0.5 at% Li doped NiO, (3) $\alpha q/2.3k$ for 0.1 at% Na doped NiO, (4) $\alpha q/2.3k$ for 0.5 at% Na doped NiO, (5) lg ϱ curve for 0.1 at% Na doped NiO, (6) lg ϱ curve for 0.1 at% Li doped NiO

semiconductors, it is well known that at low temperatures, the Fermi level is given by

$$E_F = E_A - kT \ln \left[\frac{N_A - N_D}{N_D}\right], \quad (4)$$

where N_A and N_D are the concentrations of the acceptors and donors, respectively. At $T = 0$, $E_F = E_A$ and $E_F - E_A$ is negative for $T > 0$. When σ_i becomes comparable to σ_h second term on the right-hand side of (3) becomes significant. From (4), (3) may be written as

$$\alpha = \frac{k}{q}\left[\frac{\sigma_h}{\sigma} - \left(\frac{E_F}{kT} + A\right) - \frac{\sigma_i}{\sigma} \ln\left(\frac{N_A - N_D}{N_D}\right)\right]. \quad (5)$$

The second term on the right-hand side of (5) increases with decreasing temperature so that α shows maximum when plotted against $1/T$.

It is interesting to note that in both Li and Na doped samples, the turn-over occurs at a lower temperature for the higher dopant concentration. This result can be explained as due to the fact that with higher dopant concentration the free-hole concentration increases and σ_h overshadows σ_i at a lower temperature so that the turn-over appears at lower temperature.

Another interesting feature specially noticeable in the 0.1 at% Li doped NiO and 0.1 at% Na doped NiO is the similar temperature dependence of the lg ϱ and $\alpha q/2.3k$ curves after the turn-over in α on the high temperature side. α in this range can be written as

$$\alpha = 2.3 \frac{k}{q} \lg\left(\frac{N_v e^A}{p}\right). \quad (6)$$

where p is the free hole concentration and N_v is the density of states. Also

$$\lg \varrho = \lg \frac{1}{pq\mu}; \quad (7)$$

μ being the mobility. If the mobility of holes increases exponentially with temperature, the temperature dependence of ϱ must be stronger than that of α. This is not observed experimentally and in fact, the reverse is the case for 0.5 at% Na doped NiO. Thus as also indicated by the resistivity versus $10^3/T$ curves (Fig. 1), the temperature dependence of mobility is indeed smaller which contradicts the hopping model and indicates band conduction.

Another feature is the gradual disappearance of the temperature dependence of α in the 0.5 at% Li doped NiO which is also shown slightly for 0.5 at% Na doped NiO. In the temperature range where this occurs, the lg ϱ curve does not indicate exhaustion of the majority centres. Thus the effect can be explained by assuming that N_v also increases with temperature which again indicates the band conduction mechanism in NiO.

The α values for Na doped samples are much greater than that of Li doped samples. This is expected considering the fact that the free hole concentration in Na doped samples is less than that for corresponding Li doped samples as is evident from the resistivity data.

4. Conclusions

The results obtained by us for the electrical resistivity and thermoelectric power of Li and Na doped ceramic samples of NiO indicate the band conduction mechanism in NiO. The presence of impurity conduction is evident from thermoelectric power data. Na doped samples show higher resistivity, activation energy, and α values compared to the values for Li doped NiO. These results have been explained as due to the increase in the lattice strain because of the higher ionic radius of Na than that of nickel.

References

[1] Z. M. Jarzebski, Oxide Semiconductors, Pergamon Press, 1973 (p. 150).
[2] F. J. Morin, Phys. Rev. **93**, 1199 (1954).
[3] R. R. Heikes, and W. D. Johnston, J. chem. Phys. **26**, 582 (1957).
[4] R. R. Heikes, A. A. Maradudin, and R. C. Miller, Ann. Phys. (France) **8**, 733 (1963).
[5] M. Nachman, L. N. Cojocaru, and L. V. Ribco, phys. stat. sol. **8**, 773 (1965).
[6] G. Parravano, J. chem. Phys. **23**, 5 (1954).
[7] V. P. Zhuze and A. I. Shelykh, Soviet Phys. – Solid State **5**, 1278 (1963).
[8] Ya. M. Ksendzov, L. N. Anselm, L. L. Vasileva, and V. M. Latysheva, Soviet Phys. – Solid State **5**, 1116 (1963).
[9] M. Roilos and P. Nagels, Solid State Commun. **2**, 285 (1964).
[10] H. J. van Daal and A. J. Bosman, Phys. Rev. **158**, 736 (1967.)
[11] J. B. Goodenough, J. appl. Phys. **39**, 403 (1968).
[12] I.G. Austin, A. J. Springethorpe, B. A. Smith, and C. E. Turner, Proc. Phys. Soc. **90**, 157 (1967).
[13] B. Fisher and J. B. Wagner, Jr., Phys. Letters **21**, 606 (1966).
[14] N. M. Tallan and D. S. Tannhauser, Phys. Letters A **26**, 131 (1968).
[15] J. Feinleib and D. Adler, Phys. Rev. Letters **21**, 1010 (1968).
[16] A. J. Bosman and C. Crevecoeur, Phys. Rev. **144**, 763 (1966).
[17] Ya. M. Ksendzov, B. K. Avdeenko, and V. V. Makarov, Soviet Phys. – Solid State **9**, 828 (1967).
[18] H. J. van Daal and A. J. Bosman, Phys. Rev. **158**, 736 (1967).
[19] A. J. Bosman and H. J. van Daal, Adv. Phys. **19**, 1 (1970).
[20] D. Adler and J. Feinleib, Phys. Rev. B **2**, 3112 (1970).
[21] R. A. Bari, D. Adler, and R. V. Lange, Phys. Rev. B **2**, 2898 (1970).
[22] L. B. Valdes, Proc. IRE **42**, 420 (1954).
[23] J. B. Goodenough, D. G. Wickham, and W. J. Croft, J. Phys. Chem. Solids **5**, 107 (1958).
[24] S. Koide, J. Phys. Soc. Japan **20**, 123 (1965).

(Received February 13, 1981)

phys. stat. sol. (a) 65, 371 (1981)

Subject classification: 14.4.1 and 20; 22.8.1

Lvov State Franko University[1])

Electrooptic and Electrogyration Properties of the Solid Solutions on the Basis of Lead Germanate

By

O. G. VLOKH, L. A. LAŻKO, and YA. I. SHOPA

The peculiarities and anomalies of electrooptic and electrogyration properties in solid solutions on the basis of lead germanate crystals $Pb_5Ge_3O_{11}$ with partial iso- and heterovalent substitution $Pb^{2+} \rightarrow Ba^{2+}$; $Pb^{2+} \rightarrow Bi^{3+}$; $Ge^{4+} \rightarrow Si^{4+}$ are studied. The experimental investigation results including spontaneous and induced effects in the vicinity of the phase transition are mainly illustrated with the data for solutions containing silicon. The electrooptic and electrogyration coefficients are determined.

В твердых растворах на основе германата свинца $Pb_5Ge_3O_{11}$ с частичным изо- и гетеровалентным замещением $Pb^{2+} \rightarrow Ba^{2+}$, $Pb^{2+} \rightarrow Bi^{3+}$, $Ge^{4+} \rightarrow Si^{4+}$ изучены особенности и аномалии электрооптических и электрогирационных свойств. Результаты экспериментальных исследований, которые включают спонтанные и индуцированные эффекты в окрестности фазового перехода, главным образом иллюстрируются на примере растворов, содержащих кремний. Определены электрооптические и электрогирационные коэффициенты.

1. Introduction

Lead germanate crystals $Pb_5Ge_3O_{11}$ and solid solutions on their basis belong to the hexagonal symmetry class in the paraelectric phase [1 to 4] and appear to be the only representatives of the ferroelectrics with $\bar{6} \rightarrow 3$ phase transition. Electrooptic properties and optical activity of these crystals were investigated previously by Iwasaki et al. [5 to 8], Bichard et al. [9], and Miller et al. [10]. After the anomalously big induced electrogyration effect in the vicinity of the Curie point had been discovered [11], a number of solid solutions with partial iso- and heterovalent substitution $Pb^{2+} \rightarrow Ba^{2+}$, $Pb^{2+} \rightarrow Bi^{3+}$, $Ge^{4+} \rightarrow Si^{4+}$ was investigated by us [11 to 13].

In the present work the results obtained are illustrated with the data for $Pb_5(Ge_{1-x}Si_x)_3O_{11}$ solid solutions mainly. Spontaneous and induced electrooptic and electrogyration effects according to the temperature and static electric field intensity dependences of refraction index, birefringence, and optical activity were investigated. Refraction index and birefringence measurements were held at the He–Ne laser wavelength ($\lambda = 632.8$ nm) by means of prism and Senarmon compensation techniques. Electrogyration was measured by the plane-polarized light polarization plane rotation and polarization ellipse azimuth oscillations by means of a polarimetric method with photoelectrical registration.

2. Spontaneous Effects

The spontaneous electrooptic effect in the solid solutions on the basis of lead germanate according to the spontaneous polarization direction and the polar tensor form for the initial paraelectric phase is quadratic. This fact is shown to be confirmed by the

[1]) University str. 1, 290602 Lvov, USSR.

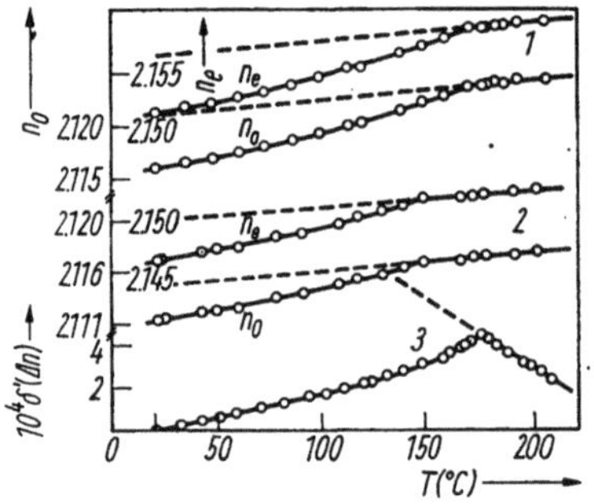

Fig. 1

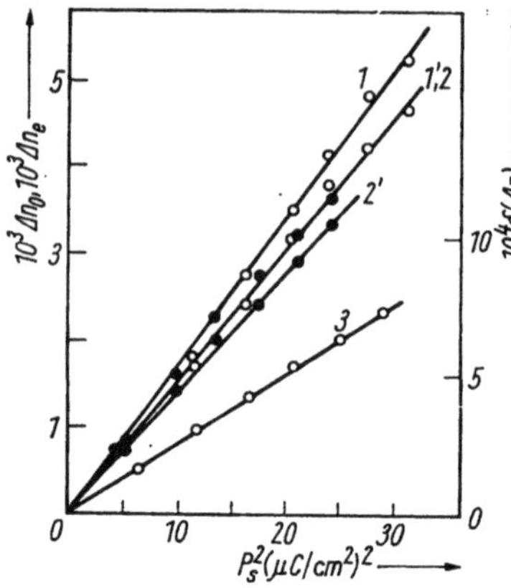

Fig. 2

Fig. 1. Temperature dependences of refraction index (1, 2) and birefringence $\delta(\Delta n_{xz})$ (3) in $Pb_5(Ge_{1-x}Si_x)_3O_{11}$ crystals. (1), (3) $x = 0$; (2) 0.10

Fig. 2. Refraction index increase Δn_e (1, 2), Δn_o (1′, 2′) and birefringence increase $\delta(\Delta n_{xz})$ (3) as functions of squared spontaneous polarization. (1), (1′), (3) $x = 0$ (2), (2′) 0.10;

refraction index and birefringence temperature dependences for $Pb_5(Ge_{1-x}Si_x)_3O_{11}$ crystals (Fig. 1). Refraction index and birefringence increases obtained from the corresponding dependence linear part extrapolation to the ferroelectric phase are proportional to the square of the spontaneous polarization (Fig. 2) and can be described by the following relations:

$$n_o = (1/2)\, n_o^3 M_{13} P_s^2 \,,$$

$$n_e = (1/2)\, n_e^3 M_{33} P_s^2 \,,$$

$$\delta(\Delta n) = \delta(n_o - n_e) = (1/2)\,(n_o^3 M_{13} - n_e^3 M_{33})\, P_s^2 \,.$$

For pure lead germanate ($x = 0$) according to the data obtained from Fig. 2 $M_{13} = 0.31$, $M_{33} = 0.34$ in units of $m^4\,C^{-2}$, and for the silicon-containing crystals ($x = 0.10$) $M_{13} = 0.29$, $M_{33} = 0.31$. The electrooptic coefficients are found to be in good agreement with the results obtained in [3]. M_{13} and M_{33} coefficients decrease slightly, though Ge → Si substitution lowers the Curie point and decreases the refraction indices essentially (Fig. 3).

The spontaneous electrogyration according to the third rank axial tensor form for the $\bar{6}$ symmetry class in contrast to the spontaneous electrooptic effect is linear. The spontaneous electrogyration temperature dependence is measured along the optical axis of single-domain crystals and the phase transition uncertainty observed

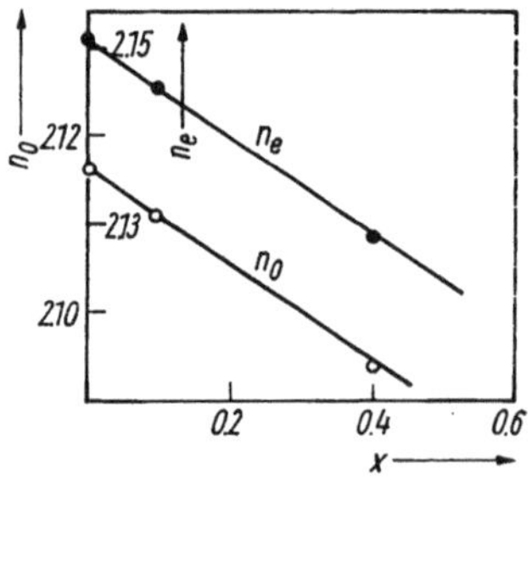

Fig. 3. Concentration dependence of refraction index in $Pb_5(Ge_{1-x}Si_x)_3O_{11}$ crystals

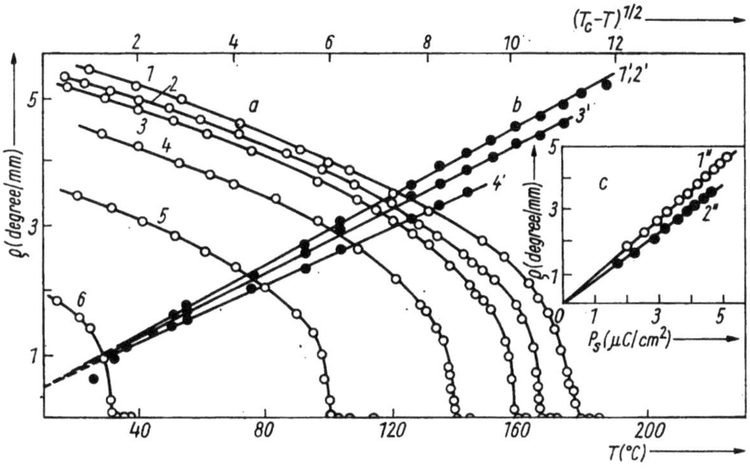

Fig. 4. Dependence of polarization plane specific rotation on temperature (a), $(T_c - T)^{1/2}$ magnitude (b), and spontaneous polarization (c) in $Pb_5(Ge_{1-x}Si_x)_3O_{11}$ crystals. (1) $x = 0$, (2), (1′) 0.03; (3), (2′) 0.05; (4), (3′), (1″) 0.10; (5), (4′), (2″) 0.20; (6) 0.40

in $(Pb_{1-x}Ba_x)_5Ge_3O_{11}$ and $(Pb_{1-x}Bi_x)_5Ge_3O_{11}$ crystals [11, 12] is not found (Fig. 4). This circumstance indicates the Ge → Si substitution regularity. Fig. 4b shows the optical activity just in the vicinity of the phase transition not to be described by the relation $\varrho \sim (T_c - T)^{1/2}$ being the consequence of the thermodynamic potential expansion in gyration powers [14]. But the linear character of the spontaneous electrogyration is confirmed (Fig. 4c). As far as the specific rotation of the polarization plane equals

$$\varrho = \frac{\pi}{\lambda n_o} \gamma_{33}^* P_s ,$$

the γ_{33}^* coefficient can be obtained. Its magnitude for $Pb_5(Ge_{1-x}Si_x)_3O_{11}$ crystals ($x = 0, 0.10, 0.20$) in units of 10^{-4} m²/C equals 8.1, 6.8, 5.6, respectively, i.e. γ_{33}^* linearly decreases with the Si concentration increase.

The optical activity in the $Pb_5Ge_3O_{11}$ crystals takes place also in the direction normal to the optical axis at the phase transition. The relation between the gyration tensor component and the spontaneous polarization in the ferroelectric phase is expressed by the following formula:

$$g_{11} = \gamma_{13}^* P_s ,$$

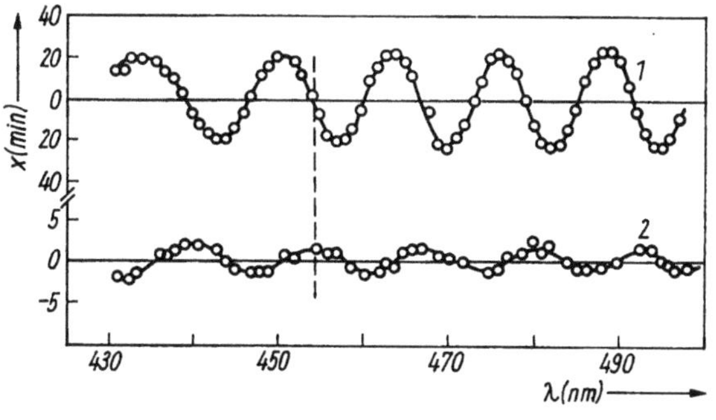

Fig. 5. Polarization ellipse azimuth oscillations in $Pb_5Ge_3O_{11}$ crystals. (1) in the case of 1° crystal shift from the extinction position, (2) for the crystal placed in the extinction position

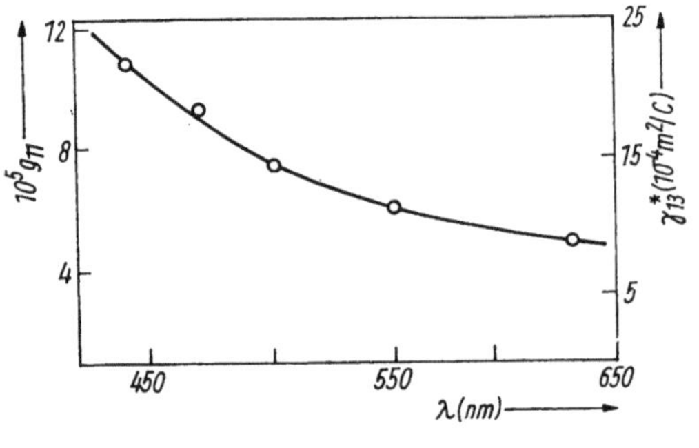

Fig. 6. Spontaneous electrogyration coefficient γ_{13}^* and corresponding dispersion of the gyration tensor component g_{11}

where $g_{11} = \chi_o(n_o^2 - n_e^2)$, χ_o is the polarization ellipse oscillation amplitude in the case of plane-polarized monochromatic light propagation through the crystal being placed into the extinction position previously. Thus, the g_{11} magnitude dispersion being obtained the γ_{13}^* spontaneous electrogyration coefficient dispersion may be calculated. Corresponding measurement results are given in Fig. 5 and 6. Their reliability is confirmed by the fact of a quarter period shift of the oscillations observed for the crystal accurately placed in extinction position relative to the oscillations obtained for the crystal moving away from the extinction position. Diagrams 1 and 2 in Fig. 5 are described by cosine and sine of the phase difference of the elliptically polarized waves in the crystal [15]. It is interesting to note that the γ_{13}^* and γ_{33}^* components are close in lead germanate.

3. Induced Effects

Induced effects at different experimental geometries in the vicinity of the phase transition, both in paraelectric and ferroelectric phases, were investigated. With the electric field directed along the spontaneous polarization direction the electrooptic and electrogyration effects are known to exhibit hysteresis features. The electrogyration and electrooptic hysteresis loops obtained for light propagation along the optical axis and the y-axis in $Pb_5(Ge_{1-x}Si_x)_3O_{11}$ crystals for $x = 0, 0.40$ are shown in Fig. 7. In $Pb_5(Ge_{1-x}Si_x)_3O_{11}$ crystals at fixed temperature the specific rotation of the polarization plane decreases almost linearly with Si impurity concentration; $\Delta\varrho \approx \approx 0.1$ degree/mm per mol%. The coercive field in thick samples increases at the same time. The same tendencies are observed in the case of Ba and Bi impurities

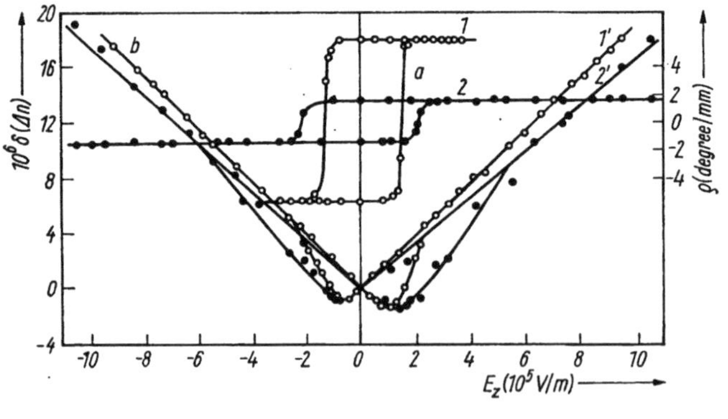

Fig. 7. Optical activity (a) and birefringence increase (b) hysteresis loops at $T = 20$ °C in $Pb_5(Ge_{1-x}Si_x)_3O_{11}$ crystals. (1), (1′) $x = 0$, (2), (2′) 0.40

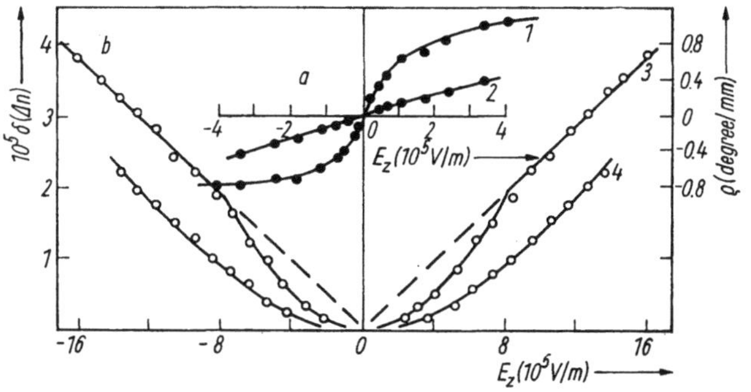

Fig. 8. Polarization plane specific rotation ϱ (a) and birefringence increase $\delta(\Delta n_{xz})$ (b) as functions of E_z electric field intensity in $Pb_5(Ge_{1-x}Si_x)_3O_{11}$ crystals. (a) $x = 0.10$, (b) 0.40; (1) $T_c + 1.1$ K, (2) $T_c + 6.5$ K, (3) $T_c + 4.3$ K, (4) $T_c + 12.0$ K

[11, 12]. The coercive field shows some peculiarities, during several first repolarization cycles its magnitude decreases and only after this becomes constant. This fixed value is considered as the coercive field magnitude.

The electrooptic loops are quadratic in contrast to the electrogyration loops. The difference observed does not contradict the symmetry restrictions of the noncentrosymmetric paraelectric phase but stays in complete accordance with spontaneous effect features. As the spontaneous electrooptic effect is quadratic a quadratic loop takes place. The total effect in the ferroelectric phase is defined both by spontaneous and induced polarization. Therefore, the birefringence increase along the y-axis is equal to

$$\delta(\Delta n)_{xz} \sim (P_s + P_i)^2 \approx P_s^2 + 2\varkappa P_s E .$$

Consequently, the induced birefringence sign does not change at fields greater than the coercive field, as far as sign reversal of the field causes a simultaneous spontaneous polarization inversion. Sign change of the birefringence increase occurs only at fields lower than the coercive one.

Other interesting features of the electrogyration and electrooptic effects in the crystals investigated are to be noted. Electrogyration both in paraelectric and ferroelectric phases is linear (Fig. 8a). But at larger fields saturation is observed and the effect magnitude depends on field sign. At temperatures close to the phase transition temperature the effect asymmetry is significant. The electrooptic effect in the way described has several distinctive features. Thus, in the ferroelectric phase the induced electrooptic effect is linear and may be described by the relation

$$\delta(\Delta n)_{xz} = (1/2)\,(n_o^2 r_{13} - n_e^2 r_{33})\,E_z .$$

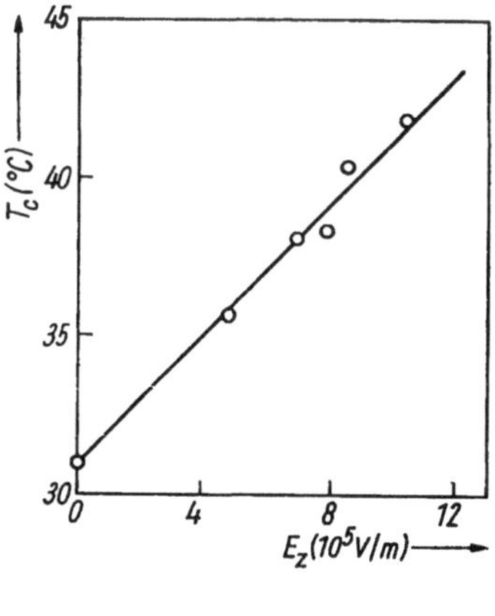

Fig. 9. Dependence of Curie temperature shift on E_z electric field in $Pb_5(Ge_{0.60}Si_{0.40})_3O_{11}$ crystals

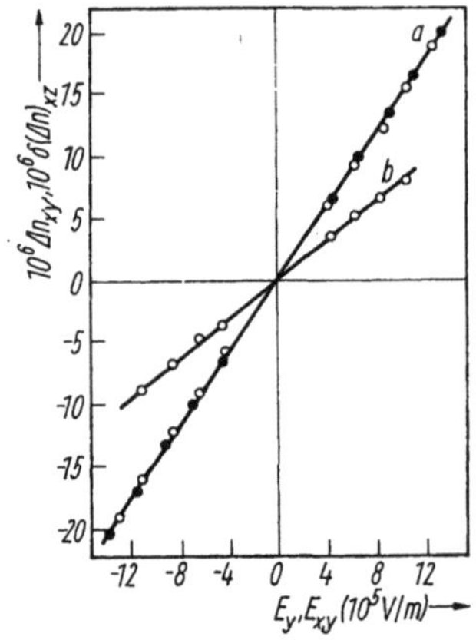

Fig. 10

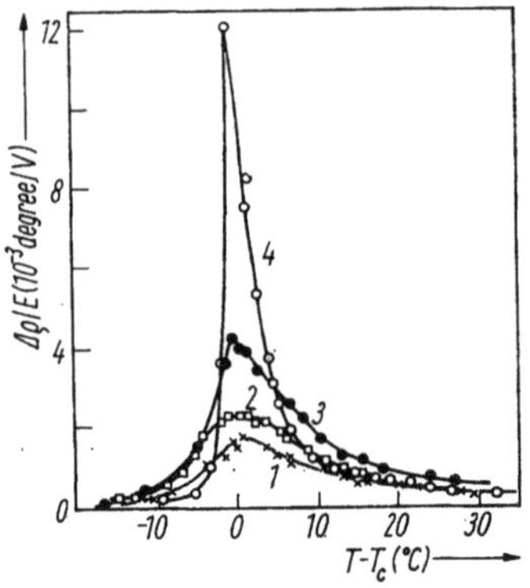

Fig. 11

Fig. 10. Dependence of birefringence (a) and birefringence increase (b) on E_x and E_y electric field intensities in $Pb_5Ge_3O_{11}$ crystals at $T = 20$ °C. (a) $\Delta n_{x'y'}$, (b) $\delta(\Delta n)_{xz}$; ● for E_x, ○ for E_y

Fig. 11. Temperature dependence of induced electrogyration in pure and impure lead germanate crystals. (1) $Pb_5Ge_3O_{11}$, (2) $(Pb_{0.98}Ba_{0.02})_5Ge_3O_{11}$, (3) $(Pb_{0.9945}Bi_{0.0055})_5Ge_3O_{11}$, (4) $Pb_5(Ge_{0.80}Si_{0.20})_3O_{11}$

According to the data of Fig. 7b for $Pb_5(Ge_{1-x}Si_x)_3O_{11}$ crystals, where $x = 0$, 0.40, at the temperature of 20 °C $n_o^3 r_{13} - n_e^3 r_{33} = 3.8 \times 10^{-11}$ and 3.4×10^{-11} m/V, respectively. The electrooptic effect under the same conditions in the paraelectric phase is quadratic (Fig. 8b). However, the phase transition being induced by an electric field of a certain value the effect is linearized. The shift of the Curie temperature depends linearly on the electric field (Fig. 9).

The electrooptic effect was measured in the case of light propagation along the optical axis on polydomain crystal to avoid the influence of optical activity. In the case of electric field E_x or E_y the lead germanate crystal becomes biaxial, the optical indicatrix rotates round the z-axis, the rotation angle being independent of the electric field magnitude. The rotation angle is defined from the $\operatorname{tg} 2\xi_{3,x} = -r_{22}/r_{11}$ relation, the electric field being directed along x-axis, and from the $\operatorname{tg} 2\xi_{3,y} = r_{11}/r_{22}$ relation in the case of E_y. Thus, the induced optical indicatrix rotation being opposite in the cases of E_x and E_y, the rotation angles are totally equal 45°. In the $Pb_5Ge_3O_{11}$ crystals $\xi_{3,x} = 1.5°$ and $\xi_{3,y} = 43.5°$ for E_x and E_y, respectively. According to this the electrooptic coefficient ratio is obtained as $r_{21}/r_{11} = 22 \pm 2$.

In the case of fields along x- and y-axes the birefringence in polydomain specimens along the z-axis depends linearly on the field (Fig. 10b). In this case the crystal is to be placed in diagonal position according to the indicatrix rotation, and then

$$\Delta n_{x'y'} = n_{x'} - n_{y'} = n_o^3 \sqrt{r_{11}^2 + r_{22}^2}\, E_{x,y} .$$

The birefringence magnitude does not depend on the field direction (along x- or y-axes) and changes sign with field sign reverse. For the longitudinal electrooptic effect measured along the y-axis, the same relation is observed. Using the relsults obtained the r_{11} and r_{22} coefficients can be calculated. For $Pb_5Ge_3O_{11}$ crystals $r_{11} = (0.7 \pm 0.1) \times \times 10^{-12}$ m/V and $r_{22} = (1.51 \pm 0.02) \times 10^{-12}$ m/V.

The temperature anomalies of the electrooptic and electrogyration coefficients in the case of field action along the optical axis are observed in the vicinity of the phase

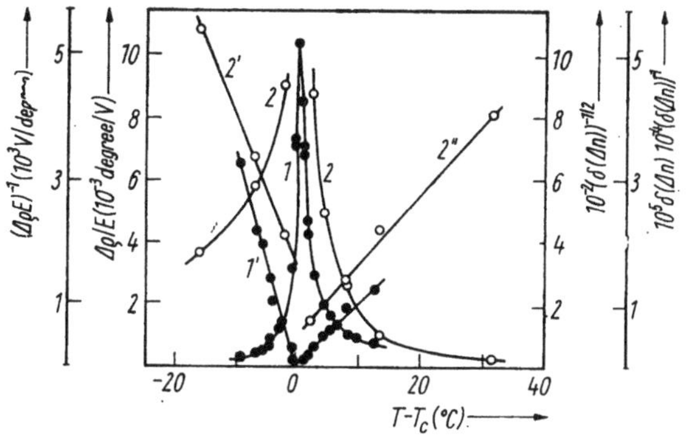

Fig. 12. Temperature dependence of induced electrogyration $\Delta\varrho/E$ (1), electrooptic effect $\delta(\Delta n)$ (2) at $E = 10^6$ V/m, their inverse values $(\Delta\varrho/E)^{-1}$ (1'), $[\delta(\Delta n)]^{-1}$ (2'), and $[\delta(\Delta n)]^{-1/2}$ (2'') in $Pb_5(Ge_{1-x}Si_5)_3O_{11}$ crystals. (1), (1') $x = 0.10$; (2), (2'), (2'') 0.40

transition. The electrogyration temperature anomalies are similar to permittivity anomalies (Fig. 11). The induced optical activity maximum increasing with impurity concentration, in contrast to the permittivity maximum, the electrogyration anomalies cannot be completely explained by the permitivity nones [11 to 13].

The temperature dependences of the induced electrooptics were measured on the crystal with 40% silicium content both above and below the phase transition temperature (Fig. 12). In the ferroelectric phase the electrooptic coefficient is proportional to $(T_c - T)^{-1}$. However, as it was noted above the effect being quadratic in the paraelectric phase is described by the relation $\delta(\Delta n) \sim (T - T_c)^{-2}$. In the case of electrogyration (Fig. 12) the $\Delta\varrho/E$ magnitude is proportional to $(T - T_c)^{-1}$ both in the ferroelectric and paraelectric phases.

4. Summary

Investigations of the electrooptic and electrogyration properties of lead germanate-type crystals provide valuable results concerning the ferroelectric phase transition. Without external electric field the spontaneous electrooptic and electrogyration methods are preferable to electric methods. Though electrooptic and electrogyration properties are complementary, the electrogyration method in some cases is prefered. In particular, the spontaneous electrogyration describes the phase transition more exactly than the spontaneous electrooptic effect.

The present investigations show the properties of solid solutions on the basis of lead germanate to be very sensitive to substitution of the ion component and to an external electric field. These factors cause an essential uncertainty and shift of the phase transition temperature. Only the isovalent substitution Ge → Si does not influence the transition character. Although the induced and spontaneous electrooptic and electrogyration effects are in good accordance with symmetry conditions, the classical thermodynamic theory is insufficient for the description of electrogyration in the phase transition region. Tendency to saturation is considered to be the most significant feature of the electrogyration effect. This feature may account for two factors: the electrical conductivity increase in the phase transition region and the transition induction by an external field. The structure mechanism of electrogyration in these crystals is defined by torsional movements of the $(GeO_4)^{4-}$ tetrahedral group.

Acknowledgements

The authors are grateful to E. V. Sinyakov and V. G. Monya for supplying the single crystals for investigations.

References

[1] K. SUGII, H. IWASAKI, and S. MIYAZAWA, Mater. Res. Bull. **6**, 503 (1971).
[2] T. YAMADA, H. IWASAKI, and N. NIIZEKI, J. appl. Phys. **43**, 771 (1972).
[3] Y. IWATA, H. KOIZUMI, N. KOYAMA, I. SHIBUYA, and N. NIIZEKI, J. Phys. Soc. Japan **35**, 314 (1973).
[4] Y. IWATA, N. KOYAMA, and I. SHIBUYA, J. Phys. Soc. Japan. **35**, 1269 (1973).
[5] H. IWASAKI, S. MIYAZAWA, H. KOIZUMI, K. SUGII, and N. NIIZEKI, J. appl. Phys. **43**, 4907 (1972).
[6] H. IWASAKI and K. SUGII, Appl. Phys. Letters **19**, 92 (1971).
[7] H. IWASAKI, K. SUGII, N. NIIZEKI, and H. TOYODA, Ferroelectrics **3**, 157 (1972).
[8] N. UCHIDA, T. SAKU, H. IWASAKI, and K. ONUKI, J. appl. Phys. **43**, 4933 (1972).
[9] V. M. BICHARD, P. H. DAVIES, K. F. HULME, Y. K. JONES, and D. S. ROBERTSON, J. Phys. D **5**, 2124 (1972).
[10] R. C. MILLER, W. A. NORDLAND, and A. A. BALLMAN, Ferroelectrics **7**, 109 (1974).
[11] O. G. VLOKH, E. V. SINYAKOV, L. A. LAŻKO, V. G. MONYA, and I. M. KLIMOV, Fiz. tverd. Tela **19**, 1031 (1977).
[12] O. G. VLOKH, E. V. SINYAKOV, L. A. LAŻKO, and V. G. MONYA, Fiz. tverd. Tela **20**, 2098 (1978).
[13] O. G. VLOKH, E. V. SINYAKOV, L. A. LAŻKO, Y. I. SHOPA, and A. Y. KREICHERECK, Fiz. tverd. Tela **22**, 227 (1980).
[14] K. AIZU, Phys. Rev. **133**, A1584 (1964).
[15] O. G. VLOKH, I. M. KLIMOV, and V. B. KOBYLYANSKY, Ukr. fiz. Zh. **18**, 2060 (1973).

(Received January 12, 1981)

phys. stat. sol. (a) 65, 379 (1981)

Subject classification: 11 and 13.4; 14.3.3; 22.1.2

Research and Development Institute for Electronic Components, Bucharest[1])

Radiation Effects of Fast Neutrons in Silicon Diffused Detectors

By

D. SACHELARIE, M. DRAGAN, M. STOICA, and M. SACHELARIE

With a new method, the ratio of the thermal emission probabilities of the carriers at the acceptor level E_3 ($E_c - 0.39$ eV) induced in silicon by fast neutron irradiation, $e_n/e_p = 500$ at 300 K, and the defect introduction rate, $\eta = 3.2$ cm^{-1}, are determined. These results are used for designing a silicon detector with p$^+$–n diffused junction used for the spectrometry and dosimetry of fast neutrons for medical and biological purposes.

Новым методом определяются соотношение вероятностей тепловой эмиссии носителей на акцеиторном уровне E_3 ($E_c - 0,39$ eV), введенным излучением быстрыми нейтронами, $e_n/e_p = 500$ при 300 K, и скорость введения дефектов $\eta = 3.2$ cm^{-1}. Эти результаты использовались для проектирования кремниевого диффузионного p$^+$–n детектора, употребленным для спектрометрии и дозиметрии быстрых нейтронов в медико-биологических прикладных задачах.

1. Introduction

The use of fast neutrons for medical and biological purposes requires semiconductor detectors of small area, good resolution and high reliability [1]. The surface-barrier detectors are not useful in such applications keeping in mind their low reliability. On the other hand, the silicon diffused-junction detectors give rise to problems connected with resolution and the thickness of the entrance window. By a careful design these detectors may, however, achieve high performances. Due to the permanent exposure of the active area to a nuclear radiation field, in the silicon crystalline lattice there appear structural defects, which are producing important modifications in the charge carrier recombination kinetics of the device. Thus the nuclear characteristics of the detector are altered. The optimization of the compromise reliability-nuclear performances represents the key of designing and obtaining semiconductor detectors for medical uses.

The defects introduced by fast neutron irradiation are defect clusters [2] generated by a cascade process initiated by the primary recoil atom. The energy levels introduced by this type of defects in the forbidden gap of silicon were measured by different methods [3 to 5], the DLTS method applied in [5], being probably the most correct one. In n-type silicon irradiated with fast neutrons three acceptor levels were found, $E_1(E_c - 0.15$ eV), $E_2(E_c - 0.21$ eV), $E_3(E_c - 0.39$ eV) with an introduction rate of 4.7, 1.4, 3.2 cm^{-1}, respectively [5]. In accordance with the recombination statistics [6], the generation–recombination processes in a p$^+$–n junction will be controlled by the deepest energy level, i.e. by E_3. The thermal emission probability e_n of the electrons at this level has the value 1.78×10^4 s^{-1}. The methods used allow only the determination of the sum $e_n + e_p$, i.e., in fact, the determination of the highest of the emission coefficients. However, the knowledge of the second coefficient e_p is essential in order to estimate the influence of the fast neutron irradiation on the generation

[1]) Str. Erou Iancu Nicolae 32B, Bucharest, Romania.

current in the depletion region and implicitly on the energy resolution of the p^+–n silicon detectors.

A method for the determination of this coefficient is proposed in the present work. This method is described in Section 2.1, and is based on the recombination parameter m determination, according to the forward bias applied to the diode at a constant temperature. In the next step, the introduction rate of E_3 defects was determined by using the reverse characteristics of a gate-controlled diode and the results were compared with those previously published. The influence of this rate is essential for the trapping and compensation processes in silicon, this rate dictating practically the reliability of the detectors in the fast neutron irradiation field. The conclusions were used in designing a p^+–n detector, that has a 20 keV resolution for 2.73 MeV tritons at a maximum irradiation fluence of 3.5×10^{13} fast neutrons/cm^2.

2. Theoretical Basis

2.1 Determination of the thermal emission probability ratio

In order to calculate the recombination current of a p–n junction forwardly biased, one may use the Sah-Noyce-Shockley theory [7], with the following assumptions: 1. the junction is linearly graded; 2. the trap concentration N_T is uniform; 3. the variation of the electrostatic potential in the transition region is linear, 4. the traps have only one energy level E_T, 5. the forward voltage U is moderate, $U < U_\text{D}/2$, where U_D is the built-in voltage, 6. in the transition region the quasiequilibrium condition is valid, 7. the width of the transition region is given by the Sah model [8],

$$W = \frac{4n_\text{i}}{a} \sinh \frac{q(U_\text{D} - U)}{2kT} \exp \frac{qU}{2kT}, \tag{1}$$

where a is the doping gradient of the junction, 8. between the capture coefficients $c_\text{n}(c_\text{p})$ and the thermal emission probabilities $e_\text{n}(e_\text{p})$ of the electrons (holes) there exist the equilibrium relations [6]

$$\left.\begin{aligned} e_\text{n} &= c_\text{n} n_\text{i} \exp \frac{E_\text{T} - E_\text{i}}{kT}, \\ e_\text{p} &= c_\text{p} n_\text{i} \exp \frac{E_\text{i} - E_\text{T}}{kT}. \end{aligned}\right\} \tag{2}$$

Thus, one obtains, with a good approximation, the expression of the recombination current

$$I = I_0 \frac{\sinh \dfrac{q(U_\text{D} - U)}{2kT}}{\dfrac{q(U_\text{D} - U)}{2kT}} \left[\exp\left(\frac{qU}{kT}\right) - 1\right] f(b), \tag{3}$$

where

$$\left.\begin{aligned} f(b) &= \frac{\arccos b}{\sqrt{1 - b^2}}, & b &< 1, \\ f(b) &= 1 & b &= 1, \\ f(b) &= \frac{\ln (b + \sqrt{b^2 - 1})}{\sqrt{b^2 - 1}}, & b &> 1. \end{aligned}\right\} \tag{4}$$

The parameter b is given by

$$b = \frac{e_n + e_p}{2\sqrt{e_n e_p}} \exp\left(-\frac{qU}{2kT}\right). \quad (5)$$

I_0 is a constant that is independent of the applied voltage,

$$I_0 = 2qn_i A_j N_T \sqrt{e_n e_p}/a\,, \quad (6)$$

where A_j is the junction area.

One can see that in this form the recombination current value depends on the emission probabilities only through their ratio e_n/e_p.

Normally, the recombination current is expressed by a relation of the form

$$I = I_1 \exp\left(\frac{qU}{mkT}\right). \quad (7)$$

By a derivation of the logarithm from equations (3) and (7) to the voltage and then by identification one gets the value of the recombination parameter m,

$$\left.\begin{aligned} m &= \left[A + \frac{b}{2\sqrt{1-b^2}}\frac{1}{\arccos b} - \frac{b^2}{2(1-b^2)}\right]^{-1}, & b<1\,,\\ m &= A^{-1}\,, & b=1\,,\\ m &= \left[A - \frac{b}{2\sqrt{b^2-1}}\frac{1}{\ln(b+\sqrt{b^2-1})} + \frac{b^2}{2(b^2-1)}\right]^{-1}, & b>1\,,\end{aligned}\right\} \quad (8)$$

where

$$A = \frac{kT}{q(U_D - U)} - \frac{1}{2}\operatorname{ctgh}\frac{q(U_D-U)}{2kT} + \frac{1}{1-\exp\left(\frac{-qU}{kT}\right)}. \quad (9)$$

Using these relations, in Fig. 1 there is shown, with the help of a HP 9825A computer, the variation of the recombination parameter m depending on the forward bias U, at different ratios e_n/e_p and for a built-in voltage $U_D = 700$ mV. By a fitting procedure between such theoretical curves and experimentally obtained curves, one may thus obtain the ratio of the emission probabilities of the carriers to the traps which dominate the recombination processes at the junction. Such a procedure is safer and easier than the method of measuring the variation of the recombination

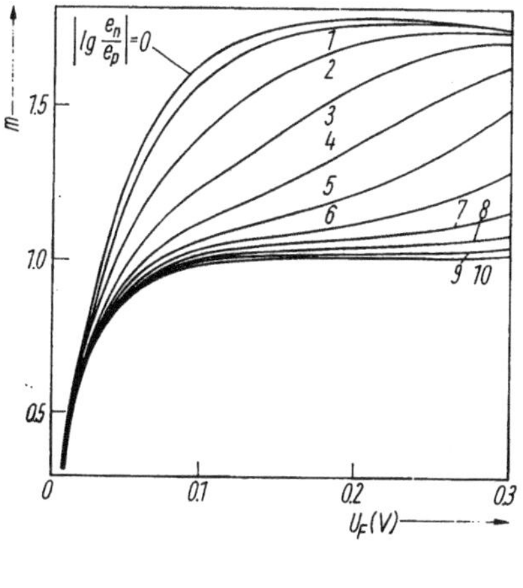

Fig. 1. Theoretical curves of the recombination parameter m against forward voltage computed from the Sah-Noyce-Shockley theory for various values of e_n/e_p. $U_D = 0.7$ V, $T = 300$ K

parameter m with temperature, used in other works [9, 10]. Knowing also the sum $e_n + e_p$ (e.g. from the measurement of the transition current), one can thus determine the thermal emission probabilities of the electrons and holes at the deep trap.

2.2 Determination of the defect introduction rate

The reverse characteristic of a gate-controlled diode [11] allows the determination of trap concentration. For this purpose one can use the difference between the value of the reverse current in the inversion regime I_{inv} and the value in the accumulation regime I_{ac},

$$I_{inv} - I_{ac} = qA_G x_d \frac{e_n e_p}{e_n + e_p} N_T , \tag{10}$$

where A_G and x_d are the gate area and the maximum width of the surface depletion region of the semiconductor. One can determine the x_d value precisely from the high frequency MOS C–U curve. With a known value of N_T one can calculate in the next step the defects introduction rate

$$\eta = \frac{N_T}{\Phi} , \tag{11}$$

where Φ is the irradiation fluence.

These methods for the determination of the ratio e_n/e_p and of the parameter η are applicable to the p–n junctions at which the generation–recombination processes in the transition region are controlled by defects with only one dominant energy level. This is the case of n-type silicon irradiated with fast neutrons, where the level E_3 (E_c – 0.39 eV) dominates, a level by far deeper than the other two levels E_1 (E_c – – 0.15 eV), E_2 (E_c – 0.21 eV) and with a comparable defect introduction rate.

3. Experimental Procedure

In order to characterize the traps introduced by fast neutron irradiation in n-type silicon, a testing device DG 10 was used, consisting of a p^+n gate-controlled diode, with the junction area 0.71×10^{-3} cm^2, the gate area 2.26×10^{-3} cm^2, and a MOS capacitor with an area of 1.51×10^{-3} cm^2. The technological characteristics of this device were epitaxial layer concentration 10^{16} cm^{-3}, junction depth 1.9 μm, surface concentration of diffused p^+-region 9×10^{19} cm^{-3}, oxide thickness 5000 Å. The diode has a second gate connected to the case, for the inversion surface delimitation.

The first variant of the detector (DJ 90) was realized on n-type planar silicon of 1.5×10^{14} cm^{-3}. The p^+ diffused zone area was 5×10^{-3} cm^2, and the junction depth 5 μm. The oxide thickness was 10000 Å. The technology of the second variant (DJ 100) of the p^+–n detectors was optimized for improvement of the nuclear characteristics.

All devices were irradiated in groups of five, at room temperature and unbiased, in the IBR30 reactor (Dubna, USSR). To eliminate the thermal neutron influence during the irradiation, the devices were wrapped in cadmium. The DG 10 diodes were irradiated at the following fast neutron fluences: 2.4×10^{13}, 4.8×10^{13}, 3.8×10^{14}, and 8.3×10^{14} neutrons/cm^2. The irradiation fluences of the DJ 90 detectors were: 2.4×10^{13}, 7.3×10^{13}, 1.5×10^{14}, 2.4×10^{14}, and 3.3×10^{14} neutrons/cm^2, and of the DJ 100 detectors: 3.4×10^{13} and 1.6×10^{14} neutrons/cm^2. Neutron fluences were determined by a proton recoil method.

The electrical characteristics of the irradiated devices were measured after stabilization at room temperature for 50 days. The current–voltage variations were measured with a Keithley picoammeter, and the capacitance–voltage characteristic (1 MHz) was obtained with the help of the Hewlett-Packard automatic measurement system.

The nuclear characterization of the detectors was done by measuring the resulting spectrum after the reaction of thermal neutrons with the ^{6}Li converter,

$$^6_3\text{Li} + ^1_0\text{n} \rightarrow ^3_1\text{H} + ^4_2\text{He} + 4.78\ \text{MeV}\,. \qquad (12)$$

The measurements were done in vacuum and the bias voltage on the detectors was 100 V.

4. Experimental Results and Discussion

4.1 Gate-controlled diodes

Fig. 2a shows the direct characteristic $I_F = f(U_F)$ of the p$^+$–n gate-controlled diodes DG 10, in the accumulation regime ($U_G = +20$ V). The measurements were done in 10 mV voltage steps. There can be observed an increase of the direct current with the irradiation dose as a result of the increase of trap concentration E_3 ($E_c - 0.39$ eV). The corresponding variation of the recombination parameter m is shown in Fig. 2b.

In the voltage range 0 to 150 mV, the recombination component dominates at the control diodes as well as at the irradiated diodes. The recombination parameter is, however, different in the two cases, evidencing different energy characteristics of the traps induced by irradiation as compared with those induced by technological processes. By a computer fitting of these experimental results and the theory developed in Section 2.1., one gets $e_n/e_p = 500$. Thus, the ratio of the thermal emission probabilities at 300 K of the carriers at the traps E_3 ($E_c - 0.39$ eV) introduced in n-silicon by fast neutron irradiation was obtained. Considering [5] $e_n = 1.78 \times 10^4$ s^{-1}, the value of the emission probability of the holes is obtained as $e_p = 35.6$ s^{-1}.

It is to be noted that at high biases (500 mV), the unirradiated diode direct current is practically formed only by the diffusion component ($m = 1$). At the irradiated diodes the recombination processes are important even at high voltages, the direct current at 500 mV being a combination between the diffusion component and the recombination component, the m parameter tending to 1.5.

Now one can calculate the trap concentration corresponding to each fast neutron fluence. Fig. 3 illustrates the variation of the reverse current I_R of the DG10 diode

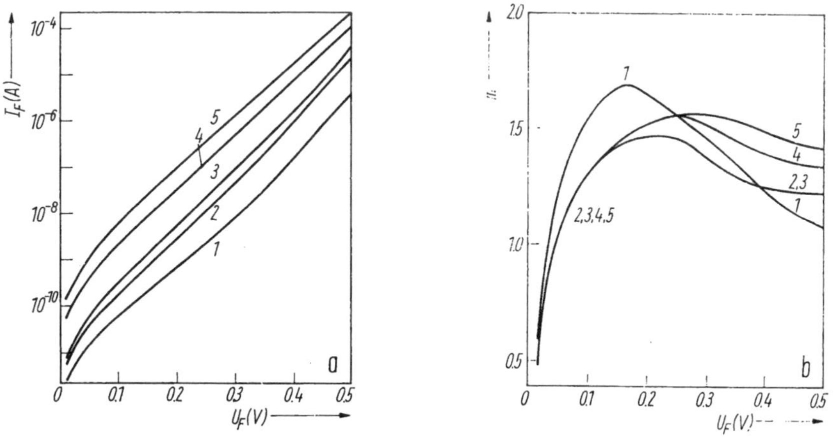

Fig. 2. a) Experimental forward I–U characteristics of the p$^+$–n gate-controlled diodes (DG 10, $U_G = +20$ V, $T = 300$ K) after various fast neutron fluences. b) Recombination parameter m as a function of applied bias: (1) before irradiation, (2) $\Phi = 2.4 \times 10^{13}$, (3) 4.8×10^{13}, (4) 3.8×10^{14}, (5) 8.3×10^{14} neutrons/cm^2

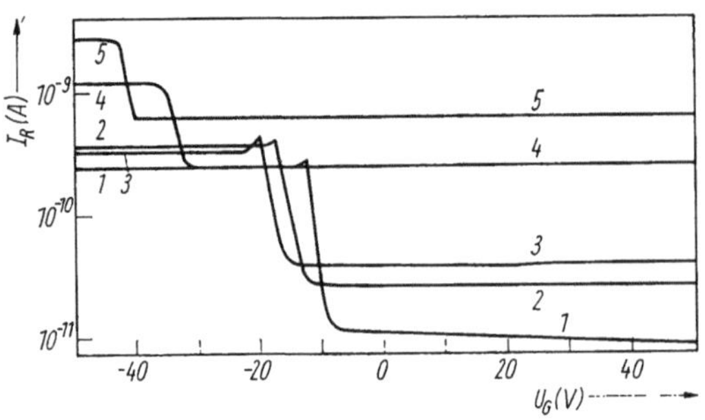

Fig. 3

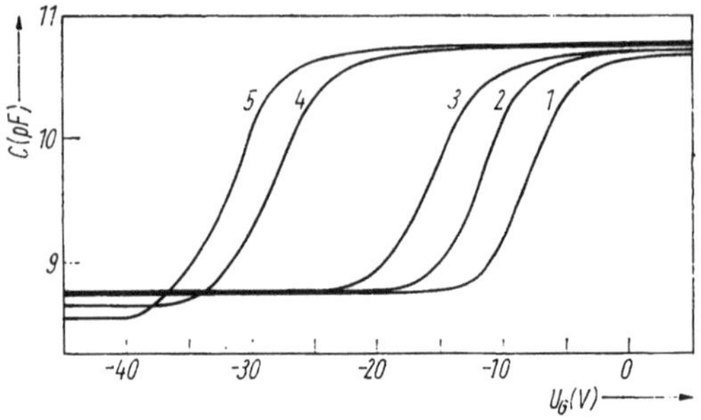

Fig. 4

Fig. 3. Reverse current as a function of gate voltage with fast neutron fluences as parameter for the p$^+$–n gate-controlled diodes (DG 10; $U_R = 1$ V, $T = 300$ K). The numbers correspond to the conditions in Fig. 2

Fig. 4. Experimental high frequency C–U characteristics of the MOS structure ($x_{\text{oxide}} = 0.5$ μm) for different fluences of fast neutron irradiation (DG 10; $T = 300$ K, $f = 1$ MHz). The numbers correspond to the conditions in Fig. 2

versus the gate voltage U_G, at a reverse applied voltage $U_R = 1$ V, and Fig. 4 shows the high frequency (1 MHz) C–U curves for the corresponding MOS capacitor.

The generation current in the metallurgical junction area as well as the current in the induced junction region increase monotonously with fast neutron fluence as a result of the increase of the E_3 traps concentration induced by irradiation. The C–U MOS curves evidence a very strong shift of the threshold voltage to negative values due to the build-up of a positive charge in the oxide. This charge is the result of a preferential trapping process [12] of the holes which appear in the oxide as a result of the ionization produced by irradiation. It results that at a p$^+$–n diode irradiated with fast neutrons, the residual current is given principally by generation of carriers at the induced traps in the region of the metallurgical junction. The irradiation decreases the importance of the generation component at the silicon surface, this region entering a high accumulation regime.

From the variation of the minimum capacitance of the MOS system one sees that the compensation effects induced by irradiation start to manifest at higher fast neutron fluences. From these C–U curves, the effective doping concentration at the silicon surface [13] and then the maximum width of the depletion region for a $U_R = 1$ V bias were determined. The knowledge of the emission probabilities of the traps as well as of the x_d value allow the determination of trap concentration by application of the method described in Section 2.2. Nevertheless, it is necessary that the generation processes in the semiconductor are controlled predominantly by the irradiation defects. Due to this relation (10) was applied only for high fluences (3.8×10^{14} and $8.3 \times$ $\times\ 10^{14}$ neutrons/cm^2). The results are given in Table 1.

Table 1

fast neutron fluence Φ (neutrons/cm^2)	control	2.4×10^{13}	4.8×10^{13}	3.8×10^{14}	8.3×10^{14}
effective doping concentration, N_B (cm^{-3})	10^{16}	10^{16}	10^{16}	7.5×10^{15}	6.5×10^{15}
trap concentration, N_T (cm^{-3})	—	7.7×10^{13}	1.5×10^{14}	1.2×10^{15}	2.6×10^{15}

By calculating the ratio between the trap concentration thus determined and the irradiation fluence, one gets the introduction rate of defects of E_3 ($E_c - 0.39$ eV) type in n-silicon by fast neutron irradiation $\eta = 3.2\ \text{cm}^{-1}$. This value is identical with that found by Tokuda et al. [5], who used the DLTS method to determine it. However, the introduction rate observed here is different from that reported by Saks [3], $\eta = 7\ \text{cm}^{-1}$.

With this value the other trap concentrations were determined corresponding to the first two irradiation fluences. As the effective doping concentration is given, in a first approximation, by the relation

$$N_B \approx N_{B0} - N_T\,, \tag{13}$$

where N_{B0} is the doping concentration before irradiation, the results obtained in the determination of the N_T concentration from the variation $I_R = f(U_G)$ of the gate-controlled diode are in accordance with the compensating effects made evident by the C–U MOS curves.

4.2 Diffused junction detectors

The p$^+$–n planar diodes may be used as nuclear detectors for the spectrometry of charged particles. The main nuclear parameters, energy resolution and collection efficiency, are dictated by the technological characteristics of the diode and undergo changes as a result of irradiation [14 to 17]. In order to investigate the fast neutron irradiation effects on the nuclear characteristics of the silicon p$^+$–n detectors, initially the DJ 90 detector, that is described in Section 3, was used. A relatively deep p$^+$–n junction (5 μm) was chosen to evidence also the importance of the entrance window of the diffused region. The electrical measurements comprised the direct characteristic current–voltage (Fig. 5) and the high frequency characteristics (1 MHz) capacitance–voltage (Fig. 6). The increase of trap concentration introduced by irradiation produces the increase of the recombination component of the direct current, a fact observable

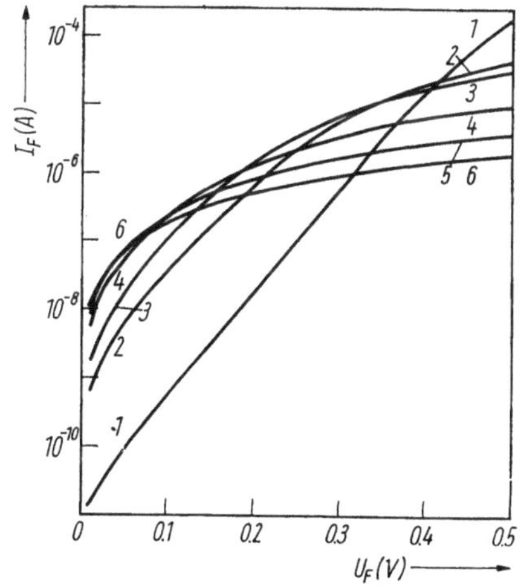

Fig. 5

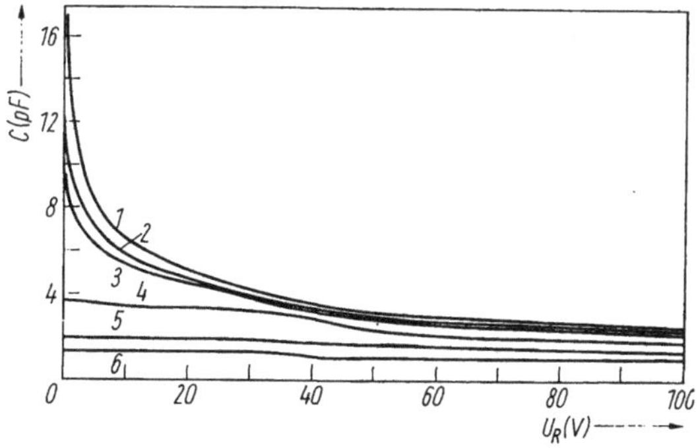

Fig. 6

Fig. 5. Forward current–voltage characteristics of the DJ 90 ($T = 300$ K) detectors after various fast neutron fluences: (1) before irradiation, (2) $\Phi = 2.4 \times 10^{13}$, (3) 7.3×10^{13}, (4) 1.5×10^{14}, (5) 2.4×10^{14}, (6) 3.3×10^{14} neutrons/cm^2

Fig. 6. C–U characteristics of the DJ 90 ($T = 300$ K) detectors irradiated by fast neutrons. The numbers correspond to the condition in Fig. 5

in the 0 to 100 mV voltage range. For higher voltages the effect of silicon compensation appears. The presence of traps increases the resistivity of the n region, the voltage on the series resistance increases and, correspondingly, the junction bias decreases. Thus, the diode current (e.g. at a 500 mV bias) decreases with the increase of the irradiation dose. The compensation effect is also reflected by the variation of the junction capacitance [18], which decreases as a result of the increase of the transition region W.

The fast neutron irradiation effects on the nuclear characteristics are made evident (Fig. 7) by the measurement of the spectrum which results from the thermal neutron reaction with the ^{6}Li converter. With the unirradiated detector, the triton energy (2.73 MeV) is correctly measured, while the α particles (2.05 MeV) are registered at a lower energy. This fact is explained by the energy loss of the α particles in the p^+ diffused region, which is relatively deep. The energy losses are important for α particles with a 8 μm range in silicon, and very small for tritons with a much larger range, 45 μm. Between the two main peaks (α and T) there are secondary peaks of smaller area. By calculating the area ratio of the diffused region and of the metallization contact, it appears that these peaks represent the response of the junction zone under the metallization. In this region the particles undergo a supplementary stopping due to metallization.

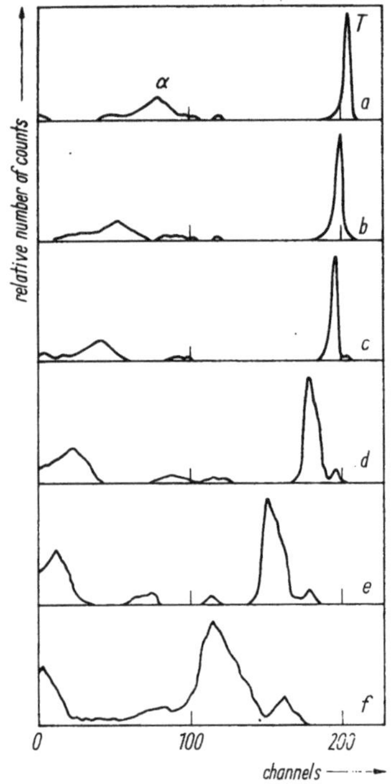

Fig. 7

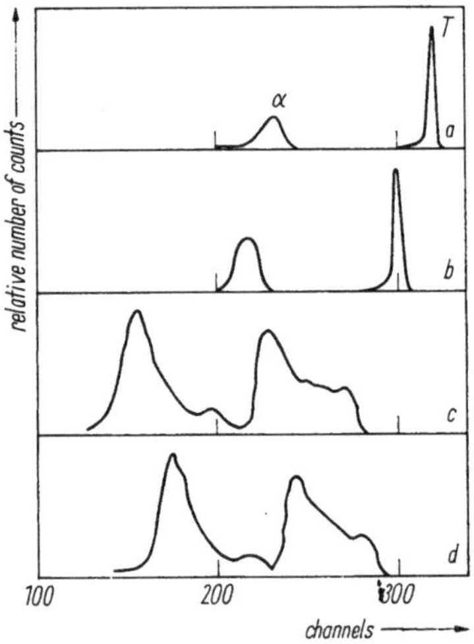

Fig. 8

Fig. 7. ^{6}Li (n, α)T spectra of the DJ 90 detectors for different fluences of fast neutron irradiation. $U_R = 100$ V, triton energy 2.73 MeV, and 1 channel = 13.4 keV. a) before irradiation, b) $\Phi = 2.4 \times 10^{13}$, c) 7.3×10^{13}, d) 1.5×10^{14}, e) 2.4×10^{14}, f) 3.3×10^{14} neutrons/cm²

Fig. 8. ^{6}Li(n, α)T spectra of the DJ 100 detectors under various fast neutron fluences. $U_R = 100$ V and 1 channel = 8.53 keV. a) before irradiation, b) $\Phi = 3.4 \times 10^{13}$ neutrons/cm², c) 1.6×10^{14} neutrons/cm² after 4 h, d) 1.6×10^{14} neutrons/cm² after 28 h

In the irradiated detectors, the increase of the fast neutron fluence produces a monotonous displacement of the spectrum towards lower energies. The displacement is due to an incomplete charge collection [19], part of it being lost by recombination. At higher fluences, the carrier lifetime

$$\tau = \frac{1}{c_p \eta \Phi} \tag{14}$$

decreases, and their transit time through the space charge region

$$t_r = \frac{W^2}{\mu U_R}, \tag{15}$$

increases. Thus, the recombination probability of the carriers generated in the transition region by a particle of energy E increases with fast neutron fluence. The energy resolution of the detectors degrades itself at fluences higher than 1.5×10^{14} neutrons/cm^2 due to the excessive increase of the generation current in the depletion region. At high fluences there appears a supplemental peak, on the right side of the triton main peak. This is the lateral depletion region response, the extension area of which increases due to the compensation of the n-type silicon. This fact was rendered evident by visualization of the space charge region with the help of a He–Ne laser scanning system.

All these observations were used in the design of a diffused junction detector, with superior nuclear characteristics and high reliability. In Fig. 8 there is shown the spectrum obtained with the new type of detector, DJ 100. It has a resolution of 17 keV for tritons and 60 keV for α-particles. In spite of the fact that is has a diffused junction, its entrance window produces only a 50 keV displacement of the α-peak. In order to check that the improvements done have not worsened the reliability, the DJ100 detectors were subjected to two strong fast neutron irradiations. After a 3.4×10^{13} neutrons/cm^2 irradiation, the displacement of the spectrum is not very important (200 keV), and the energy resolution remains good (21 keV for tritons and 60 keV for α-particles). But a 1.6×10^{14} neutrons/cm^2 irradiation induces serious modifications. The measurement of the spectrum 4 and 28 h after the irradiation shows an annealing of the fast neutron introduced defects.

5. Conclusions

The fast neutron irradiation induces in n-silicon deep level defects E_3 (E_c − 0.39 eV), which control the recombination processes in a p$^+$–n junction. The introduction rate of these defects is $\eta = 3.2$ cm^{-1}. The thermal emission probabilities ratio of the carriers to the E_3 traps is $e_n/e_p = 500$ at 300 K. The value of the minimum emission probability is $e_p = 35.6$ s^{-1}. The reverse current variation of a p$^+$–n nuclear detector irradiated with fast neutrons, is given by

$$I_R = qA_j W e_p N_T = 114\, \Phi q A_j W\,, \tag{16}$$

and the compensation degree of irradiated n-silicon is given by

$$N_B \approx N_{B0} - 3.2\Phi\,. \tag{17}$$

These two effects produce in a p$^+$–n detector irradiated with fast neutrons a degradation of the resolution, a decrease of the collection efficiency, and the appearance of the secondary peaks. p$^+$–n detectors with good nuclear characteristics and safe to fast neutron irradiation can be obtained by optimizing the diffusion profile, the oxide, and the device geometry. The DJ 100 detector described in the present work has an entrance window and an energy resolution comparable with the corresponding values of a silicon surface barrier detector, but is resistant to about 10^3 times higher fast neutron doses. Thus the silicon diffused junction detectors can be used for medical and biological purposes, for the dosimetry or spectrometry of fast neutrons in the continuous mode for about 2 to 3 years.

Acknowledgement

The authors would like to express their thanks to V. M. Nazarov of JINR, Dubna for neutron irradiation.

References

[1] K. KOVACEVIC, N. STIPCIC, G. PAIC, I. SLAUS, B. EMAN, V. PECAR, and M. ANTIC, Nuclear Instrum. and Methods **148**, 291 (1978).
[2] B. R. GOSSICK, J. appl. Phys. **30**, 1214 (1959).
[3] N. S. SAKS, IEEE Trans. Nuclear Sci. **24**, 2153 (1977).
[4] Y. TOKUDA and A. USAMI, J. appl. Phys. **47**, 4952 (1976); **48**, 1668 (1977); **49**, 603 (1978).
[5] Y. TOKUDA, N. SHIMIZU, and A. USAMI, Japan. J. appl. Phys. **18**, 309 (1979).
[6] C. T. SAH, Proc. IEEE **55**, 654 (1967).
[7] C. T. SAH, R. N. NOYCE, and W. SHOCKLEY, Proc. IRE **45**, 1228 (1957).
[8] C. T. SAH, Proc. IRE **49**, 603 (1961).
[9] C. Y. BARTHOLOMEW, IEEE Trans. Electron Devices **14**, 452 (1967).
[10] M. C. CHOW, J. L. AZAREWICZ, and C. A. GOBEN, IEEE Trans. Nuclear Sci. **15**, 88 (1968).
[11] A. S. GROVE and D. J. FITZGERALD, Solid State Electronics **9**, 783 (1966).
[12] G. W. HUGHES and G. J. BRUCKER, Solid State Technol. **22**, 70 (1979).
[13] K. H. ZAININGER and F. P. HEIMAN, Solid State Technol. **13**, 5 (1970).
[14] R. E. SCOTT, IEEE Trans. Nuclear Sci. **11**, 206 (1964).
[15] G. G. GEORGE and E. M. GUNNERSEN, Nuclear Instrum. and Methods **25**, 253 (1964).
[16] J. A. COLEMAN, D. P. LOVE, J. H. TRAINOR, and D. J. WILLIAMS, IEEE Trans. Nuclear Sci. **15**, 363, 482 (1968).
[17] S. NAKAMATO, T. AOKI, K. KAWABATA, K. NORISAWA, and M. SAKISAKA, Japan. J. appl., Phys. **12**, 1493, (1973); **13**, 524 (1974).
[18] M. G. BUEHLER, IEEE Trans. Nuclear Sci. **17**, 341 (1970).
[19] G. L. MILLER and W. M. GIBSON, Nuclear Electronics, Vol. I, IAEA, Vienna 1962 (p. 477).

(Received January 13, 1981)

phys. stat. sol. (a) 65, 389 (1981)

Subject classification: 10.2 and 13.4; 14.3; 22.1.2

Zentralinstitut für Elektronenphysik der Akademie der Wissenschaften der DDR, Berlin[1])

Eigenschaften der Energieniveaus von Versetzungen in Silizium

Von

H. Lemke

Mit den Methoden DLTS und TSCa werden die Energieniveaus und Einfangkoeffizienten bei plastisch verformten Si-Kristallen mit Versetzungsdichten von 10^6 bis $10^7\ cm^{-2}$ bestimmt. Bei n- und p-Silizium werden keine Donatorterme sondern eine breite Bande von Akzeptorniveaus unterhalb von $E_V + 0{,}50$ eV beobachtet. Möglicherweise wird diese Bande durch Versetzungen mit unterschiedlichen Kernstrukturen verursacht. Bei einer elektrostatischen Energie von 0,15 eV ($f \approx 0{,}1$) folgt für die höheren Niveaus etwa $E_V + 0{,}23$ eV und $E_V + 0{,}37$ eV, während der tiefste Term unterhalb von $E_V + 0{,}20$ eV liegen muß. Die Einfangkoeffizienten der Akzeptoren liegen bei $r_c^n \approx 10^{-9}\ cm^3\ s^{-1}$ und $r_c^p \approx 10^{-8}\ cm^3\ s^{-1}$ ($T = 90$ K). Zusätzlich wird bei einigen n-Kristallen ein Akzeptorniveau bei $E_c - 0{,}39$ eV mit $r_c^n \approx 10^{-11}\ cm^3\ s^{-1}$ beobachtet. In allen Fällen wird die Einfangrate der Elektronen im Gegensatz zur Einfangrate der Löcher wesentlich durch die Höhe der Potentialbarriere beeinflußt. Die Ergebnisse der Lebensdauermessungen unterhalb von $T = 300$ K können für n- und p-Kristalle mit unterschiedlichen Werten des Besetzungsfaktors f für das Niveau $E_V + 0{,}37$ eV erklärt werden.

The energy levels and capture constants for plastically deformed silicon crystals with dislocation densities 10^6 to $10^7\ cm^{-2}$ are been measured by DLTS- and TSCa-techniques. For n- and p-type silicon a broad band of acceptor levels below $E_V + 0.50$ eV and no donor levels are observed. Possibly this band is caused by dislocations with different core structures. For an interaction energy of 0.15 eV ($f \approx 0.1$) the lowest level must lie below $E_V + 0.20$ eV and the higher levels at $E_V + 0.23$ eV and $E_V + 0.37$ eV. These acceptors have capture constants of about $r_c^n \approx 10^{-9}\ cm^3\ s^{-1}$ and $r_c^p \approx 10^{-8}\ cm^3\ s^{-1}$ ($T = 90$ K). Besides this in some n-type crystals an acceptor level at $E_c - 0.39$ eV with $r_c^p \approx 10^{-11}\ cm^3\ s^{-1}$ is found. In all cases the capture rate of electrons contrary to the hole capture is strongly influenced by the height of the potential barrier. The data of lifetime measurements below $T = 300$ K can be explained with different values of the fill factor f of the level at $E_V + 0.37$ eV for n- and p-type crystals.

1. Einleitung

Die elektrischen Eigenschaften von Versetzungen sind in den letzten Jahrzehnten insbesondere mit Halleffekt- sowie Fotoleitungsmessungen bei Ge und Si vielfach untersucht worden. Die von Shockley [1] vorausgesagte Akzeptorwirkung konnte dabei praktisch in allen Fällen beobachtet werden, wobei jedoch sehr unterschiedliche Angaben über die Lage des Energieniveaus gemacht wurden. Akzeptorterme wurden bei Ge sowohl in der oberen [2, 3] wie auch unteren [4 bis 7] Hälfte des verbotenen Bandes nachgewiesen. Auf Grund ausführlicher Messungen von Schröter [8] an p-Ge wird heute vielfach angenommen, daß die Versetzungen bei geeigneter Ferminiveaulage auch eine Donatorwirkung aufweisen, sofern das Neutralniveau ($E_V + 0{,}1$ eV) innerhalb eines Versetzungsbandes unterschritten wird.

Ein ähnliches Energietermmodell mit einem Neutralniveau um $E_V + 0{,}4$ eV wurde inzwischen auch von Weber et al. [9] sowie Grazhulis et al. [10] zur Interpretation von

[1]) Mohrenstr. 40/41, DDR-1080 Berlin, DDR.

Halleffektmessungen bei Si-Kristallen herangezogen. Für stark verformte Kristalle (Versetzungsdichte $\approx 10^9\ cm^{-2}$) wurden in [10] zwei Energieterme — ein Akzeptorniveau bei 0,43 eV unterhalb des Leitungsbandes und ein Akzeptor–Donatorterm bei 0,42 eV oberhalb des Valenzbandes — beobachtet. Ein ähnlicher Energiewert für das Neutralniveau wurde auch von Kos und Neubert [11] mit Fotoleitungsmessungen bestimmt. Bei Vorliegen eines derartigen Akzeptor–Donatorbandes müßten bei p-Si unterhalb von Zimmertemperatur positiv geladene Versetzungen mit Raumladungsbarrieren vorliegen, so daß bei n- und p-Si für die Rekombination von Ladungsträgerpaaren ein weitgehend analoges Verhalten erwartet werden kann. Untersuchungen von Glaenzer und Jordan [12] haben jedoch gezeigt, daß eine derartige Ähnlichkeit in der Lage der jeweils wirksamen Energieniveaus und der damit verbundenen Raumladungsbarrieren nicht besteht. Diese Messungen mußten bei n-Si mit einem Energieterm bei 0,50 eV und bei p-Si je nach Höhe der Dotierung mit einem „Niveau" bei 0,38 bis 0,50 eV oberhalb des Valenzbandes interpretiert werden.

Da bei eigenen Messungen der Rekombination in plastisch verformten Si-Kristallen ähnliche Diskrepanzen bezüglich des Einflusses der Raumladungsbarrieren beobachtet wurden, erschienen weitere Untersuchungen über die energetische Lage der Rekombinationszentren sowie deren Umladungscharakter wünschenswert. Hierfür wurde die Randschichtkapazitäts-Methode [13, 14] ausgewählt, da mit derartigen Messungen umladbare Energieterme innerhalb der verbotenen Zone direkt bestimmt werden können.

2. Versuchsdurchführung

Für die Messungen wurden 0,5 mm starke n- und p-Scheiben mit Trägerkonzentrationen um $10^{14}\ cm^{-3}$ in (111)- sowie (100)-Orientierung verwendet. Die plastische Verformung erfolgte bei 1000 °C (1 bis 2′) durch Biegen um eine [110]-Richtung, so daß zwei Gleitebenen wirksam wurden. Zur Vermeidung einer Überkompensation der Kristalle wurden nur Versetzungsdichten unterhalb von $10^7\ cm^{-2}$ erzeugt. Wie ätztechnisch nachgewiesen werden konnte, enthielten die Kristallproben dabei überwiegend 60°- sowie 90°-Versetzungen. Als Schottky-Kontakte und Basiskontakte wurden geeignete Metalle bei Substrattemperaturen von maximal 400 °C aufgedampft, wobei der Sperrschichtkontakt auf ein Gebiet möglichst homogener Versetzungsdichte aufgebracht wurde.

Die Bestimmung der Energieterme erfolgte mit der Methode von Sah et al. [13] sowie mit der DLTS-Technik von Lang [14] (Meßfrequenz 3,2 MHz). Für die Energietermumladungen wurden elektrische Impulse und Lichtimpulse unterschiedlicher Länge bei einer Wiederholungsfrequenz von 50 Hz verwendet. Die direkte Messung der Energietermverteilungen (DLTS) erfolgte dabei mit einem Boxcar-Integrator bei einem Zeitkonstantenfenster von 2 bis 10 ms. Als Impulslichtquellen wurden ein YAG-Laser (Impulslänge $\approx 0{,}1\ \mu s$, $\lambda = 1{,}06\ \mu m$) und eine GaAs-Diode (Impulslänge $\approx 0{,}5$ ms, $\lambda \approx 0{,}90\ \mu m$) verwendet, so daß überwiegend über Grundgitteranregung eine effektive Umladung der Energieterme erzielt werden konnte.

3. Experimentelle Ergebnisse

3.1 DLTS-Spektren

Fig. 1 zeigt eine DLTS-Messung für einen n-Kristall mit einer Versetzungsdichte von $6 \times 10^6\ cm^{-2}$. Für die angegebene Energieskala wurde für den gesamten Temperaturbereich ein Einfangkoeffizient $r_c = 10^{-8}\ cm^3\ s^{-1}$ und eine mittlere effektive Zustandsdichte von $N_c = 7 \times 10^{18}\ cm^{-3}$ vorausgesetzt, so daß mit den verwendeten Torzeiten von 2 ms, 10 ms der Zusammenhang $E \approx 23kT$ folgt. Der linke Teil des Spektrums

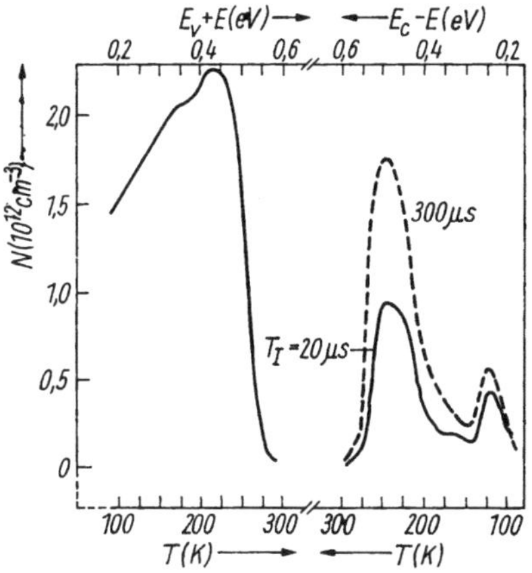

Fig 1 DLTS-Spektren für einen plastisch verformten n-Si-Kristall ($\varrho = 65\,\Omega$cm) mit Versetzungsniveaus in der oberen Bandhälfte. T_I Impulsbreite des Füllimpulses, Lichtimpulse: GaAs-Diode, $\lambda = 0{,}9\,\mu$m; $N_\mathrm{d} \approx 6 \times 10^6$ cm^{-2}

enthält die umladbaren Energieterme der unteren Bandhälfte,[2]) wobei hier die Löcherbesetzung mit Lichtimpulsen einer GaAs-Diode bewirkt wurde Man erkennt ein breites Kontinuum mit einer relativ scharfen oberen Grenze bei $E_\mathrm{v} + 0{,}50$ eV, wobei die untere Grenze wesentlich unterhalb von $E_\mathrm{v} + 0{,}20$ eV liegen muß. Diese Energieterme konnten bei allen plastisch verformten Kristallen beobachtet werden. Obgleich bei vergleichbaren Kristallen deutliche Unterschiede in der Form der Bande gefunden wurden, besteht zwischen der Gesamtkonzentration dieser Terme und der Versetzungsdichte ein eindeutiger Zusammenhang, der weiter unten angegeben werden soll.

Fig. 1 enthält im rechten Teil die Energieterme in der oberen Bandhälfte. Für die Messungen wurden Füllimpulslängen von 20 µs bzw. 300 µs verwendet, wobei deutlich eine verstärkte Umladung der Energieterme bei $E_\mathrm{c} - 0{,}50$ eV festgestellt werden kann.[3]) Auch diese Bande wird durch Versetzungszentren verursacht, da die Auffüllung der Niveaus mit Elektronen nicht mit einer exponentiellen Zeitabhängigkeit bzw. mit einer Zeitkonstanten τ_c beschrieben werden kann ($\tau_\mathrm{c} \approx 1\,\mu$s für $r_\mathrm{c} \approx$

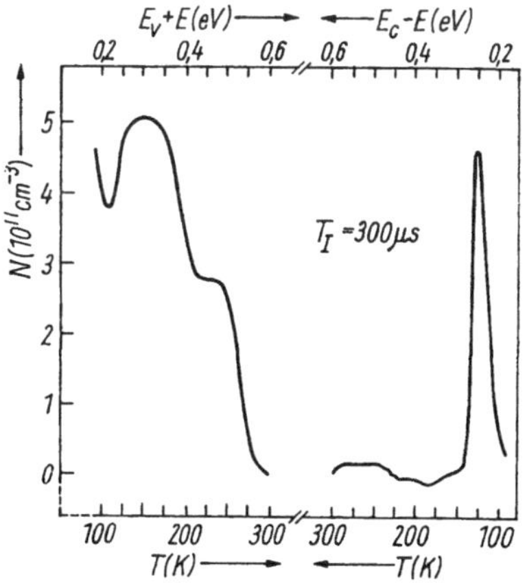

Fig. 2. DLTS-Spektrum für einen plastisch verformten n-Si-Kristall ohne Versetzungsniveaus in der oberen Bandhälfte. Lichtimpulse: YAG-Laser, $\lambda = 1{,}06\,\mu$m; $N_\mathrm{d} = 8 \times 10^5$ cm^{-2}

[2]) Hier und im folgenden ist damit das verbotene Band gemeint.

[3]) Das Niveau bei 0,20 eV wurde auch bei den Vergleichsproben beobachtet, so daß diese Terme nicht auf die Versetzungsbildung zurückgeführt werden können (Fig. 1 und 2). Die Konzentration der Fremdstörstellen lag im allgemeinen unter 5×10^{11} cm^{-3}.

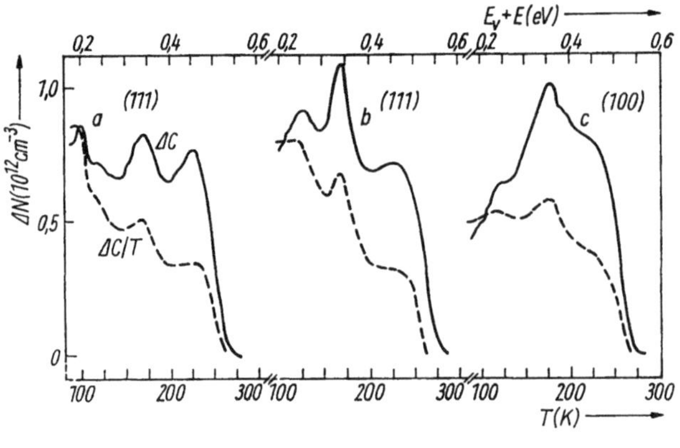

Fig. 3. DLTS-Spektren für plastisch verformte p-Si-Kristalle. a) $\varrho = 130$, $N_d = 3 \times 10^6$; b) 200, 3×10^6; c) 130 Ω cm, 3×10^6 cm^{-2}

$\approx 10^{-8}$ cm^3 s^{-1}). Der Elektroneneinfang wird hier durch Raumladungsbarrieren begrenzt, wobei die Höhe der Potentialbarriere vom Besetzungsgrad der Zentren abhängt. Für die maximal mögliche Füllung derartiger Zentren sind sehr große Zeiten erforderlich, so daß die DLTS-Spektroskopie für die Aufnahme der Energietermverteilung in diesem Fall nicht geeignet ist. Diese Bande soll daher weiter unten durch TSCa-Messungen näher untersucht werden.

Das Versetzungsniveau in der oberen Bandhälfte wurde nicht bei jedem verformten n-Kristall beobachtet. Ein Beispiel dafür zeigt Fig. 2. Während die Energieterme in der unteren Bandhälfte wieder relativ stark vertreten sind, kann das Niveau bei $E_c - 0{,}50$ eV praktisch nicht nachgewiesen werden. Charakteristisch für die Energieterme in der unteren Bandhälfte ist bei dieser Probe die Stufe bei etwa $E_v + 0{,}37$ eV, während der Verlauf bei $E_v + 0{,}20$ eV durch das Störstellenniveau in der oberen Bandhälfte verursacht wird.

Die Energieterme in der unteren Bandhälfte können bei plastisch verformten p-Kristallen direkt untersucht werden, wobei der Verlauf der Energietermbanden zuverlässig ermittelt werden kann. Fig. 3 enthält die DLTS-Spektren für drei Kristalle. Auch hier liegen breite Banden vor, die charakteristische Eigenschaften wie die obere Grenze bei $E_v + 0{,}50$ eV oder die in einigen Fällen auftretende Stufe bei $E_v + 0{,}37$ eV erkennen lassen.[4])

3.2 Bestimmung der Einfangkoeffizienten

Für die Ermittlung der genauen energetischen Lagen der Versetzungsniveaus müssen die Einfangkoeffizienten r_c^n, r_c^p für Elektronen und Löcher bekannt sein. Bei „normalen" Störstellenzentren erfolgt die Auffüllung der Niveaus nach einem exponentiellen Zeitgesetz mit der Zeitkonstanten $\tau_c = (r_c n_0)^{-1}$ (n_0 thermische Trägerkonzentration). Wie Fig. 4 zeigt, ist diese Eigenschaft für die Elektronenbesetzung der Versetzungszentren ersichtlich nicht vertreten. Für die Aufnahme des Umladediagramms wurde der n-Kristall der Fig. 1 unter angelegter Vorspannung auf 90 K abgekühlt, wobei anschließend durch elektrische Impulse variabler Breite bzw. durch optische Erzeugung von Elektron–Lochpaaren ($\lambda = 1$ µm) charakteristische Zentrenumladungen bewirkt wurden. Man erkennt, daß die Kapazität etwa um gleiche Beträge verkleinert wird, sofern die Impulsdauern jeweils verzehnfacht werden.[5]) Das Versetzungsniveau in der oberen Bandhälfte (siehe Fig. 1) wird daher nach einem

[4]) Die Spitze bei 0,20 eV (Fig. 3a) wird durch ein „normales" Störstellenniveau verursacht.

[5]) Die bei Abschalten der Sperrspannung auftretende starke Kapazitätserhöhung wurde in dem Diagramm nicht wiedergegeben.

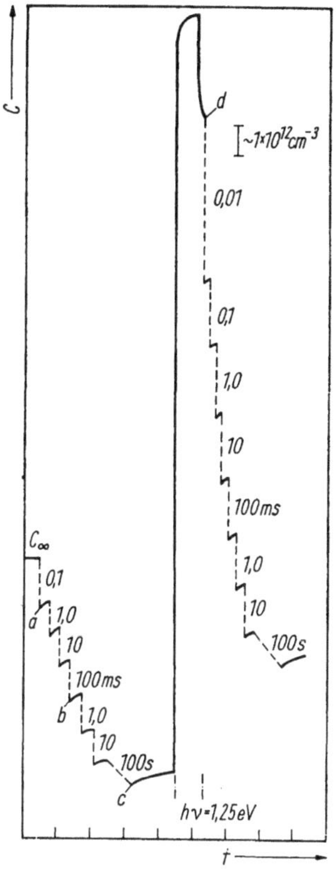

Fig. 4. Kapazitäts-Umladediagramm für den Kristall der Fig. 1 bei $T = 90$ K. t in 100 s/Skalenteil

logarithmischen Zeitgesetz mit Elektronen gefüllt. Diese Abhängigkeit kann im Rahmen eines einfachen Modells leicht abgeleitet werden. Solange die Reemission vernachlässigbar ist, gilt für die Konzentration N^- der besetzten Zentren

$$\frac{dN^-}{dt} = r_c^n n_0 (N - N^-) \exp\left(-\frac{\Phi}{kT}\right) \tag{1}$$

(N Gesamtkonzentration der wirksamen Zentren, Φ Höhe der Potentialbarriere, kT thermische Energie). Setzt man eine lineare Abhängigkeit $(\Phi/kT) = (N^-/N_0^-)$ voraus, so folgt mit $N^- \ll N$ und $N^-(0) = 0$

$$N^-(t) = N_0^- \ln\left(\frac{r_c^n n_0 N}{N_0^-} t + 1\right). \tag{2}$$

Fig. 5 zeigt eine Darstellung der Meßwerte $N^-(t)$ über $\ln t$ für das Niveau in der oberen Bandhälfte (Kurve a). Man erkennt, daß die Beziehung (2) über fünf Größenordnungen in der Zeit die experimentellen Ergebnisse richtig beschreibt. Diese Eigenschaft ist auch für die Zentren in der unteren Bandhälfte (Kurve b) gut erfüllt. Für die Messung dieser Werte wurden zunächst durch Erzeugung von Elektron–Lochpaaren die Zentren mit Löchern besetzt. Dabei geht die Kapazität (siehe Fig. 4) in den Zustand d über. Durch Einschalten von Füllimpulsen kann dann der Besetzungsgrad der Zentren mit Elektronen zeitlich verfolgt werden. Hierbei muß jedoch beachtet werden, daß in diesem Fall die Beiträge des ersten Niveaus mit enthalten sind. Die Kurve b in Fig. 5 wurde unter Berücksichtigung dieses Anteils ermittelt. Aus dem Verlauf der Geraden können die charakteristischen Größen N_0^- und t_0 bestimmt werden, so daß für den Einfangkoeffizienten

$$r_c^n = \frac{N_0^-}{n_0 N t_0} \tag{3}$$

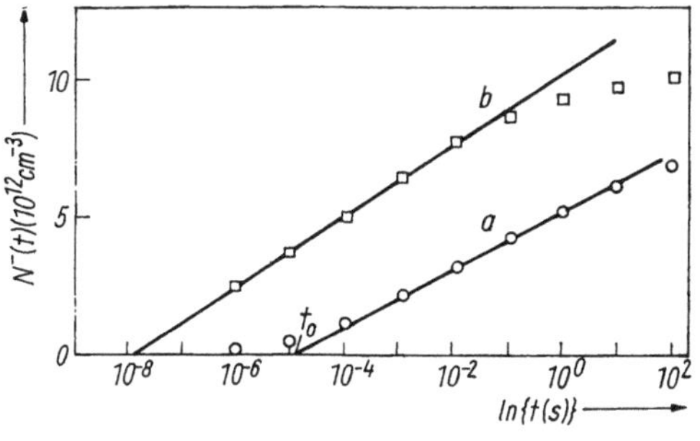

Fig. 5. Abhängigkeit der Elektronenbesetzung der Akzeptorniveaus von der Zeit bei $T = 90$ K

folgt. Setzt man für den mittleren Abstand der Versetzungszentren $a = 3 \times 10^{-8}$ cm und $N = (N_d/a) = 2 \times 10^{14}$ cm^{-3}, so folgt mit $n_0 = 7 \times 10^{13}$ cm^{-3} für die Einfangkoeffizienten der Niveaus in der oberen bzw. unteren Bandhälfte $r_c^n \approx 3 \times 10^{-12}$ cm^{3}s^{-1} bzw. 3×10^{-9} cm^3 s^{-1}. Da jedoch der überwiegende Anteil der Zentren in der unteren Bandhälfte vorliegt, entspricht die gesamte Ätzgrubendichte von 6×10^6 cm^{-2} sehr wahrscheinlich nicht der Versetzungsdichte für das Niveau in der oberen Bandhälfte, so daß im folgenden $r_c^n \approx 10^{-11}$ cm^3 s^{-1} entsprechend $N_d = 2 \times 10^6$ cm^{-2} vorausgesetzt werden soll. Diese Unsicherheit bei der Bestimmung von Parametern wird auch in einigen Fällen weiter unten vorliegen, da die gesamte Zentrenkonzentration der Versetzungen mit diesen Messungen nicht ermittelt werden kann. Unabhängig hiervon kann jedoch die effektive Barrierenhöhe wegen $\Phi/kT = N^-/N_0^-$ angegeben werden. Für die Füllzeit $t = 100$ s würde z. B. $\Phi = 0{,}15$ eV für das obere Versetzungsniveau folgen. Bei einer Versetzungsdichte von 2×10^6 cm^{-2} wäre dann $f \approx 0{,}1$, so daß die Voraussetzung $N^- \ll N$ erfüllt ist.

Die Zeitabhängigkeit für die Elektronenbesetzung der Versetzungsniveaus in der unteren Bandhälfte beweist, daß auch diese Zentren Akzeptorcharakter aufweisen. Diese Eigenschaft konnte durch Untersuchungen der Umladungen mit Löchern bei p-Kristallen bestätigt werden.

Das erste Umladediagramm in Fig. 6 enthält eine derartige Messung für den p-Kristall der Fig. 3a. Man erkennt, daß nach einer Zeit von 2 μs die Zentren in der unteren Bandhälfte praktisch vollständig mit Löchern umgeladen wurden. Im Gegensatz zu Fig. 4 liegt ersichtlich keine logarithmische Zeitabhängigkeit vor, woraus auf den Akzeptorcharakter der Zentren geschlossen werden kann.[6]) Die Umladung erfolgt näherungsweise exponentiell, so daß aus der Zeitkonstanten der Einfangkoeffizient zu $r_c^p \approx 1{,}5 \times 10^{-8}$ cm^3 s^{-1} unmittelbar bestimmt werden kann.

3.3 Bestimmung der Energieniveaus

Für die Untersuchung der Eigenschaften des Akzeptorniveaus in der oberen Bandhälfte ist die TSCa-Spektroskopie besonders geeignet, da hier große Umladezeiten leicht realisiert werden können. Fig. 7 enthält die ΔC-Verläufe in Abhängigkeit von der Temperatur für verschiedene Umladezeiten. Für diese Verläufe wurden die Ableitungen $d(\Delta C)/dT$ näherungsweise ermittelt, und als gestrichelte Kurven in den Diagrammen dargestellt. Man erkennt, daß die Konzentration der „Energieterme" bei kleinen Energien mit zunehmender Füllung des Akzeptorniveaus erhöht wird.[7])

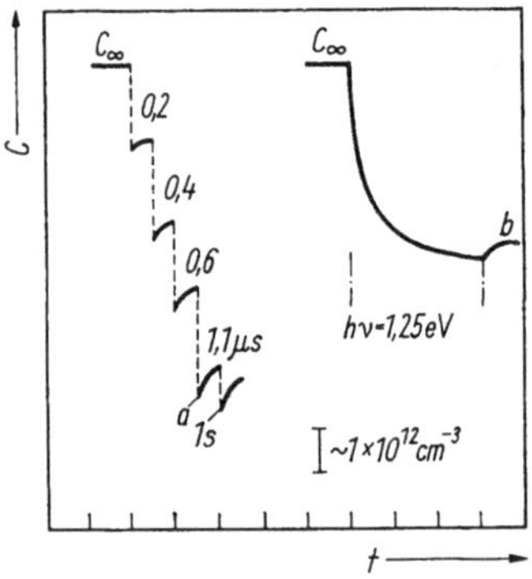

Fig. 6. Kapazitäts-Umladediagramm für den Kristall der Fig. 3a bei $T = 90$ K. t in 50 s/Skalenteil

[6]) Donatorzentren würden eine positive Aufladung der Versetzungslinien und damit den Aufbau von Barrieren bewirken.

[7]) Die Energieskala wurde für die Aufheizrate $\beta = 0{,}5$ K s^{-1} und $r_c N_c \approx 10^{11}$ s^{-1} berechnet.

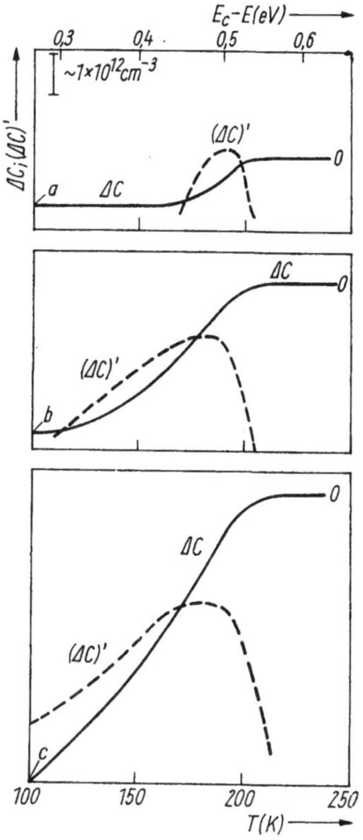

Fig. 7. TSCa-Verläufe für die Umladezustände a, b, c des n-Si-Kristalls der Fig. 4 ($\Delta C = C_\infty - C$ [13])

Diese Eigenschaft wird durch die negative Aufladung der Versetzungen und den damit zusammenhängenden elektrostatischen Energieanteil $E_e(N^-)$ verursacht. Bedeutet E_0 die Lage des Akzeptorterms unterhalb der Leitungsbandkante für $N^- \to 0$, so folgt für die Emissionsrate

$$\frac{\mathrm{d}N^-}{\mathrm{d}t} = -r_c^n N_c N^- \exp\left[-\frac{E_0 - E_e(N^-)}{kT}\right]. \tag{4}$$

Für die Berechnung der Funktion $\Delta C(T)$ muß wegen $-\Delta C \sim N^-$ die Beziehung (4) über T ($T = \beta t$) integriert werden. Da auch für den einfachen Fall $E_e \sim N^-$ keine analytische Lösung bekannt ist, wurde eine Auswertung direkt mit der experimentellen Kurve c der Fig. 7 durchgeführt. Für die Energie der Akzeptorelektronen folgt aus (4)

$$E_0 - E_e(N^-) = kT \ln\left[\frac{r_c^n N_c N^-}{\beta(-dN^-/dT)}\right]. \tag{4a}$$

Verwendet man den oben bestimmten Wert $r_c^n \approx 10^{-11}$ cm^3 s^{-1} unter Berücksichtigung eines Temperaturganges von $r_c^n N_c \sim T^2$, so erhält man den in Fig. 8 dargestellten Verlauf der Energie $E(N^-)$. Die beste Anpassung der Meßwerte wurde für $E_e \sim N^-$ erreicht, wobei für die Lage des Akzeptorniveaus $E_0 = 0{,}39$ eV bei $N^- \to 0$ folgt. Die gestrichelte Gerade zeigt den Verlauf der Energie für einen Einfangkoeffizienten von 10^{-10} cm^3 s^{-1}. Da in diesem Fall jedoch $N \approx 7 \times 10^{12}$ cm^{-3} gelten müßte, würde hier die schon nicht mehr sinnvolle Aussage $f \approx 1$ folgen. Für die maximale Besetzung im ersten Fall erhält man dagegen $f \approx 0{,}1$, wobei die elektrostatische Energie bei 0,17 eV liegt. Der elektrostatische Energieanteil ist auch für die Interpretation der Energietermverteilung in der unteren Bandhälfte von Bedeutung. Nach einer Füllung der Zentren mit Löchern (siehe Fig. 6) erfolgt die Reemission wegen $(N^x)' = -(N^-)'$

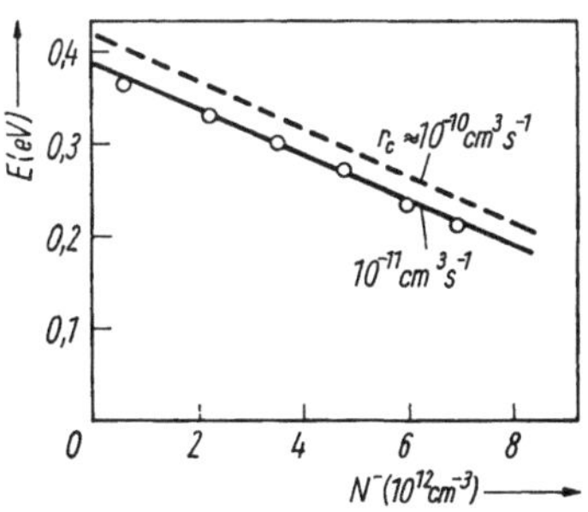

Fig. 8. Abhängigkeit der energetischen Lage des Akzeptorniveaus in der oberen Bandhälfte von der Elektronenbesetzung

gemäß der Gleichung

$$\frac{\mathrm{d}N^-}{\mathrm{d}t} = r_c^p N_v (N - N^-) \exp\left[-\frac{E_0 + E_e}{kT}\right]. \tag{5}$$

Für die TSCa-Verläufe kann die Beziehung (5), ähnlich wie oben angegeben, direkt für die Auswertung verwendet werden. Wegen der schnellen Umladbarkeit der Terme besitzen jedoch auch die DLTS-Spektren (siehe Fig. 3) einen eindeutigen Charakter der im folgenden interpretiert werden soll.

Setzt man $E_e = \alpha N^-$ in erster Näherung voraus, so erhält man mit $N^- \ll N$ eine Gleichung, die für konstantes T integriert werden kann:

$$N^-(t) = \frac{kT}{\alpha} \ln\left(\frac{\alpha N t}{kT\tau_e} + 1\right) \tag{6}$$

mit $N^-(0) = 0$ und $\tau_e = (1/r_c^p N_v) \exp(E_0/kT)$.

Für den Kristall der Fig. 3b wurden die Meßwerte $N^-(t)$ über $\ln t$ für zwei Temperaturen in Fig. 9 dargestellt. Man erkennt, daß die Beziehung (6) die experimentellen Ergebnisse über mehrere Größenordnungen in der Zeit richtig beschreibt. Jedoch zeigt eine Ablesung der charakteristischen Zeit $t_0 \approx 3 \times 10^{-5}$ s, für die entsprechend (6) $t_0 \sim \tau_e$ gelten sollte, daß die Konzentration nicht vollständig bestimmt wurde. Diese Eigenschaft wird durch das begrenzte zeitliche Auflösungsvermögen der Kapazitätsmessung verursacht, bei der Umladevorgänge unterhalb von 10^{-5} s nicht mehr erfaßt werden können. Unabhängig hiervon kann jedoch der Parameter α aus den Anstiegen der Geraden bestimmt werden, wobei Werte von 2,5 bzw. $3{,}0 \times 10^{-14}$ eV cm^3 erhalten wurden.

Für die Versetzungsdichte 3×10^6 cm^{-2} des Kristalls würde bei einem Abstand der Zentren $a = 3 \times 10^{-8}$ cm für $N = 1{,}0 \times 10^{14}$ cm^{-3} folgen, so daß $\alpha N \approx 3$ eV für $f = 1$ gilt. Mit diesem Wert kann nun leicht der theoretische Verlauf des DLTS-Spektrums für ein Versetzungsniveau ermittelt werden.

Gemäß (6) bewirkt die Löcheremission eine wesentliche Umladung, sobald der erste Summand in der Klammer größer als 1 wird. In diesem Bereich gilt für das DLTS-Signal ΔC

$$\Delta C \sim \Delta N^- = \frac{kT}{\alpha} \ln\left(\frac{t_2}{t_1}\right) \tag{6a}$$

(t_1, t_2 Torzeiten). Für ein Versetzungsniveau existiert daher ein Temperaturbereich, in dem das Signal linear mit T ansteigt und von der Emissionszeit τ_e unabhängig ist.

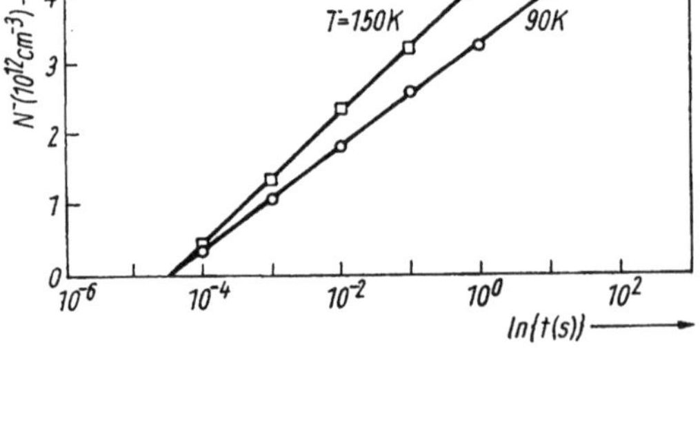

Fig. 9. Zeitliche Abhängigkeit der emittierten Löcherkonzentration für einen plastisch verformten p-Si-Kristall

Die untere Temperaturgrenze T_u charakterisiert dabei die Energie E_0 des unbesetzten Versetzungsniveaus oberhalb der Valenzbandkante, wobei näherungsweise

$$\ln (\eta_u t_2/\tau_e) \approx 0 \quad \text{oder} \quad E_0 \approx kT_u \ln (\eta_u r_c^p N_v t_2) \tag{7a}$$

gilt $(\eta_u = \alpha N/(kT_u))$.

Setzt man für $T > 100$ K $\eta_u \approx 2 \times 10^2$ und $r_c^p N_v \approx 10^{11}$ s^{-1}, so folgt mit $t_2 = 10^{-2}$ s für $E_0 \approx 26kT_u$.

Die rechte Flanke des DLTS-Spektrums bei der oberen Grenztemperatur T_0 wird durch den maximal möglichen Besetzungsgrad des Versetzungsniveaus bestimmt. Extrapoliert man die Beziehung (6) bis zu dieser Grenze, so gilt analog

$$E_0 + \alpha N_{max}^- \approx kT_0 \ln (\eta_0 r_c^p N_v t_1) \tag{7b}$$

oder $E \approx 24kT_0(\eta_0 \approx \eta_u, t_1 = 2 \times 10^{-3}$ s).

Setzt man z. B. $f_{max} = 0{,}15$, so folgt für ein Versetzungsniveau mit $E_0 = 0{,}35$ eV als DLTS-Spektrum ein trapezförmiger Verlauf zwischen 170 und 340 K.

Wegen (7a), (7b) können die charakteristischen Energiewerte der Versetzungsniveaus mit der Energieskala der Fig. 3 a bis c hinreichend genau bestimmt werden.

Dividiert man das DLTS-Signal durch T, so erhält man die in Fig. 3a bis c dargestellten gestrichelten Verläufe. Auf Grund der vorliegenden Struktur kann erwartet werden, daß die Signale eine Überlagerung von drei Beiträgen charakterisieren. Bei den deutlich erkennbaren Flanken bei etwa 130, 175 und 250 K sollte es sich entsprechend den theoretischen Überlegungen um die oberen Besetzungsgrenzen von drei Versetzungsniveaus handeln, wobei das flachste Niveau wesentlich unterhalb von $E_v + 0{,}2$ eV liegen muß. Im Bereich um 110 und 160 K scheint ein Überschneiden der oberen Grenze mit der unteren Grenze von je zwei benachbarten Banden vorzuliegen. Sofern keine Fremdstörstellen das Spektrum beeinflussen, kann aus der Höhe der umgerechneten Signale in den erkennbaren Plateaugebieten die relative Versetzungsdichte bestimmt werden. Bezeichnet man mit c den mittleren Ladungsabstand auf der Versetzungslinie, so gilt im Rahmen der linearen Näherung $E_e = \varepsilon/c$, wobei ε eine vom Versetzungstyp unabhängige Konstante bedeutet. Mit $f = a/c = N^-/N$ folgt dann

$$1/\alpha = N_d/\varepsilon \quad \text{und} \quad \alpha N = \varepsilon/a\,. \tag{8}$$

In Fig. 3 a and b ist daher der Versetzungstyp für das flachste Versetzungsniveau wesentlich häufiger als der des tiefsten Niveaus vertreten. Für den (100)-Kristall der Fig. 3c ist dagegen die Verteilung deutlich anders. Verwendet man die Ätzgrubendichte als integrale Versetzungsliniendichte, so kann aus Darstellungen entsprechend Fig. 9 der Parameter α und bei Berücksichtigung der relativen Häufigkeit von N_d die Konstante ε ermittelt werden. Mehrere Messungen bei unterschiedlichen Temperaturen ergaben dabei für die Kristalle der Fig. 3 Werte von ε im Bereich von 3,5 bis 4,5 × × 10^{-8} eV cm.

Die energetischen Lagen der tieferen Versetzungsniveaus konnten unter Verwendung der Beziehung (5) ähnlich wie in Fig. 8 mit Hilfe der experimentellen TSCa-Verläufe zu $E_0 = E_v + 0{,}23$ eV bzw. $E_v + 0{,}37$ eV ermittelt werden. Die maximale Verschiebung infolge der Energie E_e lag in beiden Fällen bei etwa 0,15 eV, so daß mit $a = 3 \times 10^{-8}$ cm und $\varepsilon \approx 4 \times 10^{-8}$ eV cm für den Besetzungsfaktor $f \approx 0{,}12$ folgt. Setzt man ähnliche Verhältnisse auch für das flachste Niveau voraus, so würde man für die Energie $E_0 \approx E_v + 0{,}1$ eV erhalten.

Für die hier untersuchten Kristalle existieren daher vier charakteristische Energiewerte bei $E_c - 0{,}39$ eV (E_1), $E_v + 0{,}37$ eV (E_2), $E_v + 0{,}23$ eV (E_3) und $E_v + 0{,}1$ eV (E_4). Die Existenz von Donatorzuständen unmittelbar unterhalb dieser Niveaus kann

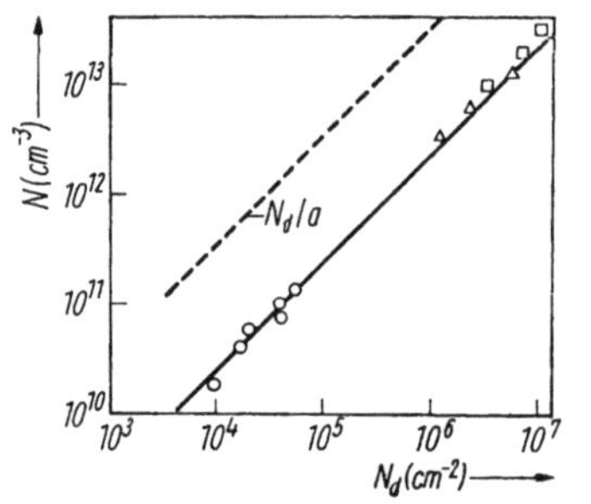

Fig. 10. Abhängigkeit der Gesamtkonzentration der besetzten Akzeptorniveaus in der unteren Bandhälfte von der Ätzgrubendichte; ○ tiegelfrei gezogene n-Kristalle, △ plastisch verformte n-Kristalle, □ plastisch verformte p-Kristalle; $a = 3 \times 10^{-8}$ cm

mit Sicherheit für E_2 sowie E_3 und wahrscheinlich auch für E_1 ausgeschlossen werden, so daß diese Energiewerte nicht die Bedeutung von Neutralniveaus halbbesetzter Energiebänder haben können. Aussagen über das Auftreten von Donatorzuständen nahe der Valenzbandkante für E_4 oder für andere voll besetzte Bänder können z. Z. noch nicht gemacht werden, da hierfür Untersuchungen bei wesentlich tieferen Temperaturen erforderlich wären. Insbesondere zwischen der Gesamtkonzentration der Akzeptorterme E_2 und E_3 und der auf {111}-Flächen bestimmten Ätzgrubendichte besteht ein linearer Zusammenhang (siehe Fig. 10), der darauf hinweist, daß diese Terme durch wichtige Versetzungstypen verursacht werden.[8])

Da nach elektronenmikroskopischen Untersuchungen [15] alle für das Gleitebenensystem {111} wichtigen Versetzungstypen [16] praktisch vollständig als Partialversetzungen vom 30°-, 60°- sowie 90°-Typ vorliegen, die grundsätzlich noch unterschiedliche Kernstrukturen aufweisen können [17], wird eine mögliche Interpretation der beobachteten Energieniveaus und Zurückführung auf bestimmte Versetzungstypen stark erschwert. Neuere theoretische Arbeiten von Marklund [18, 19] über die Energiebänder der 30°- und 90°-Partial versetzungen stehen jedoch in Übereinstimmung mit einem Teil der vorliegenden Ergebnisse. Hiernach sollte die 30°-Versetzung ohne ungepaarte Bindungen ein leeres Band in der oberen Bandhälfte und ein besetztes Band innerhalb des Valenzbandes aufweisen, so daß der Term E_1 möglicherweise auf diesen Versetzungstyp zurückgeführt werden könnte. Die 90°-Partialversetzung dagegen besitzt ungepaarte Bindungen mit zwei Energiebändern, wobei das besetzte Band eine obere Grenze von etwa $E_v + 0{,}1$ eV aufweist, während das leere Akzeptorband unmittelbar oberhalb dieser Energie liegt. Das hier beobachtete Niveau E_4 könnte daher mit diesem Versetzungstyp zusammenhängen. Bei einer derartigen Zuordnung müßten dann die häufig vertretenen Niveaus E_2 oder E_3 mit den 60°-Partialversetzungen in Verbindung gebracht werden. Da dieser Versetzungstyp bei der Dissoziation der vollständigen 90°-Versetzung [17] gebildet wird, muß man auch hier eine besonders häufig vertretene Zentrenart erwarten.[9]) Für diese Versetzung liegen zur Zeit noch keine Berechnungen der Bandstruktur vor, so daß eine Zuordnung noch nicht möglich ist. Für eine sinnvolle Deutung der experimentellen Ergebnisse müßten jedoch auch hier die besetzten Zustände entweder im Valenzband oder nahe der Valenzbandkante liegen.

---

[8]) Die Meßwerte für niedrige Versetzungsdichten wurden bei tiegelfrei gezogenen n-Kristallen bestimmt. Da die Terme bei allen n-Kristallen optisch umgeladen wurden, dürften die realen Konzentrationswerte etwa um den Faktor 2 höher liegen, wie aus einem Vergleich zwischen elektrischer und optischer Umladung in Fig. 6 ersichtlich ist (Umladezustände der Kapazität a, b).

[9]) Auf Grund der vorliegenden Ätzgrubenstrukturen kann auf Versetzungslinien-Richtungen ⟨110⟩ sowie ⟨211⟩ geschlossen werden, wobei die ⟨211⟩-Achsen den 60°-Versetzungen entsprechen könnten.

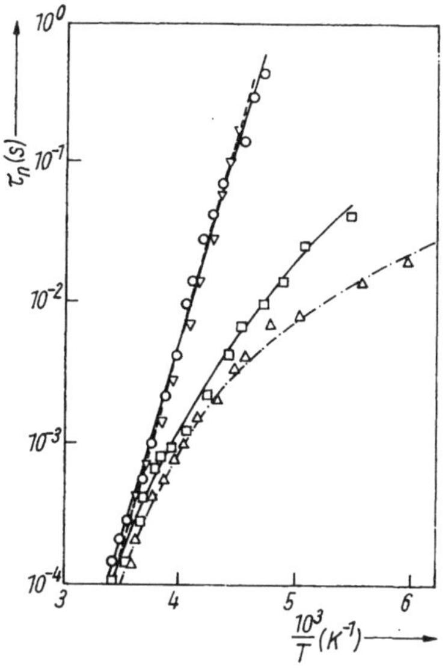

Fig. 11. Temperaturabhängigkeit der Rekombinationslebensdauer bei niedriger Anregung für pastisch verformte n-Si- und p-Si-Kristalle. ○ n, $\varrho = 500\,\Omega$ cm, $N_d \approx 3 \times 10^6$ cm^{-2}; ▽ n, $\varrho = 90\Omega$cm, $N_d \approx 2 \times 10^7$ cm^{-2}; △ p, $\varrho = 130\,\Omega$ cm, $N_d \approx 2 \times 10^6$ cm^{-2}; □ p, $\varrho = 130\,\Omega$ cm, $N_d \approx 8 \times 10^6$ cm^{-2}

3.4 Rekombinationslebensdauern

Auf Grund des Akzeptorcharakters der Versetzungsniveaus hat man ein stark unterschiedliches Verhalten der Fotoleitung bei n- und p-Kristallen zu erwarten. Diese Eigenschaft konnte durch Messung der Lebensdauern bei niedriger Anregung in Abhängigkeit von der Temperatur bestätigt werden. Fig. 11 enthält einige typische Beispiele, wobei für n-Kristalle exponentielle Abhängigkeiten mit einer Aktivierungsenergie um 0,5 eV erhalten wurden, während für p-Kristalle stark gekrümmte Verläufe vorliegen. In beiden Fällen ist wegen der Potentialbarrieren die Elektronen-Lebensdauer τ_n für die Abnahme der Fotoleitung maßgebend. Entsprechend den Ergebnissen von Figielski [20] gilt für n-Kristalle $\tau_n \sim \exp\,[E_{CF}/(kT)]$ (E_{CF} energetischer Abstand zwischen der oberen Besetzungsgrenze des Akzeptorniveaus und dem Leitungsband).

Hieraus folgt, daß in dem untersuchten Temperaturbereich das Niveau in der oberen Bandhälfte für die Rekombination noch nicht maßgebend sein kann. Dagegen wird das Niveau $E_v + 0{,}37$ eV bei den n-Kristallen so stark aufgeladen, daß man Werte von $f \approx 0{,}2$ oder $E_e \approx 0{,}2$ und damit $E_{CF} \approx 0{,}5$ eV erwarten kann.

Völlig andere Verhältnisse liegen dagegen im Fall der p-Kristalle vor. Hier sind schon bei Zimmertemperatur die Niveaus schwächer mit Elektronen besetzt, so daß auch die Potentialbarrieren niedriger als bei den n-Kristallen vorliegen. Mit fallender Temperatur nimmt der Besetzungsgrad weiter ab, wobei zusätzlich die Abschirmwirkung der freien Löcher verstärkt wird. Infolge dieser Abschirmung sind die Potentialbarrieren schmal, so daß wahrscheinlich im gesamten Temperaturbereich Tunnelströme für die Überwindung der Barrieren maßgebend sind. Entsprechend diesen Bedingungen muß man in Übereinstimmung mit den Meßergebnissen einen gekrümmten Verlauf in Abhängigkeit von T und eine Annäherung der Lebensdauern für beide Kristallarten bei $T \gtrapprox 300$ K erwarten.

4. Zusammenfassung

Für Versetzungsdichten von 10^6 bis 10^7 cm^{-2} konnte im Gegensatz zu Ergebnissen anderer Autoren [9, 10, 21] keine Donatorwirkung der Versetzungen im Temperaturbereich > 90 K beobachtet werden. In allen Fällen wurden Akzeptorniveaus in der unteren Hälfte der verbotenen Zone nachgewiesen. Aus dem Verlauf der DLTS-Spektren kann abgeleitet werden, daß hier die Überlagerung von Beiträgen mehrerer Versetzungsniveaus vorliegt. Die obere Besetzungsgrenze liegt bei $E_v + 0{,}50$ eV für einen Akzeptorterm mit $E_0 = E_v + 0{,}37$ eV. Für die energetische Lage der übrigen Niveaus ergaben erste Auswertungen $E_0 = E_v + 0{,}23$ eV und $E_0 \approx E_v + 0{,}10$ eV.

Bei einigen n-Kristallen konnte zusätzlich ein Akzeptorniveau in der oberen Bandhälfte bei $E_c - 0.39$ eV nachgewiesen werden. Die Auffüllung der Akzeptorniveaus mit Elektronen erfolgt infolge der Barrierenwirkung nach einem logarithmischen Zeitgesetz. Erste Auswertungen ergaben für den Einfangkoeffizienten des Niveaus in der oberen Bandhälfte $r_c^n \approx 10^{-11}$ cm^3 s^{-1}, während für die übrigen Niveaus ein Wert um 10^{-9} cm^3 s^{-1} bestimmt wurde ($T = 90$ K). Der Einfang der Löcher in die Akzeptorniveaus in der unteren Bandhälfte erfolgt dagegen nach einem exponentiellen Zeitgesetz, wobei für den Einfangquerschnitt ein Wert von $r_c^p \approx 10^{-8}$ cm^3 s^{-1} ermittelt werden konnte.

Unter Verwendung der Ätzgrubendichte als Maß für die integrale Zentrenkonzentration wurde der elektrostatische Energieanteil für einen Abstand c der Ladungen zu $E_e \approx 4 \times 10^{-8}(1/c)$ eV bestimmt. Der Besetzungsgrad der Akzeptorniveaus in der unteren Bandhälfte liegt innerhalb von Schottky-Randschichten bei $f \approx 0{,}1$, wodurch eine energetische Niveauverschiebung von etwa 0,15 eV verursacht wird.

Insbesondere für die Akzeptorterme $E_c - 0{,}39$ eV und $E_v + 0{,}37$ eV konnte kein konstantes Konzentrationsverhältnis beobachtet werden. Dieses Ergebnis steht im Widerspruch zur Interpretation von Grazhulis et al. [10], wonach die mit Halleffekt-Messungen bestimmten Niveaus bei $E_c - 0{,}42$ eV und $E_v + 0{,}43$ eV auf ein Hubbard-Band zurückgeführt werden. Die energetische Lage des Akzeptorterms bei $E_v + {}$ $+ 0{,}37$ eV stimmt weitgehend mit den aus Halleffekt-Messungen ermittelten Werten um 0,4 eV [10, 21] überein. Auch die von anderen Autoren beobachteten Energieterme in der oberen Bandhälfte um 0,4 eV [10] und 0,5 eV [22] entsprechen wahrscheinlich dem Akzeptorniveau bei $E_c - 0{,}39$ eV.

Die Lebensdauermessungen bei niedriger Anregung zeigten für n- und p-Kristalle unterhalb von 300 K ein Verhalten, das mit dem Akzeptorcharakter der Versetzungen qualitativ erklärt werden kann. Die in Übereinstimmung mit Glaenzer und Jordan [12] für n-Kristalle beobachtete Aktivierungsenergie von 0,5 eV entspricht dem Abstand vom Leitungsband zur oberen Besetzungsgrenze des Niveaus bei $E_v + 0{,}37$ eV für einen elektrostatischen Energieanteil um 0,2 eV. Der bei p-Kristallen auftretende Verlauf in Abhängigkeit von der Temperatur kann qualitativ mit der starken Verschiebung der Besetzungsgrenze der Akzeptorniveaus in der unteren Bandhälfte erklärt werden.

Für die Unterstützung bei der Herstellung der Meßproben möchte ich Frau N. Bloch und Frau D. Patzek danken.

Literatur

[1] W. Shockley, Phys. Rev. **91**, 228 (1953).
[2] R. A. Logan, G. L. Pearson und D. A. Kleinman, J. appl. Phys. **30**, 885 (1959).
[3] J. H. P. van Weeren, G. Koopmans und J. Blok, phys. stat. sol. **27**, 219 (1968).
[4] R. M. Broudy, Adv. Phys. **12**, 135 (1963).
[5] R. Labusch und R. Schettler, phys. stat. sol. (a) **9**, 455 (1972).
[6] M. Jastrzebska und T. Figielski, phys. stat. sol. **14**, 381 (1966).
[7] D. Mergel und R. Labusch, phys. stat. sol. (a) **41**, 431 (1977).
[8] W. Schröter, phys. stat. sol. **21**, 211 (1967).
[9] H. Weber, W. Schröter und P. Haasen, Helv. phys. Acta **41**, 1255 (1968).
[10] V. A. Grazhulis, V. V. Kveder und V. Yu. Mukhina, phys. stat. sol. (a) **43**, 407 (1977).
[11] H.-J. Kos und D. Neubert, phys. stat. sol. (a) **44**, 259 (1977).

[12] R. GLAENZER und A. JORDAN, Solid State Electronics **12**, 247 (1969).
[13] C. T. SAH, L. FORBES, L. L. ROSIER und A. F. TASCH, JR., Solid State Electronics **13**, 759 (1970).
[14] D. V. LANG, J. appl. Phys. **45**, 3023 (1974).
[15] I. L. R. RAY und D. J. H. COCKAYNE, Proc. Roy. Soc. **A325**, 534 (1971).
[16] J. HORNSTRA, J. Phys. Chem. Solids **5**, 129 (1958).
[17] P. B. HIRSCH, J. Physique **40**, 27 (1979).
[18] S. MARKLUND, phys. stat. sol. (b) **92**, 83 (1979).
[19] S. MARKLUND, phys. stat. sol. (b) **100**, 77 (1980).
[20] T. FIGIELSKI, phys. stat. sol. **6**, 429 (1964).
[21] D. NEUBERT, K. HOFFMANN, H. TEICHMANN und R. SCHLIEF, Solid State Electronics **21**, 1445 (1978).
[22] J. R. PATEL und L. C. KIMERLING, J. Physique **40**, 66 (1979).

(Received February 4, 1981)

Erratum

To the paper by N. E. Korsunskaya, I. V. Markevich, T. V. Torchinskaya, and M. K. Sheinkman

Photosensitivity Degradation Mechanism
in CdS:Cu Single Crystals

phys. stat. sol. (a) **60**, 565 (1980)

the authors request the following rectification:

On page 570, 7th line of Section 6 the inequality determining the limits of E_r must read correctly $E_d \leqq E_r \leqq E_{II} + E_d$ instead of $E_{II} \leqq E_r \leqq E_{II} + E_d$.

See Short Note by T. Mohammed-Brahim

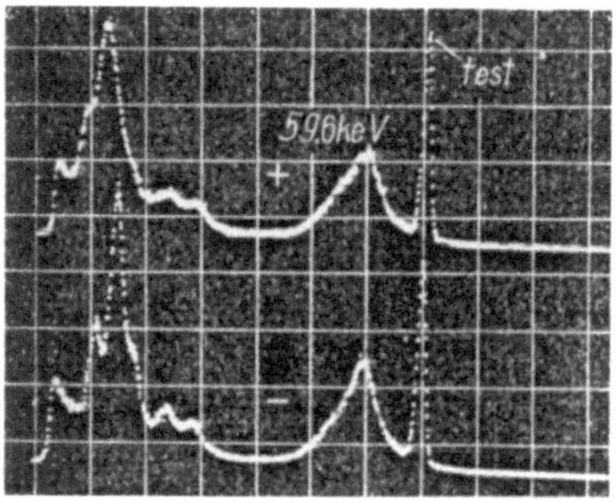

Fig. 3. ^{241}Am spectrum obtained with a sample from LETI showing a 170 K TSC peak before illumination irradiated through the negative (−) and positive + electrodes

phys. stat. sol. (a) 65, K1 (1981)
Subject classification: 13.4 and 14.3; 22.5

S.A.I.-S.E.S. C.E.N. Saclay, Gif-Sur-Yvette[1)]

Defect Levels in HgI_2 and Illumination Effect

By
T. MOHAMMED-BRAHIM[2)]

Introduction When investigating semiconductor crystals to be used for the detection of nuclear radiation at normal temperature, one has to take into account some basic criteria such as a high atomic number Z, an appropriate band-gap, and a high mobility-lifetime product.

Mercuric iodide is a high Z (80.53) material and its energy band-gap is sufficiently large (2.1 eV at normal temperature). However, its lifetime is too small; also, a great effort should be made to find out the lifetime killer defects.

The goal of this paper is to help in this effort by a systematic investigation of defect levels in samples coming from various laboratories (EG & G, University of Strasbourg, and LETI (CEA) of Grenoble) and prepared by the two growth methods of HgI_2 currently used (solution and vapor phase).

This thermally stimulated current investigation method will provide us with a comprehensive table of HgI_2 defect levels currently existing in mercuric iodide. A number of these levels have already been reported in literature /1 to 6/.

In a further step, a not yet observed effect in HgI_2 related to a prolonged illumination, will be described.

Experimental procedure The thermally stimulated current method experimental procedure is known. Because of the reactivity of mercury with most metals and the impossibility of heating HgI_2, the contacts have been achieved either with carbon lacquer deposit, or with a palladium layer deposited in an ionic vacuum unit. These contacts are injecting.

---

1) Gif-Sur-Yvette-91, France.

2) Present address: L.C.C.M.-C.S.I.N., 2 Bd. Frantz Fanon, BP 1017 Alger-Gare, Algeria.

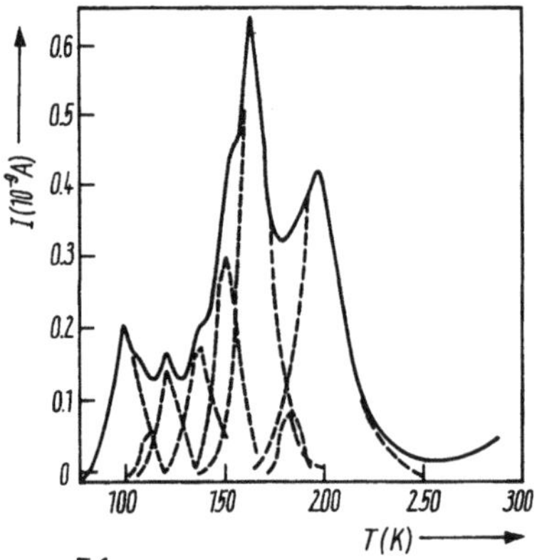

Fig1

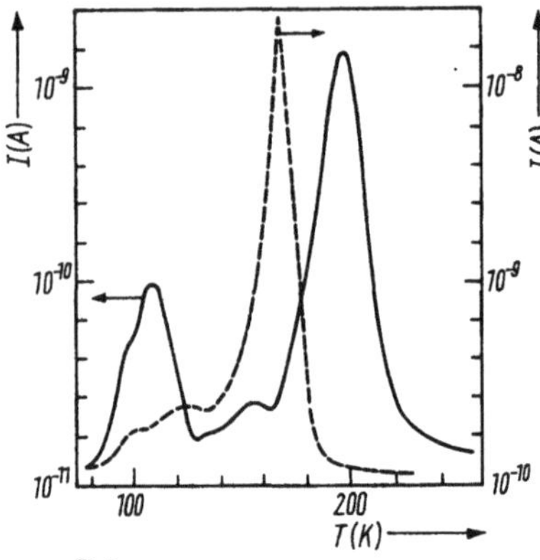

Fig.2

Fig. 1. Example of the numerical method of analysis

Fig. 2. Illumination effect on TSC spectra of a solution grown sample; —— before, - - - after illumination

All the obtained TSC spectra show several non-resolved peaks; therefore the theoretical equations of Randall and Wilkins /7/ and of Garlick and Gibson /8/ cannot be applied immediately.

To separate the peaks, two methods are used, an experimental one described in /6/ and in an internal report of CEA, and another one based on numerical methods. In the latter it is assumed that one peak at least is formed of some experimental values only related to the peak considered. With a numerical treatment of these values, a curve fitting one of the two TSC theoretical equations is obtained. The thus obtained difference between the experimental curve and the theoretical curve of the considered peak gives a curve of which some experimental values are only due to the next peak. The repeated operation of the numerical treatment generates all the spectra peaks of the curves. An example of this numerical method is shown in Fig. 1.

Measurement results TSC spectra All the studied samples (whatever their origin) show common peaks at temperature 96 to 98, 112 to 115, 125, 140, 155 to 160, 185, and 204 K with activation energy of 0.10 to 0.13, 0.16 to 0.19, 0.20 to 0.26, 0.25 to 0.30, 0.40 to 0.44, and 0.43 eV. Some samples show peaks at higher temperatures: the first solution grown sample from LETI

T a b l e 1

Temperatures of the TSC peaks (T_m) and defect levels (E).
β heating rate

Bube /1/			129 0.34	142 0.38		171 0.45	196 0.51		222 0.59		267 0.67	T_m(K) E(eV)	
Saura /2/				148 0.37			196 0.39					T_m(K) E(eV)	
Martin et al. /3/				148 0.37							280 1.05	T_m(K) E(eV)	
Stuck et al. /4/		114 0.19		145 0.22					217 0.45			T_m(K) E(eV)	β = 0.5 K/s
Hyder /5/					160 0.26				225 0.73			T_m(K) E(eV)	β = 0.56 K/s
Gelbart et al. /6/	95 0.15 to 0.23	116 0.19 to 0.23	132 0.12 to 0.29	144 0.27 to 0.35		168 0.36 to 0.49	178 0.43 to 0.51	200 0.49 to 0.54				T_m(K) E(eV)	β = 0.2 to 0.8 K/s
present work	96 to 98 0.10 to 0.13	112 to 115 0.16 to 0.19	125 0.20 to 0.23	140 0.20 to 0.26	155 to 160 0.25 to 0.30	170 0.50 to 0.59	185 0.40 to 0.44	204 0.43		252	275	T_m(K) E(eV)	β = 0.4 K/s

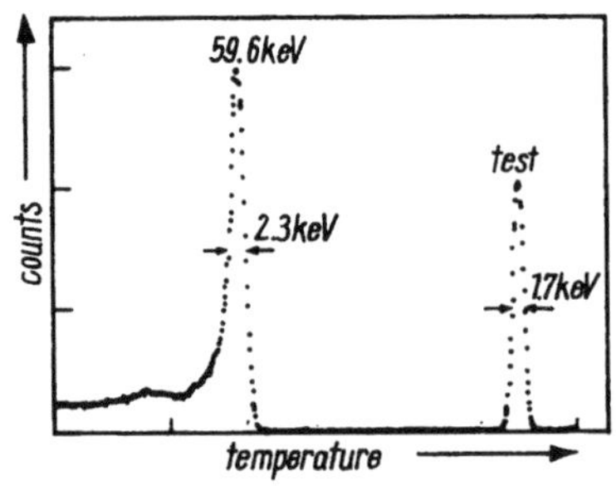

Fig. 4. ^{241}Am spectrum recorded at 59.6 keV with a sample from LETI showing a 170 K TSC peak only after illumination, irradiated through the negative electrode

shows a 250 K peak. The vapor phase grown samples from the University of Strasbourg show a 275 K peak. Finally, some solution grown samples and all vapor phase ones show another peak at 170 K with an activation energy of 0.45 to 0.50 eV.

Illumination effect An effect due to a 5900 Å prolonged (≈ 24 h) monochromatic illumination at normal temperature, corresponding to the energy band-gap, has been observed. The illumination effect on TSC spectra of a solution grown sample from LETI is shown in Fig. 2.

All the peaks show an increase in amplitude except the 203 K one which decreases until it vanishes completely. Finally, the most significant phenomenon is the emergence of a very high amplitude peak at 170 K which is far above the others. This can be seen in all the samples whatever their origin. However, it is far more significant in the solution grown samples.

Furthermore, this effect appears only if the carriers created by the light can move into the illuminated sample by either an applied field or even by the internal field when the sample is short circuited, though it is less important in this last case. Then, it decreases after storage in darkness but does not completely disappear, so it is still visible after 5 months stored in darkness.

Discussion Data of the various levels found in the literature and in this study are given in Table 1. A great similarity between levels found by Gelbart et al. and ours can be observed. These have also been found in the vapor phase crystals by Gelbart et al. The ones found by us are visible in the crystals grown in vapor phase as well as in solution except for the 170 K peak which appears in most solution grown crystals only after illumination.

The studied samples have been tested for detecting nuclear radiation at normal temperature with a ^{241}Am source. Those that show a 170 K peak in the TSC spectrum exhibit the 59.6 keV peak. An example of spectra obtained with a 200 μm x 7 mm^2 sample from LETI is shown in Fig. 3 (see on the photo pages before the Short Notes part). All other samples which do not exhibit a 170 K

Table 2

Example of the effect of an applied field on TSC spectra

applied field (V/cm)	peak temperatures (K)			
400	-	111	-	202
4000	96	114	≈170	202

TSC peak, do not show the 59.6 keV peak. After illumination, the latter show a TSC peak at 170 K and in the meantime they exhibit the 59.6 keV peak. An example of spectra obtained with a 500 μm x 20 mm^2 sample from LETI is shown in Fig. 4.

Therefore, it should be thought of a direct bond between the 170 K TSC peak and the improvement of the nuclear detection. At this stage, efforts are being made to master this bond and to eventually explain it.

The 252 K and 275 K peaks have been found respectively in the first solution sample grown in LETI and in those from the University of Strasbourg grown in vapor phase.

The following crystals from LETI grown with an excess of iodine do not show peaks at high temperature. Therefore, it must be possible to link the lack of iodine to peaks seen at high temperature.

The small number of levels found by Bube, Saura, Martin, Stuck, and Hyder is observed in Table 1. Can we conclude that the crystals used by these authors were purer? We do not think so. We rather believe in the existence of the levels found by Gelbart et al. and ourselves in all crystals whatever their origin. The only difference would be that levels can be filled up more or less easily according to the crystals. The emergence of corresponding TSC peaks would be linked with experimental procedure (amplitude of the applied field, wavelength of excitation, its duration).

Examples of the field effect and the effect of the excitation wavelength can be seen in Table 2 and 3, respectively.

Conclusion Due to the great number of thermally stimulated current spectra done on samples from various laboratories and prepared by different methods,

Table 3

Example of the effect of the excitation wavelength on two TSC peaks

wavelength (nm)	600	575	565	560	555	540	500	450
peak	206	204	204	204	204	204	207	209
temperature (K)	-	-	≈162	159	158	159	-	-

this experimental work has enabled us to build a table of defect levels existing in mercuric iodide.

It has been possible to link the lack of iodine with defects having a thermally stimulated current peak at high temperature. However, these TSC spectra specially the amplitude of the peaks and their number strongly depend on experimental conditions and on previous treatment undergone by the samples. Therefore, it would be aventure to draw a definite conclusion on the characterization of the defect levels in HgI_2.

The study of a prolonged illumination effect never seen so far has enabled us to observe an improvement of the nuclear detection by the HgI_2 crystals after illumination. This improvement has been linked to the amplitude of the 170 K TSC peak with an energy level of 0.45 to 0.50 eV.

Efforts are undertaken in order to understand more deeply the present results.

References

/1/ R.H. BUBE, Phys. Rev. 106, 703 (1957).

/2/ J. SAURA, Thesis, University of Cuyo (Brazil).

/3/ G.M. MARTIN, N.T. PHUOC, and E. FABRE, Proc. Workshop on HgI_2, Strasbourg, June 1975.

/4/ R. STUCK, J.C. MULLER, J.P. PONPON, C. SCHARAGER, C. SCHWAB, and P. SIFFERT, J. appl. Phys. 47, 1545 (1976).

/5/ S.B. HYDER, J. appl. Phys. 48, 313 (1977).

/6/ U. GELBART, Y. YACOBY, I. BEINGLASS, and A. HOLZER, IEEE Trans. Nuclear Sci. 24, 135 (1977).

/7/ J.T. RANDALL and M.H.F. WILKINS, Proc. Roy. Soc. A184, 366 (1945).

/8/ G.F. GARLICK and A.F. GIBSON, Proc. Phys. Soc. 60, 574 (1948).

(Received September 26, 1980)

phys. stat. sol. (a) 65, K7 (1981)
Subject classification: 14.1; 18.1; 18.2; 21.1

Physics Department, University of Nairobi[1)]

Temperature-Dependent Resistivity of Ni-Base Alloys

By
J. MWANJE and E.M. AYIERA

Yao /1/ has recently shown that the temperature dependent part of the resistivity of the nickel-base alloys Ni-Ti and Ni-Cu makes a good fit to the empirical formula

$$\rho_a(T) = \rho_o + AT^n ,$$

where $\rho_a(T)$ is the resistivity of the alloy measured at a temperature T, ρ_o is the resistivity at 4.2 K, A is a constant with respect to temperature. The reported fit was limited to temperatures above 300 K.

We have observed that data on Ni-V and Ni-Al also make an excellent fit to the empirical formula in the ferromagnetic state, even well below 300 K. Fig. 1a shows the plots for 7.5 at% V, 5.0 at% V, 2.5 at% V; while Fig. 1b shows the plots for 7.5 at% Al and 2.0 at% Al. Fig. 1c is a plot for pure nickel. The Curie temperatures for 7.5 at% V, 5.0 at% V, and 2.5 at% V were found to be (300 $\pm$ 5)K, (385 $\pm$ 5)K, and (543 $\pm$ 5)K, respectively. Those for 7.5 at% Al, 2.0 at% Al, and pure nickel were (403 $\pm$ 5)K, (593 $\pm$ 5)K, and (628 $\pm$ 5)K, respectively.

From these plots, the index n is found to be 1.50 $\pm$ 0.04 for all the alloys below their Curie temperatures. Therefore within the range of concentration studied, this parameter is independent of concentration and type of solute. Furthermore, the constant A is also found in this range to be independent of concentration and solute type. We find ln A = -6.30 $\pm$ 0.05. As can be seen in the figures, above the Curie temperatures a downward curvature develops in the plots and any attempt to fit the empirical formula to this paramagnetic region will produce a variable value of the indeed n. For pure nickel in the paramagnetic state we obtain a value of about 1.2 for n; which is much bigger than 0.90 reported by Yao.

---

1) P.O. Box 30197, Nairobi, Kenya.

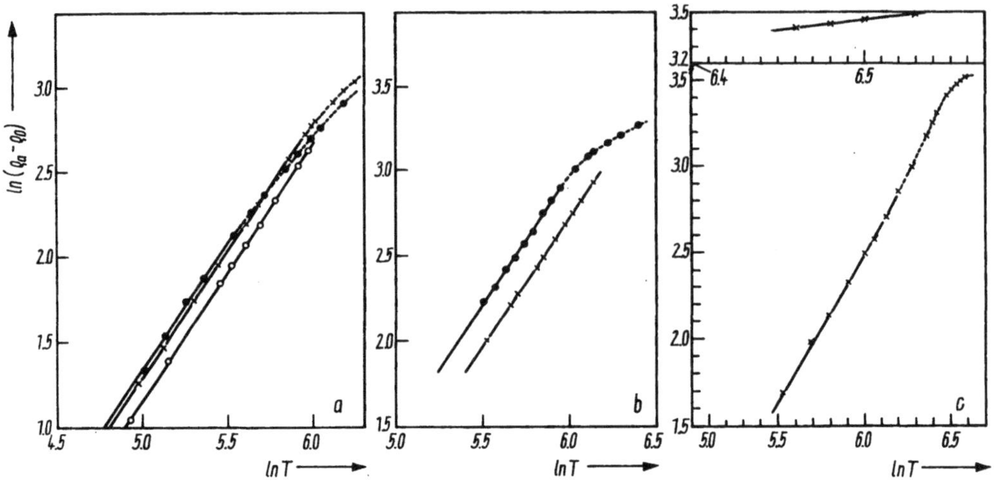

Fig. 1. a) $\ln(\rho_a - \rho_0)$ against lnT for Ni-V alloys; o Ni-7.5 at% V, × Ni-5.0 at% V, • Ni-2.5 at% V. The resistivity values are in μΩcm. b) $\ln(\rho_a - \rho_0)$ against lnT for Ni-Al alloys; o Ni-7.5 at% Al, × Ni-2.0 at% Al. The resistivity values are in μΩcm. c) $\ln(\rho - \rho_0)$ against lnT for pure nickel. The inset is an expanded region above the Curie temperature. The resistivity values are in μΩcm

The authors are grateful to the Department of Solid State Physics of the Royal Institute of Technology, Stockholm where the samples were prepared and the low temperature and some of the high temperature data were measured by one of us (E.M. Ayiera) who was on a fellowship of the International Seminars in Physics and Chemistry.

References

/1/ Y.D. YAO, phys. stat. sol. (b) 94, K137 (1979).

(Received January 5, 1981)

phys. stat. sol. (a) 65, K9 (1981)

Subject classification: 10.2 and 20.1; 3; 22.2.2

Institute of Semiconductor Physics,
Academy of Sciences of the USSR, Sibirian Branch, Novosibirsk[1)]

Depth Distribution of Growth Defects in GaP Films on Sapphire Synthesized in Vacuum

By

V.D. AKHMETOV, I.S. BZINKOVSKAYA, V.V. BOLOTOV, and R.N. LOVYAGIN

Perfect single-crystal GaP films on α-sapphire or spinel substrates are required for some applications in optoelectronic devices /1/. Of special importance here is the spatial distribution of defects that cause changes in the optical characteristics of films.

The presence of various defects in $A^{III}B^{V}$ crystals is known to result in band tailing and in an additional near-edge absorption in this region /2, 3/. Thus, in order to obtain the depth distribution of growth defects in GaP films on sapphire we measured the spectral dependence of the absorption coefficient near the fundamental absorption edge of gallium phosphide in combination with layer-by-layer etching.

In order to reduce the influence of active components of a residual medium, to prepare an atomically clean surface of the substrate and, hence, to obtain more perfect films /4, 5/ we employ the method of condensation in a high vacuum for synthesizing gallium phosphide. The arrangement for synthesis is described in /6/. The growth rate was 6×10^{-8} cm/s, the temperature 500 to 600 °C.

Gallium phosphide films on the (0001) face of single-crystal sapphire have been obtained by us for the first time. An analysis of reflection electron diffraction patterns has shown that the films are single-crystalline with microtwins. The roughness of the growth surface did not exceed several hundreds Å. Absorption spectra in the 1.6 to 2.8 eV range were measured in a two-beam spectrophotometer "Specord-UV VIS" by using differential methods.

To obtain the distribution of the absorption coefficient over the film depth we used the following relation:

---

1) 630090 Novosibirsk-90, USSR.

$$T(x_S) = (1 - R_1)(1 - R_2)\exp(-\int_0^{x_S} \alpha(x)dx) \quad , \tag{1}$$

where $T(x_S)$ is the measured transmission, R_1 and R_2 are reflection coefficients at the boundaries air-GaP and GaP-sapphire, respectively, x_S is the thickness of the GaP layer decreased by etching, $\alpha(x)$ is the absorption coefficient at a distance x from the substrate. If R_1 is assumed to be constant in the process of etching, it follows from (1) after taking the logarithm

$$-\ln T(x_S) = \int_0^{x_S} \alpha(x)dx + \text{const} \tag{2}$$

and after differentiation we obtain

$$-\frac{d}{d\,x_S} \ln T(x_S) = \alpha(x_S) \ . \tag{3}$$

The quantity $\ln T(x_S)$ was measured directly in the experiment, then a graphic differentiation of the obtained curves was performed.

The GaP films were etched in HNO_3:HCl = 1:3 etchant, the rate of etching was 0.3 μm/min, the etching step did not exceed 0.2 μm. The thickness of the etched layer was controlled by measurements in an interference microscope. Absorption spectra were measured after each etching step.

Fig. 1 shows typical absorption spectra for layers at different distances from the surface of the grown films. A considerable smearing of the fundamental absorption edge caused by growth imperfections may be seen, which is decreased as the measured layers move away from the surface (however, these layers are still far from the GaP-sapphire interface region). Since, as has been said above, the growth of the number of imperfections (radiative, dislocations, etc.) in $A^{III}B^{V}$ compounds results in increasing the near-edge absorption, it may be assumed that the amount of additional absorption is proportional to the

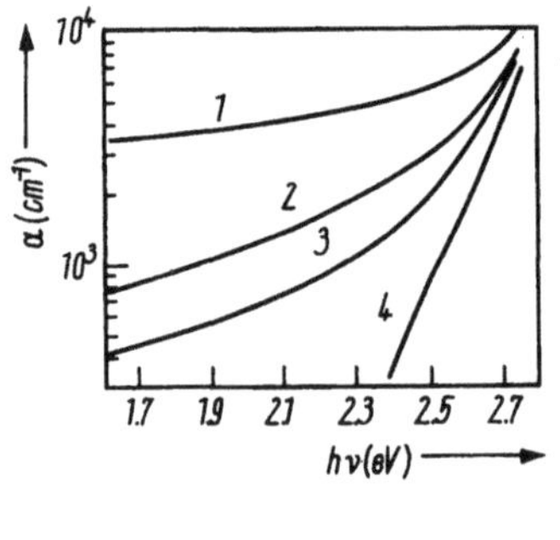

Fig. 1. The absorption edge of gallium phosphide on sapphire. The film has been grown at T = 530 °C thickness 2.5 μm, distance from the surface (1) 0.65, (2) 1.35, (3) 1.65 μm, (4) section of the spectral dependence of the absorption coefficient for a perfect single crystal /3/

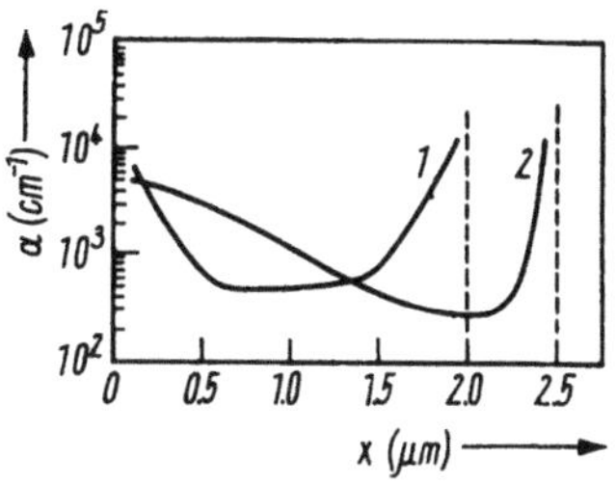

Fig. 2. The depth distribution of the absorption coefficient. Growth temperature (1) 660 °C, GaP film thickness 2 μm, (2) 530 °C, GaP film thickness 2.5 μm. The absorption coefficient is given for $h\nu = 2.48$ eV. The dashed lines mark the film-substrate interface

number of imperfections in the layer. Fig. 2 shows the depth distribution of the absorption coefficient in the films grown at 530 and 660 °C. Two most imperfect regions of films are revealed on the surface and at the gallium phosphide-sapphire interface between which there is a sufficiently perfect single-crystal GaP (minimum values of the absorption coefficient (α) in Fig. 2 agree quite well with α for a perfect single crystal /3/).

The imperfection of films is likely to be due to the deviation of composition from stoichiometry, which is caused by the deficit of phosphorus atoms in the layers. For instance, an increase of the growth temperature lowers the probability of accomodation of phosphorus to the substrate /7/ and, consequently, results in an increase in the size of the interface region up to 0.6 μm. The formation of a imperfect surface layer may be due to the evaporation of phosphorus from the surface of the grown film, as an extra vapour pressure of the elements of group V is not produced.

Thus, it is reasonable to decrease the temperature of epitaxy for reducing the imperfection of the GaP-sapphire interface region and its size.

The authors express their thanks to Prof. L.N. Aleksandrov for valuable discussion of the results.

References

/1/ J.S. VERMAK, Trans. South African Inst. Electr. Engng. 70, 2 (1979).

/2/ R. COATS and E.W.J. MITCHELL, Proc. Internat. Conf. Defects in Semicond., Reading 1972 (p. 96).

/3/ R.K. WILLARDSON and A.C. BEER (Ed.), Semiconductors and Semimetals, Vol. 3, Optical Properties of III-V Compounds, Academic Press, New York/London 1967.

/4/ A. LOPEZ-OTERO, Thin Solid Films 49, 3 (1978).

/5/ S. GONDA, Y. MATSUSHIMA, S. MUKAI, Y. MAKITA, and

O. IGARASHI, Japan. J. appl. Phys. 17, 1043 (1978).

/6/ I.S. BZINKOVSKAYA and R.N. LOVYAGIN, Study of the Conditions for Synthesis of Thin Monocrystalline Gallium Phosphide Films, Preprint 47-80, Inst. Semicond. Phys., Acad. Sci. of the USSR, Novosibirsk 1980 (p. 45).

/7/ P. HUDOCK, Trans.MS AIME 239, 338 (1967).

(Received March 30, 1981)

phys. stat. sol. (a) 65, K13 (1981)
Subject classification: 10.2 and 11; 22.4.1; 22.4.2; 22.4.3

Institute of Semiconductors,
Academy of Sciences of the Ukrainian SSR, Kiev[1)]

On Specific Features of the Defect Formation Mechanism in Cadmium Chalcogenides under Thermal Neutron Irradiation

By

I.B. ERMOLOVICH and V.V. GORBUNOV

Introduction The irradiation of semiconductors with thermal neutrons of a nuclear reactor may be an effective method for intrinsic defect production, this has been pointed out for the first time in /1, 2/. In this case defect production is not caused by elastic collisions between colliding neutrons and lattice atoms, because the kinetic energy of thermal neutrons is small (some hundreds of an eV). This process is caused by thermal neutron capturing by an atomic nucleus with mass M forming a nucleus with mass M + 1 in a highly excited state. The transition of this nucleus into the ground state is accompanied by emission of γ-quanta with energy up to ≈ 10 MeV. This is the so-called (n, γ) reaction:

$$(M,\ z) + n = (M + 1,\ z) + \gamma , \tag{1}$$

where M is the mass of the nucleus capturing the neutron, z is its charge, γ is the energy of the γ-quantum.

According to momentum conservation, the atomic nucleus having captured the thermal neutron, imparts a recoil energy E_r which in some cases (as it will be shown in this paper) is sufficient not only for the "jumping out" of this atom from its lattice site into an interstitial position (primary defect, PD) but for its motion in the lattice and secondary defect (SD) production through collisions with other atoms of both sublattices as well.

The E_r values are determined as /3/

$$E_r = \frac{537}{M + 1} (h\nu)^2 , \tag{2}$$

where $h\nu$ is the energy (in MeV) of emitted γ-quanta, M the mass (in g) of the nucleus having captured the neutron.

---

1) Prospekt Nauki 115, 252650 Kiev, USSR.

The concentration of activated nuclei is defined as /4/

$$[(M+1,\ z)] = K D\,, \qquad K = a\,\rho\,\frac{N_A}{A}\,\sigma_{th}\,, \tag{3}$$

where D is the dose of thermal neutrons (cm^{-2}), a is the percentage of isotopes, capturing neutrons, N_A is the Avogadro number, A is the molecular weight (in g) of the solid under irradiation, σ_{th} is the capture cross section of the thermal neutron (cm^2) by the isotopes.

Thus, the efficiency of defect production in semiconductors and the defect concentration under a definite dose of thermal neutron exposure depend on the parameters a and σ_{th} and on the spectrum of γ-quanta, which determines the spectrum of the recoil energy of atoms having captured neutrons, and besides on the threshold energies for displacing atoms forming the crystal lattice.

The high efficiency of defect production in semiconductors, containing Cd, as has been mentioned in /5 to 8/, is conditioned by a huge cross section σ_{th} (2×10^4 barn) of the isotope Cd^{113} /9/. The transition of the Cd^{114} nucleus from the excited state into the ground state as a result of the reaction Cd^{113} (n, γ)Cd^{114} is accompanied by irradiation of energy up to 9 MeV. The spectrum of γ-quanta for this reaction is given in /10/.

In /5 to 8/, devoted to studies of thermal neutron exposure of Cd-containing compounds, attempts have been made to estimate the concentration of introduced defects. The authors of /5/ supposed that all γ-quanta have their momentum in the same direction and are emitted simultaneously. However, in /6/ it has been stated that this is not correct as the lifetime of Cd^{114} excited states is much longer than that of defect production. Indeed, the former is 10^{-11} to 10^{-12} s/8/, whereas at E_r = 10 to 20 eV the latter (the time between two collisions) is shorter than 10^{-14} s. Therefore the authors /6/ have applied for calculations the cascade model of Kinchin and Pease /11/, which consists in the following: The Cd^{114} atom after emission of one or simultaneously some γ-quanta has time enough to traverse the lattice over a distance of 100 to 1000 Å, to lose its energy by elastic collisions with lattice atoms, and to stop on an interstitial position before emitting further γ-quanta. Then the PD irradiates γ-quanta again, begins to move, colliding and producing its next collision cascade, independent of the first one, etc. up to complete loss of the excited energy. However, in /6/ only 2 MeV γ-quanta were taken into consideration, which correspond to the maximum of the γ-quanta energy distribution curve for the $Cd^{113}(n, \gamma)Cd^{114}$ reaction. Besides the authors /6/ suggested without any reason that two 2 MeV γ-quanta are irradiated simultaneously. Judging from the values of the recoil energy, calculated by (2), they have concluded that secondary defects in CdTe crystals are not produced under thermal neutron irradiation. Consequently, in the mentioned crystals primary defects exist in the Cd sublattice only.

Watkins /12/ stated that the arguments of /6/ are not convincing and their basic conclusion is doubtful. In this paper calculations have been carried out on the concentration of defects caused by the $Cd^{113}(n, \gamma)Cd^{114}$ reaction in CdS, CdSe, and CdTe, taking into consideration the complete spectrum of emitted γ-quanta and using the cascade theory of Kinchin and Pease /11/ for calculation of the secondary defect concentration. The defect distribution on the anion and cation sublattices was determined using the results of Baroody /13/.

Calculation results To simplify the calculations one γ-quantum is supposed to be irradiated in one act. There are no definite data in literature on the combinations of γ-quanta emitted in the $Cd^{113}(n, \gamma)Cd^{114}$ reaction. Our supposition means that, if two or more γ-quanta are emitted simultaneously, their energies are additive, i.e. the direction of momenta of these two γ-quanta is supposed to be parallel. As the momenta of emitted γ-quanta are not correlated and are distributed over all possible directions, our supposition will lead to somewhat higher values of defect concentration.

In accordance with the cascade theory of Kinchin and Pease /11/ the number of secondary defects produced by PD after emission of one γ-quantum with definite energy is calculated by

$$\nu = \frac{E_r - E_d}{2\bar{E}_d}, \quad (4)$$

where E_d for compound binary semiconductors is the threshold energy for displacement of atoms, which has captured the neutron, $\bar{E}_d$ is the average of threshold energies of both component atoms (see Table 1). The total number of secondary defects, which are produced due to emission of γ-quanta with energies in a definite interval, is the product of the average number of γ-quanta emitted in this interval multiplied by

Table 1

Values of threshold energies /12/ and of the parameters A, ϱ and of calculated K and λ values for CdS, CdSe, and CdTe

compound	E_d of cation (eV)	E_d of anion (eV)	$\bar{E}_d$ (eV)	ϱ (g/cm^3)	A (g)	K (cm^{-1})	λ
CdS	7.3	8.7	8.0	4.82	114.6	50	0.695
CdSe	6.0	11.6	8.8	5.81	191.36	44	0.975
CdTe	5.6	7.8	6.7	6.20	240.00	38	1.000

$$N_d = \bar{N}_\gamma \, \nu = \bar{N}_\gamma \frac{E_r - E_d}{2 \bar{E}_d} . \qquad (5)$$

Table 2 contains the numbers of γ-quanta in the energy interval from 0 to 9 MeV taken from /10/, the values of average energies in every γ-quanta interval, calculated by (2), and the number of secondary defects, creating after collisions of PD with lattice atoms, which have been calculated by (4) for 100 acts of thermal neutron capture.

The values of parameters of (3) for the compounds mentioned above are cited in Table 1, too. The numerical values of the coefficient K in (3) are given by the coefficients of thermal neutron absorption (in cm^{-1}).

As can be seen, the depth of thermal neutrons absorption is 2000 to 2500 nm.

Quite obviously, the probability of collisions between the chaotically moving PD with atoms of either sublattice is the same. However, the energies transmitted to every atom and, consequently, the probability of defect production in one or another sublattice should depend on the correlation between the masses of atoms forming both sublattices. Harris was the first to have examined this problem /14/. Later Baroody /13/ has simplified and verified his calculation by introducing threshold energies of displacement damage in accordance with the Kinchin and Pease model /11/. From Fig. 5 in /13/ it may be seen that if the values of the parameter $\lambda = 4 M_1 M_2 (M_1 + M_2)^{-2} \geqq 0.5$ (M_1 is the mass of cadmium and M_2 is the atom mass) the correlation between the concentration of secondary defects in both sublattices is $N_2 (N_1 + N_2)^{-1} = 0.45$ ($N_{1,2}$ are the

Table 2

Theoretical results for secondary defect production in CdS, CdSe, and CdTe as a result of the Cd^{113} (n, γ)Cd^{114} reaction for 100 thermal neutron capture acts

intervals of γ-quantum energies (MeV)	average number of γ-quanta, /10/ N_γ	average recoil energies, calculated by (2) E_r (eV)	number of secondary defects, calculated by (5), for		
			CdS	CdSe	CdTe
0 to 1	132	1.9	-	-	-
1 to 2	92	10.6	-	-	-
2 to 3	96	30	135	261	350
3 to 5	73	75	308	572	755
5 to 7	17	170	172	316	416
7 to 9	1	300	18.4	33.4	44
9	0.1	-	-	-	-

Complete number of secondary defects: 633.4, 1182.4, 1565.

defect concentrations in cadmium and chalcogen sublattices, respectively), i.e. the atoms of both sublattices behave approximately identical. λ values are given in Table 1, too.

Thus, as mentioned above, the concentration of Cd atoms in chalcogenide compounds, which have captured thermal neutrons at dose D, is K D, where K values for CdS, CdSe, and CdTe are compiled in Table 1. For CdS, K = 50. Therefore, the concentration of secondary defects for CdS is $630 \cdot 50 \cdot D/100 = 315$ D, among them $315\ D \cdot 0.55 = 170$ D defects being in the Cd sublattice, and $315\ D \cdot 0.45 = 140$ D defects in the S sublattice. Thus, the total concentration of Frenkel defects in the Cd sublattice of CdS irradiated by thermal neutrons with dose D is $170\ D + 50\ D = 220$ D, and in the S sublattice it is 140 D. Similar calculations for CdSe and CdTe lead to the following results: for CdSe: 330 D in the Cd sublattice and 240 D in the Se sublattice; for CdTe: 370 D and 270 D, respectively.

Thus, taking into consideration both the complete spectrum of γ-quanta resulting from the Cd^{113}(n, γ)Cd^{114} reaction and the correlation of masses of atoms, forming the lattice of the compound semiconductors CdS, CdSe, and

CdTe, we come to the conclusion that due to thermal neutron irradiation of these semiconductors the Frenkel defects are produced in both sublattices in approximately equal concentrations (not differing by more than 35%) practically randomly over a depth of 2000 to 2500 nm of the crystals. This conclusion is very important for a correct interpretation of changes in the electrical, photoelectrical, optical, luminescent, and other properties induced by thermal neutron irradiation of cadmium chalcogenides.

References

/1/ H. SCHWEINLER, J. appl. Phys. 30, 1125 (1959).

/2/ M.V. CHUKICHEV and V.S. VAVILIV, Soviet Phys. - Solid State 3, 1522 (1961).

/3/ V.S. VAVILIV, Irradiation Effect on Semiconductors, Fizmatgiz, Moscow 1963 (p. 256).

/4/ BEKURTS and VIRTS, Neutron Physics, Atomizdat, Moscow 1968 (in Russian).

/5/ R. OSWALD and C. KICKUCHI, Nuclear Sci. Engng. 23, 354 (1965).

/6/ C. BARNES and C. KICKUCHI, Nuclear Sci. Engng. 31, 513 (1968).

/7/ L.K. VODOPIYANOV and A.A. ABRAMOV, in: Collection Cadmium Telluride, Nauka, Moscow 1968 (p. 110).

/8/ C. KICKUCHI, Proc. Internat. Conf. Radiation Effects in Semicond., 1972 (p. 317).

/9/ I.V. GORDEEV, D.A. KARDASHEV, and A.V. MALISHEV, in: Reference Book Nuclear-Physical Constants, Atomizdat, Moscow 1963 (pp. 17, 391).

/10/ A.V. GROSHEV, A.M. DEMIDOV, V.N. LUCHENKO, and V.I. PELECHOV, Izv. Akad. Nauk USSR, Ser. fiz. 26, 979 (1962).

/11/ G.H. KINCHIN and R.S. PEASE, Rep. Progr. Phys. 18, 1 (1955); Uspekhi fiz. Nauk 60, 590 (1956).

/12/ G.D. WATKINS, Inst. Phys. Conf. Ser. 31, 95 (1977).

/13/ E.M. BAROODY, Phys. Rev. 112, 1571 (1958).

/14/ E.G. HARRIS, Phys. Rev. 98, 1151 (1955).

(Received December 15, 1980)

phys. stat. sol. (a) 65, K19 (1981)

Subject classification: 1.3; 21.4

Laboratoire de Cristallographie[1] (a), Laboratoire de Diffraction Neutronique, C.E.N. (b) et Centre de Recherches sur les Très Basses Températures, C.N.R.S.[1] (c), Grenoble

Etude par diffraction neutronique de $CeRu_2D_5$

Par

D. FRUCHART (a), F. VAILLANT (a), E. ROUDAUT (b), A. NEMOZ (c) et X.G. TESSEMA (c)

Introduction Les phases de Laves de formule AB_2 adopant l'un des trois types de structures $MgCu_2$ - cubique - Fm3m - C15, $MgZn_2$ - hexagonal - $P6_3/mmc$ - C14, $MgNi_2$ - hexagonal - $P6_3/mmc$ - C36. Ils peuvent se décrire comme des empilements compacts de trois variétés de tétraédres: 2A-2B, A-3B, et 4B, la cohésion des composés faisant intervenir le facteur stérique $R_A/R_B \approx 1,22$, et la sélection du type de structure dépendant d'un facteur électronique /1/.

La stabilité relative de ces phases et la grande variété d'éléments pouvant être envisagés pour former de tels composés, favorisent la possibilité de formation d'hydrures métalliques à forte concentration d'hydrogène. Des études expérimentales ont été menées par Shaltiel et al. /2 à 4/ où l' élément le plus électropositif A est essentiellement Zr, Ti ou un métal de terre rare. Dans ce dernier cas l'élément B entrant dans la structure vient souvent de la famille de platine. Une meilleure connaissance de la concentration en hydrogène et de la structure des matériaux est nécessaire pour préciser l'évolution des propriétés physiques de $CeRu_2H_x$ en particulier; l'occupation préférentielle des sites 2A-2B, A-3B, 4B doit dépendre de la nature des éléments A et B et de la concentration en hydrogène H_x /4, 5/.

Rappel des propriétés de $CeRu_2$ Les composés cubiques supraconductrices Ce-Ru_2 et La-Ru_2 /6, 7/ de structure phase de Laves C15 sont connus pour absorber de grandes quantités d'hydrogène /2/. Par comparaison du paramètre de maille avec ceux de la série trivalent RE-Ru_2, il apparaît que le cérium est tétravalent dans $CeRu_2$/8, 9/. Des mesures de susceptibilité magnétique, montrent la disparition de la transition supraconductrice pour les deux hydrures et, dans le cas de $CeRu_2$, l'apparition d'un pic de susceptibilité à $T \approx 1,4$ K. Des mesures d'aimantation dans un champ magnétique jusqu'à 80 kOe entre 1,4 et 4,2 K, montrent que le

---

1) 166 X, 38042 Grenoble Cédex, France.

composé hydrogéné a une aimantation du type Brillouin. Ceci peut-être expliqué par l' apparition d' un moment magnétique correspondant a un état Ce^{3+} dont la valeur issue de l' aimantation à saturation est de 0,70 μ_B. Cette valeur est en accord avec un schéma de champ cristallin de l' ion Ce^{3+}, où dans un environnement cubique le doublet Γ_7 de moment magnétique théorique 0, 71 μ_B, est le niveau fondamental. Ainsi par hydrogénation, la valence du cérium a changé de Ce^{4+} à Ce^{3+} /10/.

Etude structurale L' échantillon a été préparé au four à induction, puis scellé dans un creuset en tantale et recuit à 1100 °C pendant 10 jours, et à 1300 °C pendant quelques heures. Les mesures de diffraction aux rayons X, effectuées au préalable, montrent que la structure C15 est conservée avec l' hydruration, conduisant à la formation de l' hydrure ternaire. Cependant l' étude du paramètre de maille est rendue difficile du fait de l' élargissement important des raies de diffraction /11/. Ce phénomène peut être attribué, soit à la création de désordres atomiques, soit à une fragmentation importante de matériau, lors de la réaction très exothermique avec l' hydrogène. Nos premières études par diffraction neutronique ont également conduit à des diagrammes inexploitables.

L' utilisation d' un matériel spécifique:

- multicompteur neutronique avec visualisation en temps réel,
- four adapté aux traitements in situ (P, T) sous-flux de neutrons,
- banc d' hydruration couplé au précédent permettant sur l'échantillon une pression allant jusqu' à 100 bar (H_2 ou D_2) une température allant jusqu' à 750 °C.

Nous avons pu ainsi déterminer les conditions cinétiques de préparation d' hydrures présentant des raies de diffraction fines. Pour optimiser les mesures d' intensité nous avons utilisé le deutérium de préference à l' hydrogène permettant un meilleur rapport de diffusion $\sigma_{cohérent}/\sigma_{incohérent}$. Les diffractogrammes obtenus sont représentés sur la Fig. 1. Les paramètres de maille mesurés sont respectivement: a = (7,546 ± 0,005)Å pour l' alliage désorbé et régénéré (à 700 °C) et a = (8,383 ± 0,007)Å pour $CeRu_2D_x$, x ≈ 5.

Au cours de la transformation $CeRu_2 + \frac{x}{2} D_2 \rightleftarrows CeRu_2D_x$ sous l' action de la pression ou de la température, le système de raies de chacune des deux phases reste approximativement stationnaire en angle. La structure C15 est conservée au cours de la deutération. Sur le cliché de diffraction sont visibles quelques raies dûes a une phase hexagonale qui disparaissent à la désorption. Les paramètres de maille de cette phase qui est probablement un autre deutériure sont a = = (5,830 ± 0,005) Å; c = (9,588 ± 0,005) Å. Signalons que ce phénomène est rencontré

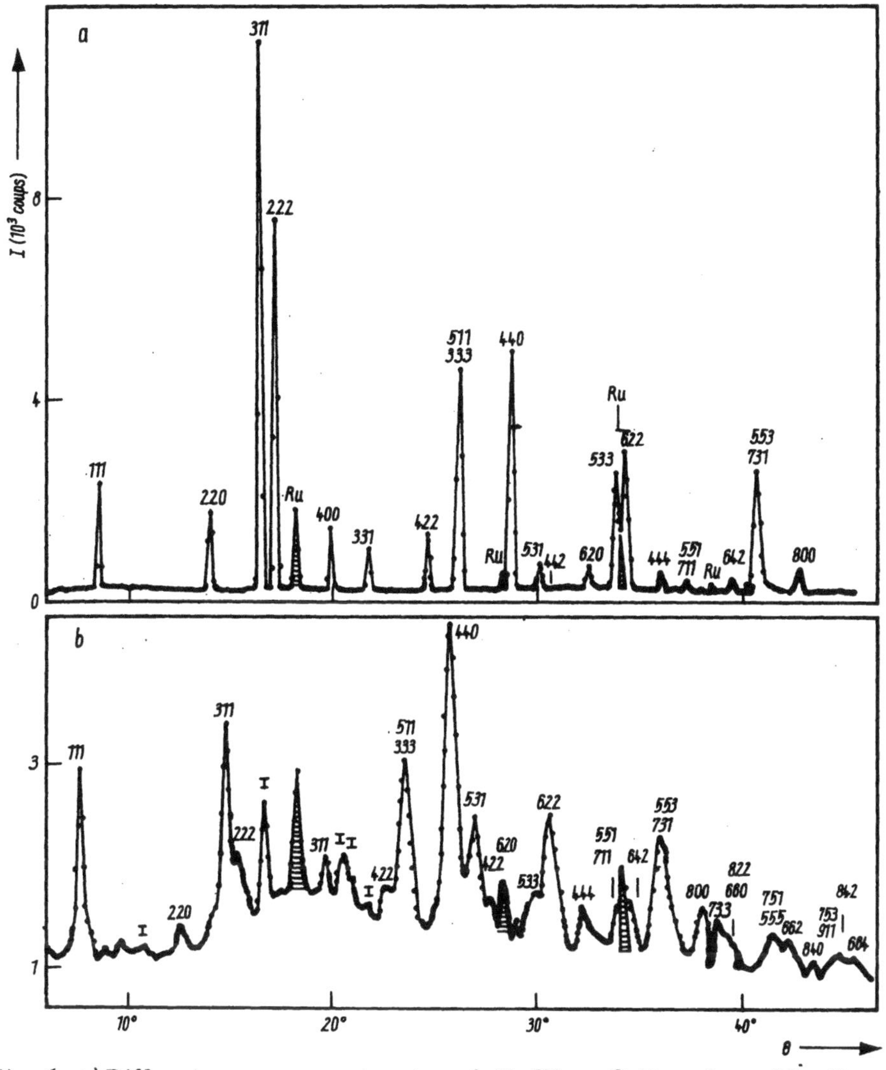

Fig. 1. a) Diffractogramme neutronique de l' alliage $CeRu_2$. Les réflections supplémentaires (R) sont celles du ruthénium libre présent après la formation de l' alliage et d' une partie démixée à la suite des cyclages d' absorption-désorption. Après plusiers de ces cycles, l' intensité des R reste constante. b) Diffractogramme neutronique du deutériure $CeRu_2D_{x\approx5}$. L' occupation des sites A_2B_2 par le deutérium produit un ordre à courte distance modulant le bruit de fond. Les quelques réflexions marquées (H) sont celles d' un deutériure hexagonal dérivant de la forme C14

avec d' autres phases de Laves /12/ où par exemple l' alliage cubique hydruré rest majoritairement cubique, avec selon le traitement thermique, apparition de traces d' hydrure hexagonale dérivé de la forme C14. En particulier avec $SmRu_2$ (C15) la transformation est alors complète puisque la structure de $SmRu_2H_{4,2}$ se rattache à l' empilement métallique C14: a = 5,84 Å; c = 9,57 Å /13/.

Le bruit de fond du diffractogramme du deutériure présente un bosse large, attribuable à un ordre à courte distance /12, 14, 15/.

L' intensité des réflexions de Bragg a été mesurée par une méthode d' interpolation gaussienne permettant d' en définir la position angulaire, les largeur et hauteur et l' intensité /16/. Un programme d' affinement dérivé de Afistr /17/, utilisant une procédure de moindres carrés, basé sur les intensités intégrées a été utilisé. La non-stoechiométrie du site 16 d du groupe Fd3m (position du ruthénium) a été mesurée à partir du diagramme de $CeRu_2$. Le produit (taux d' occupation × longueur de Fermi) du site de ruthénium correspond à une longueur de diffusion effective de $0,675 \times 10^{-12}$ cm, inférieure à celle du ruthénium ($b_{Ru} = 0,73 \times 10^{-12}$ cm) peuvant expliquer le pouvoir diffusif effectif du site 16 d. Une mesure de l' intensité diffractée par le ruthénium libre et l' hypothèse d' une substitution totale du cérium aux lacunes en 16 d conduisent à la formule globale $Ce_{0,45}Ru_{0,55}$. Cependant les analyses métallographiques réalisée après la fusion de l' alliage révélent certaines quantités de ruthénium libre.

Le diffractogramme du deutériure permet de trancher sans ambiguité entre l' occupation des sites 2A-2B, A-3B, 4B. Seul le site riche en cérium: 2Ce-2Ru, est effectivement occupé avec la formule globale $CeRu_{2-\varepsilon}D_{5,2\pm0,5}$.

Cette formula correspond bien à la capacité d' absorption volumique mesurée par ailleurs. Le Tableau 1 présente les valeurs des positions cristallographiques et taux d' occupation affinés. Rappelons qu' il existe bien 12 sites de types 2A-2B par formule et que la règle d' exclusion /5/ n' autorise qu' une occupation statistique moyenne de 1/2. Donc comme dans d' autres composés C15 our C14 hydrurés, ce sont les sites de type 2A-2B seuls, ou majoritairement occupés auquels correspond un phénomène d' ordre à courte distance /12/.

Discussion Dans $CeRu_2D_x$ ($x \approx 5$), seuls les sites 2Ce-2Ru sont occupés. Ce résultat doit être comparé à celui de $ZrV_2D_{4.9}$ où pour un volume d' hydrogène absorbé semblable, le taux d' occupation du site Zr-3V est supérieur à celui du site 2Zr-2V /15/. Plus généralement, les énergies de formation des hydrures et le degré d' occupation des sites indiquent une chaleur de formation équivalente pour les sites A-3B dès que les premiers sont relativement occupés /18/.

Ce résultat montre que le changement de valence du Ce^{4+} à Ce^{3+} favorise l' occupation de site riche en Ce, du type 2A-2B et que dans un calcul de chaleur de formation attribué aux différents sites, cette propriété devra être prise en compte. En outre, suivant l' échelle d' électronégativité utilisée par Miedema

Tableau 1

atome	position	taux d' occupation	x	y	z	b
cérium	8a	1	0	0	0	0,6 (0,3)
ruthénium	16d	0,930 (0,002)	0,5	0,5	0,5	0,55 (0,3)
cérium	8a	1	0	0	0	0,6
ruthénium	16d	0,930	0,5	0,5	0,5	0,55
deutérium	96g	0,433 (0,004)	0,338 (0,004)	0,338 (0,004)	0,157 (0,006)	2,0 (0,9)

$R = \Sigma|I_o - I_c|/\Sigma I_o$, $R_{alliage} = 0,04$, $R_{deut} = 0,07$.

La partie supérieure du tableau présente les résultats relatifs à l' alliage, la partie inférieure ceux relatifs à $CeRu_2D_{x \approx 5}$. Lorsque le paramètre a été affiné l' écart standard figure sous la valeur calculée.

et al., Ru est plus électronégatif que l' hydrogène /19/. Or la susceptibilité magnétique haute température /11/ (entre 50 et 300 K) de l' hydrure comporte un terme χ_o constant, qui s' ajoute au terme provenant du moment effectif de l' ion Ce^{3+}. χ_o est d' un ordre de grandeur supérieur à celui de $CeRu_2$ et peut s' interpréter comme provenant du remplissage de la bande du ruthénium, par des électrons venant de l' hydrogène. Ces deux mécanismes contribuent à créer un environnement, pour l' ion cérium où la charge effective de l' hydrogène serait positive. Cette interprétation conduit à un schéma de champ cristallin, où la hiérarchie des multiplets Γ_7 et Γ_8 est telle que Γ_7 est le niveau fondamental /20/ résultat cohérent avec celui déduit des mesures d' aimantation.

Les résultats expérimentaux obtenus pour $CeRu_2D_{\approx 5}$ sont à rapprocher de ceux obtenus pour l' hydrure $EuRh_2H_{5,5}$/21/. En effet, avec l' hydrogénation, l' europium change de valence, Eu^{3+} à Eu^{2+}, avec apparition d' une température d' ordre magnétique alors que Rh et Ru ont des électronégativités tout-à-fait comparables.

Une étude récente par effect Mössbauer de GdM_2 (M = Rh, Ru) et de leurs hydrures $GdM_2H_{x \approx 3}$ permet à leurs auteurs d' expliquer les variations du déplacement isomérique et de la contribution au champ hyperfin du terme dû aux électrons de conduction, observées durant l' hydruration /22/. Le modèle simple de concept d' électronégativité leur sert à justifier un déplacement de charge

$Gd \rightarrow H$, avec une diminution parallèle de la température d'ordre magnétique liée aux électrons de conduction (mécanisme RKKY). Il faut donc craindre que dans le calcul des chaleurs de formation attribuées aux différents sites tétraédriques, la nature de l'ion de terre rare ne soit un paramètre sensible.

Bibliographie

/1/ J.H. WESTBROOKE, Intermetallic Compounds, Wiley, New York 1967.

/2/ D. SHALTIEL et al., J. less-common Metals 53, 117 (1977).

/3/ I. JACOB et al., Solid State Commun. 23, 669 (1977).

/4/ D. SHALTIEL, J. less-common Metals 62, 407 (1978).

/5/ D.P. SHOEMAKER et al., J. less-common Metals 68, 43 (1979).

/6/ B. HILLENBRAND et M. WILHELM, Phys. Letters A 33, 61 (1970).

/7/ A.C. LAWSON et al., Phys. Letters A49, (1974).

/8/ K.A. GSCHNEIDNER, Rare Earth Alloys, Van Nostrand, 1961 (p. 380).

/9/ B. CORNUT, Thèse, Université de Grenoble, 1976.

/10/ G.X. TESSEMA et al., J. Physique, Lettres 40, L-105 (1979).

/11/ G.X. TESSEMA, Thèse de 3° Cycle, Université de Grenoble, 1979.

/12/ D. FRUCHART et al., J. less-common Metals 73, 396 (1980).

/13/ A.L. SHILOV et al., Soviet Phys.-J. inorg. Chem. 23, 12, 1871 (1980).

/14/ J.J. DIDISHEIM et al., Solid State Commun. 32, 1987 (1979).

/15/ J.J. DIDISHEIM et al., Solid State Commun. 32, 47 (1979).

/16/ P. WOLFERS, Système INTEGR, Institut Laue-Langevin, Grenoble.

/17/ AFISTR, Argonne Nat. Lab., App. Math. Div., Prog. Lib. 36 E7118.

/18/ I. JACOB et D. SHALTIEL, J. less-common Metals 65, 117 (1979).

/19/ A.R. MIEDEMA et al., J. less-common Metals 41, 283 (1975).

/20/ K.R. LEA et al., J. Phys. Chem Solids 23, 1381 (1967).

/21/ K.H.J. BUSCHOW et al., J. appl. Phys. 45, 5289 (1976).

/22/ I. JACOBS et al., J. Magnetism magnetic Mater. 15, 1269 (1980).

(Received February 26, 1981)

phys. stat. sol. (a) 65, K25 (1981)

Subject classification: 14.4.1; 18.4; 22.8.1

Department of Radiation Research,
Academy of Sciences of the Azerbaidzhan SSR, Baku[1)] and
Dnepropetrovsk State University[2)]

The Magnetoelectric Effect $(ME)_H$ in the Ferroelastics $NaFe(MoO_4)_2$ and $K_{0.8}Na_{0.2}Fe(MoO_4)_2$

By

I.H. ISMAILZADE (a), R.M. ISMAILOV (a), and T.M. STOLPAKOVA (b)

In /1/ it has been shown that single crystalline $NaFe(MoO_4)_2$ is ferroelastic with a Curie temperature above its melting temperature which equals ≈ 670 °C. According to /2/ the crystal $KFe(MoO_4)_2$ is also a ferroelastic with $T_c = (38 \pm 1.5)^{\circ}C$. The existence of magnetic Fe^{3+} ions in the lattice of these compounds indicates that in a definite temperature region these crystals may possess magnetic ordering. The aim of the present study is the examination of the existence of antiferromagnetic ordering in these materials and the determination of the Néel temperature in this case. This question was solved by the magnetoelectric method of T_N determination /3/ which differs from the well-known methods by its simplicity, satisfactory accuracy, and much shorter measuring time.

The single crystals of the compound $NaFe(MoO_4)$ and of $K_{0.8}Na_{0.2}Fe(MoO_4)_2$ have been grown by spontaneous crystallization. As initial components Na_2CO_3, K_2CO_3, and MoO_3 of chemical pure grade and Fe_2O_3 pure for analysis grade were used. The preparation conditions have been given in /1/. The crystals grown are in the form of platelets with ≈ 4 to 5 mm^2 area and thickness 0.1 to 0.2 mm. The (001) face is a more developed crystal face. The crystallographic [001] direction coincides with the normal of the platelets. To measure the magnetoelectric $(ME)_H$ signal, induced by the external magnetic field, the (001) faces of the crystals have been covered by silver electrodes. Thus, the direction of the applied magnetic field H coincides with the [001] axis. Owing to the very small thickness of the single crystals, measurement of the $(ME)_H$ signal P_{in} along the [100] and [010] directions was not possible. The method of P_{in} measurement will be described in /4/.

---

1) Nerimanov Prospekt 31 A, Baku-143, USSR.

2) Dnepropetrovsk, Ukrainian SSR, USSR.

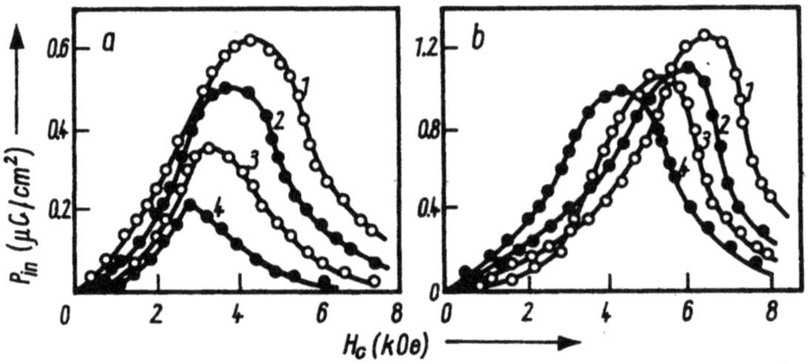

Fig. 1. The P_{in}(H) dependence: a) $NaFe(MoO_4)_2$: (1) 20, (2) 40, (3) 80, (4) 160 °C; b) $K_{0.8}Na_{0.2}Fe(MoO_4)_2$: (1) 25, (2) 48, (3) 102, (4) 150 °C

The dependence of the $(ME)_H$ signal upon the magnetic field strength P_{in}(H) has been measured at a few temperatures (Fig. 1). As can be seen, they are typical of the spin-flop phenomenon caused by the critical magnetic field H_c. Observation of the $(ME)_H$ effect shows the existence of magnetic ordering in $NaFe(MoO_4)_2$ and $K_{0.8}Na_{0.2}Fe(MoO_4)_2$ at these temperatures. Thus, these crystals belong to the 58 out of 90 magnetic symmetry classes. Spin-flop is seen as an abrupt change of the $(ME)_H$ susceptibility tensor at a definite value of H_c for a given temperature (23 °C: H_c = 4500 Oe, and 6500 Oe for $NaFe(MoO_4)_2$ and $K_{0.8}Na_{0.2}(MoO_4)_2$, respectively) due to a sudden rotation of the sublattice magnetization at H_c. Appearance of spin-flop indicates that in these crystals antiferromagnetic ordering exists. Thus, $NaFe(MoO_4)_2$ and $K_{0.8}Na_{0.2}Fe(MoO_4)_2$ are antiferromagnets with Néel temperatures above room temperature.

A considerably lower magnitude of H_c of these crystals as compared with that of the proper ferroelectric-antiferromagnetics (for compounds like $Bi_{0.90}Pb_{0.10}(Fe_{0.95}Nb_{0.05})O_3$ and $Bi_{0.85}Pb_{0.15}(Fe_{0.925}Nb_{0.075})O_3$ where H_c = 11000 and 10900 Oe, respectively /4/) attracts our attention. On the other hand, the H_c of these crystals are very close to those of antiferroelectric-antiferromagnetic compounds like $BiFeO_3$ and $Bi_{0.70}La_{0.30}FeO_3$ where H_c = 5000 and 6500 Oe, respectively /4, 5/. Therefore, our magnetoelectric data show

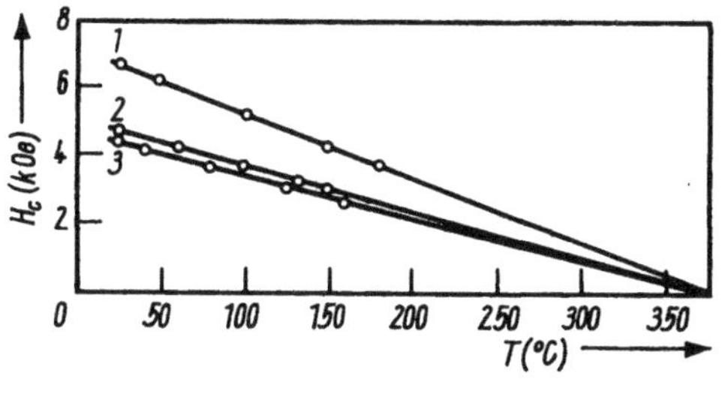

Fig. 2. The $H_c(T)$ dependence: (1) $K_{0.8}Na_{0.2}Fe(MoO_4)_2$, (2) $YFeO_3$, (3) $NaFe(MoO_4)_2$

that in the crystals studied the value of the spontaneous polarization P_s is either very small or zero, which takes place really.

The Nêel temperatures T_N of the crystals $NaFe(MoO_4)_2$ and $K_{0.8}Na_{0.2}Fe(MoO_4)_2$ have been determined by the method described in /3/. It is based on the strong linearity, established by us, of the dependence of the spin-flop magnetic field H_c upon the temperature $H_c(T)$. In this new technique four to five experimental points of $H_c(T)$ are sufficient for extrapolating H_c to 0 and thus to determine T_N within an accuracy of $\pm$ (2 to 3)°C. The $H_c(T)$ dependence for $K_{0.8}Na_{0.2}Fe(MoO_4)_2$ and $NaFe(MoO_4)_2$ and also for the ceramic sample $YFeO_3$, for which T_N = 370 to 372 °C is known /6/ are given in Fig. 2. Extrapolation of these lines to $H_c = 0$ which corresponds to the disappearance of the magnetoelectric effect at T_N, gives $T_N = (362 \pm 2)$°C for $NaFe(MoO_4)_2$, $T_N = (378 \pm 2)$°C for $K_{0.8}Na_{0.2}Fe(MoO_4)_2$, and $T_N = (370 \pm 2)$°C for $YFeO_3$. The last value well coincides with T_N given in /6/. This fact demonstrates the power of our method. On the basis of the T_N of $K_{0.8}Na_{0.2}Fe(MoO_4)_2$ and $NaFe(MoO_4)_2$ for $KFe(MoO_4)_2$ we obtain $T_N = (380 \pm 2)$°C.

Thus, the ferroelastics $NaFe(MoO_4)_2$ and $KFe(MoO_4)_2$ are antiferromagnets with Nêel temperatures $T_N = (362 \pm 2)$°C and (380 ± 2)°C, respectively.

References

/1/ E.F. DUDNIK and T.M. STOLPAKOVA, Fiz. tverd. Tela 17, 3405 (1975).

/2/ A.I. ITKO, L.N. PELICH, L.V. TOVSTYANSYI, N.M. NESTERENKO, P.S. KALININ, and A.I. ZVYAGIN, Izv. Akad. Nauk SSSR, Ser. fiz. 39, 697 (1975).

/3/ I.H. ISMAILZADE, R.M. ISMAILOV, I.M. ALIEV, A.I. ALEKPEROV, F.M. SALAEV, and A.M. HABIBOV, J. Magnetism magnetic Mater.,(1980).

/4/ I.H. ISMAILZADE, R.M. ISMAILOV, A.I. ALEKPEROV, and F.M. SALAEV, phys. stat. sol. (a) 57, 99 (1980).

/5/ I.H. ISMAILZADE, R.M. ISMAILOV, A.I. ALEKPEROV, and F.M. SALAEV, 4. EMF, Portoroz (Yugoslavia), 3. to 7. September 1979 (Abstracts, p. 276).

/6/ D. TREVERS, J. appl. Phys. 36, 1033 (1965).

(Received February 20, 1981)

phys. stat. sol. (a) 65, K29 (1981)
Subject classification: 18; 21.1; 21.4

Physics Department, Faculty of Science, Al-Fateh University, Tripoli[1)]

Magnetic Anisotropy in $ErFe_2$ Compound

By
M. HUQ

The anisotropic property of the compound $ErFe_2$ reported earlier by Clark et al. /1/ and Atzmony et al. /2/ are contradictory in the sense that the torque measurements /1/ gave the value of the first anisotropy constant K_1 larger than the second anisotropy constant K_2 whereas the Mössbauer measurements /2/ gave the value of K_2 higher than K_1. This discrepancy of the reported results prompted us to reexamine the anisotropic behaviour of the compound.

The sample of $ErFe_2$ was prepared in an argon arc furnace by melting the stoichiometric amounts of Er and Fe supplied by the Rare Earth Products and Metal Research Limited, respectively. The resulting button was inverted and remelted several times to ensure homogeneity. A part of the sample was examined by X-ray analysis which showed that the material was of single phase and the lattice parameter of 7.281 Å determined from the diffraction angle of the peak was in good agreement with the value of 7.283 Å reported by Buschow and van Stapele /3/. From the rest of the button, a disc-shaped specimen was prepared keeping the cooling direction in the plane of the disc. Like $NdCo_2$ compound /4/, it was found that the disc was anisotropic. This anisotropic behaviour arises obviously due to some preferential alignment of the crystallites along the cooling direction, the detailed mechanism of which has been described elsewhere /4/.

The torque measurements were performed on this disc only at room temperature of 297 K for four different field strengths of the external magnetic field B_a. The magnetometer used was the simplified version of the design by Aldenkamp et al. /5/. The torque curves obtained as a function of the angle between $\vec{B}_a$ and the fibre axis (cooling direction), Ψ, were of regular twofold symmetry (Fig. 1). The coefficients of sin 2Θ and sin 4Θ of the curves were obtained by Fourier analysis after making the appropriate correction for the lag angle α

---

1) P.O.Box 13569, Tripoli, Libya.

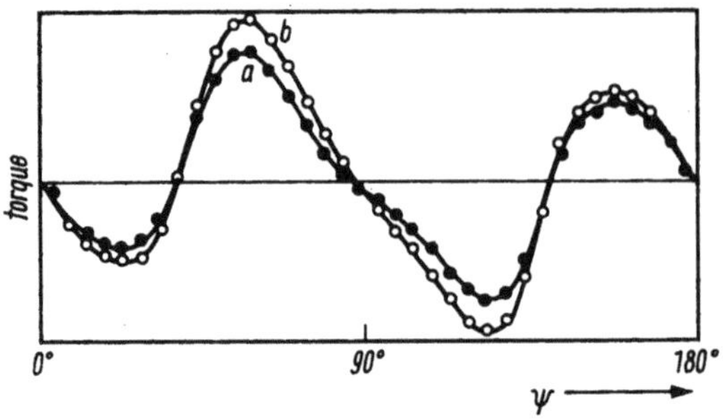

Fig. 1. Torque curves for applied magnetic field B_a of 1.4 T (curve a) and 1.9 T (curve b)

of the magnetization, $\vec{M}$, from the field direction $\vec{B}_a$, where Θ $(=\Psi-\alpha)$ is the angle between $\vec{M}$ and the fibre axis. The extracted coefficients were then plotted (Fig. 2) against $1/B_i$ ($B_i = B_a$-demagnetizing field). The value of the coefficients at $1/B_i = 0$ was taken as the value at saturation.

Due to the similarity of the preparation technique, crystal structure and preferential alignment of the RCo_2 compounds (R rare earth) with that of RFe_2 compounds, the average fibre axis of the crystallites in the compound $ErFe_2$ was taken as [110] similar to that obtained for $NdCo_2$ compound /4/ by neutron diffraction.

The values of the constants K_1 and K_2 were calculated by equating the extrapolated coefficients of sin 2Θ and sin 4Θ to the corresponding coefficients of the following torque expression for the [110] fibre axis given by Ackers /6/:

$$L_{[110]} = (K_1/32 - 3K_2/1024)\sin 2\Theta + (7K_1/64 - K_2/256)\sin 4\Theta - (39K_2/1024)\sin 6\Theta .$$

The constants thus obtained considering 100% alignment of the crystallites along [110] direction are

$$K_1 = -3.85\times10^5\ \mathrm{J/m^3} ,$$

$$K_2 = -4.38\times10^6\ \mathrm{J/m^3} ,$$

compared to the following values reported by Clark et al. /1/:

$$K_1 = -3.3\times10^5\ \mathrm{J/m^3} ,$$

$$K_2 = -9\ \times10^4\ \mathrm{J/m^3} .$$

The value of K_1 is in good agreement but the difference in K_2 is very large. Our observed magnitude of K_2 is 11.4 times larger than K_1 in contradiction to the values of Clark et al. We realize that the possibility of the fibre axis as [100] or [111] does exist. In these cases the torque expressions are /6/

$$L_{[100]} = -(K_1/8 + K_2/128)\sin 2\Theta - (7K_1/16 + K_2/32)\sin 4\Theta + (3K_2/128)\sin 6\Theta,$$

$$L_{[111]} = (K_1/12 + 16K_2/27)\sin 2\Theta + (7K_1/24 + 143K_2/432)\sin 4\Theta + (5K_2/144)\sin 6\Theta,$$

and the calculated anisotropy constants are

$$\left.\begin{aligned} K_1 &= -5.35\mathrm{x}10^5\ \mathrm{J/m^3} \\ K_2 &= 1.8\ \mathrm{x}10^7\ \mathrm{J/m^3} \end{aligned}\right\} \quad \text{for [100] fibre axis ,}$$

and

$$\left.\begin{aligned} K_1 &= 6.25\mathrm{x}10^4\ \mathrm{J/m^3} \\ K_2 &= 1.25\mathrm{x}10^4\ \mathrm{J/m^3} \end{aligned}\right\} \quad \text{for [111] fibre axis .}$$

Comparison of the above two sets of values of K_1 and K_2 with those of Clark et al. /1/ and Atzmony et al. /2/ shows that the constants differ not only in magnitude but even in sign. In view of this, we believe, the assumption of [110] as the fibre axis is correct. Thus the present measurements suggest that the magnitude of K_2 might have been underestimated in previous measurements. The argument runs as follows:

$$L_{[110]} = (K_1/32 - 3K_2/1024)\sin 2\Theta + (7K_1/64 - K_2/256)\sin 4\Theta .$$

First let us assume that $K_2 = 0$; then

$$L_{[110]} = (K_1/32)\sin 2\Theta + (7K_1/64)\sin 4\Theta .$$

This means that the coefficients of sin 2Θ and sin 4Θ must be of the same sign, which is in contradiction to the observed coefficients (Fig. 2). This implies K_2 is non-zero. For finite K_2 and to satisfy the observed conditions

$$K_1/32 - 3K_2/1024 > 0 \quad \text{and} \quad 7K_1/64 - K_2/256 < 0$$

it can be shown that K_1 and K_2 must be of the same sign and negative, and the magnitude of K_2 must lie

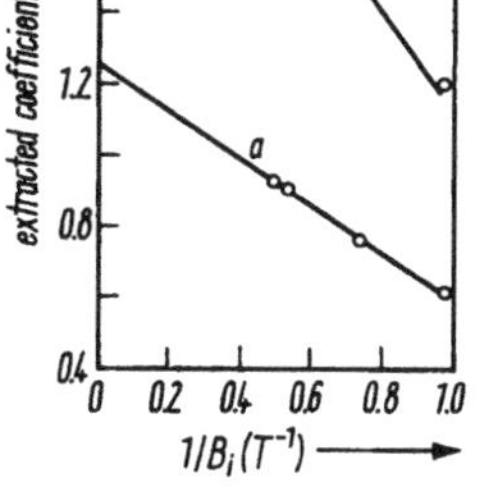

Fig. 2. Coefficients of sin 2Θ (curve a) and -sin 4Θ (curve b) of the torque curves against $1/B_i$

within the limit

$$\frac{32}{3} \left|K_1\right| < \left|K_2\right| < 28 \left|K_1\right| .$$

It can also be shown that the observed conditions cannot be simultaneously satisfied for alternate sign of K_1 and K_2.

Thus it is concluded that whatever be the error in our quoted values of K_1 and K_2, which might arise due to uncertainty of the percentage of crystallite distribution along the fibre axis, the magnitude of K_2 is certainly greater than K_1.

Thanks are due to Prof. E.W. Lee of the University of Southampton, England, for his valuable suggestions in the course of this work.

References

/1/ A.C. CLARK, H.S. BELSON, and N. TAMAGAWA, Phys. Letters A 42, 160 (1972).

/2/ U. ATZMONY, M.P. DARIEL, E.R. BAUMINGER, D. LEBENBAUM, I. NOWIK, and S. OFER, Phys. Rev. B 7, 4220 (1973).

/3/ K.H.J. BUSCHOW and R.P. VAN STAPELE, J. appl. Phys. 41, 4066 (1970).

/4/ M. HUQ, phys. stat. sol. (a) 60, 573 (1980).

/5/ A.A. ALDENKAMP, C.P. MARKS, and H. ZIJLSTRA, Rev. sci. Instrum. 31, 544 (1960).

/6/ J.E. ACKERS, Ph. D. Thesis, University of Nottingham, 1961.

(Received February 26, 1981)

phys. stat. sol. (a) 65, K33 (1981)
Subject classification: 10.2 and 20.1; 11; 13.4; 22.5.2

Faculty of Technology, Kanazawa University[1]

Post-Irradiation Annealing Effect on the Growth of the Z_3 Band in LiF:Mg Crystal

By

N. TAKEUCHI[2], K. INABE, and S. NAKAMURA

It has been reported that the thermoluminescence (TL) glow peak 5 (≈ 200 °C) in LiF:Mg is related to the Z_2 centers /1/, and that the peak 6 (≈ 250 °C) /2/, or the TL peak at 285 °C /3/, is related to the Z_3 centers. Jain /4/ and Bhuwan Chandra et al. /5/ independently have discussed that the TL peak 10 (≈ 400 °C) in LiF (TLD-100) showing that supralinearity from the beginning /6/ is related to the Z_3 centers, though their discussions are different from each other in details. The growth of the Z_2 band (310 nm) /7/ and that of the Z_3 band (225 nm) /2/ have been examined to be proportional to the dose, even in a dose range where supralinearity in TL appears. This seems in contradiction to the assumption that the Z centers play the role of trapping centers for the electrons which recombine with holes /1/. If the recombination is proportional to the TL output, the supralinear rise of the Z bands against dose should be observed. In this respect, as long as the authors are aware, the growth of the Z bands has been examined directly after irradiation at RT without heating the sample. Because TL is the phenomenon observed by heating the sample, some of the Z bands may develop during the process before the appearance of the TL. This is the purpose of the present report to examine the heating effect on the growth of Z bands.

LiF single crystals containing about 300 ppm of Mg^{2+} were cleaved in a size of 8x8x2 mm^3 and annealed at 500 °C for 0.5 h followed by cooling to RT. Because it has been reported that the growth of Z bands is correlated with the concentration of Mg^{2+}-cation vacancy dipoles (I-V dipoles) /8/, one of the above treated samples was annealed again at 80 °C (sample A) and the other at 160 °C (sample B), respectively for 1 h prior to irradiation. The concentration of the

---

1) Kodatsuno 2-40-20, Kanazawa 920, Japan.

2) To whom all correspondence should be addressed.

Table 1

Parameters used for the Gaussian analysis

peak	3.28	3.98	4.96	5.47	6.19
position (nm)	378	311	250	226	200
halfwidth (eV)	0.77	0.90	0.78	1.00	1.27

isolated I-V dipoles was determined by the dielectric loss measurement to be 65 and 30 ppm for samples A and B, respectively. The irradiation was made by using X-rays from tungsten target through an Al filter 1 mm thick, operated at 125 kV and 5 mA. The dose rate was 1.75 mC/kg s. The X-irradiated samples were pulse annealed successively at 150, 200, and 250 °C. The heating rate was 3.3 K/s and after keeping at each desired temperature for 60 s, the samples were cooled to RT. Optical absorption spectra for each step were measured at RT over the wavelength from 195 to 390 nm. The spectra were resolved into the isolated Gaussian bands by using a micro-computer. The parameters used for the Gaussian analysis are shown in Table 1.

Fig. 1a and b show the optical absorption spectra of the samples X-irradiated for 100 Gy. Both the Z_2 and Z_3 bands are larger in sample A than in sample B in agreement with Kos and Nink /8/. The growth of the F band (250 nm) seems less affected by the heat treatment before irradiation. It is estimated from the absorption intensity that the concentration of the F centers is several ten times less than that of the isolated I-V dipoles. There are some other bands peaking at 3.28 eV (378 nm) and 6.19 eV (200 nm) /1/. However, since the origins of them are not yet clear and their intensities are little affected by the pre-irradiation heat treatment, we will not treat them further.

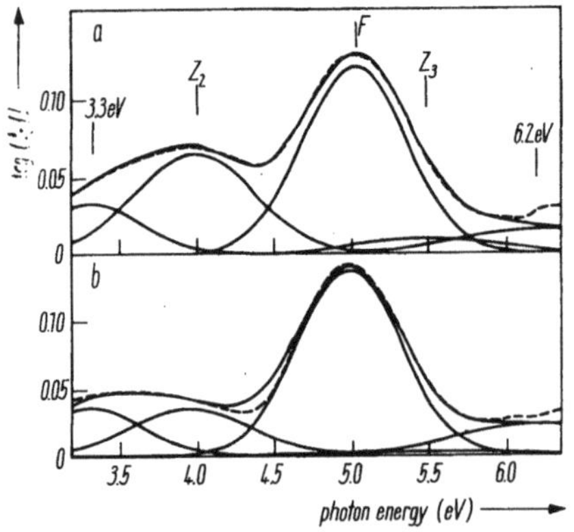

Fig. 1. Optical absorption spectra of LiF:Mg X-irradiated 100 Gy and measured at RT. The sample was pre-irradiation annealed for 1 h a) at 80 °C, and b) at 160 °C, respectively. The dashed line shows the measured spectrum. The real spectrum is the sum of each resolved Gaussian band (see Table 1)

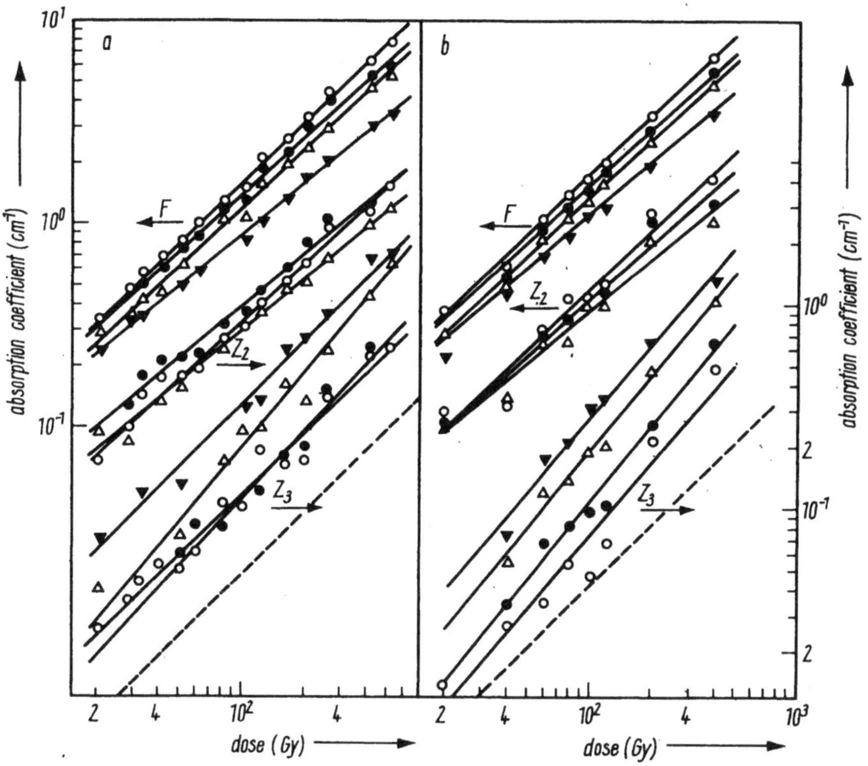

Fig. 2. Intensity of the F, Z_2, and Z_3 bands against dose. The sample was pre-irradiation annealed for 1 h a) at 80 °C, and b) at 160 °C, respectively. The white circle shows the result directly after irradiation. The black circle, white triangle, and black triangle show the results after the successive post-irradiation pulse annealings at 150 °C, 200 °C, and 250 °C, respectively. The dashed line shows the slope equal to unity

The intensities of each resolved Gaussian band for sample A and B are shown in Fig. 2a and b, respectively. The white circles indicate the result directly after the irradiation and the other marks show the results after the post-irradiation annealings. It is seen that the data points fit the straight lines up to about 400 Gy. The slopes of the lines are classified as follows: (a) All the slopes for the Z_2 and F bands are less than unity. This is not an unexpected result, because Z_2 and F centers loose electrons during the post-irradiation annealing. (b) The Z_3 band in sample A directly after irradiation is found to increase proportionally with dose. This is in accord with the result shown in /2/. According to Kos and Nink /2/, their sample was heat-treated at 480 °C for 1 h and cooled in air to RT with a mean cooling rate of 10 K/min. We consider that by this treatment a considerable amount of isolated I-V dipoles were formed. Therefore, it is considered that when the concentration of the isolated I-V dipole is high, a linear rise of the Z_3 band against dose is observed. (c) All the slopes for the Z_3 band except for sample A directly after irradiation are more than unity (i.e. supralinearity). Because the isolated I-V dipoles are closely related to the formation of the Z_2 and Z_3 centers /8/, the lower rate of formation is expected for the sample which contains more coagulated I-V dipoles, if there were no disintegration of the coagulated ones during the irradiation. However, according to Watterich and Voszka /9/, this disintegration is truly observed. Therefore, a supralinear formation for the sum of Z_2 and Z_3 centers by the irradiation is expected for a sample well annealed at temperature around 160 °C before irradiation. This may be considered at least one of the factors leading to the supralinear rise of the Z_3 band.

The reason why the supralinearity is not observed in the growth of the Z_2 band is probably due to the lower thermal stability of the Z_2 center compared with the Z_3 center even at RT.

In any way, the hypothesis that the Z_2 centers directly play the role of the electron trapping agencies for the supralinearity in TL seems very doubtful. On the contrary, the possibility that the Z_3 centers be the sources for the electrons directly leading to the supralinearity cannot be denied.

The authors are grateful to Dr. H.-J. Kos at the Physikalisch-Technische Bundesanstalt, Institut Berlin (West) for kindly supplying the single crystals studied.

References

/1/ R. NINK and H.-J. KOS, phys. stat. sol. (a) 35, 121 (1976).

/2/ H.-J. KOS and R. NINK, phys. stat. sol. (a) 56, 593 (1979).

/3/ R.K. GARTIA, phys. stat. sol. (a) 44, K21 (1977).

/4/ V.K. JAIN, phys. stat. sol. (a) 60, 351 (1980).

/5/ BHUWAN CHANDRA, A.R. LAKSHMANAN, and R.C. BHATT, phys. stat. sol. (a) 60, 593 (1980).

/6/ V.K. JAIN, S.P. KATHURIA, and A.K. GANGULY, J. Phys. C 7, 3810 (1974).

/7/ M. ROSSOTTER, D. RESS-EVANS, and S. ELLIS, J. Phys. D 3, 1816 (1970).

/8/ H.-J. KOS and R. NINK, phys. stat. sol. (a) 57, 203 (1980).

/9/ A. WATTERICH and R. VOSZKA, phys. stat. sol. (b) 93, K161 (1979).

(Received February 24, 1981)

phys. stat. sol. (a) 65, K37 (1981)
Subject classification: 17; 21.1; 21.1.1

Sektion Physik der Technischen Universität Dresden,
WB Oberflächen- und Elektronenphysik[1]

Relative Determination of L_3 Ionisation Cross Sections of Chromium, Iron, and Nickel

By
K. UHLMANN and F. STORBECK

For a quantitative analysis of measured Auger electron spectra it is important to know ionisation cross sections. Though there are many papers dealing with electron-impact ionisation cross section (see e.g. /1/), the number of measurements of L_3 ionisation cross sections of 3d transition metals is small. The calculation of these cross sections is one way to overcome this deficiency. But there are considerable differences between the electron-impact cross sections evaluated by different formulas. Furthermore, in most cases the uncertainties in Coster-Kronig transition probability data make an exact determination of total L_3 ionisation cross sections impossible.

Thus for our studies on stainless steel single crystals it was necessary to determine the total L_3 ionisation cross sections of the alloying constituents Fe, Cr, and Ni experimentally.

For our measurements we used a four-grid retarding field analyser with the primary electrons (E_p = 2 keV, I_p = 30 μA) impinging perpendicular on the sample surface and with the normal of the sample parallel to the analyser axis. All Auger spectra were recorded in derivative form (amplitude of modulation voltage 3.5 V; modulation frequency 3 kHz) and integrated numerically using the method of Antonides et al. /2/ to remove the inelastic background.

Assuming a multi-layer model with constant layer thickness and constant total atomic density in all layers, we can get from the Palmberg formula /3/

$$I_A = \sigma_A w_A r I_p \frac{\Omega T}{4\pi} \sum_{i=1}^{\infty} n_{Ai} k_{Ai} , \tag{1}$$

where I_A is the Auger electron current of a transition in element A collected by an analyser with solid angle Ω and transparence T, I_p is the primary electron

1) Mommsenstr. 13, DDR-8027 Dresden, DDR.

current, σ_A the electron-impact cross section, w_A the transition probability, and r the backscattering factor. n_{Ai} denotes the atomic density of element A within layer i and k_{Ai} is a layer contribution factor (contribution of the i-th layer to the total Auger signal of element A in a given matrix).

$$k_{Ai} = \frac{1}{\Omega} \int_{\varphi=0}^{2\pi} \int_{\vartheta_0}^{\vartheta_m} \exp\left(-\frac{(i-1)d}{\lambda_A \cos\vartheta}\right) \sin\vartheta \, d\vartheta \, d\varphi \,, \tag{2}$$

where ϑ_o and ϑ_m are the half angles of the analyser cut-out and of the analyser, d is the layer thickness (for the studied steel 0.179 nm), and λ_A is the inelastic mean free path of the Auger electrons emitted from element A (in our case we used for λ_A the values of Penn /4/ for an iron matrix).

In this simple case, the integration in (2) can be carried out analytically. The contribution factor for the first layer, k_{A1}, is always equal to one.

Introducing a constant $C = r\, I_p \Omega T/4\pi$ (r is supposed to be constant for all analysed elements) and a sensitivity factor $\beta_B = \sigma_A w_A$, (1) can be written as

$$I_A = \beta_A C \sum_{i=1}^{\infty} n_{Ai} k_{Ai} \,. \tag{3}$$

Equation (3) now shall be used for a relative estimation of the total L_3 ionisation cross sections of Cr, Fe, and Ni. Such a relative estimation is possible, if there is an overlayer with known coverage which consists of an element with known ionisation cross section. Though there are some uncertainties, too, K-shell ionisation cross sections should be taken as known.

After a thermal treatment, on the (100) surface of UR X3 CrNi 18.12 steel, nitrogen (bulk concentration lower than 0.2 wt%) is segregated, leading to the formation of a p (2x2) superstructure which can be detected by LEED /5/. In the studied temperature range (700 to 1000 K), the segregated amount of nitrogen reaches a constant value. Thus we suppose the superstructure to be nearly perfect with a nitrogen coverage of 25% of a monolayer. Since the investigation of the p (2x2) superstructure by other methods /6/ (ELS, AES fine structure, SIMS) shows results which can be explained by the formation of a chromium nitride in the outermost layer, we assume approximately that the first layer consists of 25% N and 75% Cr and that the bulk is a homogeneous mixture of the alloying constituents Cr, Fe, and Ni.

There is, however, another problem. Since in the case of an alloy the cross sections of the alloying constituents cannot be determined separately, we must assume that they can be expressed in the form

$$\sigma_{A, L_3} = K \frac{g(U)}{E^2_{A, L_3}} = K \tilde{g}_A \, , \tag{4}$$

i.e. by a common constant K, completed only by a term depending on the L_3 binding energy E_{A, L_3} (taken from /7/) and the reduced energy $U = E_p/E_{L_3}$. The energy dependence of the L_3 cross sections of iron and chromium has been proved experimentally by measuring the intensity of the L_3 series lines at various primary energies between 1.5 and 3.0 keV. We found that in the range U = 2 to 5 g(U) can be described by the energy dependent term of the Gryzinski formula /8/

$$g(U) = \frac{1}{U} \left(\frac{U-1}{U+1}\right)^{3/2} \left[1 + \frac{2}{3}\left(1 - \frac{1}{2U}\right) \ln(2.7 + \sqrt{U-1})\right] \tag{5}$$

as well as by analogous terms of other formulas (e.g. Drawin /9/).

Thus we have from (3) for the p (2x2) superstructure

$$I_N = \beta_N C \; 0.25\, n \, ,$$
$$I_{Cr} = \beta_{Cr} C \left(0.75\, n + n_{Cr} \sum_{i=2}^{\infty} k_{Cri}\right) ,$$
$$I_{Fe} = \beta_{Fe} C \, n_{Fe} \sum_{i=2}^{\infty} k_{Fei} \, ,$$
$$I_{Ni} = \beta_{Ni} C \, n_{Ni} \sum_{i=2}^{\infty} k_{Nii} \, , \tag{6}$$

where n is the total atomic density

$$n = \frac{I_N}{\beta_N C \; 0.25} = \frac{I_N}{\sigma_N w_N C \; 0.25} \tag{7}$$

and

$$n = n_{Cr} + n_{Fe} + n_{Ni} =$$

$$= \frac{1}{C K} \frac{\varkappa_{Cr}}{\varkappa_{Cr} + 0.75} \left(\frac{I_{Cr}}{\tilde{g}_{Cr} w_{Cr} \varkappa_{Cr}} + \frac{I_{Fe}}{\tilde{g}_{Fe} w_{Fe} \varkappa_{Fe}} + \frac{I_{Ni}}{\tilde{g}_{Ni} w_{Ni} \varkappa_{Ni}}\right) \tag{8}$$

with

$$\varkappa_A = \sum_{i=2}^{\infty} k_{Ai}$$

From (7) and (8) K can be calculated.

The values necessary for this calculation are given in Table 1. The ratios I_A/I_N have been determined from numerous Auger spectra. The measured ratios differ from one spectrum to another mostly by less than 10%, the differences result from the presence of impurities (S, P, Si, O) on the surface. The values in Table 1 are those from the spectrum with the smallest amount of impurities ($\Theta_S \approx 0.01$).

The transition probabilities have also been taken from measured Auger spectra by dividing the current of the transition designated in the second column by the current of the whole L_3 series. $\tilde{g}_A$ and $\varkappa_A$ are calculated values. σ_N has been calculated using the Gryzinski formula ($\sigma_N = 1.73 \times 10^{-19}$ cm^2). Since I_N is the current of the whole KLL series the transition probability w_N is equal to one.

Thus we get

$$K = 3.10 \times 10^{-13}\ \text{cm}^2\ \text{eV}^2$$

and

$$\sigma_{Cr, L_3} = 2.02 \times 10^{-19}\ \text{cm}^2\ ,$$

$$\sigma_{Fe, L_3} = 1.27 \times 10^{-19}\ \text{cm}^2\ ,$$

$$\sigma_{Ni, L_3} = 0.78 \times 10^{-19}\ \text{cm}^2 \quad (E_p = 2.0\ \text{keV})$$

with an error of about 10%.

These ionisation cross sections are in fair agreement with the cross sections of neighbouring elements Ti ($\sigma_{L_3} = 2.5 \times 10^{-19}$ cm^2) and Cu(σ_{L_3} = = 0.73×10^{-19} cm^2) (both for $U \approx 4$) given by DuCharme and Gerlach /10/ and with the cross section of Ni ($\sigma_{L_3} \doteq 0.7 \times 10^{-19}$ cm^2) calculated by Sexton et al. /11/ using the experimental data of Vrakking and Meyer /12/.

Table 1

element A	transition	I_A/I_N	w_A	$10^7\ \tilde{g}_A(\text{eV}^{-2})$	$\varkappa_A$
Cr	$L_3M_{23}M_{45}$	1.72	0.34	6.53	3.83
Fe	$L_3M_{45}M_{45}$	2.70	0.30	4.09	3.05
Ni	$L_3M_{45}M_{45}$	0.33	0.45	2.53	4.44

Simply assuming that the level which has to be ionised is filled by 6 electrons (all 2p electrons), the cross sections which can be calculated by the Gryzinski formula are about 25% greater than ours.

References

/1/ C.J. POWELL, Rev. mod. Phys. 48, 33 (1976).
/2/ E. ANTONIDES, E.C. JANSE, and G.A. SAWATZKY, Phys. Rev. B 15, 1669 (1977).
/3/ P.W. PALMBERG, Anal. Chem. 45, 549A (1973).
/4/ D.R. PENN, J. Electron Spectroscopy and Related Phenomena 9, 29 (1976).
/5/ U. FEHRMANN, H.-J. MÜSSIG, F. STORBECK, R. THIELSCH, and K. UHLMANN, Kristall und Technik 15, 455 (1980).
/6/ O. MÜLLER, private communication.
/7/ K. SIEGBAHN et al., ESCA, Atomic, Molecular and Solid State Structure Studied by Means of Electron Spectroscopy, Uppsala 1967.
/8/ M. GRYZINSKI, Phys. Rev. 138, A336 (1965).
/9/ H.-W. DRAWIN, Z. Phys. 172, 429 (1963).
/10/ A.R. DUCHARME, and R.L. GERLACH, J. Vacuum Sci. Technol. 10, 188 (1973).
/11/ B.A. SEXTON, B.G. BAKER, and N.L. ARMSTRONG, J. Vacuum Sci. Technol. 13, 1144 (1976).
/12/ J.J. VRAKKING and F. MEYER, Surface Sci. 35, 34 (1973).

(Received March 3, 1981)

phys. stat. sol. (a) 65, K43 (1981)
Subject classification: 18.2; 21.1.1

Institute of Solid State Physics,
Academy of Sciences of the USSR, Chernogolovka[1)]

Magnetization of Iron Hydride

By

V.E. ANTONOV, I.T. BELASH, E.G. PONYATOVSKII,
V.G. THIESSEN, and V.I. SHIRYAEV

Introduction At T = 523 K and P_{H_2} = 45 kbar a phase transformation occurs in iron followed by a jump-like increase in the hydrogen solubility /1/. The hydrogen content in the Fe-H samples obtained by quenching down to ≈ 95 K after hydrogenation for 10 h at P_{H_2} = 67 kbar and T = 523 K was within the H-to-metal atomic ratio n = 0.65 to 0.81. The X-ray study has shown that the hydride thus synthesized has an iron h.c.p. sublattice with the parameters a = (2.686 ± 0.003) Å, c = (4.380 ± 0.005) Å, c/a = 1.631 ± 0.003 at T ≈ 80 K and atmospheric pressure. The scattering found for the n-values (from 0.65 to 0.81) seemed to be due to the hydrogen losses from the Fe-H samples during manipulations at atmospheric pressure before a chemical analysis, since at P = 1 bar the rate of hydrogen liberation from iron hydride became noticeable even at T ≈ ≈ 150 K, and at T = 180 K the main quantity of hydrogen escaped from the samples in a few seconds. After complete liberation of hydrogen, iron recovers the initial b.c.c. crystal structure but becomes very brittle. The samples in the form of rods of 1.1 mm diameter and 5 mm length prepared from iron purified by zone melting followed by refining in hydrogen at T = 1140 K for 100 h (the content of impurities and details of the purification technique are given in /2/) were used for the measurement.

Just the similar specimens hydrogenated under the same conditions (10 h at P_{H_2} = 67 kbar and T = 523 K) were used in the present work for the investigation of the magnetization (σ) behaviour at atmospheric pressure and 4.2 ≦ T ≦ ≦ 80 K in a pulsed magnetic field H up to 320 kOe, the pulse duration being ≈ 0.01 s. Immediately after the magnetic measurement the Fe-H sample was placed in a previously cooled scaled glass vessel filled up by benzine. Then this vessel was heated to room temperature, hydrogen liberated from the sample

---

1) 142432 Chernogolovka, Moscow District, USSR.

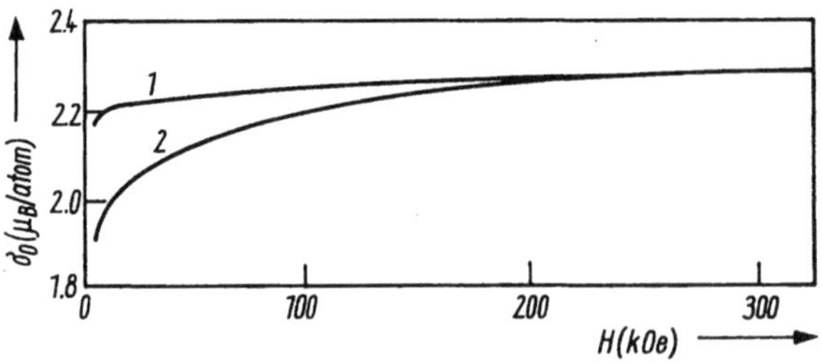

Fig. 1. Dependences of the magnetization σ upon magnetic field H at T = 4.2 K. (1) Fe, (2) Fe-H solid solution with n = 0.75 ± 0.03

and partially expelled benzine from the vessel. The content of the Fe-H sample was determined by the volume of released hydrogen with an accuracy of 4%. After a complete liberation of hydrogen, the magnetization of the iron sample was measured again in order to calibrate the measuring installation. The investigation has shown that several hydrogenation-dehydrogenation cycles do not change the σ(H) dependences for iron specimens in the studied T and H intervals, and this additional calibration permitted to attain a relative accuracy of 2% in the measurement of iron hydride magnetization.

Results and discussion A typical σ(H) dependence is shown in Fig. 1 for the Fe-H sample with n = 0.75 ± 0.03. One can see that at H < 150 kOe the σ(H) curve for iron hydride lies perceptibly below the similar curve for b.c.c. iron, the hydride magnetization being essentially dependent on H. At further H increase the hydride magnetization approaches saturation and within the experimental error coincides with that for the pure iron sample. The dependence of σ upon H observed for iron hydride at H < 150 kOe seems to be caused by a large value of the magnetic anisotropy energy that can be associated with, for example, a large magnitude of internal stresses /3/ occurring at hydrogenation. Note that the nearest analogue of the iron hydride - the ferromagnetic manganese hydride with h.c.p. manganese sublattice - also exhibits a rather large field of technical saturation exceeding 60 kOe /4/.

In the investigated temperature range 4.2 to 80 K the σ(H) dependence for iron hydride (as well as for pure b.c.c. iron) is independent of temperature within the experimental error. Hence, its Curie point $T_c \gg 80$ K and the value of spontaneous magnetization at $4.2 \leqq T \leqq 80$ K is approximately equal to σ_0 at

T = 0 K. The presence of a large region of a non-linear $\sigma(H)$ dependence for the hydride (Fig. 1) impedes a correct determination of the spontaneous magnetization. One may only state that $\sigma_0 \lessapprox 2.22\ \mu_B$/atom $\approx \sigma_0^{Fe}$, where μ_B Bohr magneton. Note that although the Fe-H sample with n = 0.75 may contain a considerable amount of pure iron phase (without hydrogen) /1/, this cannot change the above result due to the proximity of the σ_0 values for iron hydride and b.c.c. iron.

Hydrogenation leads to a strong deformation of the band structure of transition metals. The investigation of hydrogen solid solutions on the base of the f.c.c. metal sublattice has shown, however, that the behaviour of their magnetic properties may be described assuming an increase in the degree of occupation of the host metal 3d-band to be the main effect on hydrogenation, hydrogen being a donor of a fractional number of electrons $\eta < 1$ electron/proton /5, 6/. As a theoretical basis for such an approach the band structure calculations may serve to some extent. These calculations show (see, for instance, /7, 8/) that at n $\leqq$ 1 no new (hydrogen) band is formed below the Fermi energy, metal d-states change weakly whereas the energy of sp-states strongly decreases. The latter leads to an increase of the number of states below the Fermi level. Thus, a part of electrons supplied by hydrogen atoms into the conduction band is spent to fill those additional states.

Consider to what degree the description /5, 6/, which can be called a "rigid d-band" model, is useful for hydrogen solid solutions on the base of the h.c.p. metal sublattice.

The conclusion that it is an increase in the degree of occupation of the d-band which has a main effect on the magnetic properties of 3d-metal alloys at hydrogenation was based on the fact that the dependences of these properties upon hydrogen content in alloys of 3d-metals situated close to each other in the Periodic Table proved to be similar to that on changing the composition of these alloys toward an increase in the average number N^e of 3d + 4s electrons per atom /5, 6/. It is worth noting that, in its turn, the assumption of a dominating role of the degree of d-band filling for the alloys of this kind in varying the magnetic properties with composition in rather wide concentration intervals ($\Delta N^e \approx 0.5$ electrons/atom) has been confirmed while investigating hydrogen solid solutions on their base.

The values of σ_o for the f.c.c. alloys of 3d-metals closely arranged in the Periodic Table as a function of N^e fall onto a single curve (the Pauling-Slater curve). While decreasing N^e down to $\approx$ 8.8 electrons/atom σ_o increases approximately linearly with the slope $\partial\sigma_o/\partial N^e \approx -1\ \mu_B$/electron (Ni-Cu, Ni-Co, and Ni-Fe alloys), then reaches a maximum and begins decreasing. At $N^e \approx$ $\approx$ 8.35 electrons/atom the alloys become paramagnetic down to helium temperature, at further N^e decrease antiferromagnetic ordering occurs and the Néel point increases monotonously up to $T_N \approx 440$ K at $N^e = 7.65$ electrons/atom ($Fe_{65}(Ni_{1-x}Mn_x)_{35}$ alloys) /9, 5/.

The data concerning the $\sigma_o(N^e)$ dependence for the alloys with h.c.p. crystal lattice are very rare. Nevertheless, it seems to have similar character in this interval of N^e magnitudes. The h.c.p. Fe-Mn alloys (N^e = 7.71 to 7.82 electrons/atom) are antiferromagnetics with $T_N \approx 230$ K /10/. Extrapolation from Fe-Ru and Fe-Os alloys shows that at atmospheric pressure the h.c.p. iron (N^e = 8 electrons/atom) should also be antiferromagnetically ordered, but exhibit a lower Néel point $T_N \approx 100$ K /11/. Sufficiently correct data on the concentration dependence of magnetic properties for the 3d-metal h.c.p. alloys with $8 < N^e < 9$ electrons/atom are unavailable in the literature. The h.c.p. cobalt (N^e = 9 electrons/atom) is a ferromagnetic with $\sigma_o = 1.72\ \mu_B$/atom /12/ that is near the σ_o-value for f.c.c. cobalt. At further N^e increase up to $\approx$ 9.28 electrons/atom (h.c.p. Co-Ni alloys /13 to 15/) σ_o decreases approximately linearly with the slope $\partial\sigma_o/\partial N^e \approx -1\ \mu_B$/electron.

Thus, at $7.7 \lesssim N^e \lesssim 9.28$ electrons/atom the magnetic properties and their concentration dependences seem to be close for the 3d-metal alloys with f.c.c. and h.c.p. crystal structures. Investigation of hydrogen solid solutions on the base of h.c.p. cobalt has shown the hydrogenation to result in a decrease of σ_o with an initial slope $\partial\sigma_o/\partial n|_{n=0} \approx -0.6\ \mu_B$/H atom that agrees with the predictions of the "rigid d-band" model. If this model is valid for the description of the Fe-H solutions, an increase of hydrogen content in h.c.p. iron should lead (as it has been observed at the hydrogenation of the $Fe_{65}(Ni_{1-x}Mn_x)_{35}$ f.c.c. alloys /5/) to the suppression of antiferromagnetic ordering, the appearance of spontaneous magnetization at a certain value of $n = n_f$, and then to a monotonous increase of σ_o and T_c. The transformation of the antiferromagnetic h.c.p. iron into a ferromagnetic with $\sigma_o \approx 2.2\ \mu_B$/atom and $T_c \gg 80$ K at

$n \approx 0.8$ found in the present work agrees qualitatively with such a dependence of magnetic properties upon hydrogen content.

For the Fe-H system it is possible, following /5/, to introduce the coefficient of an "effective" number of electrons supplied by hydrogen into the d-band of the metal $G = \Delta N^e/\Delta n$, relating the changes in N^e and n resulting in similar changes of iron magnetic properties. By analogy with the $\sigma_o(N^e)$ dependence for the f.c.c. alloys one should expect that at $8 < N^e < 9$ electrons/atom the $\sigma_o(N^e)$ curve passes through a maximum. So, at $n \approx 0.8$ the point $\sigma_o(N^e_{Fe} + Gn) \approx 2.2\ \mu_B$/atom either falls onto the prolongation of the linear dependence $\sigma_o(N^e)$ with the slope $\varphi = \partial\sigma_o/\partial n \approx -1\ \mu_B$/electron, which is valid for h.c.p. Co-Ni alloys, or lies below this line. Hence,

$$\sigma_o^{Co} - \left[N^e_{Co} - (N^e_{Fe} + Gn)\right]\varphi \gtrsim 2.2\ \mu_B/\text{atom},$$

from whence $G \lesssim 0.6$ electrons/H atom that is a reasonable estimate.

Thus, as well as in the case of the f.c.c. alloys /5, 6/, the behaviour of iron magnetic properties at hydrogenation may be described under the assumption that these properties are mainly affected by the degree of d-band filling, and the "rigid d-band" model gives the correct sign and the order of magnitude of the observed effects.

The authors are thankful to B.K. Ponomarev who placed the equipment for magnetic measurement at their disposal and A.I. Amelin for his kind assistance in preparing the experiment.

References

/1/ V.E. ANTONOV, I.T. BELASH, V.F. DEGTYAREVA, E.G. PONYATOVSKII, and V.I. SHIRYAEV, Dokl. Akad. Nauk SSSR 252, 1384 (1980).

/2/ D.S. KAMENETSKAYA, I.B. PILETSKAYA, and V.I. SHIRYAEV, Zhelezo vysokoi stepeni chistoty, Metallurgizdat, Moskva 1978.

/3/ S.V. VONSOVSKII, Magnetizm, Izd. Nauka, Moscow 1971 (p. 836).

/4/ I.T. BELASH, B.K. PONOMAREV, V.G. THIESSEN, N.S. AFONIKOVA, V.SH. SHEKHTMAN, and E.G. PONYATOVSKII, Fiz. tverd. Tela 20, 422 (1978).

/5/ V.E. ANTONOV, I.T. BELASH, B.K. PONOMAREV, E.G. PONYATOVSKII, and V.G. THIESSEN, phys. stat. sol. (a) 52, 703 (1979).

/6/ V.E. ANTONOV, I.T. BELASH, and V.G. THIESSEN, phys. stat. sol.

(a) 59, 673 (1980).

/7/ A.C. SWITENDICK, Ber. Bunsenges. phys. Chem. 76, 535 (1972).

/8/ N.I. KULIKOV, V.N. BORZUNOV, and A.D. ZVONKOV, phys. stat. sol. (b) 86, 83 (1978).

/9/ J. CRANGLE and G.C. HALLAM, Proc. Roy. Soc. A272, 119 (1963).

/10/ H. OHNO and M. MEKATA, J. Phys. Soc. Japan 31, 102 (1971).

/11/ H. OHNO, J. Phys. Soc. Japan 31, 92 (1971).

/12/ T. WAKIYAMA and A. TŌKAIRIN, J. Phys. Soc. Japan 34, 265 (1973).

/13/ P. WEISS, R. FORRER, and F. BIRCH, C.R. Acad. Sci. (France) 189, 789 (1929).

/14/ R.M. BOZORTH, Ferromagnetism, D. van Nostrand Co., Inc., Toronto/ New York/London 1951 (p. 279).

/15/ P. WEISS and R. FORRER, Ann. Phys. (France) 12, 279 (1929).

(Received December 1, 1980)

phys. stat. sol. (a) 65, K49 (1981)

Subject classification: 18 and 20.1; 14.4.1; 22.8.1

Optics Laboratory, Institute of Physics, A. Mickiewicz University, Poznań[1]

Dispersion of the Faraday Rotation in KDP-Type Crystals by Pulse High Magnetic Field

By

M. KORALEWSKI

Magnetooptic properties of KDP-type crystals according to our knowledge have not been reported except for our earlier paper /1/. In this note we present a Faraday effect study on KDP-type crystals: KH_2PO_4 (KDP), KH_2AsO_4 (KDA), $K(D_xH_{1-x})_2AsO_4$ (DKDA), $NH_4H_2PO_4$ (ADP), $NH_4H_2AsO_4$ (ADA), and $K(D_xH_{1-x})_2PO_4$ (KDP) with various deuterations along the optical axis.

A high-intensity magnetic field, with a pulse duration of the order 0.5 ms was generated by discharging a bank of capacitors through a helicoidal solenoid of Be-Cu alloy. The system allowed us to reach magnetic fields with a peak intensity of the order of 40 T. The accuracy obtained was of $\pm 2.5\%$ at the peak. The measuring devices and methods were published in our papers /1, 2/. In the present work the light source consisted of a He-Ne laser (λ = 632.8 nm) and a high-pressure mercury lamp. The temperature dependent measurements of Faraday rotation were carried out with the use of a glass cryostat, the temperature stabilization was ± 0.5 K.

From the well known relation describing the Faraday rotation

$$\Theta = VLB , \tag{1}$$

where V is the Verdet constant, L effective length of the crystal acted on by the pulsed magnetic induction B, the Verdet constants of all examined crystals were determined by the least-squares method. The values of the Verdet constant V for λ = 632.8 nm are listed in Table 1 and dispersion curves are illustrated in Fig. 1. The relative error of the Verdet constant amounts to $\pm 6\%$. The Verdet constant was also measured for three degrees of deuterization of DKDP crystals, and it was found that with increasing deuterization the Verdet constant increased (Table 2). We found that all examined crystals possessed a positive sign of the Verdet constant and the Verdet law was obeyed for magnetic

1) ul. Grunwaldzka, 60-780 Poznań, Polska.

Table 1

Optical and magnetooptical constants of KDP-type crystals; $\vec{k} \parallel [001]$, T = 298 K

crystal	λ_c (nm)	S_c (10^{14} m^{-2})	γ	A (10^3 deg m^{-1} T^{-1})	V(deg m^{-1} T^{-1}) at λ = 632.8 nm
KDP[a)]	105.4	1.114	0.513	10.7	207
DKDP[b)]	100.5	1.219	0.666	13.8	241
ADP[a)]	106.3	1.131	0.508	10.9	230
KDA[c)]	108.2	1.202	0.856	20.2	396
DKDA[c)]	124.5	0.889	0.669	15.5	408
ADA[d)]	126.1	0.893	0.624	14.9	406

Sellmeier oscillator parameters λ_c, S_c (see /6/, equation (6)) calculated using the refractive index taken from the papers a) /13/, b) /14/, c) /15/, d) /16/.

induction up to 36.4 T at room temperature as well as in the vicinity of the Curie temperature. This is in good agreement with quantum-mechanical results of Druzhinin and Tatsenko /3/.

The dependence of the Verdet constant on temperature proved that the constant is not temperature dependent up to the Curie temperature. The results obtained for DKDP and ADA are shown on Fig. 2. So, the Verdet constant describes the so-called diamagnetic Faraday rotation only. Thus it is easy to show /2/ that like in some other dielectric crystals /4, 5/

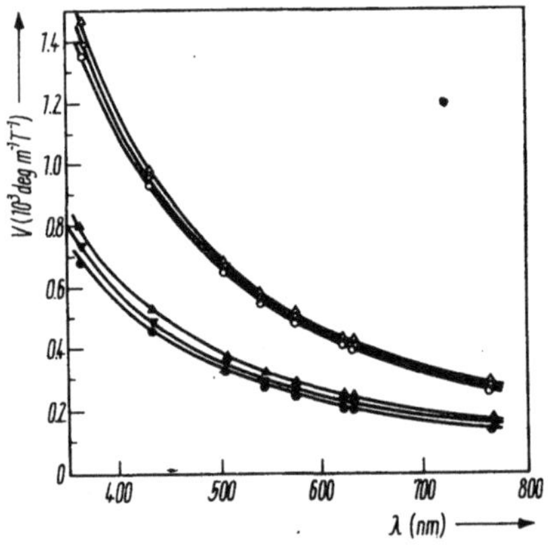

Fig. 1. Dispersion of the Verdet constant for KDP-type crystals. The solid line corresponds to the best fit to the experimental points calculated from the dispersion function (2); T = 298 K; △ DKDA, ○ KDA, ▽ ADA, ▲ DKDP, ● KDP, ▼ ADP

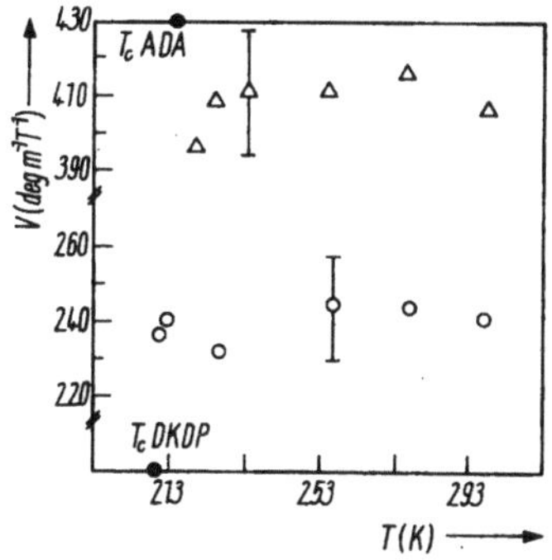

Fig. 2. Verdet constant versus temperature for DKDP (o) and ADA (Δ) crystals, λ = 632.8 nm. The bars represent an estimated error

the normal diamagnetic Verdet constant dispersion can be described by the model of a harmonic oscillator by the following single-term equation:

$$nV = A \frac{x^2}{(1-x^2)^2} , \tag{2}$$

where $x = \lambda_0/\lambda = \hbar\omega/\hbar\omega_0$; A is a constant value, n is the light refractive index, ω_0 and ω are oscillator frequency and light frequency, respectively. The constant A can be calculated as $A = (90^{\circ} e S_c \gamma_c^2/\pi mc)\,\gamma$, γ is the average magnetic anomaly given by the ratio of the effective charge and mass: $(e/m)_{eff} = \gamma\, e/m$ of the dispersing electrons, c is the speed of light in vacuum. Numerical values of λ_c, S_c, γ, and A are given in Table 1. This equation fits the experimental data points (Fig. 1) with an accuracy of $\pm 3\%$ which is enough for interpolation purposes.

The data included in Tables 1 and 2 and Fig. 1 indicate that the Verdet constant increases when potassium is replaced by an ammonium group as well as with deuterization. On the other hand, the change is known to be related with an

Table 2

Curie temperature T_c, deuterization X, and Verdet constant of DKDP and DKDA crystals

crystal	T_c(K)	X	V (deg m^{-1} T^{-1}) at λ = 632.8 nm
DKDP			
sample I	195	≈0.80	237
sample II	207	≈0.85	241
sample III	219	≈0.95	247
DKDA	159	≈0.95	408

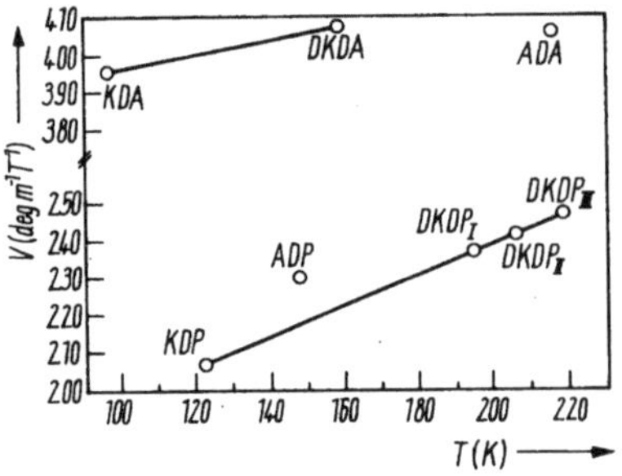

Fig. 3. Dependence of the Verdet constant for KDP-type crystals versus Curie temperature. λ = 632.8 nm, T = 298 K

increase in the Curie temperature of the crystal. For the KDP type crystals studied we found a linear relation between V and T_c for each group of crystals (Fig. 3). The diagram presented in Fig. 3 allows to determine the Verdet constant for unexamined KDP-type crystals if only its Curie temperature is known.

The diagrams in Fig. 3 and 1 imply that the influence of cations on the nonlinear part of the electron polarization is much smaller than the influence of anions. So, the Faraday effect in KDP-type crystals may be related with the electron polarization of the groups PO_4^{3-} and AsO_4^{3-} while the distribution of the electron density is greatly influenced by a hydrogen bond. The appearance of a new bond N-H...O in the ammonium crystals ADP and ADA causes a considerable change in the value of the Verdet constant. Similar conclusions concerning the groups PO_4^{3-} and AsO_4^{3-} in the Faraday effect of the investigated crystals follow from the fact that the λ_c value for KDP-type crystals is the same as for the C band. The main role in the process of electron transitions from the C band is played by PO_4^{3-} ions, and K^+ or NH_4^+ cations are of minor importance as was shown by Baldini et al. /7/ for KDP and ADP, and by Saito et al. /8/ for KDP, ADP, DKDP, and KDA. Also, as follows from Fig. 2, the process of ordering in $H_2PO_4^-$ and $H_2AsO_4^-$ dipoles to one of the two configurations of Slater-Takagi type /9, 10/ in which anions occur in the ferroelectric phase is not important in the Faraday effect. However, this problem requires some further experimental analysis.

Finally, I would like to mention the possibility of applying KDP-type crystals in the UV region, to magnetooptic devices and to express my supposition that KDP-type crystals are expected to be interesting objects as to the observation of nonlinear optical effects of higher order /11, 12/.

The author is much indebted to Prof. Marian Surma for fruitful discussions and remarks. The author expresses his thanks to Dr. B. Hilczer for his kind offer of KDA and DKDA crystals and for comments, to Dr. J. Karniewicz for DKDP crystals, and to Dr. B. Błaszkiewicz for ADA crystal.

References

/1/ M. KORALEWSKI and M. SURMA, Acta phys. Polon. A49, 803 (1976).
/2/ M. KORALEWSKI, Ph. D. Thesis, A. Mickiewicz University, Poznań 1979.
/3/ V.V. DRUZHININ and O.M. TATSENKO, Optika i Spektroskopiya 36, 733 (1974).
/4/ M. KORALEWSKI, S. HABRYŁO, and M. MRÓZ, Acta phys. Polon. A56, 419 (1979).
/5/ M. KORALEWSKI, phys. stat. sol. (a) 61, K151 (1980).
/6/ M. DI DOMENICO, Jr. and S.H. WEMPLE, J. appl. Phys. 40, 720 (1969).
/7/ G. BALDINI, M. COTTINI, and E. GRILLI, Solid State Commun. 11, 1257 (1972).
/8/ S. SAITO, K. WADA, and R. ONAKA, J. Phys. Soc. Japan 37, 711 (1974).
/9/ J.C. SLATER, J. chem. Phys. 9, 16 (1941).
/10/ Y. TAKAGI, J. Phys. Soc. Japan 3, 271 (1948).
/11/ W. JAMROZ and M. KORALEWSKI, Opt. Quantum Electronics 12, 103 (1980).
/12/ S. KIELICH and R. ZAWODNY, Optic Acta 20, 867 (1973) and references therein.
/13/ F. ZERNIKE, Jr., J. Opt. Soc. Amer. 54, 1215 (1964).
/14/ R.A. PHILLIPS, J. Opt. Soc. Amer. 56, 629 (1966).
/15/ R.S. ADHAV, J. appl. Phys. 39, 4091 (1968).
/16/ H. TOPSOE and C. CHRITIANSEN, Ann. Phys. Chem. 6, 499 (1874).

(Received February 17, 1981)

phys. stat. sol. (a) 65, K55 (1981)

Subject classification: 16 and 20.1; 14.3.4; 22.3

Institute of Physics, Academy of Sciences of the Azerbaidzhan SSR, Baku[1)]

Interference Oscillation of Photocurrent in GaSe:Sn and $GaSe_{0.95}Te_{0.05}$:Sn Single Crystals

By

E.F. BAGIRZADE, B.G. TAGIEV, G.M. MAMEDOV, I.M. ASKEROV, and F.N. SULTANOV

The oscillations of photocurrent in impurity and intrinsic regions and at $h\nu > E_g$ are primarily related to the interaction between the current carriers and and optical phonons /1, 2/. Similar photocurrent oscillations depending on the wavelength may take place by interference modulation of the light intensity inside a semiconductor, which also gives rise to photocurrent oscillations /3, 4/. The present paper deals with a study of light interference-stipulated oscillations in the photocurrent, electroabsorption, and transmission spectra of GaSe: Sn and $GaSe_{0.95}Te_{0.05}$: Sn single crystals. Owing to the laminated structure of these single crystals there were no problems in producing thin (d $\gtrsim$ 1 μm) parallel-plane photosensitive specimens in the form of Fabry-Pérot resonators. Natural and polarized light ($\vec{E} \perp \vec{C}$) passed along the C-axis of the crystal. The photoconductivity was studied according to the technique described elsewhere /3/, and the photocurrent oscillations were recorded with the use of a time-responsive H306 recorder. The electroabsorption and transmission spectra and the electrooptical effect were studied by means of an electroabsorption set-up /5/.

Fig. 1 and 2 give spectral dependences of photocurrent (curves 1) for 2 μm and 12 μm thick GaSe: Sn single crystals at 300 K and at light intensity corresponding to the linear lux-ampere characteristic. As the specimens were thin, photocurrent and its oscillations were observed in the exciton-absorption region and above. Below the exciton-absorption edge impurity photocurrent and its oscillations were observed in thick specimens. In this case the oscillation pattern showed a drastic deterioration. In the photoconductivity spectrum the principal maxima at λ = 619 nm in GaSe: Sn and λ = 624 nm in $GaSe_{0.95}Te_{0.95}$:Sn

---

1) Nazimanova 33, Baku 370/43, USSR.

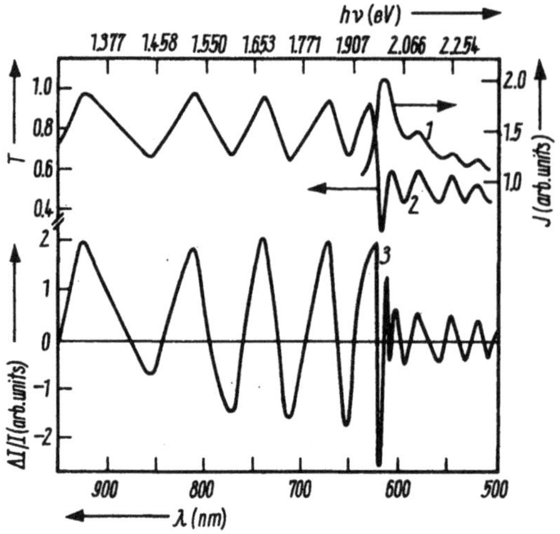

Fig. 1. Spectral dependence of photocurrent (curve 1), transmission (curve 2), and electroabsorption (curve 3) for a GaSe: Sn specimen with a thickness of d = 2 μm at 300 K

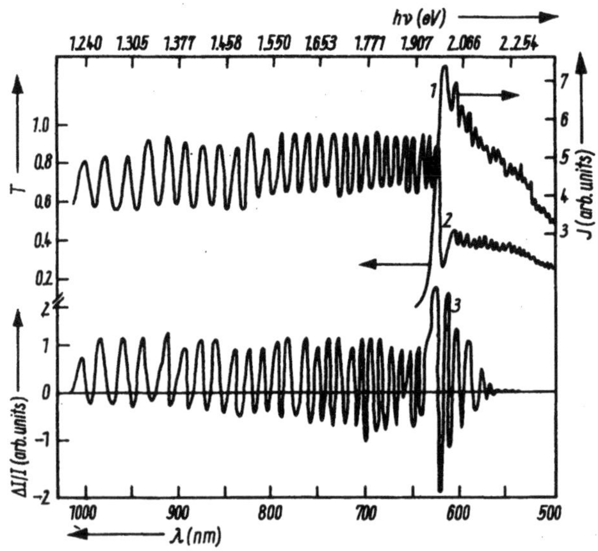

Fig. 2. Spectral dependence of photocurrent (curve 1), transmission (curve 2) and electroabsorption (curve 3) for a GaSe: Sn specimen with a thickness of d = 12 μm at 300 K

are associated with the exciton photoconductivity. This is evidenced by the agreement between the energetic positions of these maxima and the exciton minima in the transmission spectra (Fig. 1 and 2, curves 2) and the deep negative minima in the electroabsorption spectra (Fig. 1 and 2, curves 3). The differential electroabsorption spectra (Fig. 1 and 2, curves 3) coincide with those obtained earlier /5/, and a number of other oscillations are observed. Investigations of photocurrent spectra for different specimen thicknesses show that except for the exciton minima, the energetic positions of the other extrema vary. With the same specimen thicknesses the oscillation period increased, depending on wavelength. Photoconductivity spectra of nonparallel-plane specimens showed no oscillations. Experimental facts suggest that the photocurrent oscillations observed in GaSe: Sn and $GaSe_{0.95}Te_{0.05}$: Sn are not associated with the interaction between the current carriers and optical phonons. Studies on the photoconductivity, transmission, and electroabsorption spectra of these crystals show that the oscillations in these spectra are associated with the interference phenomenon. At normal light incidence upon a parallel-plane specimen under the condition $d \ll l_{coh}$ the rays subjected to multiple reflection have the phase difference $\delta = 4\pi (nd/\lambda)$ and their interference takes place /6/. As a result, the light intensity at the excit of and inside the specimen will be an oscillating function of wavelength, which is observed in the transmission spectrum (Fig. 1 and 2, curves 2). In this case the photocurrent linearly depending on the light intensity will also oscillate with the same phase as the transmission. The results obtained show that the photocurrent oscillation depth in GaSe: Sn and $GaSe_{0.95}Te_{0.05}$: Sn is defined by the visibility of the interference pattern. From the interference pattern in the transmission spectrum we determined the value of the refraction index n for $GaSe_{0.95}Te_{0.05}$: Sn according to the technique described in /6, 7/, which proved to be equal to 2.8 at $\lambda =$ $= 630$ nm.

Photocurrent oscillations were also observed in such parallel-plane specimens in strong electric fields. As shown in /3, 4/, this is associated with the interference modulation of light inside the specimen, due to a variation of the refraction index Δn. Investigations on the electrooptical effect in GaSe: Sn and $GaSe_{0.95}Te_{0.05}$: Sn, similar to those outlined in /8, 9/, show that there is a linear dependence of Δn (Fig. 3, curve 2) on the electric field. It has been

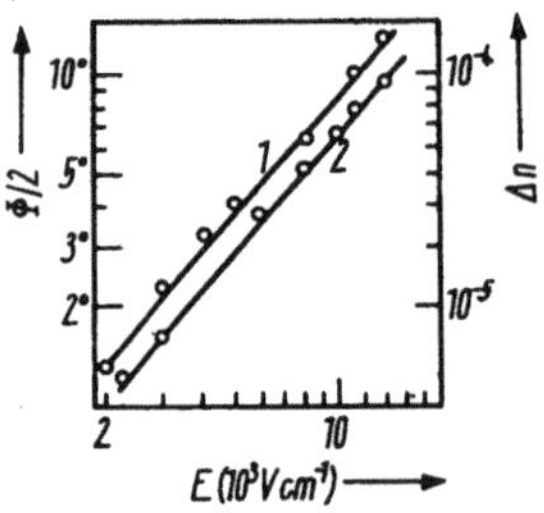

Fig. 3. Dependence of the polarization-plane rotation angle (curve 1) and the refraction index variation (curve 2) on the electric field for $GaSe_{0.95}Te_{0.05}$: Sn

found that in $GaSe_{0.95}Te_{0.05}$: Sn at λ = 630 nm the electrooptical constant has a value of 5.85×10^{-9} cm V^{-1}. Consequently, a variation of the refraction index ($\Delta n \approx 10^{-4}$) with strong electric field gives rise to a variation in the phase difference of multiple reflected rays as well as to a shift in the interference pattern. This fact is responsible for the change occurring in the transmission of parallel-plane specimens /10, 11/. This situation is responsible for the additional oscillations observed by us in the electroabsorption spectra of GaSe: Sn and $GaSe_{0.95}Te_{0.05}$: Sn single crystals (Fig. 1, and 2, curves 3).

References

/1/ KH.YU. FEN, Proc. IXth Internat. Conf. Phys. Semicond., Moscow 1968, Izd. Nauka, Leningrad 1969.

/2/ V.L. ALPEROVICH, A.F. KRAVCHENKO, N.A. PAKHANOV, and A.S. TEREKHOV, Zh. eksper. teor. Fiz., Pisma 28, 551 (1978).

/3/ B.S. RYVKIN, Fiz. tverd. Tela 13, 367 (1979).

/4/ B.S. RYVKIN, Zh. tekh. Fiz., Pisma 5, 65 (1979); 5, 586 (1979).

/5/ V.A. GAIEV, V.I. SOKOLOV, V.K. SUBASHIEV, and B.G. TAGIEV, Fiz. tverd. Tela 12, 1061 (1970).

/6/ YU.I. UKHANOV, Optical Properties of Semiconductors, Izd. Nauka, Moscow 1977.

/7/ G. ANTONIOLI, D. BIANCHI, V. CANEVARI, U. EMILIANI, and P. PODINI, phys. stat. sol. (b) 81, 665 (1977).

/8/ V.I. SOKOLOV and V.K. SUBASHIEV, Fiz. tverd. Tela 14, 222 (1972).

/9/ A. CINGOLAMI, M. FERRARA, and M. LUGARO, Solid State Commun. 29, 677 (1979).

/10/ P.I. PEROV, L.A. AVDEEVA, M.L. ELINSON, and G.V. STEPANOV, Fiz. Tekh. Poluprov. 3, 183 (1969).

/11/ D.B. KUSHEV, V.I. SOKOLOV, and V.K. SUBASHIEV, phys. stat. sol. (a) 29, 251 (1975).

(Received March 23, 1981)

phys. stat. sol. (a) 65, K59 (1981)
Subject classification: 14.3; 13.1; 22

Department of Physics (a) and Department of Earth Sciences (b), University of Roorkee[1])

Temperature Dependent Electrical Conductivity of Natural Galena Crystals

By
N.M. RAVINDRA (a), V.K. SRIVASTAVA (a), and B. MISHRA (b)

Introduction It is well known that galena is the primary ore mineral of lead. In fact it contains 80.6% lead /1/. However, it is relatively unstable /2/. The main feature of the compounds of the PbS family (PbS, PbSe, and PbTe) which makes them of practical interest is that they all have unusually small direct energy gaps in the infrared /3/. Hence they are all useful as infrared detectors. Their electrical conductivities are relatively low.

In nature, lead occurs in chalcophile phase (sulphide phase) and is generally associated with Ag, Fe, Zn, Cu, and Sb /2/.

We present here our studies of the temperature dependence of the electrical conductivity of natural galena. Also included is a study of its chemical analyses.

Experimental The samples of galena used in these studies have been taken from Agnigundala (Andhra Pradesh, India). The samples have been confirmed to be single crystals by X-ray diffraction patterns. In Table 1, we present our results of the trace element analyses in galena. Using an IL751 atomic absorption spectrophotometer, Sb, Bi, and Ni have been estimated. The estimation of Cd, Mn, Cu, Ag, and Ni have been done using a Pye Unichem atomic absorption spectrophotometer. The results of the estimation of Ni from both the spectrophotometric analyses have been found to be the same. Incidentally, we have also carried out chemical analyses of many samples of galena. Except for tra-

Table 1

Results of trace element analyses of galena crystals (in ppm)

Mn	-	6
Cd	28	59
Cu	6	12
Ag	176	175
Ni	7	17
Sb	250	198
Bi	120	50

1) Roorkee 247672, India.

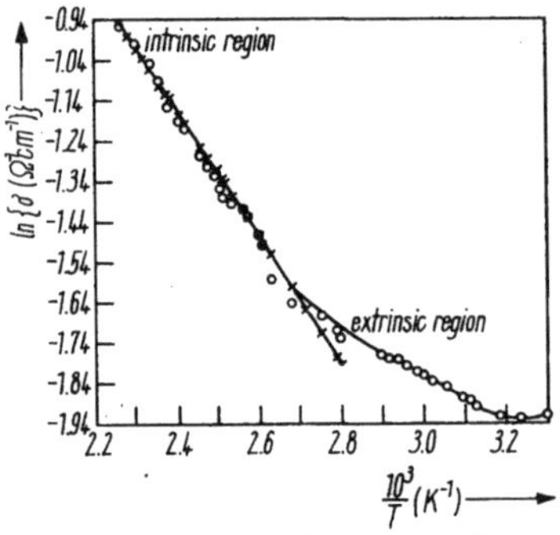

Fig. 1. Experimental (o) and theoretical (×) plots of ln σ_T versus 1/T for natural galena

ces of Mn, all the samples had almost the same constituents. Ag was generally around 175 ppm. Because of the well known geochemical affinity of galena for Sn and Au, we believe that these two should have been present but only in traces. The impurities in the galena samples are not significantly important in our studies reported here which have been carried out in the high temperature intrinsic range. We have given the above chemical analyses for completeness.

The electrical conductivity measurements have been made using the standard four-probe method. The current and voltage have been measured using an Aplab digital multimeter Model 1005 and an ECIL digital multimeter DM 6101, respectively.

Results and discussion In Fig. 1, we plot lnσ_T versus 1/T. As can be seen in the figure, the intrinsic region starts at high temperatures. In this region, the temperature dependence of the electrical conductivity is given by

$$\sigma_T = \sigma_o \exp(-E_{GT}/2kT) , \qquad (1)$$

where σ_T is the electrical conductivity at temperature T, σ_o is the high temperature limit of the electrical conductivity and is a constant, and E_{GT} is the energy gap of the semiconductor. However, in expression (1), E_{GT} is itself temperature dependent.

Our earlier studies /4, 5/ of the temperature dependence of the energy gap in PbS, PbSe, and PbTe have shown that E_{GT}, the energy gap of the semiconductor at any temperature T, is given by

$$E_{GT} = E_{Go} + \frac{\alpha T^2}{T+\beta} , \qquad (2)$$

where E_{Go} is the energy gap of the semiconductor at 0 K, the constants α and β are a measure of the electron-phonon interaction and the Debye temperature, respectively. For PbS, $\alpha = 5\times10^{-4}$ eV K^{-1}, $\beta = 40$ K, and $E_{Go} = 0.29$ eV /4, 5/. Using (2) in (1) we get

$$\sigma_T = \sigma_o \exp\left[-\frac{E_{Go}(T+\beta) + \alpha T^2}{2kT(T+\beta)}\right] . \qquad (3)$$

For our sample under consideration, the extrapolated value of σ_o = = 250.5 $(\Omega\,cm)^{-1}$. Using this value of σ_o as also the above mentioned values of E_{Go}, α, and β, we calculate the values of σ_T for various temperatures using (3). In Fig. 1, we plot the calculated values of $\ln \sigma_T$ as a function of 1/T for various temperatures. As can be seen in the figure, the calculated values of σ_T are in very good agreement with the experimental values. In fact the deviation is only $\pm 6\%$. Considering a constant value of $E_G (= E_{Go}$ = 0.29 eV), the values of σ_T have been calculated. They vary from 2.252 $(\Omega\,cm)^{-1}$ (at 357 K) to 5.717 $(\Omega\,cm)^{-1}$ (at 445 K). These calculated values of σ_T with constant E_G are, however, much higher than the experimental values. Thus from these calculations, the temperature dependence of E_G becomes very clear.

Conclusion Studies of the temperature dependence of the electrical conductivity of natural galena crystals have been presented above. Trace element analyses of the galena samples have also been reported. An interpretation for the temperature dependence of the electrical conductivity as a function of temperature has been given in terms of our earlier reported theoretical studies of the temperature dependence of the energy gap in PbS, PbSe, and PbTe.

The authors are thankful to the Directors of Wadia Institute of Himalyan Geology, Dehradun, and the University Services Instrumentation Centre for enabling them to conduct the trace element analyses. One of the authors (N.M.R.) thankfully acknowledges the financial support of C.S.I.R. (India).

References

/1/ M. KUZIN and N. EGOROV, Field Manual of Minerals, MIR Publishers, Moscow 1976.

/2/ H.E. HAWKES and J.S. WEBB, Geochemistry in Mineral Exploration, Harper and Row, New York 1965.

/3/ J.C. PHILLIPS, Bonds and Bands in Semiconductors, Academic Press, New York 1973.

/4/ N.M. RAVINDRA, SUSHIL AULUCK, and V.K. SRIVASTAVA, phys. stat. sol. (a) 52, K151 (1979).

/5/ N.M. RAVINDRA and V.K. SRIVASTAVA, phys. stat. sol. (a) 58, 311 (1980).

(Received March 17, 1981)

phys. stat. sol. (a) 65, K63 (1981)
Subject classification: 10.1 and 12.1; 21; 21.1

Physics Department, Faculty of Science, Cairo University[1)]

Internal Friction Spectra of Some Al-Cu Alloys

By

S.A. MAHMOUD, M.A. SEMARY, and Z. FARID

Introduction Internal friction measurements are one of the indirect methods to explore the relaxation processes induced by the dissolved foreign atoms or the incoherent precipitates formed in the dilute alloys. It was reported that reheating after quenching of the Al-Cu alloy revealed the presence of vacancy condensation pits containing Cu. The geometrical size and the density of the pits depend differently on the Cu concentration /1/. Ageing the Al-Cu alloy at different temperatures for different times produces different phases of precipitates /2/. Ageing at temperatures within the $\alpha + \Theta'$ field should create an approximately random distribution of large Θ' platelets /3/. Such microstructures have been illustrated in the studies of the Θ' phase where samples were water quenched and then aged for long times in the $\alpha + \Theta'$ field. The precipitates that formed at relatively low temperatures are either coherent or semicoherent with the matrix. The coherent precipitates, though increasing the background, produce no peaks in the internal friction spectra /4/. At higher temperatures the equilibrium precipitates are fitted semicoherently to the matrix structure, and peaks of relaxation character can be produced.

In this work the annealing spectra of some Al-Cu alloys are isochronally tested in the temperature range of 25 to 500 °C by observing the associated changes in internal friction. The internal friction peaks are analyzed in terms of the different precipitates formed during the successive ageing of the samples.

Experimental Each sample was brought to a metastable state at room temperature by heating above the solvus and quenching in cold water. The quenching rate is of the order of 10^4 K/s. The annealing spectrum of the samples is traced by subjecting them to rectangular heat pulses using a high heat capacity furnace maintained at the annealing temperature. The specimens are raised to various temperatures in steps of 20 °C and aged there for 15 min

---

1) Giza, Cairo, Egypt.

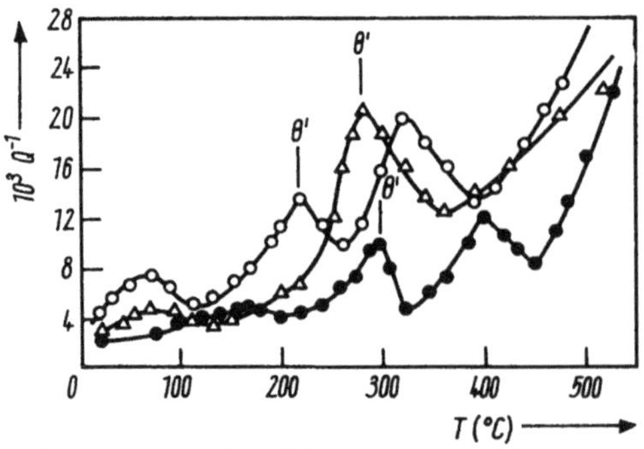

Fig. 1. Internal friction spectra of different homogenized Al-Cu alloys: △ 1% Cu at frequency 113.9 Hz, o 2% Cu at frequency 81.0 Hz, and ● 4% Cu at frequency 68.9 Hz

before the observations are made. The internal friction spectra of Al-1% Cu, Al-2% Cu, and Al-4% Cu alloy are obtained by means of the resonance curve method. A vibration detector /5/ is used to detect the amplitude of a clamped-free wire sample which vibrates transversely at low frequencies.

Results and discussions Fig. 1 shows the internal friction spectra of different homogenized Al-Cu alloys above the solvus for 6 h. From Fig. 1 we can see the presence of three annealing bands appearing as three successive peaks in the internal friction (for 1% Cu concentration, only two bands exist). The peaks are superimposed on a background of exponential temperature behaviour. Use is made of the peak maximum temperature on changing the frequency to calculate the activation energy. The energy E associated with a relaxation process is determined on the assumption that the relaxation frequency ν follows an Arrhenius equation of the type

$$\nu = \nu_o \exp(-E/kT) ,$$

where T is the absolute temperature at which the peak occurs. A plot of ln ν versus 1/T yields the value of the energy E; this is shown in Fig. 2. The three bands located in the temperature ranges 60 to 180, 200 to 315, 315 to 440 °C are thermally activated relaxation processes of energies 0.35, 1.00, and 1.4 eV for 2 and 4% Cu concentration. In the same temperature range, the 1% Cu concentration has only two bands which are characterized by activation energies of 0.35 and 1.00 eV. These energies are almost independent of Cu concentration in the matrix.

The activation energy of 0.35 eV associated with the first band is less than the migration energy of vacancies (0.5 eV) in the matrix. The vacancy-Cu impurity pair has an activation energy of 0.35 eV for its migration in the matrix. It is then reasonable to assume that the motion of these entities behind dislocation is a process responsible for the G.P. zone formation. The peak occurring

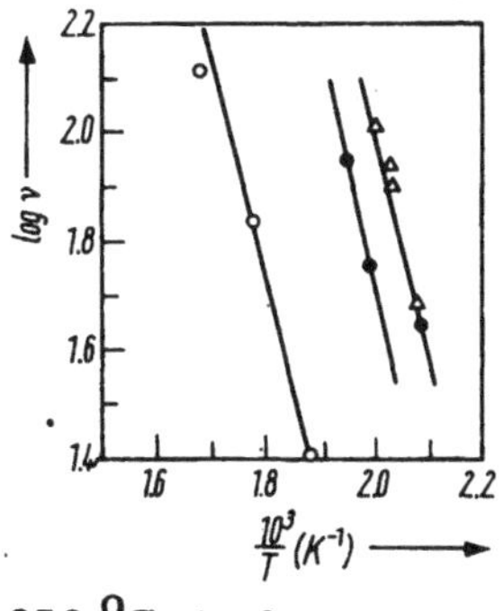

Fig. 2. A plot of log ν versus $10^3/T$ for Θ' phase, $\tau_o = 10^{-12.5}$ s; E = 1.0 eV. ● Al - 1 wt% Cu, △ Al - 2 wt% Cu, o Al - 4 wt% Cu

around 150 °C may be attributed to the formation of the Θ" phase.

In the presence of the Θ' phase, the peak near 250 °C at a frequency of about 100 Hz may be due to the misfit dislocations around the semicoherent precipitates. The Θ' precipitates have a structure with the composition Al_2Cu, they are flat discs, and have semicoherent interfaces /6/. The activation energy of this relaxation is 1 eV with a relaxation time τ_o of $10^{-12.5}$ s, in accordance with the reported values /6/. To make a double check on the determination of the activation energy associated with the Θ' phase, the internal friction peak shape method is used, using the following equation /7/:

$$\Delta T = 2.63\, k\, T_m^2/E \quad ,$$

ΔT being the difference in temperature corresponding to half the maximum of the relaxational internal friction, and T_m the peak temperature, the values obtained are in good agreement with the activation energy found from the peak shift method.

To test the coherency of Θ' precipitates a plastic deformation is done on the three different alloys. The peak height is found to increase by small plastic deformation (1.6%), which enhances the coherency loss in the interfaces /8/ (the effect of plastic deformation on Θ' phase will be published later). This proves that the investigated peak is due to the motion of misfit dislocations.

The energy of 1.4 eV associated with the peak located around 350 °C (for Cu concentrations 2% and 4%) may be attributed to a relaxation process responsible for the precipitation of the Θ phase. The value of this energy is close to the diffusion activation energy of Cu atoms in the Al matrix /9/.

References

/1/ R.H. BUKALIL, C. ROQUES-CARMES, R. TIXIER, M. AUCOUTURIER, and P. LACOMBE, Metal. Sic. J. 8, 387 (1974).

/2/ K. OSAMMA, S. FURNICHI, K. KANBAYASHI, S. TAKAMUKU, and Y. MURAKAMI, J. Japan. Inst. Metals 40, 1228 (1976).

/3/ T.J. HEADLEY and J.J. HREN, J. Mater. Sci. 11, 1867 (1976).

/4/ G. SCHOECK, phys. stat. sol. 32, 651 (1969).

/5/ R. KAMEL, J. appl. Phys. 24, 1308 (1963).

/6/ S. KISS, F.J. KEDVES, and I.Z. HARANGOZO, Nuovo Cimento 33B, 380 (1976).

/7/ R. DE BATIST, Internal Friction of Structural Defects in Crystalline Solids, North-Holland Publ. Co., 1972 (p. 61).

/8/ M. MONDINO and G. SCHOECK, phys. stat. sol. (a) 6, 665 (1971).

/9/ A.S. NOWICK and B.S. BERRY, Anelastic Relaxation in Crystalline Solids, Academic Press, New York 1972 (p. 260).

(Received November 25, 1980)

phys. stat. sol. (a) 65, K67 (1981)

Subject classification: 10.2 and 20.3; 22.5.2

Department of Chemistry, Indian Institute of Technology, Bombay[1] (a) and Health Physics Division, Bhabha Atomic Research Centre, Bombay[2] (b)

Thermoluminescence Glow Curves of Ca^{2+} and Sr^{2+} Doped LiCl Single Crystals

By

M. SHARON[3] (a), R.R. PRADHANANGA (a), and C.M. SUNTA (b)

Introduction The thermoluminescence of LiCl single crystals has been investigated by us and the analysis of glow curves for the determination of trapping parameters and recombination kinetics is reported elsewhere /1, 2/. Recently considerable interest has been focused on the study of the effect of alkaline earth dopants on the TL glow curves of alkali halides /4 to 11/. In the present investigation the effects of Ca^{2+} and Sr^{2+} dopants on the TL glow curves of LiCl crystals have been studied. Due to the bigger ionic size of Ca^{2+} and Sr^{2+} ions compared to the Li^+ host ion, it is expected that the Sr^{2+} ion will distort the lattice to a greater extent (as compared to the Ca^{2+} ion) and it is expected to produce greater changes in the TL glow curve (as compared to the TL glow curve of Ca^{2+} doped LiCl).

Experimental procedure Single crystals of pure and doped LiCl were grown by the Bridgman technique in a HCl atmosphere. For Ca^{2+} and Sr^{2+} doped crystals, appropriate cations in the form of chloride were added to the LiCl solution and the resulting solutions were evaporated in vacuum and dehydrated by treating with anhydrous HCl at high temperature, followed by vacuum dehydration. Then the single crystals were grown from thus treated materials. LiCl is highly hygroscopic, so the LiCl chips of about 3x3x0.5 mm^3 size were cleaved in dry nitrogen atmosphere. The portion adjacent to the crystal chips selected for TL was used for the determination of Ca^{2+} and Sr^{2+} concentrations by atomic absorption spectrometry. Prior to taking TL reading, all these crystals were heated to 350 °C for 1 h and then quickly cooled by removing the crystal from the furnace to the dessicator. The crystals were then

---

1) Bombay 400076, India.

2) Bombay 400083, India.

3) To whom all correspondence should be sent.

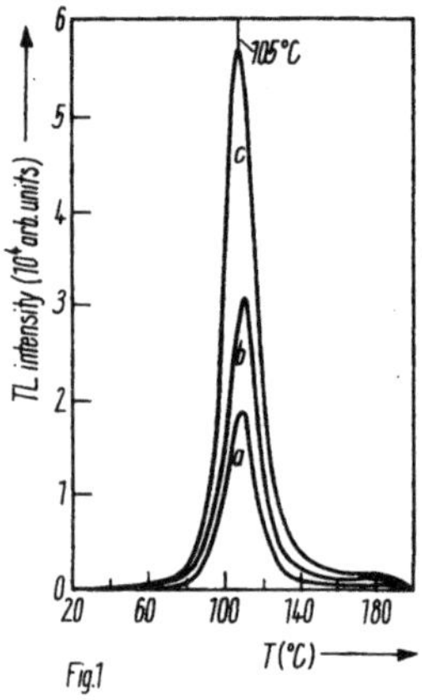

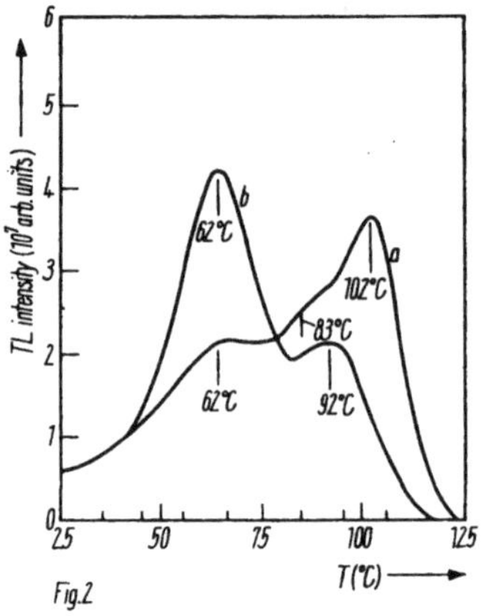

Fig. 1. Thermoluminescence glow curve of LiCl single crystal; γ-dose (a) 1000, (b) 1500, and (c) 3000 rad

Fig. 2. Thermoluminescence glow curves of LiCl crystals doped with Ca^{2+}; (a) 207×10^{-6} and (b) 929×10^{-2} mole fraction Ca^{2+}

irradiated either by ^{60}Co (γ-source) or by ^{90}Sr (β- source) for the required dose. The TL was read out in the TL reader described elsewhere /3/. The light detector was an EMI 9515S photomultiplier tube.

Results and discussion The TL glow curves of pure LiCl, γ-irradiated to different doses are shown in Fig. 1. The glow curve consists of a prominent glow peak at 105 °C, and a very weak peak at 190 °C.

Fig. 2 shows the glow curve of a β-irradiated Ca^{2+} doped crystal with a β-dose of 750 rad. With the incorporation of Ca^{2+} in the LiCl lattice a new glow peak at 62 °C was observed with a slight shift of the main glow peak of 105 °C. With the increase in Ca^{2+} dopant concentration, the prominent glow peak was suppressed while the impurity peak enhanced.

Fig. 3 shows the glow curve of a β-irradiated Sr^{2+} doped crystal with a β-dose of 750 rad. In Sr^{2+} doped crystals, in contrast to the Ca^{2+} doped crystal, only one peak at 75 °C was observed, when the Sr^{2+} dopant concentration was increased to 1716×10^{-6} mole fraction three overlapping glow peaks at 67, 83, and 97 °C were observed.

The prominent glow peaks observed at 105 °C in the case of pure LiCl can be reasonable assumed to be due to the annealing of F centers, which were

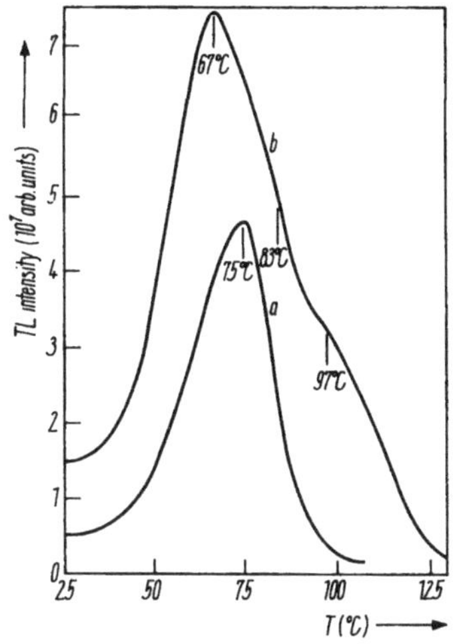

Fig. 3. Thermoluminescence glow curves of LiCl crystals doped with Sr^{2+}; (a) 251×10^{-6} and (b) 1716×10^{-6} mole fraction Sr^{2+}

formed during irradiation. The glow peaks at 62 °C in Ca^{2+} doped crystals and at 75 °C in Sr^{2+} doped crystals are clearly due to the incorporation of divalent impurities in the LiCl lattice. When divalent impurities go substitutionally in the alkali halide lattice, cation vacancies are created. These cation vacancies may be present in the lattice as a free vacancy or associated with divalent cations forming an impurity-vacancy dipole (I-V dipole). These I-V dipoles can act as a trapping center related to a TL glow peak. The glow peak at 62 °C in a Ca^{2+} doped crystal and at 75 °C in a Sr^{2+} doped crystal must be due to a release of electrons or holes from these dipoles and radiative recombination with electrons or holes. As the concentration of Ca^{2+} increased, the I-V dipoles are also expected to increase and so the intensity of the impurity peak (62 °C) should increase. This is what has been observed. With increase in the Ca^{2+} dopant concentration, the prominent peak shifts more towards low temperature. This shift is attributed to the perturbation of the F center by the aggregated center surrounding the F center. The aggregated center perhaps changes the configuration coordinate of the F center which results in a shift of the glow peak. Inabe and Takeuchi /4/ also observed such a shift of the glow peak in KCl crystals doped with Sr^{2+}. These authors explained the shift with a new model, but they also agreed that the shift in the glow peak temperature is the result of the dispersed state of divalent ion-vacancy aggregated complexes.

The complete absence of a glow peak at 105 °C and the appearance of a single glow peak at 75 °C in case of a 251×10^{-6} mole fraction Sr^{2+} doped crystal is interesting. Due to the bigger ionic size of the Sr^{2+} ion, a greater degree of association between the Sr^{2+} ion and the cation vacancy is expected to occur, which drastically reduces the free cation vacancies. Due to the abundant presence of aggregated centers (which act as a dominant trapping center responsible for the 75 °C glow peak), it must have suppressed the formation of

F-centers which are perturbed to such an extent that they are no longer dominant in the thermoluminescence emission. The other interesting aspect of Sr^{2+} doped crystals is the appearance of three overlapping peaks at 67, 83, and 97 °C. Such kinds of overlapping peaks were observed by Heckelsberg and Daniel /12/ in unpurified LiCl crystals. This might be due to the presence of background impurities in their crystals. These overlapping peaks can be reasonably assumed to be due to the stepwise release of trapped charged carriers at different aggregates of I-V dipoles.

One of us (R.R.P.) is thankful to U.G.C. and C.S.I.R. for the award of a fellowship, and Tribhuvan University, Nepal, for granting the leave to carry out this work. The authors are thankful to Dr. A.K. Ganguly for permitting them to utilize facilities of the B.A.R.C. for this work.

References

/1/ M. SHARON, R.R. PRADHANANGA, and C.M. SUNTA, J. Phys. C 13, 3795 (1980).

/2/ M. SHARON, R.R. PRADHANANGA, and C.M. SUNTA, phys. stat. sol. (a) 59, K103 (1980).

/3/ C.M. SUNTA, Ph.D. Thesis, Agra University, 1972.

/4/ K. INABE and N. TAKEUCHI, Japan. J. appl. Phys. 9, 1949 (1978).

/5/ I. KATZ, B. CHENFOUX, and N. KRISTIANPOLLER, phys. stat. sol. (a) 12, 307 (1972).

/6/ V. AUSIN and J. ALVAREZ RIVAS, J. Phys. C 5, 82 (1972).

/7/ K. INABE and N. TAKEUCHI, J. Phys. 10, 3023 (1977).

/8/ K.N. REDDY, M.L. RAO, and V.M. BABU, Indian J. pure appl. Phys. (1978).

/9/ R.V. JOSHI and N.L. KOKAN, Indian J. pure appl. Phys. 12, 79 (1974).

/10/ S.C. JAIN and P.C. MEHENDRU, Phys. Rev. 143, 957 (1965).

/11/ P.C. MEHENDRU and S. RADHAKRISHNA, J. Phys. C 2, 769 (1969).

/12/ L.F. HECKELSBERG and F. DANIEL, J. phys. Chem. 61, 414 (1957).

(Received March 5, 1981)

phys. stat. sol. (a) 65, K71 (1981)

Subject classification: 10.2 and 20.3; 22.5.2

Department of Physics, Indian Institute of Technology, Kharagpur[1] (a) and Physics Division, JNU Centre of PG Studies, Imphal[2] (b)

Emission Processes in the Thermoluminescence of Li-Doped KCl Single Crystals

By

S. P. KUILA (a), R. K. GARTIA (b), and V. V. RATNAM (a)

In an earlier paper /1/ we reported results on the thermoluminescence (TL) (hereafter termed thermally stimulated luminescence (TSL) to distinguish this phenomenon from tunneling recombination luminescence (TRL)) of F_A centres in room temperature (RT) X-irradiated Li-doped KCl crystals. There we showed that the TL emission spectra consisted of two emission bands occurring around 450 and 510 nm. These experiments were performed on comparatively thick (thickness of the order of 1 mm) crystals. Our recent experiments on thinner samples (thickness of the order of 0.5 mm) of the same crystals performed with greater care and refinement in recording reveal new features which have led to a better understanding of the recombination luminescence processes during the thermal annihilation of the X-ray induced defect centres in these crystals. Only the salient features are reported here. Detailed work on two similar systems viz., KBr:Li and KCl:Na will be reported separately.

Unlike that observed earlier /1/ where a rather broad band with an ill defined shoulder occurred we observed resolution in the TL emission spectrum in the present case. The TL emission spectra of the 90 and 210 °C glow peaks are shown in Fig. 1. As reported earlier the emission spectra of the TL peaks are the same but for the fact that we now get three emission bands around 435, 460, and 510 nm. It, therefore, appears that the 450 nm band observed as a single emission band earlier is in fact a superposition of the 435 and 460 nm emission bands (Fig. 1, curves a and c). The reason for this is mentioned later in this note.

The other important result we want to report is the change in the relative strengths of the different TL emission bands when the irradiated crystal is

1) Kharagpur 721302, India.

2) Imphal 795003, India.

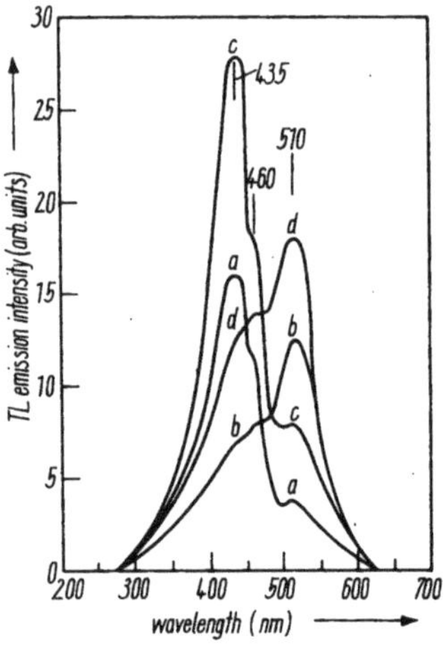

Fig. 1. TL emission spectra of a 15 min RT X-irradiated Li-doped KCl crystal. Curves a and c are of as-irradiated samples for 90 and 210 °C TL peaks, respectively, whereas b and d are those of F-bleached samples

bleached with F-light for different periods of time. The effect of 10 min F-bleaching (with 560 nm light from a Hg lamp through an interference filter) on the TL emission spectra at 90 and 210 °C is shown in Fig. 1 (curves b and d). F-bleaching is chosen because upto this time the $F \rightarrow F_A$ conversion is the dominant process /1/ and this is also the time upto which colour centre annihilation by F-bleaching is minimal as revealed by the unchanged area under the glow curve. We would like to point out that in the as-irradiated condition the 90 °C glow peak is less intense than the 210 °C glow peak whereas the spectral composition of the 90 °C peak shows abundant emission at 435 nm with lower 460 and 510 nm emissions. With different extents of F-bleaching the 435 nm band decreases while the 460 and 510 nm emissions build up and for 10 min F-bleaching the 510 nm emission is the strongest. The emission curve reported earlier is for an intermediate bleaching time of 4 min. It is to be noted from Fig. 1 (curves b and d) that F-bleaching increases the 510 nm emission at the expense of the 435 and 460 nm emissions which go down in intensity.

In the light of the results of Ikeya and Crawford /2/ on the monovalent cation doped alkali halides where the temperature dependence and variation of the X-ray luminescence spectra have been utilized to understand and explain the recombination luminescence processes, our results on the emission behaviour (during TL) can be understood.

Following the assignment of the 435 nm emission to ($[Cl_2^-]_A$ + e) recombination in which the Cl_2^- (V_k-centre) is trapped by a monovalent cation (Li in this case) and the 450 nm emission as that of the (Cl_2^- + e) recombination in which the exciton is perturbed by the monovalent cation we see that similar processes occur in TL also, performed from RT onwards. And it is a known fact that in this system (KCl) the V_k-centres necessary for these processes to occur are not available as such at these temperatures. One would naturally ask where these V_k-centres come from. The answer to this lies in the fact that though V_k-centres as such are not present at these temperatures, they are made available - produced as transients out of the higher aggregates of V-centres and interstitial aggregate hole centres which dissociate and/or also realeased from dislocation loops /3/ during the process of thermal annealing of the sample for recording the TL spectrum. The model suggested by us /3/ earlier to explain TL emission spectra will explain the present result as well.

Further the occurrence of the same wavelengths in the TL emission as in X-ray fluorescence indicates the similarity of the recombination luminescence emission processes in the two cases; i.e., in the case of X-ray fluorescence these V_k-centres which are the products of irradiation get directly involved in the luminescence processes (as proceed) whereas in the case of TL they are transients released from higher aggregates of hole centres and in different environments.

Changes in the relative strengths of the recombination luminescence emission consequent to optical (F-light) bleaching indicate preferential redistribution of the F-electrons (in the form of F_A-centres) resulting in an enhancement of the 510 nm emission. A similar preferential redistribution has been noticed in Ag-doped KCl /4/ wherein it was shown that under certain conditions of X-irradiation and extent of F-bleaching the preferential recombination is found to occur at other sites, though the thermally annealed F-centre electrons are not directly involved in the emission process. That the emission occurring around 510 nm in this system is due to the tunneling of F-electrons is in consonance with the finding of Grabovskis and Vitols /5/ who have attributed the emission occurring at 500 nm to tunneling recombination luminescence between the ground states of F-band V_k-centres, though, however, the site of recombination is not yet clear. F-bleaching of RT X-irradiated samples thus favours the increase of tunneling recombination luminescence at the cost of the extrinsic luminescence.

Preliminary results on KBr:Li and KCl:Na in progress in our laboratory are in support of the above findings.

We would like to thank Dr. R.C. Bhuniya for helpful discussion.

References

/1/ R.K. GARTIA, S.P. KUILA, and V.V. RATNAM, phys. stat. sol. (a) 59, 271 (1980).

/2/ M. IKEYA and J.H. CRAWFORD, Jr., phys. stat. sol. (b) 79, 559 (1977).

/3/ R.K. GARTIA and V.V. RATNAM, J. Phys. Chem. Solids 40, 331 (1979).

/4/ P. SRIDARAN, R.C. BHUNIYA, and V.V. RATNAM, phys. stat. sol. (b) 104, 459 (1981).

/5/ V.J. GRABOVSKIS and I.K. VITOLS, J. Lum. 20, 337 (1979).

(Received February 20, 1981)

phys. stat. sol. (a) 65, K75 (1981)

Subject classification: 13.4 and 20.3; 22.5.3

Institute for Foreign Students, Sofia[1] (a) and
Chemistry Department of Sofia University (b)

Absorption Spectra of Lead Centers in Barium Chloride

By

V. TABAKOVA (a) and K. KYNEV (b)

Lead-activated alkali halide phosphors have been extensively studied /1 to 13/. It was shown that the dipole centers (Pb^{2+}-V_k), which most often have a tetragonal symmetry, play an important rôle in the coloration processes. Depending on the activator concentration and the conditions of preparation, more complicated centers (Pb^{2+}-V_k)n, n = 2, 3, 4 can be formed in the crystal phosphors. We extended the study of the spectral characteristics of Pb^{2+} as an activator in $BaCl_2$. Because of the equal electric charges of Pb^{2+} and Ba^{2+} in the latter, it is not necessary for a cation vacancy to exist near the activator which leads to the above mentioned complicated centers of low symmetry. It is to be expected for the Pb^{2+} centers in $BaCl_2$ to have the symmetry of the host lattice. No detailed investigations of the crystal phosphor $BaCl_2$ have been known in literature so far. Some data concerning polycrystalline samples of $BaCl_2$ with 5 mol% $PbCl_2$ are given in /14/.

The preparation of the phosphor by the Bridgman method in vacuum was described in our earlier work /15/. The crystals were annealed for 24 h at 850 °C until the complete transformation of $BaCl_2$ into a stable rhombic modification. The investigations were carried out on plates cleaved parallel to the crystallographic axes b and c (deviation 3° to 5°) which was checked by X-rays. The activator concentration was determined by means of atomic absorption analysis and was 26 ppm. The measurements were done in a metal cryostat by means of a Cary 14 R spectrophotometer. Polarization filters 105 UV Polacoat were used for light polarization. $BaCl_2$:Pb crystals as well as non-activated $BaCl_2$ crystals were studied for comparison. Before measuring the samples were quenched.

The absorption spectra measured at three different temperatures are shown in Fig. 1. Five well separated bands with maxima at 4 K: 214, 232.5, 257, 288,

1) ul. Asen Belchev 27, Sofia, Bulgaria.

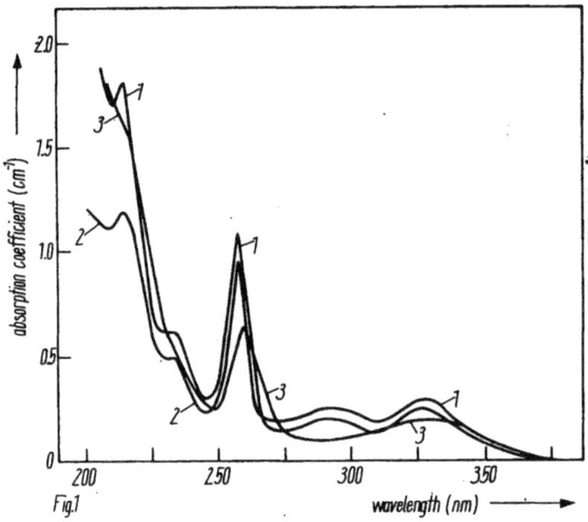

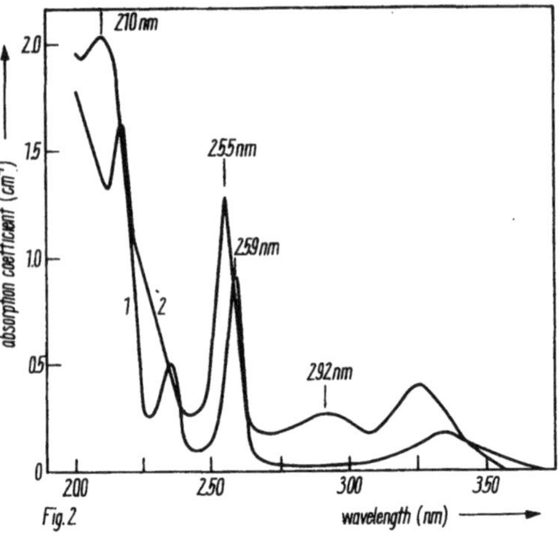

Fig. 1. Absorption spectra of $BaCl_2$:Pb at (1) 4 K, (2) 77 K, (3) 300 K

Fig. 2. Absorption spectra obtained by using polarized light at 4 K: (1) parallel to the b axis, (2) parallel to the c axis

and 326 nm can be observed in the spectrum. The strong increase of absorption at about 200 nm cannot be due only to absorption of $BaCl_2$. We suppose that below 200 nm there exists at least one more absorption band connected with the activator absorption.

The increase of temperature from 4 to 77 K leads to a strong reduction of absorption for the band at 214 nm. In the neighbouring region of 218 to 225 nm the reduction is relatively small. Deformation of the spectral curve can be observed which is more strongly manifested at 300 K. The unusual shape of the curve at 300 K can be explained if we assume the existence of one more band in the region of 218 to 225 nm.

Fig. 2 shows the absorption spectra at 4 K obtained by using polarized light. We assume that the disappearance of the structure in the region 220 to 230 nm is due to the existence of a band which is strongest for light polarized parallel to the c axis. The band at 210 nm in the same spectrum can be interpreted as a result of superposition of the band at 217 nm and the large band at 205 nm. The latter is clear in the spectrum measured at 300 K.

For the band at 257 nm in the absorption spectrum with non-polarized light (Fig. 1, curve 1) the dependence $(\ln \varkappa_0/\varkappa)^{1/2}$ versus photon energy was plotted.

The results show that the band is not Gaussian. We suggest it to be due to the fact that it is a result of superposing the bands at 255 and 259 nm which manifest themselves separately in the absorption spectra with polarized light (Fig. 2) and are described comparatively well by a Gaussian. Our assumptions were confirmed by the superposition of these bands made by us in /16/.

In discussing the results obtained we start from the assumption that Pb^{2+} isomorphously substitutes Ba^{2+} because of the equal electric charge and the proximity of the ionic radii. It can be assumed that at small activator concentration one type of centers is formed which keeps the rhombic symmetry of the host lattice. Because of the strong spin-orbit interaction in the "heavy" activator ion, the electronic transitions proceed to the mixed states $|B\rangle$: $-\nu|^1P_1\rangle + \mu|^3P_1\rangle$; $|C\rangle : \mu|^1P_1\rangle + \nu|^3P_1\rangle$; $|A\rangle$: $|^1S_0\rangle$.

In compliance with Bethe's theory /17/ for S^2-ions in a field of rhombic symmetry three absorption bands can be expected in the longer wavelength region, which correspond to the A band in alkali halides and are due to the transition $|A\rangle \rightarrow |B\rangle$. Such three bands can actually be observed in the spectrum studied with polarized light (Fig. 2) at 255, 259, and 292 nm. It must be noted that in $BaCl_2$:Pb the splitting is more strongly expressed than in the Pb-activated alkali halides (with KCl even separate A bands have not been formed). For the moment, conclusions cannot be drawn about the nature of the 326 nm band (Fig. 2). It is logical to connect the most intensive group of bands in the region of 200 to 220 nm and 217 nm with transitions to components of the $|C\rangle$ state. They are similar to the C bands in the alkali halides. We assume that below 200 nm there is another component of the same group of bands.

Although it is generally accepted that the intensity of the B band increases with increasing temperature, there are experimental data for Pb^{2+} /9, 10/ which show that this does not hold for all the components of the B band observed at low temperature. Thus one can assume that the bands at 235 nm and in the region of 225 to 230 nm are connected with a transition to the state. Additional investigations are necessary to describe the exact electronic structure of the absorption spectra of $BaCl_2$:Pb.

References

/1/ J.H. SCHULMAN, C.C. KLICK, and R.J. GINTHER, J. Opt. Soc. Amer. 40, 854 (1950).

/2/ Y. KAIFU, J. Phys. Soc. Japan 16, 1605 (1961).

/3/ A. FUKUDA, Sci. Light (Japan) 13, 3 (1964).

/4/ S.G. ZAZUBOVICH and T.A. KUKETAEV, Trudy Inst. Fiz. i Astron., Akad. Nauk Est. SSR 31, 190 (1966).

/5/ J.A. PARFIANOVICH, E.J. SHURALEVA, P.A. BABIN, and P.S. IVAKCHNENKO, Izv. Akad. Nauk SSSR, Ser. fiz. 29, 427 (1965).

/6/ R. ONAKA and A. FUKUDA, J. Phys. Soc. Japan 20, 466 (1965).

/7/ P. KOEZE and J. VOLGER, Physica 37, 467 (1967).

/8/ J. DRYDEN and G.G. HARVEY, J. Phys. C 2, 603 (1969).

/9/ S.B.S. SASTRY, V. VISWANATHAN, and C. RAMASASTRY, J. Phys. Soc. Japan 35, 508 (1973).

/10/ T. TSUBOI, Y. NAKAI, K. OYAMA, and P.W.M. JACOBS, Phys. Rev. B 8, 1968 (1973).

/11/ J.L. PASCUAL, J.M. CABRERA, and F. AGULLO-LOPEZ, Solid State Commun. 19, 917 (1976).

/12/ L. MĂRCULESCU and C. GHIŢĂ, phys. stat. sol. (b) 75, 375 (1976).

/13/ N.E. LUSHCHIK, Trudy Inst. Fiz. i Astron., Akad. Nauk Est. SSR 7, 119 (1958).

/14/ C. BETTINALI and G. FERRARESSO, Z. phys. Chem. (N.F.) 67, 1 (1969).

/15/ V. TABAKOVA, and K. KYNEV, Z. Naturf. 35a, 308 (1980).

/16/ D. GOCHEV and V. TABAKOVA, Proc. IX. Nat. Conf. Molec. Spectr. 1980, Ann. Higher Inst. Mech. Electr. Engng. 28, 73 (1981).

/17/ H. BETHE, Ann. Phys. 3, 133 (1929).

(Received December 5, 1980)

phys. stat. sol. (a) 65, K79 (1981)

Subject classification: 14.4.1 and 18.4; 22.8.1

Solid State Spectroscopy Laboratory,
Institute of Physics, A. Mickiewicz University, Poznań[1] (a) and
Institute of Nuclear Research, Świerk[2] (b)

Magnetic Ordering in $Pb(Fe_{0.5}Nb_{0.5})O_3$ at 4.2 K

By

J. PIETRZAK (a), A. MARYANOWSKA (a), and J. LECIEJEWICZ (b)

Perovskite type ferroelectrics containing Pb ions attracted large interest in past /1/. Among them lead niobate substituted by iron ions was studied by neutron diffraction at LNT /2/. As an extension of an EPR investigation /3/ we have performed a neutron diffraction study of $Pb(Fe_{0.5}Nb_{0.5})O_3$ at LHT. Apart from this, susceptibility measurements were carried out from LNT to RT.

The sample was obtained by sintering of stoichiometric amounts of PbO, Fe_2O_3, and Nb_2O_5 (analytical grade) in platinum crucible. The mixture was heated up to 1070 K and left at this temperature for 3 h. The final heating followed at 1370 K for 1/2 h.

Magnetic susceptibility measurements on polycrystalline $Pb(Fe_{0.5}Nb_{0.5})O_3$ were performed with a magnetic balance, based on the Faraday method. The magnetic susceptibility value for erbium oxide was used as a standard. The balance for measuring χ_m over considerable ranges of temperature was achieved by letting the samples hang in suitable cryostats. The temperature could be held with a stability of ± 0.1 K in the range from 77 to 300 K. A magnetic field of 1 T was applied to the sample and no dependence of χ_m on B was observed at room temperature.

Measurements of the static susceptibility χ_m were performed as a function of temperature between 77 and 300 K. The results are shown in Fig. 1. At (155 ± 2)K the magnetic susceptibility shows a pronounced anomaly. Below this temperature it increases with decreasing temperature. Above 155 K the susceptibility depends on temperature in accordance to the Curie-Weiss law $\chi_m = C/(T - \Theta)$ where C is the Curie constant. The Curie-Weiss temperature Θ amounts to -118 K.

---

1) 60-780 Poznań, Polska.

2) 05-400 Świerk-Otwock, Polska.

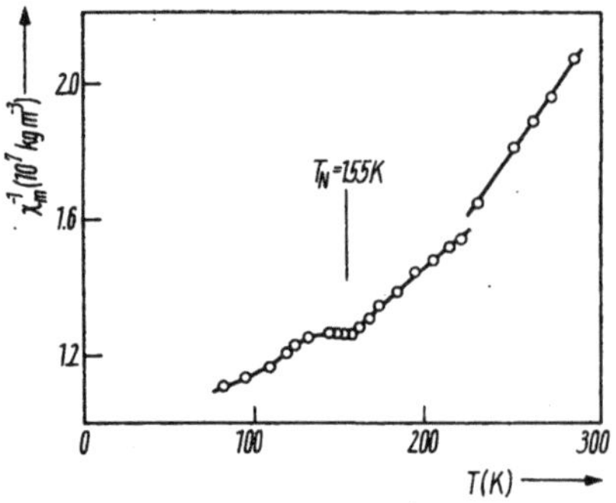

Fig. 1. The reciprocal of magnetic susceptibility of $Pb(Fe_{0.5}Nb_{0.5})O_3$ as a function of temperature; M(1/2, 1/2, 1/2)

For higher temperatures (T > 220 K) a dependence of χ_m on temperature is observed in form of the Curie law $\chi_m = C/T$, C amounts to $4.4 \times 10^{-6} m^3 K\ mol^{-1}$. The number of Bohr magnetons, μ_B, per ion is deduced from the Langevin equation. This quantity is known as

$$\mu_{eff} = 2.828(C)^{1/2} \quad \text{Bohr magnetons}$$

and its value as obtained by us, $\mu_{eff} = 5.93$, agrees well with that reported in /4/.

Neutron diffraction patterns at LHT and RT were obtained on the DN-500 neutron diffractometer at the reactor EWA in Świerk. A monochromatic beam of $\lambda = (1.324 \pm 0.001) \times 10^{-10}$ m was used. The sample in a vanadium container was placed in a liquid helium cryostat.

Six nuclear reflections were observed in the Bragg angle range from 15° to 50° covered in the experiment. All of them were indexed on a perovskite type unit cell with a lattice constant of 4.016×10^{-10} m. In fact, X-ray results reported previously /5/ show that the unit cell of $Pb(Fe_{0.5}Nb_{0.5})O_3$ is rhombohedral (space group R3) at temperatures below 387 K with the following lattice parameters: $a_o = 4.016 \times 10^{-10}$ m and $\alpha = 89^o50'$.

The space group is R3 /2/:

with

1 Pb	in	1(a):	x,	x,	x;		x = 0
1/2Fe+1/2Nb at random	in	1(a):	u,	u,	u;		
3 O	in	3(b):	x_o,	y_o,	z_o;	y_o, z_o, x_o;	
			z_o,	x_o,	y_o.		

The basic assumption in this model is a random distribution of Fe^{3+} and Nb^{5+} ions in 1(a) site. It turns out to be reasonable, since the R factor for the six reflections measured at LHT is as low as 6.6% if we take in addition $u = 0.5_o$,

Table 1

Observed and calculated intensities for $Pb(Fe_{0.5}Nb_{0.5})O_3$ at 4.2 K

hkl	I_{obs}	I_{calc}
M(1/2 1/2 1/2)	0.61	0.61
(1 1 0)	17.74	16.99
(1 1 1)	24.28	27.38
(2 0 0)	71.98	73.92
(2 1 0)	5.89	5.30
(2 1 1)	36.72	33.98

$x_o = 0.5_o$, $y_o = 0.5_o$, $z_o = 0$. This follows from Table 1. Neutron scattering lengths for lead, iron, niobium, and oxygen were taken as 0.94, 0.95, 0.71, and 0.580 (in 10^{-14} m), respectively /6/.

Apart from the above six nuclear peaks, the LHT pattern reveals only one, rather small in intensity, reflection of magnetic origin which was readily indexed on a unit cell doublet in three directions as compared to the chemical one. Its indices are M(1/2, 1/2, 1/2) in terms of the chemical unit cell or M(1 1 1) when indexing is performed on the magnetic cell. The intensities listed in Table 1 are set on absolute scale using the scaling factor for the chemical unit cell.

The magnetic unit cell is rhombohedral with $a = 2a_o$ and $\alpha = 89^\circ 50'$. The presence of the magnetic peak M(1 1 1) is an indication that the magnetic moments of iron are aligned normally to the [1 1 1] direction /7/. Due to known limitations of the neutron diffraction method for powdered crystals it was impossible to deduce uniquely moment directions within the planes normal to the unique axis i.e. [1 1 1] in this case /7/. The coupling of moments within the (1 1 1) planes is ferromagnetic but it is antiferromagnetic between the adjacent planes. The magnetic ordering scheme observed in $Pb(Fe_{0.5}Nb_{0.5})O_3$ is quite common among perovskite type magnetics and is commonly denoted as G-type /8/.

The magnitude of magnetic moment per Fe^{3+} ion at LHT was obtained from the intensity of the M(1/2, 1/2, 1/2) reflection after setting it on absolute

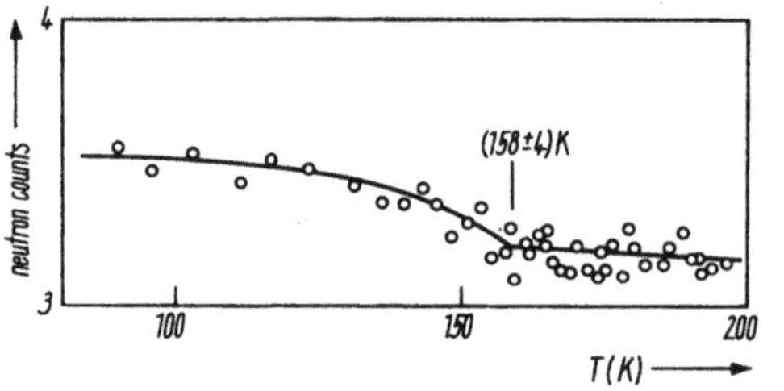

Fig. 2. Temperature dependence of the magnetic peak intensity

scale by use of a scaling factor derived from comparison of the observed and calculated nuclear intensities. The moment $\mu = 2\,S$ amounts to (2.55 ± 0.1) Bohr magnetons in ordered state at 4.2 K.

Measurements of the height of M(1/2, 1/2, 1/2) reflection as the cryostat was heated up slowly gave the Néel point at (158 ± 4)K (Fig. 2) in reasonable agreement with that one determined from susceptibility measurements (155 K).

Our neutron diffraction study confirms the presence of G-type of magnetic ordering at 4.2 K with a magnetic moment of (2.55 ± 0.10) Bohr magnetons per Fe^{3+} ion. As reported in /2/ μ is (0.80 ± 0.16) Bohr magnetons at LNT.

Random distribution of Fe^{3+} and Nb^{5+} ions in 1(a) site undoubtedly brings about weakening of magnetic interactions between neighbouring Fe^{3+} ions, but still they are strong enough to maintain the Néel point as high as 158 K. The magnetic ordering scheme as deduced from neutron data indicates also that the antiferromagnetic coupling between ions belonging to the adjacent planes prevails and makes the Curie-Weiss temperature Θ negative rather than positive.

References

/1/ G.A. SMOLENSKII and N.N. KRAINIK, Uspekhi fiz. Nauk 97, 657 (1969).

/2/ G.M. DRABKIN, E.I. MALTSEV, and V.P. PLAKHTII, Fiz. tverd. Tela 7, 1241 (1965).

/3/ A.I. KASHLINSKII, YU.N. VENEVTSEV, and V.I. CHECHERNIKOV, Proc. Internat. Meeting Ferroelectricity, Praha, Vol. 2, 1966 (p. 373).

/4/ V.A. BOKOV, J.E. MYLNIKOVA, and G.A. SMOLENSKII, Zh. eksper. teor. Fiz. 42, 643 (1962).

/5/ G.A. PLATONOV, L.A. DROBYSHEV, YU.YA. TOMASHPOLSKII, and YU.N. VENEVTSEV, Kristallografiya 14, 800 (1969).

/6/ Neutron Diffraction Commission IU Cryst., Acta cryst. A 28, 357 (1972).

/7/ G. SHIRANE, Acta cryst. 12, 282 (1959).

/8/ E.O. WOLLAN and W.C. KOEHLER, Phys. Rev. 100, 545 (1955).

(Received April 6, 1981)

phys. stat. sol. (a) 65, K83 (1981)

Subject classification: 20.3; 22.4

Department of Electrical Engineering (a) and Electronics (b), Kanazawa Institute of Technology[1)], and Department of Electrical Engineering, University of Osaka Prefecture (c)

Temperature Dependence of Electro- and Photoluminescence in ZnO Single Crystals

By

S. TAKATA (a), T. MINAMI (a), H. NANTO (b), and T. KAWAMURA (c)

Electroluminescence has been observed in ZnO single crystals, an n-type material with a direct band gap of 3.2 eV, when it was incorporated into a metal-insulator-semiconductor (MIS) /1 to 3/ and a metal-semiconductor (MS) diode /4/. In our previous publications we have reported that the electro (EL)- and photoluminescence (PL) spectra of non-intentionally doped ZnO single crystals at room temperature (RT) consist of three emissions: near ultraviolet (UV) emission, blue-green (B-G) emission, and near infrared (IR) emission. The nature of these emissions except for the near UV emission is still poorly understood. In this note we report results of EL (MS diode) and PL studies in the temperature range from liquid nitrogen temperature (LNT) to RT.

The samples used were prepared from a hydrothermally grown ZnO single crystal which has a resistivity of about 10 Ωcm and an electron mobility of about 100 cm^2/Vs. The MS diode preparation has already been described in our previous paper /4/. To measure EL, a pulse voltage was applied to the MS diodes. The pulse width and repetition rate were 300 μs and 100 Hz, respectively. To measure PL, the ZnO single crystal was excited with a pulsed nitrogen laser. The details of the instruments used have also been reported earlier /4/.

The EL and PL spectra of the same ZnO single crystal are shown in Fig. 1a and b, respectively. Compared to RT curves drawn by solid lines, curves drawn by dashed lines at LNT are sharpened. The EL spectrum which exhibits the whitish luminescence covering the visible region consists of two dominant emissions; (i) the B-G emission ($\approx$ 5000 Å) and (ii) the near IR emission ($\approx$ 9500 Å). The PL spectrum except for the UV edge emission has only an emission around 5000 Å which coincides with the B-G emission of the

---

1) 7-1, Oogigaoka, Nonoichi-machi, P.O. Kanazawa-South, Ishikawa, Japan.

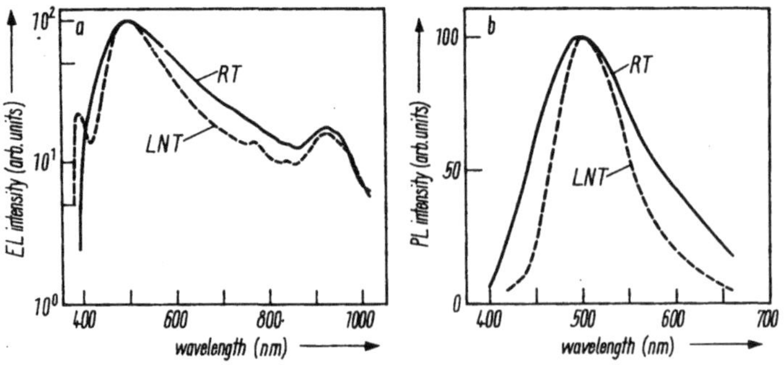

Fig. 1. Normalized EL and PL spectra at RT (solid line) and LNT (dashed line). a) EL spectra for In-ZnO-Al (In driven positive) MS diode for current of 150 mA. b) PL spectra for excitation power of about 25 kW/cm^2

EL spectrum. The peak photon energies of the B-G and near IR emissions in EL and the B-G emission in PL are little dependent on the temperature in the range from LNT to RT. The gradual decrease of the halfwidth as the temperature decreases is found in the B-G band in both EL and PL. The temperature dependences of the EL and PL intensities from LNT to RT for these emissions are shown in Fig. 2. In the cases of the B-G emission the change of the EL intensity is not simple compared with that in the PL intensity. Especially, for the EL intensity, a strong peak is seen at about 230 K. However, in the higher temperature region the B-G emissions in both EL and PL are found to be thermally quenched with the same activation energy of about 0.16 eV. The above results strongly suggest that the origin of the B-G emission is similar in both EL and PL. While, in the case of the near IR emission the change of the EL intensity is simple. The activation energy of the thermal quenching in the case is about 0.03 eV in the higher temperature region.

We have already suggested that the near UV emission in the MIS diode is attributed to excitonic recombination with the simultaneous emission of two and three LO phonons /2/. For EL an the MS diode a weak near UV emission was also observed, well resolved in the case at LNT. Although the center responsible for the B-G emission has been extensively studied, it is still subject of much controversy. For the present, two presumable interpretations for the B-G emission exist: (i) One possibility for the center is a stoichiometric zinc which may be incorporated either in the interstitial positions or at the normal lattice sites with the formation of an equivalent amount of oxygen vacancies /5, 6/.

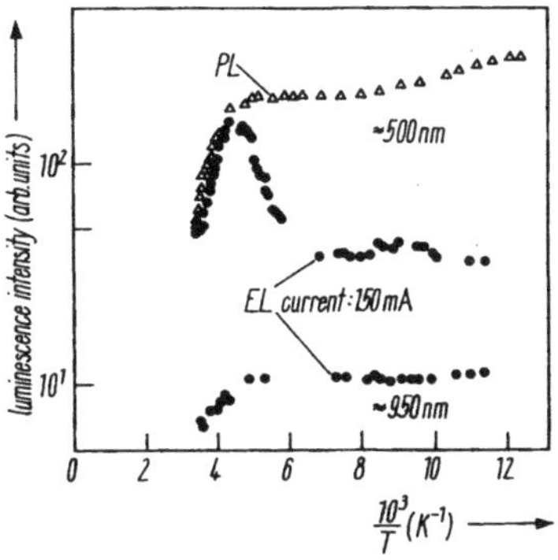

Fig. 2. Temperature dependence of the EL and PL intensities under the same experimental conditions as shown in Fig. 1

However, the detailed transition mechanism has been not fully understood. (ii) The other possibility for the center is a substitutional Cu impurity at a regular Zn lattice site. Dingle has suggested a transition mechanism that a hole recombines with an electron bound at a substitutional Cu^{2+} ion /7/. As for the B-G emission in the present sample, we shall believe that the possibility of (i) as described above is right. Recently we have suggested the possibility that the B-G emission is attributed to the transition within the self-activated (SA) center /8/. That is, some associated centers are formed by the doubly ionized zinc vacancy defect V_{Zn}^{2-} and the ionized interstitial zinc donor Zn_{in}^{+} of one and/or two at nearest neighbor interstitial sites. For the emission process, one can ascribe the B-G emission to the transition between an electron in an ionized interstitial zinc donor and a hole trapped at a doubly ionized zinc vacancy /9/. Thus the activation energy of 0.16 eV mentioned above is considered to be ascribed to the binding energy of a hole which is trapped at a doubly ionized zinc vacancy. As for EL in MS diodes we have already suggested that the excitation mechanism is due to impact ionization by free electrons which are accelerated in the high field region /4/. Further, the mechanism is supported by the exponential voltage relation ($L \propto \exp(-bV^{-1/2})$) as shown in Fig. 3. For the EL case, it is possible to consider that in the low temperature region the interstitial zinc donor in the SA center is impact-ionized to the conduction band and that at temperature above 160 K it is filled by an electron which is thermally relaxed from the conduction band. Therefore the difference of the temperature dependence between the EL and PL intensities is attributed to the difference of the excitation processes in the SA center. On the other hand, the near IR emission is nearly always observed for the EL in ZnO single crystal MS diodes, but

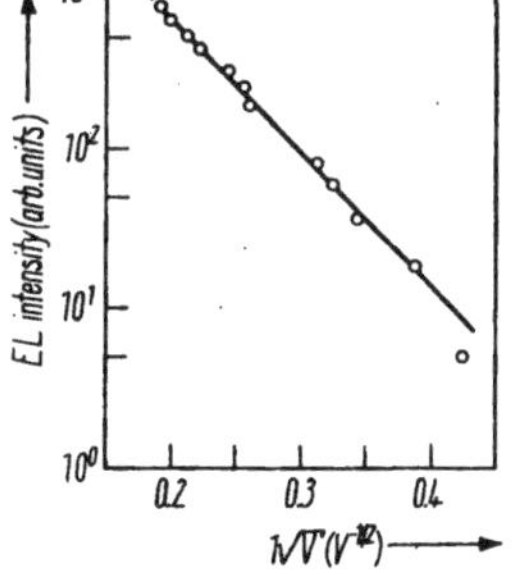

Fig. 3. EL intensity(L)-voltage ($V^{-1/2}$) characteristics of In-ZnO-Al (In driven positive) MS diode at RT

the transition mechanism is not yet known.

From the temperature dependence of EL and PL intensities the B-G emission is considered to be ascribed to a transition within the SA center. The difference of the temperature dependence between EL and PL is responsible for the excitation process.

References

/1/ B.W. THOMAS and D. WALSH, Electronics Letters 9, 362 (1973).
/2/ T. MINAMI, M. TANIGAWA, M. YAMANISHI, and T. KAWAMURA, Japan. J. appl. Phys. 13, 1475 (1974).
/3/ A. SHIMIZU, M. KANBARA, M. HADA, and M. KASUGA, Japan. J. appl. Phys. 17, 1435 (1978).
/4/ T. MINAMI, S. TAKATA, M. YAMANISHI, and T. KAWAMURA, Japan. J. appl. Phys. 18, 1617 (1979).
/5/ See, for example, F. VAN CRAEYNEST, W. MAENHOUT-VAN DER VORST, and W. DEKEYSER, phys. stat. sol. 8, 841 (1965).
/6/ S. BHUSHAN, A.N. PANDEY, and B.R. KAZA, J. Lum. 20, 29 (1979).
/7/ R. DINGLE, Phys. Rev. Letters 23, 579 (1969).
/8/ T. MINAMI, H. NANTO, and S. TAKATA, to be published.
/9/ F. WILLIAMS, J. Lum. 7, 35 (1973).

(Received January 14, 1981)

phys. stat. sol. (a) 65, K87 (1981)

Subject classification: 12.1; 21.1.1

Physics Department, University of Timişoara[1] (a)
and Institute of Physics and Chemistry of Metals,
Silesian University, Katowice[2] (b)

The Relaxation Times and the Relaxed Elastic Modulus of Prestressed B.C.C. Metals with Interstitial Impurities

By

O.F.G. ACZEL (a), G. HANECZOK (b), and J.W. MOROŃ (b)

In a previous paper /1/ the influence of a static elastic stress on the height and the temperature of the Snoek peak in b.c.c. metals was established. For a static tensile stress superposed on a variable tensile stress, both in the [001] direction of a single crystal, the following equation was found:

$$\left(\frac{\Gamma}{M_I} + \lambda B\right)\sigma + \frac{\dot{\sigma}}{M_I} = \dot{e} + \Gamma e + \text{const} \quad ,$$

all the notations being those from /1/. Comparing this equation with the equation for the standard linear body in its usual form,

$$\sigma + \tau_e \dot{\sigma} = M_R \tau_\sigma \dot{e} + M_R e + \text{const} \quad ,$$

one obtains

$$\tau_\sigma^{-1} = \nu_o \left[2 \exp(-b_1\sigma') + \exp(-b_2\sigma')\right] , \tag{1}$$

$$\tau_e^{-1} = \nu_o(\Gamma + \lambda M_I B) = \nu_o \left[2 \exp(-b_1 \sigma') + \exp(-b_2\sigma') + \frac{2C_o M_I v_o \lambda^2}{kT} \frac{1}{\exp(b_1\sigma') + 2 \exp(b_2\sigma')}\right] . \tag{2}$$

τ_σ exhibits a peak for

$$\sigma_1' = \frac{kT \ln(2\lambda_1/-\lambda_2)}{v_o(\lambda_1 - \lambda_2)} . \tag{3}$$

---

1) Bv. V. Pârvan 4, 1900 Timişoara, Romania.

2) Ul. Bankowa 12, 40-007, Katowice, Poland.

The supplementary term in (2) is small compared with Γ, so that τ_e presents also a peak near σ'_1.

M_I is considered independent of the concentration of the interstitial atoms (I-atoms) /2, 3/, and it is equal to M_I for the metal without I-atoms.

From /4/

$$\frac{M_I}{M_R} = \frac{\tau_\sigma}{\tau_e}$$

results that

$$M_R^{-1} M_I = 1 + \frac{2C_o M_I v_o \lambda^2}{kT} \frac{1}{(\exp b_1\sigma' + 2 \exp b_2\sigma')\left[2 \exp(-b_1\sigma') + \exp(-b_2\sigma')\right]}. \tag{4}$$

M_R exhibits a minimum for

$$\sigma'_2 = \frac{kT \ln 2}{v_o(\lambda_1 - \lambda_2)} \tag{5}$$

which coincides evidently with the position of the Q^{-1} peak in /1/.

The variation with temperature of the relaxation times does not present any extreme values, but only a strong decrease when the temperature increases. But M_R has a minimum for

$$T_m = \frac{v_o \lambda}{k \ln 2} . \tag{6}$$

The extreme values of M_R and Q^{-1} as a function of σ' can be explained by the fact that the equilibrium concentration of the I-atoms,

$$\bar{C}_z = C_o \frac{1}{1 + 2 \exp\left(- \frac{v_o \lambda}{kT} \sigma'\right)} \tag{7}$$

has a bending point for $\sigma' = \sigma'_2$.

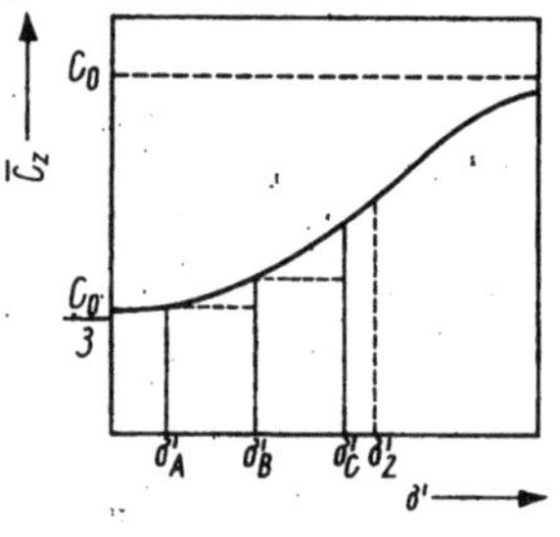

Fig. 1. Equilibrium concentration of interstitial atoms in the z-sites versus static tensile stress

From Fig. 1 it is seen that for $\sigma' < \sigma'_2$, an increase of the stress with $\sigma'_B - \sigma'_A$ leads to an increase of the equilibrium concentration, which is smaller than the increase for $\sigma'_C - \sigma'_B = \sigma'_B - \sigma'_A$, when $\sigma'_B > \sigma'_A$. Thus in the second case the relaxed deformation will be greater than in the first one, because the additional relaxation is due to the I-atoms which arrive supplementary in the z-sites. A greater relaxed deformation for the same $\Delta\sigma'$ is equivalent to a smaller modulus M_R. When $\sigma' > \sigma'_2$, the situation is inversed and M_R begins to increase when in the x- and y-sites the number of I-atoms becomes too small to produce a considerable rise of the relaxation by jumping in the z-sites.

As in /1/, the α-Fe-C solid solution will be considered for which $\lambda_1 = 0.86$, $\lambda_2 = -0.08$ /5/, $E_o = 1.333 \times 10^{-19}$ J /6/, $M_I = 1.3 \times 10^{11}$ N/m^2 /3/, $v_o = 1.2 \times 10^{-29}$ m^3. We choose the temperature of the Snoek peak at a frequency of 1 Hz: 313 K, and obtain $\sigma'_1 = 1.175 \times 10^9$ N/m^2, which exceeds the limit of elasticity. The relative variation of τ_σ at $\sigma' = 10^7$ N/m^2 compared with $\sigma' = 0$ is about 1.5%. For the same conditions we find that $\sigma'_2 = 2.65 \times 10^8$ N/m^2, while the relative variation of the relaxed modulus for $C_o = 4.65 \times 10^{-4}$ (= 0.01 wt%) and for $\sigma' = 10^7$ N/m^2 is -0.03%. Even for the maximum solubility of carbon in α-Fe (C_o = 0.02 wt%), the relative variation is only -0.05%.

Regarding the variation of τ_σ and τ_e as a function of temperature, only a small interval of temperature is of interest, i.e. that corresponding to the extreme frequencies which are used in a torsional pendulum: 292 K (0.1 Hz), and 339 K (10 Hz). The differences between the variations of τ_σ and τ_e are small and for both one obtains

$$\frac{\Delta\tau_\sigma}{\tau_\sigma} \approx \frac{\Delta\tau_e}{\tau_e} = 1.62\% \text{ at } 292 \text{ K}, \quad 1.53\% \text{ at } 313 \text{ K}, \quad 1.41\% \text{ at } 339 \text{ K}.$$

References

/1/ O.F.G. ACZEL, G. HANECZOK, and J.W. MOROŃ, phys. stat. sol. (a) 58, 297 (1980).

/2/ D. POLDER, Philips Res. Rep. 1, 5 (1945).

/3/ W. PITSCH and K. LÜCKE, Z. Phys. 139, 413 (1954).

/4/ A.S. NOWICK and B.S. BERRY, Anelastic Relaxation in Crystalline Solids, Academic Press, New York 1972 (p. 50).

/5/ J.C. SWARTZ, J.W. SHILLING, and A.J. SCHWOEBLE, Acta metall. 16, 1359 (1968).

/6/ A.E. LORD and D.N. BESHERS, Acta metall. 14, 1659 (1966).

(Received March 30, 1981)

phys. stat. sol. (a) 65, K91 (1981)

Subject classification: 18.2 and 18.4; 21.1.1

School of Engineering and Applied Sciences,
The University of Sussex, Falmer, Brighton[1]

The Saturation Magnetization of Some Substituted Fe-Si-B Metallic Glasses

By
K. HOSELITZ

Amorphous Fe-Si-B alloys have been much studied recently /1 to 3/. These alloys have interesting technical properties and could have important practical applications.

One important property, both theoretically and for their practical use, is the saturation magnetization and its variation with temperature, and this also has been investigated /1, 4, 5/.

Previous studies were much concerned with the effects of variations of the metalloid content, but the present work deals with the substitution of iron by other elements of the 3d transition metal group.

Alloys in which iron was partly replaced by V, Cr, Mn, Cu, and Zn were made by spin quenching, resulting in ribbons 1.5 mm wide by about 30 μm thick. All alloys were examined by X-ray diffractometry and no trace of a crystalline phase could be detected in any of them. Hysteresis loops were also examined, and in all ribbons the value of the coercivity was low (about 50 mOe), a very sensitive test for the absence of incipient crystallinity.

The saturation magnetization, σ, was measured to an accuracy of $\pm 0.75\%$ at 77 and 293 K by a ballistic method /6/.

Table 1 gives the composition of the first series of alloys and the results of the saturation measurements.

It can be seen that for alloys with Cu (and Zn)[2] the value of σ_{293} actually increases, whilst for the other substitutions the parameter decreases quite sharply.

The measurements allowed an extrapolation to 0 K, assuming a $(T/\Theta)^{3/2}$ law, which, according to previous authors /5/ applies in alloys of this type up

1) Brighton, Sussex BNI 9QT, England.

2) The Zn substituted results are very uncertain because of the difficulty in preparing alloys in which the Zn content could be estimated.

Table 1

1 alloy	2 composition*) (at%)	3 σ_{293} (e.m.u./g)	4 σ_{77} (e.m.u./g)	5 σ_0 (e.m.u./g)	6**) n_{μ_B} (per Fe atom)	7 Θ_c(est) (K)
0	80Fe	166	186	188	2.0	690
Cu1/2	79.5Fe 0.5Cu	173	189	191	2.04	825
Cu1	79Fe 1Cu	170	197	200	2.15	585
Mn2	78Fe 2Mn	157	183	187	2.09	565
Mn4	76Fe 4Mn	145	176	181	2.02	490
Cr1	79Fe 1Cr	164	186	190	2.02	635
Cr2	78Fe 2Cr	150	177	182	1.98	535
Cr4	76Fe 4Cr	136	162	166	1.85	530
V1	79Fe 1V	163	182	185	1.99	690
V2	78Fe 2V	156	182	186	2.02	565
V4	76Fe 4V	137	165	170	1.89	495
Zn1/2	79.5Fe 0.5Zn	177	195	197	2.04	740

*) All alloys had 5% Si and 15% B.
**) Number of Bohr magnetons per iron atom.

to room temperature.

The table shows the extrapolated moment σ_0 (column 5) and thus the value of $n_{\mu_B, Fe}$ (Bohr magnetons per iron atom) could be calculated (column 6).

The moment of the iron atom increases with Cu (and Zn) substitutions, seems to remain constant for small Mn contents, and decreases for the other metals.

This is the reverse of what would be expected if the magnetism was entirely due to the imbalance of spins in the 3d shell.

It does seem that the theory by Mizoguchi /7/ is not applicable here; however, the exact variation of the magnetic moment of the iron atom cannot yet be determined, although it seems that Mn, Cr, and V in quantities in excess of 1% cause antiferromagnetism of the iron.

Table 2

1	2				3	4	5	6	7
alloy	composition (at%)				σ_{293}	σ_{77}	σ_o	$n\mu_B$	$\Theta_{c(est)}$
	Fe	Si	B	Cu	(e.m.u./g)	(e.m.u./g)	(e.m.u./g)	(per Fe atom)	(K)
C0	80	0	20	0	172	190	191	1.99	780
CCu.4	79.6	0	20	0.4	173	192	194	2.03	720
CCu.8	79.2	0	20	0.8	175	194	196	2.07	720
D0	82	5	13	0	171	200	205	2.16	550
DCu.4	81.6	5	13	0.4	171	196	200	2.12	600
DCu.8	81.2	5	13	0.8	173	200	203	2.17	585
E0	81	2.5	16.5	0	174	197	200	2.1	645
ECu.4	80.6	2.5	16.5	0.4	176	196	199	2.1	690
ECu.8	80.2	2.5	16.5	0.8	173	197	200	2.12	615

It was of interest to examine the trend of the Curie temperature Θ_c as estimated from these measurements. The saturation should vary as

$$\sigma_T = \sigma_o(1 - \beta\,(T/\Theta)^{3/2}) \ .$$

Taking the value of $\beta = 0.43$ as found for $Fe_{80}B_{15}Si_5$ we can estimate the Curie temperature $\Theta_{c(est)}$ and this is shown in the tables (column 7). Here it is most surprising to find that for Cu1/2, and Zn1/2, $\Theta_{c(est)}$ is higher than that of alloy 0. Now this method of estimating Θ_c is not reliable but the slow fall of σ between 77 and 203 K for some of these alloys does indicate a higher Θ_c. $\Theta_{c(est)}$ is very sensitive to variation in β, a constant which is known to vary for different alloys /see e.g. 4, 8, 9/. Within a range of $\beta = 0.43 \pm 0.03$, the estimated values of Θ_c could vary by about $\pm 40^o$. The trends shown in Table 1 would be preserved.

The results so far reported show strong evidence that small substitutions of 3d elements, especially Cu, for iron in amorphous iron-boron-silicon alloys cause an increase in σ at 293 K. This was confirmed by experiments with alloys of other compositions, as shown in Table 2. Small substitutions of Cu at worst do not reduce the value of the absolute saturation, they decrease the low temperature slope of the σ-T curve, thus leading to higher values of saturation magnetization at room temperature and probably higher Curie temperatures.

This work was supported by the Science Research Council.

References

/1/ K. HOSELITZ, J. Magnetism magnetic Mater. 20, 201 (1980).
/2/ F.E. LUBORSKY, J.J. BECKER, J.L. WALTER, and H.H. LIEBERMANN, IEEE Trans. Magnetics 15, 1146 (1979).
/3/ F.E. LUBORSKY, P.G. FRISCHMANN, and L.A. JOHNSON, J. Magnetism magnetic Mater. 19, 130 (1980).
/4/ F.E. LUBORSKY, J.L. WALTER, H.H. LIEBERMANN, and E.P. WOHLFARTH, J. Magnetism magnetic Mater. 15/18, Part 3, 1357 (1980).
/5/ F.E. LUBORSKY, J.L. WALTER, and E.P. WOHLFARTH, J. Phys. F 10, 959 (1980).
/6/ A.W. SIMPSON and R.D. BRAMBLEY, J. Phys. E 3, 472 (1970).
/7/ T. MIZOGUCHI, Sci. Rep. Res. Inst. Tôhoku-Univ., Suppl. Apl1 7, 1978.
/8/ S. HATTA and T. EGAMI, J. appl. Phys. 50, 1589 (1979).
/9/ N. KAZAMA, T. MITERA, and T. MASUMOTO, Rapidly Quenched Metals III, Vol. 2, The Metals Soc., London 1978 (p. 164).

(Received April 16,1981)

phys. stat. sol. (a) 65, K95 (1981)
Subject classification: 14.3.3; 16; 22.3

Institute of Physics, Technical University of Wrocław[1)]

Minority-Carrier Diffusion Length in Zn_3P_2

By
K. NAUKA and J. MISIEWICZ

Zinc phosphide Zn_3P_2 was, until quite recently, a relatively poorly known semiconducting compound of the $A^{II}B^{V}$ type. Not long ago, a Zn_3P_2 based infrared photoconvertor was designed which enhanced our knowledge of its photoelectric parameters /1/, such as diffusion length of minority carriers. This problem was studied in /2/, and the aim of this paper is to present the results of diffusion-length measurements obtained in another way and using other samples.

It was shown in /3/ that the parameter I_{sc} (short-circuit current) for an illuminated p-n junction is given by

$$I_{sc} = \frac{e\,\Phi_0 \alpha L(1-R)}{\alpha^2 L^2 - 1}\left[\alpha L \exp(-\alpha t) - \exp(-\alpha t)\coth\left(\frac{d-t}{L}\right) + \right.$$

$$\left. + \exp(-\alpha d)\operatorname{cosech}\left(\frac{d-t}{L}\right)\right] \quad , \qquad (1)$$

where t is the sample thickness, d the junction width (i.e. width of the non-homogeneity region), L the minority-carrier diffusion length in the base region, Φ_0 the photon flux the irradiated sample, R the reflectivity coefficient, e the elementary charge, and α the absorption coefficient of the semiconductor. Taking into account the assumptions discussed in /4/, i.e. $d \gg 2L$ and $\alpha d \gg 1$ (this means that the sample should be relatively thick), the term $\exp(-\alpha d)\operatorname{cosech}((d-t)/L)$ can be ignored and $\coth((d-t)/L) \approx 1$. Now, expression (1) can be rewritten as

$$I_{sc} = \frac{e\Phi_0 L(1-R)}{L + \alpha^{-1}} \exp(-\alpha t) \quad . \qquad (2)$$

Using the well-known formula for the photon flux Φ which flows out of the sample, one can write

$$\Phi = K(L + \alpha^{-1}) \quad , \qquad (3)$$

---

1) St. Wyspiańskiego 27, 50-370 Wrocław, Polska.

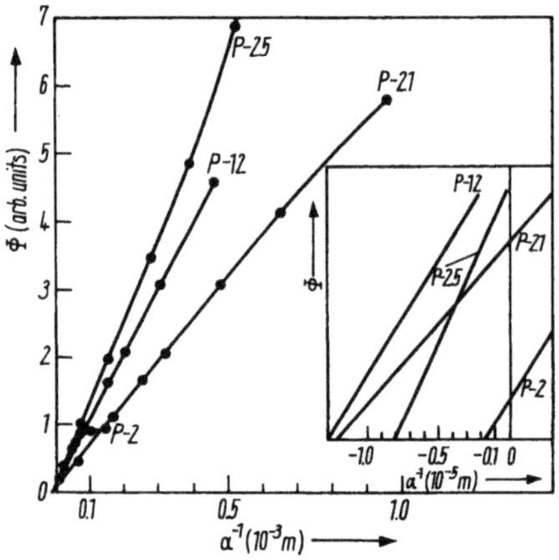

Fig. 1. Dependence of output photon flux Φ versus reciprocal of the absorption coefficient α. Inset shows the lowest part of plots

where K = const, given by physical constants and measurement conditions, see /4, 5/. Expression (3) can also be written for a metal-semiconductor junction (see /5/), Φ versus α^{-1} obeys a linear dependence, which makes it possible to determine the diffusion length of minority carriers in the base region of a metal-semiconductor junction. The linear approximation of the junction $\Phi(\alpha^{-1})$ to the point for which $\Phi = 0$ makes it possible to estimate the value of α^{-1} which satisfies the condition $|\alpha^{-1}| = L$ (see Fig. 1).

This method was used to determine the diffusion length of minority carriers in Si and GaSe, see /4, 5/, respectively.

In this work the diffusion length of electrons in the bulk region of a metal-p-type Zn_3P_2 structure was determined on the basis of the above described method. Sample preparations, their properties, and details of measurements are described in /6, 7/. Spectral characteristics of the light source were also taken into account in these calculations. The values of the absorption and transmission coefficients taken from spectral measurements of four Zn_3P_2 samples at 300 K were used to estimate the values of L. The estimation results are shown in Fig. 1. The obtained values are listed in Table 1.

Table 1

sample No.	L (μm)
P-2	1.7
P-25	8
P-21	12
P-12	13

As it is easy to show, the value of L for sample P-2 is about one order of magnitude lower than for the other samples. Because the P-2 sample was one of the first specimens made and measured in our laboratory and it was a relatively non-homogeneous sample, then the L values of 8 to 13 μm are probably more adequate for relatively pure Zn_3P_2 samples. These results are in a good agreement with results obtained by Convers Wyeth and Catalano in paper /2/, in which mean values of L were equal to 5 to 10 μm. In conclusion, one can state that the diffusion length of minority electrons in the Zn_3P_2 single crystals (which corresponds to the bulk region of metal-Zn_3P_2 structures) is of the order of several μm.

The authors are greatly indebted to Doc. Dr. J.M. Pawlikowski for helpful discussions. The financial support of the Institute of Physics of the Polish Academy of Sciences is also acknowledged.

References

/1/ J.M. PAWLIKOWSKI, Infrared Phys. 21 (1981), in the press.

/2/ N. CONVERS WYETH and A. CATALANO, J. appl. Phys. 50, 1403 (1979).

/3/ T.S. MOSS, G.J. BARREL, and B. ELLIS, Semiconductor Optoelectronics, Butterworths, 1973.

/4/ E.D. STOKES and T.L. CHU, Appl. Phys. Letters 30, 425 (1977).

/5/ V. CAPOZZI, M. DI GIULIO, S. GALASSINI, C. MANFREDOTTI, and A. TEPORE, Proc. IV. Internat. Yamada Conf. PCLM, Sendai 1980, Physica B, in the press.

/6/ J.M. PAWLIKOWSKI, N. MIROWSKA, P. BECLA, and F. KROLICKI, Solid State Electronics 23, 755 (1980).

/7/ J.M. PAWLIKOWSKI, J. MISIEWICZ, and N. MIROWSKA, J. Phys. Chem. Solids 40, 1027 (1979).

(Received January 21, 1981)

phys. stat. sol. (a) 65, K99 (1981)

Subject classification: 20.3; 10.2; 22.1.2

Institut für Angewandte Physik der Technischen Universität Clausthal[1] and SFB 126 Göttingen-Clausthal

Photoluminescence in Plastically Twisted Silicon

By

D. GWINNER[2] and R. LABUSCH

We have measured photoluminescence spectra of twisted silicon single crystals, containing a high density of screw dislocations /1/.

Luminescence was excited by a cw YAG laser with an output power of up to 1 W and analyzed with a grating monochromator (f = 220 mm). A cooled photomultiplier (RCA 7102) and a cooled PbS cell (SBRC) were used for photon energies above and below 0.95 eV, respectively.

Fig. 1 shows the results for a twisted and annealed p-type sample. 70 % of the dislocations (density $4x10^9$ cm^{-2}) are pure screws or very close to the screw direction (within 10°). So far, luminescence spectra of dislocated Si reported in the literature were obtained from bent specimens /2, 3/ and from laser annealed specimens with a high dislocation density below the surface /4/. These spectra show four rather sharp lines (D_1 to D_4).

Of these signals only D_1 is found unambiguously in our specimens, however not as a sharp line (1.535 μm). Other peaks much broader than the lines D_1 to

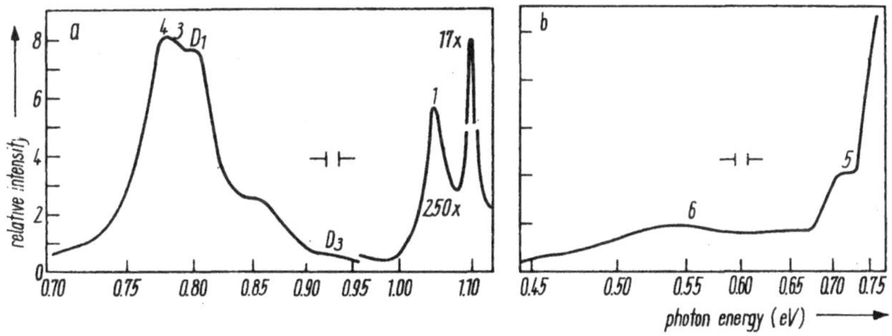

Fig. 1. Photoluminescence of twisted (90', 1000^oC) p-type Si at 12 K, recorded with a) 600 L mm^{-1} gratings and b) 300 L mm^{-1} gratings. The spectra were only corrected for the detector sensivities. Annealing was at 1150 ^oC for 4 h, $N \approx 4x10^9$ cm^{-2}

---

1) Arnold-Sommer Str. 6, D-3392 Clausthal-Zellerfeld, BRD.

2) In partial fulfillment of dissertation.

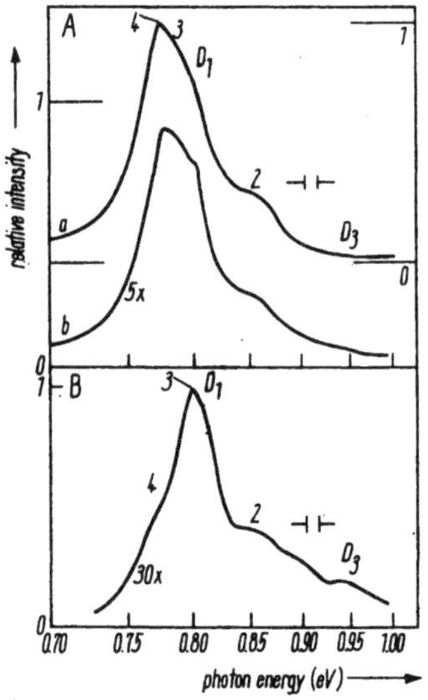

Fig. 2. Photoluminescence of twisted (Si-20) and of compressed (S/3) Si. A) p-Si, 5×10^{14} B/cm^3; (a) twisted 60', 1000 °C, (b) annealed 4 h at 1150 °C; N ≈ 10^9 cm^{-2}. B) n-Si, 8×10^{14} P/cm^3, compressed 10', 900 °C; N 3.5x10^7 cm^{-2}

to D_4 were found instead (1 to 6).

Peak 1 (1.19 μm) occurred also in non-deformed but identically heat treated samples and is therefore ruled out as dislocation induced. Peak 2 is situated at about 1.45 μm; peaks 3 and 4 at 1.56 and 1.59 μm, respectively, can only be resolved in a higher resolution record (not shown here). At lower photon energies two weak signals 5 (1.75 μm) and 6 (2.30 μm) occurred.

Curve a in Fig. 2A shows the spectrum of a twisted specimen that was not subjected to the heat treatment (4 h, 1150 °C) establishing the screw dislocation networks /1/. Curve b in Fig. 2A represents the same specimen after this heat treatment. We conclude that the annealing only lowers the amplitude of all signals, but does not alter the characteristic feature of the spectrum.

The spectrum of the compressed specimen is also given in Fig. 2B. The overall intensity is only 1/30 of that of the as-twisted sample, and peak 4 is further lowered by a factor of two. This extra factor can be removed by a 90 min heat treatment at 1000 °C.

DLTS measurements of compressed Si /5, 6/ yield an electron trap at 0.38 eV below the conduction band, which cannot be annealed out above 900 °C, while all other signals vanished at this temperature.

We therefore suggest that the strong photoluminescence around 0.8 eV (D_1 and/or 3) corresponds to a radiative transition from this trap to the valence band.

Peak 4 which appears only after heating at 1000 °C must be associated either with a dislocation-defect complex or with a particular high temperature established dislocation structure.

Finally we have to explain the absence of the sharp lines at 1.42 μm and 1.53 μm in our spectra. We propose that these lines are associated with the specimen surface in combination with dislocations and appear only if the excitation light has a smaller penetration depth than the YAG laser line. First results from the same specimen with the 647 nm line of a Kr laser as an excitation source seem to confirm this hypothesis /7/.

References

/1/ D. GWINNER and G. PACKEISER, Phil. Mag. A42, 645 (1980).

/2/ N.A. DROZDOV, A.A. PATRIN, and V.D. TKACHEV, Soviet Phys.- J. exper. theor. Phys. 23, 597 (1976).

/3/ N.A. DROZDOV, A.A. PATRIN, and V.D. TKACHEV, phys. stat. sol. (b) 83, K137 (1977).

/4/ R.H. UEBBING, P. WAGNER, H. BAUMGART, and H.J. QUEISSER, Appl. Phys. Letters 37, 1078 (1980).

/5/ J.R. PATEL and L.C. KIMERLING, J. Physique C6, 67 (1979).

/6/ L.C. KIMERLING and J.R. PATEL, Appl. Phys. Letters 34, 73 (1979).

/7/ in co-operation with Physikalisches Institut, Teil 4, Universität Stuttgart, to be published.

(Received April 24, 1981)

Pre-Printed Titles

Papers to be published
in "physica status solidi (a) - applied research"
Vol. 65, No. 2

Original Papers

H. PFEIFFER, Jena
A Simple Approach to Reactive Processes on the Interface during Crystal Growth

S. ACHARYA and O.N. SRIVASTAVA, Varanasi
Electronic Behaviour of SnS_2 Crystals

V.S. BALLANDOVICH, YU.M. TAIROV, and G.N. VIOLINA, Leningrad
Acceptor Boron in α-SiC (6H): Investigation by the Photocapacitance Method

R.P. KHARE and M. KHARE, Pilani
A New Method of Analysis of Traps Taking Part in the Electroluminescence Process

C.M. SRIVASTAVA, S.N. SHRINGI, and M.V. BABU, Bombay
Mössbauer Study of the Low-Temperature Phase of Magnetite

A.V. GORBUNOV, E.M. NADGORNYI, and S.N. VALKOVSKII, Moscow
Laser Pulse Induced Dislocation Structure in Ionic Crystals (I)

B. JENICHEN and R. KÖHLER, Berlin
Experimental and Simulated X-Ray Contrast of Striations for Nearly Plane Waves in the Laue Case

I.A. TALE, Riga
Trap Spectroscopy by the Fractional Glow Technique

T.C. LESLIE and J.W. ALLEN, St. Andrews
Thermal Quenching of Photoluminescence in ZnS:Mn and ZnSe:Mn

B. TUCK and A.J.N. HOUGHTON, Nottingham
Shallow Diffusions of Zinc in GaAs at 700 °C

Papers to be published
in "physica status solidi (b) - basic research"
Vol. 105, No. 2

Original Papers

G. AMBRAZEVIČIUS, G. BABONAS, YU.V. RUD, and A. ŠILEIKA, Vilnius
The Fundamental Absorption Edge of $MgSiP_2$

D.G. KANHERE, R.M. SINGRU, and R. HARTHOORN, Pune and Kanpur
Electron Momentum Distribution and Compton Profiles of Ag

A.K. SAXENA, Sheffield
Non-Γ Deep Levels and the Conduction Band Structure of $Ga_{1-x}Al_xAs$ Alloys

V.K. JAIN, Haryana
Host Spin-Lattice Relaxation Narrowing in $Sm_2Co_3(NO_3)_{12}24H_2O:Mn^{2+}$, Gd^{3+} Single Crystals

R.J. SAYER, R.H. PRINCE, and W.W. DULEY, Toronto
Site Effects on N and O Luminescence in Electron-Irradiated Neon Films

V.M. TALANOV, Novocherkassk
Calculation of Structural Parameters of Spinels

E.P. DOMASHEVSKAYA, L.N. MARSHAKOVA, V.A. TEREKHOV, A.N. LUKIN, YA.A. UGAI, V.E. NEFEDOV, and YA.V. SALYN, Voronezh and Moscow
Role of Noble Metals d-States in the Formation of the Electron Structure of Ternary Sulphides

C.E. VAN DIJKUM and A. LODDER, Amsterdam
The Local Density of States of Embedded Cu and Cu(Ni) Clusters

L.B. DUBOVSKII, Moscow
Density of States in Multilayer Sandwiches

W. KRAAK, R. KUHL, F. THOM, H. HAEFNER, and P. SCHMIDT, Berlin
Size Effect and Temperature Dependence of the Electrical Resistance in Bismuth Single Crystals of High Perfection

G. NEUMANN, Berlin
On the Defect Structure of Zinc-Doped Zinc Oxide

D. BONNIN et S. MULLER, Paris
Etude du gradient de champ électrique dans la muscovite par spectrométrie Mössbauer du fer

J-P. BLANC, M. DUGAY, H. ROBERT, and C. THIBAUD, Aubière
EPR Study of a Triplet Fine Structure in a 1d Alternating System: $Cs_2(TCNQ)_3$

D.D. ELEY, S. KINNEAR, and M.R. WILLIS, Nottingham
Real and Apparent Photoconduction in Single Crystals of Methyltriphenyl-phosphonium $(TCNQ)_2^-$

A.V. MALAKHOVSKII, Krasnoyarsk
Influence of Crystal Field on the Magnetic Circular Birefringence

N. PISLEWSKI and L.T.H. FERRIS, Poznań and London
Investigation of the Molecular Motions in $Cd(NH_3)_6Cl_2$ by NMR Spectroscopy

O. GOEDE, W. HEIMBRODT, and R. MÜLLER, Berlin
$CdS_{1-x}Te_x$ as Persistence-Type Semiconductor Mixed Crystals

19. Magnetic Resonance

20. Optical Properties
    20.1 Spectra. Optical Constants (X-Ray Spectra See 20)
    20.2 Lasers
    20.3 Luminescence (X-Ray Spectra See 20)

21. Metals. Alloys
    21.1 Transition Metals (Group Sc to Zn) and Their Alloys
        21.1.1 Fe and Fe Alloys
    21.2 Alkali Metals
    21.3 Alkali-Earth Metals
    21.4 Rare-Earth Metals
    21.5 Actinides
    21.6 Noble Metals
    21.7 Semi-Metals (Semiconducting Alloys 22.7)

22. Semiconductors and Ionic Crystals
    22.1 Elements
        22.1.1 Germanium
        22.1.2 Silicon
        22.1.3 Group VI Elements
    22.2 III–V Compounds
        22.2.1 Arsenides
        22.2.2 Phosphides
        22.2.3 Antimonides
        22.2.4 Mixed Crystals
    22.3 IV–IV, II–V, and III–VI Compounds
        22.3.1 Mixed Crystals
    22.4 II–VI Compounds
        22.4.1 Sulphides
        22.4.2 Selenides
        22.4.3 Tellurides
        22.4.4 Mixed Crystals
    22.5 Halides
        22.5.1 Silver Halides
        22.5.2 Alkali Halides
        22.5.3 Alkali-Earth Halides
        22.5.4 Mixed Crystals
    22.6 Simple Oxides (CdO, ZnO See 22.4)
        22.6.1 Mixed Crystals
    22.7 Semiconducting Intermetallic Compounds
        22.7.1 Mixed Crystals
    22.8 Tertiary and Higher Compounds
        22.8.1 Salts of Oxy-Acids (Carbonates, Phosphates, Silicates Including Mica etc.)
        22.8.2 Spinels, Garnets, Ferrites
    22.9 Organic Semiconductors

23. Solidified Gases

Attention

The preceding classification scheme is closely akin to the scheme used by Physics Abstracts to classify the various fields of solid state physics. A paper will either be classified under one or more of the subsections or, if none of these are relevant, it will just appear under the general section heading (i. e. sections and subsections are equivalent in this sense).

The *subject classification* given at the head of each paper begins with the main subject matter of the paper, and this is followed, if necessary, by figures referring to subsidiary subject matter and the type of solids investigated.

The papers published in this issue are listed in the order of their subject classification at the end of the table of contents.

82914

Manuscripts and letters for physica status solidi (b) — basic research and physica status solidi (a) — applied research

should be sent to

DDR-1020 Berlin, Neue Schönhauser Str. 20

or

Internationaler Buchversand GmbH, 1000 Berlin 19, Eschenallee 21.

The object of "physica status solidi (b)" and "physica status solidi (a)" us to publish rapidly important and topical results concerning fundamental research in the field of solid state physics and aspects of application and experimental techniques of fundamental solid state research, respectively. The issues of "physica status solidi (b)" and "physica status solidi (a)" will be published on the 1st and the 16th day of each month, respectively. Either series will contain the following categories:

1. Review articles written by arrangement with the editors. They will be remunerated.
2. Original papers selected from those which have not been submitted for publication elsewhere, after being reviewed. (The latest date for acceptance of contributions is the 8th and the 23rd day of each month for "physica status solidi (b)" and "physica status solidi (a)", respectively, 50 days before the delivery of the journal.)
3. Short notes published in the off-set part of the journal. They are, if necessary after being reviewed, selected by the editorial staff. (The latest date for acceptance of manuscripts for this part is the 12th and the 27th day of each month for "physica status solidi (b)" and "physica status solidi (a)", respectively, 16 days before the delivery of the journal.)
4. Pre-printed titles of papers to be published in the next issue of both series.

Editorial Information

Manuscripts should be submitted preferentially in English. Besides, papers in German, French, and Russian are admitted. The final decision on the incorporation of the papers in the series (b) or (a) is taken by the editorial staff after consultation of the referee. Authors are asked to observe the following rules:

1. Manuscripts have to be written as concisely as possible. As most, 20 and 4 pages (28 lines per page; abstracts, tables, references, figures, and captions of figures included, where 3 figures are counted as about one page) are admitted for original papers and short notes, respectively. All manuscripts should be type-written on one side only. Two copies are required. Badly arranged manuscripts are returned.
2. The author's name, initials, and institution should be given.
3. One abstract of about ten lines in English and one in German, French, or Russian are required for original papers.
4. Because of lack of space conclusions (if any) must not have the character of a summary (i.e. substantially a repetition of the contents of the abstract).
5. Formulae should not be written too small and not with pencil. Separate lines for formulae are desirable. Indistinct writing will lead to a return of the manuscript.
6. Drawings should be enclosed on separate sheets of paper, not exceeding the maximum and minimum sizes of 20×30 and 10×15 cm^2, respectively. Photographs should be submitted, on white paper, only for review articles and original papers. Duplicates are required.
7. Captions for the figures should be submitted on separate sheets of paper, and special instructions with regard to the size of illustrations should be given on these.
8. References should be numbered in order of appearance and should be given, on a separate sheet, in the form used in physica status solidi.
9. SI-units should be used.
10. Each paper should contain a proposal for the subject classification according to the classification scheme appearing on the second and third cover of each issue.

The authors will be sent galley proofs of review articles and original papers. Proof reading of short notes and pre-printed titles will be done by the editorial staff.

For each original paper and short note a total of 75 reprints is available free of charge. The number of reprints free of charge per review article is 25. Requests for additional reprints (against payment) should be given together with the submitted manuscript. **The deadline for orders of additional reprints is the 15th and the 30th of each month** for "physica status solidi (b)" and "physica status solidi (a)", respectively, two weeks before the delivery of the journal.